Progress in HPLC-HPCE Volume 7

Progress in HPLC-HPCE

Series Editors: H. Parvez and S. Parvez

Previously published in this series:

Volume 1
Gel Permeation and Ion-Exchange Chromatography of Proteins and Peptides

Volume 2
Electrochemical Detection in Medicine and Chemistry

Volume 3
Flow Through Radioactivity Detection in HPLC

Volume 4
Supercritical Fluid Chromatography and Micro-HPLC

Volume 5
Capillary Electrophoresis in Biotechnology and Environmental Analysis

Volume 6
Coulometric Electrode Array Detectors for HPLC

Progress in HPLC-HPCE Volume 7

Neurochemical markers of degenerative nervous diseases and drug addiction

Editors

G.A. Qureshi, H. Parvez
P. Caudy and S. Parvez

///VSP///

Utrecht, The Netherlands, 1998

VSP BV
P.O. Box 346
3700 AH Zeist
The Netherlands

First published in 1998

ISBN 90-6764-277-0

Printed in The Netherlands by Ridderprint bv, Ridderkerk.

CONTENTS

PREFACE

The heterogeneity in clinical features of various neurodegenerative diseases as well as drug addiction highlights the importance of neurochemical markers as a diagnostic tool. Even more important is the possibility of following a disease through its different stages and to understand the metabolic aspects of the disease identified by this marker.

Establishing a primary role for a neurochemical marker in any neurodegenerative disorder or in drug addiction is an important strategy in the design of new therapeutic agents for effective management of these clinical conditions in the future. Collecting data on neurochemical markers can provide vital information on the state of the disease in question and thereby assist immensely in the future task of designing new pharmacological approaches towards its management. Meanwhile two questions require answers. First, do these neurochemical markers reflect a primary neurophysiological or neurochemical event in the disease process? Secondly, does it reflect secondary events that appeared at the time of disease diagnosis? Only future research can provide clarification to this puzzle.

The main aim of this book is to present as much detailed information as possible that is currently available on various neurochemical markers and their metabolic pathways. It is our hope that this information could form the basis for future diagnostic support in the various neurodegenerative disorders as well as in tackling the huge problem of drug addiction. Further, the book is designed to give research students and clinicians advanced information on these diseases and the biochemical mechanisms involved.

We would like to extend our gratitude to all participants for their valuable contributions and for providing up to the minute chapters in some of the most exciting new developments in the study of neurochemical markers. To all those who contributed their efforts in organising this book, especially Mr Per-Anders Frändberg and Professor Fred Nyborg both from Uppsala Biomedical Center of Uppsala University, who rendered much assistance, we are most grateful.

G.A. Qureshi
H. Parvez
P. Caudy
S. Parvez

Progress in HPLC-HPCE, Vol. 7, pp. 3-56
G. Ali Qureshi *et al.* (Eds)

THE USE OF HPLC WITH ELECTROCHEMICAL DETECTION TO MONITOR REACTIVE OXYGEN AND NITROGEN SPECIES, MARKERS OF OXIDATIVE DAMAGE AND ANTIOXIDANTS: APPLICATION TO THE NEUROSCIENCES.

Ian N. Acworth[1,2], Bruce A. Bailey[1] and Timothy J. Maher[2]

[1]*ESA, Inc., 22 Alpha Road, Chelmsford, MA 01824-4171.*
[2]*Massachusetts College of Pharmacy, 179 Longwood Avenue, Boston, MA 02115.*

INTRODUCTION

It is an intriguing paradox that the very molecule absolutely necessary for aerobic life, oxygen, is also the precursor to the formation of a variety of harmful reduced oxygen molecules, usually misnamed[1] the reactive oxygen species (ROS) (for reviews see Acworth *et al* (1997a), Halliwell and Gutteridge (1989), Halliwell (1994)). The ROS shown in Table 1 not only include oxygen-centered free radicals (e.g., hydroxyl radical, hydroperoxyl radical, superoxide anion radical, and peroxyl radical), but also covalent, non-radical species (e.g., oxygen, and hydrogen peroxide) and excited oxygen molecules (singlet oxygen). Currently there is increasing interest in other harmful species, especially those based on nitrogen, including free radicals (e.g., nitric oxide) and covalent molecules (e.g., peroxynitrite, nitrocarbonate, oxonitrate anion and nitrosyl cation), which are collectively called the reactive nitrogen species (RNS) (see Table 1). Both the hydroxyl free radical and peroxynitrite are so energetic that they react instantly with any neighboring molecule encountered (e.g., DNA, RNA, carbohydrate, lipids or proteins), causing oxidative damage. Oxidative damage includes protein oxidation, lipid peroxidation and DNA damage (e.g., adduct formation, strand cessation).

In order to exist in an oxidizing environment, organisms have evolved a number of defenses designed to minimize the consequences of oxidative damage. These protective mechanisms are varied and include the prevention of ROS/RNS formation, destruction of ROS/RNS when produced, and removal and elimination of molecules damaged by ROS/RNS. Furthermore, organisms use a selection of low molecular weight antioxidants to help maintain the cellular redox status in a strongly reducing

---

[1] The ROS show a wide range of reactivities with the hydroxyl free radical being the most aggressive (typical rate constants of 10^9 to 10^{11} $M^{-1}s^{-1}$) while hydrogen peroxide is only mildly reactive (typical rate constants of 10^1 to 10^3 $M^{-1}s^{-1}$) (see Acworth *et al.* (1997a)).

environment and to react with free radicals and other pro-oxidant species. Normally, a fine balance exists between ROS/RNS production and the antioxidant defenses. However, under certain conditions (e.g., a loss of antioxidants or protective enzymes, an excess of ROS/RNS, or both) the antioxidant defenses can be overwhelmed and the cell is said to be in "oxidative stress" (Sies (1985)). Oxidative stress can be produced by a variety of endogenous and exogenous effectors including an organism's genetic predisposition (e.g., iron storage diseases and other inborn errors of metabolism) and exposure to various neurotoxins (e.g., tetrahydroisoquinolines, 6-hydroxydopamine, methamphetamine, β-carbolines and α-guanidinoglutaric acid). If oxidative stress is uncontrolled, it may eventually cause a variety of diseases including cancer and, as discussed in greater detail below, neurodegenerative diseases such as Parkinson's and Alzheimer's diseases (Beal (1997), Kehrer (1993), Kontos (1989),Youdim and Riederer (1997)).

At a fundamental level, the vast majority of chemical reactions involving the ROS/RNS can be considered as redox reactions where electrons are transferred from one species (e.g., an electron-rich antioxidant) to another (e.g., the electron poor free radical). In an analogous fashion, an electrochemical detector measures the flow of electrons (current) when an electron-rich compound loses electrons to the working electrode's surface while it itself undergoes oxidation (conversely, electron-poor compounds accept electrons from the working electrode's surface while undergoing reduction). It should come as no surprise therefore, that the electrochemical detector is one of the most versatile analytical instruments and can be used to study the many different aspects of oxidative metabolism. This chapter reviews the use of HPLC-electrochemical detection (ECD) to monitor ROS/RNS levels, measure markers of oxidative damage, examine metabolism of antioxidants, and study the role of oxidative metabolism in a variety of neurological diseases.

ELECTROCHEMICAL DETECTION

Electrochemical detection is one of the most sensitive and selective techniques available for use with HPLC. The theory behind it has been extensively reviewed elsewhere (Acworth and Bowers (1997) and references therein; Acworth *et al.* (1997b)). Although the first commercially available amperometric detector offered better sensitivity than spectrophotometric ones, this came with a price: unreliability and considerable down-time. This was due in part to its thin-layer sensor design, low reaction efficiency, wet reference electrode and susceptibility to noise from external electromagnetic interference due to insufficient grounding. Consequently, electrochemical detection quickly gained a reputation in the scientific community as a difficult to use detector which unjustly continues to persist to this day. It is indeed fortunate that not all electrochemical sensors are the same. One specialized amperometric electrode, whose reaction efficiency approaches 100%, is the flow-through coulometric sensor. The coulometric sensor offers many advantages over the thin-layer design including superior sensitivity and reliability, with little or no maintenance (Acworth and Gamache (1996), Acworth *et al.* (1997b)). But the true power of the coulometric electrode is perhaps better appreciated when it is used in a serial array. Using a coulometric electrode array, analytes can be characterized by their voltammetric profile (hydrodynamic voltammetric behavior) which can be used to qualify analytes and detect possible co-elutions (Acworth *et al.* (1997b), Svendsen (1993)).[2] Such voltammetric resolution is only feasible with the coulometric sensor and is impossible with an array of amperometric thin-layer electrodes due to their inherent low electrochemical efficiency (see Figure 1).

As discussed above, electrochemical detection has great applicability to the study of oxidative metabolism and has been used to measure many aspects of redox biology. This includes the measurement of: ROS/RNS (e.g., $HO^{\bullet}$ production can be inferred from endogenous levels of tyrosine isomers or from levels of dihydroxybenzoic acids following salicylate infusion; $NO^{\bullet}$ release can be estimated from nitrite production; 3-nitrotyrosine can be used as an indicator of peroxynitrite production); the reaction products of ROS/RNS attack (8-hydroxy-2'-deoxyguanosine as a marker of DNA damage; lipid peroxides as a marker of lipid peroxidation); and antioxidants (ascorbate, glutathione, α-tocopherol) (see Acworth and Bailey, 1995). Of particular interest is the use of the coulometric electrode array to unequivocally identify low levels of analytes in complex biological matrices e.g., 8-hydroxy-2'-deoxyguanosine in urine (Acworth *et al.* (1997a)) and to study various biomarkers associated with neurodegenerative diseases.

THE REACTIVE OXYGEN SPECIES

During aerobic respiration molecular oxygen can undergo a series of single electron reduction reactions culminating in the formation of water. The relationship between oxygen and the ROS is presented in Figure 2. One of the major sources of ROS formation *in vivo* is cytochrome oxidase, the terminal enzyme of the respiratory chain, located in the inner membrane of mitochondria. This enzyme reduces oxygen to water in a four electron reaction. Unfortunately, this reaction is not 100% efficient, and ROS are produced as byproducts. Another ROS, singlet oxygen, is not a reduced oxygen species but is best regarded as an excited form of oxygen, produced when molecular oxygen absorbs energy.

The hydroxyl free radical

The hydroxyl free radical ($HO^{\bullet}$) is extremely aggressive, reacting at diffusion controlled rates with any molecule in close proximity. It has a half-life of 10^{-10}s and only diffuses a few angstroms from its site of production before reacting (Antunes *et al.* (1996), Pryor (1986)). There are several pathways leading to $HO^{\bullet}$ formation including the homolytic fission of water induced by ionizing radiation: the reaction of hydrogen peroxide with oxidizable metals (e.g., iron, copper) [the Fenton and Haber-Weiss reactions] (Halliwell and Gutteridge (1990, 1992)) and the reaction between superoxide anion radical and hypochlorous acid (Candeias *et al.* (1993)). Furthermore, peroxynitrous acid oxidizes substrates in a similar fashion to $HO^{\bullet}$ but without direct formation of $HO^{\bullet}$ (Prior and Squadrito (1995)) (see below).

$HO^{\bullet}$ is far too reactive to be controlled enzymatically. Instead the body attempts to prevent its formation (by chelating iron and copper in non-reactive protein bound

[2] In an analogous way the photodiode array detector uses spectral information to qualify analytes, but lacks the sensitivity of the coulometric electrode array.

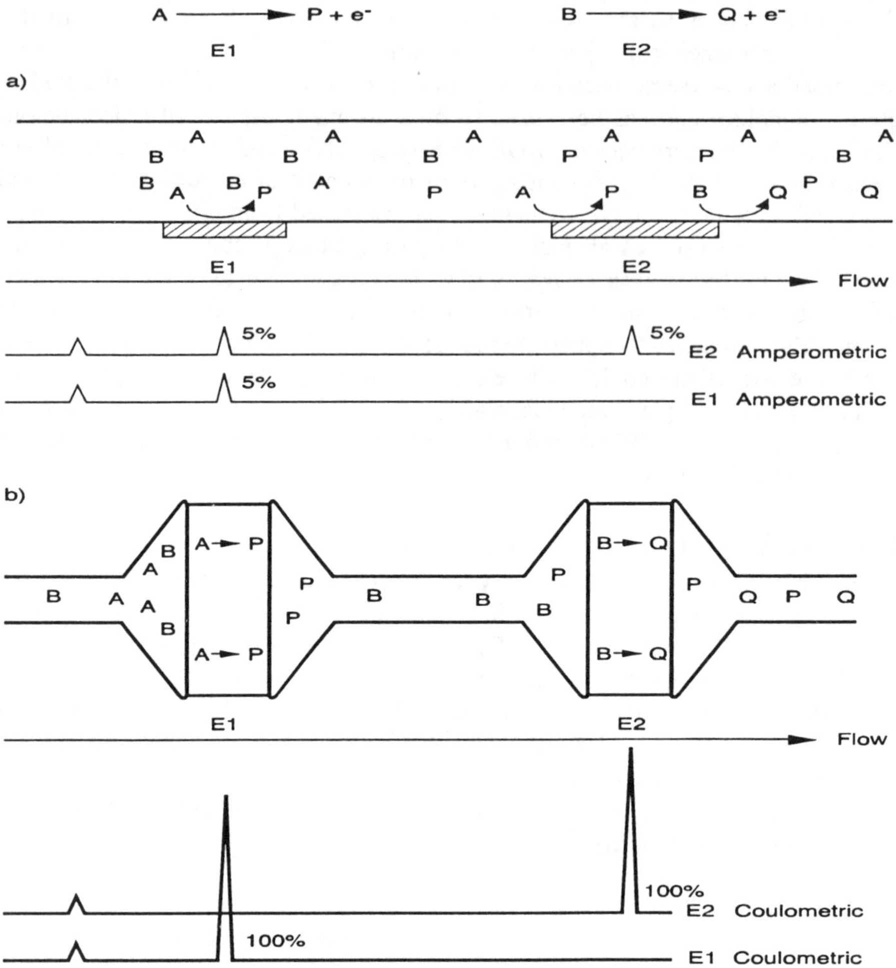

Figure 1. Resolving co-elutions. This figure illustrates differences between dual amperometric and dual coulometric electrodes for the measurement of compound A (oxidizing at 200mV) and compound B (oxidizing at 600mV). Figure 1a shows that with the inefficient amperometric electrodes, signal from compound A "spills over" onto the downstream electrode. If these compounds were poorly resolved chromatographically then measurement of compound B would be compromised. Figure 1b shows that with coulometric electrodes all of compound A is measured on the upstream electrode thereby effectively removing it from interfering at the downstream electrode. By such screening the coulometric approach overcomes poor chromatographic resolution. Notice that with the coulometric measurement the signal is twenty-fold that of the amperometric sensor. Reproduced with permission of VS Press, The Netherlands.

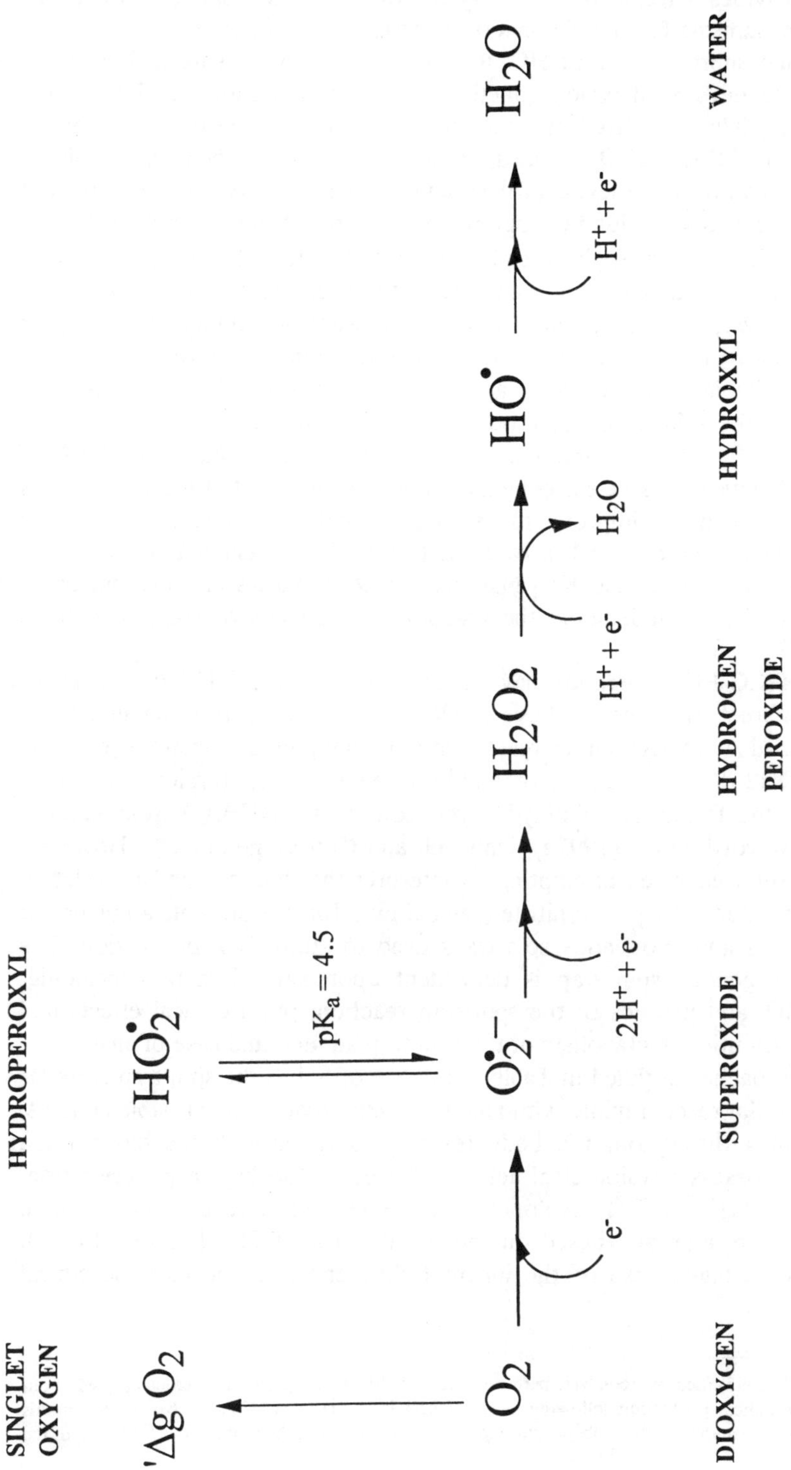

Figure 2. The chemical relationship between the different ROS.

forms). It also provides a tremendous variety of water and fat soluble antioxidants, and possesses mechanisms for the elimination of reaction products (see below). Due to its reactivity and short half-life of HO• it is impractical to measure it directly, so methods have been developed which quantify the reaction products of HO• attack (Acworth *et* al. (1997a), Halliwell and Gutteridge (1989)). Electron paramagnetic resonance (EPR) and HPLC-ECD are the approaches which show the most promise.

EPR approaches can measure levels of free radicals as low as 10^{-10}M (*in vitro*) but only if the free radical survives long enough to be measured (Halliwell and Gutteridge (1989)). The use of spin traps (such as α-phenyl-tert-butylnitrone) prolongs the life of the HO• (and other free radicals) and can extend the usefulness of EPR to *in vivo* applications.[3] However, there are several issues to be aware of including quenching of the spin trap adduct *in vivo* by cellular reducing agents such as ascorbic acid (Floyd (1983)), tissue availability of the spin trap (e.g., passage through the blood brain barrier), toxicity of the spin trap, and production of free radicals from the spin trap adduct (Floyd (1983), Halliwell and Gutteridge (1989), Takemura *et al.* (1992)). Many spin trap adducts are also electrochemically active and HPLC-ECD recently has been used in an attempt to improve the sensitivity and overcome the quenching problems of EPR (Floyd *et al.* (1984a), Iwahashi (1996), Motohashi and Mori (1989), Stronks *et al.* (1984), Towel and Kalyanaraman (1991)). Fast-scan voltammetry is also being used to better understand the reaction mechanisms of some spin traps (Baur *et al.* (1996)).

A variety of HPLC-ECD methods have been used to study HO• levels. These approaches are indirect, with the synthesis of HO• being inferred from the abundance of a product formed when HO• either reacts with an endogenous substrate (e.g., the formation of 8-OH2'dG from DNA) or administered spin trap (typically aromatic compounds, e.g., the formation of dihydroxybenzoic acids (DHBAs) from salicylic acid (Sal)) (see Acworth *et al.* (1997a), Halliwell and Gutteridge (1989)). However, care should be exercised when attempting to interpret the data as similar oxidation products can be produced by peroxynitrite (see below). Table 2 presents a number of endogenous markers and exogenous spin traps used to study HO• production. The best choice of exogenous spin trap is dependent upon several factors including: reactivity with HO• and kinetics of the spin trap reaction; physiological effects and toxicity; tissue distribution; metabolism of the spin trap adduct; and ease of analysis.

Of all of the approaches outlined in Table 2, the use of Sal as the spin trap is by far the most common. Sal reacts rapidly with HO• producing readily quantifiable adducts, it rapidly distributes throughout the body (even passing through the blood brain barrier), but may possess physiological activity if used at too high a concentration. Recently, McCabe *et al.* (1997b) developed an extremely sensitive and selective dual coulometric electrode approach based on the initial work of Floyd *et al.* (1984b). McCabe's technique makes use of the inherent differences in the electrochemical

[3] Recently, Dugan *et al.* (1995) used microdialysis perfusion with 5,5-dimethyl-1-pyrroline N-oxide coupled to on-line EPR to study free radical production following focal cerebral ischemia-reperfusion *in vivo*. Although this approach shows promise, the authors caution about spin trap stability, toxicity and the spontaneous formation of the spin trap-HO• adduct.

behavior between the DHBAs and Sal.[4] The upstream coulometric electrode is set at a low potential to completely oxidize both 2,3- and 2,5-DHBA isomers while the downstream coulometric electrode is set at a higher potential for the oxidation of Sal (Figure 3). Consequently, as the DHBAs are measured at a low potential there is less noise (a marked improvement in the signal/noise ratio) and less chance for chromatographic interference. A major advantage of this enhanced sensitivity is that less Sal has to be administered to the animal, thereby minimizing possible physiological side effects. Furthermore, this approach offers the ability to measure the 2,3-DHBA isomer which is often too low in concentration to be measured by amperometric techniques. An alternative, but less suitable, approach uses a single amperometric electrode to simultaneously measure the DHBAs and Sal. As expected, a high potential has to be applied to the amperometric working electrode in order to measure Sal. This results in increased noise, poor signal/noise ratios for the DHBAs and more complex chromatograms with a greater chance of coelutions. Yet a different approach uses HPLC-ECD to measure DHBAs in combination with UV absorbance for the measurement of Sal. Fortunately, McCabe's coulometric method makes this complicated approach unnecessary.

Not all methods measure Sal and erroneously assume that changes in the levels of DHBAs always reflect changes in $HO^{\bullet}$ production. It is important to remember that altered levels of DHBA, at least when Sal is administered peripherally, could be due to a number of factors other than $HO^{\bullet}$ production including: differences in Sal metabolism, variability in Sal's tissue distribution and differences in the site of Sal administration. It is therefore important that DHBA data be "normalized". Many papers report the use of DHBA/Sal ratios. One exception is central microdialysis when Sal is perfused through the microdialysis probe at a constant flow-rate and therefore is unaffected by peripheral metabolism. However, a growing number of papers now report a significant level of exogenous formation of DHBAs when Sal is dissolved in the microdialysis perfusion medium and this situation can be further exacerbated by metals in the perfusion system (e.g., the syringe needle, metal in the probe itself) (Acworth *et al.* (1997a), McCabe *et al.* (1997b), Montgomery *et al.* (1995)). Whether conditions do exist where Sal can be effectively used as a spin trap, while coupled to microdialysis perfusion, are yet to be fully explored.

Recently, the Sal/DHBA approach has been extended to simultaneously measure a variety of monoamines and metabolites and this has been used for the analysis of brain and CSF tissues (Patthy *et al.* (1995), Sloot *et al.* (1995)). This is a very important development with great applicability to the field of neurodegeneration. However, during the development of such methods it is vitally important to fully characterize each eluting peak (both chromatographically and voltammetrically) to ensure its authenticity. Unfortunately, such approaches are incredibly tedious and time consuming. All too often, a peak in a sample is identified solely upon whether its elution time matches that of an authentic standard. Based upon this criteria, it should come as no surprise that, all too often, peaks are misrepresented or are a product of

---

[4] This technique permits the analysis of both 2,3-DHBA and 2,5-DHBA. The former is currently thought to be a better marker of $HO^{\bullet}$ production, as the 2,5-DHBA isomer is also produced *in vivo* by cytochrome P450-dependent hydroxylation (see McCabe *et al.* (1997b) and references therein).

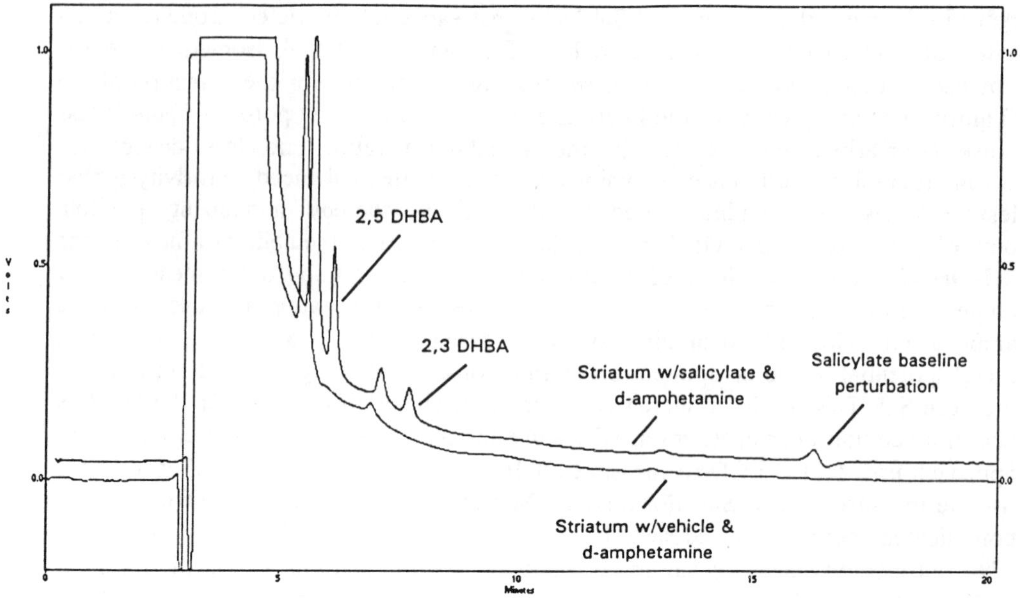

B

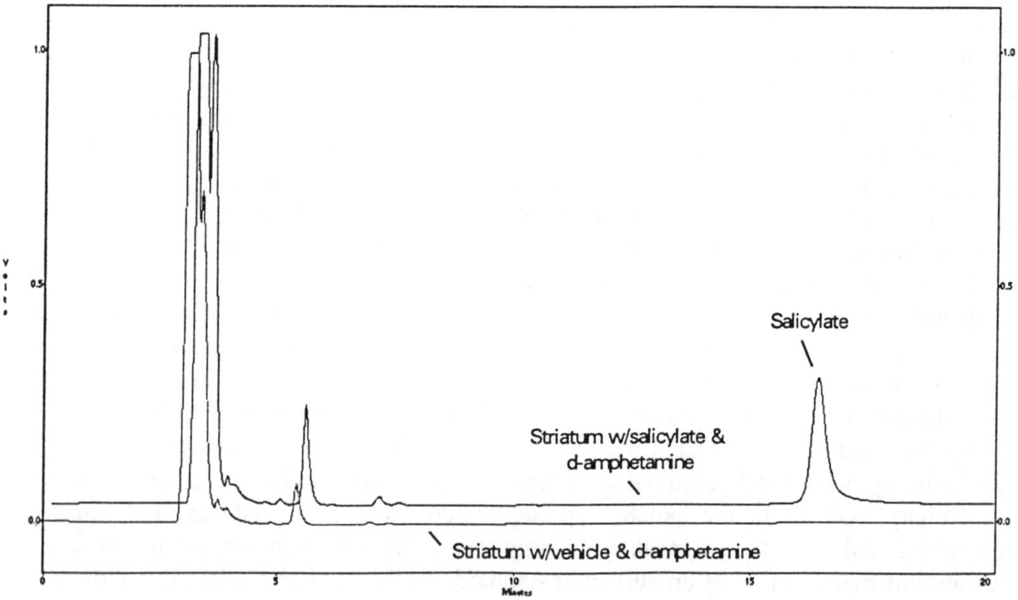

Figure 3. The simultaneous measurement of DHBA isomers and salicylate using an HPLC dual-coulometric electrode approach. The chromatograms show stimulation of DHBA production following administration of salicylate. Figure 3a presents the measurement of DHBA isomers on the upstream electrode at +250mV (5nA gain). Figure 3b presents the measurement of salicylate on the downstream electrode at +750mV (5μA gain). See McCabe *et al.* (1997b) for greater details.

coelution. Coulometric electrode array analysis produces instantaneous voltammetric data for each analyte eluting from the column. With this approach an analyte is identified by both its chromatographic and voltammetric behavior, thereby greatly minimizing misidentification (Figure 4). Beal *et al.* (1995) recently modified their gradient coulometric array method (capable of accurately measuring more than twenty-four neurochemicals and related analytes) to include Sal, the 2,3- and 2,5-DHBA isomers, 3-nitrotyrosine and 3-aminotyrosine simultaneously (Beal *et al.* 1990). They used this method to study the effects of the mitochondrial toxin 3-nitropropionic acid on the metabolism of copper/zinc superoxide dismutase in transgenic mice and concluded that both ROS and peroxynitrite play important roles in the pathogenesis of this toxin. Sontag *et al.* (1997) used another coulometric array method to examine the endogenous levels of tyrosine isomers and related their levels in tissues to the degree of exposure to ionizing radiation.

Other ROS

Although there are many HPLC-ECD methods for the analysis of $HO^{\bullet}$ there are relatively few for the other ROS, namely superoxide ($O_2^{\bullet -}$), hydrogen peroxide (H_2O_2) and singlet oxygen ($^1\Delta O_2$).

$O_2^{\bullet -}$ is formed by the addition of one electron to oxygen (Figure 2). It is produced in biological systems by exposing oxygen to ionizing radiation, as part of normal metabolism (e.g., as a byproduct when xanthine is converted to uric acid by xanthine oxidase), from leaky electron transport chains located in the inner mitochondrial membrane, and is one of the "microbicides" produced during the respiratory burst of the body's immune system. $O_2^{\bullet -}$ is also produced by autoxidation of catecholamines, folates, neurotoxins (see below), drugs and bipyridylium herbicides. Although $O_2^{\bullet -}$ can act as both a mild oxidizing agent (e.g., oxidation of catecholamines into the semiquinone) and reducing agent (e.g., reducing Fe^{3+} to Fe^{2+} as part of the Haber-Weiss reaction) and may appear to be relatively innocuous, it also has a more sinister side as it can react with $NO^{\bullet}$ producing peroxynitrite, one of the most reactive RNS known (see below). Levels of $O_2^{\bullet -}$ are mainly controlled by spontaneous and enzymatic dismutation catalyzed by superoxide dismutase. The resulting H_2O_2 is a possible precursor of $HO^{\bullet}$. $O_2^{\bullet -}$ is a weak acid and under physiological conditions about 1% will be in the form of the perhydroxyl radical ($HO_2^{\bullet}$). The abundance of $HO_2^{\bullet}$ increases markedly under acidic conditions such as in the microenvironment associated with cellular membranes or in lysosomes. $HO_2^{\bullet}$ is chemically more reactive than $O_2^{\bullet -}$, can pass through biological membranes and may play a role in lipid peroxidation. There are no HPLC-ECD methods currently capable of measuring $O_2^{\bullet -}$ even though it is EC active and can be measured using either a cytcohrome c/gold bonded voltammetric electrode (McNeil, *et al.* (1992) or one composed of glassy carbon covered with polypyrrole/horseradish peroxidase and superoxide dismutase (Lvovich and Scheeline (1997)).

H_2O_2 is produced *in vivo* by several different enzymes (e.g., superoxide dismutase, D-amino acid oxidase, monoamine oxidase), by most systems producing $O_2^{\bullet -}$ (e.g., electron transport chain of the mitochondrion, respiratory burst of immune system), as a byproduct of the cytochrome P450 system, and from redox cycling of

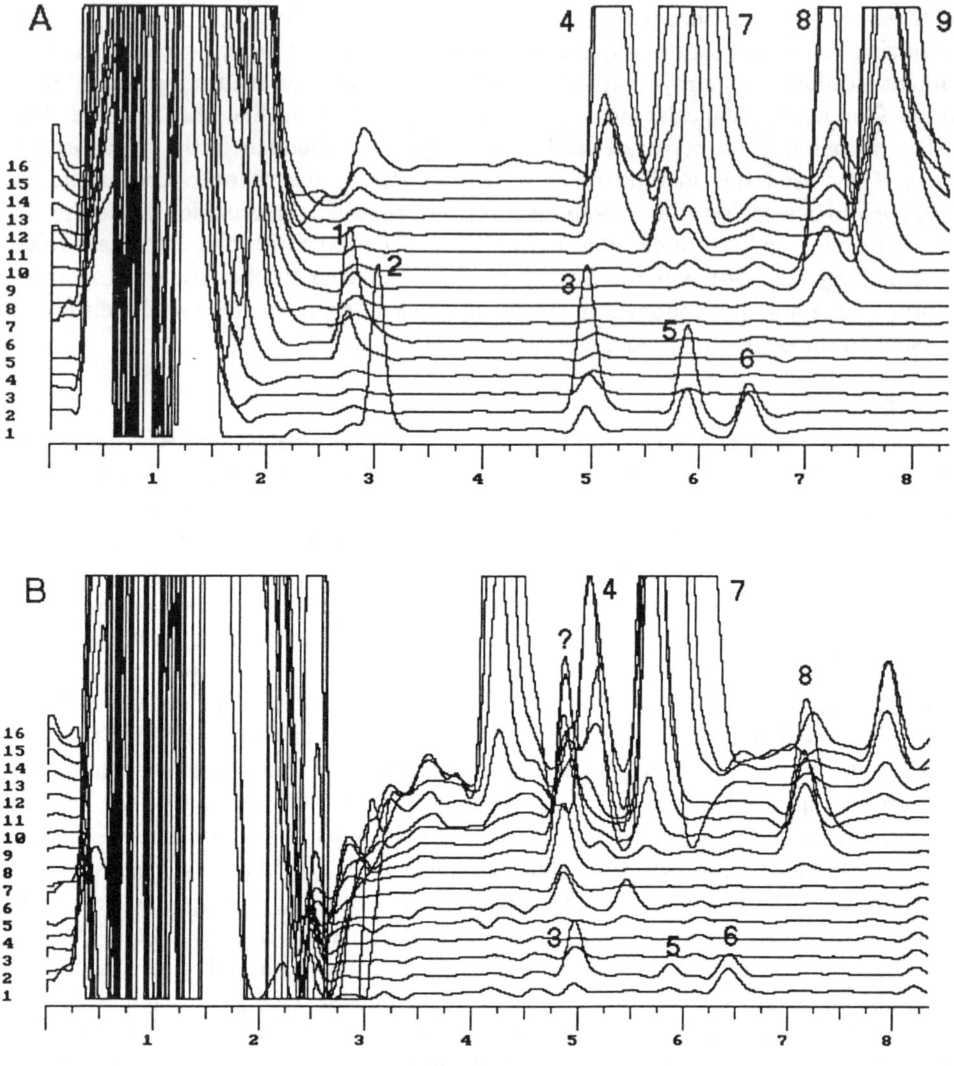

Figure 4. Coulometric array chromatogram of standard mixture (A) and rat posterior cortex homogenate (B). Analytes were resolved using a methanol gradient and sixteen channel array (0 to 900mV in 60mV increments). Analytes measured were: 1 - MHPG, 2 - Homogentisic acid, 3 - Norepinephrine (NE) , 4 - Guanosine (GN), 5 - Dihydroxyphenylacetic acid, 6 - L-Dihydroxyphenylalanine, 7 - Methionine, 8 - 4-Hydroxyphenyllactic acid, 9 - 4-hydroxybenzoic acid. Although NE and GN co-eluted they were well resolved voltammetrically (A). However, in the tissue homogenate an unknown peak (?) also co-eluted with NE and GN. Ratio accuracies for NE were 0.960 (preceding/dominant; channel 1:2) and 0.640 (following/dominant; channel 3:2). The low value for the second ratio accuracy was due to signal from ? corrupting the HDV for NE. The coulometric electrode array can immediately inform the user of such co-elutions. A single electrode HPLC system would not be able to measure this co-elution leading to the overestimation of NE tissue levels. For further details see Acworth and Bowers (1997). Reproduced with permission of VS Press, The Netherlands.

catecholamines and drugs. Like $O_2^{\bullet -}$, H_2O_2 acts as both a weak oxidizing and reducing agent. It does react with thiols, and if these are located in the active site of an enzyme its activity may be disrupted (e.g., glyceraldehyde-3-phosphate dehydrogenase). H_2O_2 can readily pass through membranes and if it encounters "reactive" (non-chelated) iron will produce the extremely aggressive $HO^{\bullet}$ (see above). The level of H_2O_2 is mainly controlled enzymatically brought about by catalase and glutathione peroxidase (the latter is a selenium dependent enzyme using reduced glutathione as a cofactor). H_2O_2 is electrochemically active and can be measured on metal working electrodes. For example, HPLC-ECD can be used to indirectly measure choline and acetylcholine based upon the amount of H_2O_2 produced when choline is oxidized by choline oxidase (Greaney *et al.* (1993)). However, the authors are unaware of any direct HPLC-ECD method capable of measuring H_2O_2 in biological samples.

Oxygen molecules can absorb energy to form an excited species, a singlet oxygen state represented as $^1\Delta g\ O_2$. This excited species can be produced in biological systems from the reaction between H_2O_2 and hypochlorite, formed during the respiratory burst of the immune system (see Figure 2). When light reacts with the pigment rhodopsin in the eye, $^1\Delta g\ O_2$ can also be formed. $^1\Delta g\ O_2$ reacts with a variety of metabolites including amino acids (e.g., methionine produces methionine sulfoxide, tryptophan produces N-formylkynurenine, cysteine forms cystine), lipids (producing the hydroperoxide), phenols, carotenoids, cholesterol and bilirubin (Halliwell and Gutteridge, (1989)). *In vivo* $^1\Delta g\ O_2$ is usually quenched by its reactions with β-carotene (and other carotenoids) or α-tocopherol. $^1\Delta g\ O_2$ production may be inferred from levels of its reaction products which can be measured using HPLC-ECD, e.g., N-formylkynurenine (Matson *et al.* (1987) used gradient HPLC with coulometric array to measure N-formylkynurenine, kynurenine and other tryptophan metabolites), cholesterol hydroperoxide (Arai *et al.* (1996), Korytowski *et al.* (1991)) and possibly lipid hydroperoxides (O'Gara *at al.* (1989) used a novel glutathione-linked coulometric assay). However, whether levels of these metabolites solely reflect the action of $^1\Delta g\ O_2$ remains unclear. $^1\Delta g\ O_2$ can be determined by measuring the light emitted when it decays back to its ground state but, when using this approach, care should be exercised as luminescence can come from many biological sources (Halliwell and Gutteridge (1989)). As discussed above, $^1\Delta g\ O_2$ can also be measured using HPLC-ECD as a spin trap adduct (Motohashi and Mori (1989)).

THE REACTIVE NITROGEN SPECIES

As shown in Table 1, a number of RNS (both free radicals and non radicals) exist and many have been shown to occur *in vivo*. The relationship between the different RNS is presented in Figure 5 and between ROS, RNS and other reactive species in Figure 6. Nitric oxide and the other RNS have recently been the subject of several books (Feelisch and Stamler (1996), Ignarro and Murad (1995), Lancaster (1996), Packer (1996a,b)).

Nitric oxide

$NO^{\bullet}$ is produced *in vivo* by the action of NO-synthase on the amino acid L-arginine (L-citrulline is a byproduct).[5] In biological systems $NO^{\bullet}$ plays a variety of important roles including the regulation of blood pressure (Moncada *et al.* (1991)), it is a precursor to nitrosyl chloride (NOCl) which is used by the immune system to destroy pathogenic organisms (Koppenol (1994)), and it acts as a retrograde neurotransmitter which has been implicated in the formation of long term memory (Feldman *et al.* (1993), Snyder and Bredt (1992)). However, $NO^{\bullet}$ possesses a sinister side too, reflective of its character as a major environmental pollutant and, not surprisingly, it has been implicated in the pathophysiology of certain diseases (Gross and Wolin (1995), Dawson and Dawson (1994, 1996)). $NO^{\bullet}$ directly damages iron-sulfur proteins and other metalloproteins (Radi (1996)); it reacts with both tryptophyl and tyrosyl free radicals which act as intermediates in a variety of enzymatic reactions (Eiserich *et al.* (1995)); it reacts with free and protein-bound thiols (in a reaction possibly initiated by $^{\bullet}NO_2$, NO^+ or peroxynitrite) producing nitrosothiols[6] (Upchurch *et al.* (1995)); it directly attacks DNA (Nguyen *et al.* (1992), Wink *et al.* (1991)) which can result in ADP-ribosylation of proteins (Zhang *et al.* (1994)), and eventually to cell death. Recently, d'Ischia and Costantini (1995) have reported that under aerobic conditions, $NO^{\bullet}$ can react with catecholamines to produce potentially neurotoxic nitrocatecholamine adducts. Whether this reaction *per se* occurs *in vivo* has yet to be proven. The formation of nitrocatecholamine adducts is also likely to occur through the reaction of catecholamines with another RNS, peroxynitrous acid (see below).

Although $NO^{\bullet}$ reacts with O_2 and is often regarded as a beneficial regulator of $O_2^{\bullet-}$'s levels, the reaction product, peroxynitrite, is an incredibly aggressive species attacking many biomolecules (see below). $NO^{\bullet}$ can also react with $HO^{\bullet}$ producing nitrous acid. However, for this reaction to take place, unphysiologically high $NO^{\bullet}$ levels would have to be present in order to compete with the plethora of other molecules capable of reacting with $HO^{\bullet}$. For this reason, the reaction between $NO^{\bullet}$ and $HO^{\bullet}$ is thought to be biologically unimportant.

A variety of different approaches have been used to measure $NO^{\bullet}$ levels and these have been reviewed in detail elsewhere (Acworth *et al.* (1997a)). $NO^{\bullet}$ is electrochemically active and can be directly measured *in vivo* using a variety of modified voltammetric electrodes (Friedemann *et al.* (1996), Malinski and Taha (1992), Maskus *et al.* (1996)). However, the reader should exercise caution when exploring this approach as the various voltammetric methods differ considerably in their sensitivity, selectivity and reaction kinetics with $NO^{\bullet}$, and therefore their applicability to biological assays. Unfortunately, $NO^{\bullet}$'s short biological half-life (<1s

---

[5] Arginine is also the precursor to a variety of polyamines (e.g., putrescine, spermidine and spermine) which have been shown to be involved in a variety of cellular processes involving DNA replication, transcription and translation. The polyamines can be readily measured as their *o*-phthaldialdehyde derivatives using HPLC and coulometric array technology (Achilli *et al.* (1993)).

[6] Nitrosothiols are electrochemically active and have been measured using HPLC-ECD on a gold amalgam working electrode, although with such rapid elution and instability of the working electrode this method may not be practical for the routine analysis of biological tissues (Stamler and Feelisch (1996)). Detection on coulometric graphite electrodes may be a viable alternative - see measurement of thiols below.

in blood; <6-10s in tissue) ensures that its direct measurement by HPLC-ECD is virtually impossible. Several indirect HPLC-ECD approaches have been used to quantitate the level of $NO^{\bullet}$ in biological samples. Wang and Maher (1992) used HPLC with coulometric electrochemical detection with precolumn o-phthaldialdehyde derivatization to study changes in the levels of arginine and citrulline in rat hippocampal microdialysis perfusates. The level of citrulline was found to increase under conditions where levels of $NO^{\bullet}$ are known to be elevated (e.g., following administration of the excitatory amino acid agonist, kainic acid) while the level of citrulline was decreased under conditions where levels of $NO^{\bullet}$ are known to be reduced (e.g., following treatment with the nitric oxide synthase inhibitor, N-nitro-L-arginine). Another approach makes use of the fact that $NO^{\bullet}$ is readily oxidized *in vivo* to nitrite (NO_2^-). Preik-Steinhoff and Kelm (1996) used a simple and sensitive HPLC-ECD method with limited sample preparation to directly measure levels of NO_2^- in human blood. Other approaches measure both NO_2^- and its product NO_3^- as markers of NO production.[7] However, these approaches may be inaccurate as the NO_3^- level is influenced by a variety of exogenous factors including intestinal bacterial NO_3^- synthesis, NO_3^- coming from the diet and inhalation of atmospheric nitrogen compounds. Furthermore, as the tissue level of NO_3^- is approximately 10 to 100 fold higher than that of NO_2^-, any significant changes in endogenous NO_2^- levels will be overwhelmed when NO_3^- is reduced during these analytical procedures.[6]

Peroxynitrite

$NO^{\bullet}$ reacts with $O_2^{\bullet-}$ at near diffusion controlled limits, and at approximately three times the dismutation rate of $O_2^{\bullet-}$ by superoxide dismutase, to produce peroxynitrite (ONO_2^-). The peroxynitrite anion is fairly stable, but with a pKa of 6.8, it is protonated at physiological pH forming the unstable, peroxynitrous acid (ONO_2H) (Crow and Beckman, (1995)). ONO_2H has a half-life of <1s and is capable of acting as both a powerful oxidizing agent (reacting similarly to $HO^{\bullet}$) and nitrating agent (e.g., producing 3-nitrotyrosine from tyrosine, 8-nitroguanine from guanine and possibly 6-nitrocatecholamines from the catecholamines). The reaction mechanisms of ONO_2H have been dealt with in detail elsewhere (Beckman *et al.* (1994), Pryor and Squadrito (1995)).

ONO_2H reacts with many important biomolecules and promotes lipid peroxidation (Radi *et al.* (1991), Rubbo *et al.* (1994)), DNA damage (Tamir *et al.* (1996)) and oxidizes and nitrates many amino acids, peptides and proteins (Ischiropoulos and Al-Mehdi (1995), van der Vliet *et al.* (1994)). It also reacts with free and protein-bound thiols, deoxyribose and ascorbic acid (see Alvarez *et al.* (1995) and references therein). As ONO_2H rapidly reacts with carbon dioxide (or bicarbonate), eventually leading to the formation of the equally reactive nitrocarbonate anion (Uppa *et al.* (1996), the reactions attributed to ONO_2H might in fact be due to involvement of nitrocarbonate.

---

[7] NO_2^- can be determined by HPLC-UV of the Greiss derivative following metal catalyzed reduction of NO_3^- to NO_2 (Yamada and Nabeshima (1997), Muscara and Nucci (1996)) or by chemiluminescence following enzymatic reduction of NO_3^- (Salter *et al.* (1996)).

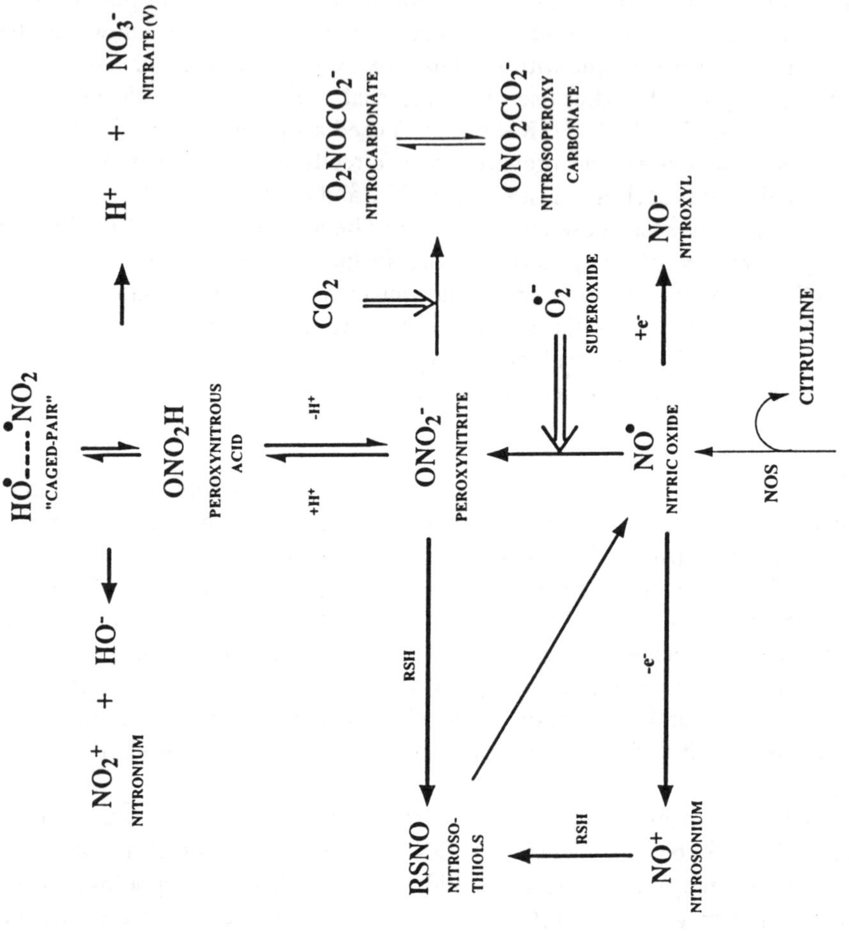

Figure 5. The chemical relationship between the different RNS.

Although ONO_2^- absorbs light at 302nm and can be measured spectrophotometrically *in vitro*, such an approach is unsuitable for *in vivo* analysis due to its inherent lack of sensitivity and selectivity. Instead, the level of ONO_2^- in biological samples is usually indirectly determined by the measurement of one of its reaction products, e.g., 3-nitrotyrosine (dealt with in greater detail below).

MARKERS OF ROS/RNS DAMAGE

All the major macromolecules found in cells can react with ROS/RNS and there is considerable interest in measuring the reaction products as markers of damage.

Amino acid and protein damage

Free amino acids, peptides and proteins can be covalently modified by reaction with the ROS/RNS. These modifications may be metabolically harmful if they were to occur in an essential area of a protein, such as the active site of an enzyme. Fortunately, protein oxidation (along with deamidation, glycation and ubiquitin conjugation) usually marks the protein for subsequent proteolysis. However, recent reports have suggested that accumulation of modified proteins have been linked to aging (Smith *et al.* (1992)) and Alzheimer's disease (Smith *et al.* (1991)).

When proteins are oxidized by $HO^•$ and ONO_2^-, several amino acids are converted to carbonyls (e.g., proline and arginine into glutamic semialdehyde, histidine into 2-oxohistidine and lysine into lysylcarbonyl). These protein carbonyls have been used as markers of oxidative stress and have been measured as their 2,4-dinitrophenylhydrazine derivatives using HPLC-UV (Ayene *et al.* (1993), Fung and Grosjean (1981), Smith *et al.* (1991)). Although improvements in sensitivity could be achieved by adaptation of an HPLC-ECD method (Chiavari *et al.* (1985)), the value of protein carbonyl measurement is unclear at present as protein carbonylation can take place by mechanisms not involving ROS/RNS (Giuivi and Davies (1994)). Other ROS/RNS induced structural modifications include oxidation of cysteine into cystine and methionine into methionine sulfoxide, as well as the conversion of proline into glutamate, histidine into asparagine and protein fragmentation.

The chemical modification of aromatic amino acids by ROS/RNS is receiving a great deal of interest because of their possible use as oxidative stress markers. As discussed above (see Table 2), the measurement of the levels of tyrosine isomers and dihydroxyphenylalanine isomers (produced from phenylalanine and tyrosine, respectively) can be used to infer production of $HO^•$ (Ishimitsu *et al.* (1986), Kaur and Halliwell (1994c), Liu (1993), Ramezanian *et al.* (1996), Sontag *et al.* (1997), Sun *et al.* (1993), van der Vliet *et al.* (1994)). However, as these markers can also be produced by reaction with ONO_2^- (van der Vliet *et al.* (1994)) care should be taken when attempting to relate changes in these markers to those of $HO^•$. Dityrosine is also used as a marker of oxidative damage to proteins. It is however relatively unspecific and can also be formed by UV irradiation and exposure to H_2O_2, $HO^•$ and ONO_2^- (Giuilivi and Davies (1994), van der Vliet (1995)). Dityrosine has been determined in protein hydrolysates by HPLC-UV (Ischiropoulos *et al.* (1992)) or HPLC-fluorescence (Giulivi and Davies (1994)). Free dityrosine is electrochemically active and has been measured simultaneously with 3-nitrotyrosine and various markers of

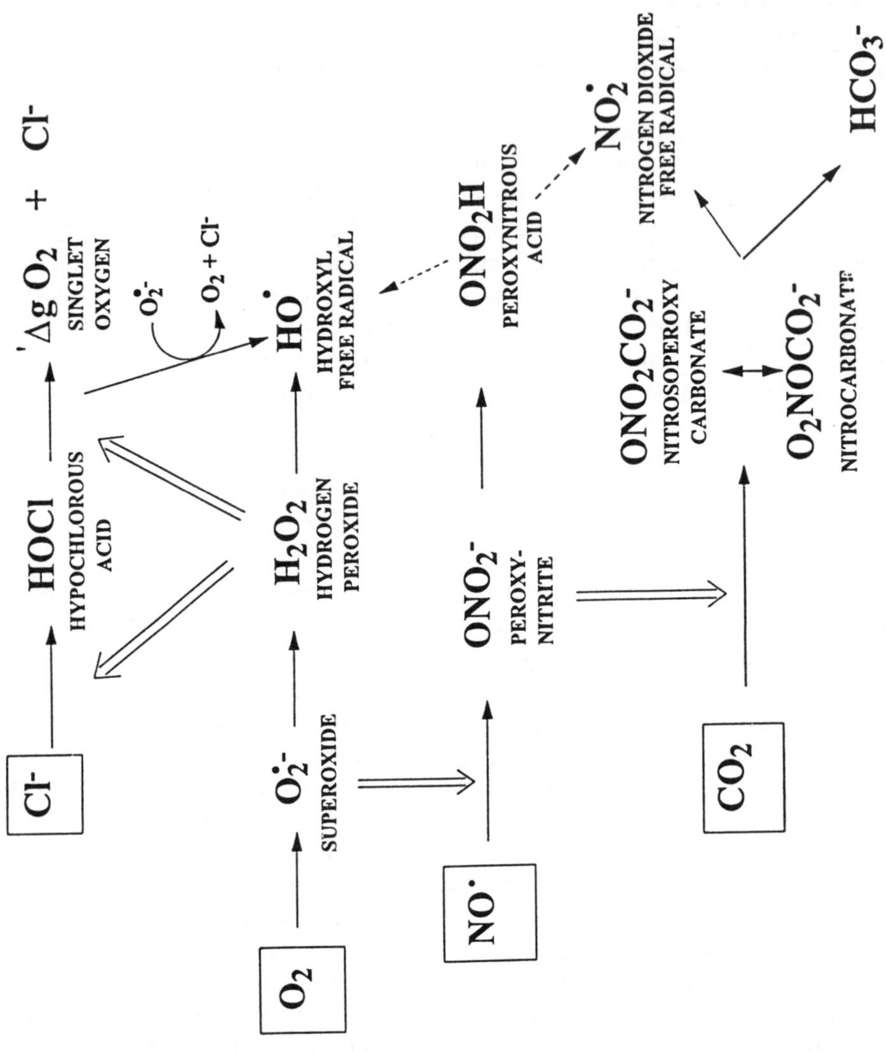

Figure 6. The chemical relationship between ROS, RNS, chloride and carbon dioxide.

DNA damage by gradient HPLC with coulometric array detection (Milbury and Matson (1995)).

In addition to oxidation, ONO_2^- can nitrate aromatic amino acids, producing 3-nitrotyrosine from tyrosine (Ischiropoulis and Al-Mehdi (1995), Oury *et al.* (1995), Salman-Tabcheh *et* al. (1995), van der Vliet *et al.* (1994, 1995)), 4-nitrophenylalanine from phenylalanine (van der Vliet *et al.* (1994), and 5- and 6-nitrotryptophan from tryptophan (Padmaja *et al.* (1996)). 3-Nitrotyrosine has been identified as the chemically stable end product of the reaction between free or protein-bound tyrosine and ONO_2^-, however, the involvement of other RNS such as $^{\bullet}NO_2$ (van der Vliet *et al.* (1995)), N_2O_3 (Kharitonov *et al.* (1995)) and NO_2Cl (Eiserich *et al.* (1996)) cannot be discounted. Metabolically, free 3-nitrotyrosine can undergo enzymatic conversion to 3-nitro-4-hydroxyphenylethylamine, and 3-nitro-4-hydroxyphenylacetic acid and these too can be used as markers. The level of 3-nitrotyrosine has been shown to be altered in lung injury (Haddad *et al.* (1994), Kooy *et al.* (1995)), atherosclerosis (Beckman *et al* (1994), Kaur and Halliwell (1994)), in patients suffering from both renal failure and septic shock (Fukuyama *et al.* (1997)) and may also be associated with cancer (Liu and Hotchkiss (1995)). Various detection techniques have been used to detect 3-nitrotyrosine in biological samples including immunochemical approaches and methods based on HPLC. Although monoclonal and polyclonal antibodies raised against 3-nitrotyrosine-containing proteins have been used in numerous studies (e.g., Crow and Beckman (1995), Ye *et al.* (1996)), these methods are less quantitative than chromatographic approaches. Free 3-nitrotyrosine and 3-nitrotyrosine obtained from hydrolyzed proteins have been measured using HPLC-UV (Crow and Ischiropoulos (1996), Kaur and Halliwell (1994c), van der Vliet *et al.* (1995, 1996)). Although HPLC-fluorescence generally offers improved limits of detection, unfortunately the electron withdrawing effect of the nitro group renders 3-nitrotyrosine only weakly fluorescent. Therefore HPLC-fluorescence is normally only used following derivatization of 3-nitrotyrosine with a fluorogenic agent such as phenylisothiocyanate (Ischiropoulos and Al-Mehdi (1995)) or following the reduction of 3-nitrotyrosine to 3-aminotyrosine by dithionite (Crow and Ischiropoulos (1996)). Recently extremely low levels of free 3-nitrotyrosine and tyrosine have been accurately determined using HPLC with dual coulometric electrochemical detection (Maruyama *et al.* (1996)) and gradient HPLC with coulometric array detection (Beal *et al.* (1995), Maruyama *et al.* (1996), Schulz *et al.* (1995)). Using the latter approach Beal *et al* (1996) reported increased concentrations of 3-nitrotyrosine and one of its metabolites, 3-nitro-4-hydroxyphenylacetic acid, in samples obtained from both sporadic and familial amyotrophic lateral sclerotic patients. Freeto (1997) has recently developed a coulometric array method capable of simultaneously quantitating many of the products of the ONO_2^-/tyrosine reaction including the tyrosine isomers, 3-nitrotyrosine and its chemically reduced analog, 3-aminotyrosine (see Figure 7). The advantage of measuring 3-aminotyrosine is that it electrochemically oxidizes at a lower potential (+70mV vs. Pd) than 3-nitrotyrosine (+600mV vs. Pd) and can thus be measured with improved limits of detection due to less system noise.

The use of HPLC to determine levels of 3-nitrotyrosine in proteins is more complex as it first has to be liberated from proteins before analysis. Acid hydrolysis of proteins may artificially alter 3-nitrotyrosine levels (see Shigenaga *et al.* (1997)). Shigenaga *et al.* (1997) have recently reported an HPLC-coulometric electrochemical method capable of measuring 3-nitrotyrosine in protein hydrolysates. Here, following pronase E protein hydrolysis and extensive sample preparation, 3-nitrotyrosine was determined at low sensitivity as its N-acetyl-3-aminotyrosine analog.

DNA damage

DNA is a relatively fragile molecule and can be readily damaged by ionizing radiation and its interaction with free radicals resulting in destruction of bases and deoxyribose, and single and double strand breaks (Breen and Murphy (1995), Marnet and Burcham (1993), Steenken (1989), von Sonntag (1991)). Such damage has been postulated to underlie aging and a variety of pathological states such as cancer, mutagenesis, carcinogenesis and cell death (Halliwell and Gutteridge (1989), Wiseman and Halliwell (1996)). DNA is readily attacked by $HO^{\bullet}$ (Pryor (1988)) and a major product of 2'-deoxyguanosine damage, 8-hydroxy-2'-deoxyguanosine (8-OH'2dG), is one of the most mutagenic lesions known (Ohtsuka and Nishimura (1987), Shibutani *et al.* (1991)). DNA is also susceptible to attack by alkylating agents (forming various methylated products, e.g., 3-methyl-, 7-methyl- and O-6-methyl-2'deoxyguanosine), lipid peroxidation products such as 4-hydroxynonenal and malondialdehyde (Douki and Ames (1994), Goda and Marnett (1991)) and compounds capable of undergoing redox cycling.[8] The reaction of DNA with ONO_2^- produces a variety of modified species, including 8-OH2'dG (Douki *et al.* (1996)).[9]

DNA adducts have been measured using a variety of methods including immunochemical, GC-MS, GC-MS-MS, ^{32}P- post labeling, HPLC-UV and HPLC-^{32}P post labeling. These have been reviewed elsewhere (see Acworth *et al.* (1997a)). Many DNA adducts (especially those based on guanosine) are electrochemically active and can be measured easily with HPLC-ECD.[10] This technique permits sensitive and selective measurement of DNA adducts without the need for extensive sample preparation, the cost of expensive analytical instrumentation or the possible safety hazards of dealing with radioisotopes. HPLC-ECD has been used to measure a variety of DNA adducts including: 8-OH2'dG (Acworth *et al.* (1997a), Adachi *et al.* (1993, 1995), Beechler *et al.* (1992), Brown *et al.* (1995), Claycamp and Ho (1993), Conaway *et al.* (1991), Culp *et al.* (1989), Degan *et al.* (1991), Denda *et al.* (1994), Faux *et al.* (1992), Fischer-Nielson *et al.* (1992, 1994), Floyd *et al.* (1986), Fraga *et al.* (1990), Frenkel *et al.* (1991), Germadnik *et al.* (1997), Han and Liehr (1994), Hattori-Nakakuki *et al.* (1994), Haegele *et al.* (1994), Herbert *et al.* (1996), Kelly *et al.* (1992), Lagorio *et al.* (1994), Liu and Wells (1995), Loft *et al.* (1995), Lunec *et al.* (1994), McCabe *et al.* (1997a), Mecocci *et al.* (1993), Nicotera *et al.* (1994), Park

---

[8] The catecholestrogens (the hydroxylated metabolites of mammalian estrogens) are capable of undergoing redox cycling producing reactive intermediates which can form adducts with purines. For example, 4-hydroxyestrone reacts with guanine producing 7-[hydroxyestron-1(α,β)-yl]guanine which has been measured using HPLC with coulometric electrode array detection (Acworth (1996)).

[9] Yermilov *et al.* (1995) used HPLC-UV to demonstrate that 8-nitroguanine is also produced when DNA reacts with ONO_2^-. Improvements in their detection limit will be achievable with HPLC-ECD.

[10] Considerable variation in the levels of guanine adducts appear in the literature probably due to the differences in laboratory methodologies and/or the extent of DNA repair mechanisms (Adachi *et al.* (1995)).

et al. (1989, 1992), Shigenaga *et al.* (1989, 1990, 1994), Tagesson *et al.* (1994), Takeuchi and Morimoto (1993)); alkylguanines (Douki and Ames (1994), Kaur and Halliwell (1996), Tan *et al.* (1990), van Delft *et al* (1993)); malondialdehyde-DNA adducts (Goda and Marnett (1991)); and cytidine adducts (Wagner *et al.* (1992)). Some researchers have tried to quantitate several DNA adducts simultaneously but have been hampered by the lack of selectivity of amperometric electrodes (Kaur and Halliwell (1996)) and are often limited to the measurement of a single adduct in chemically-treated non-biological samples (Xie and Duda (1994)). McCabe *et al.* (1997a) have recently developed a gradient HPLC-coulometric electrode array method capable of measuring both precursors and adducts simultaneously and with low pg sensitivity (see Figure 8). Mecocci *et al.* (1993) have used a similar coulometric array approach to show that levels of 8-OH2'dG were not only markedly higher in mitochondrial DNA obtained from different human brain regions, but that the mitochondrial adduct was more abundant in aged brain than younger control samples. They concluded that such mitochondrial DNA damage may contribute to age-dependent increases in the incidence of neurodegenerative diseases.

Lipid peroxidation

ROS/RNS induced lipid damage has been linked to aging and a variety of diseases including: atherosclerosis, stroke and reperfusion injury, and neurodegenerative disorders (Beal (1995), Jialal and Devaraj (1996), Spiteller (1996)). Polyunsaturated fatty acids (PUFAs) such as arachidonic, linoleic and linolenic acids, which occur naturally as either the free acid or as part of more complex lipid (e.g., triglycerides, phospholipids), are exquisitely sensitive to ROS/RNS attack. Abstraction of a proton from PUFAs by $HO^{\bullet}$ or ONO_2^- produces a lipid radical which is thought to be the principle initiating mechanism of lipid peroxidation ($HO_2^{\bullet}$ and $\Delta g\ O_2$ may only play minor roles) (Halliwell and Chiroco (1993)). The lipid radical can react with molecular oxygen producing a lipid peroxyl radical, which can then abstract a proton from another PUFA to form a lipid peroxide, thereby propagating the chain reaction. Chain breaking antioxidants (CBAs) such as α-tocopherol can readily donate a proton to the lipid peroxyl radical thus terminating the chain reaction (see below).

Damage to the lipid bilayer can have major ramifications especially for neurons which contain high levels of PUFAs. First, lipid peroxidation will result in altered membrane fluidity which can affect membrane dependent processes such as receptor function, enzymatic activity, ion channels and signaling processes. Second, the lipid peroxyl radical can react and damage important macromolecules such as intrinsic membrane proteins and DNA (Park and Floyd (1992)). Third, lipid peroxides and many of their reaction products (e.g., malondialdehyde, 4-hydroxynonenal) are cytotoxic (Esterbauer *et al.* (1991)) (see Figure 9). For example, 4-hydroxynonenal can react with thiol and amine groups, malondialdehyde can cross link proteins (Smith *et al.* (1996)) and both can damage DNA (Douki and Ames (1994), Goda and Marnett (1991)).

A variety of approaches have been used to study lipid peroxidation processes (e.g., UV absorbance, colorimetric assay, chemiluminescence, GC-MS, HPLC with UV and fluorescence, iodine liberation) and these have been reviewed elsewhere (Esterbauer

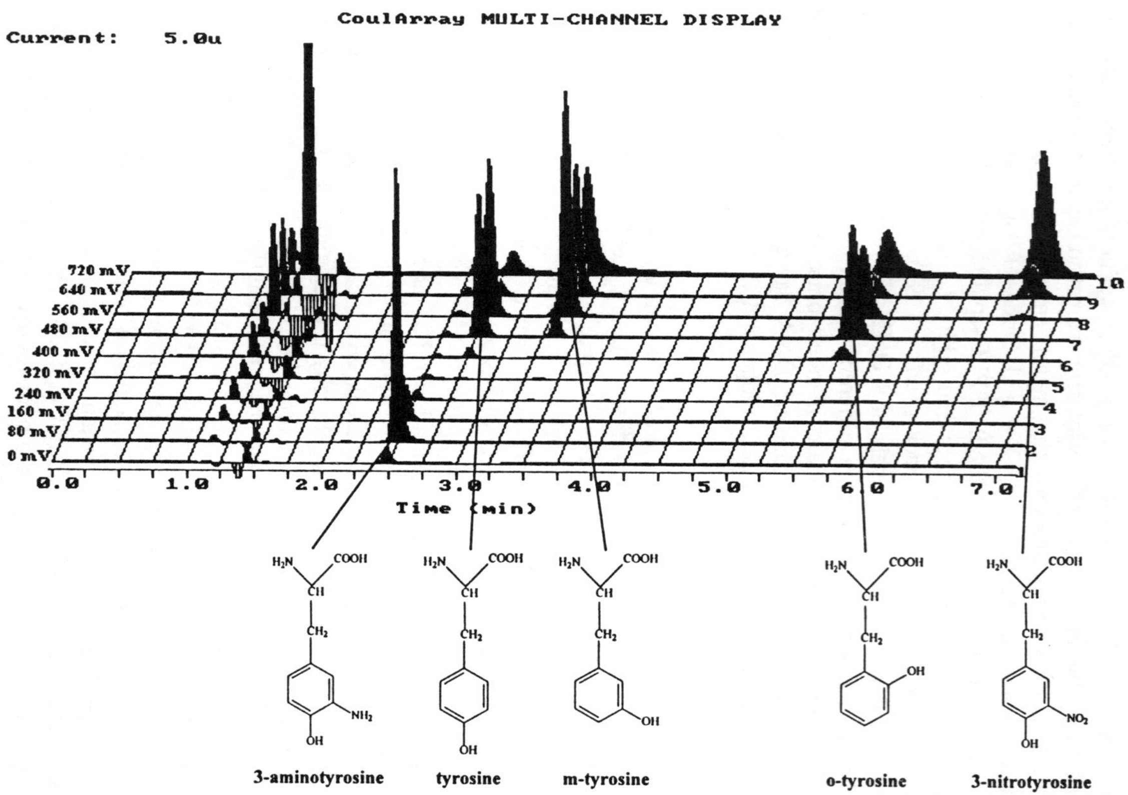

Figure 7. The simultaneous measurement of 3-aminotyrosine, 3-nitrotyrosine and tyrosine isomers (5ng on column) using HPLC with coulometric array detection. See Freeto (1997) for further details.

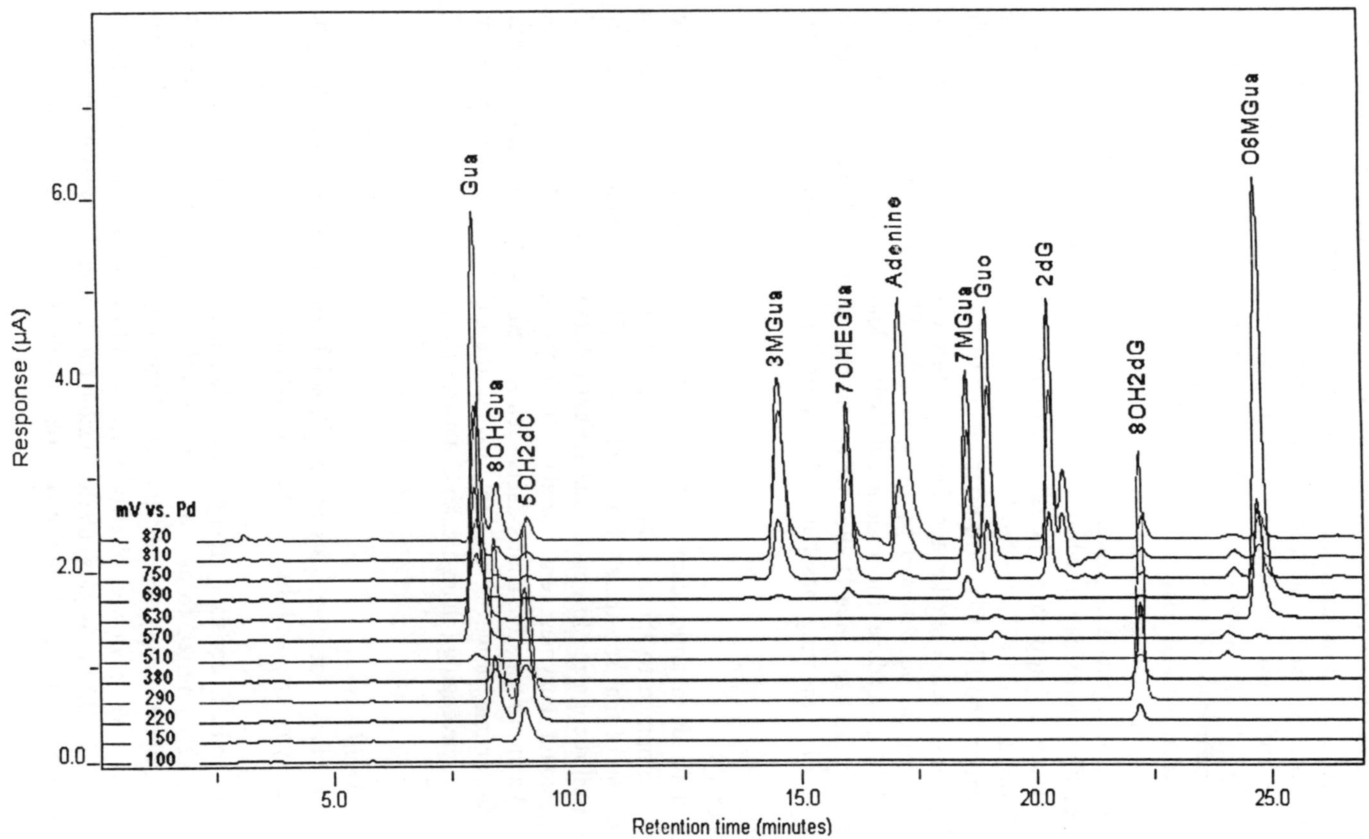

Figure 8. The simultaneous measurement of DNA bases and adducts using HPLC with coulometric array detection. See Gamache *et al.* (1997) for further details. Gua - guanine, 8OHGua - 8-hydroxyguanine, 5OH2dC - 5-hydroxy-2'-deoxycytidine, 3MGua - 3-methylguanine, 7OHEGua - 7-hydroxy-ethylguanine, 7MGua - 7-methylguanine, Guo - guanosine, 2dG - 2'-deoxyguanosine, 8OH2dG - 8-hydroxyguanosine, O6MGua - O-6-methylguanine.

et al. (1991), Halliwell and Chirico (1993)). Lipid peroxides have also been measured by HPLC-ECD using both direct (Mulchandani and Rudolph (1995)) and indirect (O'Gara *et al.* (1989)) methods. As shown in Table 3, a variety of HPLC-ECD methods have been used to measure lipids and peroxidation products.

Considerable interest has been focused on two lipid peroxidation products, malondialdehyde and 4-hydroxynonenal, due to their reactivity and toxicity (e.g., Esterbauer *et al.* (1991)). Malondialdehyde has been measured using a variety of techniques, but one method, the thiobarbituric acid assay, is used most often, probably because it is inexpensive and easy to perform. Although it is said to be specific for malondialdehyde, in fact this assay lacks specificity and can react with any number of aldehydes. Furthermore, the assay conditions can promote malondialdehyde formation and also induce lipid peroxidation which may lead to false positive results. A better and more selective approach uses precolumn derivatization with 2,4-dinitrophenylhydrazine (DNPH) and HPLC separation. Ekstrom *et al.* (1988) used DNPH derivatization and HPLC-UV to measure urinary malondialdehyde levels while Cordis *et al.* (1994) used a similar approach to measure it in heart tissue. Perhaps better sensitivity would be obtained by using HPLC-ECD. Although there are no reports of such a procedure for malondialdehyde, Goldring *et al.* (1993) have developed an extremely sensitive DNPH derivatization, HPLC-coulometric detection method capable of measuring basal liver levels of 4-hydroxynonenal and in cell cultures.

ANTIOXIDANTS

Halliwell has defined an antioxidant as "any substance that, when present at low concentrations compared to those of an oxidizable substrate, significantly delays or prevents oxidation of that substrate" (Halliwell (1994)). This definition therefore not only includes well recognized low molecular weight substances such as ascorbic acid, but also antioxidant enzymes (e.g., catalase, glutathione peroxidase, superoxide dismutase, etc.) and preventative antioxidants such as metal-binding proteins (ferretin, transferrin). This article, however, will only review the electrochemically active low molecular weight scavenging (chain breaking) antioxidants.

Fat-soluble antioxidants

Several different classes of chemicals act as fat-soluble antioxidants including the tocopherols, coenzyme Q_{10}, and the carotenoids/retinoids.

Tocopherols. Vitamin E consists of eight, essential, naturally occurring, lipophilic molecules collectively referred to as the tocopherols.[11] The importance of vitamin E to the human central nervous system and the neurological disorders which result when it is deficient has been reviewed elsewhere (Sokol (1989)). α-Tocopherol has a plasma level of 20-30μM but is particularly abundant in biological membranes where it acts as a chain-breaking antioxidant and is responsible for the prevention of lipid

[11] α-Tocopheryl quinone, the oxidized form of α-tocopherol, is a potent anticoagulant that may be responsible for the reported beneficial effects of α-tocopherol in preventing heart attacks and stroke (Dowd and Zheng (1995)).

peroxidation. α-Tocopherol can intercept a lipid peroxyl radical forming a lipid hydroperoxide and the much less reactive, α-tocopheryl radical. α-Tocopherol can then be regenerated from the α-tocopheryl radical in reactions involving ascorbic acid (Beyer (1994), Chan (1993)), ubiquinone (Beyer (1994)), reduced glutathione (Sies and Murphy (1991)) and/or lipoic acid (Kagan *et al.* (1992)). Recent research has suggested that α-tocopherol and γ-tocopherol may have different biological roles (Christen *et al.* (1997), Moore and Papas (1996)). For example, α-tocopherol is the form which is the more potent antioxidant against the ROS, whereas the γ-form effectively reacts with RNS and is nitrated in the process.

Several HPLC-ECD methods exist for the measurement of the tocopherols either alone or with other lipophilic antioxidants (Table 4). In general, coulometric detection offers superior sensitivity and selectivity over amperometric approaches. Furthermore, only with coulometric detection can both tocopherols and tocopherylquinones be effectively measured electrochemically and overcome the need for chemical manipulation of the sample (also see coenzyme Q_{10} below) (Murphy and Kehrer (1989)). Finckh *et al.* (1995) developed an HPLC-dual coulometric electrode method capable of simultaneously measuring twelve lipophilic analytes including tocopherols, carotenoids, ubiquinone and ubiquinol in neonatal plasma microsamples. This approach was recently extended by the use of gradient HPLC with coulometric array detection, and is capable of measuring both α- and γ-tocopherol as well as carotenoids, retinoids, CoQ_{10} and vitamin K simultaneously (Gamache *et al.* (1997)) (Figure 10). This method has now been expanded to include retinoic acid, the D vitamins and metabolites, and the carotenoids astaxanthin, cryptoxanthin and xeoxanthin (Gamache (1997)).

Coenzyme Q_{10}. Coenzyme Q_{10} (CoQ_{10}; ubiquinone; 2,3-dimethoxy-5- methyl-6-decaprenylbenzo-quinone) is a lipophilic compound with a plasma level of <2.0μM but is found at higher levels in membranes. It is an essential component of the mitochondrial respiratory chain (where it shuttles electrons between NADH-Q reductase and cytochrome reductase) and helps to stabilize and control membrane fluidity. CoQ10 is also thought to play a major role in preventing membrane lipid peroxidation, possibly by the regeneration of the tocopheryl radical (Beyer (1994), Frei *et al.* (1990)). The biological importance of CoQ_{10}, its role in disease and its possible use in the treatment of neurodegenerative disorders has recently been reviewed (Beal (1995)).

HPLC-ECD has been used to measure reduced CoQ_{10} directly (Lang and Packer (1987) or the total (reduced and oxidized) amount following chemical reduction with sodium borohydride (Edlund (1988), Okamoto *et al.* (1988)). Poddo *et al.* (1996) used HPLC-ECD to measure ubiquinols and HPLC-UV for the ubiquinones. Coulometric approaches overcome the need for chemical reduction, and the need for multiple detectors, and permit quantitation of both reduced and oxidized CoQ_{10} simultaneously. The upstream electrode can electrochemically reduce oxidized CoQ_{10} with 100% conversion efficiency leaving the downstream electrode to measure reduced CoQ_{10}. Several researchers have taken advantage of this approach including

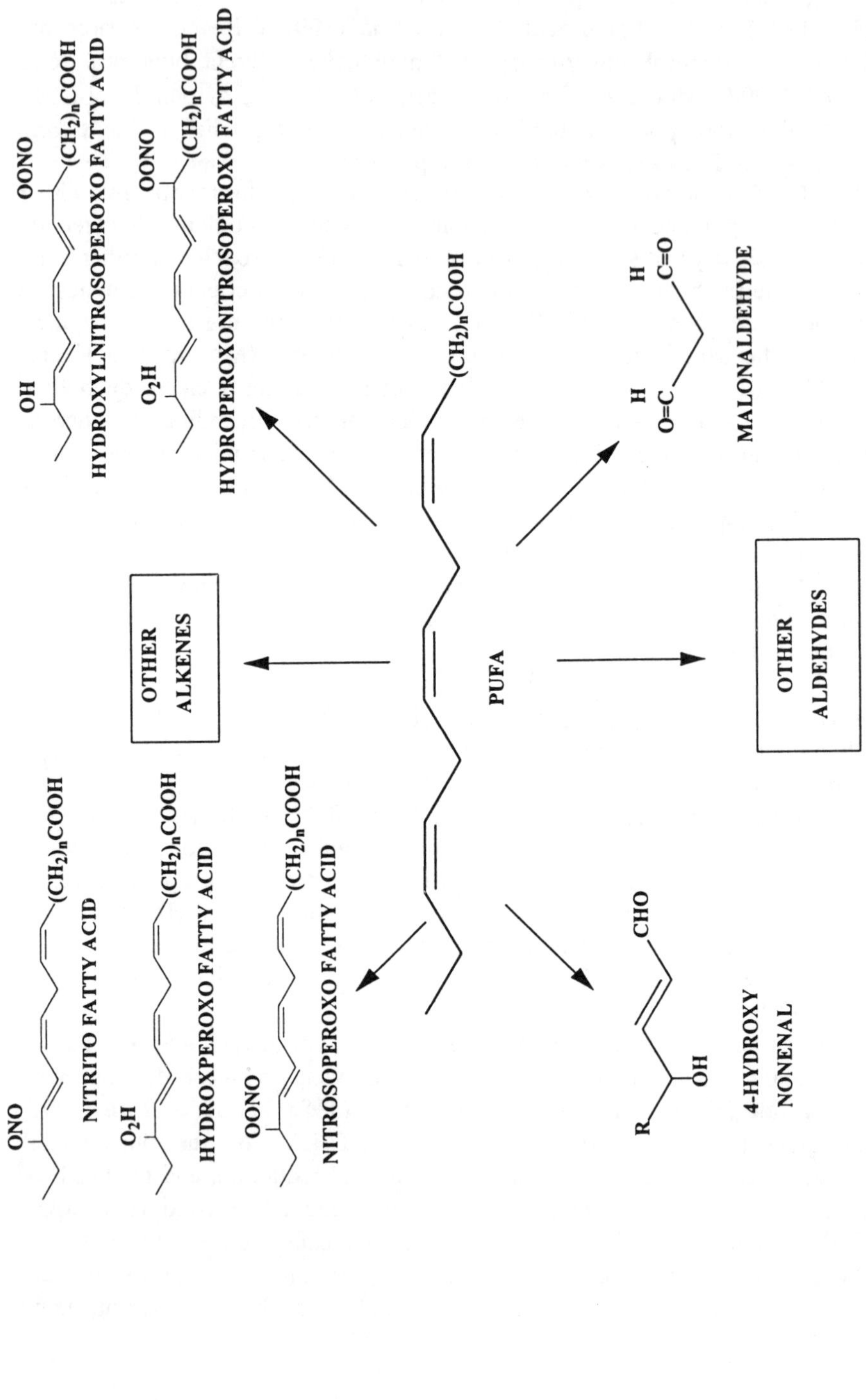

Figure 9. A variety of different potentially toxic products are formed when PUFAs undergo lipid peroxidation.

Finckh *et al.* (1995), Gamache *et al.* (1997), Grossi *et al.* (1992) and Motchnik *et al.* (1994).

Carotenoids and retinoids. The role of carotenoids in health and disease prevention has been reviewed recently (Canfield *et al.* (1993), Mayne (1996), Krinsky (1993)). Although there is considerable evidence that carotenoids act as chain breaking antioxidants, scavengers and quenchers of $^1\Delta g$ O_2 *in vitro*, there is only indirect evidence implicating them as antioxidants *in vivo* (Krinsky (1993), Miller *et al.* (1996)). Retinol (vitamin A), is formed from its major precursor β-carotene in the intestine. Retinol's principle function is photoreception in the retina, however it can also sensitize oxygen forming $^1\Delta g$ O_2. It should come as no surprise that the eye contains a plethora of antioxidants, particularly glutathione, ascorbate, and tocopherol capable of handling $^1\Delta g$ O_2 produced during phototransduction. Although the importance of β-carotene has been known for some time, recent research has concentrated on several more obscure carotenes (lycopene) and xanthophyls (astaxanthin, β-cryptoxanthin, lutein and zeaxanthin) (Miller *et al.* (1996)).

Several HPLC-ECD methods have been used to measure carotenoids (MacCrehan and Schonberger (1987a) and retinoids (Bryan *et al.* (1991), Motchnik *et al.* (1994), Wring and Hart (1989), Wring *et al.* (1988)). Using a gradient HPLC coulometric array method Gamache *et al.* (1997) reported measurement of a variety of carotenoids and retinoids (Figure 10). This method is now being expanded to include retinoic acid, astaxanthin and zeaxanthin.

Water-soluble antioxidants

Several different classes of compounds are used as water-soluble antioxidants including ascorbic acid, thiols/disulfides and uric acid.

Ascorbic acid. Ascorbic acid (vitamin C) is an essential (in man) water soluble compound with circulating levels of 25-100μM and its chemistry has recently been reviewed (Levine *et al.* (1996)). It plays several major metabolic roles such as being a cofactor or the synthesis of collagen (Levine (1986)). It also is important in the central nervous system where it acts as a cofactor for dopamine-β-hydroxylase (responsible for the synthesis of norepinephrine from dopamine) and may act as a neuromodulator (Grunewald (1993)). Along with albumin, glutathione and uric acid, ascorbic acid is a major antioxidant reacting with $O_2^{\bullet-}$, $HO_2^{\bullet}$, $HO^{\bullet}$, $NO^{\bullet}$, $NO_2^{\bullet}$, $RS^{\bullet}$, $^1\Delta g$ O_2 and hypochlorous acid (see Acworth *et al.* (1997a) and references therein). It has been suggested that one of ascorbic acid's principle functions is in lipid peroxidation where it is responsible for the regeneration of α-tocopherol from the α-tocopheryl radical at the cytosol/membrane interface (Beyer (1994), Chan (1993)). Under *in vitro* conditions, ascorbic acid may also act as a pro-oxidant in the presence of iron, but whether this is biologically significant still remains unclear (Halliwell (1996)).

The unstable nature of ascorbic acid makes its analysis difficult and paying special attention to sample preparation and storage is critical. Several analytical approaches have been used to measure ascorbic acid (sometimes in conjunction with one of its

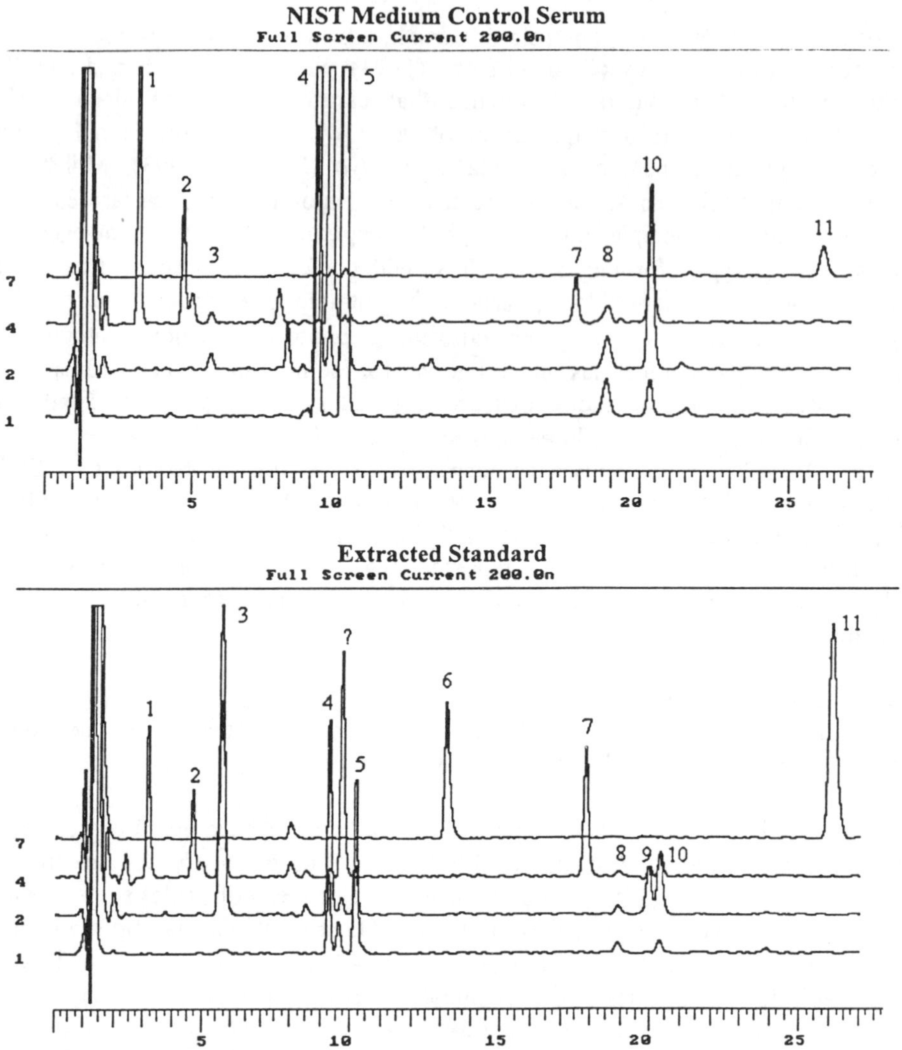

Figure 10. The simultaneous analysis of different fat soluble vitamins using gradient HPLC with coulometric array detection. Only four channels are shown for clarity. Channel 6 is set at -1000mV to reduce compounds prior to their oxidation and measurement on channel 7 (+200mV). This redox approach allows the measurement of vitamin K1 and coenzyme Q_{10}. The upper chromatogram presents the analysis of NIST medium control serum following hexane extraction, evaporation and reconstitution. This chromatogram is presented at a gain of 200nA. The lower chromatogram is for extracted standards (1μg/mL) and is presented at a gain of 200nA. 1- retinol; 2- retinyl acetate; 3- lutein; 4- γ-tocopherol; 5- α-tocopherol; 6- vitamin K1; 7- retinyl palmitate; 8- lycopene; 9- α-carotene; 10- β-Carotene; 11- coenzyme Q10. See Gamache *et al.* (1997) for further details. Reproduced with permission of VS Press, The Netherlands.

metabolites, dehydroascorbic acid) and these have been critically reviewed. (Lykkesfeldt *et al.* (1995), Washko *et al.* (1992)). These authors examined the use of HPLC-ECD in detail and concluded that coulometric detection was more reliable and offered marked improvement in sensitivity over amperometric approaches. Several groups have used HPLC-coulometric detection to measure ascorbate either alone (Lykkesfeldt *et al.* (1995), Schell and Bode (1993), Xu and Wells (1996)) or with other antioxidants (e.g., reduced glutathione, oxidized glutathione, uric acid) simultaneously (Rose and Bode (1995), Sofic *et al.* (1991)).

Thiols and disulfides. Thiols and their respective disulfides act as redox couples helping to keep biological tissues in a reducing environment:

$$2RSH + OXIDANT \rightarrow RSSR + REDUCED\text{-}OXIDANT$$

Several thiol/disulfide couples exist including glutathione (GSH)/oxidized glutathione (GSSG), cysteine/cystine, homocysteine/homocystine and dihydrolipoic acid/lipoic acid. The relative importance of each of these couples is very much location specific. For example, in plasma the total thiol concentration is 400-500μM primarily resulting from plasma proteins, mainly albumin, with only a little contribution from GSH (5μM); whereas the intracellular thiol concentration is derived primarily from GSH (2mM in erythrocytes).

The tripeptide GSH (γ-glutamyl-cysteinyl-glycine) has many important biological activities (see Acworth *et al.* (1997a) for greater details). It readily acts as an antioxidant directly reducing a variety of ROS (e.g., $HO^{\bullet}$, $RO^{\bullet}$, $RO_2^{\bullet}$) forming a thiyl free radical which can undergo numerous reactions (D'Aquino *et al.* (1994)). GSH is also the cofactor for glutathione peroxidase which helps to regulate ROS activity by reducing H_2O_2 and organic peroxides into water and alcohols, respectively. GSH is also thought to play a role in regenerating α-tocopherol from its radical either directly or through the regeneration of ascorbic acid from its radical (Packer (1996c), Sies and Murphy (1991)). GSH is an essential biochemical for maintaining a functional redox status within the mitochondria. Mitochondrial GSH, which represents only 10% of the total cellular levels, totally relies on an effective transport system to keep these levels constant due to the lack of synthetic capabilities within the mitochondria. When neurons, which contain lower GSH levels than other tissues, experience a high peroxidative situation they become very susceptible to lower mitochondrial GSH levels. This effects the energy output and produces dysfunctional mitochondria which can lead to a vicious cycle of cellular degradation (Benzi and Moretti (1995)). It is currently thought that mitochondrial dysfunction may be a major deleterious mechanism related to several neurodegenerative diseases (Beal (1995), Benzi and Moretti (1995), Floyd (1996)).

Several HPLC-ECD methods have been developed to measure GSH and GSSG and these usually either use a mercury working electrode (either as a mercury drop electrode or gold amalgam) or coulometric electrode (see Table 5). In general, methods using coulometric electrodes offer good sensitivity and reliability without having to use expensive and toxic polarographic electrodes or unstable, noisy, short-lived and toxic gold amalgam working electrodes.

Another very interesting thiol antioxidant is α-lipoic acid (1,2-dithiolane-3-valeric acid) which has been known for decades to function as a cofactor (lipoamide) in many dehydrogenase complexes. Unlike GSH which has only one thiol group, α-lipoic acid contains two thiols that can form an intramolecular disulfide bond when oxidized. The redox potential of dihydrolipoic acid/α-lipoic acid is -0.32V (compared to -0.24V for GSH/GSSG) making this a strongly reducing couple (Packer (1996c)). Apart from being an antioxidant in its own right reacting with many ROS and RNS (Acworth *et al.* (1997b), Packer (1996c), Scott *et al.* (1994)), α-lipoic acid is also thought to be intimately involved in the recycling of vitamin E during prevention of lipid peroxidation (Kagan *et al.* (1992), Packer (1996c), Packer *et al.* (1997)). Exogenous dihydrolipoic acid/α-lipoic acid may be useful in treating oxidative stress conditions and may play a role in preventing brain disorders (e.g., it protects against cerebral-ischemia reperfusion injury and reduces the extent of *N*-methyl-D-aspartate-induced and malonic acid-induced striatal lesions) (Packer (1996c) and references therein)). α-Lipoic acid has been measured using HPLC-ECD (Han *et al.* (1995), Kamata and Akiyama (1990), Teichert and Preiss (1992, 1995)).

Homocysteine, its free oxidized conjugated forms (homocystine, homocysteine-cysteine mixed disulfides) and its protein bound oxidized form occur in plasma at a total concentration of ~10μM and are thought to be important contributors to the plasma redox thiol status (Ueland (1995)). A variety of HPLC-based methods have been used to measure plasma homocysteine levels which can be used clinically as a marker for folate and vitamin B12 status (Jacobsen *et al.* (1994)). Free homocysteine levels have been measured using HPLC-coulometric electrochemical detection (Acworth and Bailey (1995)), while total homocysteine has been determined following sodium borohydride reduction (Malinow *et al.* (1989)).

Uric acid. Uric acid is synthesized from hypoxanthine by xanthine oxidase with $O_2^{\bullet-}$ being produced as a byproduct. In man uric acid is considered to be the final product of purine metabolism as humans lack the necessary enzymes to metabolize it further. Some regard uric acid to be a waste product even though destruction of purines is physiologically costly and >90% of uric acid is reabsorbed from the urine resulting in plasma levels of 200-450μM, close to the solubility limit of this metabolite. Others have suggested that uric acid should be considered as an important circulating antioxidant. Indeed, Ames *et al.* (1981) have suggested that an important step in human evolution was the increased plasma concentration of uric acid which, through its antioxidant capacity, enabled man to live longer and avoid the cancers commonly found in short-lived species.

Uric acid has a variety of important antioxidant roles including its ability to react with $\Delta g\ O_2$, HOCl and $RO_2^{\bullet}$, to bind iron and copper in forms that do not react with H_2O_2, and to protect against damage from heme intermediates containing iron in +4 and +5 valencies. The fact that uric acid can react with $HO^{\bullet}$ or peroxyl radicals to produce potentially damaging urate radicals led Benzie and Strain (1996) to question uric acid's true biological importance. Fortunately, the reactive uric acid radical can be destroyed by its ready reaction with ascorbic acid.

Uric acid can be measured using HPLC-UV absorbance (Ames *et al.* (1981), Tang-Liu and Riegelman (1982)) or with better selectivity and sensitivity, by HPLC-ECD. HPLC-ECD approaches either measure it alone (Aoki *et al.* (1984), Iwamoto *et al.* (1983), Roch-Ramel *et al.* (1980)) or in conjunction with ascorbate (Honegger *et al.* (1989), Irayama *et al.* (1984), Shirachi and Omaye (1992)) or other metabolites (Rose and Bode (1995)). Uric acid has also been determined as part of a global method for metabolic profiling using gradient HPLC with coulometric array detection (Rizzo *et al.* (1991)).

Although uric acid is regarded as the final enzymatically-formed metabolite in man, uric acid can react with ROS to produce a variety of potentially toxic products including allantoin, oxonic acid, parabanic acid, cyanuric acid, oxaluric acid and urea. Possible reaction mechanisms of uric acid breakdown have been studied using coulometric electrochemical detection coupled to thermospray-tandem mass spectrometry (Volk *et* al. (1989). These analytes have also been measured using HPLC-UV (Hicks *et al.* (1993), Kaur and Halliwell (1990)) and HPLC-ECD (Rose and Bode (1995)). Interestingly, Hillered and Persson (1995) recently measured parabanic acid as a indicator of oxidative stress in microdialysis samples obtained from patients with severe acute brain injuries.

NEURODEGENERATIVE DISEASES

There is considerable interest in the role that oxidative stress may play in the development and progression of neurodegenerative (Alzheimer's, Parkinson's, Huntington's, amyotrophic lateral sclerosis and cerebellar degeneration) diseases, other diseases (e.g., epilepsy) and aging. This topic has been the subject of several excellent books (Beal (1995), Packer *et al.* (1996)) and reviews (Beal (1997), Fahn and Cohen (1992), Gerlach *et al.* (1994), Kontos (1989), LeBel and Bondy (1991, 1992), Smith *et* al. (1996), Youdim (1988), Youdim and Riederer (1997)). As presented in Table 6, the brain is in an unfortunate situation. It has a great demand for oxygen and is a highly energetic tissue, yet it only possesses weak antioxidant defenses and contains many potentially toxic chemicals that can auto-oxidize, redox cycle and produce ROS/RNS. The remainder of this article examines the formation of a variety of neurotoxic metabolites formed from oxidation of the monoamines.

Monoamines, autoxidation and possible neurotoxins

The monoamines, which include the catecholamines (dopamine, norepinephrine) and the indolealkylamine (serotonin), are important neurotransmitters in the brain. Their synthesis (e.g., tyrosine hydroxylase, tryptophan hydroxylase, dopamine-β-hydroxylase) and catabolism (monoamine oxidase) require molecular oxygen. Most monoamines, their precursors and metabolites are electrochemically active and many HPLC-ECD methods have been developed for their determination (see Acworth *et al.* (1997b), Downer *et al.* (1985), Naoi *et al* (1993)).

The catecholamines. The synthesis of the catecholamines is oxygen dependent. The initial rate-limiting step in catecholamine synthesis is the conversion of tyrosine into dopamine in a reaction controlled by tyrosine hydroxylase which requires molecular oxygen and tetrahydrobiopterin as cofactors. This reaction is tightly coupled to the reduction of oxygen to water, without production of ROS. Tyrosine hydroxylase is an iron-dependent enzyme which could theoretically produce $HO^{\bullet}$ if its iron atom was to become liberated and it encounters H_2O_2, but the physiological relevance of the resulting localized damage is unclear. The major catabolic route for catecholamines is through monoamine oxidase which produces the catecholamine aldehyde (a reactive intermediate which can form potentially toxic metabolites when it forms a Schiff base with a variety of cellular amines (e.g., reaction with dopamine produces the potentially harmful salsolinol)) and H_2O_2.

Catecholamines are notoriously unstable *in vitro* and readily auto-oxidize under neutral/basic conditions to form aminochromes (also called adrenochromes). *In vivo,* catecholamines can be chemically reactive (when not protected by the antioxidant environment of storage vesicles) producing a variety of ROS as well as many toxic adducts which may be involved in neurodegeneration. Catecholamine molecules which are not repackaged in vesicles for future release have several metabolic fates (see Figure 11). They can undergo oxidative deamination by monoamine oxidase which produces H_2O_2 as a byproduct. Catecholamine molecules can also undergo non-enzymatic autoxidation leading to the formation of reactive semiquinone and quinone intermediates and ROS (Bindoli *et al.* (1992)).[12] These reactive intermediates can then undergo intramolecular cyclization to form cytotoxic aminochromes which can polymerize to form neuromelanin which is capable of reducing non-heme Fe^{3+} to Fe^{2+} (Bindoli *et al.* (1989), Graham *et al.* (1978, 1989)) or react with a variety of nucleophiles present in brain tissue producing catecholamine adducts (see below) or, most likely, undergo reduction (e.g., by ascorbic acid) back to the original catecholamine (Figure 11).

The most abundant source of cellular nucleophiles are the sulfhydryls which can form stable adducts with both the catecholamines (via the quinone intermediate) and aminochromes (Bindoli *et al.* (1992)). These adducts are toxic, can inhibit the activity of several enzymes and interfere with mitochondrial energy processes, leading some researchers to speculate that they may contribute to the neurodegeneration seen in Parkinson's disease (Bindoli *et al.* (1992)). Cysteine, which is present in cells as the free amino acid and as part of GSH and proteins, readily reacts with catecholamine-quinone to produce the 5-*S*-cysteinyl-catecholamine adduct. Fornstedt *et al.* (1990) used HPLC-ECD to measure the 5-*S*-cysteinyl-adducts of dopamine, L-DOPA, and DOPAC in dopamine-enriched brain areas of man and other mammals and warned about artifact formation during sample processing. Hastings used HPLC-ECD to investigate free and protein bound dopamine-cysteinyl adducts formed when exogenous dopamine was administered to either brain slices (Hastings and Zigmond (1994)) or intrastriatally (Hastings *et al.* (1996)). Hastings showed that dopaquinone formed from the oxidation of dopamine binds to proteins in the striatum, an event

[12] The neuronal damage produced by administration of the neurotoxin 6-hydroxydopamine is probably caused by the ROS produced when this drug undergoes redox cycling (Sachs and Johnson (1975)).

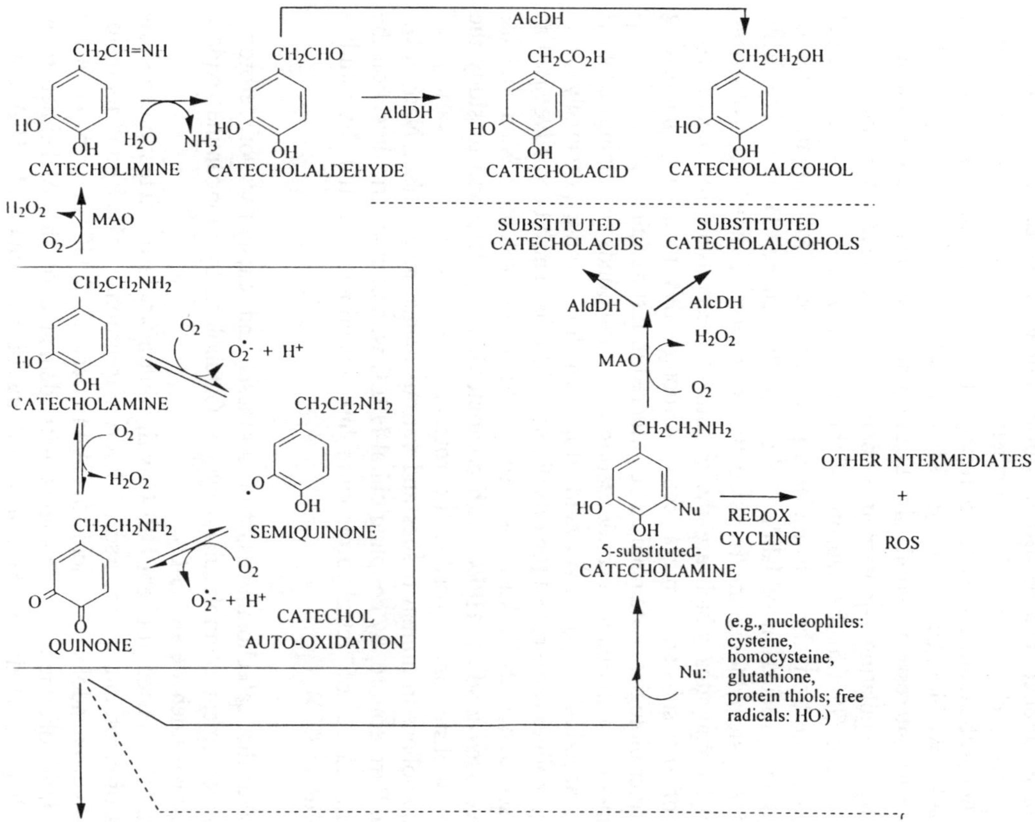

Figure 11. From catabolism, to redox cycling, to the production of possible neurotoxic intermediates, the chemistry of the catecholamines is intimately associated with the ROS.

which correlated to specific loss of dopaminergic terminals, and concluded that the selective degeneration of dopamine neurons in Parkinson's disease may be caused by an imbalance between the oxidation of dopamine and the availability of antioxidant defenses. Dryhurst used cyclic voltammetry and HPLC-UV to study the mechanisms of interaction between cysteine and dopamine (Shen and Dryhurst (1996a), Zhang and Dryhurst (1994)) or norepinephrine (Shen and Dryhurst (1996b)). Dryhurst reported that many cysteinyl adducts can be formed during dopamine oxidation which divert dopamine away from melanin formation. Furthermore, some adducts, especially those based on the dihydrobenzothiazines are extremely toxic. Combining his mechanistic studies with the known biochemical changes seen in Parkinson's disease, Dryhurst has put forward a very interesting hypothesis: 1) the brain responds to some unknown chronic insult by upregulating γ-glutamyltranspeptidase which leads to an increase in the level of cytosolic GSH in the neurons of the substantia nigra. 2) Elevated GSH levels divert dopamine from the neuromelanin pathway leading to depigmentation of these cells and the production of labile 5-*S*-cysteinyldopamine. 3) The very facile oxidation of 5-*S*-cysteinyldopamine leads to the formation of dihydrobenzothiazine and its cysteinyl conjugates which are the endotoxins responsible for the selective degeneration of the substantia nigra cells. Although this is a very interesting hypothesis, further biological data need to be gathered to test its validity.

Catecholamines can also undergo condensation with compounds containing carbonyl moieties (e.g., acetaldehyde, pyruvic acid, dopaldehyde) forming potentially toxic tetrahydroisoquinolines (also called the salsolinols) which have also been advocated as endotoxins associated with Parkinson's disease (Naoi, this volume). Naoi and his colleagues have used gradient HPLC with coulometric array detection to study the formation, metabolism and possible neurotoxic mechanisms of a variety of tetrahydroisoquinolines in human tissues and various animal models (see Naoi *et al.* (1993, 1997)). Furthermore, by coupling chiral-HPLC with coulometric detection the biological significance of tetrahydroisoquinoline enantiomers could be studies (Maruyama *et al.* (1997)).

Serotonin. Like the catecholamines, the synthesis and catabolism of serotonin requires the involvement of molecular oxygen. Catabolism by monoamine oxidase produces a reactive aldehyde and H_2O_2.

Volicer *et al.* (1985) used gradient HPLC with coulometric array detection to show that the levels of serotonin, it precursor, 5-hydroxytryptophan (5HTP) and its major metabolite, 5-hydroxyindoleacetic acid (5HIAA), were all significantly lower in the cerebrospinal fluid obtained from patients with dementia of the Alzheimer type. Interestingly, they reported that the concentrations of serotonin and 5HIAA were lower when measured with the coulometric array than with conventional amperometric detection, due to the coelution of compounds that could not be electrochemically resolved by the amperometric electrode. Furthermore, the voltammetric behavior of these contaminants across the coulometric array suggested that they were partially oxidized forms of serotonin and 5HTP. Volicer went on to identify one of the contaminants as the neurotoxic compound, tryptamine-4,5-dione (Chen *et al.* (1992), Fishman *et al.* (1991)). Wong *et al.* (1993) have suggested that tryptamine-4,5-dione is likely to be incredibly short-lived *in vivo*, as it is an electron deficient molecule and will readily react with nucleophiles such as thiols (cysteine, GSH etc.). They suggest that 7-*S*-glutathionyl-tryptamine-4,5-dione (or one of its conjugates) is more likely to be an aberrant metabolite of serotonin in Alzheimer's disease and induces toxicity by producing ROS when it undergoes redox cycling. No biological evidence for the existence of that 7-*S*-glutathionyl-tryptamine-4,5-dione was given, possibly due to the insensitivity of their HPLC-UV approach. Wrona *et al.* (1992) have suggested that another oxidized form of serotonin, 5,5'-dihydroxy-4,4'-bitryptamine, might also play an important role in neurodegenerative diseases, as it acts as a neurotoxin when administered centrally to mice.

Serotonin and other indoleamines can also undergo condensation with compounds containing carbonyl moieties (e.g., acetaldehyde, pyruvic acid) forming β-carbolines, Musshoff *et al.* (1996)). These compounds (which are structurally similar to the neurotoxin and Parkinsonmimetic, 1-methyl-4-phenyl-1,2,3,6-tetrahydropyridine), can be activated in the brain by *N*-methylation producing toxic derivatives. These *N*-methylated analogs are potent mitochondrial inhibitors and the resulting oxidative stress may promote neurodegeneration (Collins *et al.* (1992)).

CONCLUSIONS

Living in an oxygen environment is not without ramifications. Living organisms have developed a variety of antioxidant mechanisms in an attempt to control the consequences of molecular damage resulting from a variety of hostile ROS/RNS. The oxidative stress placed upon the organism when these defenses fail may eventually lead to a variety of diseases. The human brain is especially sensitive to oxidative stress which has been implicated to underlie a variety of diseases including Alzheimer's and Parkinson's disease. Although we may not know the exact biochemical nature of the initial insults in Parkinson's disease, evidence is accumulating that oxidative metabolism plays an essential role during the degenerative phase of the disease process. HPLC-ECD is a valuable tool which can be used to study the many aspects of oxidative stress including the generation of ROS/RNS, their reaction products, the antioxidant defense systems, production of endogenous toxins and the effects of disease progression on brain neurochemistry.

REFERENCES

Achilli, G., Cellerino, G. P. and Melzi d'Eril, G. V. (1994). Determination of amines in wine by high-performance liquid chromatography with electrochemical coulometric detection after precolumn derivatization. *J. Chromatogr.* **661**, 201-205.

Acworth, I. N. (1996). Catecholestrogens. *Application Note #70-2217. ESA, Inc.*

Acworth, I. N. and Bailey, B. A. (1995). *The Handbook of Oxidative Metabolism.* ESA, Inc., Chelmsford, MA.

Acworth, I. N. and Bowers, M. (1997). An introduction to HPLC-based electrochemical detection: From single electrode to multi-electrode arrays. In: *Coulometric Electrode Array Detectors for HPLC.* Progress in HPLC-HPCE (6), I. N. Acworth, M. Naoi, S. Parvez and H. Parvez (Eds.) VSP Publications, The Netherlands, pp. 1-48.

Acworth, I. N. and Gamache, P. H. (1996). The coulometric electrode array for use in HPLC analysis. *Am. Lab.* **5**, 33-37.

Acworth, I. N., McCabe, D. R. and Maher, T. J. (1997a). The analysis of free radicals, their reaction products and antioxidants. In: *Antioxidants, Oxidants and Free Radicals*, S. Baskin and H. Salem (Eds.). Taylor and Francis, Washington DC.

Acworth, I. N., Naoi, M., Parvez, S. and Parvez, H. (Eds). (1997b). *Coulometric Electrode Array Detectors for HPLC.* Progress in HPLC-HPCE (6), VSP Publications, The Netherlands.

Adachi, S., Takemoto, K., Hirosue, T., and Hosogai, Y. (1993). Spontaneous and 2-nitropropane induced levels of 8-hydroxy-2'-deoxyguanosine in liver DNA of rats fed iron-deficient or manganese- and copper-deficient diets. *Carcinogen.* **14**, 265-268.

Adachi, S., Zeisig, M. and Moller, L. (1995). Improvements in the analytical method for 8-hydroxydeoxyguanosine in nuclear DNA. *Carcinogen.* **16**, 252-258.

Allison, L. A., and Shoup, R. E. (1983). Dual electrode liquid chromatography detector for thiols and disulfides. *Anal. Chem.* **55**, 8-12.

Alvarez, B., Denicola, A. and Radi, R. (1995). Reaction between peroxynitrite and hydrogen peroxide: Formation of oxygen and slowing of peroxynitrite decomposition. *Chem. Res. Toxicl.* **8**, 859-864.

Ames, B. N., Cathcart, R., Schwiers, E. and Hochstein, P. (1981). Uric acid provides an antioxidant defense in humans against oxidant- and radical-caused aging and cancer: A hypothesis. *Proc. Natl. Acad. Sci.* **78**, 6858-6862.

Antunes, F., Salvador, A., Marhino, H. S., Alves, R. and Pinto, R. E. (1996). Lipid peroxidation in mitochondrial membranes. 1. An integrative kinetic model. *Free Radic. Biol. Med.* **21**, 917-943.

Aoki, T., Yoshiura, M., Iwamoto, T. and Iriyama, K. (1984). Postmortem changes of uric acid in various rat tissues: Determination of uric acid by reversed-phase high-performance liquid chromatography with electrochemical detection. *Anal. Biochem.* **143**, 113-118.

Arai, H., Terao, J., Abdalla, D. S. P., Suzuki, T. and Takama, K. (1996). Coulometric detection in high-performance liquid chromatographic analysis of cholesteryl ester hydroperoxides. *Free Radic. Biol. Med.* **20**: 365-371.

Ayene, S. I., Al-Mehdi, A. B. and Fisher, A. B. (1993). Inhibition of lung tissue oxidation during ischemia/reperfusion by 2-mercaptopropionylglycine. *Arch. Biochem. Biophys.* **303**, 307-312.

Baur, J. E., Wang, S., and Brandt, M. C. (1996). Fast-scan voltammetry of cyclic nitroxide free radicals. *Anal. Chem.* **68**, 3815-3821.

Beal, M. F. (1995). *Mitochondrial dysfunction and oxidative damage in neurodegenerative diseases.* Springer-Verlag, Germany.

Beal, M. F. (1997). Oxidative damage in neurodegenerative diseases. *The Neuroscientist* **3**, 21-27.

Beal, M. F., Ferrante, R.J., Henshaw, R., Matthews, R. T., Chan, P. K., Kowall, N. W., Epstein, C. J. and Schulz, J. B. (1995). 3-Nitropropionic acid neurotoxicity is attenuated in copper/zinc superoxide dismutase transgenic mice. *J. Neurochem.* **65**, 919-922.

Beal, M. F., Ferrante, R. J., Matthews, R. T. Kowall, N. W. and Brown, R. H. (1996). 3-Nitrotyrosine in ALS. *Soc. Neurosci. Abstr.* **22**, 1943.

Beal, M. F., Matson, W. R., Swartz, K. J., Gamache, P. H. and Bird, E. D. (1990). Kynurenine pathway measurements in Huntington's disease striatum: Evidence for reduced formation of kynurenic acid. *J. Neurochem.* **55**, 1327-1339.

Beckman, J. S., Chen, J., Ischiropoulos, H. and Crow, J. P. (1994). Oxidative chemistry of peroxynitrite. *Meth. Enzymol.* **233**, 229-240.

Beckman, J. S., Ye, Y. Z., Anderson, P. G., Chen, J., Accavitti, M. A., Tarpey, M. M. and White, C. R. (1994). Extensive nitration of protein tyrosines in human atherosclerosis detected by immunohistochemistry. *Biol. Chem. Hoppe-Seyler* **375**, 81-88.

Beehler, B. C., Przybyszewski, Box, H. B. and Kulesz-Martin, M. F. (1992). Formation of 8-hydroxyguanosine within DNA of mouse keratinocytes exposed in culture to UVB and H_2O_2. *Carcinogen.* **13**, 2003-2007.

Benzi, G. and Moretti, A. (1995). Age- and peroxidative stress-related modifications of the cerebral enzymatic activities linked to mitochondria and the glutathione system. *Free Radic. Biol. Med.* **19**, 77-101.

Benzie, I. F. F. and Strain, J. J. (1996). Uric acid: Friend or foe? *Redox Report* **2**, 231-234.

Beyer, R. E. (1994). The role of ascorbate in antioxidant protection of biomembranes: Interaction with vitamin E and coenzyme Q. *J. Bioenerg. Biomemb.* **26**, 349-358.

Bindoli, A., Rigobello, M. P. and Deeble, D. J. (1992). Biochemical and toxicological properties of the oxidation products of catecholamines. *Free Radic. Biol. Med.* **13**, 391-405.

Bindoli, A., Rigobello, M. P. and Galzinga, L. (1989). Toxicity of aminochromes. *Toxicol. Lett.* **48**, 3-20.

Breen, A. P. and Murphy, J. A. (1995). Reactions of the oxyl radicals with DNA. *Free Radic. Biol. Med.* **18**, 1033-1077.

Brown, R. K., McBurney, L., Lunec, J. and Kelly, F. J. (1995). Oxidative damage to DNA in patients with cystic fibrosis. *Free Radic. Biol. Med.* **18**, 801-806.

Bryan, P. D., Honigberg, I. L. and Meltzer, N. M. (1991). Electrochemical detection of retinoids using normal phase HPLC. *J. Liq. Chromatogr.* **14**, 2287-2295.

Candeias, L. P., Patel, K. B. Stratford, M. R. L. and Wardman, P. (1993). Free hydroxyl radicals are formed from the reaction between the neutrophil-derived species, superoxide anion and hypochlorous acid. *FEBS Letts.* **333**, 151-153.

Canfield, L. M., Krinsky, N. I. and Olson, J. A. (Eds.). *Carotenoids in Human Health*. New York Acad. Sci. **691**.

Carro-Ciampi, G., Hunt, P. G., Turner, C. J. and Wells, P. G. (1988). A high-performance liquid chromatographic assay for reduced and oxidized glutathione in embryonic, neonatal, and adult tissue using a porous graphite electrochemical detector. *J. Pharmacol. Meth.* **19**, 75-83.

Castle, M. C. and Cook, W. J. (1985). Measurement of vitamin E in serum and plasma by high-performance liquid chromatography and electrochemical detection. *Ther. Drug Monit.* 7, 364-368.

Chan, A. C. (1993). Partners in defense, vitamin E and vitamin C. *Can. J. Physiol. Pharmacol.* **71**, 725-731.

Chen, J-C., Schnepper, P. W., To, A. and Volicer, L. (1992). Neurochemical changes in the rat brain after intraventricular administration of tryptamine-4,5-dione. *Neuropharmacol.* **31**, 215-219.

Chiavari, G. and Bergamini, C. (1985). High-performance liquid chromatography of carbonyl compounds as 2,4-dinitrophenylhydrazones with electrochemical detection. *J. Chromatogr.* **318**, 427-432.

Christen, S., Woodall, A. A., Shigenaga, M. K., Southwell-Keely, P. T., Duncan, M. W. and Ames, B. N. (1997). γ-Tocopherol traps mutagenic electrophiles such as NO_x and complements α-tocopherol: Physiological implications. *Proc. Natl. Acad. Sci.* **94**, 3217-3222.

Claycamp, H. G. and Ho, K.-K. (1993). Background and radiation-induced 8-hydroxy-2'deoxyguanosine in γ-irradiated *Escherichia coli*. *Int. J. Radiat. Biol.* **63**, 597-607.

Collins, M. A., Neafsey, E. J., Matsubara, K., Cobuzzi, R., Albores, R., Fields, J. and Rollema, H. (1992). Indole-*N*-methylation of β-carbolines: The brain's bioactivation route to toxins in Parkinson's disease. New York, Acad. Sci., **648**, 263-265.

Conoway, C. C., Nie, G., Hussain, N. S. and Fiala, E. S. (1991). Comparison of oxidative damage to rat liver DNA and RNA by primary nitroalkanes, secondary nitroalkanes, cyclopentanone oxime, and related compounds. *Cancer Res.* **51**, 3143-3147.

Cordis, G. A., Bagchi, D., Maulik, N. and Das, D. K. (1994). High-performance liquid chromatographic method for the simultaneous detection of malonaldeyde, acetaldehyde, formaldehyde, acetone and propionaldehyde to monitor the oxidative stress in heart. *J. Chromatogr. A.* **661**, 181-191.

Crow, J. P. and Beckman, J. S. (1995). Reactions between nitric oxide, superoxide and peroxynitrite: Footprints of peroxynitrite *in vivo*. *Adv. Pharmacol.* **34**, 17-43.

Crow, J. P. and Ischiropoulos, H. (1996). Detection and quantitation of nitrotyrosine residues in proteins: *In vivo* marker of peroxynitrite. *Meth. Enzymol.* **269**, 185-194.

Culp, S. J., Cho, B. P., Kadlubar, F. F. and Evans, F. E. (1989). Structural and conformational analyses of 8-hydroxy-2'deoxyguanosine. *Chem Res. Toxicol.* **2**, 416-422.

D'Aquino, M., Bullion, C., Chopra, M., Devi, D., Devi, S., Dunster, C., James, G., Komuro, E., Kundu, S., Niki, E., Raza, F., Robertson, F., Sharma, J. and Willson, R. (1994). Sulfhydryl free radical formation enzymatically by sonolysis, by radiolysis, and thermally: Vitamin A, curcumin, muconic acid, and related conjugated olefins as references. *Meth. Enzymol.* **233**, 34-46.

Dawson, T. M. and Dawson, V. L. (1994). Nitric oxide: Actions and pathological roles. *Neuroscientist*, **Preview Issue**, 9-19.

Dawson, V. L. and Dawson, T. M. (1996). Pathologic roles of nitric oxide in the central nervous system. In: *Free Radicals in Brain Physiology and Disorders*. L. Packer, M. Hiramatsu and T. Yoshikawa (Eds.) Academic Press, New York, pp. 83-100.

Degan, P., Shigenaga, M., Park, E.M., Alperin, P. E. and Ames, B. N. (1991). Immunoaffinity isolation of urinary 8-hydroxy-2'-deoxyguanosine and 8-hydroxyguanine and quantitation of 8-hydroxy-2'-deoxyguanosine in DNA by polyclonal antibodies. *Carcinogen.* **12**, 865-871.

Denda, A., Tang, Q., Endoh, T., Tsujiuchi, T., Horiguchi, K., Noguchi, O., Mizumoto, Y., Nakae, D. and Konishi, Y. (1994). Prevention by acetylsalicylic acid of liver cirrhosis and carcinogenesis as well as generations of 8-hydroxydeoxyguanosine and thiobarbituric acid-reactive substances caused by a choline-deficient, L-amino acid-defined diet in rats. *Carcinogen.* **15**, 1279-1283.

d'Ischia, M. and Costantini, C. (1995). Nitric oxide-induced nitration of catecholamine neurotransmitters: A key to neuronal degeneration? *Bioorg. Med. Chem.* **3**, 923-927.

Douki, T. and Ames, B. N. (1994). An HPLC-ECD assay for 1,N^2-propano adducts of 2'deoxyguanosine with 4-hydroxynonenal and other α,β-unsaturated aldehydes. *Chem. Res. Toxicol.* **7**, 511-518.

Douki, T., Cadet, J. and Ames, B. N. (1996). An adduct between peroxynitrite and 2'-deoxyguanosine: 4,5-dihydro-5-hydroxy-4-(nitrosooxy)2'-deoxyguanosine. *Chem. Res. Toxicol.* **9**, 3-7.

Dowd, P. and Zheng, Z. B. (1995). On the mechanism of the anticlotting action of vitamin E quinone. *Proc. Natl. Acad. Sci.* **92**, 8171-8175.

Downer, R. G. H., Bailey, B. A. and Martin, R. J. (1985). Estimation of biogenic amines in biological tissues. In: *Neurobiology* R. Gilles and J. Balthazart (Eds.) Springer-Verlag, Berlin, pp. 248-263.

Dugan, L. A., Lin, T-S., He, Y. -Y., Hsu, C. Y. and Choi, D. W. (1995). Detection of free radicals by microdialysis/spin trapping EPR following focal cerebral ischemia-reperfusion and a cautionary note on the stability of 5,5-dimethyl-1-pyrroline N-oxide (DMPO). *Free Rad. Res.* **23**, 27-32.

During, M. J., Acworth, I. N. and Wurtman, R.J. (1988). Phenylalanine administration influences dopamine release in the rat's corpus striatum. *Neurosci. Lett.* **93**, 91-95.

Edlund, P. O. (1988). Determination of coenzyme Q10, α-tocopherol and cholesterol in biological samples by coupled-column liquid chromatography with coulometric and ultraviolet detection. *J. Chromatogr.* **425**, 87-97.

Eiserich, J. P., Butler, J., Van Der Vleit, A., Cross, C. E., and Halliwell, B. (1995). Nitric oxide rapidly scavenges tyrosine and tryptophan radicals. *Biochem. J.* **30**, 745-749.

Ekstrom, T., Garberg, P., Egestad, B. and Hogberg, J. (1988). Recovery of malondialdehyde in urine as a 2,4-dinitrophenylhydrazine derivative analyzed with high-performance liquid chromatography. *Chem. Biol. Interact.* **66**, 177-187.

Esierch, J. P., Cross, C. E., Jones, A. D., Halliwell, B. and van der Vliet, A. (1996). Formation of nitrating and chlorinating species by reaction of nitrite with hypochlorous acid. A novel mechanism for nitric oxide-mediated protein modification. *J. Biol. Chem.* **271**, 19199-19208.

Esterbauer, H., Schaur, R. J. and Zollner, H. (1991). Chemistry and biochemistry of 4-hydroxynonenal, malondialdehyde and related aldehydes. *Free Radic. Biol. Med.* **11**, 81-128.

Fahn, S. and Cohen, G. (1992). The oxidant stress hypothesis in Parkinson's disease: Evidence supporting it. *Ann. Neurol.* **32**, 804-812.

Faux, S. P., Francis, J. E., Smith, A. G. and Chipman, J. K. (1992). Induction of 8-hydroxydeoxyguanosine in *Ah*-responsive mouse liver by iron and Aroclor 1254. *Carcinogen.* **13**, 247-250.

Feelisch, M. and Stamler, J. S. (Eds.) (1996). *Methods in Nitric Oxide Research*. Wiley, New York.

Feldman, P. L., Griffith. O. W., and Stuehr, D. J. (1993). The surprising life of NO. *C and EN.* December, 26-38.

Finckh, B., Kontush, A., Commenz, J., Hubner, C., Burdelski, M. and Kohlschutter, A. (1995). Monitoring ubiquinol-10, ubiquinone-10, carotenoids, and tocopherols in neonatal plasma microsamples using high-performance liquid chromatography with coulometric electrochemical detection. *Anal. Biochem.* **232**, 210-216.

Fischer-Nielsen, A., Jeding, I. B. and Loft, S. (1994). Radiation-induced formation of 8-hydroxy-2'-deoxyguanosine and its prevention by scavengers. *Carcinogen.* **15**, 1609-1612.

Fischer-Nielsen, A., Poulsen, H. E. and Loft, S. (1992). 8-Hydroxydeoxyguanosine *in vitro*: Effects of glutathione, ascorbate and 5-aminosalicylic acid. *Free Radic. Biol. Med.* **13**, 121-126.

Fishman, J. B., Rubins, J. B., Chen, J-C., Dickey, B. F. and Volicer, L. (1991). Modification of brain guanine nucleotide-binding regulatory proteins by tryptamine-4,5-dione, a neurotoxic derivative of serotonin. *J. Neurochem.* **56**, 1851-1854.

Floyd, R. A. (1983). Hydroxyl free-radical spin-adducts in rat brain synaptosomes observations on the reduction of the nitroxide. *Biochem. Biophys. Acta.* **756**, 204-216.

Floyd, R. A. (1996). Mitochondrial damage in neurodegenerative disease. In: *Free Radicals in Brain Physiology and Disorders.* L. Packer, M. Hiramatsu, and T. Yoshikawa (Eds.). Academic Press, New York. pp. 313-329.

Floyd, R. A., Lewis, C. A. and Wong, P. K. (1984a). High-pressure liquid chromatography-electrochemical detection of oxygen free radicals. *Meth. Enzymol.* **105**, 231-327.

Floyd, R. A., Watson, J. J., Harris, J., West, M. and Wong, P. K. (1986). Formation of 8-hydroxydeoxyguanosine, hydroxyl free radical adduct of DNA in granulocytes exposed to the tumor promoter, tetradeconylphorbolacetate. *Biochem. Biophys. Res. Commun.* **137**, 841-846.

Floyd, R. A., Watson, J. J. and Wong, P. K. (1984b). Sensitive assay of hydroxyl free radical formation utilizing high pressure liquid chromatography with electrochemical detection of phenol and salicylate hydroxylation products. *J. Biochem. Biophys. Meth.* **10**, 221-235.

Floyd, R. A., Watson, J. J., Wong, P. K., Altmiller, D.H. and Rickard, R.C. (1986). Hydroxyl free radical adduct of deoxyguanosine: Sensitive detection and mechanisms of formation. *Free Rad. Res. Comm.* **1**, 163-172.

Fornstedt, B., Bergh, I., Rosengren, E. and Carlsson, A. (1990). An improved HPLC-electrochemical detection method for measuring brain levels of 5-*S*-cysteinyldopamine, 5-*S*-cysteinyl-3,4-dihydroxyphenylalanine, and 5-*S*-cysteinyl-3,4-dihydroxyphenylacetic acid. *J. Neurochem.* **54**, 578-586.

Fraga, C. G., Shigenaga, M. K., Park, J-W., Degan, P. and Ames, B. N. (1990). Oxidative damage to DNA during aging: 8-hydroxy-2'-deoxyguanosine in rat organ DNA and urine. *Proc. Natl. Acad. Sci.* **87**, 4533-4537.

Frei, B., Kim, M. C. and Ames, B. N. (1990). Ubiquinol-10 is an effective lipid-soluble antioxidant at physiological concentrations. *Proc. Natl. Acad. Sci.* **87**, 4879-4883.

Frenkel, K., Zhong, Z., Wei, H., Karkoszka, Patel, U., Rashid, K., Georgescu, M. and Solomon, J. J. (1991). Quantitative high-performance liquid chromatography analysis of DNA oxidized *in vitro* and *in vivo*. *Anal. Biochem.* **196**, 126-136.

Freeto, S. (1997). The simultaneous analysis of 3-nitrotyrosine, 3-aminotyrosine and tyrosine isomers using HPLC with coulometric array detection. ESA, Inc., Application Note. In preparation.

Friedmann, M. N., Robinson, S. W. and Gerhardt, G. (1996). o-Phenylenediamine-modified carbon fiber electrodes for the detection of nitric oxide. *Anal. Chem.* **68**, 2621-2628.

Fukuyama, N., Takebayashi, Y., Hida, M., Ishida, H., Ichimore, K. and Nakazawa, H. (1997). Clinical evidence of peroxynitrite formation in chronic renal failure patients with septic shock. *Free Radic. Biol. Med.* **22**, 771-774.

Fung, K. and Grosjean, D. (1981). Determination of nanogram amounts of carbonyls as 2,4-dintrophenylhydrazones by high-performance liquid chromatography. *Anal. Chem.* **53**, 168-171.

Gamache, P. H. (1997). Personal communications.

Gamache, P. H., McCabe, D. R., Parvez, H., Parvez, S, and Acworth, I. N. (1997). The measurement of markers of oxidative damage, anti-oxidants and related compounds using HPLC and coulometric array analysis. In: *Coulometric Electrode Array Detectors for HPLC.* Progress in HPLC-HPCE (6), I. N. Acworth, M. Naoi, S. Parvez and H. Parvez (Eds). VSP Publications, The Netherlands, pp. 99-126.

Gerlach, M., Ben-Shachar, D., Riederer, P. and Youdim, M. B. H. (1994). Altered brain metabolism of iron as a cause of neurodegenerative diseases? *J. Neurochem.* **63**, 793-807.

Germadnik, D., Pilger, A. and Rudiger, H. W. (1997). Assay for the determination of urinary 8-hydroxy-2'deoxyguanosine by high-performance liquid chromatography with electrochemical detection. *J. Chromatogr. B.* **689**, 399-403.

Giulivi, C. and Davies, K. J. A. (1994). Dityrosine: A marker for oxidatively modified proteins and selective proteolysis. *Meth. Enzymol.* **318,** 427-432.

Goda, Y. and Marnett, L. J. (1991). High-performance liquid chromatography with electrochemical detection for determination of the major malondialdehyde-guanine adduct. *Chem. Res. Toxicol.* **4**, 520-524.

Goldring, C., Casini, A. F., Maellaro, E., Bello, B. D. and Comporti, M. (1993). Determination of 4-hydroxynonenal by high-performance liquid chromatography with electrochemical detection. *Lipids*, **28**, 141-145.

Graham, D. (1978). Oxidative pathways for catecholamines in the genesis of neuromelanin and cytotoxic quinones. *Mol. Pharmacol.* **14**, 633-643.

Graham, D., Tiffany, S. M., Bell, W. R. and Gutknecht, W. F. (1978). Autoxidation versus covalent binding of quinones as mechanisms of toxicity of dopamine, 6-hydroxydopamine, and related compounds toward C1300 neuroblastoma cells *in vitro*. *Mol. Pharmacol.* **14** 644-653.

Greaney, M. D., Marshall, D. L., Bailey, B. A. and Acworth, I. N. (1993). Improved method for the routine analysis of acetylcholine release *in vivo*: Quantitation in the presence and absence of esterase inhibitors. *J. Chromatogr.* **622**, 121-135.

Gross, S. S. and Wolin, M. S. (1995). Nitric oxide: Pathophysiological mechanisms. *Annu. Rev. Physiol.* **57**, 737-769.

Grossi, G., Bargossi, A. M., Fiorella, P. L. and Piazzi, S. (1992). Improved high-performance liquid chromatographic method for the determination of coenzyme Q10 in plasma. *J. Chromatogr.* **593**, 217-226.

Grunewald, R. A. (1993). Ascorbic acid in the brain. *Brain Res. Rev.* **18**, 123-133.

Haddad, I. Y., Pataki, G., Hu, P., Galliani, C., Beckman, J. S. and Matalon, S. (1994). Quantitation of nitrotyrosine levels in lung sections of patients and animals with acute lung injury. *J. Clin. Invest.* **94**, 2407-2413.

Haegele, A. D., Briggs, S. P. and Thompson, H. J. (1993). Antioxidant status and dietary lipid unsaturation modulate oxidative DNA damage. *Free Radic. Biol. Med.* **16**, 111-115.

Halliwell, B. (1994). Free radicals and antioxidants: A personal view. *Nutr. Rev.* **52**, 253-265.

Halliwell, B. (1996). Vitamin C: Antioxidant or pro-oxidant *in vivo*? *Free Radic. Res.* **25**, 439-454.

Halliwell, B. and Chiroco, S. (1993). Lipid peroxidation: Its mechanism, measurement and significance. *Am. J. Clin. Nutr.* **57**, 715S-725S.

Halliwell, B., and Gutteridge, J. M. C. (Eds.) (1989). *Free Radicals in Biology and Medicine*. Clarendon Press, Oxford.

Halliwell, B. and Gutteridge, J. M. C. (1990). Role of free radicals and catalytic metal ions in human disease: An overview. *Meth. Enzymol.* **186**, 1-85.

Halliwell, B. and Gutteridge, J. M. C. (1992). Biologically relevant metal ion-dependent hydroxyl radical generation. *FEBS Letts.* **307**, 108-112.

Han, D., Handelman, G. J. and Packer, L. (1995). Analysis of reduced and oxidized lipoic acid in biological samples by high-performance liquid chromatography. *Meth. Enzymol.* **251**, 315-325.

Han, X. and Liehr, J. G. (1994). 8-Hydroxylation of guanine bases in kidney and liver DNA of hamsters treated with estradiol: Role of free radicals in estrogen-induced carcinogenesis. *Cancer Res.* **54**, 5515-5517.

Harvey, P. R. C., Ilson, R. G. and Strasberg, S. M. (1989). The simultaneous determination of oxidized and reduced glutathione in liver tissue by ion pairing reversed-phase high performance liquid chromatography with a coulometric electrochemical detector. *Clin. Chim. Acta* **180**, 203-212.

Hastings, T. G. and Zigmond, M. J. (1994). Identification of catechol-protein conjugates in neostriatal slices incubated with [^{3}H]dopamine: Impact of ascorbic acid and glutathione. *J. Neurochem.* **63**, 1126-1132.

Hastings, T. G., Lewis, D. A. and Zigmond, M. J. (1996). Role of oxidation of the neurotoxic effects of intrastriatal dopamine injections. *Proc. Natl. Acad. Sci.* **93**, 1956-1961.

Hattori-Nakakuki, Y., Nishigori, C., Okamoto, K., Imamura, S., Hiai, H. and Toyokuni, S. (1994). Formation of 8-hydroxy-2'deoxyguanosine in epidermis of hairless mice exposed to near-UV. *Biochem. Biophys. Res. Commun.* **201**, 1132-1139.

Herbert, K. E., Evans, M. D., Finnegan, M. T. V., Farooq, S., Mistry, N., Podmore, I. D., Farmer, P. and Lunec, J. (1996). A novel HPLC procedure for the analysis of 8-oxoguanine in DNA. *Free Radic. Biol. Med.* **20**, 467-473.

Herrman, T., Steinhilber, D. and Roth, H. J. (1987). Determination of leukotriene B_4 by high-performance liquid chromatography with electrochemical detection. *J. Chromatogr.* **416**, 170-175.

Hicks, M., Wong, L. S. and Day, R. O. (1993). Identification of products from oxidation of uric acid induced by hydroxyl radicals. *Free Rad. Res. Comms.* **18**, 337-351.

Hill., B. A., Kleiner, H. E., Ryan, E. A., Dulik, D. M., Monks, T. J. and Lau, S.S. (1993). Identification of multi-*S*-substituted conjugates of hydroquinone by HPLC-coulometric electrode array analysis and mass spectrometry. *Chem. Res. Toxicol.* **6**, 459-469.

Hillered, L. and Persson, L. (1995). Parabanic acid for monitoring of oxygen radical activity in the injured human brain. *NeuroReport* **6**, 1816-1820.

Honegger, C. G., Langeman, H., Krenger, W. and Kempf, A. (1989). Liquid chromatographic determination of common water-soluble antioxidants in human tissue. *J. Chromatogr.* **487**, 463-468.

Ignarro, L. and Murad, F. (Eds.) (1995). *Nitric Oxide. Biochemistry, Molecular Biology, and Therapeutic Implications. Advances in Pharmacol.* **34**. Academic Press, New York.

Irayama, K., Yoshiura, M., Iwamoto, T. and Ozaki, Y. (1984). Simultaneous determination of uric and ascorbic acids in human serum by reversed-phase high-performance liquid chromatography with electrochemical detection. *Anal. Biochem.* **141**, 238-243.

Ischiropoulos, H. and Al-Mehdi, A. B. (1995). Peroxynitrite-mediated oxidative protein modifications. *FEBS Letts.* **364**, 279-282.

Ischiropoulos, H., Zhu, L., Chen, J., Tsai, M., Martin, J. C., Smith, C. D. and Beckman, J. S. (1992). Peroxynitrite-mediated tyrosine nitration catalyzed by superoxide dismutase. *Arch. Biochem. Biophys.* **298**, 431-437.

Ishimitsu, S., Fujimoto, S. and Ohara, A. (1986). Determination of *m*-tyrosine and *o*-tyrosine in human serum by high-performance liquid chromatography with fluorometric detection. *J. Chromatogr.* **378**, 222-225.

Iwahashi, H. (1996). Quantitative detection of reduced, radical and oxidized forms of α-(4-pyridyl-1-oxide)-*N*-*tert*-butylnitrone radical adduct using high-performance liquid chromatography with electrochemical detection. *J. Chromatogr. A.* **753**, 235-242.

Iwamoto, T., Yoshiura, M. and Irayama, K. (1983). A simple, rapid and sensitive method for the determination of rat serum uric acid by reversed-phase high-performance liquid chromatography with electrochemical detection. *J. Chromatogr.* **278**, 156-159.

Jacobsen, D. W., Gatautis, V. J., Green, R., Robinson, K., Savon, S. R., Secic, M., Ji, J., Otto, J. M. and Taylor, L. M. (1994). Rapid HPLC determination of total homocysteine and other thiols in serum and plasma: Sex differences and correlation with cobalamin and folate concentrations. *Clin. Chem.* **40**, 873-881.

Jialal, I. and Devaraj, S. (1996). Low-density lipoprotein oxidation, antioxidants and atherosclerosis: A clinical biochemistry perspective. *Clin. Chem.* **42**, 498-506.

Kagan, V. E., Shvedova, A., Serbinova, E., Khan, S., Swanson, C., Powell, R. and Packer, L. (1992). Dihydrolipoic acid - universal antioxidant both in the membrane and in the aqueous phase: Reduction of the peroxyl, ascorbyl and chromanoxyl radicals. *Biochem. Pharmacol.* **44**, 1637-164.

Kamata, K. and Akiyama, K. (1990). High-performance liquid chromatography with electrochemical detection for the determination of thioctic acid and thioctic acid amide. *J. Pharmaceut. Biomed. Anal.* **8**, 453-456.

Kaur, H. and Halliwell, B. (1990). Action of biologically-relevant oxidizing species upon uric acid. Identification of uric acid oxidation products. *Chem.-Biol. Interacts.* **73**, 235-247.

Kaur, H. and Halliwell, B. (1994a). Aromatic hydroxylation of phenylalanine as an assay for hydroxyl radicals: Measurement of hydroxyl radical formation from ozone and in blood from premature babies using improved HPLC methodology. *Anal. Biochem.* **220**, 11-15.

Kaur, H. and Halliwell, B. (1994b). Detection of hydroxyl radicals by aromatic hydroxylation. *Meth. Enzymol.* **233**, 67-82.

Kaur, H. and Halliwell, B. (1994c). Evidence for nitric oxide-mediated oxidative damage in chronic inflammation. Nitrotyrosine in serum and synovial fluid from rheumatoid patients. *FEBS Letts.* **350**, 8-12.

Kaur, H. and Halliwell, B. (1996). Measurement of oxidized and methylated DNA bases by HPLC with electrochemical detection. *Biochem. J.* **318**, 21-23.

Kehrer, J. P. (1993). Free radicals as mediators of tissue injury. *Crit. Rev. Toxicol.* **23**, 21-48.

Kelly, J. D., Orner, G. A., Hendricks, J. D. and Williams, D. E. (1992). Dietary hydrogen peroxide enhances hepatocarcinogenesis in trout: Correlation with 8-hydroxy-2'-deoxyguanosine levels in liver DNA. *Carcinogen.* **13** 1639-1642.

Kharitonov, V. G., Sundquist, A. R. and Sharma, V. S. (1995). Kinetics of nitrosation of thiols by nitric oxide in the presence of oxygen. *J. Biol. Chem.* **270**, 28158-28164.

Knopse, J., Steinhilber, D., Herrmann, T. and Roth, H. (1988). Picomole determination of 2,4-dimethoxyanalides of prostaglandins by high-performance liquid chromatography with electrochemical detection. *J. Chromatogr.* **442**, 444-450.

Kontos, H. A. (1989). Oxygen radicals in CNS damage. *Chem. Biol. Interactions* **72**, 229-255.

Koppenol, W. H. (1994). Thermodynamic considerations on the formation of reactive species from hypochlorite, superoxide and nitrogen monoxide. *FEBS Letts.* **347,** 5-8.

Korytowski, W., Bachowski, G. and Girotti. A. W. (1991). Chromatographic separation and electrochemical determination of cholesterol hydroperoxides generated by photodynamic action. *Anal. Biochem.* **197**, 149-156.

Korytowski, W., Geiger, P. G. and Girotti, A. W. (1995). High-performance liquid chromatography with mercury cathode electrochemical detection: Application to lipid hydroperoxide analysis. *J. Chromatogr. B.* **670**, 189-197.

Krien, P. M., Margou, V. and Kermici, M. (1992). Electrochemical determination of femtomole amounts of free reduced and oxidized glutathione. *J. Chromatogr.* **576**, 255-261.

Krinsky, N. I. (1993). Actions of carotenoids in biological systems. *Annu. Rev. Nutr.* **13**, 561-587.

Lagorio, S., Tagesson, C., Forastiere, F., Iavarone, I., Axelson, O. and Carere, A. (1994). Exposure to benzene and urinary concentrations of 8-hydroxydeoxyguanosine, a biological marker of oxidative damage to DNA. *Occup. Environ. Med.* **51**, 739-743.

Lakritz, J., Plopper, C. G. and Buckpitt, A. R. (1997). Validated high-performance liquid chromatography-electrochemical method of glutathione and glutathione disulfide in small tissue samples. *Anal. Biochem.* In press.

Lancaster, J. (Ed.) (1996). *Nitric Oxide. Principles and Actions.* Academic Press, New York.

Lang, J. K. and Packer, L. (1987). Quantitative determination of vitamin E and oxidized and reduced coenzyme Q by high-performance liquid chromatography. *J. Chromatogr.* **385**, 109-117.

LeBel, C. P., and Bondy, S. C. (1991). Oxygen radicals: Common mediators of neurotoxicity. *Neurotox. Teratol.* **13**, 341-346.

LeBel, C. P., and Bondy, S. C. (1992). Oxidative damage and cerebral aging. *Prog. Neurobiol.* **38**, 601-609.

Levine, M. (1986). New concepts in the biology and biochemistry of ascorbic acid. *N. Engl. J. Med.* **314**, 892-902.

Levine, M., Rumsey, S., Wang, Y., Park, J., Kwon, O., Xu, W., and Amano, N. (1996). Vitamin C. In: *Present Knowledge in Nutrition.* Filer, L. J. and Ziegler, E. E. (Eds.) International Life Sciences. pp 146-159.

Liu, D. (1993). Generation and detection of hydroxyl radical *in vivo* in rat spinal cord by microdialysis administration of Fenton's reagents and microdialysis sampling. *J. Biochem. Biophys. Meth.* **27**, 281-291.

Liu, L. and Wells, P. G. (1995). DNA oxidation as a potential molecular mechanism mediating drug-induced birth defects: Phenytoin and structurally related teratogens initiate the formation of 8-hydroxy-2'-deoxyguanosine *in vitro* and *in vivo* in murine maternal hepatic and embryonic tissues. *Free Radic. Biol. Med.* **19**, 639-648.

Liu, R. H. and Hotchkiss, J. H. (1995). Potential genotoxicity of chronically elevated nitric oxide: A review. *Mutat. Res.* **339**, 73-89.

Loft, S., Larsen, P. N., Rasmussen, A., Fischer-Nielsen, A., Bondesen, S., Kirkegaard, P., Rasmussen, R. S., Ejlersen, E., Tornoe, K., Bergholdt, R. and Poulsen, H. E. (1995). Oxidative DNA damage after transplantation of the liver and small intestines in pigs. *Transplant.* **59**, 16-20.

Lunec, J., Herbert, K., Blount, S., Griffiths, H. R. and Emery, P. (1994). A marker of oxidative DNA damage in systemic lupus erythematosus. *FEBS Letts.* **348**, 131-138.

Lvovich, V. and Scheeline, A. (1997). Amperometric sensors for simultaneous superoxide and hydrogen peroxide detection. *Anal. Chem.* **69**, 454-462.

Lykkesfeldt, J., Loft, S. and Poulsen, H. (1995). Determination of ascorbic acid and dehydroascorbic acid in plasma by high-performance liquid chromatography with coulometric detection - are they reliable markers of oxidative stress? *Anal. Biochem.* **229**, 329-335.

MacCrehan, W. A. and Schonberger, E. (1987a). Determination of retinol, α-tocopherol and β-carotene in serum by liquid chromatography and electrochemical detection. *Clin. Chem.* **33**, 1585-1592.

MacCrehan , W. A., and Schonberger, E. (1987b). Reversed-phase high-performance liquid chromatographic separation and electrochemical detection of retinol and its isomers. *J. Chromatogr.* **417**, 65-78.

Maine, S. T. (1996). Beta-carotene, carotenoids, and disease prevention in humans. *FASEB J.* **10**, 690-701.

Malinow, M. R., Kang, S-S., Taylor, L. M., Wong, P. W. K., Coull, B. and Inahara, T. (1989). Prevalence of hyperhomocyst(e)inemia in patients with peripheral arterial occlusive disease. *Circulation* **79**, 1180-1188.

Malinski, T. and Taha, Z. (1992). Nitric oxide release from a single cell measured *in situ* by a porphyrinic-based microsensor. *Nature* **358**, 676-678.

Marnett, L. J. and Burcham, P. C. (1993). Endogenous DNA adducts: Potential and paradox. *Chem. Res. Toxicol.* **6**, 771-785.

Maruyama, W., Deng, Y., Dostert, P. and Naoi, M. (1997). Analysis of endogenous enantiomers of neurotoxins in clinical samples by a multi-electrode detection system with a chiral column. In: *Coulometric Electrode Array Detectors for HPLC*. Progress in HPLC-HPCE (6), I. N. Acworth, M. Naoi, S. Parvez and H. Parvez (Eds.) VSP Publications, The Netherlands, pp. 339-350.

Maruyama, W., Hashizume, Y., Matsubara, K. and Naoi, M. (1996). Identification of 3-nitro-L-tyrosine, a product of nitric oxide and superoxide, as an indicator of oxidative stress in the human brain. *J. Chromatogr. B.* **676**, 153-158.

Maskus, M., Parient, F., Wu, Q., Toffanin, A., Shapleigh, J. P. and Abruna, H. D. (1996). Electrocatalytic reduction of nitric oxide at electrodes modified with electropolymerized films of $[Cr(v\text{-}tpy)_2]^{3+}$ and their application to cellular NO determinations. *Anal. Chem.* **68**, 3128-3134.

Matson, W. R., Gamache, P. G., Beal, M. F. and Bird, E. D. (1987). EC array sensor concepts and data. *Life Sci.* **41**, 905-908.

McCabe, D. R., Acworth, I. N., Maidt, L. and Floyd, R. (1997a). A sensitive and selective method for the determination of tissue 8-hydroxy-2'deoxyguanosine using HPLC with electrochemical array detection. Presented at the American Association for Cancer Research meeting. San Diego.

McCabe, D. R. Maher, T. J. and Acworth, I. N. (1997b). Improved method for the estimation of hydroxyl free radical levels *in vivo* based on liquid chromatography with electrochemical detection. *J. Chromatogr. B.* In press.

McNeil, C. J., Greenough, K. R., Weeks, P. A., Self, C. H. and Cooper, J. M. (1992). Electrochemical sensors for the direct reagentless measurement of superoxide production by human neutrophils. *Free Rad. Res. Comms.* **17**, 399-406.

Mecocci, P., MacGarvey, U., Kaufman, A. E., Koontz, D., Shoffner, J. M., Wallace, D. C. and Beal, M. F. (1993). Oxidative damage to mitochondrial DNA shows marked age-dependent increases in human brain. *Ann. Neurol.* **34**, 609-616.

Milbury, P. E. and Matson, W. R. (1995). Tyrosine and purine derived indicators of oxidative stress and free radical action. Annual Meeting of The Oxygen Society, Pasadena, California, #2-104.

Miller, N. J., Sampson, J., Candeleias, L. P., Bramley, P. M. and Rice-Evans, C. A. (1996). Antioxidant activities of carotenes and xanthophyls. *FEBS Lett.* **384**, 240-242.

Mitton, K. P., and Trevithick, J. R. (1994). High-performance liquid chromatography-electrochemical detection of antioxidants in vertebrate lens: Glutathione, tocopherol, and ascorbate. *Meth. Enzymol.* **233**, 523-539.

Moncada, S., Palmer, R. M. J., and Higgs, E. A. (1991). Nitric oxide physiology, pathophysiology, and pharmacology. *Pharmacol. Rev.* **43**, 109-141.

Montgomery, J., Ste-Marie, L., Boismenu, D. and Vachon, L. (1995). Hydroxylation of aromatic compounds as indices of hydroxyl radical production: A cautionary note revisited. *Free Radic. Biol. Med.* **19**, 927-933.

Moore, A. S. and Papas, A. M. (1996). Biochemistry and health significance of vitamin E. *J. Adv. Med.* **9**, 11-29.

Motchnik, P. A., Frei, B. and Ames, B. N. (1994). Measurement of antioxidants in human blood plasma. *Meth. Enzymol.* **234**, 269-279.

Motohashi, N. and Mori, I. (1989). High-performance liquid chromatography-electrochemical detection of singlet oxygen by reaction with 2,2,6,6-tetramethyl-4-piperidone. *J. Chromatogr.* **465**, 417-421.

Mulchandani, A. and Rudolph, D. C. (1995). Amperometric determination of lipid hydroperoxides. *Anal. Biochem.* **225**, 277-282.

Murphy, M. E. and Kehrer, J. P. (1987). Simultaneous measurement of tocopherols and tocopheryl-quinones in tissue fractions using high-performance liquid chromatography with redox-cycling electrochemical detection. *J. Chromatogr.* **421**, 71-82.

Muscara, M. N. and de Nucci, G. (1996). Simultaneous determination of nitrite and nitrate anions in plasma, urine and cell culture supernatants by high-performance liquid chromatography with post-column reactions. *J. Chromatogr. B.* **686**, 157-164.

Musshoff, F., Daldrup, Th., Bonte, W., Leitner, A. and Lesch, O. M. (1996). Formaldehyde-derived tetrahydroisoquinolines and tetrahydro-β-carbolines in human urine. *J. Chromatogr. B.* **683**, 163-176.

Naoi, M., Maruyama, W., Acworth, I. N., Nakahara, D. and Parvez, H. (1993). Multi-electrode detection system for the determination of neurotransmitters. In: *Methods in Neurotransmitter and Neuropeptide Research. Techniques in the Behavioral and Neural Sciences.* S. H. Parvez, M. Naoi, T. Nagatsu and S. Parvez (Eds.) Elsevier, New York, **11**, 1-39.

Naoi, M., Maruyama, W., Nakahara, D. and Parvez, H. (1997). The application of the multi-electrode detection system to neurochemistry: *In vivo* microdialysis studies. In: *Coulometric Electrode Array Detectors for HPLC.* Progress in HPLC-HPCE (6), I. N. Acworth, M. Naoi, S. Parvez and H. Parvez (Eds). VSP Publications, The Netherlands, pp. 227-246.

Nguyen, T., Brunson, D., Crespi, C. L., Penman, B. W., Wishnok, J. S. and Tannenbaum, S. R. (1992). *Proc. Natl. Acad. Sci.* **89**, 3030-3034.

Nicotera, T. M., Munson, B. R. and Fiel, R. J. (1994). Photosensitized formation of 8-hydroxy-2'deoxyguanosine and DNA strand breakage by a cationic *meso*-substituted porphyrin. *Photochem. Photobiol.* **60**, 295-300.

O'Gara, C. Y., Maddipati, K. R. and Marnett, L. J. (1989). A sensitive electrochemical method for quantitative hydroperoxide determination. *Chem. Res. Toxicol.* **2**, 295-300.

Ohtsuka, E. and Nishimura, S. (1987). Misreading of DNA templates containing 8-hydroxydeoxyguanosine as the modified base and at adjacent residues. *Nature* **327**, 77-99.

Okamoto, T., Fukanaga, Y., Ida, Y. and Kishi, T. (1988). Determination of reduced and total ubiquinones in biological materials by liquid chromatography with electrochemical detection. *J. Chromatogr.* **430**, 11-19.

Oury, T. D., Tatro, L., Ghio, A. J. and Piantadosi, C. A. (1995). Nitration of tyrosine by hydrogen peroxide and nitrite. *Free Rad. Res.* **23**, 537-547.

Ozcimder, M., Louter, A. J. H., Lingeman, H., Voogt, W. H., Frei, R. W. and Bloemeddal, M. (1991). Determination of oxidized, reduced and protein-bound glutathione in eye lenses by high-performance liquid chromatography and electrochemical detection. *J. Chromatogr.* **570**, 19-28.

Packer, L. (Ed.) (1996a). Nitric oxide. Part A. Sources and detection of NO and NO synthase. *Meths. Enzymol.* **268**.

Packer, L. (Ed.) (1996b). Nitric oxide. Part B. Physiological and pathological processes. *Meths. Enzymol.* **269**.

Packer, L. (1996c). Prevention of free radical damage in the brain: Protection by α-lipoic acid. In: *Free Radicals in Brain Physiology and Disorders*. L. Packer, M. Hiramatsu, and T. Yoshikawa (Eds.). Academic Press, New York. pp. 19-34.

Packer, L., Hiramatsu, M. and Yoshikawa, T. (Eds.) (1996). *Free Radicals in Brain Physiology and Disease*. Academic Press, New York.

Packer, L., Tritschler, H. J. and Wessel, K. (1997). Neuroprotection by the metabolic antioxidant α-lipoic acid. *Free Radic. Biol. Med.* **22**, 359-378.

Padmaja, S., Ramazenian, M. S., Bounds, P. L. and Koppenol, W. H. (1996). Reaction of peroxynitrite with L-tryptophan. *Redox Report* **2**, 173-177.

Palumbo, G., Carlucci, G., Mazzeo, P., Frieri, G., Pimpo, M. T. and Fanini, D. (1995). Simultaneous determination of 5-aminosalicylic acid, acetyl-5-aminosalicylic acid and 2,5-dihydroxybenzoic acid in endoscopic intestinal biopsy samples in humans by high-performance liquid chromatography with electrochemical detection. *J. Pharmaceut. Biomed. Anal.* **14** 175-180.

Park, E-M., Shigenaga, M. K., Degan, P., Korn, T. S., Kitzler, J. W., Wehr, C. M., Kolachana, P. and Ames, B. N. (1992). Assay of excised oxidative DNA lesions: Isolation of 8-oxoguanine and its nucleoside derivatives from biological fluids with a monoclonal antibody column. *Proc. Natl. Acad. Sci.* **89**, 3375-3379.

Park, J-W., Cundy, K. C. and Ames, B. N. (1989). Detection of adducts by high-performance liquid chromatography with electrochemical detection. *Carcinogen.* **10**, 827-832.

Park, J-W. and Floyd, R. A. (1992). Lipid peroxidation products mediate the formation of 8-hydroxydeoxyguanosine in DNA. *Free Radic. Biol. Med.* **12**, 245-250.

Patthy, M., Kiraly, I. and Sziraki, I. (1995). Separation of dihydroxybenzoates, indicators of *in vivo* hydroxyl free radical formation, in the presence of transmitter amines and some metabolites in rodent brain, using high performance liquid chromatography with electrochemical detection. *J. Chromatogr. N.* **664**, 247-252.

Pileblad, E. and Magnusson, T. (1989). Intracerebroventricular administration of L-buthionine sulfoxamine: A method for depleting brain glutathione. *J. Neurochem.* **53**, 1878-1882.

Podda, M., Weber, C., Traber, M. G. and Packer, L. (1996). Simultaneous determination of tissue tocopherols, tocotrienols, ubiquinols and ubiquinones. *J. Lip. Res.* **37**, 893-901.

Preik-Steinhoff, H. and Kelm, M. (1996). Determination of nitrite in human blood by combination of a specific sample preparation with high-performance anion-exchange chromatography and electrochemical detection. *J. Chromatogr. B.* **685**, 348-352.

Pryor, W. A. (1986). Oxy-radicals and related species: Their formation lifetimes and reactions. *Ann. Rev. Physiol.* **48**, 657-663.

Pryor, W. A. (1988). Why is the hydroxyl radical the only radical that commonly adds to DNA? Hypothesis has a rare combination of high electrophilicity, high thermochemical reactivity and a mode of production that can occur near DNA. *Free Radic. Biol. Med.* **4**, 219-223.

Pryor, W. A. and Squadrito, G. L. (1995). The chemistry of peroxynitrite: A product from the reaction of nitric oxide with superoxide. *Am. J. Physiol.* **268**, L699-L722.

Radi, R. (1996). Reactions of nitric oxide with metalloproteins. *Chem. Res. Toxicol.* **9**, 828-835.

Radi, R., Beckman, J. S., Bush, K. M. and Freeman, B. A. (1991). Peroxynitrite-induced membrane lipid peroxidation: The cytotoxic potential of superoxide and nitric oxide. *Arch. Biochem. Biophys.* **288**, 481-487.

Radzik, D. M., Roston, D. A. and Kissinger, P. T. (1983). Determination of hydroxylated aromatic compounds produced via superoxide-dependent formation of hydroxyl radicals by liquid chromatography/electrochemistry. *Anal. Biochem.* **131**, 458-464.

Ramezainian, M. S., Padmaja, S. and Koppenol, W. H. (1996). Nitration and hydroxylation of phenolic compounds by peroxynitrite. *Chem. Res. Toxicol.* **9**, 232-240.

Rizzo, V., Melzi D'Eril, G., Achilli, G. and Cellerino, G. P. (1991). Determination of neurochemicals in biological fluids by using an automated high-performance liquid chromatographic system with a coulometric array detector. *J. Chromatogr.* **536**, 229-236.

Roch-Ramel, F., Diezi-Chomety, F., Roth, L. and Weiner, I. M. (1980). A micropuncture study of urate excretion by Cebus monkeys employing high-performance liquid chromatography with amperometric detection of urate. *Pflugers Arch.* **383**, 203-207.

Rose, R. and Bode, A. (1995). Analysis of water-soluble antioxidants by high-pressure liquid chromatography. *Biochem. J.* **306**, 101-105.

Rubbo, H., Radi, R., Trujillo, M., Telleri, R., Kalyanaraman, B., Barnes, S., Kirk, M. and Freeman, B.A. (1994). Nitric oxide regulation of superoxide and peroxynitrite-dependent lipid peroxidation. *J. Biol. Chem.* **269**, 26066-26075.

Sachs, C. H. and Johnson, G. (1975). Mechanisms of action of 6-hydroxydopamine. *Pharmacol.* **24**, 1-8.

Salter, M., Duffy, C., Garthwaite, J. and Strijbos, P. J. (1996). *Ex vivo* measurement of brain tissue nitrite and nitrate accurately reflects nitric oxide synthase activity *in vivo. J. Neurochem.* **66**, 1683-1690.

Schell, D. A. and Bode, A. M. (1993). Measurement of ascorbic acid and dehydroascorbic acid in mammalian tissue utilizing HPLC and electrochemical detection. *Biomed. Chromatogr.* **7**, 267-272.

Schneider, J. E., Browning, M. M., Zhu, X., Eneff, K. L. and Floyd, R. A. (1989). Characterization of hydroxyl mediated damage to plasmid pBR322 DNA. *Mutation Res.* **214**, 23-31.

Schultz, J. B., Matthews, R. T., Muqit, M. M. K., Browne, S. E. and Beal, M. F. (1995). Inhibition of neuronal nitric oxide synthase by 7-nitroindazole protects against MPTP-induced neurotoxicity in mice. *J. Neurochem.* **64**, 936-939.

Scott, B. C., Arouma, O. I., Evans, P. J., O'Neill, C., van der Vliet, A., Cross, C. E., Tritschler, H. and Halliwell, B. (1994). Lipoic and dihydrolipoic acids as antioxidants. A critical evaluation. *Free Rad. Res.* **20**, 119-133.

Shen, X-M. and Dryhurst, G. (1996). Further insights into the influence of L-cysteine on the oxidation chemistry of dopamine: Reaction pathways of potential relevance to Parkinson's disease. *Chem. Res. Toxicol.* **9**, 751-763.

Shen, X-M. and Dryhurst, G. (1996). Oxidation chemistry of (-)-norepinephrine in the presence of L-cysteine. *J. Med. Chem.* **39**, 2018-2029.

Shibutani, S., Takeshita, A. and Grollman, A. P. (1991). Insertion of specific bases during DNA synthesis past the oxidation-damaged 8-oxodG. *Nature* **349**, 431-434.

Shigenaga, M. K., Aboujaoude, E. N., Chen, Q. and Ames, B. N. (1994). Assays of oxidative DNA damage biomarkers 8-oxo-2'-deoxyguanosine and 8-oxoguanine in nuclear DNA and biological fluids by high-performance liquid chromatography with electrochemical detection. *Meth. Enzymol.* **234**, 16-33.

Shigenaga, M. K., Gimeno, C. J. and Ames, B. N. (1989). Urinary 8-hydroxy-2'deoxyguanosine as a biological marker of *in vivo* oxidative DNA damage. *Proc. Natl. Acad. Sci.* **86**, 9697-9701.

Shigenaga, M. K.., Lee, H., Blount, B., Christen, S., Shigeno, E. T., Yip, H. and Ames, B. N. (1997). Inflammation and NO_x-induced nitration: Assay for 3-nitrotyrosine by HPLC with electrochemical detection. *Proc. Natl. Acad. Sci.* **94**, 3211-3216.

Shigenaga, M. K., Park, J-W., Cundy, K. C., Gimeno, C. J. and Ames, B. N. (1990). *In vivo* oxidative DNA damage: Measurement of 8-hydroxy-2'deoxyguanosine in DNA and urine by high-performance liquid chromatography with electrochemical detection. *Meth. Enzymol.* **186**, 521-530.

Shirachi, D. Y. and Omaye, S. T. (1992). The simultaneous measurement of uric acid and ascorbic acid in the lateral ventricles of freely-moving rats by brain microdialysis and electrochemical detection. *Proc. West Pharmacol. Soc.* **35**, 161-163.

Sies, H. (1985). Oxidative Stress: Introductory remarks. In: *Oxidative Stress*, H. Sies (Ed.) Academic Press, London.

Sies, H. and Murphy, M. E. (1991). Role of tocopherols in the protection of biological systems against oxidative damage. *J. Photochem. Photobiol.* **8**, 211-224.

Skinner, K. A., Wang, Z. and Parks, D. A. (1996). Detection of salicylate nitration and hydroxylation products after liver preservation. The Third Annual Meeting of The Oxygen Society, Florida. Poster #2-39.

Sloot, W. N. and Gramsbergen, J. B. P. (1995). Detection of salicylate and hydroxylated adducts 2,3- and 2,5-dihydroxybenzoic acids as possible indices for *in vivo* hydroxyl radical formation in combination with catechol- and indoleamines and their metabolites in cerebrospinal fluid and brain tissue. *J. Neurosci. Meth.* **60**, 141-149.

Smith, C. D., Carney, J. M., Starke-Reed, P. E., Olive, C. N., Stadtman, E. R., Floyd, R. A. and Markesbery, W. R. (1991). Excess brain protein oxidation and enzyme dysfunction in normal aging and Alzheimer disease. *Proc. Natl. Acad. Sci.*, **88**, 10540-10543.

Smith, C. D., Carney, J. M., Tatsumo, T., Stadtman, E. R., Floyd, R. A. and Markesbery, W. R. (1992). Protein oxidation in aging brain. *Annals New York Acad. Sci.* **663**, 110-119.

Smith, M. A., Sayre, L. and Perry, G. (1996). Is Alzheimer's a disease of oxidative stress? *Alzheimer's Disease Review* **1**, 63-67.

Smith, N. C., Dunnett, M. and Mills, P. C. (1995). Simultaneous quantitation of oxidized and reduced glutathione in equine biological fluids by reversed-phase high-performance liquid chromatography using electrochemical detection. *J. Chromatogr.* **673**, 35-41.

Snyder, S. H. and Bredt, D. S. (1992). Biological roles of nitric oxide. *Scientific American,* **5**, 68-77.

Sofic, E., Lange, K. W., Jellinger, K. and Riederer, P. (1992). Reduced and oxidized glutathione in substantia nigra of patients with Parkinson's disease. *Neurosci. Letts.* **142**, 128-130.

Sofic, E., Riederer, P., Burger, R., Gsell, W. and Heuschneider, G. (1991). Determination of glutathione, glutathione disulfide, ascorbic acid and dehydroascorbic acid in tissues by reversed-phase liquid chromatography with electrochemical detection. *Fresenius J. Anal. Chem.* **339**, 258-260.

Sokol, R. J. (1989). Vitamin E and neurological function in man. *Free Radic. Biol. Med.* **6**, 189-207.

Sontag, G., Bernweiser, I. and Krach, C (1997). HPLC with electrode array detection and its application in analytical food chemistry. In: *Coulometric Electrode Array Detectors for HPLC*. Progress in HPLC-HPCE (6), I. N. Acworth, M. Naoi, S. Parvez and H. Parvez (Eds.) VSP Publications, The Netherlands, pp. 75-126.

Spiteller, G. (1996). Enzymic lipid peroxidation-consequence of cell injury? *Free Radic. Biol. Med.* **21**, 1003-1009.

Stamler, J. S. and Feelisch, M. (1996). Preparation and detection of S-nitrosothiols. In: *Methods in Nitric Oxide Research*. Felich, M., and Stamler, J. S. (Eds.). Wiley, New York. pp. 521-539.

Steenken, S. (1989). Purine bases, nucleosides and nucleotides: Aqueous solution chemistry and transformation reactions of their radical cations and e^- and OH adducts. *Chem. Rev.* **89**, 503-520.

Ste-Marie, L., Boismenu, D., Vachon, C. and Montgomery, J. (1996). Evaluation of sodium 4-hydroxybenzoate as an hydroxyl radical trap using gas chromatography-mass spectrometry and high-performance liquid chromatography with electrochemical detection. *Anal. Biochem.* **241**, 67-74.

Stronks, H. J., Janzen, E.G. and Weber, R. J. (1984). Electrochemical detection of spin adduct aminoxyls (nitroxides) separated by high performance liquid chromatography. *Anal. Letts.* **17**, 321-328.

Svendsen, C. N. (1993). Multi-electrode array detectors in high-performance liquid chromatography: A new dimension in electrochemical analysis. *Analyst*, **118**, 123-129.

Tagesson, C., Kallberg, M., Klintenberg, C. and Starkhammer, H. (1995). Determination of urinary 8-hydroxydeoxyguanosine by automated coupled-column high performance liquid chromatography: A powerful technique for assaying *in vivo* oxidative DNA damage in cancer patients. *Eur. J. Cancer* **31A**, 934-940.

Takeda, H., Shibuya, T., Yanagawa, K., Kanoh, H. and Takasaki, M. (1996). Simultaneous determination of α-tocopherol and α-tocopherolquinone by high-performance liquid chromatography and coulometric detection in the redox mode. *J. Chromatogr. A.* **722**, 287-294.

Takeuchi, T. and Morimoto, K. (1995). Increased formation of 8-hydroxydeoxyguanosine, an oxidative DNA damage, in lymphoblasts from Franconi's anemia patients due to possible catalase deficiency. *Carcinogen.* **14**, 1115-1120.

Tamir, S., Burney, S. and Tannenbaum, S. R. (1996). DNA damage by nitric oxide. *Chem. Res. Toxicol.* **9**, 821-827.

Tan, B. H., Bencsath, F. A. and Gaubatz, J. W. (1990). Steady-state levels of 7-methylguanine increase in nuclear DNA of postmitotic mouse tissues during aging. *Mutat. Res.* **237**, 229-238.

Tang-Liu, D. D-S. and Riegelman, S. (1982). An automated HPLC assay for simultaneous quantitation of methylated xanthines and uric acids in urine. *J. Chrom. Sci.* **20**, 155-159.

Teichert, J. and Preiss, R. (1992). HPLC-methods for the determination of lipoic acid and its reduced form in human plasma. *J. Clin. Pharm. Ther. Toxicol.* **30**, 511-512.

Teichert, J. and Preiss, R. (1995). Determination in human plasma by high-performance liquid chromatography with electrochemical detection. *J. Chromatogr. B.* **672**, 277-281.

Towel, J. and Kalyanaraman, B. (1991). Detection of radical adducts of 5,5-dimethyl-1-pyrroline *N*-oxide by the combined use of high-performance liquid chromatography with electrochemical detection and electron spin resonance. *Anal. Biochem.* **196**, 111-119.

Upchurch, G. R., Welch, G. N. and Loscalzo, J. (1995). S-Nitrosothiols: Chemistry, biochemistry and biological actions. In: *Nitric Oxide. Biochemistry, Molecular Biology, and Therapeutic Implications. Advances in Pharmacol.* **34**. Academic Press, New York. L. Ignarro and F. Murad. (Eds.). 343-349.

Uppa, R. M., Squadrito, G. L. and Pryor, W. A. (1996). Acceleration of peroxynitrite oxidations by carbon dioxide. *Arch. Biochem. Biophys.* **327**, 335-343.

van Delft, J. H. M., van Winden, M. J. M., van den Ende, A. M. C. and Baan, R. A. (1993). Determining N7-alkylguanine adducts by immunochemical detection: Applications in animals studies and monitoring human exposure to alkylating agents. *Environ. Health Perspectives* **99**, 25-32.

van der Vliet, A., Eiserich, J. P., Kaur, H., Cross, C. E. and Halliwell, B. (1996). Nitrotyrosine as biomarker for reactive nitrogen species. *Meth. Enzymol.* **269**, 175-184.

van der Vliet, A., Eiserich, J. P., O'Niell, C. A., Halliwell, B. and Cross, C. E. (1995). Tyrosine modification by reactive nitrogen species: A closer look. *Arch. Biochem. Biophys.* **319**, 341-349.

van der Vliet, A., O'Neill, C. A. O., Halliwell, B., Cross, C. E. and Kaur, H. (1994). Aromatic hydroxylation and nitration of phenylalanine and tyrosine by peroxynitrite. *FEBS Letts.* **339**, 89-92.

Vatassery, G. T., Smith, W. E. and Quach, H. T. (1993). A liquid chromatographic method for the simultaneous determination of α-tocopherol and tocopherolquinone in human red blood cells and other biological samples where tocopherol is easily oxidized during sample treatment. *Anal. Biochem.* **214**, 426-430.

Volicer, L., Langlais, P. J., Matson, W. R., Mark, K. A. and Gamache, P. H. (1985). Serotoninergic system in dementia of the Alzheimer type: Abnormal forms of 5-hydroxytryptophan and serotonin in cerebrospinal fluid. *Arch. Neurol.* **42**, 1158-1161.

Volk, K. J., Yost, R. A. and Brajter-Toth, A. (1989). On-line electrochemistry/thermospray/tandem mass spectrometry as a new approach to the study of redox reactions: The oxidation of uric acid. *Anal. Chem.* **61**, 1709-1717.

von Sontagg, C. (1991). The chemistry of free-radical-mediated DNA damage. In: *Physical and Chemical Mechanisms in Molecular Radiation Biology.* W. A. Glass and M. N. Varma (eds.) Plenum Press, New York. pp287-321.

Wagner, J. R., Hu, C-C. and Ames, B. N. (1992). Endogenous oxidative damage of deoxycytidine in DNA. *Proc. Natl. Acad. Sci.* **89**, 3380-3384.

Wang, S. S. and Maher, T. J. (1992). Changes in hippocampal ECF amino acid levels in anesthetized rats during kainic acid-induced seizures: A microdialysis study. *Pharmacologist*, **34**, 190.

Washko, P. W., Welch, R. W., Dhariwal, K. R., Wang, Y. and Levine, M. (1992). Ascorbic acid and dehydroascorbic acid analyses in biological samples. *Anal. Biochem.* **204**, 1-14.

Wink, D. A., Kasprzak, K. C., Maragos, C. M., Elespuru, R. K., Misra, M., Dunams, T. M., Cebula, T. A., Koch, W. H., Andrews, A. W., Allen, J. S. and Keefer, L. K. (1991). DNA deaminating ability and genotoxicity of nitric oxide and its progenitors. *Science*, **254**, 1001-1003.

Wiseman, H. and Halliwell, B. (1996). Damage to DNA by reactive oxygen and nitrogen species: Role in inflammatory disease and progression to cancer. *Biochem. J.* **313**, 17-29.

Wong, K-S., Goyal, R. N., Wrona, M. Z., Blank, C. L. and Dryhurst, G. (1993). 7-*S*-glutathionyl-tryptamine-4,5-dione: A possible aberrant metabolite of serotonin. *Biochem. Pharmacol.* **46**, 1637-1652.

Wring, S. A. and Hart, J. P. (1989). Electrochemical determination of all-trans-retinol and correlation with retinol binding protein in liver cirrhosis. *Med. Lab. Sci.* **46**, 367-369.

Wring, S. A., Hart, J. P. and Knight, D. W. (1988). Voltammetric behavior of all-trans-retinol (vitamin A1) at a glassy carbon electrode and its determination in human serum using high-performance liquid chromatography with electrochemical detection. *Analyst* **113**, 1785-1789.

Wrona, M. Z., Goyal, R. N., Turk, D. J., Blank, C. L. and Dryhurst, G. (1992). 5,5'-Dihydroxy-4,4'-bitryptamine: A potentially aberrant, neurotoxic metabolite of serotonin. *J. Neurochem.* **59**, 1392-1398.

Xie, F. and Duda, C. (1994). Determination of an oxidation product of guanine in DNA by microbore liquid chromatography. *Current Sep.* **13**, 18-21.

Yamada, K., Terao, J. and Matsushita, S. (1987). Electrochemical detection of phospholipid hydroperoxides in reversed-phase high-performance liquid chromatography. *Lipids* **22**, 125-128.

Yamada, K. and Nabeshima, T. (1997). Simultaneous measurement of nitrite and nitrate levels as indices of nitric oxide release in the cerebellum of conscious rats. *J. Neurochem.* **68**, 1234-1243.

Yamashita, G. T. and Rabenstein, D. L. (1989). Determination of penicillamine, penicillamine disulfide and penicillamine-glutathione mixed disulfide by high-performance liquid chromatography with electrochemical detection. *J. Chromatogr.* **491**, 341-354.

Yang, C-S., Tsai, P-J., Chen, W-Y., Liu, L. and Kuo, J-S. (1995). Determination of extracellular glutathione in livers of anesthetized rats by microdialysis with on-line high-performance liquid chromatography. *J. Chromatogr. B.* **667**, 41-48.

Ye, Y. Z., Strong, M., Huang, Z-Q. and Beckman, J. S. (1996). Antibodies that recognize nitrotyrosine. *Meth. Enzymol.*, **269**, 201-209.

Yermilov, V., Rubio, J., Becchi, M., Friesen, M. D., Pignatelli, B. and Ohshima, H. (1995). Formation of 8-nitroguanine by the reaction of guanine with peroxynitrite *in vitro*. *Carcinogen.* **16**, 2045-2050.

Youdim, M. B. H. (1988). Iron in the brain: Implications for Parkinson's and Alzheimer's diseases. *Mount Sinai J. Med.* **55**, 97-101.

Youdim, M. and Riederer, P. (1997). Understanding Parkinson's disease. *Scientific American*, **1**, 52-59.

Zhang, F. and Dryhurst, G. (1994). Effects of L-cysteine on the oxidation chemistry of dopamine: New reaction pathways of potential relevance to idiopathic Parkinson's disease. *J. Med. Chem.* **37**, 1084-1098.

Zhang, J., Dawson, V. L., Dawson, T. M. and Snyder, S. H. (1994). Nitric oxide activation of poly(ADP-ribose) synthetase in neurotoxicity. *Science*, **263**, 687-689.

Table 1. Some ROS, RNS and other free radicals identified in biological systems.

Reactive Oxygen Species		**Reactive Nitrogen Species**		**Miscellaneous**	
Free Radicals					
Alkoxyl	$LO^{\bullet}$	Nitric Oxide (monoxide)	$NO^{\bullet}$	Carbon-centered radicals	e.g., $CCl_3^{\bullet}$
Hydroperoxyl	$HO_2^{\bullet}$	Nitrogen Dioxide	$NO_2^{\bullet}$	Disulfide radical	$RSSR^{\bullet-}$
Hydroxyl	$HO^{\bullet}$			Hydrogen atom	$H^{\bullet}$
Peroxyl	$LO_2^{\bullet}$			Thiyl	$RS^{\bullet}$
Superoxide	$O_2^{\bullet-}$				
Non-Radicals					
Hydrogen Peroxide	H_2O_2	Alkyl Peroxynitrite	LO_2NO^-	Disulfide	RSSR
Lipid Peroxides	LO_2H	Dinitrogen Pentoxide	N_2O_5	Thiol	RSH
Oxygen	O_2	Dinitrogen Tetroxide	N_2O_4	Hypochlorous acid	HOCl
Ozone	O_3	Dinitrogen Trioxide	N_2O_3		
Singlet Oxygen	$^1\Delta g\ O_2$	Nitrate	NO_3^-		
		Nitrite	NO_2^-		
		Nitrocarbonate	$O_2NOCO_2^-$		
		Nitronium (Nitryl)	NO_2^+		
		Nitrosonium (Nitrosyl)	NO^+		
		Nitrosoperoxycarbonate	$ONO_2CO_2^-$		
		Nitrosyl chloride	NOCl		
		Nitroxyl	NO^-		
		Nitryl chloride	NO_2Cl		
		Peroxynitrite	ONO_2^-		

Table 2. HO˙ can be measured using a variety of spin traps.

ENDOGENOUS

Spin Trap	Product	Comments	References
Guanosine	8-OH2'dG	Low tissue levels; may be more reflective of H_2O_2 damage.	Schneider *et al.* (1989), Floyd *et al.* (1986).
L-Phenylalanine (L-Phe)[1]	o-, m-Tyrosine	Precursor not EC active; products occur at low levels in tissues.	Ishimitsu *et al.* (1986), Sontag *et al.* (1997).
L-Tyrosine	3,4-, 2,4-L-DOPA	Low tissue levels; 3,4-isomer formed enzymatically.	

EXOGENOUS and *IN VITRO*

Spin Trap	Product	Comments	References
5-Aminosalicylic acid (5HSA)	5-Acetyl-HSA, dihydroxy-5-aminobenzoic acid, 2,5-DHBA	Reduces leukotriene production, inhibits interleukin-1 release, inhibits prostaglandin synthetase or lipoxygenase, interferes with antibody production.	Palumbo *et al.* (1997).
Aniline	o- and p-Aminophenol	Toxic. Not practical for biological experiments.	Radzik *et al.* (1983).
DMPO (5,5-dimethyl-pyrroline-*N*-oxide)	DMPO-OH, DMPO-OH_2	HPLC-ECD overcomes the biological quenching problem associated with EPR.	Floyd *et al.* (1984a).
4-Hydroxybenzoic acid	(2,4-), 3,4-DHBA	Similar in reactivity to salicylate but with less physiological activity. Readily formed by intestinal microbes which may be problematic if gut is damaged.	Montgomery *et al.* (1995), Ste-Marie *et al.* (1996).
PBN (α-phenyl -*tert*-butylnitrone	Aminoxyl adduct	HPLC-ECD method may not be suitable for biological reactions. Attack of phenyl ring by HO˙ may result in complex chromatography.	Stronk *et al.* (1984).
Phenol	Catechol, resorcinol	May not be practical for biological experiments.	Floyd *et al.* (1984b), Radzik *et al.* (1983).
D-Phe[1]	o-, m- D-Tyrosine	Precursor not EC active; D-Phe to L-Phe isomerization; L-Phe can affect dopamine synthesis and release in brain[2].	Kaur and Halliwell (1994a,b)[1].
L-Phe[1]	o-, m-Tyrosine isomers	Precursor not EC active; L-Phe can affect dopamine synthesis and release in brain.[2] Rate of reaction slower than for Sal.	Kaur and Halliwell (1994c), Liu (1993), Ramezanian *et al.* (1996), Sontag *et al.* (1997), Sun *et al.* (1993), van der Vliet *et al.* (1994).
4-POBN (α-(4-pyrisyl-1-oxide)-*N*-*tert*-butylnitrone	4-POBN radical adduct	HPLC-ECD permits the study of reaction mechanisms but no	Iwahashi (1996).

		biological data presented.	
Salicylic acid[3]	2,3-, 2,5-DHBA	Enzymatic formation of 2,5-isomer. Physiological effects at higher concentrations.	Floyd *et al.* (1984b). McCabe *et al.* (1997b) and references therein.
L-Tyrosine	3,4-Dihydroxy-phenylalanine	*In vitro* study only.	Ramezanian *et al.* (1996).

[1]Non-HPLC-ECD methods. Although tyrosine isomers are EC-active these methods use either HPLC-UV or HPLC-fluorescence of the amino acid derivative (the former does not possess sufficient sensitivity for basal *in vivo* levels).
[2]See During *et al.* (1988).
[3]This approach has also been used to measure peroxynitrite production via the formation of 2-hydroxy-5-nitrobenzoate (Skinner *et al.* 1996).

Table 3. The use of ECD and HPLC-ECD to measure lipids and lipid peroxidation products.

Analyte	Technique	Reference
Fatty acid hydroperoxides	Ferrocene derivatization/ Voltammetry.	Mulchandani and Rudolph (1995).
Fatty acid hydroperoxides	Indirect HPLC-coulometric detection via glutathione measurement.	O'Gara *et al.* (1989).
Lipid hydroperoxides	HPLC-polography.	Korytowski *et al.* (1995).
Phospholipid hydroperoxides	HPLC-amperometric detection.	Yamada *et al.* (1987).
Cholesterol hydroperoxides	HPLC-amperometric detection.	Korytowski *et al.* (1991).
Cholesterol ester hydroperoxides	HPLC-coulometric detection.	Arai *et al.* (1996).
4-Hydroxynonenal	HPLC-coulometric detection of DNPH derivative.	Goldring *et al.* (1993).
Leukotriene	HPLC-amperometric detection.	Herrmann *et al.* (1987).
Prostaglandins	HPLC-amperometric detection of 2,4-dimethoxyanilides.	Knopse *et al.* (1988).

Table 4. Measurement of tocopherols.

Analyte	Method	Reference
Tocopherols, tocotrienols, ubiquinols, ubiquinones	HPLC-amperometric detection for tocopherols, tocotrienols, ubiquinols; HPLC-UV detection for ubiquinones.	Poddo *et al.* (1996).
Tocopherol	HPLC-dual coulometric detection.	Castle and Cook (1985).
Tocopherols/tocopherylquinones	HPLC-triple coulometric electrode detection.	Murphy and Kehrer (1987).
Tocopherols/tocopherylquinones	HPLC-dual coulometric detection.	Vatassery *et al.* (1993).
α-Tocopherol, CoQ10	HPLC-dual coulometric detection.	Edlund (1988).
Tocopherol, tocopherylquinones	HPLC-triple coulometric electrode detection.	Takeda *et al.* (1996).
Tocopherol, ubiquinol, ubiquinone	HPLC-amperometric detection for tocopherol only.	Lang and Packer (1987).
α-Tocopherol, β-carotene and retinol	HPLC-UV and amperometric detection.	MacCrehan and Schonberger (1987).
Tocopherols, β-carotene, lycopene, ubiquinol, ubiquinone	HPLC-dual coulometric electrode detection.	Motchnik *et al.* (1994).
Tocopherols, carotenoids, ubiquinol, ubiquinone	HPLC-dual coulometric electrode detection.	Finckh *et al.* (1995).
Tocopherols, CoQ10, retinoids, carotenoids, Vitamin K	Gradient HPLC-coulometric array detection.	Gamache *et al.* (1997a).

Table 5. HPLC-ECD measurement of GSH and related compounds.

Analyte	Electrode Material	Tissue	Reference
GSH, GSSG	Glassy carbon	eye lens	Ozcimder *et al.* (1991).
GSH	Gold amalgam	brain	Pileblad and Magnusson (1989).
GSH	Gold amalgam	liver microdialysis perfusion	Yang *et al.* (1995).
GSH, GSSG	Gold amalgam	blood	Allison and Shoup (1983).
GSH, GSSG	Gold amalgam	red blood cells	Yamashita and Rabenstein (1989).
GSH, GSSG	Gold amalgam	eye-lens	Mitton and Trevithick (1994).
GSH, ascorbic acid	Coulometric	human brain	Sofic *et al.* (1991).
GSH, ascorbic acid, uric acid, cysteine, allantoin	Coulometric	liver and peripheral tissues	Rose and Bode (1995).
GSH, GSSG, cysteine, homocysteine, methionine, *N*-acetylcysteine	Coulometric	plasma	Acworth and Bailey (1995).
GSH, GSSG, cystathionine, cycteine, cystine, homocysteine, homocystine, methionine	Coulometric	bronchoalveolar lavage, plasma	Smith *et al.* (1995).
GSH, GSSG	Coulometric	lung	Lakritz *et al.*(1997).
GSH, GSSG	Coulometric	human substantia nigra	Sofic *et al.* (1992).
GSH, GSSG	Coulometric	human hair	Krien *et al.* (1992).
GSH, GSSG, methionine, *N*-acetyl-cysteine	Coulometric	bile, kidney, lens, liver	Carro-ciampi *et al.* (1988).
GSH, GSSG, cysteine, homocysteine, cystathionines	Coulometric	bile, liver	Harvey *et al.* (1989).
GSH-conjugates	Coulometric array	bile, urine	Hill *et al.* (1992).

Table 6. Factors which make the brain highly susceptible to damage by oxidative stress.

- High blood flow.
- Enormous capillary bed (can readily rupture during ischemia and reperfusion, stroke, head trauma).
- High number of mitochondria.
- High oxygen utilization.
- High levels of ascorbic acid (which may act as a pro-oxidant in the presence of iron).
- High levels of iron in certain brain regions (e.g., substantia nigra).
- Lower levels of iron storage and transport antioxidants.
- Neurons have lower GSH levels than other tissues.
- Blood-brain barrier can limit antioxidant levels.
- Monoamines can auto-oxidize producing ROS and other neurotoxins.
- High $NO^{\bullet}$ production which can lead to the formation of other RNS.
- High PUFA content in membrane lipids, readily undergo lipid peroxidation.
- Unique immune system (microglia) leading to marked production of ROS, RNS and other species.
- Integrity of ionic gradients. Excitatory amino acids (e.g., glutamate) can lead to disruption of ionic gradients, increasing intracellular Ca^{2+} levels and leading to NOS activation.
- Once damaged neurons have limited regeneration capacity.
- Retina can also be affected by UV radiation which can also promote cataract formation.

Progress in HPLC-HPCE, Vol. 7, pp. 57-91
G. Ali Qureshi *et al.* (Eds)

MONITORING OF EXCITOTOXICITY IN NEUROLOGIC DISORDERS

G.Ali Qureshi
Clinical Research Center, Huddinge University Hospital, Novum,Karolinska Institute, S-141 57 Huddinge, Sweden.

INTRODUCTION

Neurodegenerative diseases are insidious and subtle in onset and run a gradually progressive inexorable course. Many major neurological disorders such as Alzheimer disease (AD), Huntington disease (HD), amyotropic lateral sclerosis (ALS), cerebrovascular disorders (CVD), Parkinson's disease (PD), bacterial or viral meningitis and multiple sclerosis (MS) present a characterstic pattern of damage, at both the gross and histological levels, which is not distributed uniformly throughout the brain. Neurological symptoms and neurodegeneration are silent features of many neuropathological conditions. These include acute metabolic and physiological insult, chronic motor and dementia disorders as well as inflammatory diseases, which are an important cause of morbidity and mortality in man. The role of toxic processes mediated by excitatory amino acids (EAA) and their analogues, and free radicals in neurodegeneration has been the subject of a major research effort for some time. Despite the mechanisms responsible for nerve cell death and neurological dysfunction are likely to be complex, never the less, rapid progress is being made in elucidating the etiology of neuronal degeneration in these diseases. In recent years, increasing numbers of studies have focused attention upon neurodegeneration mechanism as avenues for exploring human neuropatholigical conditions in which such mechanisms appear to play some important role. Techniques of neurochemical pathology developed over the last decade have allowed the investigation of abnormalities of neurotransmission which may give rise to localize changes and may thus, underlie disease pathogenesis. Among the various analytical techniques available, high-performance liquid chromatography (HPLC) has shown a remarkable growth in recent years. It has developed into a modern and powerful technique in chemical analysis. Although liquid chromatographs were available in 1969, it was not until six or seven years later that the technique gained popularity. Within the last two decades have seen the emergence of this biochemical separation technique as mature intellectual and practical endeavors. Very prominent in this continuing growth was the advent of high-quality, well-designed instruments for use in HPLC. The utility and success of HPLC as a quantitative method are due to both its rapidity and simplicity as well as to its sensitivity. New pump designs, column packings and sensitivity detectors make HPLC as a dynamic tool in clinical analysis. These developments have further been responsible for the unprecedented growth in sophistication of life sciences enabling progessively to enunciate biological phenomena more precisely in terms of their molecular origins. In the present time, biochemical separation techniques especially HPLC can clearly perceive as the power house that supplies not only the experimental methods and tools of analysis and isolation, but also the sources of such of the intellectual drive to evaluate in quantitative physicochemical terms many of the diverse biological phenomena. Transmitter and its metabolite concentrations in brain tissue are reflected by density of cells using specific transmitter in a brain region and the activity of the neurons within the area. The heterogeneity in the chemical features of various degenerative disease highlights the importance of biological markers as diagnostic tools. Even more important is the possibility to follow each disease during its different stages, and to understand the metabolic aspects of the disease identified by the biological marker. Establishing a primary role of a biochemical marker in neurodegenerative disease is an important strategy in the design of new pharmacological treatments. Quantitative analyses of biochemical constituents of cerebrospinal fluid (CSF) represents an initial strategy in attempts to link neurological and psychiatric disorders with pathology within specific neurotransmitters and/ or neuromodulator systems in the central nervous system (CNS).

In addition, oxygen free radicals may play a role in triggering neuronal destruction and nitric oxide (NO) has been implicated in a number of diverse physiological processes *(Snyder,1992)*. There is increasing evidence that NO plays an important role in pathophysiology of variety of CNS disorders including cerebral ischemia and seizures and in acute and chronic inflammation (*Koedel et al,1995; Moncada et al,1991; Dawson et al,1992)*. Endogenous NO is synthesized in neurons during the conversion of Arginine (Arg) to Citrulline (Cit) by cleavage of terminal guanidino nitrogen by nitric oxide synthase (NOS). As NO is a very labile molecule, it is difficult to determine NO directly. In aqueous solution, NO reacts rapidly with O_2 and accumulates in aqueous medium as nitrite (NO_2^-) and nitrate (NO_3^-). NO_2^- is freely permeable to membrane of cells and could be detected by HPLC with electrochemical detection. Besides, NO explosure causes cobalamin deficiency resulting from its inactivity which is known to produce impaired neurological functions. Congenital defects of cobalamin metabolism and nutritional deficiencies have been reported to produce abnormal CNS and peripheral neuronal function in humans (*Stollhoff et al, 1987; Hall, 1990*). Psychomotor impairment caused by dietary and genetic defects indicates that cobalamin is essential for normal human brain development (*Stollhoff et al, 1987; Hall, 1990*). Neuropsychiatric abnormalities have been found in cobalamin-deficient adults.

This chapter deals with CSF analysis of amino acids, nitrite (a metabolite of free radical, nitric oxide), vitamin B_{12} and homocysteine in various neurologic disorders . These levels are reported with the assumption that CSF concentration reflect brain or spinal cord concentrations and perhaps synaptic activity, since CSF is in constant exchange with the extracellular fluid of the CNS. Based on the facts that most of the transmitters in the brain are present in the CSF (*Herkenham, 1987*) and that alterations in the concentration of transmitters in the CSF can parallel alterations of the same transmitters in the brain of man (*Wester et al., 1990*), the results are compiled in patients with various neurologic disorders and compared with healthy subjects.

BLOOD-BRAIN BARRIER

The presence of the blood-brain barrier (BBB) has major implications for the passage of relatively large and hydrophilic compounds in brain. As a result , the entry of certain endogenous compounds like nutrients is restricted. Essential nutrients are transported into the brain by means of (selective) carrier mechanisms. Several transport system have been characterized varying from passive transport (diffusion) to active and energy requiring processes. Under healthy conditions, the BBB not only regulates the entry of drug or endogenous compounds into the brain, but also diminishes cellular infiltration compound to peripheral organ. However, the barrier function of BBB can change dramatically during various CNS diseases especially inflammatory reactions underlie the changes in the integrity of BBB during cerebral diseases like MS, CVD and bacterial meningitis (*Yu et al., 1994*). In some of the neurologic disorders such as MS and TBM, the common consequence is an alteration of blood brain barrier permeability (*Saez-Llorens et al., 1990)* and leakage of serum proteins and macromolecules into the CSF which results in the development of vasogenic edema. Additionaly larger numbers of leukocytes enter the subarachnoid space and release toxic substances that contribute to the production of cytotoxic edema (*Talan et al., 1988*). Antidiuretic hormone secretion also contributes to the development of cytotoxic edema by decreasing the toxicity of brain parenchymal fluid and increasing the permeability of the brain to water (*Kaplan, 1989)*. All these events if not modulated promptly cause alterations of CSF dynamics of brain metabolisms and of cerebrovascular autoregulation which results in irrevesible brain injury (*Lipton et al., 1993; Saez-Llorens et al., 1990*). In order to study these changes, CSF obtained by lumbar puncture technique is the most important and widely used diagnostic tool in evaluating the levels of neurotransmitters and their metabolites. Lumbar CSF neurochemical measurements aim to serve as indirect in vivo markers of human central neurotransmission and these markers can be used in the diagnosis of neurodegeneration process.

EXCITOTOXICITY AND NEURODEGENERATION

The etiology of neuronal death in neurodegenerative diseases remains mysterious, however, great advances in both molecular genetics and neurochemistry have improved our knowledge of fundamental processes involved in cell death including both excitotoxicity and oxidative damage. Excitotoxicity refers to neuronal cell death caused by activation of excitatory amino acid (EAA) receptors N-methyl-D-aspartate (NMDA), and other excitatory amino acid receptors have been implicated in neuronal death for sometimes. These receptors not only mediate EAA neurotransmission within the hippocampus, basal ganglia and cerebral cortex but when activated for a sufficient period of time , may also results in neurodegeneration and seizures (*Meldrum et al., 1990*). An increase body of evidence has implicated excitotoxicity as a mechanism of cell death in both acute and chronic neurological diseases (*Choi, 1988; Beal, 1992a; Coyle, 1993*). The concept of slow or weak excitotoxicity has been proposed in neurodegenerative diseases that evolve slowly over many years (*Albin et al., 1992; Beal, 1992a*). Calcium (Ca^{+2}) influx via NMDA receptors are much more effective in mediating cell death than that occurring through non-NMDA receptors or voltage dependent Ca^{+2} channels (*Tymiaski et al.*, 1993). This suggests compartmentalization of Ca^{+2}-dependent neurotoxic processes within neurons with a preferential localization in the submembrane space adjacent to NMDA receptors. Increased intracellular Ca^{+2} levels can initiate a number of deleterious processes including activation of nitric oxide synthase and free radical generation (*Beal, 1995*). NMDA receptors appear to have a critical role in excitotoxicity although non-NMDA receptors could also mediate prolonged stimulation (*Choi, et al., 1987*). NMDA receptor activation is able to trigger neuronal injury more readily than can non-NMDA receptors such as kainic acid (KA) and 3-hydroxy-5-methyl-4-isoxazoleproprionic acid (AMPA) activation due to a greater ability to induce Ca^{2+} overload within a cell. Glutamate (GLU) and Aspartate (ASP) are considered as neurotoxic when present in higher amounts in brain (*Rothman et al, 1986*). The neurotoxic effects of ASP and GLU were described for some time and these have been found responsible for neuronal death in a wide range of neurological disorders (*Rothman,1984; Farber et al,1981; Rothman et al, 1986*). An association between excitotoxic mechanisms and the slow, often insidious development of human neurodegenerative disorders such as MS, PD, AD and ALS, has undergone widespread scrutiny. And the relatively recent recognition of interaction between excitotoxicity and neuronal metabolic stress underscores the likelihood that these two biological mechanisms may act in nature to produce neuropathological conditions (*Albin et al., 1992*). Cell injury or cell death may be introduced by several processes. It has been suggested (*Choi, 1992*) that neuronal injury following brief intense exposure to GLU involves two components, distinguishable by differences in time-course and ionic dependence. The first component marked by neuronal swelling probably reflects Na^{+} ion influx with subsequent passive Cl^{-} and water entry leading to an expansion of cell volume. It occurs within the minutes and depends on the presence of extracellular Na^{+} and Cl^{-} ions and may or may not be lethal. The second and possibily more significant component may involve Ca^{+2} influx mediated by NMDA subtype of glutamate receptors (*Collingridge et al., 1989*). The process leads to a delayed neuronal disintegration (*Choi, 1992*) occurring over a period of hours after exposure to GLU in cultured cells and is dependent on the presence of extracellular Ca^{+2} ion. GLU is known to be released in excess during cerebral ischemia and GLU receptors are enriched in brain regions susceptible to ischemic injury (*Nakanishi, 1992*). Neurotoxic mechanisms mediated by both free radicals and neuronal death following hypoxic or ischemic injury to brain (*Choi, 1990 ; Floyd, 1990*) . Generation of toxic free radicals is also thought to play an important role in the pathogenesis of ischemic induced neuronal death. Recent evidence has suggested that excitotoxicity and oxidative stress (the cytotoxic effect of oxygen radicals) may be sequential and interactive mechanisms leading to neuronal degeneration (*Coyle, 1993*). The most direct evidence linking excitotoxicity to oxidative stress is the finding that NMDA explosure leads to superoxide generation in cultures of cerebellar neurons (*Lofton-Cazal et al., 1993*).

The most accepted theory dependent on various biochemical marker leading to neurodegeneration is as follows;
The cycle of neurodegeneration = Free Radicals (oxidative stress)+ Cytosolic Ca^{+2} + Mitochondrial damage + Excitotoxicity of EAA and homocysteine (HC) + Vitamin B_{12}, vitamin B_6 and folate deficiency . All these factors one way or other result into cell death process in the brain.

MATERIAL AND METHODS

Patients and healthy subjects and routine analysis

The results on various patient groups along with healthy subjects are included in this chapter. Most of the patients were recruited from the department of Neurology, Huddinge University Hospital, Sweden, except the patients with tuberculous meningitis who were recruited from the department of Neurology , the Aga Khan University Hospital, Karachi, Pakistan. The details on patient groups as well as healthy subjects are shown in Table 1. 20 patients with Parkinson's disease (PD) were included in this study. The diagnosis of PD was made according to established clinical criteria (*Lang et al., 1989*). The details on these patients is previously described (*Bednar et al., 1996*). 16 patients with ischemic stroke were included in this study. Computerized tomography (CT scan) and/or MRI confirmed the evidence of stroke and its site. All the strokes were in territory arround the middle cerebral artery.The details on these patients are published earlier (*Halawa et al., 1996*). 20 patients with definite multiple sclerosis (MS) were included in this study . MS was diagnosed according to Schaumacher criteria (*Schaumacher etal., 1965*). The detail on MS patients is previously (*Baig et al., 1995*). 18 patients with aseptic meningitis (AM) and 14 patients with tuberculous meningitis (TBM) were included in this study . The diagnosis of the disease was based on standard diagnosis criteria (*Futrell et al., 1992*). Subsequent examination of CSF was carried on the basis of clinical and therapeutic considerations.The diagnosis of tuberculous meningitis was established on clinical and CSF findings , positive cerebrospinal Gram stain and latex agglutination and positive presence of growth of pathogenic bacteria. The details on AM and TBM patients are as published previously (*Qureshi et al., 1996a*). CSF samples were also collected from 14 healthy individual with complaints of mascular tension headache. These individual were considered as healthy subjects (HS) because CSF routine analysis as well as blood complete examination, liver function test, electrolytes, ESR were within normal limits. Their general physical examination, neurological observation and CT scan of head was also normal.

The clinical data on the healthy subjects and patient groups are presented in Table 1. All values are expressed as mean ± Standard Error Mean (SEM).

10-12 ml cerebrospinal fluid (CSF) was collected from each patient and healthy subject in a sitting position at the L4-L5 levels. Blood samples were collected by venipuncture . The basic CSF analyses included : cell counting by phase-contrast microscopy (*Sörnäs, 1967*), determination of CSF/serum albumin ratio and CSF/immunoglobuline G (IgG) index (*Link et al., 1977*) as well as isoelectric focusing for detection of oligoclonal IgG band. Serum and CSF albumin and IgG were determined using Hitachi 737 Automatic Analyzer (Naka Works, Hitachi Ltd., Tokyo, Japan. CSF and serum samples were kept at -70° C if not analysed immediately.

TABLE 1. Clinical data on the healthy subjects (HS), patients with Parkinson's disease (PD), cerebrovescular disorders (CVD), multiple sclerosis (MS), asceptic (AM) and Tuberculous meningitis (TBM).

Patients	n	Females	Age (years)	CSF-albumin	CSF-IgG	IgG-index
HS	14	5	53±5	218±17	34±4	0.44±0.01
PD	20	7	72±12	332±42*	51±7**	0.43±0.01
CVD	16	6	70±10	313±31*	46±6*	0.42±0.01
MS	20	8	54±6	188±14	65±7**	1.03±0.14**
AM	18	9	46±5	412±53**	83±12***	0.68±0.06**
TBM	14	5	51±4	442±61**	88±11***	0.73±0.11**

*p<0.05,**p<0.01,***p<0.001

Strategy of chemical Analysis of amino acids, nitrite, vitamin B12 and homocysteine

Various chromatographic techniques constitute the arsenal of procedures now available as biochemical separation methods. All are based on the principle of differential migration of the bioanalytes under a defined set of conditions. The interdependence between the biochemical separation sciences and modern biology currently finds its greatest expression in the field anecdotally known as high performance liquid chromatography or HPLC. The development in the methodology of HPLC during the last two decades has resulted in making this analytical method as one of the most used techniques in neurochemical analysis of biological samples such as CSF, blood, brain tissues. The technique provide with a possibility of fast, effecient and reproducible separation with high sensitivity

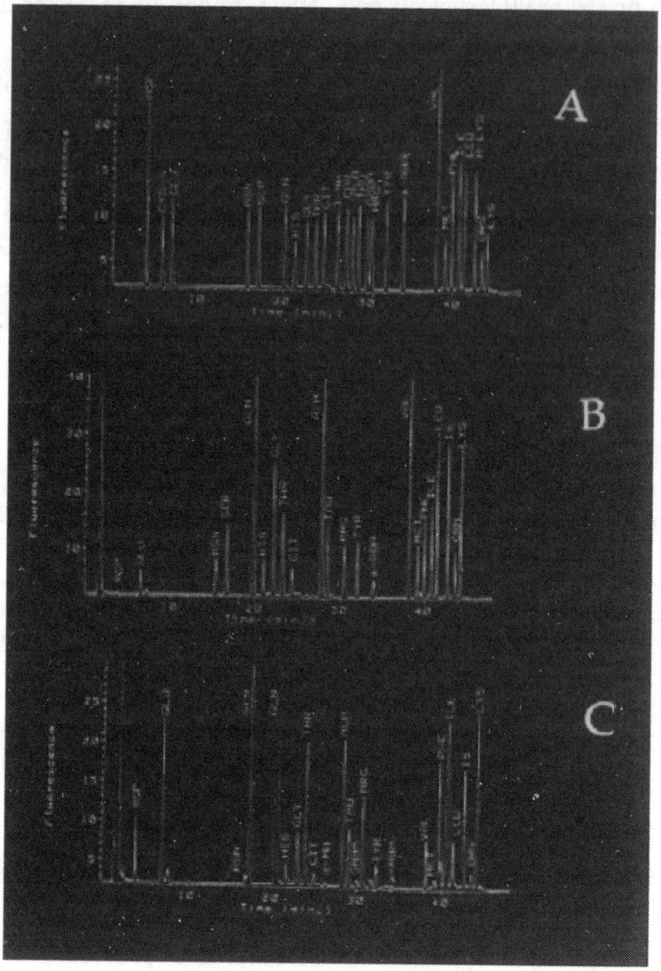

Fig. 1. HPLC separation of 26 OPA-2ME derivatized amino acids (A) standard 10 μmol/l, (B) plasma sample and (C) CSF sample under the chromatgraphic conditions (see ref. Halawa etal., 1991).
Abbrevation for amino acids: ASP=Aspartic acid, CYS-S=Cystein sulphonic acid, GLU=Glutamic acid, ASN=Aspargine, SER= Serine, GLN=Gltamine, HIS=Histidine, GLY=Glycine, THR=Theoronine, CIT=Citrulline, 3-MH=3-methylhistidine, CAR=Carnosine, ALA=Alanine, TAU=Taurine, GABA=Gamma-aminobutyric acid, ARG=Arginine, TYR=Tyrosine, a-ABA= alpha-aminobutyric acid, VAL=Valine, MET=Methionine, PHE=Phenylalanine, ILE=Isoleucine, LEU=Leucine, H-LYS=hydroxylysine, ORN=Ornithine, LYS=Lysine.

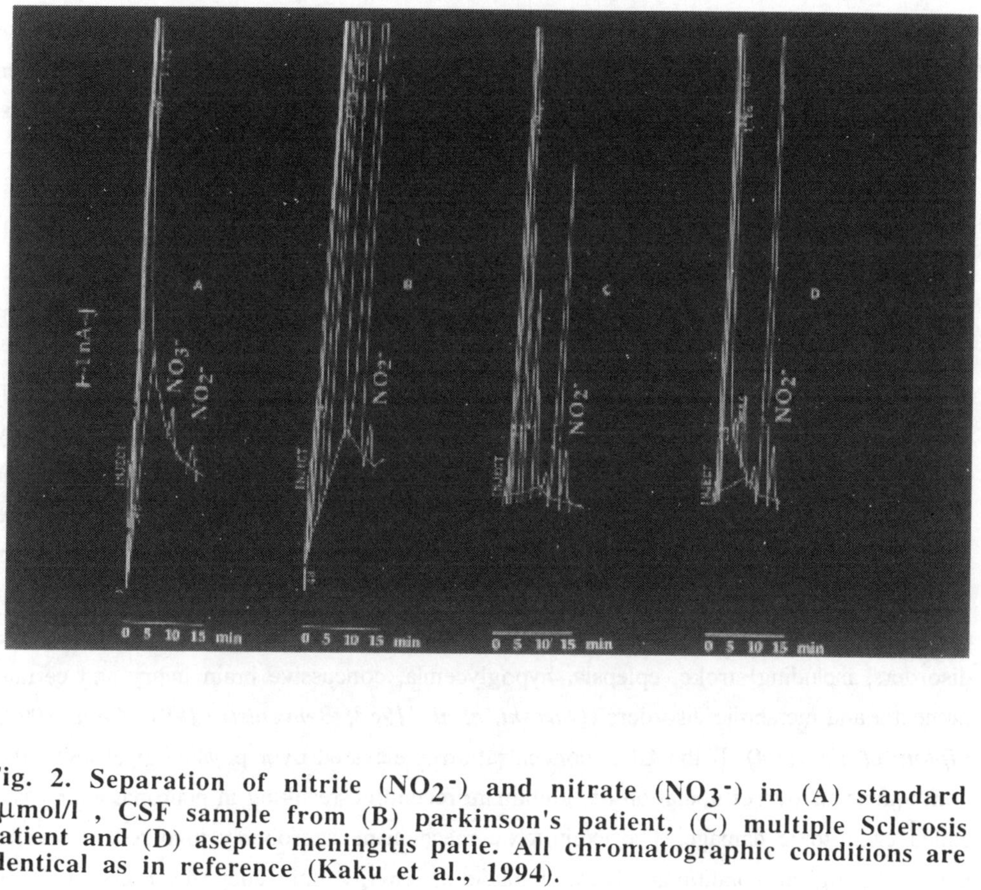

Fig. 2. Separation of nitrite (NO_2^-) and nitrate (NO_3^-) in (A) standard 1μmol/l , CSF sample from (B) parkinson's patient, (C) multiple Sclerosis patient and (D) aseptic meningitis patie. All chromatographic conditions are identical as in reference (Kaku et al., 1994).

Fig. 2 shows the separation of nitrate and nitrite in CSF. Homocysteine was quantitated using previously described HPLC method (*Hyland et al. 1992*) and vitamin B_{12} was quantitated using a competitive protein binding assay (CPBA) based on radioactivity measurement with intrinsic factor (obtain from Amershan Int., Buck., England) (*Ikeda et al., 1990*)

Data are presented as mean ±SEM differences in concentration of amino acids, nitrite, homocysteine and vitamin B12 were analyzed with ANOVA and group comparison were made with t-text. A p-value less than 0.05 was considered significant. Correlation between HC and GLU versus nitrite were submitted to a stepwise polynomial regression analysis (*Zar, 1984*).

NEUROTOXICITY

EXCITATORY AMINO ACIDS

Glutamate (GLU) is the main excitatory neurotransmitter apart from aspartate (ASP) in mammalian brain. In normal physiological conditions GLU acts as a neurotransmitter in several brain regions, regulates neuronal outgrowth and synaptogenesis during normal development and adult plasticity, and has a special function in learning and memory processes (*Monaghan et al., 1989; Lipton et al., 1989; Gasic et al., 1992)* . The concept of excitotoxicity and the observations collectively construed that excessive amounts of the intrinsic EAA specifically GLU is capable of exciting CNS neurons to such an extent as to become toxic, ultimately killing GLU sensitive neurons through over-excitation. An excitotoxic mechanism is involved in the pathogenesis of several neurodegenerative disorders, including stroke, eplepsia, hypoglycemia, concussive brain injury and certain dementia and metabolic disorders (*Qureshi, et al., 1988; Benveniste, 1991; Choi, 1992; Lipton, et al., 1994*). If the GLU concentration is elevated over physiological level, the consequence is overestimulation of glutamate receptors, resulting in neuronal injury, so-called excitotoxic damage. The excitotoxic mechanism is probably involved in several neuropathological conditions. These include a) energy deficiency disorders such as hypoxia and ischemia (*Rothman, 1984; Benveniste, 1984*), Hypoglycemia and prolonged seizures (*Schwarcz, et al., 1984*), b) metabolic disorders such as olivopontocerebellar degeneration (*Plaitakis, et al., 1984*), amyotropic lateral sclerosis (ALS) (*Plaitakis, et al., 1987*) and sulfide oxide deficiency (*Olney, et al., 1975*), and c) disorders induced by food excitotoxic such as endemically occuring neurolathyrism (*Ross, et al., 1987*). In addition, making HPLC as an important tool in various studies in the area of neurology and neurochemistry.

The classical analysis of amino acids was performed by the ion-exchange column developed nearly 4 decade ago (*Spackman et al., 1958*). The application of chromatographic techniques revolutionized the analysis of samples with biological interest and the separation chemistry in general. The samples of biological interest were applied to the column at low PH, and the elution was carried out by using series of buffers of increasing pH and/or ionic strength. Detection of amino acids was accomplished using post-column derivatization with ninhydrin to produce a product that was detected colorimetrically. Although subsequent refinements in the columns and instrument technology have decreased analysis times from days to minutes, and the sensitivity from micromoles to picomoles, this method remained unchanged and is still most widely used for the quantitation of amino acids. Except for ninhydrin, o-phthalaldehyde (OPA) and fluorescaminehave also been used for the amino acid derivatives which offers high sensitivity and capabilities. As a pre-derivatization procedure using reverse-phase columns, the hydrophobicity, seletivity and stability of the reaction products are the main criteria in selection of a derivatizing agent. At the present time, pre-column derivatization procedures are the preferred methods for amino acid analysis since these methods are more sensitive and do not require eloborate systems as compare to post-column derivatization. Many suitable reagents are introduced in the recent years increasing the application area of pre-column systems.

In our laboratory, amino acid analysis was performed based on the pre-column derivatization of OPA with amino acids as previously described (*Qureshi et.al., 1993*). The reproducibility for each amino acid was <2% with this method.Fig. 1 shows the separation of amino acids in CSF and plasma samples.

Since the short half-life time for NO, its concentration is quantitated indirectly by measuring the levels of its degraded product nitrite. Spectrophotometric quantitation of nitrite level is possible by diazo-coupling or measuring methemoglobin converted from hemeglobin by reaction with NO, however, the sensitivity is very low. On the other hand, chemiluminesence method can meet the low sensitivity requirement but the special apparatus are required. Isotropic method can be used ,however, radioactive hazard is the consequence. The better alternative is to use HPLC technique with electrochemical detection to quantitate nitrite and nitrate simultaneously. Recent HPLC method in combination with electrochemical detection is used for this purpose (*Kaku et al., 1994*).

an excitotoxic mechanism has been hypothetized to be involved in several other neuropathological conditions, including Huntington's disease (*Schwarcz et al., 1978),* Alzheimer's disease (*Geddes et al., 1986*), parkinsonism (*Sonsall et al., 1989*) and AIDS encephalopathy (*Lipton, 1992*). An elevated GLU (or related excitatory compounds) concentration is common to all of these conditions. Subsequently it was observed that another intrinsic EAA, the Glu analogue aspartate (ASP), is also capable of inducing an excitotoxic effect (*Olney, 1978*). More recently, however, it has been recognized that excitotoxic neurodegeneration considerably enhanced when intrecellular energy levels are reduced (*Zeevalk et al., 1991; McMaster et al., 1991*), and evidence now indicates that even in the face of relatively low levels of excitotoxic substances, when neuronal energy metabolism is impaired, characterstic excitotoxic neurodegeneration can be demonstrated. Recognition of the interaction between excitatory effects of EAA and reduced intracellular levels has led to studies on more slowly-evolving or chronic form of excitotoxic neurodegeneration. And these investigations have brought into sharper focus the posssibility that certain slowly-evolving neurodegenerative diseases of the CNS may in fact have a chronic excitotoxic neurodegenerative basis or component.

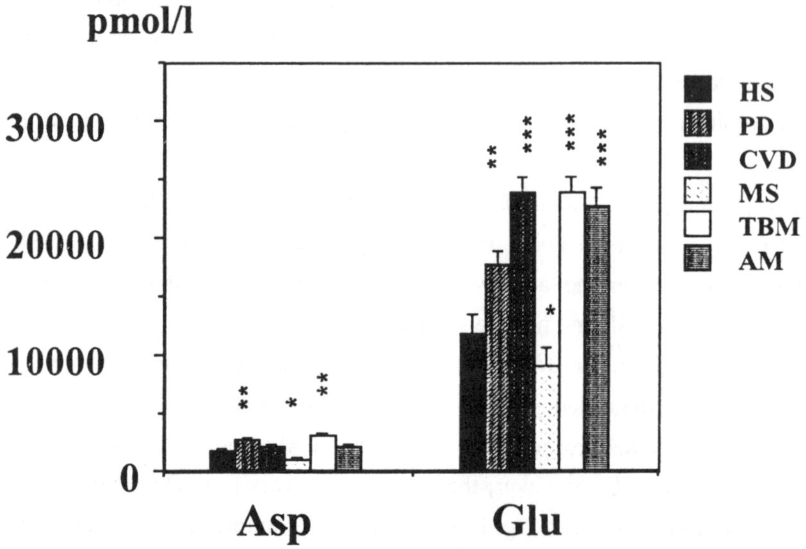

Fig. 3. CSF levels of glutamate and aspartate in various neurologic patients and healthy subjects . All the values are expressed as mean ± SEM. * p<0.05, **p<0.01, ***p<0.001.

Evidence which came to light a decade ago showed that various types of iGLU-receptors responded differentialy when stimulated by their respective ligands so that different subtypes of iGLU-receptors are selective in allowing influx of Na^{+}, Cl^{-}, and Ca^{+2} . These differential responses bear significantly on iGLU-receptor function in both normal and pathological conditions in the CNS (*Schwarcz et al., 1978; Monaghan etal., 1989*). Cell death could result from distrubances of the metabolism or physiology that maintains EAA concentrations at sub-toxic levels. These may include defective catabolism or uptake or excessive production of EAA particularly GLU (*Dodd et al., 1994*). A number of excitotoxic hypotheses have been formed to account for regional variations in the pathology of different orders. It has also been proposed (*Albin et al., 1992*) that neurons in circumscribed brain areas could possess altered receptor subtypes which could make them more susceptible to the excitotoxic effects of endogenous EAA.

Activation of EAA receptor leads to increased intracellular Ca^{+2} followed by activation of protein kinases, phospholipases, nitric oxide synthase, impaired mitochondrial function and the generation of free radicals (*Beal, 1992b*). Our results (*Qureshi, 1997b*) showed the levels of ASP and GLU were increased in the CSF of PD, CVD, TBM and AM patients , however, decreased levels of ASP and GLU were found in MS patients (Fig. 3). In PD, we have shown previously that when comparing the CSF levels of GLU and dopamine (DA), there is an evidence that the interaction between DA and GLU in human brain. And our findings are consistent with the possibility that GLU secretion is enhanced in PD and that DAergic treatment normalizes not only DA but also GLU level (*Bednar et al., 1996*). It has also been suggested that GLU release and interact with neighbouring catecholamine systems to cause catecholamine release via a mechanism which requires production of nitric oxide (NO) (*Montague et al., 1994*). In PD, with the death of DA cells, these interactions are functionally disrupted and GLU may act as a neurotoxin (*Olney, 1990*). Enhanced GLU release increases NO production, causes DNA damage, energy depletion in neurons and neuronal death as occurs in PD (*Zhang et al., 1994*). This hypothesis is also supported by our recent study (*Qureshi et al., 1995*) which showed high level of arginine and nitrite (metabolite of NO) in CSF of PD patients. Similar to PD patients, CVD patients also show elevated levels of GLU and ASP. This is in agreement with the previous results showing that during experimental hypoxia or isechemia , there is a rapid accumulation of extracellular GLU and ASP (*Goldberg et al., 1988*). This also supports the hypothesis that EAA are implicated in the pathophysiology of stroke

(*Rothman, 1984;Farber et al, 1981)*. Local application of GLU or ASP on experimental animal neurons causes spreading depression (*Murphy et al, 1989*). This process is blocked by GLU antagonist (*Siesjö et al,1989*). In diseases with acute neuronal damage caused by ischemia,hypoxia or seizures, it has been suggested that EAA are released in excess and after specific receptor stimulation, EAA can promote or intensify neuronal destruction (*Molennen,1983* ; *Buchan,1990*). Both of these EAA are considered to be responsible for fast excitatory postsynaptic potentials (EPSP) in brain and spinal cord. A major impetus in understanding how the excitotoxic potential of endogenous GLU may be unleashed to cause neuronal degeneration came with the discovery (*Benveniste et al,1984*) that conditions such as hypoxia/ ischemia and head trauma cause a marked outpouring of GLU and ASP from the intracellular to extracellular compartment of brain where they interact with EAA receptors to trigger an excitotoxic cascade. It was also demonstrated that cerebral ischemia causes marked elevation in the extracellular concentration of the endogenous excitotoxins GLU and ASP in rat hippocampus, and in vivo hippocampel cell culture (*Kochhar et al, 1988*) , EAA antagonists can prevent anoxic neuronal degeneration. The latter finding was confirmed suggesting that NMDA antagonist can protect from hypoxic/ischemic brain damage (*Buchan,1990; Rothman et al,1995*). The excitatory action of GLU is mediated by membrane receptors resulting in increase sodium permeability of neurons. Tissue culture studies had also shown that postsynaptic amino acid antagonist 6-glutamylglycine which blocks GLU and ASP receptors reduced EPSP and Asp and GLU responses. These studies further indicate that GLU- and ASP-antagonists blocked neural death when culture was exposed to anoxic condition. Although the detailed molecular basis of GLU neurotoxicity is not known, however, it appears that Ca^{+2} influx plays a critical role (*Siesjö et al,1990; Goldberg et al,1988*).

Unlike PD (*Bednar et al., 1996*) , TBM , AM and CVD (*Qureshi et al., 1996a; 1996b ; Halawa et al, 1996; Qureshi et al., 1997a; Qureshi, 1997b*) , MS patients do not show elevated levels of GLU and ASP in CSF (Fig.3), however, elevated levels are recorded in the serum of MS patients (*Qureshi et al., 1993)*.

A distrubance in transmitter concentration is usually interpreted morphologically i.e. a decrease in transmitter concentration reflects a loss of neurons using that particular transmitter. These levels of CSF GLU and ASP indicate that in brain the neurons containing ASP and GLU are subjected to death process in MS patients.

Similar to Parkinson's disease (PD) (*Bednar et al., 1996*) and brain stroke patients (*Halawa et al., 1996*) , both AM and TBM patients show elevated levels of ASP and

GLU in CSF (Fig. 3) . Hence, the changes in CSF amino acid levels in TBM as well as in AM are assumed to be a consequece of an additional parameter other than individual concentration variation in blood and blood-CSF barrier function. It is also reasonable to suggest that inflammatory changes in choroid plexuses or meninges or the combination of the two may produce distrubance in amino acid transport across the blood-brain barrier in these patients.

GLU, ASP and related EAA account for most of the excitatory synaptic activity in the CNS are released via a Ca^{2+}-dependent mechanism by approximately 40% of all synapses (*Fonnum, 1984*). A transient intense influx of Ca^{+2} may lead to uncontrolled activation of one or more of these potentially lethal processes. It has been demonstrated that acute excitotoxicity experimentally induced by excessive amount of GLU or GLU-analogues can produce neuropathological changes of acute CNS infarction like those seen with ischemia,hypoxia or head trauma (*Meldrum et al., 1990; Choi et al., 1988; Katayama et al., 1988*). It is also established that excessive release of excitatory neurotransmitter GLU and sustained activation of GLU receptors may also be responsible for neuronal degeneration associated with epilepsy, cerebral ischemia, hypoglycemia and other neurodegenerative diseases (*Rothman et al., 1987; Choi, 1988; Meldrum et al., 1990*). It is now emerging that free radical formation and GLU receptor activation may act in concert , cooperating in the genesis and propagation of neuronal damage (*Bondy et al., 1993; Chio, 1990; Coyle et al., 1993).*

OTHER AMINO ACIDS

Apart from GLU, GABA is the brain's commonest transmitter and between GLU and GABA, at least 50 to 80% of CNS synapses (*Walker, 1983*) are used . GABA is derived from GLU via a single enzymatic step (glutamate decarboxylase) and its inhibitory action directly opposes GLU's excitatory action. Both excitatory and inhibitory receptors frequently coexist on the same post synaptic cell. CSF GABA level in nearly all patient groups was increased similar to GLU except in MS where decreased level of GABA was observed similar to GLU (Fig. 4). Among other amino acids, glycine (GLY), and tryptophan (TRP) are elevated in all patients except in MS patients. Phenylalanine (PHE) is increased in TBM and decreased in MS patients (Fig.5). Taurine (TAU) is increased only in CVD where as decreased levels are observed in MS,PD and TBM patients (Fig. 5). Among these amino acids, GLY is an essential intermediate in the metabolism of

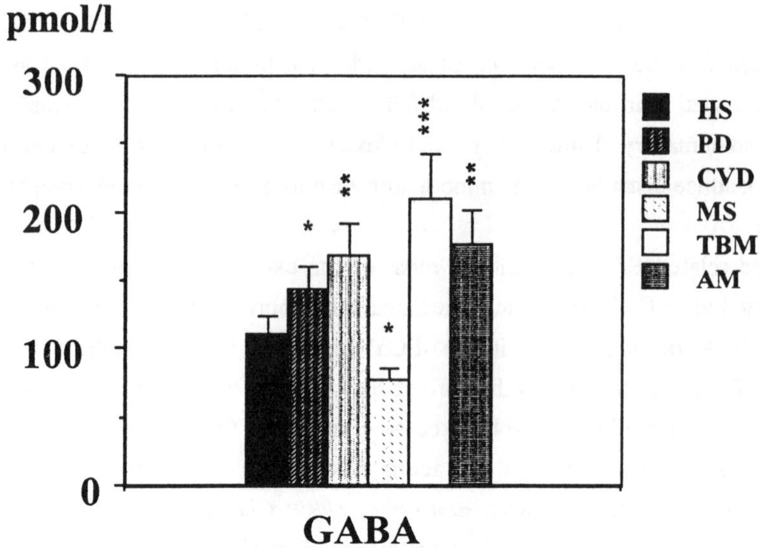

Fig. 4. CSF level of GABA in various neurologic patients and healthy subjects. All the values are expressed as mean ± SEM. * p<0.05, **p<0.01, ***p<0.001.

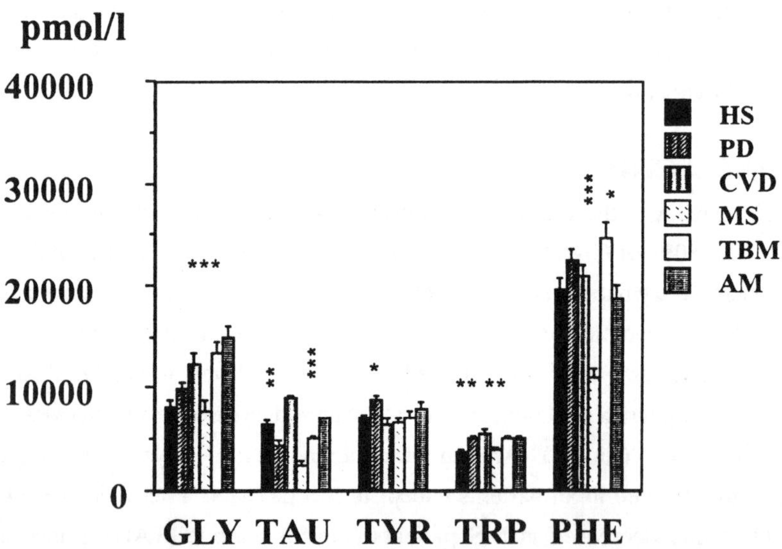

Fig. 5. CSF levels of glycine,taurine,tyrosine,tryptophan and phenylalanine in various neurologic patients and healthy subjects. All values are expressed as mean ± SEM. *p<0.05, ***p<0.001.

protein, peptides, one-carbon fragments, nucleic acids, porphyrin and bile salts. It is also beleived to play a role as a neurotransmitter in CNS. GLY is also a neurotransmitter with cortical hypothalamic projection and its presence in CSF originates from glycinergic neurons in the spinal cord. Hence, the increased GLY levels as observed in most of these patients is suggestive of increased release from spinal cord and brain or decreased in outflow from CSF (*Halawa et al., 1991*). Increased levels of PHE as well as TRP are found in all these patients except in MS where significant decreased levels are noted. These increased and decreased levels of PHE, as well as TRP are of immence importance since both are percursors for catecholamines and serotonin respectively.

FREE RADICALS

A free radical is any species that contains one or more unpaired electrons. Most of the free radicals are unstable species that can extract an electron from neighbouring molecules leading to oxidative demage. Although , there are number of intracellular sources of free radicals, the mitochondria are thought to be the most important. The most reliable and robust risk factor for neurodegenerative disease is normal aging since there is a substantial evidence that mitochondrial function declines with age (*Beal, 1995*). A great deal of interest has been focused on the possibility that oxidative demage plays an important role in the pathogenesis of PD with increased levels of malondialdehyde and lipid peroxide in substantia nigra (*Dexter et al., 1994*), HD patients with significant increase levels of lactate in occipital cortex (*Koroshetz et al., 1994*) and reduction in glutamine synthetas activity which is particularly vulnerable to oxidative stress in AD patients (*Smith et al., 1991*). Hence, the role of free radical in cell death induced by activation of EAA receptors is an area of expanding interest (*Kontos, 1989; Coyle et al., 1993; Halliwell, 1992*). The initial report linking free radicals to excitotoxicity (*Dykens et al., 1987*) showed that kianate induced damage to cerebellar neurons could be attenuated by superoxide dismutase, allopurinol and hydroxyl radical scavengers such as mannitol. The brain is known to have high oxygen consumption needs and is rich in oxidizable substrates especially unsaturated lipids and catecholamines. It is a highly oxygenated organ, consuming almost one fifth of the body's total oxygen, and it derives most of its energy from oxidative metabolism of mitochondrial respiratory chain. Oxygen-centred free radicals are the main types of radicals formed in cells as normal attribute of aerobic life, either as accidental by-products of metabolism or as selectively generated species.

For example, the oxidative burst of neutrophils and macrophages during inflammation relies on de novo synthesis of reactive oxygen species (ROS). Furthermore, ROS are also involved in intracellular signalling processes, modifying the activity of proteins at the post-translational level and regulating gene expression (*Sies, 1991; Halliwell et al. , 1989*). As a result, it is not surprizing that several CNS disorders are related to oxygen radicals either as primary process or a secondary one. In this latter case, secondary free radical formation may result in increased damage over above that is caused by the primary disease process since the brain is sensitive to hyperbaric oxygen exposure. Although the fundamental mechanism is not fully understood, however, excessive oxygen radical production is suggested by the protective action of liposome entrapped superoxide dismutase and catalase (*Yusa et al., 1984*). One of the prerequisites of aerobic life is a competent defence system against ROS (*Yu, 1994*). As with atherosclerosis and acute myocardial infarction, oxygen radicals are probably equally important in ischemic brain injury. In addition to these conditions, a various studies, have shown increased radical formation and lipid peroxidation in most of the neurologic and neuropsychiatric disorders including schizopherenia, MS and AD (*Hunter et al., 1985; Qureshi, 1997b; Halawa et al., 1996*). Oxygen-derived free radicals are very important mediators of cell injury and death. Not only these highly reactive chemical species are important in the aging process but directly or indirectly these species are involved in a wide variety of clinical disorders (*Knight, 1995; Floyd, 1990; Hunter et al., 1985*). In addition, these species play an important role in chronic granulmatous diseases and act as secondary sources of cellular injury in chronic inflammatory processes. Biological systems have developed a comprehensive array of defence mechanisms to protect against free radicals. These includes enzymes to decompose peroxides, proteins to sequester transition metal ions and range of compounds to scavenge free radicals. The term "oxidative stress" is caused by accumulation of free radicals, and is defined as a distrubance in prooxidant-antioxidant balance in favour of former, leading to potentially damaging reactions with biological molecules (*Sies, 1985*) . It has been suggested that oxidative stress is involved in the etiology of various biological processes such as inflammation, aging, carcinogenesis and ischaemia/reperfusion damage (*Sies,, 1991; Halliwell et al., 1989*). However, establishing the involvement of ROS in the pathogenesis of a disease is extremely difficult due to the short lifetimes of these species. The detoxication of ROS is one of the critical prerequisites of all aerobic live. In order to maintain cellular integrity, organism have evolved an array of antioxidant defence which can operate at several different levels within cell. These include;

a. the prevention of radical formation

b. the interception of formed radical

c. the repair of oxidative damage

d. increased elimination of damage molecules

e. non-repair-recognition of excessive damaged molecules in order to prevent the fixation of mutations.

The generation of ROS in vitro or in vivo, in sufficient amounts to overwhelm normal defence mechanisms, can result in serious cell and tissue damage. All of the major classes of biological macromolecules may be attacked, but membrane lipids are probably the most susceptible. The overall effect of lipid peroxidation is to decrease membrane fluidity, thus, destabilising membrane proteins such as receptors. Lipid peroxidation has been implicated in a wide range of tissue injuries and diseases (*Hallliwell et al., 1989; Cross et al., 1987*). On the other hand, the protective effects of iron chelation in several disorders, superoxide dismutase and catalase in ischemia reperfusion and the protection by various vitamins and other oxidants under variety of experimental conditions, along with other examples clearly indicate a key role for free radicals in many, if not most of these disorders (*Halliwell et al, 1985; McCord, 1985*). Although free radicals can damage cell directly through oxidation of biological molecules, they can also inhibit mitochondrial activity, induce the release of EAA and promote a rise in the level of cytosolic free Ca^{2+}, thus promoting a cascade of events leading to neuronal degeneration (*Olanow et al., 1994*). The two most actively investigated oxygen radicals at present are the superoxide anion($O_2 \cdot$) and nitric oxide ($NO \cdot$). The mammalian brain may be exceptionally vulnerable to oxidative stress through attack of these radicals (*Dawson et al., 1996; Cross et al., 1987*).

The free radical nitric oxide (NO) has been implicated in a range of neurodegenerative diseases (*Dodd et al.,1992; Qureshi et al., 1995; Halawa et al., 1996; Qureshi, et al., 1996a, 1996b*). ARG the precursor of NO and nitrite (a metabolite of NO) are significantly increased in all patient groups (*Qureshi et al., 1995; Qureshi et al., 1996a; Halawa et al., 1996*) (Figs. 6 and 7).

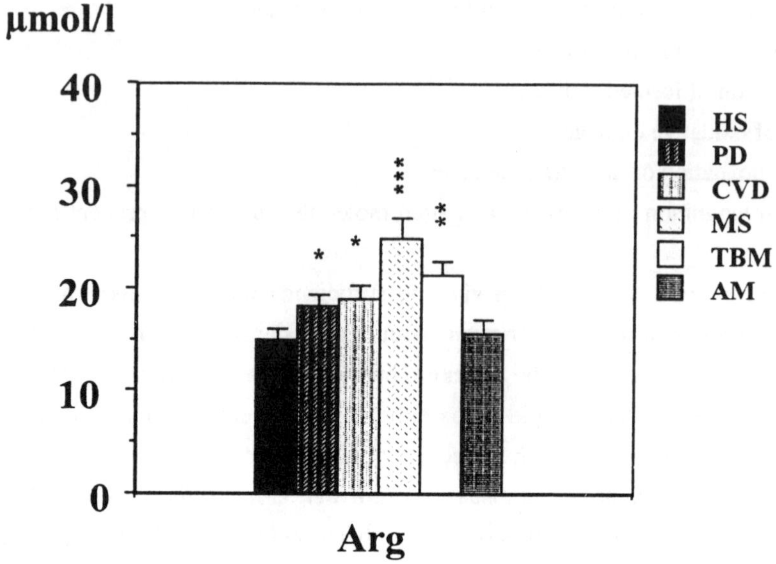

Fig. 6. CSF level of arganine in various neurologic patients and healthy subjects. The value is expressed as mean ± SEM. *p<0.05, **p<0.01.

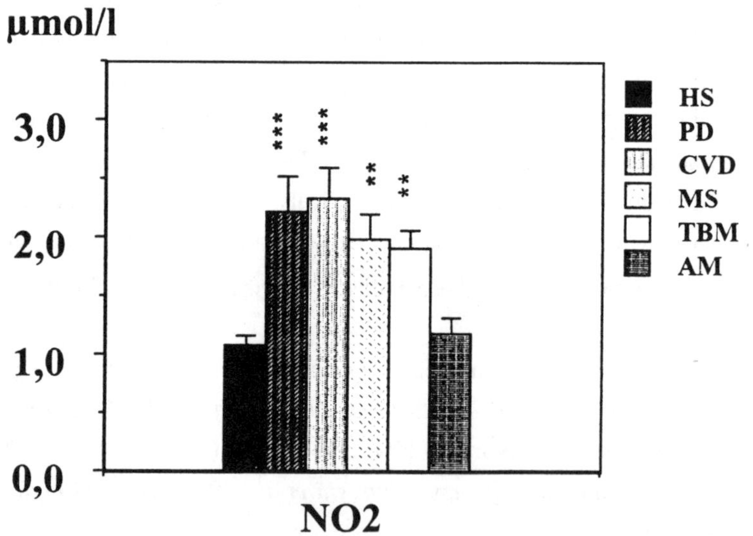

Fig. 7. CSF levels of nitrite (metabolite of nitric oxide) in various neurological patients and healthy subjects. The value is expressed as mean ± SEM. **p<0.01, ***p<0.001.

It has been previously shown (*Qureshi et al., 1996b*) that there is a linear relationship between ARG vs nitrite , GLU vs ARG and GLU vs nitrite in TBM, CVD and MS patients. There is probably no single intracellular pathway that is responsible for neurotoxicity, however, this shows that these neurotoxic generating processes based on GLU and free radical NO are interrelated in these patients (*Qureshi etal.,1997a, 1997c*). According to the exitotoxic theory of ischemia-hypoxic neuronal damage (*Choi 1988, Bondy et al., 1993*) formation of reactive oxygen species may be one consequence of excessive GLU receptor activation, mediated by increased cytosolic Ca^{+2} level within the neuron after GLU exposure (*Pellegini-Giampietro 1990; Volterra et al 1994*). Accumulation of reactive oxygen species in cerebral tissues may be prevented by agents that have antioxidant activity (*Rice -Evans et al., 1993*).

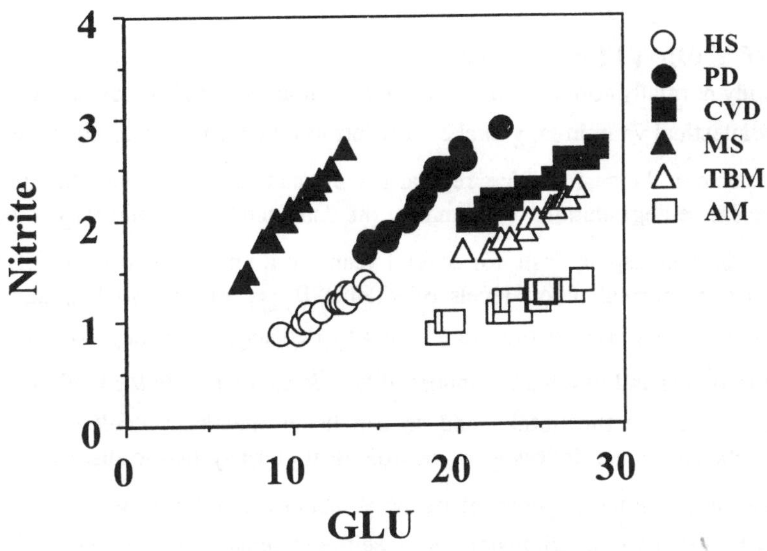

Fig. 8. Relationship between nitrite and Glu in CSF from healthy subjects and patients with various neurologic disorders.

NO function as neuronal messanger in nervous system and in situation of excessive production , it may act as a neurotoxic. It has been clearly demonstrated that GLU and related amino acids such as NMDA markedly stimulate NO synthase (NOS) activity in cerebeller slices monitored by the conversion of Arginine (ARG) to Citrulline (CIT). NMDA receptor activation by EAA and Glu is known as a major trigger for NO formation in several areas of the brain (*Garthwaite, 1991; Dawson et al., 1992*). NO shows enhanced production in post-synaptic cell when NMDA receptors are excessively stimulated (*Garthwaite, 1991*) and promotes the toxic effects of GLU on cells in culture (*Dawson et al., 1992*). Our results clearly show the linear interrelationship between CSF nitrite and GLU in healthy subjects as well as in all patient groups (Fig. 8) as reported ealier (*Qureshi, 1997b*).

An abundance of experimental evidence supports the concept that an acute event causing sudden focal, regional or global loss of oxygen and energy in the CNS can induce a state of neuronal membrane depolarization which allows ion and water shift characterstic of excitoxicity. While these changes appear to be quite consistent with the immediate effects of ion and water influx, the compounding effects of mitochondrial demage due to direct Ca^{+2} toxicity and impairment of mitochondrial function must also be taken into account.

DEFICIENCY OF VITAMIN B12

The human body normally contains ca. 5-10 mg of vitamin B_{12} and the deficiency can be caused by low intestinal vitamin B_{12} intake , low intrinsic factor in the stomach, deficiency of hydrochloric acid in the gastric juice, regular use of laxative or medicine , low uptake in the CNS or excessive degradation. In humans , the daily need of vitamin B_{12} is very low and its storage in liver usually lasts for several years even under malnutrition. However, even in presence of normal blood levels of vitamin B_{12} , an abnormal uptake and/or utilization of the vitamin B_{12} in the brain, leads to deficiency and may contribute to a number of neurological and psychiatric abnormalities (*Ikeda et al., 1990*). Lack of Ca^{+2} in the food can also reduce the uptake and so can heavy metals (*Shevell et al., 1992*). Vitamin B_{12} deficiencies are followed by neurological and psychiatric disorders such as disturbed sense of coordination, loss of memory, abnormal reflexes, weakness, loss of muscle strength, exhaustion, confusion, low self-confidence, incontinence, impaired vision, abnormal gait, frequent need to pass water and physiological deviance (*Metz, 1992*). Although the mechanisms responsible for the neurological lesion of vitamin B_{12} deficiencies are less well understood, however, vitamin B_{12} analogues including methylcobalamin (MCbl) have been widely used in therapy of neurological diseases (*Chanarin etal., 1988*). Vitamin B_{12} was shown to improve memory, emmotional function and communication ability in Alzheimer's patients (*Ikeda etal., 1992*). MCbl is an active coenzyme of vitamin B_{12} analogue that is essential for cell growth and replication (*Hall, 1990*) . Vitamin B_{12} is known to be required in two biochemical reactions in humans. The coenzyme forms of vitamin B_{12} differs in two reactions (a) vitamin B_{12} in the form of adenosyl cobalamin (Ad Cbl) is required as cofactor for methylmolonyl CoA (MMCoA) which is essential for the isomerization of of MMCoA to succinyl CoA - the final step in the metabolism of propionate to succinyl CoA, and (b) Cbl in the form of MCbl is required as cofactor for the enzyme methionine synthetase which catalyzes the recycling of homocysteine to methionine. Folate in the form of methyltetrahydrofolate is also required in this latter reaction (*Lindenbaum etal.,1990*). Less than 5% of vitamin B_{12} is free and the rest of intracellular vitamin B_{12} is bound to the two intracellular enzymes i.e. methionine synthase and MMCoA mutase. Psychomotor impairment caused by dietary and genetic defects indicate that Cbl is essential for normal human brain development (*Stollhoff et al., 1987; Hall, 1990*). Neuropsychiatric abnormalities have been found in Cbl-deficient adult and vitamin B_{12} administeration produced improvement in these patients and partially reversed their elevate serum methylmalonate and homocystein (HC) levels.

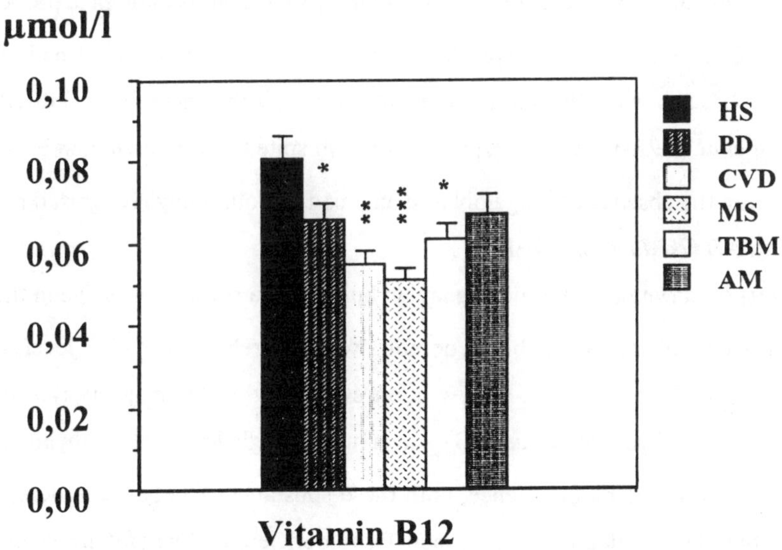

Fig. 9. CSF levels of Vitamin B12 in various neurologic patients and healthy subjects. the value is expressed as mean ± SEM. *p<0.05, **p<0.01, ***p<0.001.

HC accumulates in methylcobalamin deficiency is thought to be neurotoxic for CNS (*Watkin etal., 1989*). Recently, a study (*Reynold et al., 1992*) showed lower level of serum vitamin B_{12} and elevation of unsaturated R-binder capacity in MS patients.

The patients with vitamin B_{12} deficiency manifest as pernicious anemia and macrocytosis of red cells is observed. It is also known (*Crellin et al., 1990;*) that mild macrocytosis which is relatively common in MS patients as compared to the other neurological diseases . Besides, vitamin B_{12} plays an important role in the peripheral and CNS through demethylation and this reaction occuring in the CNS has been regarded as similar to the one in the peripheral nerve system (*Kang et al.,1987*). Compared with other types of vitamin B_{12}, MCbl is more smoothly transferred to nerve tissues (*Imada et al., 1981*) and therefore, it can promote the repair of demage nerves by enhancing the MET synthesis reaction. It is suggested (*Rochelle, 1995*) that in presence of heavy metals the cobalt atom in vitamin B_{12} oxidized from Co^{+2} to Co^{+3} , and with the result the properties of the cobalamin are hypothetically changed and vitamin B_{12} has lost its biological properties.

The reduced CSF cobalamin has been shown in Alzheimer's disease and alcholic dementia (*Regland, 1991*), however, this reduction is also seen in other neurological diseases such as MS (*Baig et al., 1995*). Vitamin B_{12} deficiency in spite of normal vitamin B_{12} levels in serum and/or in the absence of megaloblastic anemia is a commonly recognized problem in patients with AD (*Ikeda et al. 1991*).

In fact, detection of vitamin B_{12} deficiency in humans is a problem to solve in the clinical area , because data on its level in blood do not always correlates with CSF levels (*Ikeda et al., 1990; 1992*). However, there is now good evidence that the majority (>%90) of the patients who are deficient in vitamin B_{12} accumulate high levels of methylmalonic acid and homocysteine in the blood. Hence, both these substances can be considered a better marker for the clinical diagnosis of vitamin B_{12}- deficiency (*Chandefaux et al., 1994*). Vitamin B_{12} exerts a control on one-carbon metabolism by its double role as an acceptor of a methyl group from MTHF. A distrubance in the integrated mechanism between these vitamin-cofactors is expected to cause significant changes in the synthesis and availibility of a number of compounds directly and indirectly linked to vitamin B_{12} / folate metabolism.

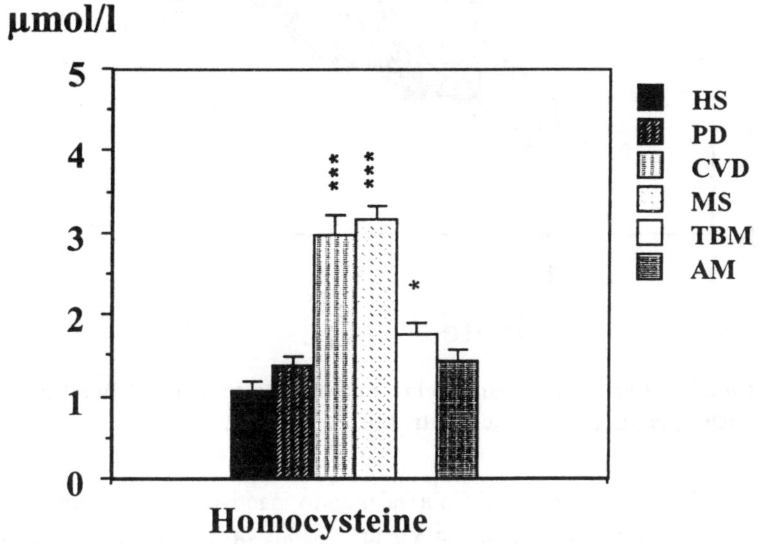

Fig. 10. CSF level of homocysteine in various neurologic patients and healthy subjects. the value is expressed as mean ± SEM. *p<0.05, **p<0.01, ***p<0.001.

It is suggested that such an abnormality produces a severe derangement of folate metabolism which may contribute to etiology of the disorder (*Regland et al., 1992*). NO has been associated with decreased fertility , an increased incidence of spontaneous abortions and the appearance of neurological problems and other disorders (*Henry, 1992*). NO toxicity due to its interaction with vitamin B_{12}-dependent enzyme i.e. methionine synthase produces haematological and less frequently neurological symptoms resembling those of pernicious anaemia (*Yagiela, 1991*). In mammals, the activity of these cobalamin-dependent enzymes is thought to depend on the availibility of vitamin B_{12}. It is also shown that NO inactivates this enzyme. The simultaneous estimation of HC is considered to diagnose cobalamin deficiency because elevated HC occurs due to failure to convert HC to MET, which is a byproduct of functional impairment of methionine synthase due to inadequate cobalamin cofactor synthesis (*Lindebaum et al., 1990*).

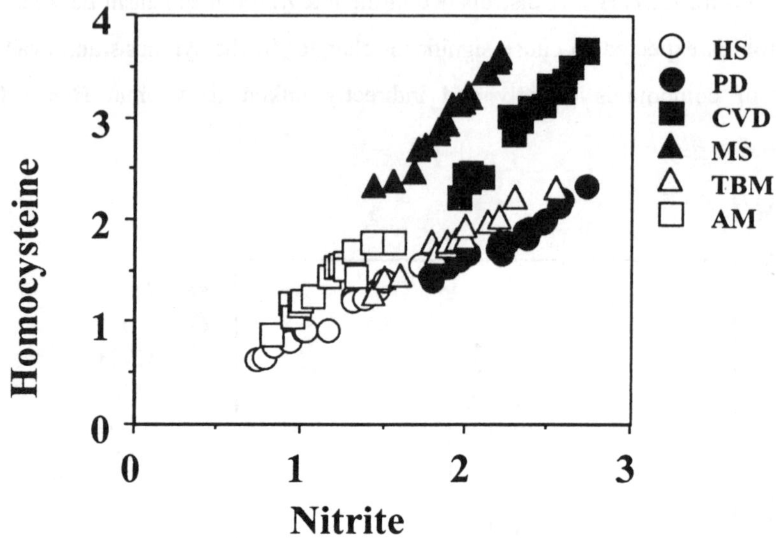

Fig. 11. Relationship between homocysteine and nitrite in CSF of healthy subjects and patients with various neurologic syndromes.

NO is known to oxidize active reduced Cob(I)alamin to inactive Cob(III)alamin (*Banks et al., 1968*). As Cbl in the form of reduced MCbl is required as cofactor for methionine synthetase, exposure to NO causes rapid inactivation of this enzyme (*Deacon et al., 1978*). The inactive CBl is excreted, so that repeated exposure to NO results in depletion of body Cbl stores, with reduced Ado Cbl, so that the activity of the Ado Cbl dependent enzyme MMCoA is also affected (*Deacon et al., 1978*). Long-term Cbl deficiency resulting from its inactivation following NO exposure can produced impaired neurological function. Congenital defects of Cbl metabolism and nutritional deficiencies have also been reported to produce abnormal CNS and peripheral neuronal function in humans. NO interaction with cobalamin results in the liberation of hydroxy free radicals responsible for inactivation of methionine synthase (*Drummond et al., 1994*).

Our results also shows reduced level of vitamin B12 and elevated level of HC in all patient groups except AM patients (Figs 9 & 10). This may be due to the high level of nitrite possibily, suggesting the inter-relationship between nitrite and Vitamin B_{12} in these patients whereas unchanged levels of Vitamin B_{12} and HC are found in AM patients. There is a linear relationship beween CSF nitrite and HC (Fig. 11) in all patient groups, suggesting that both the CSF levels of nitrite and HC are interconnected in the process leading to neurodegeneration . It also suggest the involvement of free radical NO, deficiency of Vitamin B_{12} and increased in level of HC could add up to multifactorial process in the cell death in some of the neurologic disorders such as TBM, PD, MS and CVD (*Qureshi et al., 1997c*).

And there is a clear evidence from our previous results (*Qureshi et al., 1997b*) showing that there is a linear correlation between the level of GLU and nitrite in PD, CVD and TBM patients.

ACKNOWLEDGEMENT

The author is thankful to Dr. Shahid Baig of the Aga University Hospital for providing with the CSF samples from TBM patients.

CONCLUSIONS

There is probably no single intracellular pathway that is responsible for the cell death of neurological diseases (*Rothman et al,1995*). However, hyperactivity of the gluergic innervation, the associated activation of NMDA-receptor ion channels selective for Ca^{+2} and ensuing increase in the intracellular concentration of Ca^{+2} may contribute to the neurodegenerative process (*Rothman et al,1995)* . Besides, oxygen-free radicals, nitric oxide (NO) in particular is thought to mediate NMDA-toxicity (*Dawson et al,1996*). The evidence is accumulating to support the view that besides its beneficial actions, NO production from Arg could contribute and even initiate the cytotoxic effect and brain cell death . The mechanism by which NO kills brain cells is unknown, however, free radical formation has been generally implicated in various forms of neurotoxicity (*Whetsell, 1996*). Our results do indicate increased accumulation of ARG and increased production of nitrite in all patients groups except AM. And there is a clear evidence from our previous studies that there is a linear correlation between the level of GLU and nitrite in CVD, TBM patients (*Halawa et al., 1996; Qureshi, 1997a, 1997b, 1997c; Qureshi et al., 1996*). Besides, the deficiency of Vitamin B_{12} and the elevated level of homocysteine, and the correlation between HC and NO do suggest the possibility of their involvement in neuronal death in these patients. If such a scenario indeed occurs, then a variety of therapeutic approach are possible. This include glutamate release inhibitors, EAA antagonists, agents to improve mitochondrial function, free radical scavengers and neurotropic factors . It is possible that combinations of therapies with vitamin B_{12} that act at sequential steps in the degenerative process may have improved efficacy.

REFERENCES

Albin,R.L. and Greenamyre, J.T. (1992) Alternative excitatoxic hypotheses. *Neurology* **42**, 733-738.

Baig, S.M. and Qureshi, G.A. (1995) Homocysteine and Vitamin B12 in Multiple Sclerosis. *Biogenic Amine* **11(6)**, 479-485.

Bank,R.G.S., Henderson, J.R. and Pratt, J.M. (1968) Reactions of gases in solution. III Some reactions of Nitric oxide with transition metal complexes. *J.Chem.Soc.* **(A) 3**, 2886-90.

Beal, M.F. (1992a) Does impairment of energy metabolism result in excitotoxic neuronal death in neurodegenerative illness ?. *Ann. Neurol* **31**, 119-130.

Beal M.F. (1992b) Mechanism of excitotoxicity in neurological disease, *FASEB J.* **6**, 3338-3344.

Beal, M.F. (1995) Aging, Energy and oxidative stress in neurodegenerative disease. *Ann.Neurol.* **38**, 357-366.

Bednar, I., Baig, S., Silberring, J., Stenfors, C., Södersten, P., Siden, Å. and Qureshi, G.A. (1996) Excitatory amino acids, monoamines, neuropeptide Y and Cholecystokinin in cerebrospinal fluid of Parkinson patients. *Biogenic Amine* **12(4)**, 275-284.

Benveniste, H., Drejer, J., Schousboe, A. and Diemer, N.M. (1984) Elevation of the extracellular concentrations of glutamate and aspartate in hippocampus during transient cerebral ischemia monitored by intracerebral microdialysis.*J.Neurochem.* **43**, 1369-1374

Benveniste, H. (1991) The excitotoxin hypothesis in relation to cerebral ischemia.*Cerebrovasc.Brain metab.Rev.* **3**, 213-245.

Bondy,S.C. and LeBele,C.P. (1993) The relationship between excitotoxicity and oxidative stress in the central nervous system. *Free Rad. Biol.Med.* **14**,633-642.

Buchan, A.M. (1990) Do antagonist protect against cerebral ischemia . Are clinical trails warranted ? *Cerebrovasc.Brain Metab.Rev.* **2**, 1-26.

Chanarin I., Deacon R., Lumb M., Muir M. and Perry J. (1988). Cobalamin folate interrations : A Critical Review. *Blood* **66**,479-482.

Chandefaux, B., Cooper, B.A., Gilfix, B.M., Lue-Shing, H., Carson, W., Gavsie, A. and Rosenbladt, D.S. (1994) Homocysteine: relationship to serum cobalamin, serum folate, erythrocyte folate, and lobation of neutrophils, *Clin.Invest.Med* **17**, 540-550.

Choi,D.W. and Rothman,S.M. (1987) The role of glutamate neurotoxicity in hypoxic-ischemic neuronal death. *Ann.Rev.Neurosci*, **13**, 171-182.

Choi,D.W. (1988) Glutamate neurotoxicity and diseases of nervous system. *Neuron* **1**, 623-634.

Choi, D.W. (1990). Cerebral hypoxia: Some new approaches and unanswered question. *J.Neurosci.* **10**, 2493-2496.

Choi, D.W. (1992). Excitoxic Cell death. *J.Neurobiol.* **23,** 1261-1276.

Collingridge, G.L. and Lester, R.A.J. (1989). Excitatory amino acid receptors in the vertebrate central nervous sytem. *Pharmac.Rev.* **40**, 143-210.

Coyle, J.T. and Puttfarcken, P. (1993) Oxidative stress, glutamate and neurodegenerative disorders. *Science* **262**, 689-694.

Crellin R.F., Bottiglieri T. and Reynolds E.H. (1990). Multiple sclerosis and macrocytosis. *Acta. Neurol. Scand* . **81** ,388-91.

Cross,C.E.,Halliwell,B.,Borish,E.T.,Pryor,W.A.,Ames,B.N.,Saul, R.L.,McCord, J.M. and Harman, D. (1987) Oxygen Radicals and Human Disease, *Annal Int.Med.* **107**, 526-545.

Dawson, T.M., Dawson, V.L. and Snyder, S.H. (1992) A novel neuronal messager molecule in brain : The free radical, nitric oxide. *Ann.Neurol.* **32**, 297-311.

Dawson,V.L. and Dawson, T.M. (1996) Free radicals and neuronal cell death. *Cell Death and Differentiation* **3**, 71-78.

Deacon,R.,Lamb,M. and Perry,J. (1978) Selective inactivation of vitamin B12 in rats by nitrous oxide. *Lancet* **2**, 1023-24.

Dexter, D.T., Holley, A.E. and Flitter, W.D. (1994). increased levels of lipid hydroperoxides in Parkinsonian substantia nigra : An HPLC and ESR study. *Mov.Disord.* **9**, 92-97.

Dodd,P.R.,Williams,S.H.,Gundlach,A.L.,Harper,P.A.W.,Healy,P.J.,Dennis,J.A. and Johnston,G.A.R. (1992) Glutamate and GABA neurotransmitter systems in the acute phase of Maple Syrup urine disease and citrullinemia encephalopathies in new born calves. *J.Neurochem.* **59**, 582-590.

Dodd,P.R.,Scott,H.L. and Wesrphalen,R.I. (1994) Excitotoxic mechanism in the pathogenesis of dementia. *Neurochem.Int.* **25**, 203-219.

Drummond, J.T. and Methews, R.G. (1994) Nitrious oxide degradation by cobalamin-dependent synthase : Characterization of reactants and products in the inactivation reaction. *Biochemistry* **33**, 3732-41.

Dykens, J.A., Stern, A. and Trenkner, E. (1987) Mechanisms of kainate toxicity tocerebellar neurons in vitro is analogue to reperfusion tissue injury.

J.Neurochem. **49**, 1222-1228.
Farber, J.I., Chein, K.R. and Mittnachi, S. (1981) Thepathogenesis of irreversible cell injury in ischemia. *Am.J.Pathol*. **102**, 271-281.
Floyd, R.A. (1990). Role of oxygen free radicals in brain ischemia. *FASEB J*. **4**, 2587-2592.
Fonnum, F. (1984) Glutamate : a neurotransmitters in mammalian brain.*J.Neurochem*. **42**, 1-6.
Futrell, N., Schultz, L.R. and Malliken, C. (1992) Central nervous system disease in patients with systemic lupus erythematosus. *Neurology* **42**, 1649-1657.
Garthwaite,J. (1991) Glutamate,Nitric oxide and cell-cell signalling in the nervous system. *Trends Neurosci*. **14**, 60-67.
Gasic, G.P. and Hollmann, M. (1992) molecular neurobiology of glutamate receptors. *Ann.Rev. Physiol*. **54**, 507-536.
Geddes, J.M., Chang-Chiu, J., Cooper, S.M., Lott, I.T. and Cotman, C.W. (1986) Density and distribution of NMDA receptors in the human hippacampus in Alzheimer's disease . *Brain Res*. **3-9**, 156-161.
Goldberg, M.P., Monyer, H. and Chio, D.W. (1988) Hypoxic neuronal injury in vitro depends on extracellular glutamine. *Neurosci.Lett*. **94**, 52-57.
Halawa, I., Baig, S. and Qureshi, G.A. (1991) Application of HPLC in defining the abnormalities in patients with aseptic meningitis. *J. Biomed. Chromatogr.*, **5**, 216-222.
Halawa, A., Baig, S. and Qureshi, G.A. (1996) Amino acids and nitrite in ischemic brain stroke. *Biogenic Amines* **12 (1)**, 27-36.
Hall, C.A. (1990) Function of Vitamin B12 in the central nervous system as revealed by congenital defects. *Amr.J.Hemalt*. **34**, 121-127.
Halliwell, B. and Gutteridge, J.M.C. (1985) The importance of free radicals and catalytic metal ions in human diseases. *Molec.Aspects Med*. **8**, 189-193.
Halliwell, B. and Gutteridge, J.M.C. (1989) Free radicals in biology and medicine, Clarendon Press, Oxfod.
Halliwell, B. (1992) Reactive oxygen species and the central nervous system. *J.Neurochem*. **59**, 1609-1614.
Henry, R.J. (1992) Assessing environmental health concerns associated with nitric oxide. *J.Amr.Dent.Assoc*. **123**, 41-47.
Herkenham, M. (1987) Mismatches between neurotransmitter and receptor localizations in brain: observations and implications.*Neuroscience*, **23**, 1-38.
Hunter, M.I.S., Niemadim, B.C. and Davidson, D.L.W. (1985). Lipid peroxidation products and antioxidant proteins in plasma and CSF from multiple sclerosis patients. *Neurochem.Res*. **10**, 1645-1652.
Hyland, K. and Bottiglieri, T. (1992) Measurement of total plasma and CSF homocysteine by fluorescence detection followed HPLC and pre-column derivatization with OPA. *J.Chromatogr*. **579**, 55-62.
Ikeda T., Funikawa Y., Mashimoto T., Takahashi K. and Yamada M. (1990). Vitamin B12 levels in serum and CSF of people with

Alzheimer disease. *Acta. Psychiatr. Scand* . **82**, 327-329.

Ikeda, T., Yamamoto, K., Takahashi, K., Kaku, Y., Uchiyama,M., Sugiyama, K. and Yamada; M. (1992) Treatment of Alzheimer-type dementia with intravenous mecobalamin. *Clin.Therapeutics* **14**, 426-437.

Imada, M., Katou T., Kameyam M. and Toyoshima M .(1981). Adsorption and transport of methylcobalamin. In (Kuroiwa G. ed) Nerve System and methyl B12. (pp 23-29), Kyowa Kikatu Tsushin, Tokyo Japan .

Kaku, S., Tanaka,, M., Muramatsu, M. and Otomo, S. (1994) Determination of nitrite by HPLC with EC : Measurement of Nitric oxide synthase activity in rat cerebellum cytosol. *J.Biomed.Chromatogr.* **8**, 24-28.

Kang , S.S. and Wong P.W.K. (1987). Norusis M. Homocysteinuri due to folate deficiency. *Metabolism* **36**, 458-462.

Kaplan, S.L. (1989) Recent Advances in bacterial meningitis. *Adv. Pediatr.Infect.Dis.* **4**, 83-110.

Knight,J.A. (1995) Diseases related to oxygen-derived free radicals. *Annal Clin.Lab.Sci.* **25(2)** ,111-121.

Kochhar, A., Zivin, J.A., Lyden, P.D. and Mazzarella, V. (1988) Glutamate antagonist theraphy reduces neurologic deficit produced by focal nervous sytem ischemia. *Arch.Neurol.* **45**, 148-153.

Koedel, U., Bernatowics, A., Paul, R., Frei, K., Fontana, A. and Pfister, H-W. (1995). Expermental pneumococcal Menigitis : Cerebrovascular alternations, Brain Edema and meningal inflammation are linked to the production of NO. *Annal.Neurology* **37**, 313-323.

Kontos, H.A. (1989) Oxygen radicals in CNS damage. *Chem.Biol.Interact.* **72**, 229-236.

Koroshetz, W.J., Jenkins, B. and Rosen, B. (1994) Evidence from metabolic disorder in Huntington's disease. *Neurology* **44** (suppl.2), A338.

Lang, A.E. and Fahn, S. (1989) Assessment of Parkinson's disease. In: Munsat, T.L. (ed), *Quantification of NeurologicalDeficit*. Butterworths, Boston, pp. 285-309.

Lindenbaum,J.,Healton,E.B. and Savage,D.G. (1988) Neuropsychiatric disorders caused by cobalamin deficiency in the absence of anemia or macrocytosis. *J.Eng.J.Med.* **318**, 1720-1728.

Link, H. and Tibbling, G. (1977) Priciples of albumin and IgG analyses in neurological disorders, III. Evaluation of IgG synthesis within the central nervous system in multiple sclerosis. *Scand.J.Clin.Lab.Invest.* **37**, 297-301

Lipton, S.A. (1992) Models of neuronal injury in AIDS: another role for the NMDA receptor ? *Trends Neurosci.* **15**, 75-79.

Lipton, S.A. and Kater, S.B. (1989) neurotransmitter regulation of neuronal outgrowth, plasticity and survival. *Trends Neurosci.* **12**, 265-270.

Lipton, J.D. and Schafermeyer, R.W. (1993) Evolving concepts in pediatric bacterial meningitis : Part1-pathophysiology and diagnosis. *Ann.Emerg.Med.* **22** ,1602-1615.

Lipton, S.A. and Rosenberg, P.A. (1994) Excitatory amino acids as a final

common pathway for neurologic disorders. *N.Engl.J.Med.* **330**, 613-622.
Lofton-Cazal, M., Pietri, S., Culcasi, M. and Bockaert, J. (1993) NMDA-dependent superoxide production and neurotoxicity. *Nature* **364**, 535- 537.
McCord,J.M. (1985) Oxygen-derived free radicals in post ischemic tissue injury. *N.Engl.J.Med.* **321**, 159-163.
McMaster, O.G., Du, F., French, E.D. and Schwarcz, R. (1991). Focal injection of aminooxyacetic acid produces seizures and lesions in rat hippocampus: Evidence for mediation by NMDA receptors. *Exp.Neurol.* **113**, 378-385.
Meldrum,B.and Garthwaite,J.(1990) Excitatory amino acid neurotoxicity and neurodegeneration disease, *Trends in Pharmacollogical Science* **11**, 379-86.
Metz, J. (1992) Cobalamin deficiency and the pathogensis of nervous system disease. *Ann.Ren.Nub.* **12**, 59-79.
Monaghan,D.R., Bridges,R,J. and Cotman, C.W. (1989) The excitatory amino acid receptors : Their classes, pharmacology and distinct properties in the function of central nervous system.*Ann.Rev.Pharmacol.Toxicol.* **29**, 365-402.
Moncada, S., Plamer, R.M.J. and Higgs, E.A. (1991). Nitric oxide : Physiology, phathophysiology and pharmacology. *Pharmacol. Rev.* **43**, 109-142.
Montague, P.R., Cancayco, C., Winn, M.J., Marchase, R.B. and Friedlander, M.J. (1994) Role of NO production in NMDA receptor-mediated neurotransmitter release in cerebral cortex. *Science,* **263,** 973-977.
Murphy, T.H., Miyamoto, M., Saste, A., Schnaar,, R.L. and Coyle, J.T. (1989) Glutamate toxicity in neuronal cell line involves inhibition of cystine transport leading to oxidative stress. *Neuron* **2**, 1547-1548.
Nakanishi, S. (1992). Molecular diversity of Glutamate receptors and implications for brain function. *Science* **258**, 597-64.
Olney; J.W., Misra, C.H. and deGubareff, T. (1975) Cysteine-S-Sulfate: Brain damaging metabolite in sulfite oxidasedefiency.J.*Neuropathol. Exp.Neurol.* **34**, 167-176.
Olney,J.W. (1978) Neurotoxicity of excitatory amino acids In: McGeer,E.G.,Olney,J.W. and McGeer,P.L. (eds) Kainic acid as a tool in neurobiology. pp 37-69 New York, Raven Press .
Olney, J.W. (1990) Excitatory amino acids and neuropsychiatric disorders. *Annu. Rev. Pharmacol. Toxicol.,* **30,** 47-71.
Pellegrini-Giampietro,D.E.,Cherici,G.,Alesiani,M.,Carla,V.,Moroni,F. (1990) Excitatory amino acid release and free radical formation may cooperate in the genesis ofischemia-induced neuronal damage. *J.Neurosci.* **10**, 1035-1041.
Plaitakis, A. and Caroscio, J.T. (1987) Abnormal glutamate metabolism in amyotropic lateral sclerosis. *Ann.Neurol.* **22**, 575-579.
Plaitakis;A., Berl, S. and Yahr, M.D. (1984) Neurobiological disorders associatedwith deficiency of glutamate dehydrogenase. *Ann.Neurol.* **7**, 297-303.

Qureshi, G.A. and Baig, S.M. (1988) Quantitation of amino acids in biological samples by HPLC. *J.Chromatogr.* **459**, 237-244.

Qureshi, G.A. and Baig, S. (1993) Role of neurotransmitters amino acids in multiple sclerosis in exacerbation, remission, and progressive course. *Biogenic Amines* **10**, 39-48.

Qureshi,G.A., Baig,S., Bednar, I., Södersten, P., Forsberg, G. and Siden, Å. (1995) Increased cerebrospinal fluid concentration of nitrite in Parkinson's disease. *NeuroReport* **6**, 1642-1644.

Qureshi, G.A., Halawa, A. and Baig, S. (1996a) Multiple Sclerosis and Neurotransmission, *Biogenic Amines* **12(5)**, 353-376.

Qureshi, G.A. Baig, S,M. and Parvez, S.H. (1996b) Biochemical markers in tuberculosis meningitis. *Biogenic Amines* **12(6)**, 499-519.

Qureshi, G.A., Halawa, A,, Baig, S.M., Bednar, I. and Parvez, S.H. (1997a) The neurochemical markers in cerebrospinal fluid to differentiate between viral and bacterial meningitis *J.Neurochem.* (in press)

Qureshi, G.A. (1997b) Neuroxicity and the role of aspartic acid, glutamic acid and GABA in neurologic disorders. *Biochem.Mol.Biol.Int.* (in press).

Qureshi, G.A. and Baig, S.M. (1997c) The relationship between the deficiency of Vitamin B12 and neurotoxicity of homocysteine with nitrite level in cerebrospinal fluid of neurological patients. *Biochem.Mol.Biol.Int.* (in press).

Regland, B. (1991) Vitamin B12 defeciency in dementia disorders, Ph.D. thesis, Gothenborg University, Gothenborg , Sweden.

Reynolds E.H., Linnell J.C. and Faludy J.E. (1991). Multiple sclerosis associated with vitamin B12 deficiency. *Arch. Neurol.* **48** , 808-811.

Reynold E.H., Bottiglieri T., Laundy M., Crellin R.F., Kirker S.G.(1992). Vitamin B12 metabolism in multiple sclerosis. *Arch. Neurol.* **49** , 649- 652.

Rice-Evans,C.A. and Diplock,A.T. (1993) Current status of antioxidant therapy. *Free Rad.Biol.Med.* **15,** 77-96.

Rochelle,L.G., Morana, S.J., Kruszyna,H., Russel, M.A., Wilcox, D.E. and Smith, R.P. (1995). Interactions between hydroxocobalamin and nitric oxide: Evidence for redox reaction between nitric oxide and reduced cobalamin and reversible NO binding to oxidized cobalamin.*J.Pharm & Exp. Ther.***275**,48-52.

Ross, S.M., Seelig, M. and Spencer, P.S. (1987) Specific antagonism of excitotoxic action of uncommon amino acids assayed in organotypic mouse cortical cultures. *Brain Res.* **425**, 120-127.

Rothman, S.M. (1984) Synaptic release of excitatory amino acids mediates anoxic neuronal death. *J.Neurosci.* **4**, 1884-1891.

Rothman, S.M. and Olney, J.W. (1986) Glutamate and pathophysiology of Hypoxic-Ischemic brain damage . *Ann.Neurol.* **19**, 105-111.

Rothman,S.M. and Olney, J.W. (1987) Excitotoxicity and the NMDA receptor. *Trends Neurosci*, **10**, 299-303

Rothman, S.M- and Olney, J.W. (1995) Excitatory and NMDA receptors- still lethal after eight years. *Trends in Neurosci.* **18,** 57-58.

Saez-Llorens, X., Ramilo, O. and Mustafa, M.M. (1990) Molecular Phthophysiology of bacterial meningitis : Current concepts and therpeutic implications, *J.Pediatr.* **116**, 671-684.

Schumacher G.A., Becke G., Kibler R.E., Kurland L., Kurtzke J.,McDowell F.,Nogler B., Sibley W., Tourtellotte W.W. and Willmon T.(1965). Problem on the experimental trails of therapy in Multiple Sclerosis : report by a panel on the evaluation of experimental trails of therapy in Multiple Sclerosis.*Ann. N.Y. Acad. Sci.* **122**, 552-558.

Schwarcz,R.,Scholz,D. and Coyle,J.T. (1978) Structure-activity relations for the neurotoxicity of kainic acid derivatives and glutamate analogues. *Neurophamacology* **17**, 145-151.

Schwarcz, R., Foster, A.C., French, E.D., Whetsell, W.O.Jr. and Köhler, C. (1984) Excitotoxic models for neurodegenerative disorders. *Life Sci.* **35**, 19-32.

Shevell M.I. and Rosenblatt D.S. (1992). The Neurology of Cobalamin. *Can. J. Neurol.Sci.* **19**, 472-486.

Sies, H. (1985) *Oxidative Stress*, Academic Press Inc, Ltd., London.

Sies, H. (1991) *Oxidative Stress: Oxidants and Antioxidants*, Academic Press, London.

Siesjö, B.K. and Bengtsson, F. (1989) Calcium fluxes, calcium antagonists and calcium related pathology in brain ischemia, hypoglycemia and spreading depression, a unifying hypothesis. *Cereb.Blood Flow Metab.* **9**, 127-140.

Smith, C.D.,Carney, J.M. and Strake-Reed, P.E. (1991). Excess brain protein oxidation and enzyme dysfunction in normal aging and Alzehamier's disease.*Proc.Natl.Acad.Sci.USA* **88**, 10540-10543.

Snyder, S.H. and Bredt, D.S. (1992) Biological roles of nitric oxide. *Science* **299**, 68-71.

Sonsall, P.K., Niklas, W.J. and Heikkila, R.E. (1989) Role excitatory amino acids in metamphetamine-induced nigrostriatal dopaminergic toxicity. *Science* **243**,398-400.

Sörnäs, R. (1967) A new method for the cytological examination of the cerebrospinal fluid. *J.Neurol.Neurosurg.Psych.* **30**, 568-577.

Spackman, D., Stein, W.H. and Moore, S. (1958) Automatic recording apparatus for use in chromatography of amino acids. *Anal.Chem.* **30,** 1190-1195.

Stollhoff,K. and Schulte. F.J. (1987) Vitamin B12 and brain development. *Eur.J.Paediatr.* **146**, 201-205.

Talan, D.A., Guterman, J.J. and Overturf, G.D. (1988) Influenza of granulocytes of brain edema, intracranial pressure and CSF concentrations of Lactate and protein in experimental meningitis. *J.Infect.Dis.* **157**, 456-459.

Tymiaski,M., Charlton, M.P. and Carlen, P.L. (1993) Source specificity of early Calcium neurotoxicity in cultured embryonic spinal neurons. *J.Neurosci.***13**,2085-2104.

Volterra,A.,Trotti,D.,Tromba,C.,Floridi,S. and Racagni,G. (1994) Glutamate uptake inhibition by oxygen free radicals in rat cortical astrocytes.*J.Neurosci.***14**,2924-32.

Walker, J.E. (1983) Glutamate, GABA and CNS diseases. *Neurochem.Res.* **8**, 521-550.

Wester, P., Bergström, U. and Eriksson, A. (1990) Venticular cerebrospinal fluid monoamine transmitter and metabolite concentrations refect human brain neurochemistry in autopsy cases. *J.Neurochem.* **54**, 1148-1156.

Whetsell, W.O. (1996) Current Concept of excitotoxicity. *J.Neuropath.Expt.Neurology* **55**, 1-13

Yagiela, J.A. (1991) health Hazards and nitric oxide : a time for reappriasal. *Anesth.Prog.* **38**, 1-11.

Yu,B.P.(1994) cellular defences against damage from reactive oxygen species. *Physiol. Rev.* **74**, 139-162.

Yusa,T., Crapo, J.D. and Freeman, B.A. (1984) Lipaso mediated augmentation of brain SOD and catalase inhibits CNS oxygen toxicity. *J.Physiol.* **57**, 1674-1681.

Zar, J.H. (1984) *Biostatistical Analysis*. Englewood Cliffs, Prentice Hall, London.

Zeevalk, G.D. and Nicklas, W.J. (1991) Mechanism underlying initiation of excitatoxicity associated with metabolic inhibition. *J.Pharmacol.Exp.Ther.* **257**, 870-878.

Zhang, J., Dawson, V.L., Dawson, T.M. and Snyder, S.H. (1994) Nitric oxide activation of poly (ADP-ribose) synthetasein neurotoxicity. *Science*,687-89.

Progress in HPLC-HPCE, Vol. 7, pp. 93-110
G. Ali Qureshi *et al.* (Eds)

OXIDATIVE DAMAGE: AN EARLY AND CENTRAL EVENT IN THE PATHOGENESIS OF ALZHEIMER DISEASE

George Perry and Mark A. Smith

Institute of Pathology, Case Western Reserve University, 2085 Adelbert Road, Cleveland, Ohio 44106

INTRODUCTION

Studies over the past five years have established oxidative stress and damage not only in the lesions of Alzheimer disease but also in neurons at risk of death (Smith et al., 1994a,b, 1995a-c, 1996a,b, 1997a,b; Sayre et al., 1997a). Current studies are establishing how oxidative stress is related to other possible causes of Alzheimer disease as well as whether oxidative stress is an initiator or is instead a result of the disease process (Figure 1). The distinction of where oxidative stress lies in this scheme is more than academic since it is essential to know whether reducing oxidative stress will have therapeutic value to protect brain function.

Here we present evidence that oxidative damage is the earliest cytopathological marker of neuronal dysfunction in Alzheimer disease and, moreover, impinges on all proposed pathogenic risk factors and mechanisms implicated in Alzheimer disease.

The pathological presentation of Alzheimer disease, the leading cause of senile dementia, involves regionalized neuronal death and an accumulation of intraneuronal and extracellular lesions termed neurofibrillary tangles and senile plaques, respectively (reviewed in Smith, 1997). Several independent hypotheses have been proposed to link the pathological lesions and neuronal cytopathology with, among

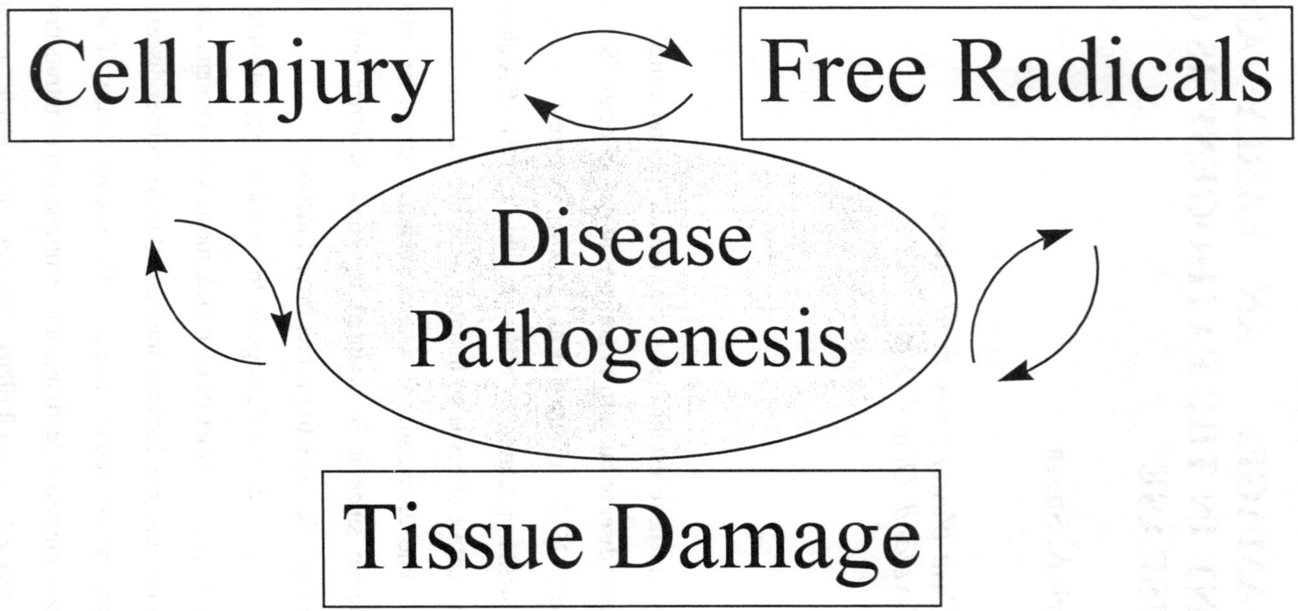

Figure 1: Free radicals are associated with a variety of diseases involving cell injury and tissue damage. Understanding whether free radicals are a primary and early factor in the disease process is critical to determining whether efforts to reduce cellular damage by antioxidant therapy might be efficacious.

others, apolipoprotein E genotype (Corder et al., 1993; Roses, 1995); hyperphosphorylation of cytoskeletal proteins (Trojanowski et al., 1993a), and amyloid-ß metabolism (Selkoe, 1997). However, not one of these theories alone is sufficient to explain the diversity of biochemical and pathological abnormalities of Alzheimer disease which involves a multitude of cellular and biochemical changes. Furthermore, attempts to mimic the disease by a perturbation of one of these elements using cell or animal models, including transgenic animals, do not result in the same spectrum of pathological alterations. The most striking case is that while amyloid-ß plaques are deposited in some transgenic rodent models overexpressing ß-protein precursor, there is no neuronal loss - a seminal feature of Alzheimer disease.

What many of these theories have failed to incorporate is that Alzheimer disease is a disease of aging (Katzman, 1986) (Figure 2). Importantly, this holds true even in individuals with a genetic predisposition, i.e., those individuals with an autosomal dominant inheritance of Alzheimer disease or in individuals with Down's syndrome who develop the pathology of Alzheimer disease. Therefore, age is a clear contributor in 100% of Alzheimer disease cases, whatever the genetic background. The free radical theory of aging (Harman, 1956), hypothesizes that the aging process is associated with (i) an increase in the adventitious production of oxygen-derived radicals, i.e., reactive oxygen species (ROS), together with (ii) a concurrent decrease in the ability to defend against such ROS that leads to the accumulation of oxidatively-modified macromolecules. The decrease in ROS buffering capacity also leads to a compromised ability to deal with abnormal sources of ROS such as those associated with genetic predisposition and/or disease status.

Oxygen Radical Generation and Oxidative Stress in Alzheimer Disease

ROS production occurs as a ubiquitous byproduct of both oxidative phosphorylation and the myriad of oxidases necessary to support aerobic metabolism. In Alzheimer disease, in addition to this background level of ROS, there are a number of additional contributory sources that are thought to play an important role in the disease process: (1) Iron, in a redox-active state, is increased in neurofibrillary tangles as well as in amyloid-ß deposits (Good et al., 1992; Smith et al, 1997a). Iron catalyzes the formation of •OH from H_2O_2 as well as the formation of advanced glycation end products. Furthermore, aluminum, which also accumulates in neurofibrillary tangle-containing neurons (Good et al., 1992), stimulates iron-induced lipid peroxidation (Oteiza, 1994). (2) Activated microglia, such as those that surround most senile plaques (Cras et al., 1990), are a source of NO and O_2^- (Colton and Gilbert, 1987) which can react to form peroxynitrite, leaving nitrotyrosine as an identifiable marker (Good et al., 1996; Smith et al., 1997b). (3) Amyloid-ß itself has been directly implicated in ROS formation through peptidyl radicals (Butterfield et al., 1994; Hensley et al., 1994; Sayre et al., 1997b). (4) Advanced glycation end products in the presence of transition metals (see above) can undergo redox cycling with consequent ROS production (Baynes, 1991; Yan et al., 1994, 1995). Additionally, advanced glycation end products and amyloid-ß activate specific receptors such as the receptor for advanced glycation end products (RAGE) and the class A scavenger-receptor to increase ROS production (Yan et al., 1996; El Khoury et al., 1996). (5) Abnormalities in mitochondrial metabolism, such as deficiencies in key enzyme function, resulting in part from detection of the mitochondrial genome, suggest the mitochondria may be the major and possible initiating source of ROS (Davis et al., 1997). An exact determination of the contribution of each source of oxidative stress is complicated if for no other reason than that most sources have positive feedback. Nonetheless, the overall result is damage including advanced glycation end products (Smith et al., 1994a), nitration (Smith et al., 1997b), lipid peroxidation adduction products (Sayre et al., 1997a) as well as carbonyl-modified neurofilament protein and free carbonyls (Smith et al., 1991; Smith et al., 1994a, 1995b, 1996a; Ledesma et al., 1994; Vitek et al., 1994; Yan et al., 1994; Montine et al., 1996a; Sayre et al., 1997a) with the involvement extending beyond the lesions to neurons not displaying obvious degenerative change. Oxidative crosslinking makes proteins not only insoluble (reviewed in

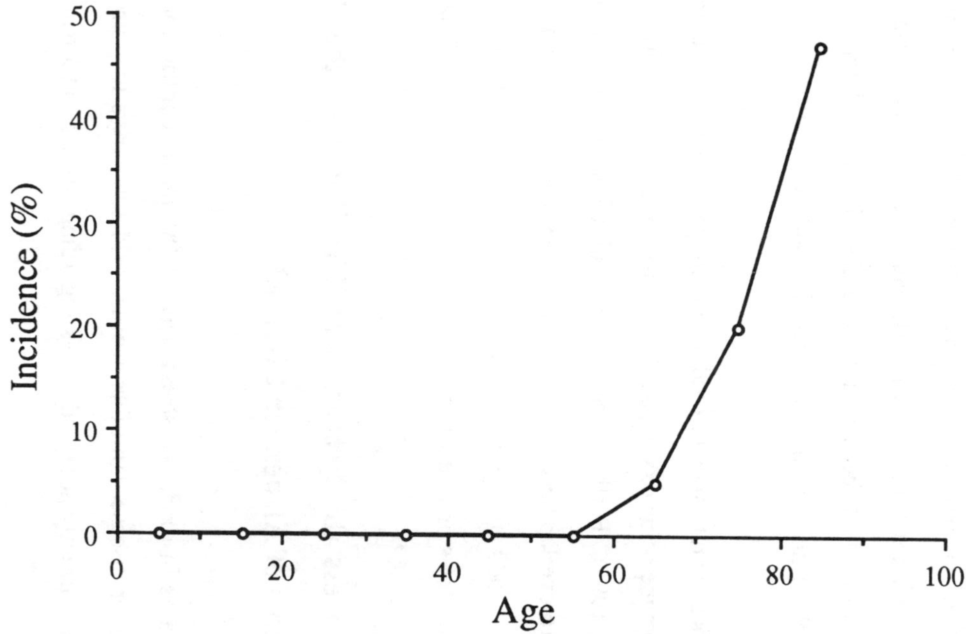

Figure 2: The prevalence of Alzheimer disease increases sharply with age. Adapted from statistics from the National Institute of Aging and the Alzheimer Association.

Smith et al., 1995a; Smith et al., 1996b) but also resistant to proteolytic removal (Cras et al., 1995) by competitively inhibiting the proteosome (Friguet et al., 1994). Therefore, oxidative crosslinking may be significant in the accumulation of ubiquitin conjugates in neurofibrillary tangles (Perry et al., 1987) in the face of numerous proteolytic activities which are highly active against abnormal proteins (Smith and Perry, 1995). Indeed, it may not be coincidental that fibrillary inclusions found in neurodegenerative diseases other than Alzheimer disease are also extensively ubiquinated, e.g., Lewy/Pick bodies and Rosenthal fibers (Galloway et al., 1988; Manetto et al., 1989) and are also oxidatively modified (Castellani et al., 1995, 1996, 1997).

The cytopathological significance of oxidative damage is seen by the upregulation of the antioxidant enzyme heme oxygenase-1 in neurons with NFT (Smith et al., 1994b; Premkumar et al., 1995). Furthermore, in quantitative immunocytochemical studies, there is a complete overlap between heme oxygenase-1 and Alz50, an early marker of τ abnormalities, indicating that cytoskeletal abnormalities are associated with increased oxidative stress or vice versa (Smith and Perry, unpublished observation).

Is Oxidative Stress the Process that Ties Together the Seemingly Separable Events of Alzheimer Disease?

Age-related increases in oxidative stress may be the major contributor toward all aspects of Alzheimer disease. In other words, oxidative stress may provide a link between all of the currently accepted views regarding disease pathogenesis (Figure 3).

A. Oxidative Stress And Cytoskeletal Phosphorylation

The mechanisms by which normal soluble cytoskeletal elements, such as τ and neurofilaments, are transformed into insoluble paired helical filaments is an important issue (Selkoe et al., 1982; Smith et al., 1996b). Insolubility has been linked to the most well known posttranslational change of τ, abnormal

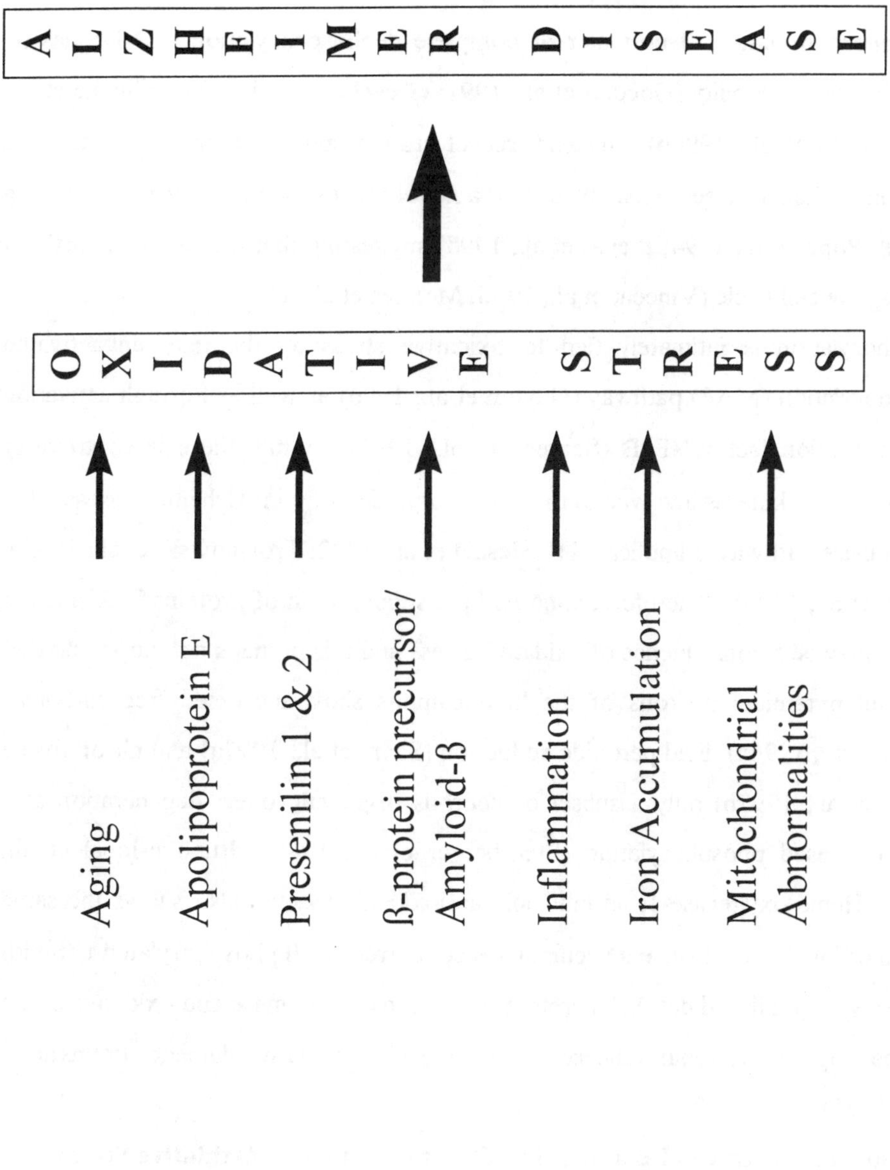

Figure 3: Oxidative stress may be the central mechanism linking the various risk factors of Alzheimer disease into a common pathology.

phosphorylation (Goedert et al., 1991; Greenberg et al., 1992) and a number of specific kinases and phosphatases have been implicated (reviewed in Trojanowski et al., 1993a). However, while increased phosphorylation decreases microtubule stability, a salient feature of the pathology of Alzheimer disease (Perry et al., 1991; Alonso et al., 1994, 1996; Iqbal et al., 1994; Praprotnik et al., 1996a,b), NFT insolubility is not mediated by phosphorylation (Smith et al., 1996b). Indeed, *in vitro* phosphorylation of normal τ or complete dephosphorylation of NFT has no effect on their solubility (Goedert et al., 1991; Greenberg et al., 1992; Gustke et al., 1992; Smith et al., 1996b). Instead, recent studies suggest τ phosphorylation as found in Alzheimer disease may be part of a novel process similar to that seen during mitosis (Pope et al., 1994; Preuss et al., 1995) suggesting that it might be abortively entering the cell cycle (Vincent et al., 1996; McShea et al., 1997).

Phosphorylation is intimately tied to oxidative stress by the mitogen activated phosphorylation (MAP) pathway (Guyton et al., 1996) as well as through activation of transcription factor NFκB (Schreck et al., 1991). While there is controversy concerning the kinases involved in the phosphorylation of τ in Alzheimer disease, the MAP kinase pathway is implicated (Ledesma et al., 1992; Trojanowski et al., 1993b; Hyman et al., 1994). Therefore, abnormal phosphorylation of proteins in Alzheimer disease may be a consequence of oxidative stress and it is perhaps not surprising that while all pyramidal neurons of the hippocampus show increased free carbonyls (Smith et al., 1996a), lipid peroxide adduction (Sayre et al., 1997a) and nitrotyrosine (Smith et al., 1997b) only a subset of neurons displaying overt degeneration also show increased phosphorylation (Sternberger et al., 1985; Grundke-Iqbal et al., 1986). Heme oxygenase-1, an inducible antioxidant enzyme, also shows the same pattern of involvement since it accumulates concurrent with phosphorylated τ (Smith and Perry, unpublished data). Therefore, induction of the kinase and oxidative stress responses appears to require the same critical level of oxidative damage (Premkumar et al., 1995).

B. Influences of Amyloid-ß and Other Genetic Factors on Oxidative Stress

A number of mechanisms have been invoked for the neurotoxicity of amyloid-ß (Yankner et al., 1990; reviewed in Iversen et al., 1995; Sayre et al., 1997b) including membrane depolarization (Carette et al., 1993), increased sensitivity to excitotoxins (Koh et al., 1990), and alterations in calcium homeostasis (Mattson et al., 1992). However, the leading hypothesis is that neuronal damage by amyloid-ß is mediated by free radicals and, as such, can be attenuated using antioxidants such as vitamin E (Behl et al., 1992, 1994) or catalase (Zhang et al., 1996) although there has been some controversy regarding whether the protection from the latter involves its antioxidant activity (Lockhart et al., 1994). Critically addressing whether amyloid-ß initiates oxidative damage in Alzheimer disease requires careful study of the relationship of amyloid-ß deposition and increased oxidative damage. In this regard, it will be extremely interesting to determine the oxidative status of the recently reported transgenic rodent models of amyloid-ß deposition (Games et al., 1995; Hsiao et al., 1996).

Mutations in ßPP are the cause of Alzheimer disease in a small number of familial cases where the mutations lead to increased amyloid-ß deposition (reviewed in Selkoe, 1996, 1997). Transgenic rodents overexpressing ßPP, which display many of the neuropathological correlates of Alzheimer disease, support the important role of amyloid-ß (Games et al., 1995; Hsiao et al., 1996). It is interesting to consider that ßPP/amyloid-ß may be related to oxidative stress. Indeed, amyloid-ß is reported to spontaneously generate peptidyl radicals (Butterfield et al., 1994; Goodman et al., 1994; Harris et al., 1995; Prehn et al., 1996). Further, mutations in ßPP are associated with increased DNA fragmentation, possibly involving oxidative mechanisms (see below).

Presenilins 1 and 2 (Sherrington et al., 1995; Selkoe, 1997) are genetic factors where the biological mechanism, although not established, may also involve oxidative damage. Increased presenilin 2, expression increases DNA fragmentation and apoptotic changes (Wolozin et al., 1996), both important consequences of oxidative damage. Apolipoprotein E, in brains and cerebrospinal fluid, is found adducted with the highly reactive lipid peroxidation product, hydroxynonenal

(Montine et al., 1996b). Furthermore, apolipoprotein E is a strong chelator of copper and iron, important redox-active transition metals (Miyata and Smith, 1996). Finally, interaction of apolipoprotein E with amyloid-ß only occurs in the presence of oxygen (Strittmatter et al., 1993).

In programmed cell death, i.e., apoptosis, cells are digested within their own membrane by proteases and nucleases as well as by increased ROS. However, without the full range of morphological changes, it is unclear whether DNA fragmentation is apoptotic or is, instead, mediated solely by oxidative stress (Berlin and Haseltine, 1981). While the relative contribution of oxidative stress and apoptosis related DNA fragmentation is unresolved, the relative infrequency of apoptosis defined by morphology (Su et al., 1994; Cotman and Su, 1996) and the broad findings of fragmentation in all cells in cases of Alzheimer disease argues for widespread oxidative DNA damage. Certainly, this interpretation is consistent with the oxidative nuclear damage in all cells of the brain in areas affected in Alzheimer disease (Smith et al, 1996a, 1997a).

Therapeutic Value

An important question in discerning whether reducing oxidative stress may have therapeutic value is whether it is a primary or secondary event in disease pathogenesis (Mattson et al., 1995; Smith et al., 1995c) (Figure 2). Recent evidence supports oxidative damage as the earliest cytopathological and biochemical change of Alzheimer disease (Ledesma et al., 1994; Smith et al., 1994a, 1995a, 1996a, 1997b; Vitek et al., 1994; Yan et al., 1994; Sayre et al., 1997a).

Agents that inhibit free radical formation, reduce both the incidence and the progression of Alzheimer disease (McGeer and Rogers, 1992; Rogers et al., 1993; Rich et al., 1995; Stewart et al., 1997). For the latter, several studies have found epidemiological relationships (Breitner et al., 1994; McGeer and Rogers, 1992) that, together with the results of metal chelation treatment (McLachlan et al., 1991), strongly suggest that oxidative stress precedes cell and tissue damage and therefore agents that prevent oxidative damage show promise in the treatment of Alzheimer disease.

Summary

Oxidative stress may underline all of the commonly accepted notions on Alzheimer disease pathogenesis, including hyperphosphorylation, apolipoprotein E genotype, and mutations of the ß-protein precursor. Further studies, to examine the types and extent of oxidative damage, will undoubtedly identify which antioxidant agents will prove most efficacious in the treatment of Alzheimer disease.

Acknowledgments

This work was supported through grants from the National Institutes of Health (AG09287) and the American Health Assistance Foundation (AHAF). M.A.S. is a Daland fellow of the American Philosophical Society.

References

Alonso, A. C., Zaidi, T., Grundke-Iqbal, I. and Iqbal, K. (1994). Role of abnormally phosphorylated tau in the breakdown of microtubules in Alzheimer disease. *Proc. Natl. Acad. Sci. USA* **91**, 5562-5566.

Alonso, A. C., Grundke-Iqbal, I. and Iqbal, K. (1996). Alzheimer's disease hyperphosphorylated tau sequesters normal tau into tangles of filaments and disassembles microtubules. *Nature Med.* **2**, 783-787.

Baynes, J. W. (1991). Role of oxidative stress in development of complications in diabetes. *Diabetes* **40**, 405-412.

Behl, C., Davis, J., Cole, G. M. and Schubert, D. (1992). Vitamin E protects nerve cells from amyloid-ß protein toxicity. *Biochem. Biophys. Res. Commun.* **186**, 944-950.

Behl, C., Davis, J. B., Lesley, R. and Schubert, D. (1994). Hydrogen peroxide mediates amyloid ß protein toxicity. *Cell* **77**, 817-827.

Berlin, V. and Haseltine, W. A. (1981). Reduction of adriamycin to a semiquinone-free radical by NADPH cytochrome P-450 reductase produces DNA cleavage in a reaction mediated by molecular oxygen. *J. Biol. Chem.* **256**, 4747-4756.

Breitner, J. C. S., Gau, B. A., Welsh, K. A., Plassman, B. L., McDonald, W. M., Helms, M. J. and Anthony, J. C. (1994). Inverse association of anti-inflammatory treatments and Alzheimer's disease: initial results of a co-twin control study. *Neurology* **44**, 227-232.

Butterfield, D. A., Hensley, K., Harris, M., Mattson, M. and Carney, J. (1994). ß-amyloid peptide free radical fragments initiate synaptosomal lipoperoxidation in a sequence-specific fashion: implications to Alzheimer's disease. *Biochem. Biophys. Res. Commun.* **200**, 710-715.

Carette, B., Poulain, P. and Delacourte, A. (1993). Electrophysiological effects of 25-35 amyloid- ß-protein on guinea-pig lateral septal neurons. *Neurosci. Lett.* **151**, 111-114.

Castellani, R., Smith, M. A., Richey, P. L., Kalaria, R., Gambetti, P. and Perry, G. (1995). Evidence for oxidative stress in Pick disease and corticobasal degeneration. *Brain Res.* **696**, 268-271.

Castellani, R., Smith, M. A., Richey, P. L. and Perry, G. (1996). Glycoxidation and oxidative stress in Parkinson disease and diffuse Lewy body disease. *Brain Res.* **737**, 195-200.

Castellani, R. J., Perry, G., Harris, P. L. R., Monnier, V. M., Cohen, M. L. and Smith, M. A. (1997). Advanced glycation modification of Rosenthal fibers in Alexander disease. *Brain Res.* (in press).

Colton, C. A. and Gilbert, D. L. (1987). Production of superoxide anions by a CNS macrophage, the microglia. *FEBS Lett.* **223**, 284-288.

Corder, E. H., Saunders, A. M., Strittmatter, W. J., Schmechel, D. E., Gaskell, P. C., Small, G. W., Roses, A. D., Haines, J. L. and Pericak-Vance, M. A. (1993). Gene dose of apolipoprotein E type 4 allele and the risk of Alzheimer's disease in late onset families. *Science* **261**, 921-923.

Cotman, C. W. and Su, J. H. (1996). Mechanisms of neuronal death in Alzheimer's disease. *Brain Pathol.* **6**, 493-506.

Cras, P., Kawai, M., Siedlak, S., Mulvihill, P., Gambetti, P., Lowery, D., Gonzalez-DeWhitt, P., Greenberg, B. and Perry, G. (1990). Neuronal and microglial involvement in ß-amyloid protein deposition in Alzheimer's disease. *Am. J. Pathol.* **137**, 241-246.

Cras, P., Smith, M. A., Richey, P. L., Siedlak, S. L., Mulvihill, P. and Perry, G. (1995). Extracellular neurofibrillary tangles reflect neuronal loss and provide further evidence of extensive protein cross-linking in Alzheimer disease. *Acta Neuropathol.* **89**, 291-295.

Davis, R. E., Miller, S., Herrnstadt, C., Ghosh, S. S., Fahy, E., Shinobu, L. A., Galasko, D., Thal, L. J., Beal, M. F., Howell, N. and Parker, W. D. Jr. (1997). Mutations in mitochondrial cytochrome c oxidase genes segregate with late-onset Alzheimer disease. *Proc. Natl. Acad. Sci. USA* **94**, 4526-4531.

El Khoury, J., Hickman, S. E., Thomas, C. A., Cao, L., Silverstein, S. C. and Loike, J. D. (1996). Scavenger receptor-mediated adhesion of microglia to ß-amyloid fibrils. *Nature* **382**, 716-719.

Friguet, B., Stadtman, E. R. and Szweda, L. I. (1994). Modification of glucose-6-phosphate dehydrogenase by 4-hydroxy-2-nonenal. Formation of cross-linked protein that inhibits the multicatalytic protease. *J. Biol. Chem.* **269**, 21639-21643.

Galloway, P. G., Grundke-Iqbal, I., Iqbal, K. and Perry, G. (1988). Lewy bodies contain epitopes both shared and distinct from Alzheimer neurofibrillary tangles. *J Neuropathol Exp Neurol* **47**, 654-663.

Games, D., Adams, D., Alessandrini, R., Barbour, R., Berthelette, P., Blackwell, C., Carr, T., Clemens, J., Donaldson, T., Gillespie, F., Guido, T., Hagopian, S., Johnson-Wood, K., Khan, K., Lee, M., Leibowitz, P., Lieberburg, I., Little, S., Masliah, E., McConlogue, L., Montoya-Zavala, M., Mucke, L., Paganini, L., Penniman, E., Power, M., Schenk, D., Seubert, P., Snyer, B., Soriano, F., Tan, H., Vitale, J., Wadsworth, S., Wolozin, B. and Zhao, J. (1995). Alzheimer-type neuropathology in transgenic mice overexpressing V717F ß-amyloid precursor protein. *Nature* **373**, 523-527.

Goedert, M., Sisodia, S. S. and Price, D. L. (1991). Neurofibrillary tangles and beta-amyloid deposits in Alzheimer's disease. *Curr. Opin. Neurobiol.* **1**, 441-447.

Good, P. F., Perl, D. P., Bierer, L. M. and Schmeidler, J. (1992) Selective accumulation of aluminum and iron in the neurofibrillary tangles of Alzheimer's disease: a laser microprobe (LAMMA) study. *Ann. Neurol.* **31**, 286-292.

Good, P. F., Werner, P., Hsu, A., Olanow, C. W. and Perl, D. P. (1996). Evidence of neuronal oxidative damage in Alzheimer's disease. *Am. J. Pathol.* **149**, 21-28.

Goodman, Y., Steiner, M. R., Steiner, S. M. and Mattson, M. P. (1994). Nordihydroguaiaretic acid protects hippocampal neurons against amyloid beta-peptide toxicity, and attenuates free radical and calcium accumulation. *Brain Res.* **654**, 171-176.

Greenberg, S. G., Davies, P., Schein, J. D. and Binder, L. I. (1992). Hydrofluoric acid-treated tau PHF proteins display the same biochemical properties as normal tau. *J. Biol. Chem.* **267**, 564-569.

Grundke-Iqbal, I., Iqbal, K., Tung, Y. C., Quinlan, M., Wisniewski, H. M. and Binder, L. I. (1986). Abnormal phosphorylation of the microtubule-associated protein τ (tau) in Alzheimer cytoskeletal pathology. *Proc. Natl. Acad. Sci. USA* **83**, 4913-4917.

Gustke, N., Steiner, B., Mandelkow, E. M., Biernat, J., Meyer, H. E., Goedert, M. and Mandelkow, E. (1992). The Alzheimer-like phosphorylation of tau protein reduces microtubule binding and involves Ser-Pro and Thr-Pro motifs. *FEBS Lett.* **307**, 199-205.

Guyton, K. Z., Liu, Y., Gorospe, M., Xu, Q. and Holbrook, N. J. (1996). Activation of mitogen-activated protein kinase by H_2O_2. Role in cell survival following oxidant injury. *J. Biol. Chem.* **271**, 4138-4142.

Harman, D. (1956). Ageing: a theory based on free radical and radiation chemistry. *J. Gerontol.* **11**, 298-300.

Harris, M. E., Carney, J. M., Cole, P. S., Hensley, K., Howard, B. J., Martin, L., Bummer, P., Wang, Y., Pedigo, N. W. J. and Butterfield, D. A. (1995). ß-amyloid peptide-derived, oxygen-dependent free radicals inhibit glutamate uptake in cultured astrocytes: implications for Alzheimer's disease. *Neuroreport* **6**, 1875-1879.

Hensley, K., Carney, J. M., Mattson, M. P., Aksenova, M., Harris, M., Wu, J. F., Floyd, R. A., and Butterfield, D. A. (1994). A model for ß-amyloid aggregation and neurotoxicity based on free radical generation by the peptide: relevance to Alzheimer disease. *Proc. Natl. Acad. Sci. USA* **91**, 3270-3274.

Hsiao, K., Chapman, P., Nilsen, S., Eckman, C., Harigaya, Y., Younkin, S., Yang, F. and Cole, G. (1996). Correlative memory deficits, Aß elevation, and amyloid plaques in transgenic mice. *Science* **274**, 99-102.

Hyman, B. T., Elvhage, T. E. and Reiter, J. (1994). Extracellular signal regulated kinases. Localization of protein and mRNA in the human hippocampal formation in Alzheimer's disease. *Am. J. Pathol.* **144**, 565-572.

Iqbal, K., Zaidi, T., Bancher, C. and Grundke-Iqbal, I. (1994). Alzheimer paired helical filaments. Restoration of the biological activity by dephosphorylation. *FEBS Lett.* **349**, 104-108.

Iversen, L. L., Mortishire-Smith, R. J., Pollack, S. J. and Shearman, M. S. (1995). The toxicity in vitro of ß-amyloid protein. *Biochem. J.* **331**, 1-16.

Katzman, R. (1986). Alzheimer's disease. *N. Engl. J. Med.* **314**, 964-973.

Koh, J. Y., Yang, L. L. and Cotman, C. W. (1990). ß-amyloid protein increases the vulnerability of cultured cortical neurons to excitotoxic damage. *Brain Res.* **533**, 315-320.

Ledesma, M. D., Correas, I., Avila, J. and Diaz-Nido, J. (1992). Implication of brain cdc2 and MAP2 kinases in the phosphorylation of tau protein in Alzheimer's disease. *FEBS Lett.* **308**, 218-224.

Ledesma, M. D., Bonay, P., Colaco, C. and Avila, J. (1994). Analysis of microtubule-associated protein tau glycation in paired helical filaments. *J. Biol. Chem.* **269**, 21614-21619.

Lockhart, B. P., Benicourt, C., Junien, J. L. and Privat, A. (1994). Inhibitors of free radical formation fail to attenuate direct ß-amyloid$_{25-35}$ peptide-mediated neurotoxicity in rat hippocampal cultures. *J. Neurosci. Res.* **39**, 494-505.

Manetto, V., Abdul-Karim, F. W., Perry, G., Tabaton, M., Autilio-Gambetti, L. and Gambetti, P. (1989). Selective presence of ubiquitin in intracellular inclusions. *Am J Pathol* **134**, 505-513.

Mattson, M. P., Cheng, B., Davis, D., Bryant, K., Lieberburg, I. and Rydel, R. E. (1992). ß-amyloid peptides destabilize calcium homeostasis and render human cortical neurons vulnerable to excitotoxicity. *J. Neurosci.* **12**, 376-389.

Mattson, M. P., Carney, J. W. and Butterfield, D. A. (1995). A tombstone in Alzheimer's? *Nature* **373**, 481.

McGeer, P. L. and Rogers, J. (1992). Anti-inflammatory agents as a therapeutic approach to Alzheimer's disease. *Neurology* **42**, 447-449.

McLachlan, D. R., Kruck, T. P., Lukiw, W. J. and Krishnan, S. S. (1991). Would decreased aluminum ingestion reduce the incidence of Alzheimer's disease? *Can. Med. Assoc. J.* **145**, 793-804.

McShea, A., Harris, P. L. R., Webster, K. R., Wahl, A., and Smith, M. A. (1997). Abnormal expression of the cell cycle regulators P16 and CDK4 in Alzheimer's disease. *Am. J. Pathol.* **150**, 1933-1939.

Miyata, M. and Smith, J. D. (1996). Apolipoprotein E allelle-specific antioxidant activity and effects on cytotoxicity by oxidative insults and ß-amyloid peptides. *Nature Genetics* **14**, 55-61.

Montine, T. J., Amarnath, V., Martin, M. E., Strittmatter, W. J. and Graham, D. G. (1996a). E-4-hydroxy-2-nonenal is cytotoxic and cross-links cytoskeletal proteins in P19 neuroglial cultures. *Am. J. Pathol.* **148**, 89-93.

Montine, T. J., Huang, D. Y., Valentine, W. M., Amarnath, V., Saunders, A., Weisgraber, K. H., Graham, D. G., and Strittmatter, W. J. (1996b). Crosslinking of apolipoprotein E by products of lipid peroxidation. *J. Neuropathol. Exp. Neurol.* **55**, 202-210.

Oteiza, P. I. (1994). A mechanism for the stimulatory effect of aluminum on iron-induced lipid peroxidation. *Arch. Biochem. Biophys.* **308**, 374-379.

Perry, G., Friedman, R., Shaw, G. and Chau, V. (1987). Ubiquitin is detected in neurofibrillary tangles and senile plaque neurites of Alzheimer disease brains. *Proc. Natl. Acad. Sci. USA* **81**, 3033-3036.

Perry, G., Kawai, M., Tabaton, M., Onorato, M., Mulvihill, P., Richey, P., Morandi, A., Connolly, J. A. and Gambetti, P. (1991). Neuropil threads of Alzheimer's disease show a marked alteration of the normal cytoskeleton. *J. Neurosci.* **11**, 1748-1755.

Pope, W. B., Lambert, M. P., Leypold, B., Seupaul, R., Sletten, L., Krafft, G., and Klein, W. L. (1994). Microtubule-associated protein tau is hyperphosphorylated during mitosis in the human neuroblastoma cell line SH-SY5Y. *Exp. Neurol.* **126**, 185-194.

Praprotnik, D., Smith, M. A., Richey, P. L., Vinters, H. V. and Perry, G. (1996a). Plasma membrane fragility in dystrophic neurites in senile plaques of Alzheimer's disease: an index of oxidative stress. *Acta Neuropathol.* **91**, 1-5.

Praprotnik, D., Smith, M. A., Richey, P. L., Vinters, H. V. and Perry, G. (1996b). Filament heterogeneity within the dystrophic neurites of senile plaques suggests blockage of fast axonal transport in Alzheimer's disease. *Acta Neuropathol.* **91**, 226-235.

Prehn, J. H., Bindokas, V. P., Jordan, J., Galindo, M. F., Ghadge, G. D., Roos, R. P., Boise, L. H., Thompson, C. B., Krajewski, S., Reed, J. C. and Miller, R. J. (1996). Protective effect of transforming growth factor-ß 1 on ß-amyloid neurotoxicity in rat hippocampal neurons. *Mol. Pharmacol.* **49**, 319-328.

Premkumar, D. R. D., Smith, M. A., Richey, P. L., Petersen, R. B., Castellani, R., Kutty, R. K., Wiggert, B., Perry, G. and Kalaria, R. N. (1995). Induction of heme oxygenase-1 mRNA and protein in neocortex and cerebral vessels in Alzheimer's disease. *J. Neurochem.* **65**, 1399-1402.

Preuss, U., Doring, F., Illenberger, S. and Mandelkow, E. M. (1995). Cell cycle-dependent phosphorylation and microtubule binding of tau protein stably transfected into Chinese hamster ovary cells. *Mol. Biol. Cell* **6**, 1397-1410.

Rich, J. B., Rasmusson, D. X., Folstein, M. F., Carson, K. A., Kawas, C. and Brandt, J. (1995). Nonsteroidal anti-inflammatory drugs in Alzheimer's disease. *Neurology* **45**, 51-55.

Rogers, J., Kirby, L. C., Hempelman, S. R., Berry, D. L., McGeer, P. L., Kaszniak, A. W., Zalinski, J., Cofield, M., Mansukhani, L., Willson, P. and Kogan, F. (1993). Clinical trial of indomethacin in Alzheimer's disease. *Neurology* **43**, 1609-1611.

Roses, A. D. (1995). On the metabolism of apolipoprotein E and the Alzheimer diseases. *Exp. Neurol.* **132**, 149-156.

Sayre, L. M., Zelasko, D. A., Harris, P. L. R., Perry, G., Salomon, R. G. and Smith, M. A. (1997a). 4-Hydroxynonenal-derived advanced lipid peroxidation end products are increased in Alzheimer's disease. *J. Neurochem.* **68**, 2092-2097.

Sayre, L. M., Zagorski, M. G., Surewicz, W. K., Krafft, G. A. and Perry, G. (1997b). Mechanisms of neurotoxicity associated with amyloid β deposition and the role of free radicals in the pathogenesis of Alzheimer's disease. A critical appraisal. *Chem. Res. Toxicol.* **10**, 518-526.

Schreck, R., Rieber, P. and Baeuerle, P. A. (1991). Reactive oxygen intermediates as apparently widely used messengers in the activation of the NF-κB transcription factor and HIV-1. *EMBO J.* **10**, 2247-2258.

Selkoe, D. J., Ihara, Y. and Salazar, F. J. (1982). Alzheimer's disease: insolubility of partially purified paired helical filaments in sodium dodecyl sulfate and urea. *Science* **215**, 1243-1245.

Selkoe, D. J. (1996). Amyloid beta-protein and the genetics of Alzheimer's disease. *J. Biol. Chem.* **271**, 18295-18298.

Selkoe, D. J. (1997). Alzheimer's disease: genotypes, phenotypes, and treatments. *Science* **275**, 630-631.

Sherrington, R., Rogaev, E. I., Liang, Y., Rogaeva, E. A., Levesque, G., Ikeda, M., Chi, H., Lin, C., Li, G., Holman, K., Tsuda, T., Mar, L., Foncin, J. -F., Bruni, A. C., Montesi, M. P., Sorbi, S., Rainero, I., Pinessi, L., Polinsky, R. J., Wasco, W., Da Silva, H. A. R., Haines, J. L., Pericak-Vance, M. A., Tanzi, R. E., Roses, A. D., Fraser, P. E., Rommens, J. M. and St. George-Hyslop, P. H. (1995). Cloning of a gene bearing missense mutations in early-onset familial Alzheimer's disease. *Nature* **375**, 754-760.

Smith, C. D., Carney, J. M., Starke-Reed, P. E., Oliver, C. N., Stadtman, E. R., Floyd, R. A. and Marksberry, W. R. (1991). Excess brain protein oxidation and enzyme dysfunction in normal aging and in Alzheimer disease. *Proc. Natl. Acad. Sci. USA* **88**, 10540-10543.

Smith, M. A., Taneda, S., Richey, P. L., Miyata, S., Yan, S.-D., Stern, D., Sayre, L. M., Monnier, V. M. and Perry, G. (1994a). Advanced Maillard reaction products are associated with Alzheimer disease pathology. *Proc. Natl. Acad. Sci. USA* **91**, 5710-5714.

Smith, M. A., Kutty, R. K., Richey, P. L., Yan, S.-D., Stern, D., Chader, G. J., Wiggert, B., Petersen, R. B. and Perry, G. (1994b). Heme oxygenase-1 is associated with the neurofibrillary pathology of Alzheimer's disease. *Am. J. Pathol.* **145**, 42-47.

Smith, M. A., Sayre, L. M., Monnier, V. M. and Perry, G. (1995a). Radical AGEing in Alzheimer's disease. *Trends Neurosci.* **18**, 172-176.

Smith, M. A., Rudnicka-Nawrot, M., Richey, P. L., Praprotnik, D., Mulvihill, P., Miller, C. A., Sayre, L. M. and Perry, G. (1995b). Carbonyl-related posttranslational modification of neurofilament protein in the neurofibrillary pathology of Alzheimer's disease. *J. Neurochem.* **64**, 2660-2666.

Smith, M. A., Sayre, L. M., Vitek, M. P., Monnier, V. M. and Perry, G. (1995c). Early AGEing and Alzheimer's. *Nature* **374**, 316.

Smith, M. A., Perry, G., Richey, P. L., Sayre, L. M., Anderson, V. E., Beal, M. F. and Kowall, N. (1996a). Oxidative damage in Alzheimer's. *Nature* **382**, 120-121.

Smith, M. A., Siedlak, S. L., Richey, P. L., Nagaraj, R. H., Elhammer, A. and Perry, G. (1996b). Quantitative solubilization and analysis of insoluble paired helical filaments from Alzheimer disease. *Brain Res.* **717**, 99-108.

Smith, M. A., Harris, P.L.R., Sayre, L.M. and Perry, G. (1997a). Redox-active iron is associated with the pathological lesions in Alzheimer disease. *J. Neuropath. Exp. Neurol.* **56**, 608.

Smith, M. A., Harris, P. L. R., Sayre, L. M., Beckman, J. S. and Perry, G. (1997b). Widespread peroxynitrite-mediated damage in Alzheimer's disease. *J. Neurosci.* **17**, 2653-2657.

Smith, M. A. (1997). Alzheimer Disease. In: *International Review of Neurobiology*, R. J. Bradley and R. A. Harris (Eds.). Academic Press, San Diego. In press.

Smith and Perry (1995) ?

Sternberger, N. H., Sternberger, L. A. and Ulrich, J. (1985). Aberrant neurofilament phosphorylation in Alzheimer disease. *Proc. Natl. Acad. Sci. USA* **82**, 4274-4276.

Stewart, W. F., Kawas, C., Corrada, M. and Metter, E. J. (1997). Risk of Alzheimer's disease and duration of NSAID use. *Neurology* **48**, 626-631.

Strittmatter, W. J., Weisgraber, K. H., Huang, D. Y., Dong, L. M., Salvesen, G. S., Pericak-Vance, M., Schmechel, D., Saunders, A. M., Goldgaber, D. and Roses, A. D. (1993). Binding of human apolipoprotein E to synthetic amyloid ß peptide: isoform-specific effects and implications for late-onset Alzheimer disease. *Proc. Natl. Acad. Sci. USA* **90**, 8098-8102.

Su, J. H., Anderson, A. J., Cummings, B. J. and Cotman, C. W. (1994). Immunohistochemical evidence for apoptosis in Alzheimer's disease. *Neuroreport* **5**, 2529-2533.

Trojanowski, J. Q., Schmidt, M. L., Shin, R.-W., Bramblett, G. T., Goedert, M. and Lee, V. M.-Y. (1993a). PHF-τ (A68): From pathological marker to potential mediator of neuronal dysfunction and degeneration in Alzheimer's disease. *Clin. Neurosci.* **1**, 184-191.

Trojanowski, J.Q., Mawal-Dewan, M., Schmidt, M.L., Martin, J. and Lee, V.M. (1993b) Localization of the mitogen activated protein kinase ERK2 in Alzheimer's disease neurofibrillary tangles and senile plaque neurites. *Brain Res.* **618**, 333-337.

Vincent, I., Rosado, M. and Davies, P. (1996). Mitotic mechanisms in Alzheimer's disease? *J. Cell Biol.* **132**, 413-425.

Vitek, M. P., Bhattacharya, K., Glendening, J. M., Stopa, E., Vlassara, H., Bucala, R., Manogue, K. and Cerami, A. (1994). Advanced glycation end products contribute to amyloidosis in Alzheimer disease. *Proc. Natl. Acad. Sci. USA* **91**, 4766-4770.

Wolozin, B., Iwasaki, K., Vito, P., Ganjei, J. K., Lacana, E., Sunderland, T., Zhao, B., Kusiak, J. W., Wasco, W., and D'Adamio, L. (1996). Participation of presenilin 2 in apoptosis: enhanced basal activity conferred by an Alzheimer mutation. *Science* **274**, 1710-1713.

Yan, S. -D., Chen, X., Schmidt, A. -M., Brett, J., Godman, G., Zou, Y. -S., Scott, C. W., Caputo, C., Frappier, T., Smith, M. A., Perry, G., Yen, S. -H., and Stern, D. (1994). Glycated tau protein in Alzheimer disease: a mechanism for induction of oxidant stress. *Proc. Natl. Acad. Sci. USA* **91**, 7787-7791.

Yan, S.D., Yan, S.F., Chen, X., Fu, J., Chen, M., Kuppusamy, P., Smith, M.A., Perry, G., Godman, G.C., Nawroth, P., Zweier, J.L. and Stern, D. (1995). Non-enzymatically glycated tau in Alzheimer's disease induces neuronal oxidant stress resulting in cytokine gene expression and release of amyloid ß-peptide. *Nature Medicine* **1**, 693-699.

Yan, S. D., Chen, X., Fu, J., Chen, M., Zhu, H., Roher, A., Slattery, T., Zhao, L., Nagashima, M., Morser, J., Migheli, A., Nawroth, P., Stern, D. and Schmidt, A. M. (1996). RAGE and amyloid-ß peptide neurotoxicity in Alzheimer's disease. *Nature* **382**, 685-691.

Yankner, B. A., Duffy, L. K. and Kirschner, D. A. (1990). Neurotrophic and neurotoxic effects of amyloid beta protein: reversal by tachykinin neuropeptides. *Science* **250**, 279-282.

Zhang, Z., Rydel, R. E., Drzewiecki, G. J., Fuson, K., Wright, S., Wogulis, M., Audia, J. E., May, P. C. and Hyslop, P. A. (1996). Amyloid beta-mediated oxidative and metabolic stress in rat cortical neurons: no direct evidence for a role for H_2O_2 generation. *J. Neurochem.* **67**, 1595-1606.

Progress in HPLC-HPCE, Vol. 7, pp. 111-131
G. Ali Qureshi *et al.* (Eds)

Multiple Cerebral Infarction in Rats : Pathological and Pathophysiological Studies

HIDEYA SAITO[1)], KEN MORII[2)], HIROKO TOGASHI[2)], MITSUHIRO YOSHIOKA[2)], NISHIO NAKAMURA[3)] AND SHINICHI KIMURA[1)].

1) Department of Clinical Pharmacology and Toxicology, Faculty of Pharmaceutical Sciences, Health Sciences University of Hokkaido, Ishikari-Tobetsu 061-02, 2) 1st Department of Pharmacology, Hokkaido University School of Medicine, Sapporo 060, 3) Department of Laboratory Technology, College of Medical Technology, Hokkaido University, Sapporo 064, Japan.

INTRODUCTION

Of all organs of the body, the brain is the most sensitive to decreases in blood flow. This is because a large amount of oxygen and glucose are necessary for brain function; and because brain function is only maintained under aerobic metabolism. In other words, brain function can not be maintained during anaerobic metabolism. Very small amounts of oxygen and glucose accumulation are essential for brain metabolism and a continuous supply of these substances are necessary to maintain the brain function mediated by cerebral circulation. It has been reported that insufficient amounts of oxygen and glucose supplies led to the rapid development of permanent organic impairment of the brain (Sokoloff, 1961). With regard to the cerebral cell death due to ischemia-induced energy metabolism impairment, Evidence has accumulated that cell groups vulnerable to ischemia existed within the cerebral cells. After transient ischemia, these vulnerable cell groups died, in spite of the recovery of both cerebral blood flow and brain energy metabolism (post-ischemic neuronal death) (Kirino, 1982; Onodera et al., 1986). Although the mechanism of this post-ischemic neuronal death is still obscure, it has been recognized that selective vulnerable cell groups are distributed both in the limbic system and the basal ganglia, and that these cell groups receive nerve fibers which are mediated by glutamate, an excitatory neurotransmitter (Araki et al., 1989).

Cerebral infarction is caused by organic cerebral impairment and is thought to be the cause of cerebral vascular dementia (Tomlinson et al., 1970; Rogers et al., 1986; Hachinski et al., 1974). Numerous investigators have studied the pathology of cerebral infarction using various experimental animal models of cerebral ischemia, which were classified into two categories, global ischemia (hemispheric ischemia) or focal ischemia. The following rat models have been reported: (1) Focal cerebral ischemia following middle cerebral artery occlusion (Tamura et al.,

1981; Tamura et al., 1981), (2) cerebral infarction produced by homologous, blood clot emboli (Kudo et al., 1982), (3) cerebral ischemia induced by bilateral carotid artery ligation (Eklof et al., 1972), (4) cerebral ischemia initially induced by bilateral vertebral artery ligation and followed by bilateral carotid artery ligation (Pulsinelli et al., 1979), and (5) cerebral ischemia following the injection of carbon-microspheres into the internal carotid artery (Kogure et al., 1974; Kogure et al., 1975; Kogure et al., 1980). In order to clarify the pathology and pathophysiology of cerebral ischemia, cerebral blood flow, EEG, ECoG, evoked potential, energy metabolism, monoamine metabolism have been scrutinized as criteria of brain function (Branston et al., 1974; Momma et al., 1988; Chen et al., 1986).

Multiple cerebral infarction which were classified into global ischemia is thought to be an important cause of cerebrovascular dementia. However, pathological and pathophysiological studies are need to establish an animal model for multiple cerebral infarction. In this study we used carbon-microspheres (50 μm in diameter) to develop a novel type of cerebral ischemia in rats. During the acute stage (within 1 hour after the development of cerebral ischemia) the changes in electrocorticography (ECoG) and cerebral cortical blood flow were studied. Seven days after the development of cerebral ischemia, the passive avoidance test was performed to evaluate the functions of learning and memory (Yamazaki et al.,1977; Kiyota et al., 1985; Chorover et al., 1965; Rigter et al., 1974). Pathological studies were also done 5-and 7-days after the development of cerebral ischemia in this model. In this manuscript we evaluate whether this new experimental ischemic model is appropriate for the study of multiple cerebral infarction or not.

METHODS

1. Generals.

Male Wistar rats ranging from 350 to 400 grams and from 10-12 weeks were used. Rats were acclimatized to a light and dark alternation cycle (12 hours light and 12 hours dark). Rats were anesthetized intraperitoneally with a dose of 100 mg/kg ketamine hydrochloride. Using artificial respiration, ventilation volumes were set for a final CO_2 expiration concentration 4.5 ±0.5%. The right femoral artery and vein were cannulated. In order to record electromyogram for computer analysis of ECoG frequency, ketamine hydrochloride at the dose calculated by pharmacokinetic parameters (Benet et al., 1990) was continuously infused into the right femoral vein with an infusion pump. Simultaneous arterial pressure and heart rates were continuously measured from the femoral artery (Figure 1).

2. Creation of Experimental Cerebral Ischemic Rats.

A. Acute Cerebral Ischemic Rats.

Experimental ischemic rats were created by modifying the method of Kogure (Kogure et al., 1974) and Kiyota et al (Kiyota et al., 1985). As shown in Figure 2, bifurcation of the common carotid artery was exposed. After ligation of the right occipital artery, the right cranial thyroid artery, and the external carotid artery, a PE-10 polyethylene cannula (I.D. 0.28 x O.D. 0.61 mm) was inserted into the right external carotid artery retrogressively. The cannula tip was fixed at the centrifugal site 1 mm from the bifurcation of the internal and external carotid arteries. Two thousands carbon-microspheres (50 μm in diameter) were injected into internal carotid artery during one minute (100 μl suspension in 20% dextran) to bring about embolization of the arterioles.

B. Cerebral Ischemic Rats 7 days after microsphère injection.

Using the same method as for the acute cerebral ischemic rats, microspheres were injected into the right internal carotid artery. Under spontaneous respiration, a PE-10 polyethylene cannula was inserted into the right external carotid artery as quickly as possible to the rat anesthetized with ketamine. Rats were bred to the same experimental conditions as the pre-operation period for 7 days. Passive avoidance tests were then performed to clarify learning and memory impairment as described later.

3. Histopathological examination.

Five hours after the injection of carbon microspheres, the acute ischemic rats were anesthetized. Whole brains of them were isolated and fixed in 10% formalin. The fixed brains were cut into slices in coronal sections at 2 mm in thickness and two slices were chosen in each brain at the two levels; 1. through the heads of caudate nuclei, 2. through the basal ganglia and thalami. The slices were embedded in paraffin blockes and serially sectioned by a microtome at 5 μm in thickness. Two hundreds of unstained specimens were cut from each block, which corresponded to 1 mm in thickness, however, the neighbouring 2 specimens in every 10 of serial sections were chosen for staining. They were stained by hematoxylin and eosin (H.E.) and klüver-Barrera (K.B.) methods. It is estimated that 20 H.E. stains were prepared at 50 μm-intervals in 1 mm thickness as well as 20 K.B. preparations were made from a block respectively.

Under the microscope the microspheres with 50 μm in diameter were detected while foci of infarction were examined. The distribution of microspheres and foci of infarction were simultaneously plotted and illustrated on the figures which were constructed from the inspection of the specemens.

4. Electrocorticography (ECoG)

As shown in Figure 3, a stereotaxic apparatus was use to expose the cranium. Two bipolar electrodes were inserted into the cerebral cortex at 3 mm caudate from the Bregma, 2 mm lateral right and left, at a depth of 1 mm from the brain surface. Electrodes were made from insulated stainless steel, the tip of which had less than 0.2 mm insulation. As shown in Figure 1, ECoG derived from the bilateral cerebral cortex was amplified with a time constant of 0.3 with the high cut filter off. Amplifiers were connected directly to the recorder and computer analyzing

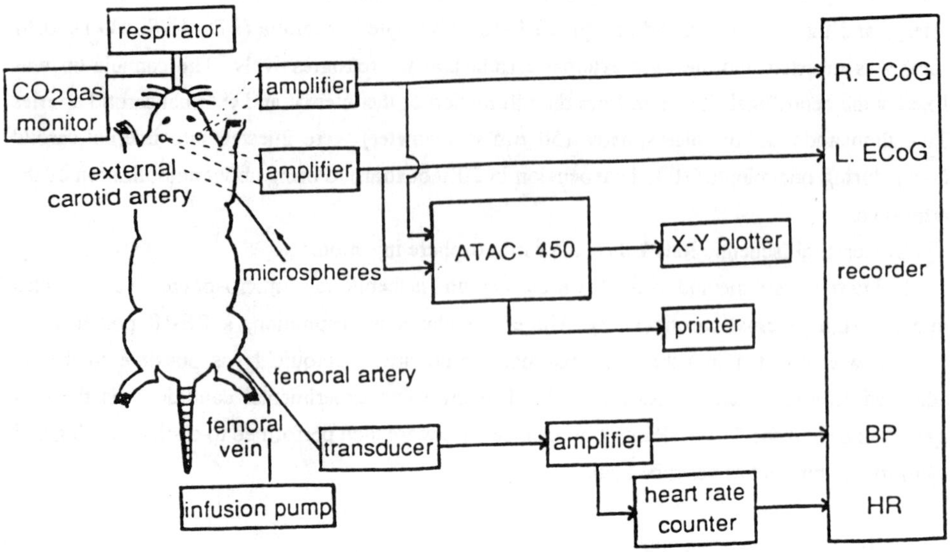

Fig. 1 Blockdiagram of the recording and analyzing system for the electrocorticogram (ECoG).

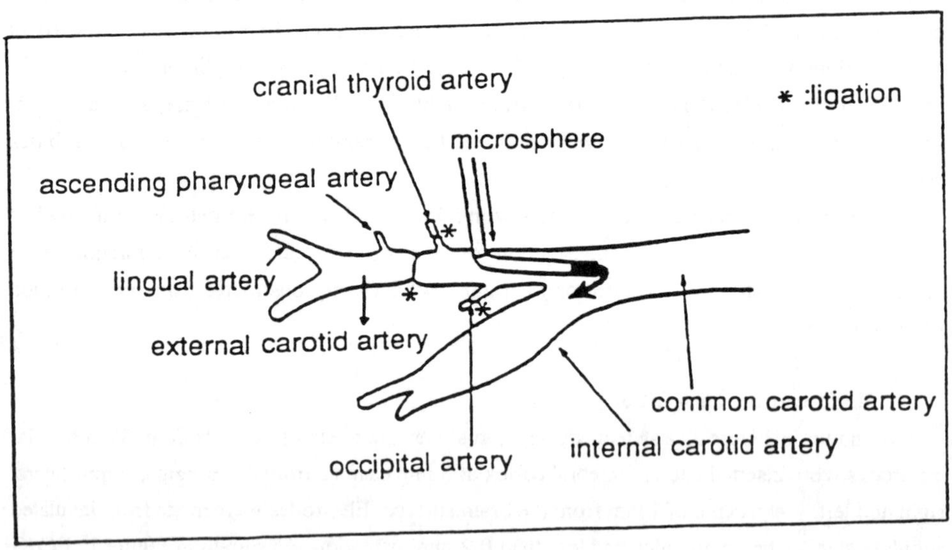

Fig. 2 Schematic illustration showing a cannulation procedure * cranial thyroid artery, occipital artery and external carotid artery are ligated.

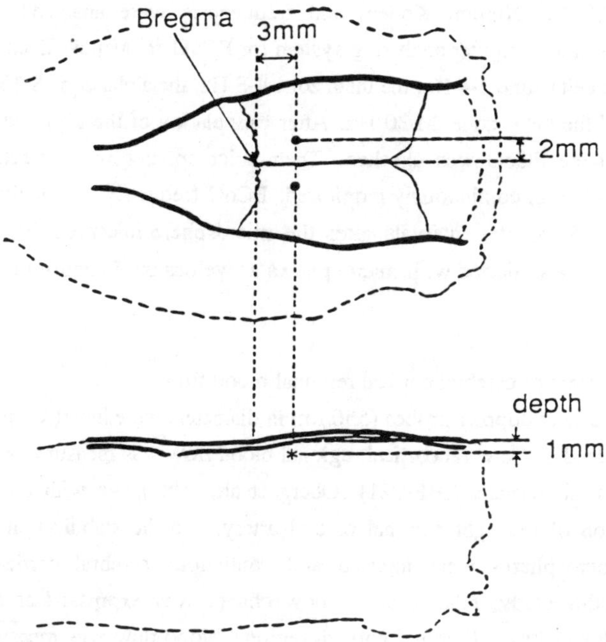

Fig. 3 Schematic illustration showing the localization of the ECoG electrodes and the laser doppler probe.

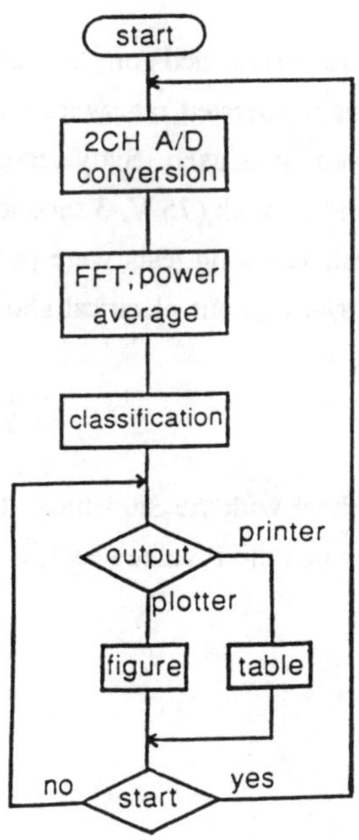

Fig. 4 Blockdiagram of the computer analyzing system for the ECoG frequency.

system (ATAC-450, Nippon Koden) and frequencies were analyzed. Figure 4 shows a blockdiagram of the computer analyzing system for ECoG frequency. Each frequency zone was as follows: the delta zone 2-4 Hz; the theta zone 4-8 Hz; the alpha zone 8-13 Hz; the $beta_1$ zone 13-20 Hz; and the $beta_2$ zone 20-30 Hz. After cannulation of the right external carotid artery, the stability of the ECoG was checked. Then, microspheres were injected over a 1 minute period and ECoG was continuously monitored. ECoG frequency was analyzed at 1 minute, 5 minutes and at 5 minutes intervals after the microsphere injection. ECoG monitoring was performed on 10 rats injected with microspheres as well as on 7 control rats that received saline injections.

5. Measurement of cerebral cortical regional blood flow.

Needle type laser doppler probes (550 μm in diameter) were inserted into close proximity of the ECoG electrodes. Cerebral cortical regional blood flow was measured with a laser type flow meter (Biomedical Science, LBF-221) (Oberg et al., 1984). As with the ECoG monitoring, after cannulation of the right external carotid artery, and the stabilization of blood flow was confirmed, microspheres were injected and continuous cerebral cortical blood flow was measured. In this study, relative blood flow change was expressed as a percentage of the preinjected value (100%). Cerebral cortical regional blood flow was measured both 5 rats that were injected with microspheres and in 6 control rats that received injections of saline.

6. Passive avoidance test.

Passive avoidance tests were performed on 8 cerebral ischemic rats 7 days after microsphere injection. And, 7 sham-operated rats were studied as a control. A step-through type passive avoidance test apparatus was used. Sixty seconds after acclimatization in the light chamber, rats were given an electrical shock (75 V, 3 seconds) on the grid of the dark chamber. Five days after the acquisition trial, retention tests were performed for 5 days. Retention tests were identical to the acquisition trial without electrical shocks. Latency of the response (sec) was measured up to 600 seconds.

7. Statistical analysis

Statistical analysis was done either with the Studentϛs t-test or the Cochran test. Differences were considered to be significant when the P value was less than 0.05.

RESULTS

1. Pathological and pathophysiological studies of acute cerebral ischemia in rats.

A. Results of histopathological studies.

The intracranial distribution of microspheres (50 μm in diameter) was inspected on the specimens produced from the materials taken at 5 hours after the injection of the emboli into the right internal carotid artery. Most of the microspheres were observed on the injected side (ipsilateral side) and only a small number of microspheres were on the contralateral side. Many microspheres were captured in the branches of the right anterior and middle cerebral arteries. Several microspheres were also detected in the parenchyma of the frontal and parietal cortices, the basal ganglia, the thalamus and hippocampus on the right side. On the other hand, only a few microspheres were observed in the branches of the left anterior cerebral artery and in the superficial area of their territories. Figure 5A and 5B illustrate the distribution of microspheres in two coronal sections of a representative case. They were plotted and summarized from the 20 specimens which were stained at 50 μm-intervals in a distance of 1 mm in a block. Since the distance of the interval was same as the diameter of the microspheres, it is indicated that the numbers of the microspheres themselves within the thickness of 1 mm in the block were accurately calculated without exception. The drawings of the slices at the levels through the frontal lobe (Figure 5A) and through the basal ganglia and thalami (Figure 5B) show that 75-80 % of the microspheres were located on the ipsilateral (right) side, while 10 % of them were on the contralateral (left) side, and about 10-15% of them were in the blood vessels on the midline, where the right and left anterior cerebral arteries are intermingled.

More than half of the microspheres on the ipsilateral side remained in cortical vessels, however, some of them reached deeply into the parenchyma. Only a few microspheres were seen in the cortex in the territory of the left anterior cerebral artery (contralateral side).

An inspection was also made to clarify the distribution of ischemic changes and/or acute cerebral infarction. Hyperchromatism of the nuclei and/or shrinkage of perikarion with dilatation of pericellular spaces are caused by acute ischemia and considered to be a finding of acute infarction in very early stage (Figure 10). Depopulation of nerve cells with reduced stainability of neuropile is also another finding of acute infarction in the next stage. Then proliferation of macrophages may occur in the subacute stage. Proliferation of astrocytes is not observed in the lesion of acute infarction, although it may lately occur in subacute or chronic stages. The microscopical examination revealed multiple foci of acute infarction in the frontal and parietal cortices, basal ganglia and thalamus on the right side (Figures 5C, 5D, 6, 7, 8, 10). Some small foci seemed to unite themselves into larger ones. The foci of infarction were always accompanied by the emboli of microspheres in the neighbourhood (Figures 6, 7, 8, 10). Foci of acute infarction were also observed in the frontal and parietal cortices on the left (contralateral) side, however, the area was very small and restricted to the territory of superficial branches of the anterior cerebral artery (Figure 9). Distribution of these foci of acute infarction in each case was respectedly noted on the figure shown as in Figures 5C and 5D.

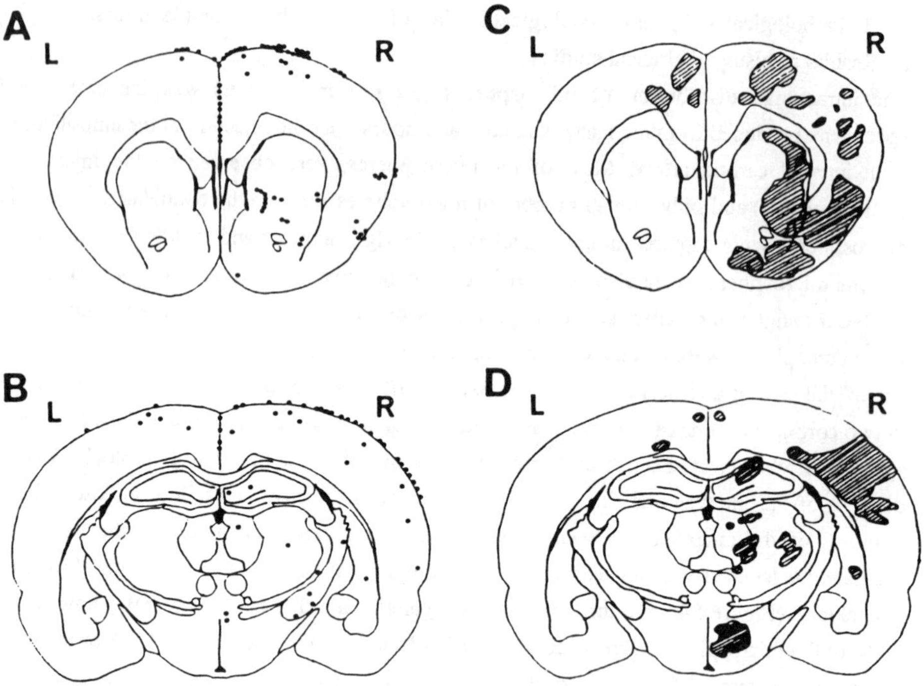

Fig. 5 Diagramatic representation of distribution of carbon microspheres and foci of infarction. Coronal sections at the levels through the frontal lobe and through the basal ganglia and thalami are presented. The location of microspheres are indicated by black dots. Two hundred slices a block of the paraffin-embedded brain were serially cut at 5μm in thickness. A set of two specimens for H.E. and K.B. stains were chosen every 10 out of them. It means that 20 specimens with H.E. stain at 50μm-intervals were prepared in the distance (thickness) of 1mm in a block as well as another set prepared with K.B. stain. Every particle at 50μm in diameter is not inspected twice in the series of 20 specimens and the numbers of the microspheres are accurately plotted into the drawings. Foci of infarction are indicated by the areas filled with diagonal lines. A, B, C, D; Distribution of microspheres (A, B) and foci of infarction (C, D) in the frontal lobe (A, C) and in the basal ganglia, thalamus and parietal cortex (B, D).

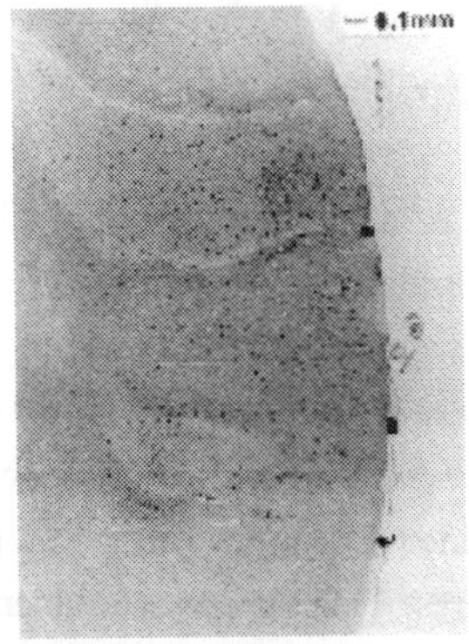

Fig. 6 Micrograph of coronal section at the right (ipsilateral) parietal cortex (H.E. stain). Note that 3 microspheres are seen in cortical branches of the anterior cerebral artery. Some nuclei of nerve cells are darkly stained and some show shrinkage of cell bodies with dilatation of pericellular spaces, which indicate acute ischemic change in the cerebral cortex. The area where the stainability of the parenchym reduced indicate foci of acute infarction. Some of them may be composed of fusion of smaller ones. Thus multifocal infarctions are seen in the cortex.

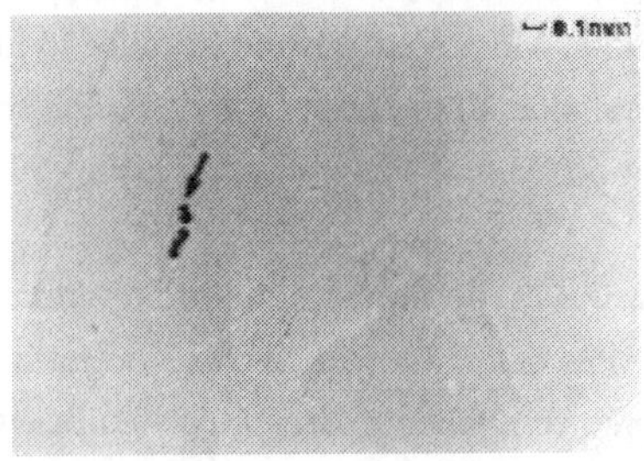

Fig. 7 Micrograph at the right (ipsilateral) basal ganglia (H.E. stain) showing foci of acute infarction where the stainability is obviously reduced. An arrow indicates a rosary-like aggregate of microspheres in a penetrating branch of the right anterior cerebral artery.

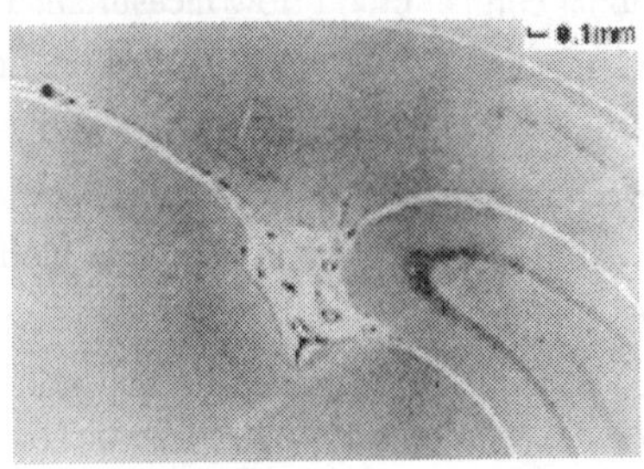

Fig. 8 Micrograph at the right (ipsilateral) hippocampus (H.E. stain). A solitary focal infarction is observed in CA1 area of the hippocampus (right hand side, top). A microsphere which may be corresponding to this focus is seen in a branch of the anterior cerebral artery at the hippocampal fissure.

B. Changes in electrocorticography.

(1) Case report of ECoG.

Figure 11 shows the ECoG of a rat that received an injection of 2000 microspheres into its right internal carotid artery. One minute and 5 minutes after the injection, the ECoG turned into a slow wave. This slow wave was recognized synchronously both at ipsilateral (injected) site and the contralateral site. These changes were most intense during the 5 minutes after microsphere injection. The general slow wave disappeared 30 minutes after microsphere injection. But, the ECoG retained a slow wave as compared with that recorded before the administration of the microspheres. This slow wave tendency is remarkable in the ipsilateral site which 60 minutes after injection remained, whereas in the contralateral site, the ECoG wave appeared to recover to its pattern before microsphere administration.

(2) Quantitative time-course observation of ECoG.

With regard to the time-course observation of the powers of each frequency zone of ECoG, comparisons were performed between the microsphere-injected group and saline injected group (Figure 12). Our results showed that the delta wave increased immediately after the microsphere injection, its power increased maximally and then gradually decreased. These changes were remarkable in the ipsilateral site. Moreover, power continued to increase remarkably 60 minutes after injection in the ipsilateral site. In the contralateral site, on the other hand, only a transient and statistically insignificant difference was observed during 60 minutes, although a tendency toward increase was noted.

The power of the alpha waves decreased significantly immediately after injection in both ipsilateral and contralateral sites and both sites showed decreasing tendencies thereafter. $Beta_1$ waves, one of the fastest wave components, decreased significantly immediately after injection, but other changes were not observed. Both $beta_2$ and theta waves did not show significant differences.

C. Changes in cerebral cortical regional blood flow.

(1) A case report of regional cortical blood flow measurement.

Figure 13 shows the changes in cerebral cortical regional blood flow when 2000 microspheres were injected into the right internal carotid artery. Relative blood flow changes were calculated as the percentage of blood flow value before microsphere administration (100%). Immediately after microsphere injection, regional blood flow in both ipsilateral and contralateral sites decreased. Ipsilateral site blood flow decreased markedly as compared with that of the contralateral site. Blood flow in both sites had not recovered 60 minutes after the injection of microspheres.

(2) Time course observation of cortical regional blood flow.

The regional blood flows of the microsphere-injected group and the saline-injected group were compared (Figure 14). In the microsphere-injected group, blood flow in the ipsilateral site decreased approximately 50% as compared with that in the contralateral site. Blood flow

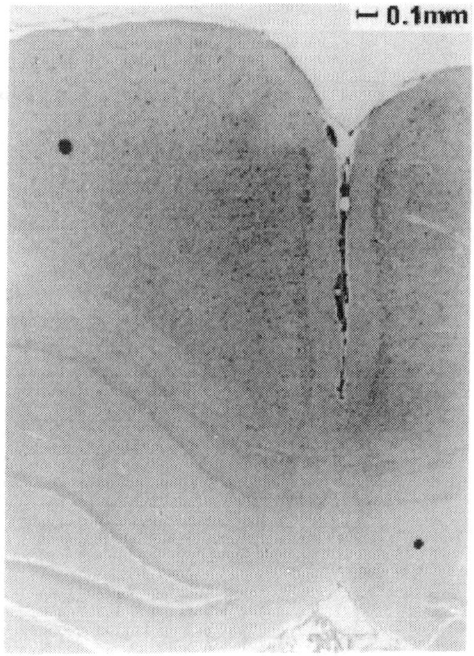

Fig. 9 Micrograph of coronal section at the paramedian portion of cerebrum (H.E. stain). A microsphere and small foci of infarction are seen in the parietal cortex on the left (contralateral) side as well as the one seen in the hippocampus on the right (ipsilateral) side.

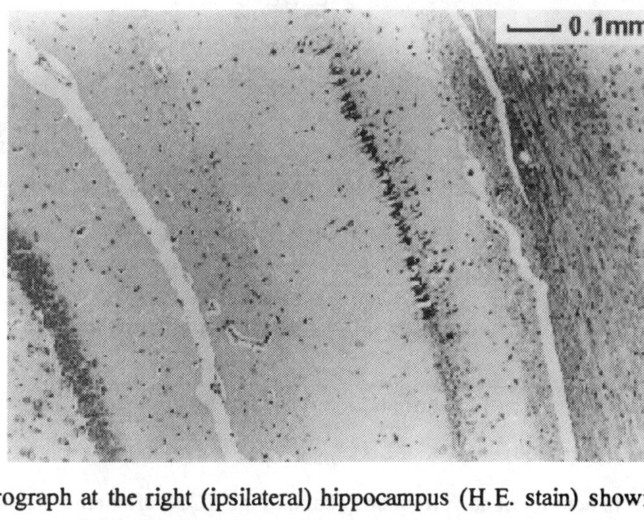

Fig. 10 Micrograph at the right (ipsilateral) hippocampus (H.E. stain) showing a focus of very acute infarction. Hyperchromatism of nuclei and shrinkage of perikaryon of nerve cells are noted and stainability of neuropils is somewhat reduced in the area.

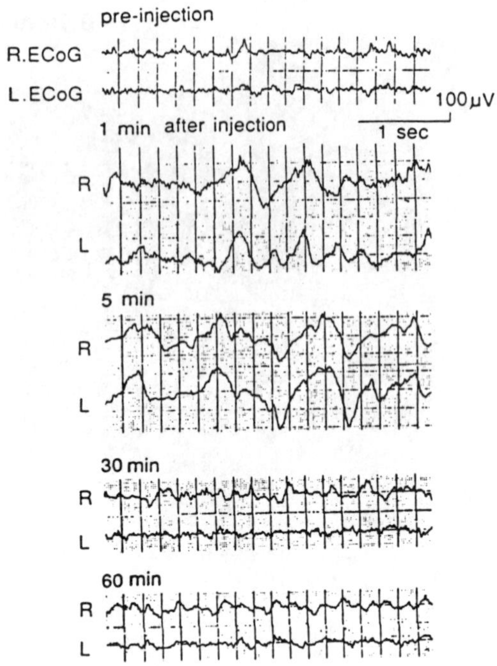

Fig 11 Actual recordings of the ECoG in the embolized rat. R; right (ipsilateral) hemisphere, and L; left (contralateral) hemisphere. The ECoG was recorded before (pre-injection) and 1 min , 5 min, 60 min after injection of carbon- microspheres.

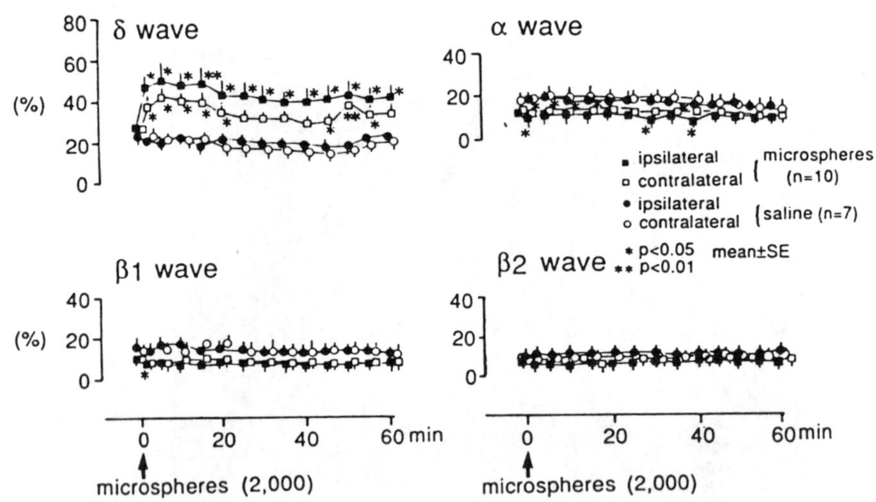

Fig. 12 Serial changes in activity of each frequency band in the ECoG after embolization. The activity of theta wave which did not change significantly , is not shown in this figure. Each value indicates mean ±S.E. (%). * P <0.05, ** p < 0.01 vs. saline-treated control.

decrease continued until 60 minutes after injection. In the contralateral site, blood flow tended to decrease up to 20 minutes after injection. However, no significant difference was noted between the contralateral site and the control group. Thereafter, a significant decrease in blood flow of approximately 30% occurred. The blood flow decrease continued 60 minutes after the injection of microspheres.

2. Pathological and pathophysiological findings 7 days after microsphere injection.

A. Passive avoidance test.

Seven days after 2000 microspheres were injected into the right carotid artery, a comparison between brain-ischemic rats and sham-operated rats was undertaken with respect to the time required for acquisition trials. No significant difference was demonstrated between brain-ischemic rats and sham-operated rats. Seven days after the injection of microspheres, retention trials were done for 5 days after the acquisition trial in the brain-ischemia group. For 3 days after the acquisition trials, the response latency of the microsphere-injected rats shortened significantly as compared with that of the sham-operated rats (Figure 15).

B. Pathological studies.

Eight weeks after injection of microsphere, pathological examinations were undertaken for brain-ischemic rats on which the passive avoidance test had been performed. Histopathological examination revealed multiple foci of infarction. Shrinkage of cell bodies and depopulation of nerve cells with proliferation astrocytes were observed in these lesions. In the white matter, pallor in staining of myelin (K.B. stain) were seen in and around the foci of infarction. In some cases, deposition of hemosiderin granules was observed as the evidence of focal hemorrhagic infarction (Figure 16). Also in some cases, cystic changes in the frontal and parietal cortices were seen, which indicated that foci of infarction might have united into a cystic cavity in the chronic stage.

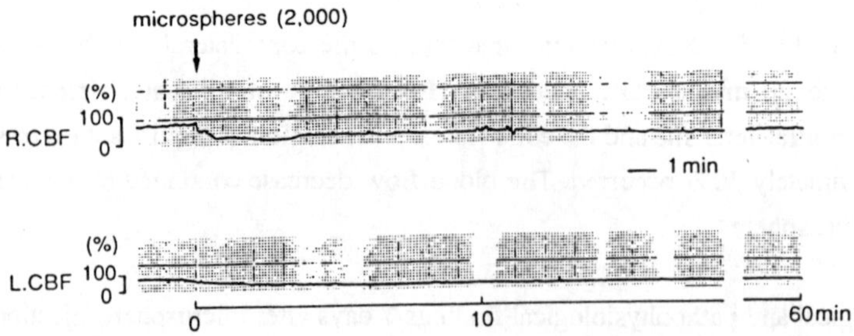

Fig. 13 Actual recordings of the regional cortical blood flow (CBF) in the embolized rat. R; right (ipsilateral) hemisphere, and L; left (contralateral) hemisphere.

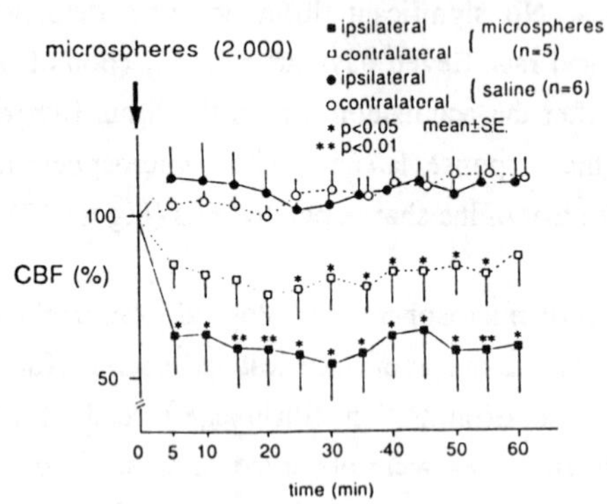

Fig. 14 Serial changes in the CBF after embolization. The CBF obtained after embolization is presented as a relative CBF by defining the value before embolization as 100 %. Each value indicates mean ±S.E. (%). * P <0.05, ** p < 0.01 vs. saline-treated control.

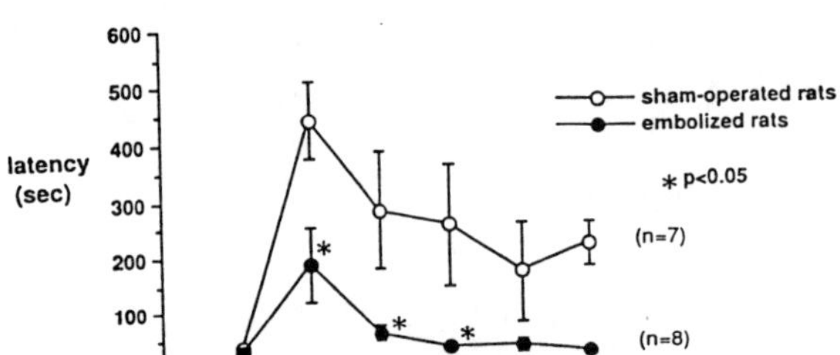

Fig. 15 Effect of embolization on passive avoidance test in rats. The acquisition trial (electrical shock, 75 V for 3 sec) was performed 7 days after embolization (shown as -1 day). The first retention test was carried out 24 hours after the acquisition trial, followed by four further retention tests. Vertical; the mean latency of the response (sec). Abscissa ; the retention trial. Each value indicates the mean ± S.E. (sec). * p < 0.05 vs. sham -operated control.

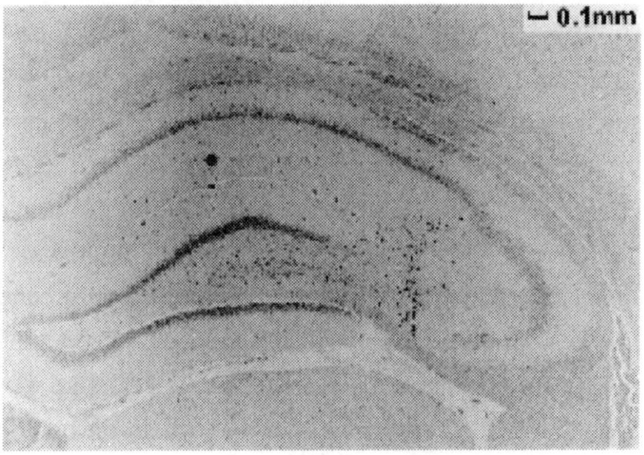

Fig. 16 Micrograph at the right (ipsilateral) hippocampus (K.B. stain). A focus of chronic infarction with deposition of hemosiderin granules is observed in the hippocampus 8 weeks after embolization. A microsphere deeply reached into the parenchym is also seen near to this lesion.

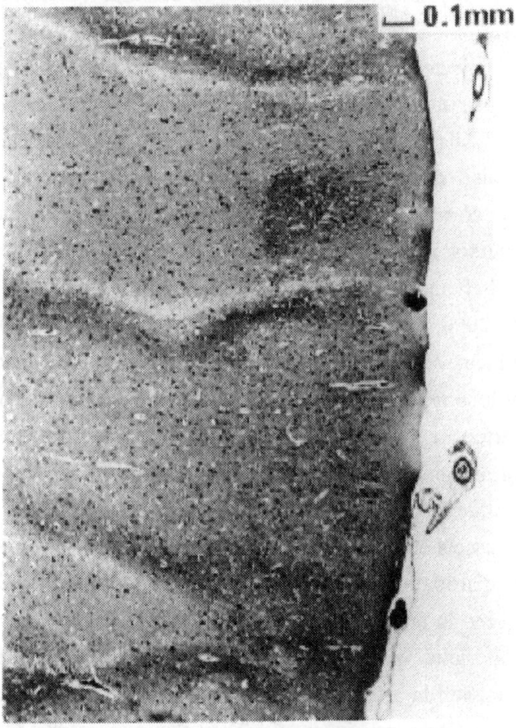

Fig. 17 Micrograph of coronal section at the surface of the right parietal cortex (H.E. stain). Microspheres stay in the lumen at 50μm or less in diameter of the cortical arteries. The lumen are sometimes vacant where the vessel might be larger then 50μm in diameter.

DISCUSSION

In order to clarify the pathology and pathophysiology of multiple cerebral infarction, microspheres were used to create a cerebral ischemic rat model. The pathology and pathophysiology of this animal model were studied both during the acute period and 7 days after microsphere injection. In this experiment, 2000 microspheres (50 μm in diameter) were injected into the right internal carotid artery. Large or small multiple cerebral infarctions were demonstrated in the right cerebral hemisphere. Small cerebral infarctions were also observed in the contralateral hemisphere. There is no report in the literature concerning this kind of multiple cerebral infarctions in rats. Experimentally-induced cerebral infarctions caused by microspheres injected into the ipsilateral internal carotid artery were previously studied by Kogure et al. They reported that microspheres over 70 μm in diameter produced multifocal microinfarctions. The diameter and number of microspheres, therefore are important factors with respect to producing induced cerebral infarctions. In this experiment, we used 50 μm carbon-microspheres in diameter which were injected into the right internal carotid artery.

1. Pathology and pathophysiological studies of acute cerebral-ischemic rats.

A Pathological studies.

A. Kishikawa et al. injected 0.14 mg microspheres (35 μm in diameter) into the right carotid artery. Through pathological examination, they reported that most microspheres were distributed mainly in the ipsilateral site and a small amount of microspheres were found at the contralateral site. They also detected cerebral infarctions in the ipsilateral hippocampus, the thalamus and the cerebral cortex. The infarctions in the cerebral cortex were distributed mainly in the middle cerebral artery region. They proposed that their model displayed either focal ischemia or hemispheric ischemia of the middle cerebral artery region.

In this experiment, 2000 microspheres (50 μm in diameter) produced small cerebral infarctions, some of which were fused to each other that were distributed in the ipsilateral frontal lobe, the parietal lobe and the basal ganglia. These cerebral infarctions were not limited to the middle cerebral artery region. Approximate 75-85% of microspheres was noted in the territories of the anterior and middle cerebral arteries on the ipsilateral side. About 10% of the microspheres were distributed contralaterally. Also, about 10-15% of the microspheres were distributed in the blood vessels on the median line.

We examined the distribution of intravascular microspheres on the brain surface. Microspheres were not seen in arteries that were larger than their own diameter, but tended to be captured mainly in arterioles with diameters less than 50 μm. This finding suggests that microspheres produced emboli in arterioles that were 50 μm in diameter and that these emboli induced cerebral infarctions (Refer to Figure 17). They also suggest that injected microspheres did not produce large infarctions mediated by the interaction with various coagulation factors in the blood. Therefore, we did not consider that large emboli would form in large arteries. Next, we examined the histology of cerebral infarction. We found that a focus of cerebral infarction was caused by small infarctions, some of them united each other. These small infarctions were produced at the peripheral portion of blood vessels, so that collateral circulation could not take place. Therefore, it was presumed that changes in cerebral neuronal cells might appear within a short period. No reports are available in the literature concerning such multiple cerebral infarctions. However, because these multiple cerebral infarctions occurred within a short period, we should consider the suitability of this model for patients in the clinical course. It is not yet known whether the course of human cerebral infarction coincides with that of this experimentally induced cerebral infarction in the rat. Further study is necessary to develop an optimal animal model for patients with multiple cerebral infarctions.

B. Changes in electrocorticography (ECoG) and regional cortical blood flow (CBF).

In order to detect the activity of cerebral cell groups, ECoG was measured. Computer analysis was performed for ECoG frequency to evaluate ECoG changes quantitatively. Moreover, a Laser type flowmeter was used for real time measurement of regional cortical blood flow (Cohen et al., 1976; Matousek et al., 1979). As a result, 1 minute after the ECoG of both ipsilateral and contralateral sites could be observed. Slow wave changes and these slow wave changes continued up to 60 minutes after microsphere injection. The fact that generalized slow waves appeared 1 minute and 5 minutes after microsphere injection suggests that after the occurrence of abrupt cessation of blood supply (ischemia) a possible loss of homeostasis in the brain may induce compensatory activity. Thereafter, the generalized slow wave disappeared. The disappearance of this slow wave, might relate to cerebral neuronal death, causing a new homeostasis to occur. The precise mechanism is still obscure. In terms of changes in regional cortical blood flow, a significant blood flow decrease occurred ipsilaterally after microsphere injection which continued up to 60 minutes. In the contralateral site, blood flow tended to decrease up to 20 minutes after injection of microspheres. Thereafter, a significant decrease in blood flow occurred up to 60 minutes after microsphere injection. With regard to the relationship between ECoG changes and CBF changes, many reports indicated that parallel changes occurred in ECoG and CBF. However, the present study demonstrated that a general slow wave appeared at the time of the onset of the beginning of CBF decrease. The ECoG changes are not necessarily related to CBF decrease. Moreover, although delta wave increases occurred markedly during the 20 minutes after microsphere injection, a marked CBF decrease did not occur during the same period. Therefore, changes in CBF can not completely explain the changes in ECoG. The present experiment revealed that both ECoG and CBF did not recover to pretreatment levels for 60 minutes after injection of microspheres.

2. Pathological findings 7 days after microsphere injection.

It has been realized that, from the viewpoint of histopathology, multiple cerebral infarctions may be associated with patients with cerebrovascular dementia. In this experiment, we examine the brains 7 days after microsphere injection. It was presumed that the disappearance of cerebral neuronal cells which occur during the chronic stage of multiple cerebral infarction is relevant to the defective intellectual state in patients with vascular dementia. Using the passive avoidance test, an attempt was made to clarify whether these experimental ischemic rats show impairment of learning and memory.

A. Passive avoidance test.

One week after microsphere injection into the right carotid artery, passive avoidance tests were performed. The cerebral ischemic rat group revealed significant shortened response latency of retention trials three days after the acquisition trial as compared with those in sham-operated rats. This finding suggests the impairment of learning and memory.

B. Histopathological examination.

Eight weeks after microsphere injection, histopathological examination revealed loss of neuronal cells and proliferation of glia cells. Loss of neuronal cells might indicate the deterioration of cerebral function. The fact that hemosiderin-laden cells were observed in the hippocampus appears to be associated with focal bleeding at the site of infarction.

In this experiment, the injection of 2000 microspheres (50 μm in diameter) into the right carotid artery produced cerebral ischemic rats. These rats demonstrated both slow wave changes in the ECoG and a CBF decrease. This CBF decrease did not recover to pretreatment values until 60 minutes after microsphere injection. In these cerebral ischemic rats, slow wave changes in the ECoG which indicate the activity of cerebral cortical neuronal cells did not recover within 60 minutes. In contralateral sites, both slow wave changes of the ECoG and a CBF decrease continued. This study revealed that ECoG might change in parallel with CBF changes. In other word, cerebral neuronal cell changes induced by CBF decrease might reflect ECoG changes. However, both CBF and ECoG changes do not necessarily coincide. Therefore, with respect to neuronal cell changes, hyperirritability and glutamate accumulation should be considered as a possible mechanism for neuronal cell changes. Seven days after the microsphere injection, cerebral ischemic rats showed impairment of learning and memory. And 8 weeks after the microsphere injection, histopathological study revealed loss of neuronal cells which is a characteristic sign of multiple cerebral infarction. These findings indicate that this rat model of cerebral ischemia may be useful for the pathophysiological study of multiple cerebral infarctions.

SUMMARY

1. In this study, carbon microspheres (50 μm in diameter) which were injected into the right carotid artery produced multiple cerebral infarction in rats.

2. Histopathological examination revealed that multiple cerebral infarctions were distributed mainly in the frontal lobe, the parietal lobe, the basal ganglia, the hippocampus, the thalamus and the hypothalamus on the ipsilateral side, and a small amount of microspheres were distributed in the frontal lobe on the contralateral side.

3. Acute cerebral ischemia induced by microsphere injection demonstrated that slow wave changes in ECoG initially occurred in parallel with CBF decrease. However, at the onset of CBF decrease, ECoG changes could not be explained only by the changes in CBF.

4. This type of ischemic animal model shows multiple cerebral infarction foci which are induced by emboli at the periphery of the cerebral artery. In this cerebral ischemic rat model, collateral circulation might not take place. The fact that both regional cortical blood flow (CBF) and electrocorticography (ECoG) did not recover up to 60 minutes after microsphere injection was demonstrated.

5. Seven days after injection of microspheres, this cerebral ischemia may produce an impairment in learning and memory. Therefore, this animal model will be useful for the pathophysiological study of multiple cerebral infarctions.

REFERENCE

Araki, T., Kato, H. and Kogure, K. (1989). Selective neuronal vulnerability following transient cerebral ischemia in the gerbil: distribution and time course. *Acta Neurol. Scand.*, 80: 548-553.

Benet, L.Z. and Williams, R.L. (1990) Design and optimization of dosage regimens: Pharmacokinetic data. *In* " Goodman and Gilman's The Pharmacological Basis of Therapeutics. 8th ed." Gilman, Goodman, L.C.,Rall,T.W. and Murad, F., Macmillan (Eds), New York, Toront, London, pp1650-1735.

Branston, N.M., Symon, L., Crockard, H.A. and Pasztor, E. (1974). Relationship between the cortical evoked potential and local cortical blood flow following acute middle cerebral artery occlusion in the baboon. *Exp. Neurol.,* 45: 195- 208.

Chen, S.T., Hsu, C.Y., Hogan, E.L., Maricq, H. and Balentine, J.D. (1986). A model of focal ischemic stroke in the rat: Reproducible extensive cortical infarction. *Stroke*, 17: 738-743.

Chorover, S.L. and Schiller, P.H. (1965). Short-term retrograde amnesia in rats. *J. Comp. Physiol. Psychol.*, 59: 73-78.

Cohen, B.A., Bravo-fernandez, E.J. and Sances Jr, A. (1976). Quantification of computer analyzed serial EEGs from stroke patients. *Electroenceph. Clin. Neurophysiol.,* 41:397-386.

Eklof, B. and Sisjo, B.K. (1972). The effect of bilateral cartid artery ligation upon the blood flow and energy state of the rat brains. *Acta Phsiol. Scand.*, 86:155-165.

Hachinski, V.C., Lassen, N.A. and Marshall, J. (1974). Multi-infarct dementia, a cause of mental deterioration in the elderly. *The Lancet*, 2: 207-210.

Kirino,T. (1982). Delayed neuronal death in the gerbil hippocampus following ischemia. *Brain Res.*, 239:57-69.

Kiyota, Y., Hamajo, K., Miyamoto, M. and Nagaoka, A. (1985). Effect of Idebenon (CV-2619) on memory impairment observed in passive avoidance task in rats with cerebral embolization. *Japan. J. Pharmacol.*, 37: 300-302.

Kogure, K., Busto, R., Schwartzmann, B.R.J. and Scheinberg, P. (1980). The dissociation of cerebral blood flow, metabolism, and function in the early stages of developing cerebral infarction. *Ann. Neurol.*, 8: 278-290.

Kogure, K., Busto, R., Scheinberg, P. and Reinmuth, O.M. (1974). Energy metabolites and water content in rat brain during the early stage of development of cerebral infarction. *Brain*, 97: 103-114.

Kogure, K., Scheinberg, P., Matsumoto, A., Busto, R. and Reinmuth, O.M. (1975). Catecholamines in experimental brain ischemia. *Arch. Neurol.*, 32: 21-24.

Kudo, M., Aoyama, A., Ichimori, S. and Fukunaga, N. (1982). An animal model of cerebral infarction, homologous blood clot emboli in rats. *Stroke.*, 13: 505-508.

Matousek, M., Arvidsson, A. and Friberg, S. (1979). Serial quantitative electroencephalography. *Electroenceph. Clin. Neurophysiol.*, 47: 614-622.

Momma, F., Sabin, H.M., Branston, H.I., Braston, N.M.and Symon, L. (1988). The effect of experimental ischaemia on the direct cortical response of the motor cortex in primates. *Electroenceph. Clin. Neurophysiol.*, 71: 304- 309.

Oberg, P.A., Tenland, T. and Nilsson, G.E. (1984). Laser-Dopper flowmetry---a non-invasive and continuous method for blood flow evaluation in microvascular studies. *Acta.Med.Scand.* (suppl), 687:17-24.

Onodera, H., Iijima, K. and Kogure, K. (1986). Mononucleotide metabolism in the rat brain after transient ischemia. *J. Neurochem.*, 46: 1704-1710.

Pulsinelli, W.A. and Brierley, J.B. (1979). A new model of bilateral hemispheric ischemia in the unanesthetized rat. *Stroke*, 10: 267-272.

Rigter, H., van Riezen, H. and de Wied, D. (1974). The effects of ACTH- and vasopressin- analogues on CO2-induced retrograde amnesia in rats. *Physiol.Behav.*, 13: 381-388.

Rogers, R.L.,Meyer,J.S., Mortel, K.F.,Mahurin, R.K. and Judd, B.W. (1986). Decreased cerebral blood flow precedes multiinfarc dementia, but follows senile dementia of Alzheimer type. *Neurology*, 36: 1-6.

Sokoloff, L. (1961). Aspcets of cerebral circulatory physiology of relevance to cerebrovascular disease. *Neurology*, 11 (part 2) : 34-40.

Tamura, A., Graham, D.I., McCulloch, J. and Teasdale, G.M. (1981). Focal cerebral ischaemia in the rat. 2. Regional cerebral blood flow determined by [14C] iodoantipyrine autoradiography following middle cerebral artery occlusion. *J. Cereb. Blood Flow Metab.*, 1.

Tamura, A., Graham, D.I., McCulloch, J. and Teasdale, G.M. (1981). Focal cerebral ischaemia in the rat: 1. Description of the technique and early neuropathological consequences following middle cerebral artery occlusion. *J. Cereb. Blood Flow Metab.,* 1: 53-60.

Tomlinson,B.E.,Blessed,G. and Roth, M. (1970). Observations on the brains of demented old people. *J. Neurol. Sci.*, 11: 205-242.

Yamazaki, N., Niihama, K. and Imada, H. (1977). Effects of electroconvulsive shock on the retention of avoidance behavior. *Japan. J. Psychol.*, 43: 303-306.

Progress in HPLC-HPCE, Vol. 7, pp. 133-170
G. Ali Qureshi *et al.* (Eds)

Prospects of recombinase-mediated HPRT Gene Therapy for Lesch-Nyhan disease

BERTOLOTTI, R.

CNRS, Molecular Genetics, Department of Hepato-Gastroenterology, Faculty of Medicine, University of Nice Sophia Antipolis, 06107 Nice, France

1. INTRODUCTION

An approach to gene therapy based on pre-made DNA-recombinase filaments

In december 1991, we devised a new approach to Gene Therapy based on pre-made DNA-recombinase filaments (Bertolotti R., Grant application to French MRT, 1991; Bertolotti, 1996a, 1996b and 1997). The idea is to generate, *in vitro*, human recombinase-DNA complexes analogous to the bacterial RecA presynaptic filament and use them as vectors to promote, upon transfection, targeting homologous recombination between exogenous DNA and chromosomal cognate DNA. Such a directed homologous recombination process should culminate in the substitution of chromosomal DNA by homologous genomic DNA from wild type (gene repair) or mutant (gene inactivation) origin (Bertolotti, 1996a, 1996b and 1997).

Recombinase-mediated gene repair/inactivation as an alternate to the current minigene expression therapy

Unlike our recombinase-mediated gene repair/inactivation strategy, current protocols for gene therapy are based on transfer and expression of minigenes (Blaese *et al.*, 1995; Mulligan, 1993; Roemer and Friedmann, 1992; Verma and Somia, 1997), *i.e.* DNA corresponding to mature messenger RNA (cDNA) under various gene expression cassettes. This has been our first approach to Gene Therapy as a collaborative preliminary projet on the Gaucher disease with Dr. Edward Ginns at the National Institutes of Health, Bethesda, USA (McKinney C., Mellouk S., LaMarca M., Bertolotti R. and Ginns E., 1990-1991, unpublished). The strategy is to enforce appropriate cells from a patient to synthesize an active protein upon introduction of the corresponding cDNA and its driving expression cassette.

As illustrated in figure 1, typical minigenes comprise 1) a viral or mammalian promoter for transcription initiation, 2) a cDNA encoding the protein to be produced, 3) a DNA sequence for efficient transcript

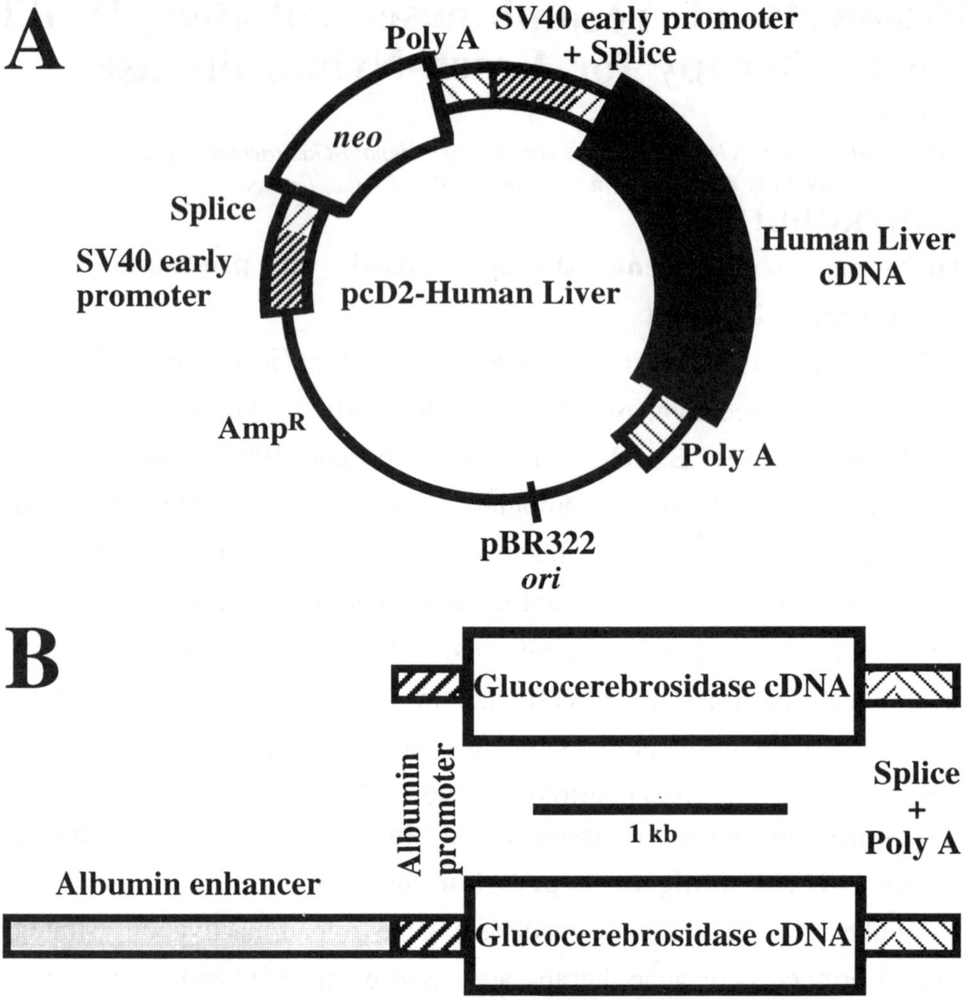

Figure 1. Minigenes from our liver expression programs (Bertolotti,1997). **A.**The viral SV40 early promoter ( ) is active both in mammalian cells (Mulligan and Berg, 1981a) including differentiated hepatoma lines (Lutfalla *et al.*, 1985) and in the fission yeast (Lee and Nurse, 1987). It is used to drive transcription of full-length cDNA in polyvalent pcD-libraries (Okayama and Berg, 1983). The pcD2 vector (Chen and Okayama, 1987) we have used to construct human liver cDNA expression libraries (Bertolotti *et al.*, 1995) comprises two SV40-based expression cassettes driving the bacterial *neo* "gene" for transfectant selection (left) and library cDNA encoding human liver proteins (right), respectively. In addition to the SV40 promoter for transcription initiation, both cassettes comprise SV40 DNA sequences providing for efficient transcript termination and maturation (Poly A, Splice; Mulligan and Berg, 1981a). On the other hand, plasmid pcD contains the pBR322 sequences ( ) required for self-replication (*ori*) and selection (Amp^{R}) in bacteria (library amplification). **B.** Unlike its SV40 counterpart, the strong albumin promoter ( ) is tissue-specific (Gorski *et al.*, 1986) and is used for liver-restricted expression. However, high-level transcription is achieved only when the *cis*-acting albumin enhancer is present (Pinkert *et al.*, 1987). Both cassettes have been used for the construction of gluco-cerebrosidase minigenes (McKinley C., Bertolotti R. and Ginns E., 1990-91, unpublished).

termination and polyadenylation (Poly A) and 4) a short intervening sequence for mRNA splicing (Splice) either on the 5' (Fig. 1A) or 3' end (Fig. 1B) of the cDNA. In addition, promoter strength can be improved upon insertion of a *cis*-acting enhancer as shown in figure 1B, providing occasionally for position-indepent expression of integrated minigenes (Stief *et al.*, 1989). Unlike most of their normal counterparts, such "synthetic" genes are short enough to be handle in viral vectors or bacterial plasmids (Fig. 1). Indeed, as illustrated below, mammalian genes are usually large (20-100 kb); their coding sequence is interrupted by non-coding intervening sequences (introns) that are removed upon splicing of nascent mRNA (Breathnach and Chambon, 1981; Leff *et al.*, 1986). Depending on tissues, mature mRNA have an average size of about 1.5 kb, ranging from about 0.2 to 14 kb. Interestingly enough, most large mRNA have long 5' or 3' non-coding ends that can eventually be deleted in their cDNA counterparts.

The main drawback of this approach is that minigenes are usually not under complete or normal physiological genetic regulations. In addition to missing non-coding sequences, they behave as true foreign cellular elements, either free or "random-integrated" into host chromosomes. Therefore, they cannot fully substitute for wild-type genes. This is why, in december 1991, we devised a new approach to Gene Therapy aimed at repairing mutant genes (inherited diseases) and at inactivating deleterious or vital viral genes (cancer, chronic viral infections such as AIDS and hepatitides). As indicated above, our strategy is based on premade DNA-recombinase filaments (Bertolotti R., Grant application to French MRT, 1991; Bertolotti, 1996a, 1996b, 1997).

HPRT gene: target of Lesch-Nyhan mutations and model for gene therapy

Complete or virtually complete deficiency in hypoxanthine-guanine phosphoribosyl transferase (HPRT) enzyme activity results in Lesch-Nyhan syndrome (Seegmiller *et al.*, 1967) with hyperuricemia, choreoathetosis, compulsive self-mutilation and mental retardation (Lesch and Nyhan, 1964; Rossiter and Caskey, 1995). Lesch-Nyhan mutations are well-characterized at the molecular level (Gibbs *et al.*, 1989; Cariello and Skopek, 1993; Rossiter and Caskey, 1995). They were instrumental to design a series of DNA-backbones for nucleoprotein filaments aimed at inactivating or repairing the human HPRT gene (Bertolotti, 1996b). Recombinase-mediated gene therapy strategies (Bertolotti R., Grant application to French MRT, 1991; Bertolotti, 1996a and 1997) based on these Lesch-Nyhan mutants and on human RAD51 nucleoprotein filaments (Bertolotti, 1996b and 1997) are detailed below. In addition, prospects are presented both for *ex vivo* and *in vivo* gene therapy of the Lesch-Nyhan disease as well as of other inherited diseases, cancer and chronic viral infections. They are discussed in light of the current minigene expression therapy approach (Blaese *et al.*, 1995; Kohn and Parkman, 1997; Verma and Somia, 1997) as well as of emerging RNA-DNA oligonucleotide technology (Cole-Strauss *et al.*, 1996).

2. SELECTION AND TRANSFECTION OF HPRT$^-$ CELLS

The HPRT gene as a ''single-copy'' X-linked selective marker

The HPRT gene is a classic marker of somatic cell genetics (Ephrussi, 1972; Szybalski, 1992). It is expressed in all cells and, to a notably higher level in the brain (Rossiter and Caskey, 1995; Jiralespong and Patel, 1996). It provides both for positive and negative selections of cultured cells (Szybalska and Szybalski, 1962; Ephrussi, 1972; Szybalski, 1992). In addition, it is a "single-copy" X-linked gene, *i.e.* it is hemizygous in males and it has only a single functional copy in adult females (X chromosome inactivation; Lyon, 1996). It is therefore an ideal target for gene repair/inactivation experiments where each targeting event can be readily scored upon positive or negative selection of transfectant cells in HPRT-specific growth media.

Selection of HPRT-deficient cells *in vitro*

HPRT (inosine monophosphate : pyrophosphatase phosphoribosyl transferase; EC 2.4.2.8) is a key purine salvage enzyme that catalyzes the transfer of the ribose phosphate moiety from phosphoribosyl pyrophosphate to hypoxanthine or guanine (Fig. 2). It thus recycles catabolic purine bases yielding inosine monophosphate (IMP) or guanosine monophosphate (GMP), respectively.

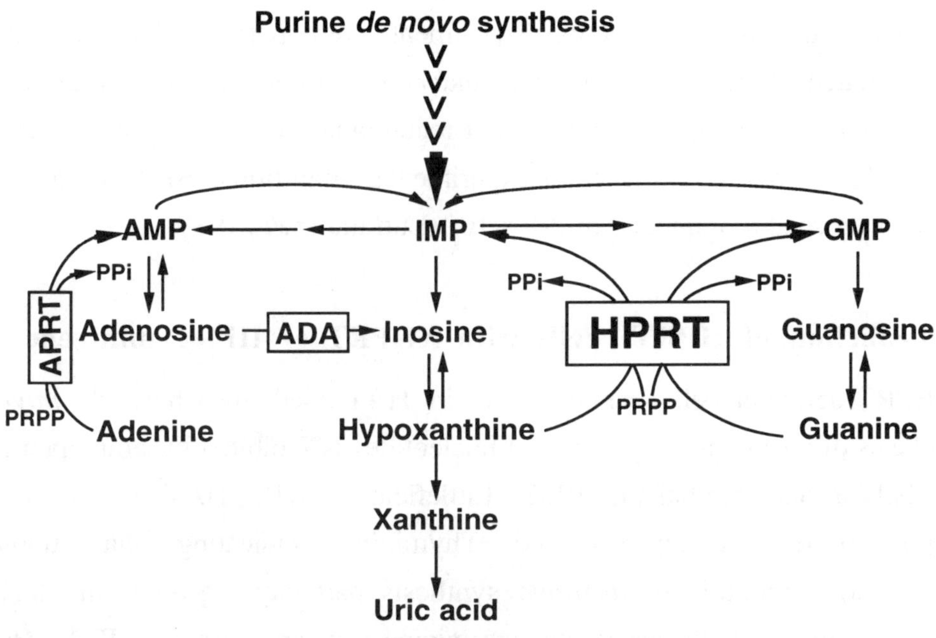

Figure 2. Simplified metabolic pathways for *de novo* synthesis and salvage of purines (based on Rossiter and Caskey, 1995; Nyhan and Wong, 1996; and Simmonds *et al.*, 1995). Purine nucleotides (IMP, AMP and GMP) are produced upon either *de novo* purine synthesis or recycling of catabolic purines. The recycling of catabolic or exogenous purine bases is mediated by HPRT and adenine phosphoribosyl transferase (APRT) upon transfer of the phosphoribosyl moiety from 5-phosphoribosyl-1-pyrophosphate (PRPP) to hypoxanthine, guanine and adenine, respectively. PPi denotes inorganic pyrophosphate, IMP inosine monophosphate, AMP adenosine monophosphate and GMP guanosine monophosphate. Adenosine deaminase (ADA) deficiency results in severe combined immunodeficiency (SCID) syndrome (see below).

In addition, HPRT binds and phosphoribosylates toxic purine analogues such as azahypoxanthine, azaguanine or thioguanine (Szybalski, 1992;

Szybalska and Szybalski, 1962). This ability has been used to isolate HPRT-deficient mutant cells upon selection in drug-containing media (Szybalska and Szybalski, 1962; Szybalski, 1992). Under these conditions, HPRT-positive cells die (counter-selection) while HPRT-negative mutants survive because they are unable to metabolize toxic purine analogs (negative selection).

Growth medium containing 15µg/ml of thioguanine provided for efficient selection of HPRT-deficient variants from the human hepatic HepG2 line (Lutfalla *et al.*, 1989). Selection experiments were performed on a well-differentiated HepG2 subclone (Lutfalla *et al.*, 1989; Armbruster *et al.*, 1992), an essential target of our current recombinase-mediated gene therapy approach (see below). Under our experimental conditions, HPRT-deficient variants arose at a frequency of 2-5 x 10^{-6} (Lutfalla *et al.*, 1989).

Transfection of HPRT$^-$ cells with XGPRT or HPRT minigenes

HPRT-deficient cells cannot grow in HAT medium where *de novo* synthesis of purine (and pyrimidine) nucleotides is inhibited by aminopterin (Szybalska and Szybalski, 1962; Littlefield, 1964). HAT stands for Hypoxanthine, Aminopterin and Thymidine, indicating that upon aminopterin inhibition of *de novo* synthesis pathways, growth in HAT medium relies on both purine and pyrimidine salvage enzymes: HPRT for phosphoribosylation of hypoxanthine and thymidine kinase for phosphorylation of thymidine (Szybalska and Szybalski, 1962; Littlefield, 1964; Szybalski, 1992).

Efficient survival of Lesch-Nyhan cells in HAT medium (positive selection) was first achieved upon transfection with XGPRT minigenes where the *Escherichia coli* DNA sequence encoding bacterial xanthine-guanine phosphoribosyltransferase (XGPRT) was placed under the control of an SV40 expression cassette (Mulligan and Berg, 1980; see also Fig. 1). Unlike its mammalian HPRT counterpart, XGPRT phosphoribosylates both xanthine and hypoxanthine; it was therefore used as a universal bacterial selective marker for mammalian cells either with HAT medium for HPRT mutants or with xanthine selective medium for normal cells (Mulligan and Berg, 1980 and 1981b). Full-length cDNA encoding mammalian HPRT were subsequently isolated and introduced as selectable minigenes into HPRT-deficient cells (Brennand *et al.*, 1983; Jolly *et al.*, 1983).

We have used XGPRT minigenes in our early transfection experiments with rat, mouse and human hepatoma cells (Bertolotti and Lutfalla, 1983; Lutfalla, Blanc and Bertolotti, 1985). As illustrated in figure 3, growth of XGPRT transfectants in HAT medium was directly correlated to the production of active bacterial XGPRT enzyme in mammalian hepatoma cells (Lutfalla *et al.*, 1985).

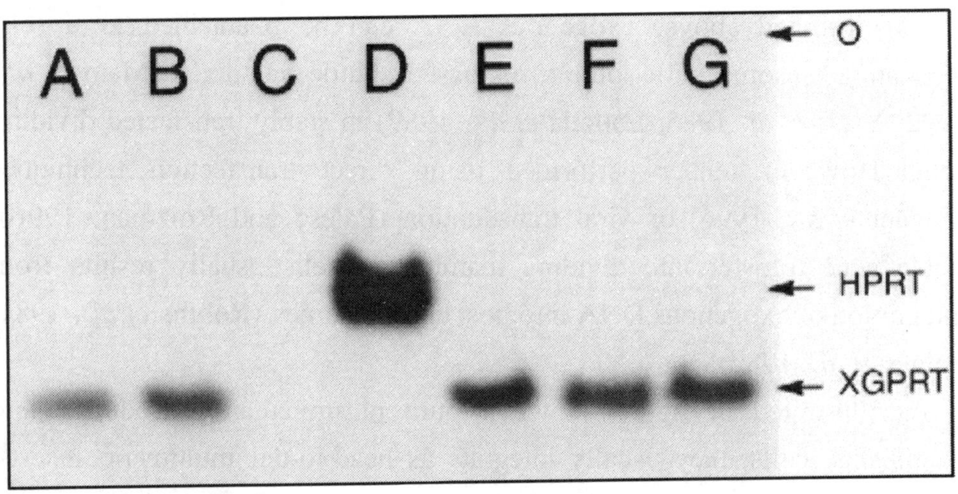

Figure 3. XGPRT and HPRT zymograms of transfected and control cells. Rat and mouse HPRT-deficient hepatoma cells were transfected with plasmids comprising a XGPRT minigene (Bertolotti and Lutfalla, 1983; Lutfalla *et al.*, 1985). Transfectants were selected in HAT medium and their XGPRT-HPRT content analyzed upon electrophoresis of cellular protein extracts on cellulose acetate gels (Shin *et al.*, 1971) and subsequent fluorography of purine phosphoribosyltransferase activity (Chasin and Urlaub, 1976; Lutfalla *et al.*, 1985). Unlike HPRT-deficient (lane C) and wild-type rat controls (lane D), rat (lanes A, B and F) and mouse (lanes E and G) transfectant cells expressed bacterial XGPRT enzyme as expected (Lutfalla *et al.*, 1985).

We have then introduced an XGPRT minigene in a plasmid comprising the origin of replication of the Epstein-Barr virus and shown that it was maintained as a self-replicating episome when HepG2 cells were engineered to produce trans-acting Epstein-Barr nuclear antigen 1 (EBNA) (Lutfalla *et al.*, 1989). Stable XGPRT transfectants arose in HAT medium at a frequency of 2-3 x 10^{-2} with EBNA-producing HepG2 cells instead of 10^{-4} with standard HPRT-deficient HepG2 cells. Such a dramatic 200-fold increase in transfection efficiency was associated to episomal maintenance of the XGPRT vector, suggesting that Epstein-Barr shuttle vectors may be useful to gene therapy protocols based on minigene expression.

3. RECOMBINASE-DNA NUCLEOPROTEIN VECTORS

Integration of exogenous DNA in stable transfectant cells

As illustrated above, exogenous DNA can be maintained as a self-replicating episome with some virus-based shuttle vectors (DiMaio *et al.*, 1982; Yates *et al*, 1985; Lutfalla *et al.*, 1989) in stably transfected dividing cells. However, either performed using direct transfection techniques (Zauner *et al.*, 1996) or viral transduction (Palese and Roizman, 1996), stable gene transfer into dividing mammalian cells usually results from integration of exogenous DNA into host chromosomes (Robins *et al.*, 1981; Folger *et al.*, 1982).

As illustrated in figure 4, when pure plasmids are introduced into mammalian cells, they usually integrate as head-to-tail multimeric inserts. Such structures are generated by homologous recombination (Folger *et al.*, 1982). Indeed, in most transfected cells, exogenous DNA molecules are suject to an intense homologous recombination activity (Capecchi, 1989; Folger *et al.*, 1982; Miller and Temin 1983; Lutfalla *et al.*, 1985). Such a homologous recombination activity coincides with the early transfection phase when newly introduced DNA is not yet packaged into chromatin (Capecchi 1989). It is transient and, in association with ligation and end-

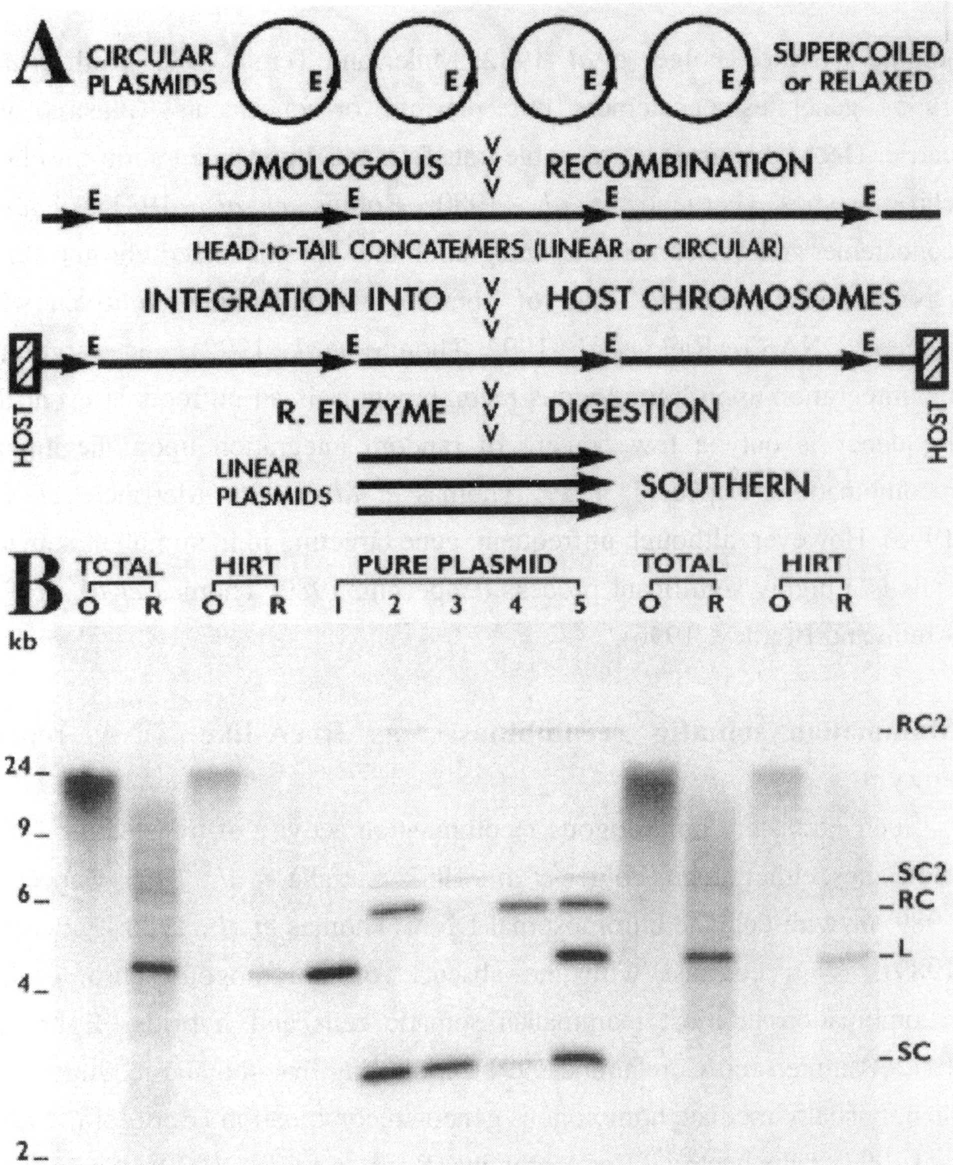

Figure 4 (Bertolotti,1997). **A.** Formation and integration of head-to-tail concatemers upon mammalian cell transfection with pure plasmids. Digestion of circular plasmids or head-to-tail concatemers with restriction enzyme R that cuts the plasmid in a single point (E) gives rise to full-length linear plasmids. **B.** Southern blot analysis of the plasmid content of stable transfectants.Low-molecular weight (Hirt) and total DNA were extracted and digested either with enzyme O that does not cut the plasmid or with R. Plasmid-specific probes showed that plasmid DNA was associated to host chromosomal DNA (O lanes) ; no free plasmid either supercoiled (SC and dimer SC2) or relaxed circular (RC and dimer RC2) was ever found (see pure plasmid combinations). Digestion with R generated essentially full-length linear plasmids (L; R lanes) indicative of head-to-tail multimeric inserts (Lutfalla *et al.*, 1985).

joining activities (Folger *et al.* 1982; Miller and Temin 1983; Roth *et al.*, 1995), generates concatemers (pure plasmid) or concatenates (plasmid and carrier DNA) that give rise to stable transformants upon integration into host chromosomes (Perucho *et al.*, 1980; Robins *et al.*, 1981). Unlike concatemer genesis, concatemer integration into chromatinized chromosomal DNA is usually not the result of homologous recombination. Even with isogenic DNA (Te Riele *et al.*, 1992; Thomas *et al.*, 1992), gene targeting, *i.e.* integration upon homologous recombination, is an unfrequent event; its frequency is only a few percent of random integration upon illegitimate recombination (Capecchi, 1989; Thomas *et al.*, 1986; Merrihew *et al.*, 1996). However, although unfrequent, gene targeting in mammalian somatic cells is a highly significant process (Capecchi, 1989; Thomas *et al.*, 1992; Abuin and Bradley, 1996).

Mammalian somatic recombinases as RecA-like DNA repair enzymes

Such a transient homologous recombination activity of transfecting DNA molecules, either alone (Folger *et al.*, 1982; Lutfalla *et al.*, 1985; Capecchi, 1989) or with cognate chromosomal DNA (Thomas *et al.*, 1986; Capecchi, 1989), is in contrast with the absence of homologous chromosome recombination in most mammalian somatic cells and hybrids (Ephrussi, 1972; Panthier and Condamine 1991), indicating that somatic recombinases do not usually mediate homologous genetic recombination (Bertolotti, 1996a and 1997). Like bacterial RecA proteins (Roca and Cox, 1990), they appear to be mainly involved in DNA repair (Bertolotti, 1996a and 1997). Therefore, in december 1991, our new approach to gene therapy (Bertolotti R., Grant application to French MRT, 1991) started with the search for such a mammalian homolog of RecA (Borchiellini *et al.*, 1997) that could be used to produce, *in vitro*, with single-stranded DNA (ssDNA), nucleoprotein complexes analogous to the bacterial RecA presynaptic filament (Fig. 5).

Upon transfection into mammalian cells, premade filaments should promote efficient spooling of exogenous DNA into chromatinized chromosomal DNA, culminating in directed homologous recombination (Bertolotti, 1996a, 1996b and 1997). Direct transfer of active nucleoprotein filaments may thus increase homologous recombination efficiency between transfecting DNA and cognate chromosomal DNA to a level compatible with a selection-free processing of transfected cells (Bertolotti R., Grant application to French MRT, 1991; Bertolotti, 1996a, 1996b and 1997). Therefore, by contrast to present gene targeting procedures (Capecchi, 1989; Te Riele *et al*, 1992; Thomas *et al.*, 1992; Abuin and Bradley, 1996), our approach could possibly be amenable both to *ex vivo* and *in vivo* gene therapy where drastic selective growth of transfectant cells is impossible.

Recombinase-DNA nucleoprotein filaments as gene targeting vectors

In bacteria, the active presynaptic filament able to mediate *in vitro* the key reaction from homologous recombination, *i.e.* homologous pairing-strand exchange with double-stranded DNA (dsDNA), is a right-handed helical ssDNA-nucleoprotein (Fig. 5) that comprises 6 RecA monomers and 18-19 nucleotides per helical turn (Kowalczykowski *et al.*, 1994; Roca and Cox, 1990). It is formed, in the presence of ATP, upon RecA protein polymerization onto ssDNA at a stochiometry of 3 bases per monomer (Kowalczykowski *et al.*, 1994; Roca and Cox, 1990). Homologous pairing and strand exchange reactions occur within the nucleoprotein filament upon synapsis with dsDNA (Howard-Flanders *et al.*, 1984; Roca and Cox, 1990; Radding, 1991; Stasiak and Egelman, 1994). The mammalian recombinase-based counterpart of such an active nucleoproteic structure should thus provide transfecting ssDNA with protection, nuclear tagging and synapsing ability with mammalian chromosomal dsDNA (Bertolotti, 1996a, 1996b and 1997). Indeed, in its recombinase protein sheath, transfecting ssDNA should be transiently protected from cellular nuclease degradation, from illegitimate

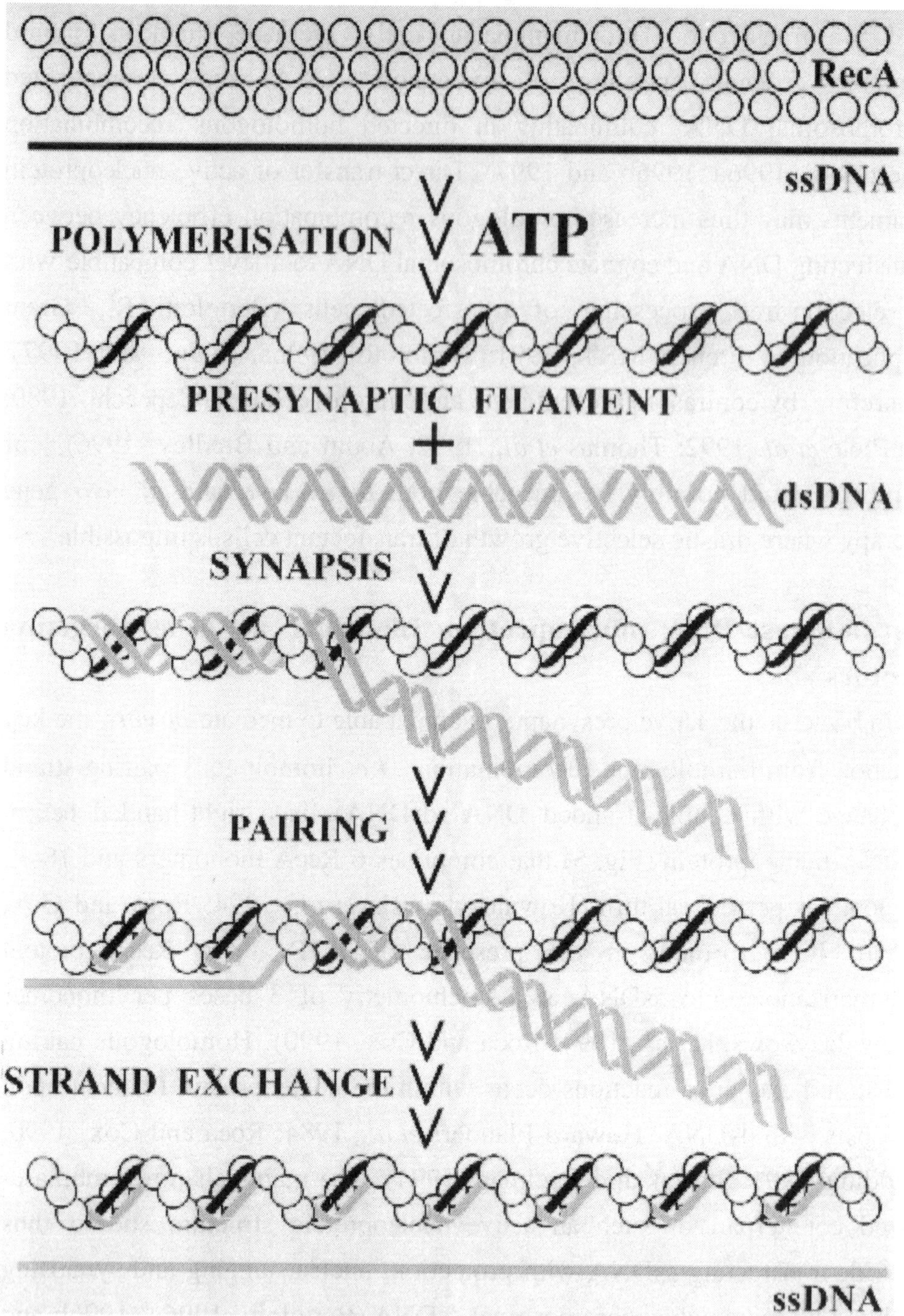

Figure 5. Homologous pairing and strand exchange mediated by RecA *in vitro* (based on Roca and Cox, 1990; Bertolotti 1997). DNA helicity in RecA filaments (18.6 bases/turn of the helix) differs from Watson-Crick B-form (dsDNA: 10 bases/turn): the elongation factor is 50-60%.

recombination and random integration into host chromosomes (Merrihew *et al.*, 1996; Roth *et al.*, 1995) and, possibly, from a variety of DNA-binding proteins that could hamper homologous recombination. In addition, with or without the intermediate formation of recombinosomes (Hays *et al.*, 1995), human recombinase coating should provide for nuclear delivery (Haaf *et al.*, 1995) and putative protein-protein interaction sites which are absent in RecA protein and which are supposed to be essential for an efficient spooling and processing of exogenous ssDNA into chromatinized mammalian chromosomal dsDNA (Bertolotti, 1996b and 1997).

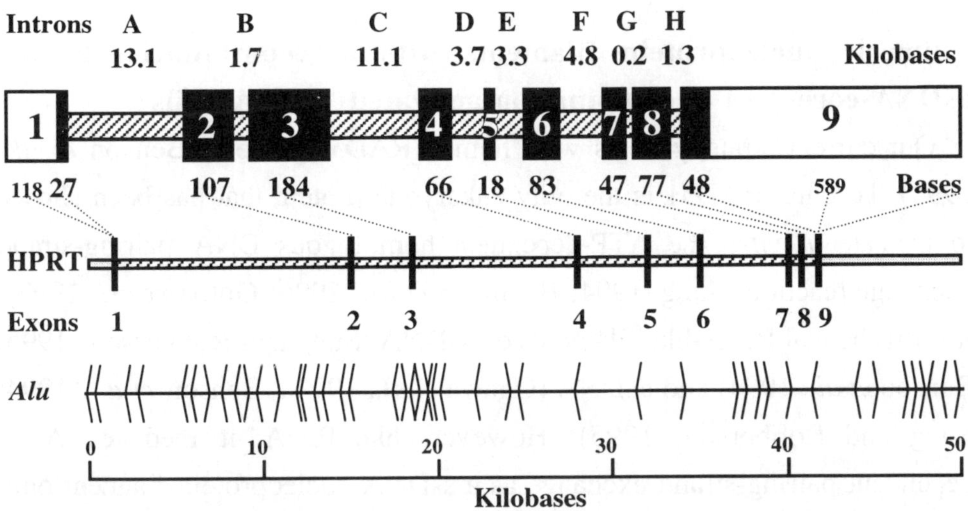

Figure 6. Structure of the human HPRT gene (based on Edwards *et al.*, 1990, and on EMBL/GeneBank file No. M26434; Bertolotti, 1997). The HPRT gene comprises 9 exons (boxes) and 8 introns (hatched areas). Introns and flanking sequences (stippled areas) contain repetitive elements such as *Alu* (46 repeats, positioned with orientation) and LINE (not shown). Enlarged at the upper part of the figure, exons, not in scale with introns, span about 1364 base pairs (bp) out of a ~40 kilobases (kb) genomic sequence. The coding sequence is 657 bp long (filled boxes). Exons 1 and 9 comprise 5' and 3' non-coding regions (empty boxes), respectively.

Upon transfection into human cells, human recombinase presynaptic filaments should therefore be able of carrying out genome-wide search on chromatinized human DNA (Bertolotti, 1996b and 1997) in the same way *E. coli* RecA filaments do it *in vitro* on naked human DNA (Ferrin and

Camerini-Otero, 1991; Camerini-Otero and Hsieh, 1993). The ensuing directed homologous recombination should thus generate specific alterations to endogenous dsDNA sequences resulting in efficient gene inactivation or repair (Bertolotti, 1996b and 1997). However, as illustrated with the human HPRT gene, mammalian genomic DNA mainly comprises introns with repetitive elements such as *Alu* and LINE (Fig. 6). Therefore, in order to avoid unspecific targeting to repetitive chromosomal elements and to bypass potential arrests of the homologous DNA pairing-strand exchange reaction by therapeutic heterologous sequences, we have devised a strategy in which presynaptic filaments may comprise a dsDNA core (Bertolotti, 1996b).

Chimeric nucleoprotein filaments with a recombination-locked dsDNA-core and recombination-activated ssDNA tails

Our current strategy deals with human RAD51 protein (Benson *et al.*, 1994). To date, RAD51 is the only eukaryotic protein that has been shown to mediate *in vitro* the ATP-dependent homologous DNA pairing-strand exchange reaction (Sung, 1994; Baumann *et al.*, 1996; Gutpa *et al.*, 1997). It forms helical RecA-like filaments on ssDNA (Sung and Robberson, 1995; Benson *et al.*, 1994) and dsDNA (Ogawa *et al.*, 1993a; Benson *et al.*, 1994; Sung and Robberson, 1995). However, like RecA, it mediates ATP-dependent pairing-strand exchange as a ssDNA-nucleoprotein filament only (Sung and Robberson, 1995). Efficient genesis of such an active filament depends on the presence of replication protein A (RPA) (Sung, 1994; Sung and Robberson, 1995). RPA is part of a putative recombinosome (Hays *et al.*, 1995) comprising proteins like RAD52 (Ogawa *et al.*, 1993b) that may interact *in vivo* with premade filaments (see discussion in Bertolotti, 1997).

Importantly enough, as illustrated in figure 7, coating dsDNA with RAD51 prevents duplex DNA from participating in the pairing reaction but does not affect the ability of uncoated concurrent dsDNA to react with RAD51-ssDNA filaments (Sung and Robberson, 1995). The idea is thus to devise chimeric targeting filaments in which therapeutic heterologous bases

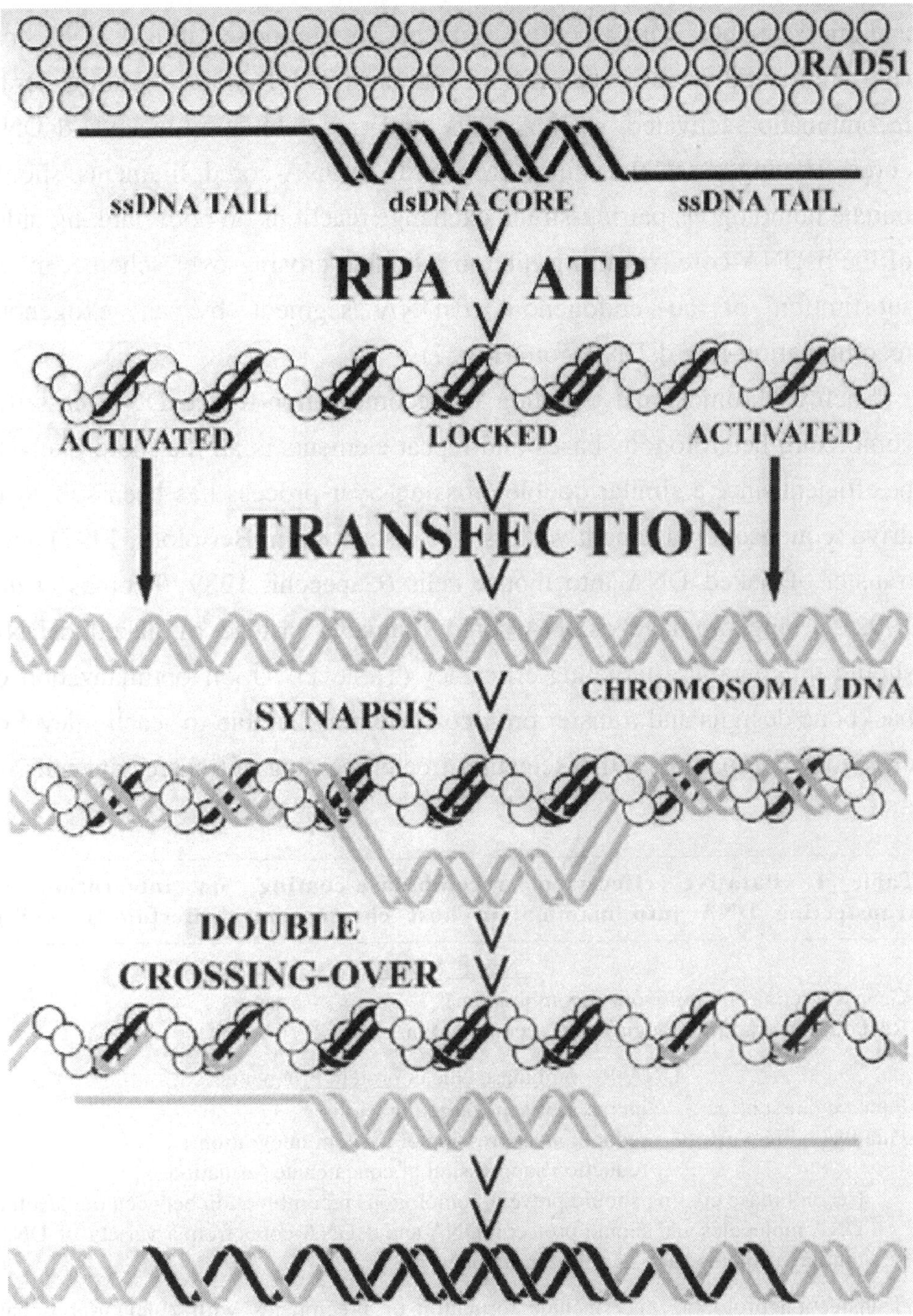

Figure 7. Targeting recombination-locked dsDNA cassettes to chromosomal DNA with dsDNA-cored RAD51 filaments (Chromatinian structures are not shown; Bertolotti, 1997)

and any sequence from repetitive elements are comprised in a dsDNA core (Fig. 7). Indeed, coating such DNA backbones with RAD51 should generate recombination-activated ssDNA tails and recombination-locked dsDNA cores. Having a ssDNA tail at each end, dsDNA-cored filaments should initiate homologous pairing-strand exchange reactions on both flanking sides of the dsDNA core, resulting, like in a double crosing-over scheme, in the substitution of an endogenous dsDNA segment by an exogenous recombination-free dsDNA core (Fig. 7).

Such a chromosomal targeting of recombination-free dsDNA cassettes comprising heterologous bases and repeat elements is all the more likely to be efficient since a similar double crossing-over process has been shown to drive sequence replacement vectors (see discussion in Bertolotti, 1997) upon transfer of naked DNA into mouse cells (Capecchi, 1989; Thomas *et al.*, 1992). Using premade dsDNA-cored filaments instead of naked dsDNA should thus increase targeting efficiency (Table 1). Upon optimalization of backbone designs and transfer protocols, we may be able to reach a level of efficiency compatible with a selection-free processing of transfectant cells.

Table 1. Putative effects of recombinase-coating on integration of transfecting DNA into mammalian host chromosomes (Bertolotti, 1997).

"Naked"* DNA

-Gene Targeting (homologous recombination):	rare
-Random Integration ("illegitimate" recombination):	frequent

DNA-Recombinase Nucleoprotein Filaments

-Putative direct effect:	increase of gene targeting
-Putative indirect effects:	reduction/suppression of random integration
	reduction/suppression of concatenate formation

(recombinase coating should prevent homologous recombination between transfecting DNA molecules and should protect ssDNA and dsDNA-cores from a variety of DNA-binding proteins such as end-joining ones**.)

*Transfection protocols may include formation of precipitates with calcium-phosphate (Graham and Van der Eb, 1973) or of various complexes (Zauner *et al.*, 1996).

**End-to-end ligation described for linear dsDNA (Baumann *et al.*, 1996) should be impossible with standard or dsDNA-cored ssDNA filaments.

DNA backbones for standard and dsDNA-cored RAD51 filaments

A series of DNA-backbones for nucleoprotein filaments aimed at inactivating then repairing the HPRT gene from human cells were designed (Bertolotti, 1996b) using EMBL/GenBank Data file No. M26434. This file comprises the nucleotide sequence from the human HPRT gene, its structure and its molecular alterations in a series of Lesch-Nyhan mutants (Edwards *et al.*, 1990). Some DNA backbones corresponding to two such mutants are shown (Fig. 8). Other selected mutations involve either introns or exons.

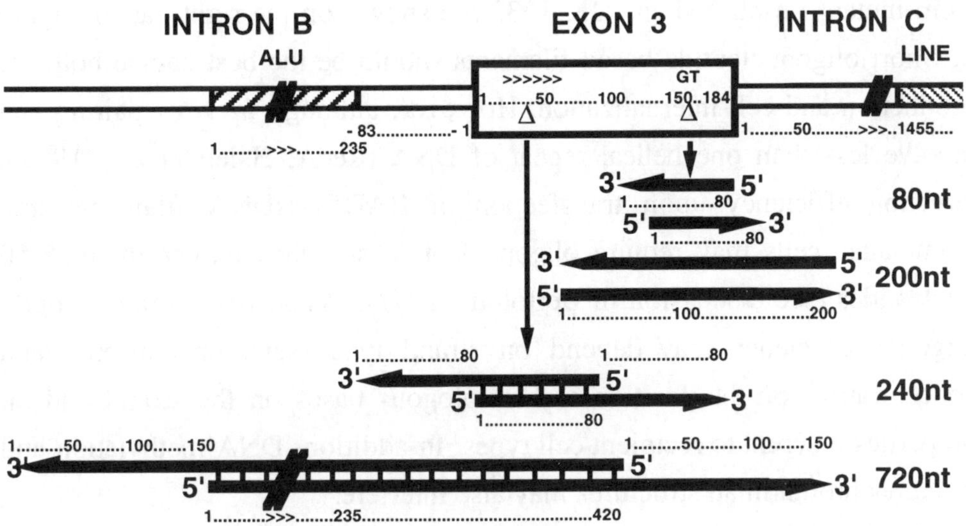

Figure 8. Designing HPRT DNA-backbones for human RAD51 filaments (Bertolotti, 1996b). Mutations Lesch-Nyhan RJK 2108 and 1332 are deletions (Δ) of either 40 base-pairs (bp) or 2 bp (GT/CA) from exon 3. Flanking exon 3 (184 bp), the 3' end of intron B (1.7 kb) and the 5' end of intron C (11.1 kb) are shown with their respective terminal *Alu* and proximal LINE repeat sequences. ssDNA stretches (thick horizontal arrows) are either derived from mutant (inactivating filaments) or wild-type DNA (reconstructing filaments). Small arabic numbers refer to nucleotide (nt) length.

As illustrated in figure 8, the idea is to use standard recombinase-ssDNA filaments when dealing with point-mutations or very short lesions (GT/CA deletion on the 3' end of exon 3) from non-repetitive elements, and to use chimeric filaments with a dsDNA core comprising heterologous base clusters and any sequence from repetitive elements when required (40 bp deletion on the 5' end of exon 3). RecA protein-mediated exchange of DNA strands can accomodate quite important DNA lesions and mismatched base pairs (Kowalczykowski *et al.*, 1994). Such informations are missing for RAD51. Therefore, the strategy should rely on results obtained with both approaches, *i.e.* standard presynaptic filament versus dsDNA-cored one. On the other hand, dsDNA-cored filaments may prove to be more efficient at sequence transfer than standard presynaptic ones, even for point-mutations and short lesions. They may therefore develop into a universal gene targeting device.

Two sets of sens and antisens strands designed for the GT /CA deletion from mutant Lesch-Nyhan RJK 1332 are shown on the right part of figure 8. Short oligonucleotide-based filaments should be the best choice both for production and cell-internalization. However, although *in vitro* pairing can involve less than one helical repeat of DNA (RecA; Hsieh *et al.*, 1992), targeting efficiency upon transfection of RAD51-ssDNA filaments into mammalian cells may require oligonucleotide stretches longer than 25-50 nucleotides (see discussion in Bertolotti, 1997). Aside from strand length, targeting efficiency may depend on strand type (sens or antisens) and concentration, on the position of heterologous bases on the strand and on properties inherent to recipient cell types. In addition, DNA methylation and discrete chromatinian structures may also interfere.

The DNA backbones of two dsDNA-cored filaments corresponding to mutant Lesch-Nyhan RJK 2108 are also presented in figure 8. The first one is 240 nt long. It comprises an 80 bp dsDNA core including a 40 bp cluster corresponding to deletion RJK 2108 and two 80 nt ssDNA tails. The second one is 720 nt long. It comprises a 420 bp dsDNA core including a 235 bp *Alu* repeat and the aforementioned 40 bp cluster; its tails are 150 nt long. In these cases, targeting efficiency may depend on the total length of the nucleoprotein filament as well as on the length of each component, *i.e.* tails and core (see discussion in Bertolotti, 1997). Aside from filament concentration, methylation, chromatinian structures and cell type, additional potential parameters include single-stranded tail type. Importantly enough, actual complexes are based on human RAD51 protein, the yeast form of which has been shown to mediate *in vitro* ATP-dependent homologous DNA pairing-strand exchange reaction with a polarity opposite to that of RecA (Sung and Robberson, 1995). The designs of the dsDNA-cored filaments presented on the figure are thus based on the assumption that, in addition to a structural conservation (Shinohara *et al.*, 1993; Yoshimura *et al.*, 1993; Benson *et al.*, 1994) with its yeast homolog (Aboussekhra *et al.*, 1992; Basile *et al.*, 1992; Shinohara *et al.*, 1992; Ogawa *et al.*, 1993a), human RAD51 protein also catalyzes ATP-dependent homologous pairing and strand exchange *via* a similar mechanism. Thus, although unlikely[1], human RAD51-ssDNA filaments could initiate the pairing-strand exchange reaction on the 3' side of the complementary strand of the duplex partner, necessitating 5' ssDNA-tailed filaments instead of the 3' yeast-type ones from figure 8.

Completion of the strand-exchange reaction upon transfection

Although central to homologous recombination, homologous pairing and DNA strand exchange require additional activities in order to have the exchange process achieved. As shown in figure 7, outgoing DNA must be cleaved from chromosomal DNA and replaced upon ligation of incoming DNA. Such reactions are typical from gene targeting with replacement vectors (Thomas *et al.*, 1986; Thomas and Capecchi, 1987; Thomas *et al.*, 1992) where crossovers often occur near the ends of the vectors (Deng *et al.*, 1993). The enzymatic machinery able to carry out the double crossingover reaction is thus present in somatic cells. Indeed, with our premade dsDNA-cored filaments where DNA is protected by a RAD51 protein sheath and ssDNA tails ready for synapsis, we should be close from

---

[1] Recent experiments suggest that the polarity of strand transfer promoted by human RAD51 is indeed the same as yeast RAD51 one (Baumann *et al.*, 1996).

the aforedescribed *in vitro* conditions (Bertolotti, 1997) where homologous pairing and strand-exchange are mediated by RAD51 alone (Sung and Robberson, 1995). Efficient spooling of recombination-activated ssDNA tails into chromatinian structures should promote straightforward scanning of chromosomal DNA culminating with homologous pairing of ssDNA tails with chromosomal dsDNA counterparts. The ensuing DNA strand exchange should be restricted to invading ssDNA tails; branch migration should not involve recombination-locked dsDNA cores from invading filaments and may thus be completed without the need of mammalian counterparts of bacterial specialized proteins (RuvA-B and RecG; Eggleston and West, 1996). Putative Holliday junctions may thus form at the edges of dsDNA cores and can be cleaved upon somatic resolvase (Hyde *et al.*, 1994) processing (Bertolotti, 1997).

Tissue-specific delivery with recombinase-DNA nucleoprotein filaments

The ability of vectors to deliver therapeutic DNA to the correct cells *in vivo* is an important feature for gene therapy. In this respect, in addition to gene targeting, recombinase protein may provide for cell and tissue-specific delivery upon partial glycosylation or fusion with cell-binding peptides (Bertolotti, 1997). Indeed, receptor-mediated gene delivery was first accomplished on hepatocytic cells both *in vitro* (HepG2 cells; Wu and Wu, 1987) and *in vivo* (hepatocytes; Wu and Wu, 1988) *via* the liver-specific asialoglycoprotein receptor (Schwartz, 1984). This receptor binds and internalizes galactose-terminal glycoproteins such as asialoorosomucoid (AsOR) and asialofetuin (Schwartz, 1984). It provides for hepatocyte-specific endocytosis of toxins (Simpson *et al.*, 1982; Dodeur M., Gros P. and Bertolotti R., 1983, unpublished), drugs (Wu, 1988) and DNA (Wu and Wu, 1987) when they are fixed to a galactose-terminal glycoprotein or complex.

The original DNA carrier system for gene delivery consisted in AsOR covalently linked to DNA-binding poly-L-lysine (Wu and Wu, 1987 and 1988). The DNA carrier was subsequently improved upon substituing lactosylated poly-L-lysine to AsOR/poly-L-lysine conjugate (Midoux *et al.*, 1993). Introducing some lactosylated recombinase protein into recombinase-DNA nucleoprotein filaments could thus provide for hepatocyte targeting without altering homologous recombination capability.

Cell-type targeting of poly-L-lysine can be changed using other glycosylations such as mannosylation (Midoux *et al.*, 1993). Differential glycosylation of recombinase protein could thus provide for tissue-specific delivery of nucleoprotein filaments. However, we believe that the genesis of a true tissue-specific repertoire for DNA delivery most likely relies on the potential availability of short random peptides that bind and enter cells in a tissue-specific way (Barry *et al.*, 1996). Fusing such small peptides to the carboxy or amino terminus of recombinase protein may not impair 1) polymerisation onto ssDNA with wild-type recombinase and 2) homologous recombination activity upon co-polymerisation into mainly wild-type nucleoprotein filaments. Interestingly enough, *E. coli* RecA protein has been shown to retain its biological activity when fused to the nuclear location signal peptide of SV40 large T-antigen (Kido *et al.*, 1992).

Efficiency of gene delivery mediated by asialoglycoprotein (Wu and Wu, 1987) or transferrin (Wagner *et al.*, 1990) receptors is depending on lysosomotrophic (Cotten *et al.*, 1990; Midoux *et al.*, 1993) or endosome-disrupting co-factors (Curiel *et al.*, 1991; Wagner *et al.*, 1992; Cristiano *et al.*, 1993; *Midoux et al.*, 1993). These ancillary factors inhibit endosomial-lysosomial degradation (Cotten *et al.*, 1990; Seglen, 1983; Yoshimori *et al.*, 1991) and promote cytoplasmic release of endocytosed therapeutic DNA (Curiel *et al.*, 1991), respectively. Endosome-disruption is the essential step. It has been achieved using adenoviral capsides or fusigenic peptides (Curiel *et al.*, 1991; Wagner *et al.*, 1992; Midoux *et al.*, 1993). Providing an endosome-disrupting factor *in vivo* with recombinase-DNA nucleoprotein filaments is feasible but may have side-effects. However, lysosomotrophic and endosome-disrupting co-factors have been shown to be ineffective when DNA delivery is mediated by synthetic RGD integrin-binding peptides (Hart *et al.,* 1995) suggesting that random cell-binding peptides may provide, like fusigenic Sendai-virus liposomes (Dzau *et al.*, 1996), for direct introduction of DNA into cell cytoplasm.

The RGD integrin-binding motif does not confer tissue-specificity (Hart *et al.*, 1995; Barry *et al.*, 1996) but stands as a means of assessing the potential use of recombinase-DNA nucleoprotein filaments as direct *in vivo* transfection vehicles. On the other hand, encapsulation of native recombinase-DNA nucleoprotein filaments into new emerging liposomes such as fusigenic sendai-virus ones (Dzau *et al.*, 1996) or polyvalent vesosomes (Walker *et al.*, 1997) is an other way to approach *in vivo* DNA delivery. In this case, tissue-specificity can be achieved upon exposing cell-binding molecules at the surface of the liposomes (Remy *et al.*, 1995). Another advantage of leakproof liposomes is to circumvent putative immunological reactions against native or modified recombinase protein. Indeed, as expected, human RAD51 is an intracellular protein (Haaf *et al.*, 1995) and may therefore be an autoimmune disease hazard for gene therapy patients.

4. RAD51-MEDIATED GENE THERAPY

An HPRT-target/OTC-control set of ''single-copy'' genes as an *in vitro* model system for Gene Therapy

We have previously used *in vivo* homologous recombination as a shuttle driving mechanism for integrating vectors from hepatoma cells to *E. coli* (Lutfalla *et al.*, 1985) and as a hybrid-shuttle processor for expression cloning in fission yeast (Bertolotti *et al.*, 1995). Using a set of selectable "single copy" X-linked genes, HPRT as a target and ornithine transcarbamylase (OTC) as a control, we wish now to direct homologous recombination upon transfection of human cells with RAD51 nucleoprotein filaments (Bertolotti, 1996a and 1996b).

Both HPRT and OTC genes have similar structures: they are large (about 40 and 70 kb respectively), comprise 9 and 10 exons respectively, and encode a ~1.5 kb mature mRNA (Edwards *et al.*, 1990; Horwich *et al.*, 1984; Hata *et al.*, 1988). In addition, mutations that inactivate these human genes have been well-characterized (Cariello and Skopek, 1993; Rossiter and Caskey, 1995; Brusilow and Horwich, 1995). Importantly enough, we have isolated well-differentiated variants from the human HepG2 line upon selection in liver-specific growth-media (Lutfalla *et al.*, 1989). These variants are able to grow in arginine-free medium containing ornithine and, as expected, do express OTC (Armbruster *et al.*, 1992). These OTC-positive cells can be efficiently killed with tritiated ornithine (Armbruster and Bertolotti, unpublished) as described for rat cells (Goss, 1984; Bertolotti, unpublished). Therefore, using either thioguanine or HAT (Szybalski, 1992) for HPRT, and arginine-free medium with normal or tritiated ornithine for OTC, we can select or counter-select cells that express either HPRT or OTC. Thus, upon transfecting human cells with RAD51-HPRT nucleoprotein filaments or with control RAD51-OTC ones, we can evaluate homologous recombination targeting efficiencies and compare them to the rates of unspecific insertional mutagenesis (Bertolotti, 1996b).

Table 2. HPRT and OTC genes.

	Features
-"Single" copy (X-linked)	
-Similar structure:	large (~40 and 70 kb, respectively)
	9 and 10 exons, respectively
	~1.5 kb mRNA
-Well-characterized mutations	
-Selection and counter-selection for both genes	
	Strategy
-Target::	HPRT (Targeting efficiencies)
-Control:	OTC (Rates of unspecific insertional mutagenesis)

From transfecting cell lines *in vitro* toward targeting specific somatic cells *in vivo*

Producing active nucleoprotein filaments with human RAD51 (Benson *et al.*, 1994) and RPA (Henricksen *et al.*, 1994) recombinant proteins, and demonstrating their ability at improving gene targeting upon electroporation (Neumann *et al.*, 1982) of HepG2 and control cells is our present concern. Reducing potential unspecific insertional mutagenic effects of RAD51 nucleoprotein filaments and increasing homologous recombination efficiency to a level compatible with a selection-free processing of transfectant cells should be the next achievement. Protocols could then be devised for *ex vivo* gene therapy (Blaese *et al.*, 1995; Ponder *et al.*, 1991; Groves *et al.*, 1993) when appliable and, most importantly, for *in vivo* tissue-specific delivery (Wu and Wu, 1988; Wilson *et al.*, 1992; Barry *et al.*, 1996) of RAD51-DNA nucleoprotein filaments in any other cases.

From transfecting cell lines *in vitro* towards targeting specific somatic cells *in vivo*, our ultimate goal would be to develop a recombinase-mediated gene therapy that would substitute to the present minigene replacement approach (Blaese *et al.*, 1995; Mulligan, 1993; Roemer and Friedmann, 1992; Verma and Somia, 1997). Repairing mutant genes and inactivating deleterious or vital viral genes could be an effective way to cure inherited diseases, cancer and chronic viral infections such as AIDS and hepatitides. Interestingly, chimeric RNA-DNA oligonucleotides were recently devised that appear now as a potential partial alternate to our recombinase-DNA nucleoprotein filaments. Indeed, RNA-DNA oligonucleotides were first designed to inactivate key viral proteins upon mutating viral episomes (Yoon *et al.*, 1996) and have been shown to be also effective at genomic targeting (Cole-Strauss *et al.*, 1996). However, unlike this new emerging method, our approach is not restricted to point mutations and is thus more appropriate for the potential treatment of inherited diseases.

5. PROSPECTS FOR THE LESCH-NYHAN DISEASE

Hyperuricemia and neurologic syndrome of the disease

The main metabolic feature of Lesch-Nyhan patients is hyperuricemia with uricosuria, tophaceous gouty arthritis, urinary tract calculi and urate nephropathy (Lesch and Nyhan, 1964; Rossiter and Caskey, 1995). It results from HPRT enzyme deficiency (Seegmiller *et al.*, 1967) and a subsequent increase in the rate of *de novo* purine synthesis (Lesch and Nyhan, 1964; Nyhan *et al.*, 1965) that leads to an overproduction of hypoxanthine upon HPRT recycling deficiency (Rossiter and Caskey, 1995). Hypoxanthine is subsequently converted to xanthine and uric acid mainly by liver xanthine oxidase (Rossiter and Caskey, 1995). Allopurinol, a xanthine oxidase inhibitor, is an effective medication for Lesch-Nyhan hyperuricemia and associated symptoms (Rossiter and Caskey, 1995). Unfortunatelly, allopurinol has no effect on the neurologic symptoms of the disease. This is consistent with the fact that these neurologic symptoms are not casually related to uric acid since its concentration in the cerebrosidal fluid was shown to be the same in patients and control subjects (Rosenbloom *et al.*, 1967). Interestingly enough, the central nervous system is protected from circulating uric acid by the blood-brain barrier; with a low concentration in the cerebrosidal fluid, uric acid is indeed a sensitive marker of dysfunction of the blood-brain barrier (Hallgren *et al.*, 1983). In fact, there is not yet a cure to the neurologic symptoms of the Lesch-Nyhan syndrome (Nyhan and Wong, 1996; Rossiter and Caskey, 1995). They are still open to HPRT gene therapy as first proposed more than 10 years ago (Friedmann, 1985).

Brain dopaminergic deficit and gene therapy timing

Unlike typical neurodegenerative diseases such as Parkinsonism and Alzheimer‘s disease which primarily affect the elderly, the Lesch-Nyhan disease appears to be developmental in origin. Its neurological features, which include delayed motor development, severe spasticity, choreoathetosis, compulsive self-mutilation and mental retardation, appear to be a consequence of alterations in brain dopamine systems (Lloyd *et al.*, 1981; Wong *et al.*, 1996; Ernst *et al.*, 1996).

Lesch-Nyhan patients have a reduced number of dopaminergic nerve terminal and cell bodies (Wong *et al.*, 1996; Ernst *et al.*, 1996). Such a dopaminergic deficit has been described in HPRT-deficient transgenic mice where it is mainly localized in basal ganglia and emerges during the first two months of postnatal development (Jinnah *et al.*, 1994). In Lesch-Nyhan patients, the dopaminergic deficit appears to involve all dopaminergic pathways (Ernst *et al.*, 1996) although it was first identified in basal ganglia (Lloyd *et al.*, 1981); it is pervasive and appears to be stable with age (patients were 10 to 20 years old; Ernst *et al.*, 1996). Such a stability implies that we are not dealing with a life-long degenerative process (Ernst *et al.*, 1996). It is consistent with the clinical course of the Lesch-Nyhan neurobehavioural syndrome that is fully expressed and constant by the age of 2 years (Mizuno, 1986; Anderson *et al.*, 1992). The neural syndrome of the Lesch-Nyhan disease appears thus to be developmental in origin (*Ernst et al.,* 1996). Whether the aforedescribed dopaminergic deficit results from miss-proliferation/differentiation processes or from true developmental degeneration of dopaminergic neurones is not yet known. In any case, it is not likely to be reversed upon restauration of neuronal or glial HPRT activity of Lesch-Nyhan patients. Therefore, HPRT gene repair should be attempted before the onset of the neurologic pathology. Gene therapy of the Lesch-Nyhan disease may thus succeed on newborn children or late-stage fetuses.

Pathology of the disease and choice of target cells

Choosing the right target cells is another important clue for gene therapy. It will depend on what is still to be learned about the pathology of the Lesch-Nyhan disease. Indeed, the neurologic dysfunction of Lesch-Nyhan patients is supposed to result either from the primary absence of HPRT activity in brain cells or from the secondary increased concentration of circulating brain purines or related metabolites (Rosenbloom *et al.*, 1967; Rossiter and Caskey, 1995). In the first case, although in an unknown way, HPRT activity of neurones or glial cells may be important for the efficient genesis of dopaminergic pathways. In the second case, overproduction of circulating purines or related metabolites by brain, liver and other organs may induce developmental alterations in the brain dopamine systems. In both cases, targeting the brain may be sufficient to prevent the onset of the neurologic features of the disease. In the second case, targeting liver cells or circulating lymphocytes may be the simplest and most efficient way to control purine overproduction and subsequent alterations of dopaminergic pathways. On the one hand, liver is a large organ and appears as a good candidate for future protocols, both for *ex vivo* (Ponder *et al.,* 1991; Roy Chowdhury *et al.*, 1991) and *in vivo* (Wilson *et al.*, 1992; Koeberl *et al.*, 1997; Snyder *et al.*, 1997) gene therapy. On the other hand, T lymphocytes are already involved in successful minigene expression therapy trials (Blaese *et al.*, 1995; Kohn *et al.*, 1995; Kohn and Parkman, 1997; Bonini *et al.*, 1997). In addition, $HPRT^+$ lymphocytes appear to have a selective survival advantage over $HPRT^-$ ones in human heterozygotes (Dancis *et al.*, 1968; Nyhan *et al.*, 1970; McDonald and Kelley, 1972) indicating that HPRT-engineered lymphocytes may accumulate to therapeutic levels upon *ex vivo* gene therapy. Consistent with this idea, normal HPRT activity in peripheral blood cells has been obtained in an adult Lesch-Nyhan patient upon bone marrow transplantation (Nyhan *et al.*, 1986).

Neonatal and fetal gene therapy

The delivery of recombinase-HPRT nucleoprotein vectors to cells and tissues is discussed above in the tissue-specific delivery section. It deserves some additional comments regarding neonatal and fetal gene therapy. The proliferating capabilities of most neonatal and late-stage fetal tissues is an important parameter for gene targeting. Indeed, integration of transfecting

DNA essentially occurs in dividing cells (Capecchi, 1989). The same holds true for retroviral vectors with the exception of lentiviral ones that can also integrate into the chromosomes of resting cells (Naldini *et al.*, 1996). Therefore, targeting lymphocytes or organs such as liver and brain that undergo significant late-fetal and postnatal growth offers a real promise of success. Indeed, in conjunction with intramuscular injection of adenosine deaminase (ADA) enzyme conjugated to polyethylene glycol, autologous umbilical cord blood lymphocytes have been used to treat ADA-deficiency on three newborn children in the first such trial based on minigene expression therapy (Kohn *et al.*, 1995). So far, all three children have recovered from their severe combined immune deficiency (SCID) syndrome (Kohn and Parkman, 1997). As a matter of fact, neonatal and fetal gene therapy approaches are increasingly attractive (Kohn and Parkman, 1997; Coutelle *et al.*, 1995). They raise moral, ethical and safety questions that may not be easy to solve for brain gene therapy.

REFERENCES

Aboussekhra, A., Chanet, R., Adjiri, A. and Fabre, F. (1992). Semidominant suppressors of Srs2 helicase mutations of *S. cerevisiae* map in the *RAD51* gene, whose sequence predicts a protein with similarities to procaryotic RecA proteins. *Mol. Cell. Biol.*, 12: 3224-3234.

Abuin, A. and Bradley, A. (1996). Recycling selectable markers in mouse embryonic stem cells. *Mol. Cell. Biol.*, 16: 1851-1856.

Anderson, L. T., Ernst, M. and Davis, S. V. (1992). Cognitive abilities of patients with Lesch-Nyhan disease. *J. Autism Dev. Disord.*, 22: 189-203.

Armbruster, L., Cavard, C., Briand, P. and Bertolotti, R. (1992). Selection of variant hepatoma cells in liver-specific growth media: regulation at the mRNA level. *Differentiation*, 50: 25-33.

Barry, M. A., Dower, W. J. and Johnston, S. A. (1996). Toward cell-targeting gene therapy vectors: selection of cell-binding peptides from random peptide-presenting phage libraries. *Nature Med.* , 2: 299-305.

Basile, G., Aker, M. and Mortimer, R. K. (1992). Nucleotide sequence and transcriptional regulation of the yeast recombinational repair gene *RAD51*. *Mol. Cell. Biol.*, 12: 3235-3246.

Baumann, P., Benson, F. and West, S. C. (1996). Human Rad51 protein promotes ATP-dependent homologous pairing and strand transfer reactions *in vitro*. *Cell*, 87: 757-766.

Benson, F. E., Stasiak, A. and West, S. C. (1994). Purification and characterization of the human Rad51 protein, an analogue of *E. coli* RecA. *EMBO J.*, 13: 5764-5771.

Bertolotti, R. (1996a). Recombinase-mediated gene therapy: targeting single stranded DNA to chromosomal DNA with human RAD51 presynaptic fibers. *Hepatology* , 24: 484A.

Bertolotti, R. (1996b). Recombinase-mediated gene therapy: strategies based on Lesch-Nyhan mutants for gene repair/inactivation using human RAD51 nucleoprotein filaments. *Biogenic Amines*, 12: 487-498.

Bertolotti, R. (1997). Recombinase-DNA nucleoprotein filaments as Gene Therapy vectors. *Biogenic Amines* (in the press).

Bertolotti, R., Armbruster-Hilbert, L. and Okayama, H. (1995). Liver fructose-1,6-bisphosphatase cDNA: trans-complementation of fission yeast and characterization of two human transcripts. *Differentiation*, 59: 51-60.

Bertolotti, R. and Lutfalla, G. (1983). Genes programming cell differentiation. *J. Cell. Biochem.*, 7A: 149.

Blaese, R. M., Culver, K., Miller, A. D., Carter, C., Fleisher, T., Clerici, M., Shearer, G., Chang, L., Chiang, Y., Tolstoshev, P., Greenblatt, J., Rosenberg, S., Klein, H., Berger, M., Mullen, C., Ramsey, W., Muul, L., Morgan, R. and French-Anderson, W. (1995). T lymphocyte-directed gene therapy for ADA$^-$ SCID: initial trial results after 4 years. *Science*, 270: 475-480.

Bonini, C., Ferrari, G., Verzeletti, S., Servida, P., Zappone, E., Ruggieri, L., Ponzoni, M., Rossini, S., Mavilio, F., Traversari, C. and Bordignon, C. (1997). HSV-TK gene transfer into donnor lymphocytes for control of allogenic graft-versus-leukemia. *Science*, 276: 1719-1724.

Borchiellini, P., Angulo, J. F. and Bertolotti, R. (1997). Genes encoding mammalian recombinases: cloning approach with anti-RecA antibodies. *Biogenic Amines*, 13: 195-215.

Breathnach, R. and Chambon, P. (1981). Organization and expression of eucaryotic split genes coding for proteins. *Ann. Rev. Biochem.*, 50: 349-383.

Brennand, J., Konecki, D. S. and Caskey, C. T. (1983). Expression of human and chinese hamster hypoxanthine-guanine phosphoribosyltransferase cDNA recombinants in cultured Lesch-Nyhan and chinese hamster fibroblasts. *J. Biol. Chem.*, 258: 9593-9596

Brusilow, S. and Horwich, A. (1995). Urea cycle enzymes. *In*: The metabolic and molecular bases of inherited disease, C. Scriver, A. Beauded, W. Sly and D. Valle (Eds.), McGraw-Hill, New York, pp. 1187-1232.

Camerini-Otero, R. D. and Hsieh, P. (1993). Parallel DNA triplexes, homologous recombination and other homology-dependent DNA interactions. *Cell*, 73: 217-223.

Capecchi, M. (1989). The new mouse genetics: altering the genome by gene targeting. *Trends Genet.*, 5: 70-76.

Cariello, N. and Skopek, T. (1993). Analysis of mutations occurring at the human *hprt* locus. *J. Mol. Biol.*, 231: 41-57.

Chasin, L. A. and Urlaub, G. (1976). Mutant alleles for hypoxanthine phosphoribosyl transferase: codominant expression, complementation, and segregation in hybrid Chinese hamster cells. *Somatic Cell Genet.*, 2: 453-467.

Chen, C. and Okayama, H. (1987). High-efficiency transformation of mammalian cells by plasmid DNA. *Mol. Cell. Biol.*, 7: 2745-2752.

Cole-Strauss, A., Yoon, K., Xiang, Y., Byrne, B., Rice, M., Gryn, J., Holloman, W. and Kmiec, E. (1996). Correction of the mutation responsible for sickle cell anemia by an RNA-DNA oligonucleotide. *Science*, 273: 1386-1389.

Cotten, M., Langle-Rouault, F., Kirlappos, H., Wagner, E., Mechtler, K., Zenke, M., Beug, H. and Birnstiel, M. L. (1990). Transferrin-polycation-mediated introduction of DNA into human leukemic cells: stimulation by agents that affect the survival of transfected DNA or modulate transferrin receptor levels. *Proc. Natl. Acad. Sci. USA*, 87: 4033-4037.

Coutelle, C., Douar, A.M., Colledge, W. H. and Froster, U. (1995). The challenge of fetal gene therapy. *Nature Med.*, 1: 864-866.

Cristiano, R., Smith, L. and Woo, S. (1993). Hepatic gene therapy: adenovirus enhancement of receptor-mediated gene delivery and expression in primary hepatocytes. *Proc. Natl. Acad. Sci. USA*, 90: 2122-2126.

Curiel, D., Agarwal, S., Wagner, E. and Cotten, M. (1991). Adenovirus enhancement of transferrin-polylysine-mediated gene delivery. *Proc. Natl. Acad. Sci. USA*, 88: 8850-8854.

Dancis, J., Berman, P. H., Jansen, V. and Balis, M. E. (1968). Absence of mosaicism in the lymphocyte in X-linked congenital hyperuricosuria. *Life Sci.*, 7: 587-591.

Deng, C., Thomas, K. and Capecchi, M. (1993). Location of crossovers during gene targeting with insertion and replacement vectors. *Mol. Cell. Biol.*, 13: 2134-2140.

DiMaio, D., Treisman, R. and Maniatis, T. (1982). Bovine papilloma-virus vector that propagates as a plasmid in both mouse and bacterial cells. *Proc. Natl. Acad. Sci. U.S.A.*, 79: 4030-4034.

Dzau, V., Mann, M., Morishita, R. and Kaneda, Y. (1996). Fusigenic viral liposome for gene therapy in cardiovascular diseases. *Proc. Natl. Acad. Sci. U.S.A.*, 93: 11421-11425.

Edwards, A., Voss, H., Rice, P., Civitello, A., Stegemann, J., Schwager, C., Zimmermann, J., Erfle, H., Caskey, C. and Ansorge, W. (1990). Automated DNA sequencing of the human *hprt* locus. *Genomics*, 6: 593-608.

Eggleston, A. and West, S. (1996). Exchanging partners: recombination in *E. coli*. *Trends Genet.*, 12: 20-26.

Ephrussi, B. (1972). Hybridization of somatic cells. Princeton University Press, Princeton, N. J.

Ernst, M., Zametkin, A. J., Matochik, J. A., Pascualvaca, D., Jons, P. H., Hardy, K., Hankerson, J. G., Doudet, D. J. and Cohen, R. M. (1996). Presynaptic dopaminergic deficits in Lesch-Nyhan disease. *New Eng. J. Med.*, 334: 1568-1572.

Ferrin, L. and Camerini-Otero, R. D. (1991). Selective cleavage of human DNA: RecA-assisted restriction endonuclease (RARE) cleavage. *Science*, 254: 1494-1497.

Folger, K., Wong, E., Wahl, G. and Capecchi, M. (1982). Patterns of integration od DNA microinjected into cultured mammalian cells: evidence for homologous recombination between injected plasmid DNA molecules. *Mol. Cell. Biol.*, 2: 1372-1387.

Friedmann, T. (1985). HPRT gene transfer as a model for gene therapy. *In*: Genetic Engineering-Principles and Methods, J. Setlow and A. Hollaender (Eds), Plenum Press, New York, Vol. 7 pp. 263-282.

Gibbs, R. A., Nguyen, P. N., McBride, L. J., Koepf, S. M., and Caskey, C. T. (1989). Identification of mutations leading to the Lesch-Nyhan syndrome by automated direct sequencing of *in vitro* amplified cDNA. *Proc. Natl. Acad. Sci. USA*, 86: 1919-1923.

Graham, F. and Van der Eb, J. (1973) A new technique for the assay of infectivity of human adenovirus 5 DNA. *Virology*, 52: 456-467.

Gorski, K., Carniero, M. and Schibler, U. (1986). Tissue-specific *in vitro* transcription from the mouse albumin promoter. *Cell,* 47: 767-776.

Goss, S. J. (1984). Arginine synthesis by hepatomas *in vitro*. II. Isolation and characterization of Morris hepatoma variants unable to convert ornithine to arginine, and modulation of urea-cycle enzymes by dexamethasone and cyclic-AMP. *J. Cell Sci.*, 68: 305-319.

Groves, A. K., Barnett, S., Franklin, R., Crang, A., Mayer, M., Blakemore, W. and Noble, M. (1993). Repair of demyelinated lesions by transplantation of purified O-2A progenitor cells. *Nature*, 362: 453-455.

Gupta, R., Bazemore, L. R., Golub, E. and Radding, C. (1997). Activities of human recombination protein RAD51. *Proc. Natl. Acad. Sci. USA*, 94: 463-468.

Haaf, T., Golub, E., Reddy, G., Radding, C. and Ward, D. (1995). Nuclear foci of RAD51 recombination protein in somatic cells after DNA damage and its localization in synaptonemal complexes. *Proc. Natl. Acad. Sci. U.S.A.*, 92: 2298-2302.

Hallgren, R., Niklasson, F., Terent, A., Akerblom, A. and Widerlov, E. (1983). Oxypurines in cerebrospinal fluid as indices of disturbed brain metabolism. A clinical study of ischemic brain diseases. *Stroke*, 14: 382-388.

Hart, S., Harbottle, R., Cooper, R., Miller, A., Williamson, R. and Coutelle, C. (1995). Gene delivery and expression mediated by an integrin-binding peptide. *Gene Ther.*, 2: 552-554.

Hata, A., Tsuzuki, T., Shimada, K., Takiguchi, M., Mori, M. and Matsuda, I. (1988). Structure of the human ornithine transcarbamylase gene. *J. Biochem.*, 103: 302-308.

Hays, S., Firmenich, A. and Berg, P. (1995). Complex formation in yeast double-strand break repair: participation of Rad51, Rad52, Rad55 and Rad57 proteins. *Proc. Natl. Acad. Sci. U.S.A.*, 92: 6925-6929.

Henricksen, L., Umbricht, C. and Wold, M. (1994). Recombinant replication protein A: expression, complex formation and functional characterization. *J. Biol. Chem.*, 269: 11121-11132.

Horwich, A. L., Fenton, W., Williams, K., Kalousek, F., Kraus, J. P., Doolittle, R., Konigsberg, W. and Rosenberg, L. (1984). Structure and expression of a complementary DNA for the nuclear coded precursor of human mitochondrial ornithine transcarbamylase. *Science*, 224: 1068-1074.

Howard-Flanders, P., West, S. C. and Stasiak, A. (1984). Role of RecA protein spiral filaments in genetic recombination. *Nature*, 309: 215-220.

Hsieh, P., Camerini-Otero, C. and Camerini-Otero, R. D. (1992). The synapsis event in the homologous pairing of DNAs: RecA recognizes and pairs less than one helical repeat of DNA. *Proc. Natl. Acad. Sci. U.S.A.*, 89: 6492-6496.

Hyde, H., Davies, A., Benson, F. and West, S. (1994). Resolution of recombination intermediates by a mammalian activity functionally analogous to *E. coli* RuvC resolvase. *J. Biol. Chem.*, 269: 5202-5209.

Jinnah, H. A., Wojcik, B. E., Hunt, M., Narang, N., Lee, K. Y., Goldstein, M., Wamsley, J. K., Langlais, P. J. and Friedmann, T. (1994). Dopamine deficiency in a genetic mouse model of Lesch-Nyhan disease. *J. Neurosci.*, 14: 1164-1175.

Jiralerspong, S. and Patel, P. I. (1996). Regulation of the hypoxanthine phosphoribosyl transferase gene: *in vitro* and *in vivo* approaches. *Proc. Soc. Exp. Biol. Med.*, 212: 116-127.

Jolly, D. J., Okayama, H., Berg, P., Esty, A., Filpula, D., Bohlen, P., Johnson, G. G., Shively, J. E., Hunkapiller, T. and Friedmann, T. (1983). Isolation and characterization of a full-length expressible cDNA for human hypoxanthine phosphoribosyltransferase. *Proc. Natl. Acad. Sci. U.S.A.*, 80: 477-481.

Kido, M., Yoneda, Y., Nakanishi, M., Uchida, T. and Okada, Y. (1992). *E. coli* RecA protein modified with a nuclear location signal binds to chromosomes in living mammalian cells. *Exp. Cell Res.*, 198: 107-114.

Koerberl, D. D., Alexander, I. E., Halbert, C. L., Russell, D. W. and Miller, A. D. (1997). Persistent expression of human clotting factor IX from mouse liver after intravenous injection of adeno-associated virus vectors. *Proc. Natl. Acad. Sci. U.S.A.*, 94: 1426-1431.

Kohn, D. B. and Parkman, R. (1997). Gene therapy for newborns. *FASEB J.*, 11: 635-639.

Kohn, D. B., Weinberg, K. I., Nolta, J. A., Heiss, L. N., Lenarsky, C., Crooks, G. M., Hanley, M. E., Annett, G., Brooks, J. S., El-Khoureiy, A., Lawrence, K., Wells, S., Shaw, K., Moen, R. C., Bastian, J., Williams-Herman, D. E., Elder, M., Wara, D., Bowen, T., Hershfied, M. S., Mullen, C. A., Blaese, R. M. and Parkman, R. (1995). Engraftment of gene-modified cells from umbilical cord blood in neonates with adenosine deaminase deficiency. *Nature Med.*, 1: 1017-1026.

Kowalczykowski, S., Dixon, D., Eggleston, A., Lauder, S. and Rehrauer, W. (1994). Biochemistry of homologous recombination in *E. coli*. *Microbiol. Rev.*, 58: 401-465.

Lee, M. and Nurse, P. (1987). Complementation used to clone a human homologue of the fission yeast cell cycle control gene *cdc2*. *Nature*, 327: 31-35.

Leff, S., Rosenfeld, M. and Evans, R. (1986). Complex transcriptional units: diversity in gene expression by alternative RNA processing. *Ann. Rev. Biochem.*, 55: 1091-1117.

Lesch, M. and Nyhan, W.L. (1964). A familiar disorder of uric acid metabolism and central nervous system function. *Am. J. Med.*, 36: 561-570.

Littlefied, J. (1964). Selection of hybrids from mating of fibroblasts in vitro and their presumed recombinants. *Science*, 145: 709-710.

Lloyd, K. G., Hornykiewicz, O., Davidson, L., Shannak, K., Farley, I., Goldstein, M., Shibuya, M., Kelley, W. N. and Fox, I. H. (1981). Biochemical evidence of dysfunction of brain neurotransmitters in the Lesch-Nyhan syndrome.*New Eng. J. Med.*, 305: 1106-1111.

Lutfalla, G., Armbruster, L., Dequin, S. and Bertolotti, R. (1989). Construction of an EBNA-producing line of well-differentiated human hepatoma cells and of appropriate Epstein-Barr virus-based vectors. *Gene*, 76: 27-39.

Lutfalla, G., Blanc, H. and Bertolotti, R. (1985). Shuttling of integrated vectors from mammalian cells to *E. coli* is mediated by head-to-tail multimeric inserts. *Somat. Cell Mol. Genet.*, 11: 223-238.

Lyon, M. F. (1996). X-chromosome inactivation: pinpointing the centre. *Nature*, 379: 116-117.

McDonald, J. A. and Kelley, W. N. (1972). Lesch-Nyhan syndrome: absence of the mutant enzyme in erythrocytes of a heterozygote for both normal and mutant hypoxanthine-guanine phosphoribosyltransferase. *Biochem. Genet.*, 6: 21-26.

Merrihew, R., Marburger, K., Pennington, S., Roth, D. and Wilson, J. (1996). High-frequency illegitimate integration of transfected DNA at preintegrated target sites in a mammalian genome. *Mol. Cell. Biol.*, 16: 10-18.

Midoux, P., Mendes, C., Legrand, A., Raimond, J., Mayer, R., Monsigny, M. and Roche, A. C. (1993). Specific gene transfer mediated by lactosylated poly-L-lysine into hepatoma cells. *Nucleic Acids Res.*, 21: 871-878.

Miller, C. and Temin, H. (1983). High-efficiency ligation and recombination of DNA fragments by vertebrate cells. *Science*, 220: 606-609.

Mizuno, T. (1986). Long-term follow-up of ten patients with Lesch-Nyhan syndrome. *Neuropediatrics*, 17: 158-161.

Mulligan, R. (1993). The basic science of gene therapy. *Science*, 260: 926-932.

Mulligan, R. and Berg, P. (1980). Expression of a bacterial gene in mammalian cells. *Science*, 209: 1422-1427.

Mulligan, R. and Berg, P. (1981a). Factors governing the expression of a bacterial gene in mammalian cells. *Mol. Cell. Biol.*, 1: 449-459.

Mulligan, R. and Berg, P. (1981b). Selection for animal cells that express the *Escherichia coli* gene coding for xanthine-Guanine phosphoribosyltransferase. *Proc. Natl. Acad. Sci. U.S.A.*, 78: 2072-2076.

Naldini, L., Blomer, U., Gallay, P., Ory, D., Mulligan, R., Gage, F. H., Verma, I. M. and Trono, D. (1996). *In vivo* gene delivery and stable transduction of nondividing cells by a lentiviral vector. *Science*, 272: 263-267.

Neumann, E., Schaefer-Ridder, M., Wang, Y. and Hofschneider, P. (1982). Gene transfer into mouse lyoma cells by electroporation in high electric fields.*EMBO J.*, 1: 841-845

Nyhan, W. L., Bakay, B., Connor, J. D., Marks, J. F. and Keele, D. (1970). Hemizygous expression of glucose-6-phosphate dehydrogenase in erythrocytes of heterozygotes for the Lesch-Nyhan syndrome. *Proc. Natl. Acad. Sci. U.S.A.*, 65: 214-218.

Nyhan, W. L., Olivier, W. J. and Lesch, M. (1965). A familial disorder of uric acid metabolism and central nervous system function. *J. Pediat.*, 67: 257-263.

Nyhan, W. L., Parkman, R., Page, T., Gruber, H. E., Pyati, J., Jolly, D. and Friedmann, T. (1986). Bone marrow transplantation in Lesch-Nyhan disease. *Adv. Exp. Med. Biol.*, 195A: 167-170.

Nyhan, W. L. and Wong, D. F. (1996). New approaches to understanding Lesch-Nyhan disease. *New Eng. J. Med.*, 334: 1602-1604.

Ogawa, T., Yu, X., Shinohara, A. and Egelman, E. H. (1993a). Similarity of the yeast RAD51 filament to the bacterial RecA filament. *Science*, 259: 1896-1899.

Ogawa, T., Shinohara, A., Nabetani, A., Ikeya, T., Yu, X., Egelman, E. H. and Ogawa, H. (1993b). RecA-like recombination proteins in eukaryotes: functions and structures of *RAD51* genes. *Cold Spring Harbor Symp. Quant. Biol.*, 58: 567-576.

Okayama, H. and Berg, P. (1983). A cDNA cloning vector that permits expression of cDNA inserts in mammalian cells. *Mol. Cell. Biol.*, 3: 280-289.

Palese, P. and Roizman, B. (Eds) (1996). Genetic engineering of viruses and of virus vectors. *Proc. Natl. Acad. Sci. USA*, 93: 11287-11425.

Panthier, J. and Condamine, H. (1991). Mitotic recombination in mammals. *BioEssays*, 13: 351-356.

Perucho, M., Hanahan, D. and Wigler, M. (1980). Genetic and physical linkage of exogenous sequences in transformed cells. *Cell*, 22: 309-317.

Pinkert, C., Ornitz, D., Brinster, R. and Palmiter, R. (1987). An albumin enhancer located 10 kb upstream functions along with its promoter to direct efficient, liver-specific expression in transgenic mice. *Genes Dev.*, 1: 268-276.

Ponder, K. P., Gupta, S., Leland, F., Darlington, G., Finegold, M., DeMayo, J., Ledley, F. D., Chowdhury, J. R. and Woo, S. L. (1991). Mouse hepatocytes migrate to liver parenchyma and function indefinitely after intrasplenic transplantation. *Proc. Natl. Acad. Sci. USA*, 88: 1217-1221.

Radding, C. M. (1991). Helical interactions in homologous pairing and strand exchange driven by RecA protein. *J. Biol. Chem.*, 266: 5355-5358.

Remy, J. , Kichler, A., Mordvinov, V., Schuber, F. and Behr, J. P. (1995). Targeted gene transfer into hepatoma cells with lipopolyamine-condensed DNA particles presenting galactose ligands: a stage toward artificial viruses. *Proc. Natl. Acad. Sci. U.S.A.*, 92: 1744-1748.

Robins, D., Ripley, S., Henderson, A. and Axel, R. (1981). Transforming DNA integrates into the host chromosome. *Cell*, 23: 29-39.

Roca, A. and Cox, M. (1990). The RecA protein: structure and function. *Crit. Rev. Biochem. Mol. Biol.*, 25: 415-455.

Roemer, K. and Friedmann, T. (1992). Concepts and strategies for human gene therapy. *Eur. J. Biochem.*, 208: 211-225.

Rosenbloom, F. M., Kelley, W. N., Miller, J., Henderson, J. F. and Seegmiller, J. E. (1967). Inherited disorder of purine metabolism: correlation between central nervous system dysfunction and biochemical defects. *J.A.M.A.*, 202: 175-177.

Rossiter, B. J. F. and Caskey, C. T. (1995). Hypoxanthine-Guanine Phosphoribosyl Transferase deficiency: Lesch-Nyhan syndrome and gout. *In*: The metabolic and molecular bases of inherited disease, C. Scriver, A. Beauded, W. Sly and D. Valle (Eds.), McGraw-Hill, New York, pp. 1679-1706.

Roth, D. B., Lindahl, T. and Gellert, M. (1995). How to make ends meet. *Curr. Biol.*, 5: 496-499 (1995).

Roy Chowdhury, J., Grossman, M., Gupta, S., Roy Chowdhury, N., Baker, J. R. and Wilson, J. M. (1991) Long-term improvement of hypercholesterolemia after *ex vivo* gene therapy in LDLR-deficient rabbits. *Science*, 254: 1802-1805.

Schwartz, A. L. (1984). The hepatic asialoglycoprotein receptor. *Crit. Rev. Biochem.*, 16: 207-233.

Seegmiller, J. E., Rosenbloom, F. M. and Kelley, W. N. (1967). Enzyme defect associated with a sex-linked human neurological disorder and excessive purine synthesis. *Science*, 155: 1682-1687.

Seglen, P. O. (1983). Inhibitors of lysosomal function. *Methods Enzymol.*, 96: 737-764.

Shin, S., Meera Khan, P. and Cook, P. R. (1971). Characterization of hypoxanthine-guanine phosphoribosyl transferase in man-mouse somatic cell hybrids by an improved electrophoretic method. *Biochem. Genet.*, 5: 91-99.

Shinohara, A., Ogawa, H. and Ogawa, T. (1992). Rad51 protein involved in repair and recombination in *S. cerevisiae* is a RecA-like protein. *Cell*, 69: 457-470.

Shinohara, A., Ogawa, H., Matsuda, Y., Ushio, N., Ikeo, K. and Ogawa, T. (1993). Cloning of human, mouse and fission yeast recombination genes homologous to RAD51 and recA. *Nature Genetics*, 4: 239-243.

Simmonds, H. A., Sahota, A. S. and Van Acker, K. J. (1995). Adenine phosphoribosyltransferase deficiency and 2,8-dihydroxyadenine lithiasis. *In*: The metabolic and molecular bases of inherited disease, C. Scriver, A. Beauded, W. Sly and D. Valle (Eds.), McGraw-Hill, New York, pp. 1707-1724.

Simpson, D., Cawley, D. and Herschmam, H. (1982). Killing of cultured hepatocytes by conjugates of asialofetuin and EGF linked to the A chains of ricin or diphteria toxin. *Cell*, 29: 469-473.

Snyder, R. O., Miao, C. H., Patijn, G. A., Spratt, S. K., Danos, O., Nagy, D., Gown, A. M., Winther, B., Meuse, L., Cohen, L. K., Thompson, A. R. and Kay, M. A. (1997). Persistent and therapeutic concentrations of human factor IX in mice after hepatic gene transfer of recombinant AAV vectors. *Nature Genet.*, 16: 270-276.

Stacey, A., Schnieke, A., McWhir, J., Cooper, J., Coleman, A. and Melton, D. (1994). Use of double-replacement gene targeting to replace the murine α-lactalbumin gene with its human counterpart in embryonic stem cells and mice. *Mol. Cell. Biol.*, 14: 1009-1016.

Stasiak, A. and Egelman, E. (1994). Structure and function of RecA-DNA complexes. *Experientia*, 192-203.

Stief, A., Winter, D. M., Strätling, W. H. and Sippel, A. E. (1989). A nuclear DNA attachment element mediates elevated and position-independent gene activity. *Nature*, 341: 343-345.

Sung, P. (1994). Catalysis of ATP-dependent homologous DNA pairing and strand exchange by yeast RAD51 protein. *Science*, 265: 1241-1243.

Sung, P. and Robberson, D. (1995). DNA strand exchange mediated by a RAD51-ssDNA nucleoprotein filament with polarity opposite to that of RecA. *Cell*, 82: 453-461.

Szybalska, E. and Szybalski, W. (1962). Genetics of human cell lines, IV. DNA-mediated heritable transformation of a biochemical trait. *Proc. Natl. Acad. Sci. USA*, 48: 2026-2034.

Szybalski, W. (1992). Use of the HPRT gene and the HAT selection technique in DNA-mediated transformation of mammalian cells: first step toward developing hybridoma techniques and gene therapy. *Bioessays*, 14: 495-500.

Te Riele, H., Maandag, E. and Berns, A. (1992). Highly efficient gene targeting in embryonic stem cells through homologous recombination with isogenic DNA constructs. *Proc. Natl. Acad. Sci. USA*, 89: 5128-5132.

Thomas, K., Folger, K. and Capecchi, M. (1986). High frequency targeting of genes to specific sites in the mammalian genome. *Cell*, 44: 419-428.

Thomas, K. and Capecchi, M. (1987). Site-directed mutagenesis by gene targeting in mouse embryo-derived stem cells. *Cell*, 51: 503-512.

Thomas, K., Deng, C. and Capecchi, M. (1992). High-fidelity gene targeting in embryonic stem cells by using sequence replacement vectors. *Mol. Cell. Biol.*, 12: 2919-2923.

Verma, I. M. and Somia, N. (1997). Gene therapy-promises, problems and prospects. *Nature*, 389: 239-242.

Wagner, E., Plank, C., Zatloukal, K., Cotten, M. and Birnstiel, M. L. (1992). Influenza virus hemagglutinin HA-2 N-terminal fusogenic peptides augment gene transfer by transferrin-polylysine-DNA complexes: toward a synthetic virus-like gene-transfer vehicle. *Proc. Natl. Acad. Sci. USA*, 89: 7934-7938.

Wagner, E., Zenke, M., Cotten, M., Beug, H. and Birnstiel, M. L. (1990). Transferrin-polycation conjugates as carriers for DNA uptake into cells. *Proc. Natl. Acad. Sci. USA*, 87: 3410-3414.

Walker, S. A., Kennedy, M. T. and Zasadzinski, J. (1997). Encapsulation of bilayer vesicles by self-assembly. *Nature*, 387: 61-64.

Wilson, J. M., Grossman, M., Wu, C. H., Chowdhury, N. R., Wu, G. Y. and Chowdhury, J. R. (1992). Hepatocyte-directed gene transfer *in vivo* leads to transient improvement of hypercholesterolemia in low density lipoprotein receptor-deficient rabbits. *J. Biol. Chem.*, 267: 963-967.

Wong, D. F., Harris, J. C., Naidu, S., Yokoi, F., Marenco, S., Dannals, R. F., Ravert, H. T., Yaster, M., Evans, A., Rousset, O., Bryan, R. N., Gjedde, A., Kuhar, M. J. and Breese, G. R. (1996). Dopamine transporters are markedly reduced in Lesch-Nyhan disease in vivo. *Proc. Natl. Acad. Sci. USA*, 93: 5539-5543.

Wu, G. Y. (1988). Targeting in diagnosis and therapy. *In*: The liver: biology and pathobiology, I. Arias, W. Jakoby, H. Popper, D. Schachter and D. Shafritz (Eds.), Raven Press, New York, pp. 1303-1313.

Wu, G. Y. and Wu, C. H. (1987). Receptor-mediated *in vitro* gene transformation by a soluble DNA carrier system. *J. Biol. Chem.*, 262: 4429-4432.

Wu, G. Y. and Wu, C. H. (1988). Receptor-mediated gene delivery and expression *in vivo*. *J. Biol. Chem.*, 263: 14621-14624.

Yates, J., Warren, N. and Sugden, B. (1985). Stable replication of plasmids derived from Epstein-Barr virus in various mammalian cells. *Nature*, 313: 812-815.

Yoon, K., Cole-Strauss, A. and Kmiec, E. (1996). Targeted gene correction of episomal DNA in mammalian cells mediated by a chimeric RNA-DNA oligonucleotide. *Proc. Natl. Acad. Sci. USA*, 93: 2071-2076.

Yoshimori, T., Yamamoto, A., Moriyama, Y., Futai, M. and Tashiro, Y. (1991). Bafilomycin A1, a specific inhibitor of vacuolar-type H^+-ATPase, inhibits acidification and protein degradation in lysosomes of cultured cells. *J. Biol. Chem.*, 266: 17707-17712.

Yoshimura, Y., Morita, T., Yamamoto, A. and Matsushiro, A. (1993). Cloning and sequence of the human RecA-like gene cDNA. *Nucleic Acids Res.*, 21: 1665-1665.

Zauner, W., Kichler, A., Schmidt, W., Sinski, A. and Wagner, E. (1996). Glycerol enhancement of ligand-polylysine/DNA transfection. *Biotechniques*, 20: 905-913.

7. SUMMARY

In december 1991, we devised a new approach to Gene Therapy based on premade recombinase-DNA nucleoprotein filaments. Aimed at repairing mutant genes (inherited diseases) and at inactivating deleterious or vital viral genes (cancer, hepatitides, AIDS), the idea is to generate, *in vitro*, human recombinase-DNA complexes analogous to the bacterial RecA presynaptic filament and mimic, upon transfection, recombinational DNA repair that occurs in most dividing mammalian cells. The aim is to substitute chromosomal DNA segments by exogenous homologous genomic DNA from wild type (gene repair) or mutant (gene inactivation) origin. Our actual strategies deal with human RAD51 protein that has been shown to mediate *in vitro*, in the presence of replication protein A, the ATP-dependent homologous DNA pairing-strand exchange reaction. We describe how, in order to avoid unspecific targeting to repetitive chromosomal elements and to bypass potential arrests of the homologous DNA pairing-strand exchange reaction by therapeutic heterologous bases, we have designed presynaptic filaments that comprise a double-stranded DNA core. Our current *in vitro* model system comprises a set of selectable "single copy" X-linked genes, hypoxanthine-guanine phosphoribosyl transferase (HPRT) as a target and ornithine transcarbamylase (OTC) as a control. From transfecting well-differentiated hepatic lines *in vitro* with RAD51-HPRT nucleoprotein filaments toward targeting specific somatic cells *in vivo*, our ultimate goal would be to develop a recombinase-mediated Gene Therapy that would substitute to the current minigene expression protocols. The prospects of such an approach to Gene Therapy are presented for HPRT enzyme deficiency that results in brain dopaminergic deficit and subsequent neurobehavioural Lesch-Nyhan syndrome.

Progress in HPLC-HPCE, Vol. 7, pp. 171-192
G. Ali Qureshi *et al.* (Eds)

OPIOID PEPTIDES IN DRUG DEPENDENCE AND NEUROLOGICAL DISEASES.

Ingrid Nylander* and Jerzy Silberring+

**Department of Pharmaceutical Biosciences, Division of Pharmacology, Uppsala University, Uppsala, Sweden,+ Laboratory for High Resolution Mass Spectrometry, Faculty of Chemistry, Jagiellonian University, Krakow, Poland.*

INTRODUCTION

Endorphins, enkephalins and dynorphins are all members of the growing family of endogenous opioid peptides (see Herz, 1993 for review). These peptides derive from three distinct precursors, proopiomelanocortin (Nakanishi, *et al* 1979), proenkephalin (Noda, *et al*, 1982) and prodynorphin (Kakidani *et al*, 1982). Recently two additional peptides were identified in mammalian tissue, the endomorphin-1 (Tyr-Pro-Trp-Phe-NH2) and endomorphin-2 (Tyr-Pro-Trp-Gly-NH2) (Zadina *et al*, 1997). The opioid peptides are endogenous ligands for the opiate receptors of which three types have been identified and cloned so far, the μ-, δ- and κ-receptors (see Reisine, 1995 for review). Endorphin binds both to the μ- and δ-receptor with comparable affinity, the endomorphins bind selectively to the μ-receptor, the enkephalins preferably to the δ-receptor, whereas the dynorphins bind to the κ-receptor.

The three opioid peptide systems each have a distinct pattern of distribution in the central nervous system (Mansour et al, 1995). The enkephalins, Met-enkephalin and Leu-enkephalin are widely distributed in the central nervous system and are present both in smaller interneurons as well as in longer neuronal loops (Kachaturian *et al*, 1983). β-endorphin are localised in hypothalamic neurons originating in the arcuate nucleus and projecting to the limbic system and brainstem. β-endorphin are also synthesized within the pituitary gland and are released into the circulation (Crine et al, 1978). The prodynorphin derived peptides, dynorphin A, dynorphin B, and the neoendorphins are present in distinct neuronal pathways. The striato-nigral pathway (Christensson-Nylander *et al*, 1986) and dynorphinergic pathways within the hippocampus (McGinty *et al*, 1983) are well described. Dynorphins are also localised in neurosecretory neurons with cell bodies in the hypothalamus and terminals in the neurointermediate lobe of the pituitary gland (Millan *et al*, 1983)

In the past 20 years the opioid peptides have been shown to mediate a large number of physiological effects and to modulate a variety of other neurotransmittor systems. Taken together these peptides may be regarded as critical regulators of neuronal pathways mediating pain, reward, motivation and hormone release. Here we focus on the role of opioid peptides in drug dependence mechanisms and attempt to summarise present knowledge about opioids as possible biochemical substrates for individual differences in drug taking behaviour and also review drug effects on opioid peptide systems. In addition, we give a brief survey of the involvement of opioid peptides in degenerative neurological diseases and their possible role as biochemical markers.

DRUG DEPENDENCE

The molecular basis for individual differences in disposition for drug dependence is largely unknown today and there are no valid markers that easily can be measured in, for instance, body fluids. A number of studies have been devoted to finding a marker for genetic predisposition for drug dependence. In this section we summarise research efforts concerning the opioid peptides that may lead to identification of possible future markers of drug vulnerability.

OPIOIDS AND REWARD

It is well known that opiates have rewarding properties in man and in animals. They produce place preference and are readily self-administered intravenously or directly into brain areas such as the VTA and nucleus accumbens by experimental animals (Koob and Bloom, 1988; Koob, 1992). The opioid peptides mediate effects depending on which receptor they activate (Shippenberg *et al*, 1992). The enkephalins, β-endorphin, μ- and δ-receptor agonists, and also an enkephalinase inhibitor produce conditioned place preference (Bals-Kubik *et al*, 1990; Phillips *et al*, 1982; Glimcher *et al*, 1984), and thus have the same properties as exogenously administered opiates. Opiate μ- and δ-receptor agonists are also self-administered (Devine et al, 1994). The dynorphin peptides and κ-agonists, on the other hand, produce place aversion and are generally not self-administered in a non-dependent animal (Mucha and Herz, 1985; Bals-Kubik *et al*, 1989). Activation of κ-receptors have also been associated with a dysphoric state in humans (Pfeiffer *et al*, 1986). A few reports have however shown place preference (Iwamoto, 1988) as well as self-administration of dynorphin, although the latter was found to be mediated by μ- and not κ-receptors (Stevens *et al*, 1991). Opposing dynorphin effects have also been reported depending on the state of the animal, dynorphin inhibits, for example, morphine-induced analgesia in naive animals, but potentiates morphine effects in a morphine-tolerant animal (Tulunay *et al*, 1981). Furthermore, despite the aversive effects observed in naive animals, both dynorphin and the non-opioid dynorphin A(2-17), can substitute for morphine (Hooke *et al*,

1995) and attenuate withdrawal signs in opiate dependent rats (Green *et al*, 1988), monkeys (Aceto *et al*, 1982) and humans (Wen and Ho, 1982).

The picture is thus complicated by the fact that some effects of dynorphin seem to be mediated by other opiate or even non-opiate receptors. These contrasting effects might be explained by either *in vivo* metabolism of dynorphins such as enzymatic conversion of the κ-receptor active dynorphin peptides into shorter peptides, dynorphin (1-6) and leu-enkephalin, acting on δ-receptors which mediate other physiological effects, or by enzymatic truncation of the N-terminal tyrosine residue generating dynorphin A (2-17) which can produce non-opioid effects (see Chapters 17, Nyberg, and 25, Silberring, of this issue).

INTERACTIONS WITH DOPAMINE

Dopamine transmission in the central nervous system is strongly associated with dependence mechanisms. Two main pathways exist in the brain: mesostriatal dopamine neurons originating in the substantia nigra and projecting to the dorsal striatum (caudate and putamen) and the mesolimbic pathway which projects from the VTA to the nucleus accumbens (ventral striatum). Opioid peptides interact closely with these pathways and the terminal release of dopamine are regulated by substances acting on opiate receptors (see below).

Dopamine transmission in the mesolimbic pathway is critical for the acute reinforcing effects of drugs and is regarded as the brain reward pathway (see Wise and Bozarth, 1987; Wise, 1989, for reviews). Three classes of drugs with a clear difference in mechanism of action induce dopamine release from the terminals in the nucleus accumbens, the opiates, the psychostimulants and alcohol (Di Chiara *et al*, 1988; Spanagel *et al*, 1990). Opiates activate dopamine neurons via disinhibition of GABAergic interneurons within the VTA, whereas the psychostimulants act on the dopmaine transporter in the nucleus accumbens inhibiting reuptake of transmitter (see Koob and Bloom, 1988 for review). Chronic opiate administration induces long lasting changes in the mesolimbic dopamine system and opiate withdrawal results in a marked reduction in extracellular dopamine concentrations (Acquas and Di Chiara, 1992; Spanagel *et al*, 1993) and neuronal firing (Diana *et al*, 1995). Opiate-induced dopamine release also changes during abstinence, from no response one day after withdrawal to an increased release after 3 days, before returning to a normal opiate effect seen in naive animals (Acquas and Di Chiara, 1992).

Endogenous opioid peptides interact with the mesolimbic dopamine system and recently two opposing tonically active opioid systems have been described (Spanagel *et al*, 1992). In the VTA there is a dense population of μ-receptors but a relatively sparse population of δ-receptors (Quirion *et al*, 1983; Mansour *et al*, 1987; Dilts *et al*, 1990). Enkephalins and β-endorphin are present in the VTA and the enkephalins also in the nucleus accumbens (see Schafer *et al*, 1991 for review). A μ-agonist activates dopamine neurons in the VTA by inhibiting inhibitory neurons correlating well with its reinforcing effects. Consequently, a μ-antagonist inhibits dopamine release (Spanagel *et al*, 1992). Dynorphin fibers and κ-receptors are abundant in the nucleus accumbens and sparse in the VTA (see Schafer *et al*, 1991 for review). Opposite to other opiate receptor agonists, it has been reported that a κ-agonist reduces, while a κ-antagonist increases dopamine release via an action on κ-receptors in the nucleus accumbens (Spanagel *et al*, 1992). These biochemical findings correlate to the behaviourally aversive effect of dynorphin peptides.

These findings thus imply a delicate balance between the different endogenous opioid peptides in regulating dopaminergic activity. The relative activity of κ- versus μ/δ-opioid systems controls the sensitivity of the mesolimbic pathway in a certain individual and an innate imbalance in opioid systems could thereby affect drug vulnerability. The results further led to the question whether an imbalance in opioid peptide systems could influence drug taking behaviour and also whether the opioid peptides might serve as biological substrates for genetic differences in disposition for drug dependence. κ-receptors mediate for example inhibition of dopamine release in the nucleus accumbens, as pointed out above. A low activity of the dynorphin system would thus result in less inhibition, a higher drug-induced release of dopamine and thereby a stronger reinforcing effect of a certain drug. Chronic drug administration might also induce changes in opioid peptides and disturb the physiological balance between neuronal systems, which partly could cause the strong craving reported by drug addicts.

INDIVIDUAL DIFFERENCES

Studies in the past have given evidence for the existence of genetic predisposition to alcoholism as well as possible dependence of other drugs in humans. The biological sustrate for the genetic differences is, however, still unknown and finding the determinants of dispositional factors for drug abuse would be of considerable value.

Very little is known about such biochemical factors deciding individual differences in drug-taking behaviour and vulnerability for drug dependence. Studies of individuals in "high risk" families with a history of alcoholism have revealed that they have lower basal levels of plasma β-endorphin as do abstinent alcoholics. Alcohol drinking induces a higher increase in peptide levels in "high risk" individuals which may indicate that they also experience a greater reinforcing effect of the drug than "low risk" individuals (Gianoulakis *et al*, 1989).

To study biochemical variables in individuals with a high vulnerability for drug dependence, it is common to use certain inbred animal strains with preference for alcohol and/or other drugs. Examples are the Lewis rat which self-administers drugs to a higher degree than the Fischer rat (Terwilliger *et al*, 1991) and the C57BL/6 mice which prefer alcohol and the DBA/2 mice which avoid alcohol (Schuster, 1989). Selective outbreeding has also been used to generate lines of rats with different responses to drugs, for instance the AA (Alko Alcohol) rats which prefer alcohol and the ANA (Alko Non Alcohol) rats with no alcohol intake (Eriksson, 1968)

Several studies have focused on the Lewis rat strain since these rats readily self-administers opiates, cocaine and alcohol. Biochemical analyses have revealed strain differences in tyrosine hydroxylase as well as in the cAMP pathway and G-proteins in the mesolimbic pathway (See Nestler, 1992 for review). Lewis rats resemble morphine treated Sprague-Dawley rats in this respect (see below). Differences in basal levels of opioid peptides between Lewis and Fischer rats have also been described. Lewis rats have lower levels of dynorphin peptides in several brain areas. Moreover, morphine treatment affect peptide levels differently in Lewis and Fischer rats. Lewis rats show a comparative unresponsiveness to morphine with regard to enkephalin peptides, and the morphine-induced changes in dynorphin peptide levels are strain-specific. The results also indicate a discrepancy in morphine effects on enzymatic conversion of prodynorphin-derived peptides (Nylander *et al*, 1995). Strain differences have finally been reported in opiate withdrawal behaviour (Guitart *et al*, 1993) and opioid peptide levels during withdrawal (Nylander et al, 1995).

Differences between opioid peptides have also been shown in the AA and ANA rats. The AA rats who freely drink alcohol in a two-bottle choice between alcohol and water have, for instance, lower basal levels of dynorphin in the nucleus accumbens and in the pituitary gland (Nylander *et al*, 1994). The content of β-endorphin was instead higher in the anterior part of the pituitary and in some brain areas (Gianoulakis *et al*, 1992). Alcohol drinking for four weeks induced an increase in met-enkephalinArgPhe (a marker for proenkephalin) levels in the pituitary gland and in the nucleus accumbens, whereas dynorphin levels were largely unaffected by alcohol.

Low levels of dynorphin in the nucleus accumbens may render an individual more sensitive to the reinforcing effect of the drug caused by an increased drug-induced dopamine release. These results thus indicate that the innate basal activity of opioid peptide systems could control the motivation to initiate drug intake. A low activity of the κ-opioid system would, for example, increase the motivation to take drugs. The drive to continue drinking could on the other hand depend on the μ- and/or δ-opioid system and the alcohol-induced stimulation of enkephalins and/or endorphins. The present findings of biochemical differences in mesolimbic dopamine neurons as well as in opioid peptide systems in various addiction-prone animals may be the basis of an increased knowledge of biochemical determinants for genetic differences in disposition for drug dependence and for the driving force to take drugs. The development of, and future access to, animals that lack or possibly overexpress a certain opioid system will help in breaking new ground to further study the role of various opioid systems in drug dependence mechanisms. The recent successful breeding of mice lacking the μ-opioid receptor was a breakthrough in opioid research (Matthes et al, 1996). Morphine-induced place preference, morphine withdrawal and other classical morphine effects are abolished in these animals giving evidence that the μ -opioid receptor is critical for morphine action (Matthes et al, 1996).

DRUG EFFECTS ON OPIOID PEPTIDE SYSTEMS

Numerous studies have been conducted to describe how addictive drugs interfere with opioid peptide systems. An alteration of opioid peptide function may underlie important components to addiction, withdrawal and drug seeking behaviour. Drugs may for instance affect gene expression, receptor proteins, precursor proteins and/or various enzymes participating in conversion or metabolism of peptides and thereby change opioid peptide levels.

Expression of immediate early genes such as the Fos proto-oncogene protein are often used as an anatomical marker of activated brain areas. Drugs of abuse such as cocaine (Young *et al*, 1991), amphetamine (Graybriel *et al*, 1990), morphine (Chang *et al*, 1990; Liu *et al*, 1994) and ethanol (Chang et al, 1995) have recently been shown to activate the expression of immediate early genes. Drug-induced changes in c-fos and junB and their target genes, possibly proenkephalin and prodynorphin, may cause long-term changes in neuronal transmitter systems involved in dependence mechanisms. Other drug-induced effects on opioid peptide systems are summarised below focusing on opiates, cocaine (as a representative of the psychostimulants) and alcohol.

Opiates

Early on it was suggested that exogenous administration of opiates would down-regulate endogenous opioid peptide systems leading to a deficit which in opiate withdrawal contributed to the abstinence syndrome (Hughes and Kosterlitz, 1975). It has however been difficult to demonstrate a clear effect on peptide levels or their receptors and divergent results have been published (see Loh and Smith, 1990 for review). Some of the earlier studies failed to detect any changes in opioid peptide levels, but more recent studies report a reduction in met-enkephalin, endorphin and proenkephalin mRNA levels (Gudehithly *et al*, 1991; Uhl *et al*, 1988) in CNS areas of tolerant animals. Opioid systems in the spinal cord seem to be regulated differently and an increase in β-endorphin levels has been reported after morphine treatment (Bronstein *et al*, 1993). Receptor density or affinity seem to be unaffected by chronic morphine, heroin addicts have, for instance, no change in μ-receptors (Gabilondo *et al*, 1994).

Studies of opiate administration on the dynorphin system have provided divergent results. The variability might be explained by different experimental set-ups, for instance the use of continuous (pellet implantation or infusion by a miniosmotic pump) or intermittent administration of drugs. Dynorphin peptide levels increase by chronic morphine treatment in the striatum and nucleus accumbens (Trujillo *et al*, 1989; Nylander *et al*, 1995) and in amygdala (Rattan *et al*, 1992), whereas both a reduction (Nylander *et al*, 1995) and an increase (Rattan *et al*, 1992) has been reported in the pituitary gland. Tolerant and abstinent animals have lower levels in the spinal cord (Nylander *et al*, 1991; Rattan et al, 1992). Acute systemic administration of morphine induces a long lasting increase in extracellular dynorphin B levels in the substantia nigra (You *et al*, 1996) and reduces tissue levels in striatum and hippocampus (Nylander *et al*, 1995). These results indicate a release of peptide after acute administration possibly followed by an up-regulation if the administration continues. Prodynorphin mRNA levels in several brain areas are, however, reduced in tolerant animals (Romuladi *et al*, 1991) together with the finding of higher peptide levels. This apparent contradiction may be explained by an inhibited release in the chronic state leading to accumulation of peptide and later on a decrease in peptide synthesis. A more plausible explanation would be that opiates also affect processing and/or converting enzymes as described by Persson et al. 1989 and Vlaskovska et al. 1994.

Opiates also induce changes in intracellular systems which have been shown in mesolimbic dopamine neurons and in the locus coeruleus, a brain area often used as a model system studying opiate action. In these systems it has been shown that acute opiate administration inhibits adenylate cyclase and cAMP dependent protein phosphorylation. In the tolerant state, however, all steps of the cAMP system from receptor to physiological response are up-regulated providing evidence for opiate-induced adaptive changes in intracellular mechanisms. An up-regulaton of the cAMP system has also been shown in dopaminergic neurons in the nucleus accumbens, not only after opiate, but also after cocaine administration. These results have been summarised by Nestler, 1992, who also have outlined experimental approaches for the further study of neurobiological mechanisms underlying drug addiction processes (Nestler et al, 1993).

Cocaine

Several lines of evidence have focused on a possible role of dynorphin in the response to psychostimulants. Dopamine agonists (Nylander *et al*, 1987) as well as cocaine have been shown to increase striatal dynorphin levels, the cocaine effect being attenuated by dopamine antagonists (Sivam *et al*, 1989). This has led to further investigations of interactions between opioid peptides and psychostimulants, especially cocaine. Cocaine addicts have marked reductions in the levels of proenkephalin mRNA and μ-opioid receptor-binding in the caudate and putamen, whereas levels of prodynorphin mRNA and κ-receptor binding were elevated (Hurd *et al*, 1993). These results suggest that chronic cocaine usage may cause an up-regulation of the dynorphin system which is further supported by the findings of higher peptide levels (Sivam *et al*, 1989) and prodynorphin mRNA levels (Spangler *et al*, 1993; Hurd *et al*, 1992) in animals given chronic cocaine. Higher peptide levels are thus found both after chronic cocaine and opiates while prodynorphin mRNA levels are oppositely affected. Increased dynorphin and κ-receptor levels concomitantly with decreased enkephalin and u-receptor levels after chronic cocaine administration imply a heightened responsiveness of a peptide system mediating dysphoria and an inhibition of a peptide system mediating euphoria (Hurd *et al*, 1993).

Alcohol

Several studies have given evidence for involvement of opioid peptides in alcohol reinforcement. Alcohol influences several neurotransmitter systems such as the dopamine, GABA and glutamate systems, in addition to the endogenous opioid system. Effects of ethanol administration depend on the dose, the route of administration and, in experimental animals, the strain or line of animal. Acute ethanol administration generally increase the release of endorphin peptides and increase tissue levels of enkephalin in man and animals (see Gianoulakis, 1989 for review). This observation and the finding of reduced alcohol consumption after opiate antagonist treatment have led to the use of naltrexone in pharmacological treatment of alcohol dependence (O'Malley et al, 1992; Volpicelli et al, 1995). Dynorphin peptides do not seem to be affected by acute ethanol administration.

Enkephalin and endorphin levels have been shown to be either reduced or unaffected after chronic ethanol administration. A reduction in enkephalin and endorphin levels is generally seen during withdrawal which, however, may reflect a stress-induced depletion of peptide rather than a direct effect of ethanol (Gianoulakis, 1989). Lower levels of dynorphin peptides have been shown in some brain areas such as the hypothalamus and hippocampus, whereas other areas are largely unaffected by ethanol (Seizinger *et al*, 1983).

Summary drug effects

Divergent results have thus been reported with regard to drug effects on opioid peptide systems which might be due to differences in the experimental set-up such as route of administration, dosage regimes, time schedules and the strain of animal. Peptide levels seem to be more affected by intermittent administration, which resembles drug taking behaviour by addicts, as compared to chronic continuous administration. Further studies are, however, needed to identify the effects of various drugs of abuse on the endogenous opioid peptide systems.

CONCLUSIONS AND FUTURE DIRECTIONS

Research efforts to date have provided evidence that the κ- and the μ-/δ-opioid systems, respectively, mediate different and often opposite physiological effects. An imbalance in these systems, either innate or drug-induced, could have serious consequences for drug-taking behaviour, sensitivity to drug effects and vulnerability for drug dependence in a certain individual. Further experiments are needed to evaluate the role of opioid peptides in drug dependence mechanisms and should involve detailed co-ordinated studies of receptors, gene expression, enzymatic conversion of peptides, peptide levels and intracellular mechanisms in drug dependent individuals. An effort should be made to establish reliable and sensitive assays, which are easy to use in clinical situations, to measure opioid peptide activity in humans. *In vivo* microdialysis in various brain regions has previously given valuable information about dependence mechanisms and are currently used to analyse extracellular levels of peptides. This technique will hopefully continue to generate useful data in the near future. Studies of animals with a low or high activity of a certain opioid system, which can mimic an innate or acquired imbalance of opioids in humans, will be useful when investigating drug taking behaviour and the role of opioid peptide systems in drug dependence mechanisms. Finally, it is of interest to identify possible drug-induced changes in the opioid peptide systems that persist after detoxification and withdrawal. It is very important to find markers for such adaptive changes which could contribute to various abstinence symptoms and be involved in craving mechanisms. This would help in finding new pharmacological treatments to restore drug-induced imbalances, such as deficit or excess of peptide, re-establishing physiological function and, thereby, alleviate the craving and drug seeking behaviour in drug addicts.

NEUROLOGICAL DISEASES

After the discovery of neuropeptides, particularly opioid peptides, it was suggested that they might be involved in the degenerative neurological diseases or that they could serve as markers of such diseases. The availability of human material for such biochemical studies is, however, limited. Therefore, measurements are restricted to plasma and cerebrospinal fluid (CSF). Of these two compartments, CSF is in direct contact with the nervous tissue, and thus is most commonly used as a source of samples. The general concept of peptide measurements in the body fluids and basal levels of major opioid peptides were described by Nyberg (1993). Here, we will focus on the CNS disorders where opioid peptides were reported to be significantly affected. This alteration may have several reasons, which are described in the final paragraph of this chapter.

ALZHEIMER'S DISEASE

Alzheimer's disease is associated with the deposition of amyloid in the brain tissue. It is a common dementing illness and a leading cause of mortality in the geriatric population (Terry and Katzmann, 1983). In general, the majority of laboratories report decreased levels of opioid peptides. CSF β-endorphin was markedly reduced in Alzheimer patients and the peptide levels correlated negatively with the degree of dementia (Heilig et al., 1995). In another study (Kaiya, 1983), the greatest decrease was seen in patients with pre-senile and senile dementia of Alzheimer type (SDAT). The β-endorphin immunoreactivity significantly correlated with psychological functions. In a similar study, β-endorphin and β-lipotropin were measured in patients with Alzheimer's disease, multi-infarct dementia and controls (Sulkava et al, 1985). In both the dementia groups, the mean concentration of β-endorphin was significantly lower than in controls. The mean β-lipotropin levels did not differ significantly. This altered ratio of β-endorphin and its precursor may suggest alteration in the processing pathways, generated by the prohormone convertases PC1/PC3 and PC2, which differ in specificity towards the proopiomelanocortin molecule (Benjannett et al., 1991).

CSF measures of dynorphin A have been compared in three groups; Alzheimer patients, elderly depressives, and age-matched normal controls (Sunderland et al. 1991). The Alzheimer patients revealed a 40% decrease in CSF dynorphin compared with controls. CSF dynorphin did not correlate significantly with clinical variables or other CSF measures of monoamine metabolites. One explanation might be a possible up-regulation of opiate receptors in Alzheimer's disease. The reduced levels of dynorphin did not correlate with density or regulation by G-proteins in the temporal cortex of patients with Alzheimer's disease as tested on the post-mortem tissue (Garlind et al., 1995).

In more general studies, endogenous opioid receptoractive peptides were measured in CSF of patients showing senile dementia of Alzheimer type (Muhlbauer et al., 1986). The total opioid peptide content was quantitated using labelled etorphine as a ligand. The peptide level, expressed as Met-enkephalin equivalents, was significantly higher than the corresponding level from patients with no known neurological disorders. As etorphine is a nonselective ligand, this result suggests that other opioid peptides may contribute to this elevated level.

PARKINSON'S DISEASE

Parkinson's disease is associated with a damage of dopaminergic neurons. Baronti et al. (1991) have shown that the ganglionic dopamine system interacts with Met-enkephalin and with dynorphin in the striato-nigral pathway and have also studied peptide levels in the CSF of parkinsonian patients. The concentration of a proenkephalin derivative, Met-enkephalinArgGlyLeu, was markedly reduced, as compared to controls. In contrast, dynorphin (1-8) levels remained unchanged indicating abnormalities in the enkephalin system. These results are in agreement with the studies performed in the basal ganglia regions of post-mortem parkinsonian brains (Sivam, 1991) and with the work performed by Toquet et al. (1985). As the striatal dynorphin system may be affected in parkinsonian patients, a clinical trial was performed in order to test the hypothesis that selective dynorphin agonists may improve motor symptoms (Giuffra et al., 1993). The results suggest that dynorphin replacement strategies, using spiradoline-like kappa$_1$-receptor agonists, may have limited value in the therapy of patients with Parkinson's disease. The levels of the neuropeptides Met- and Leu-enkephalin were measured (Fernandez et al., 1996) by a combined high performance liquid chromatography and radioimmunoassay method in post-mortem samples of basal ganglia from Parkinson's disease patients and incidental Lewy body disease patients (pre-symptomatic Parkinson's disease). The levels of Met-enkephalin were reduced in the caudate nucleus and putamen in patients with Parkinson's disease as well as incidental Lewy body disease. Parkinsonian patients also had lower levels in the substantia nigra. Leu-enkephalin levels were decreased in the putamen and were undetectable in the substantia nigra in patients with Parkinson's disease. Leu-enkephalin levels were unchanged in incidental Lewy body disease, although there was a tendency to a reduction in putamen. The altered enkephalin levels are thought to be an integral part of the pathology of the illness and not secondary to dopamine neuronal loss.

SEIZURES

Seizures are mostly studied using various animal models. However, a number of reports have been published on the immunocytochemistry or nuclear magnetic resonance showing substantial changes in the hippocampal mossy fibers. The presence of aberrant mossy fiber collaterals in the hippocampus has been confirmed by dynorphin immunohistochemistry indicating its relation to epilepsy (Babb et al., 1991) Experimental models, such as electroconvulsive shocks (ECS) and amygdaloid kindling, have commonly been used to study the dynamic status of hippocampal opioid peptides in response to seizures. Both ECS- and kindling-induced seizures triggered an initial large release of enkephalin and dynorphin, but produced opposite long-term effects on the biosynthesis of these two peptides, an increase of enkephalin, and a drastic decrease of dynorphin.

Dynorphin has been implicated in specific types of seizures (Simonato et al., 1996). Dynorphin appears to be released in the hippocampus and other brain areas during complex partial seizures. The released dynorphin inhibits excitatory neurotransmission at multiple synapses in the hippocampus via activation of kappa opioid receptors. On the other hand, kappa opioid receptor agonists are highly effective against limbic seizures. It is anticipated that dynorphin plays a role as an endogenous anticonvulsant in complex partial seizures and in some cases of tonic-clonic seizures. This pattern of effects may coincide with the antiseizure spectrum of selective kappa agonists.

Becker et al. (1991) reported an interesting application of the casomorphins, peptides possessing an opioid-like activity. Derivatives of beta-casomorphin (1-5) were tested for their influence on pentylenetetrazol, picrotoxin and electrically induced seizures. Tyr-Pro-Phe-D-Pro-Gly was found to facilitate pentylenetetrazol-evoked seizures, whereas desTyr derivatives Pro-Phe-D-Pro-Gly and Pro-Phe-Pyr exhibited anticonvulsant properties against those convulsions. The tripeptide was effective only 10 minutes after application.

GILLES DE LA TOURETTES SYNDROME

Tourette's Syndrome was first described over a hundred years ago by a French neurologist by the name of George Gilles de la Tourette. He described nine patients, primarily a Madame de D. who: "...at the age of 7 was afflicted by convulsive movements of the hands and arms... She was felt to be suffering from overexcitement and mischief, and... she was subjected to reprimand and punishment. It soon became clear that these movements were indeed involuntary... involved the shoulders, the neck, and the face, and resulted in contortions and extraordinary grimaces."

The main facts about Tourette's Syndrome are that it is a hereditary neurological condition characterised by recurring, compelling impulses to make movements, and usually has an onset early in childhood. The disease affects more males than females and does not get progressively worse. It is now recognised that Tourette's Syndrome is not a psychological disorder or neurosis, but rooted in biology and neurology (http://www-personal.umd.umich.edu/~infinit/tsfaq1.html).

It has been suggested that, besides a dopaminergic mechanism, deranged opioid functions may be important in the pathophysiology of Tourette's Syndrome (Sandyk, 1988). In immunocytochemical studies, Haber et al. (1986) showed that enkephalin-immunoreactive-positive woolly fibers were densely stained and of normal distribution. Dynorphin-immunoreactive staining was, however, considerably less dense than normal throughout the brain. Moreover, there was the total absence of dynorphin-positive woolly fibers in the dorsal part of the external segment of the globus pallidus. These observations indicate a decrease of dynorphin in striatal fibers projecting to the globus pallidus. In another study (Leckman et al., 1988), dynorphin A (1-8) CSF levels were, however, shown to be higher in patients with the syndrome. A possible explanation for the discrepancies in these reports has been given by Nyberg (1993), who has suggested a change in processing rate of the enzyme forming the octapeptide.

INTERPRETING NEUROPEPTIDE ALTERATIONS

Quantitation of neuropeptides by immunological techniques still remains the method of choice in the majority of laboratories. These methods have, however, some drawbacks. Opioid peptides, particularly those having relatively short sequences, are poor immunogens. Moreover, the N-terminal enkephalin sequence is common for all structures, which often causes unwanted detection of elongated molecules or, simply, measurement of epitopes present within the precursor molecules. In this respect, it is crucial to develop a proper separation strategy before the particular molecule can be detected and quantitated. Antibody-based methods, such as radioimmunoassay or ELISA, lack structural specificity causing methodological problems as mentioned above. Direct measurement of neuropeptides in tissues or body fluids may lead to severe degradation by proteinases present in the same compartment, or to artefacts if the extraction procedure is not properly designed. Often, solid-phase extraction does not completely remove proteolytic activities. Application of high performance liquid chromatography for preseparation of the extracts may partially increase selectivity of measurements. Introduction of modern mass spectrometry has greatly facilitated verification of the peptide structures. This technique has been successfully applied for the quantitation of several endogenous neuropeptides (Desiderio, 1996) and from this point of view, it seems to be a promising future technique in up-to-date neurochemistry.

The concentration of a particular peptide depends not only on its expression in the tissue, but also on the proteolytic activities responsible for release and degradation (see chapters 17, Nyberg, and 25, Silberring, of this issue). The discrepancies in peptide levels reported by various laboratories may therefore be related to the presence of unwanted proteolysis during sample withdrawal or may be caused by an incorrect extraction procedure. Therefore, it is a useful strategy to test the metabolism of the peptide in, for instance, the body fluid subjected to examination to assure that the peptide will not degrade during the entire procedure.

REFERENCES

Acquas, E., Di Chiara, G. (1992) Depression of mesolimbic dopamine transmission and sensitization to morphine during opiate abstinence. *Journal of Neurochemistry* **58**, 1620-1625.

Babb, T.L., Kupfer, W.R., Pretorius, J.K., Crandall, P.H., Levesque, M.F. (1991) Synaptic reorganization by mossy fibers in human epileptic fascia dentata. *Neuroscience* **42**, 351-363.

Bals-Kubik, R., Herz, A., Shippenberg, T.S. (1989) Evidence that the aversive effects of opioid antagonists and κ-agonists are centrally mediated. *Psychopharmacology* **98**, 203-206.

Bals-Kubik, R., Shippenberg, T.S., Herz, A. (1990) Involvement of central μ- and δ-opioid receptors in mediating the reinforcing effects of β-endorphin in the rat. *European Journal of Pharmacology* **175**, 63-69.

Baronti, F., Conant, K., Giuffra, M., Davis, T., Brughitta, G., Iadarola,M., Berrettini, W., Chase, T., Mouradian M. (1991) Opioid peptides in Parkinson's disease: effects of dopamine repletion. *Brain Research* **560**, 92-96.

Becker, A., Grecksch, G., Matthies, H. (1991) Influence of beta-casomorphin derivatives on chemically and electrically induced seizures. *Peptides* **12**, 43-45.

Benjannet, S., Rondeau, N., Day, R., Chretien, M., Seidah N. (1991) PC1 and PC2 are proprotein convertases capable of cleaving proopiomelanocortin at distinct pairs of basic residues. *Proceedings of National Academy of Science USA*. **88**, 3564-3568.

Bronstein, D.M., Gutstein, H.B., Akil, H. (1993) Effects of chronic morphine treatment on β-endorphin-related peptides in the caudate medulla and spinal cord. *Journal of Neurochemistry* **60**, 2304-2307.

Chang, S.L., Harlen, R.E. (1990) The Fos proto-oncogene protein: Regulation by morphine in the rat hypothalamus. *Life Sciences* **46**, 1825-1832.

Chang, S.L., Patel, N.A., Romero, A.A. (1995) Activation and desensitization of Fos immunoreactivity in the rat brain following ethanol administration. *Brain Research* **679**, 89-98.

Christensson-Nylander, I., Herrera-Marschitz, M., Staines, W., Hökfelt, T., Terenius, L., Ungerstedt, U., Cuello, C., Oertel, W.H., Goldstein, M. (1986) Striato-nigral dynorphin and substance P pathways in the rat. I. Biochemical and immunohistochemical studies. *Experimental Brain Research* **64**, 169-192.

Crine, P., Gianoulakis, C., Seidah, N.D., Gossard, F., Pezalla, P.D.,

Lis, M., Chrétien, M. (1978) Biosynthesis of β-endorphin from β- lipotropin and a larger molecular weight precursor in rat pars intermedia. *Proceedings of National Academy of Science* **75**, 4719-4723.

Desiderio, D.M. (1996) Mass spectrometric quantification of neuropeptides. *Methods of Molecular Biology* **61**, 57-65.

Devine, D.P., Wise, R.A. (1994) Self-administration of morphine, DAMGO, and DPDPE into the ventral tegmental area of rats. *Journal of Neuroscience* **14**, 1978-1984.

Diana, M., Pistis, M., Muntoni, A., Gessa, G. (1995) Profound decrease of mesolimbic dopaminergic neuronal activity in morphine withdrawn rats. *The Journal of Pharmacology and Experimental Therapeutics* **272**, 781-785.

Di Chiara, G., Imperato, A. (1988) Drugs abused by humans preferentially increase synaptic dopamine concentrations in the mesolimbic system of freely moving rats. *Proceedings of National Academy of Science* **85**, 5274-5278.

Dilts, R.P., Kalivas, P.W. (1990) Autoradiographic localization of delta opioid receptors within the mesocortical dopamine system using radioiodinated (2-D-penicillinamine, 5-D-penicillamine)enkephalin (125 I-DPDPE) *Synapse* **6**, 121-132.

Fernandez, A., de Ceballos, M.L., Rose, S., Jenner, P., Marsden, C.D. (1996) Alterations in peptide levels in Parkinson's disease and incidental Lewy body disease. *Brain* **119**, 823-830.

Gabilondo, A.M., Meana, J.J., Barturen, F., Sastre, M., Garcia-Sevilla, J.A. (1994) μ-opioid receptor and α_2-adrenoceptor agonist binding sites in the postmortem brain of heroin addicts.*Psychopharmacology* **11**, 135-140.

Garlind, A., Cowburn, R.F., Wiehager, B., Ravid, R., Winblad, B., Fowler, C. (1995) Preservation of kappa 1 opioid receptor recognition site density and regulation by G-proteins in the temporal cortex of patients with Alzheimer's disease. *Neuroscience Letters* **185**, 131-134.

Gianoulakis, C., Bèliveau, D., Angelogianni, P., Meaney, M., Thavundayil, J., Tawar, V., Dumas, M. (1989) Different pituitary β-endorphin and adrenal cortisol response to ethanol in individuals with high and low risk for future development of alcoholism. *Life Sciences* **45**, 1097-1109.

Gianoulakis, C. (1989) The effect of ethanol on the biosynthesis and regulation of opioid peptides. *Experientia* **45**, 428-435.

Gianoulakis, C., de Waele, J.P., Kiianmaa, K. (1992) Differences in the brain and pituitary β-endorphin system between the alcohol-preferring AA and alcohol-avoiding ANA rats. *Alcoholism: Clinical and Experimental Research* **16**, 453-459.

Giuffra, M., Mouradian, M.M., Davis, T.L., Ownby, J., Chase, T.N. (1993) Dynorphin agonist therapy of Parkinson's disease. *Clinical Neuropharmacology* **16**, 444-447.

Glimcher, P.G., Giovino, A.A., Margolin, D.H., Hoebel, B.G. (1984) Endogenous reward induced by an enkephalinase inhibitor thiorphan, injected into the ventral midbrain. *Behavioural Neuroscience* **98**, 262-268.

Graybriel, A.M., Moratalla, R., Robertson, H.A. (1990) Amphetamine and cocaine induce drug-specific activation of the c-Fos gene in striaosome-matrix compartments and limbic subdivisions of the striatum. *Proceedings of National Academy of Science* **87**, 6912-6916.

Gudehithlu, K.P., Tejwani, G.A., Bhargava, H.N. (1991) β-endorphin and methionine-enkephalin levels in discrete brain regions, spinal cord, pituitary gland and plasma of morphine tolerant-dependent and abstinent rats. *Brain Research* **553**, 284-290.

Haber, S.N., Kowall, N.W., Vonsattel, J.P., Bird, E.D., Richardson, E.P. Jr (1986) Gilles de la Tourette's syndrome. A postmortem neuropathological and immunohistochemical study. *Journal of Neurological Science* **75**, 225-241

Heilig, M., Sjögren, M., Blennow, K., Ekman, R. and Wallin, A. (1995) Cerebrospinal fluid neuropeptides in Alzheimerís disease and vascular dementia. *Biological Psychiatry* **38**, 210-216.

Herz, A. Ed. (1993) Opioids I. Springer-Verlag, New York.

Hong, J.S. (1992) Hippocampal opioid peptides and seizures. *Epilepsy Research* Suppl. **7**, 187-95.

Hooke, L.P., He, L., Lee, N.M. (1995) Dynorphin A modulates acute and chronic opioid effects. *The Journal of Pharmacology and Experimental Therapuetics* **273**, 292-297.

Hurd, Y.L., Brown, E.E., Finlay, J.M., Fibiger, H.C., Gerfen, C.R. (1992) Cocaine self-administration differentially alters mRNA expression of striatal peptides. *Molecular Brain Research* **13**, 165-170.

Hurd, Y., Herkenham, M. (1993) Molecular alterations in the neostriatum of human cocaine addicts. *Synapse* **13**, 357-369.

Iwamoto, E.T. (1988) Dynorphin A(1-17) induces "reward" in rats in the place conditioning paradigm. *Life Sciences* **43**, 503-508.

Kachaturian, H., Lewis, M., Watson, S.J. (1983) Enkephalin systems in diencephalon and brainstem of rat. *Journal of Comparative Neurology* **220**, 310-320.

Kaiya, H., Tanaka, T., Takeuchi, K., Morita, K., Adachi, S., Shirakawa, H., Ueki, H., Namba, M. (1983) Decreased level of beta-endorphin-like immunoreactivity in cerebrospinal fluid of patients with senile dementia of Alzheimer type. *Life Sciences* **33**, 1039-1043.

Koob, G.F., Bloom, F.E. (1988) Cellular and molecular mechanisms of drug dependence. *Science* **242**, 715-723.

Koob, G.F. (1992) Drugs of abuse: anatomy, pharmacology and function of reward pathways. *Trends in Pharmacological Science* **13**, 177-184.

Laatikainen, T. (1985) CSF beta-endorphin and beta-lipotropin in Alzheimer's disease and multi-infarct dementia. *Neurology* **35**, 1057-1058.

Leckman, J.F., Riddle, M.A., Berrettini, W.H., Anderson, G.M., Hardin, M., Chappell, P., Bissette, G., Nemeroff, C.B., Goodman, W,K., Cohen, D.J. (1988) Elevated CSF dynorphin A [1-8] in Tourette's syndrome. *Life Sciences* **43**, 2015-2023.

Liu, J., Nickolenko, J., Sharp, F.R. (1994) Morphine induces c-*fos* and *junB* in striatum and nucleus accumbens via D_1 and *N*-methyl-D-aspartate receptors. *Proceedings of National Academy of Science* **91**, 8537-8541.

Loh, H.H., Smith, A.P. (1990) Molecular characterization of opioid receptors. *Annual Reviews in Pharmacology and Toxicology* **30**, 123-147.

Mansour, A., Kachaturian, H., Lewis, M.E., Akil, H., Watson, S.J. (1987) Autoradiographic differentiation of mu, delta, and kappa opioid receptors in the rat forebrain and midbrain. *Journal of Neusoscience* **7**, 2445-2464.

Mansour, A., Fox, C.A., Akil, H., Watson, S.J. (1995) Opioid-receptor mRNA expression in the rat CNS: anatomical and functional implications. *Trends in Neuroscience* **18**, 22-29.

Matthes, H.W.D., Maldonado, R., Simonin, F., Valverde, O., Slowe, S.,

Kitchen, I., Befort, K., Dierich, A., Le Meur, M., Dollé P., Tzavara, E., Hanoune, J., Roques, B.P., Kieffer, B.L. (1996) Loss of morphine-induced analgesia, reward effect and withdrawal symptoms in mice lacking the μ-opioid-receptor gene. *Nature* **383**, 819-823.

McGinty, J.F., Henriksen, S.J., Goldstein, A., Terenius, L., Bloom, F.E.(1983) Dynorphin is contained within hippocampal mossy fibers: immunohistochemical alterations after kainic acid administration and colchicine-induced neurotoxicity. *Proceedings of National Academy of Science* **80**, 589-593.

Millan, M.H., Millan, M.J., Herz, A. (1983) The hypothalamic paraventricular nucleus: relationship to brain and pituitary pools of vasopressin and oxytocin as compared to dynorphin, b-endorphin and related opioid peptides in the rat. *Neuroendocrinology* **38**, 108-166.

Muhlbauer, M., Metcalf, J., Robertson, J., Fridland, G., Desiderio, D. (1986) Opioid peptides in the cerebrospinal fluid of Alzheimer patients. *Biomedical Chromatography* **1**, 155-158.

Nestler, E.J., Hope, B.T., Widnell, K.L. (1993) Drug Addiction: A model for the molecular basis of neural plasticity. *Neuron* **11**, 995-1006.

Nyberg, F. (1993) CSF opioids in pathophysiology. Handbook of Experimental Pharmacology, vol. 104/II. Opioids II (A. Hertz ed.) Springer-Verlag, Berlin, pp. 653-672.

Nylander, I., Terenius, L. (1987) Dopamine receptors mediate alterations in striato-nigral dynorphin and substance P pathways. *Neuropharmacology* **26**, 1295-1302.

Nylander, I., Sakurada, T., Le Greves P., Terenius, L. (1991) Levels of dynorphin peptides, substance P and CGRP in the spinal cord after subchronic administration of morphine in the rat. *Neuropharmacology* **30**, 1219-1223.

Nylander, I., Vlaskovska, M., Terenius, L. (1995) The effects of morphine treatment and morphine withdrawal on the dynorphin and enkephalin systems in the Sprague-Dawley rats. *Psychopharmacology* **118**, 391-400.

Nylander, I., Vlaskovska, M., Terenius, L. (1995) Brain dynorphin and enkephalin systems in Fischer and Lewis rats: effects of morphine tolerance and withdrawal. *Brain Research* **683**, 25-35.

Nylander, I., Hyytiä, P., Forsander, O., Terenius, L. (1994)

Differences between alcohol-preferring (AA) and alcohol-avoiding (ANA) rats in the prodynorphin and proenkephalin systems. *Alcoholism: Clinical and Experimental Research* **18**, 1272-1279.

O'Malley, S.S., Jaffe, A.J., Chang, G., Schottenfeld, R.S., Meyer, R.E., Rounsaville, B. (1992) Naltrexone and coping skills therapy for alcohol dependence. *Archieves of General Psychiatry* **49**, 881-887.

Persson, S., Post, C., Alari, L., Nyberg, F., Terenius, L. (1989) Increased neuropeptide-converting enzyme activities in cerebrospinal fluid of opiate-tolerant rats. *Neuroscience Letters* **107**, 318-322.

Pfeiffer, A., Brandt, V., Herz, A. (1986) Psychotomimesis mediated by kappa opiate receptors. *Science* **233**, 774-776.

Phillips, A.G., LePiane, F.G. (1982) Reward produced by microinjection of (D-Ala^{2}), (Met^{5})-enkephalinamide into the ventral tegmental area. *Behavioural Brain Research* **5**, 225-229)

Quirion, R., Zajac, J.M., Morgat, J.L., Roques, B.P. (1983) Autoradiographic distribution of mu and delta opiate receptors in rat brain using highly selective ligands. *Life Sciences* **33** suppl 1, 227-230.

Rattan, A.K., Koo, K.L., Tejwani, G.A., Bhargava, H.N. (1992) The effect of morphine tolerance dependence and abstinence on immunoreactive dynorphin (1-13) levels in discrete brain regions, spinal cord, pituitary gland and peripheral tissues of the rat. *Brain Research* **584**, 207-212.

Reisine, T. (1995) Opiate receptors. *Neuropharmacology* **34**, 463-472.

Romualdi, P., Lesa, G., Ferri, S. (1991) Chronic opiate agonists down-regulate prodynorphin gene expression in rat brain. *Brain Research* **563**, 132-136.

Romualdi, P., Lesa, G., Donatini, A., Ferri, S. (1995) Long-term exposure to opioid antagonists up-regulates prodynorphin gene expression in rat brain. *Brain Research* **672**, 42-47.

Sandyk, R., Bamford, C.R. (1988) Opioid modulation of gonadotrophin release in Tourette's syndrome. *International Journal of Neuroscience* **39**, 233-234

Schafer, M.K., Day, R., Watson, S.J., Akil, H. (1991) Distribution of opioids in brain and peripheral tissues. *In* The Neurobiology of Opioids. O.F.X. Almeida and T.S. Shippenberg, Eds. pp 53-71.

Springer-Verlag, New York.

Seizinger, B.R., Bovermann, K., Maysinger, D., Höllt, V., Herz, A. (1983) Differential effects of acute and chronic ethanol treatment on particular opioid peptide systems in discrete regions of rat brain and pituitary. *Pharmacology Biochemistry and Behaviour* **18**, suppl. 1, 361-369.

Shippenberg, T.S., Herz, A., Spanagel, R., Bals-Kubik, R., Stein, C. (1992) Conditioning of opioid reinforcement: Neuroanatomical and neurochemical substrates. *Annals New York Academy of Sciences* **654**, 347-356.

Simonato, M., and Romualdi, P. (1996) Dynorphin and epilepsy. *Progression in Neurobiology* **50**, 557-83.

Sivam, S.P. (1989) Cocaine selectively increases striatonigral dynorphin levels by a dopaminergic mechanism. *Journal of Pharmacology and Experimental Therapeutics* **250**, 818-824.

Sivam, S.P. (1991) Dopamine dependent decrease in enkephalin and substance P levels in basal ganglia regions of postmortem parkinsonian brains. *Neuropeptides* **4**, 201-207.

Sonneberg, J.L., Rauscher, F.J., Morgan, J.I., Curran, T. (1989) Regulation of proenkephalin by Fos and Jun. *Science* **246**, 1622-1625.

Spanagel, R., Herz, A., Shippenberg, T.S. (1990) The effects of opioid peptides on dopamine release in the nucleus accumbens: An in vivo microdialysis study. *Journal of Neurochemistry* **55**, 1734-1740.

Spanagel, R., Herz, A., Shippenberg, T.S. (1992) Opposing tonically active endogenous opioid systems modulate the mesolimbic dopaminergic pathway. *Proceedings of Natlional Academy of Science* **89**, 2046-2050.

Spanagel, R., Almeida, O.F.X., Shippenberg, T.S. (1993) Long lasting changes in morphine-induced mesolimbic dopamine release after chronic morphine exposure. *Synapse* **14**, 243-245.

Spangler, R., Unterwald, E.M., Kreek, M.J. (1993) "Binge" cocaine administration induces a sustained increase of prodynorphin mRNA in rat caudate-putamen. *Molecular Brain Research* **19**, 323-327.

Sulkava R, Erkinjuntti T, Laatikainen T. (1985) CSF beta-endorphin and beta-lipotropin in Alzheimer's disease and multi-infarct dementia. *Neurology* **35**, 1057-1058.

Sunderland, T., Berrettini, W.H., Molchan, S.E., Lawlor, B.A.,

Martinez, R., Vitiello, B., Tariot, P.N., Cohen, R.M. (1991) Reduced cerebrospinal fluid dynorphin A(1-8) in Alzheimer's disease. *Biological Psychiatry* **30**, 81-87.

Taquet, H., Javoy-Agid, F., Giraud, P., Legrand, J.C., Agid, Y., Cesselin, F. (1985) Dynorphin levels in parkinsonian patients: Leu5-enkephalin production from either proenkephalin A or prodynorphin in human brain. *Brain Research* **341**, 390-392.

Terry, R.D. and Katzmann, R.K. (1983) Senile dementia of the Alzheimer type. Annals in Neurology 14, 497-506.

Terwilliger, R.Z., Bradberry, C., Guitart, X., Beitner-Johnson, D., Marby, D., Kosten, T.A., Roth, R.H., Nestler, E.J. (1991) The Lewis and Fischer 344 rats and drug addiction: behavioural and biochemical correlates. *Society for Neuroscience Abstracts* **17**, 823.

Trujillo, K.A., Akil, H. (1989) Changes in prodynorphin peptide content following treatment with morphine or amphetamine: possible role in mechanisms of action of drug of abuse. *NIDA Research Monograph* **95**, 550-551.

Tulunay, F.C., Fen, M.F., Chang J.K., Loh, H.H., Lee, N.M. (1981) Possible regulatory role of dynorphin on morphine- and β-endorphin-induced analgesia. *Journal of Pharmacology and Experimental Therapeutics* **219**, 296-301.

Uhl, G.R., Ryan, J.P., Schwartz,J.P. (1988) Morphine alters preproenkephalin gene expression. *Brain Research* **459**, 391-397.

Vlaskovska, M., Geijer, T., Nylander, I., Silberring, J., Terenius, L. (1994) Changes in the prodynorphin system after chronic morphine treatments in vitro. *Regulatory Peptides* **suppl.1**, 287-288.

Volpicelli, J.R., Watson, N.T., King, C.A., Sherman, C.E., O'Brien C.P. (1995) Effect of naltrexone on alcohol "high" in alcoholics. *American Journal of Psychiatry* **152**, 613-615.

White, J.D., Gall, C.M. (1987) Differential regulation of neuropeptide and protooncogene mRNA content in the hippocampus following recurrent seizures. *Molecular Brain Research* **3**, 211-229.

Wise, R.A. (1989) Opiate reward: sites and substrates. *Neuroscience and Biobehavioural Reviews* **13**, 129-133.

Wise, R.A., Bozarth, M.A. (1987) A psychomotorstimulant theory of addiction. *Psychology Reviews* **94**, 469-492.

Young, S.T., Porrino, L.J., Iadarola, M.J. (1991) Cocaine induces striatal c-Fos-immunoreactive proteins via dopaminergic D1 receptors.*Proceedings of National Academy of Science* **88**, 1291-1295.

You, Z-B., Herrera-Marschitz, M., Nylander, I., Goiny, M., Kehr, J., Ungerstedt, U., Terenius, L. (1996) Effect of morphine on dynorphin B and GABA release in the basal ganglia of rats. *Brain Research* **710**, 241-248.

Zadina, J.E., Hackler, L., Ge, L-J., Kastin, A.J. (1997) A potent and selective endogenous agonist for the μ-opiate receptor *Nature* **386**, 499-501.

Progress in HPLC-HPCE, Vol. 7, pp. 193-221
G. Ali Qureshi *et al.* (Eds)

ALCOHOLIC INTOXICATION

J. Nikolić
Institute of Biochemistry, Faculty of Medicine, University of Niš, Yugoslavia

Introduction

Ethanol (alcohol) is the most frequently used drug in the world despite the knowledge of the social, economic and medical problems associated with its abuse. In excessive intake alcohol associates with clinical symptoms of acute intoxication, while, long-term intake of high doses leads to development of chronic ethanol intoxication (alcoholism, alcohol addiction). It is believed that alcoholism and alcohol-related derangements of variety organs develops when person consumes more than 21 units of alcohol weekly for 5 years. For woman limits is 14 units because women have low activity of gastric alcohol dehydrogenase (ADH). 1 unit (10g) of alcohol is contained in 28 ml of whisky or similar drinks, 85 ml of wine or 230 ml of beer. The nature of the beverage is not important. When alcohol is ingested in moderate amount about 20% does not enter the systematic circulation and is oxidized in the stomach. Gastric ethanol metabolism modulates its systemic effects and potential toxicity and represents a "protective barrier" against systemic effects. Gastric barrier is low in women so women have increased susceptibility to ethanol (Frezza et al. 1990). Gastric ADH is inhibited by some commonly used drugs as aspirin and cimetidine, after surgical resection and long consumption of ethanol (DiPadova et al., 1989; Roine et al., 1990; Caballeria et al., 1991).

Metabolism of ethanol

After absorption ethanol is mainly metabolized in the liver, however, extrahepatic metabolism exist also ((Hawkins and Kalant, 1972; Lieber, 1984). There are different ways in the metabolism of ethanol, predominantly they are oxidative pathways but also it can be metabolized by nonoxidative metabolism. Glucuronide and sulfate conjugation and fatty acid esterification have small contributions to the total disposition of ethanol in the body. All oxidative pathways of ethanol metabolism result in production of acetaldehyde. Acetaldehyde has been produced in different cellular compartments:

• alcohol dehydrogenase pathway is the major pathway for ethanol disposition. Alcohol dehydrogenase is citosolar enzyme which catalyzes the oxidation of ethanol to acetaldehyde by the reduction of NAD to NADH. The liver contains high concentration of alcohol dehydrogenase . Human liver ADH is a zinc metaloenzyme with five classes of multiple molecular forms that arise from the association of eight different types of subunits: α, $\beta 1$, $\beta 2$, $\beta 3$, $\gamma 1$, $\gamma 2$, π, χ into active dimmer molecules. Extrahepatic tissues contain isozymes of ADH with a much lower affinity for ethanol than the hepatic. Therefore, extrahepatic metabolism of ethanol is negligible (Li and Bosron, 1987). The presence of alcohol dehydrogenase has been detected in the brain but its activity is small so it has small contribution in ethanol oxidation in the central nervous system (Raskin, 1973, Rawat and Kuriyama, 1972).

• catalase, peroxizome enzyme, catalysis the oxidation of ethanol in presence H_2O_2- generating system.. Under physiological condition catalase play minor role in ethanol metabolism. By the action of the enzyme catalase acetaldehyde may be produced in the brain during alcohol intoxication (Gill et al. 1992).

• microsomal ethanol oxidizing system, (MEOS) with location in endoplasmic reticulum. For activity it required citochrom P-450, NADPH-cytochrome P-450 reductase and phospholipid. The reaction involves molecular oxygen, one atom of which is incorporated into ethanol and form acetaldehyde, while, the other atom is used to generate water from reduced pyridine nucleotide (NADPH). MEOS has a major role in long-term consumption of ethanol (Lieber et al., 1987). Isoenzime specific for ethanol oxidation is cytochrome P-450IIE1 (Takahashi et al., 1993). The presence of cytochrome P-450IIE1 is detected in brain microcosms (Anandatheerthavarada et al., 1993).

The major mechanism for acetaldehyde disposition is oxidation by acetaldehyde dehydrogenase (AcDH) to acetate by reduction of NAD in NADH predominantly in mitochondria. Brain mitochondria possess acetaldehyde dehydrogenase for acetaldehyde metabolism (Quintanitilla and Tampier, 1992; Zimatkin et al., 1992). Most of them are oxidized in Krebs cycle in H_2O, CO_2 and released energy (7. 1 kcal/g). Acetate can be used also in alternative metabolic pathways (Fig.1).

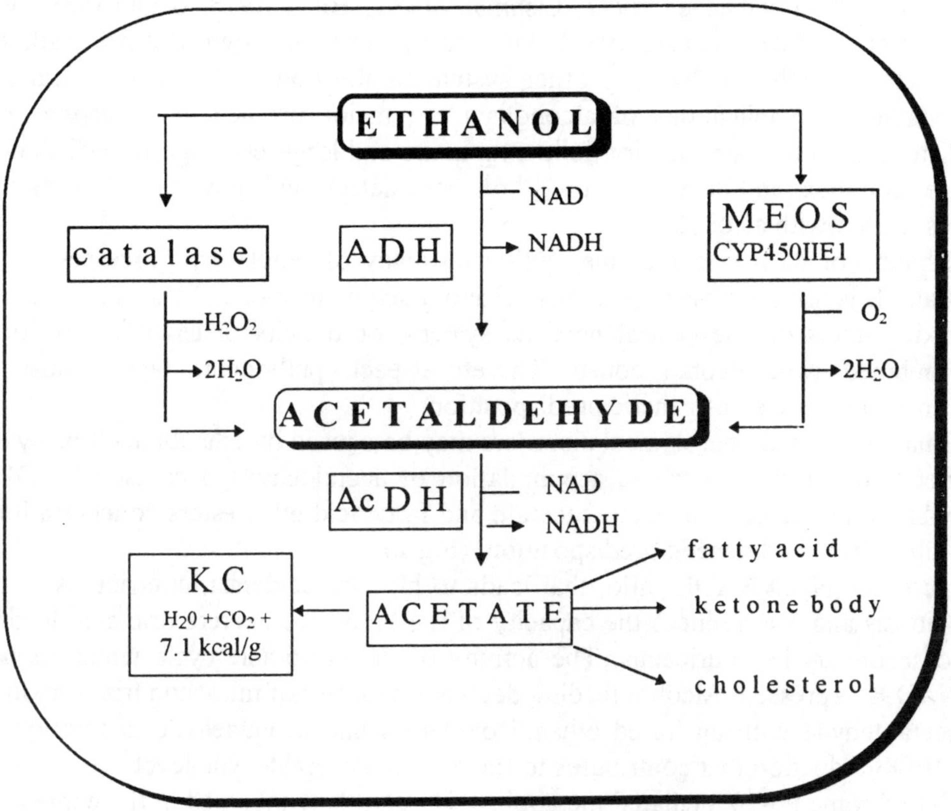

Fig. 1 Metabolism of ethanol

Neurotoxicity of ethanol

Many of epidemiological studies have shown that excessive intake of ethanol is the risk in developing of variety of organ, neuronal and metabolic disorders. Alcohol-related metabolic derangements affect almost all body organs and their functions including the liver, the heart , the central nervous system, pancreas, the endocrine cells (Lieber, 1994; Dance and Maxwell, 1986; Hillbom and Kaste,1990; Isselbacher, 1977; Lishman, 1981; Bing, 1978; Rubin,1989; Van Thiel et al, 1987; Leonard,1987). The central nervous system is more markedly affected by ethanol than any other system of the body. The neurological and behavioral complications of alcoholism are major problem in medicine. The disturbances in brain function following acute and long term ingestion of alcohol are complex and a wide variety of psychiatric and physical illnesses are associated with alcoholism.

Ethanol, or its metabolites may influence many of regulatory processes in the brain. It is not established, whether ethanol itself or its metabolites cause primary toxic effects on the central nervous system, or deficits of essential nutritiens combined with alcohol abuse. Therefore each pathogenic factor must be considered as well as genetic predisposition.

Ethanol-induced neurological disorders may be caused by ethanol itself or by its metabolic products such as accumulation of acetaldehyde, increased NADH / NAD ratio, increase in lactate, acetate and fatty acid ethyl esters concentrations, malnutrition and genetic predisposition (Fig.2).

Increased NAD/NADH ratio, that leads to hiperlactacidemia, contributes to the acidosis and also reduces the capacity of the kidney to excrete uric acid leading to secondary hyperuricemia. The activity of the citric acid cycle which require NAD is depressed. Alcohol feeding decrease capacity of mitochondria to oxidize acetaldehyde with unaltered ethanol oxidation and acetaldehyde generation by MEOS induction that contributes to the elevated acetaldehyde level.

Cytochrome P 450 mediated metabolism is potentially involved in the expression toxicological effects of many drugs and environmental chemicals upon tissue which contain this metabolic system. Chronic ethanol ingestion resulted in induction of brain CYP 450 IIE1, so chronic alcohol ingestion could enhance the sensitivity of brain to chemicals that are metabolized to more toxic derivatives by this system. Ethanol-inducibile cytohrome P-450IIEl causes metabolic tolerance to ethanol and to various commonly used drugs. Induction of microsomal pathway contributes to increase of acetaldehyde generation which promotes GSH depletion , lipid peroxidation and other toxic effects. (Mattila, 1990; Lieber et al., 1987; Lieber, 1988; Morton and Mitchell,1985).

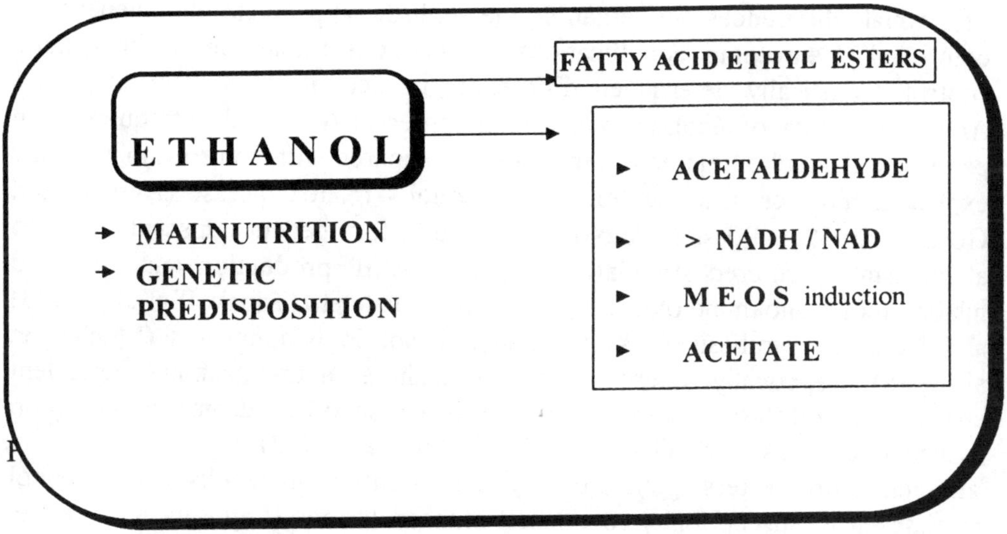

Fig.2 Mechanisms of neurotoxicity of ethanol.

Neurotoxic effect of ethanol and acetaldehyde was noted on developing brain (West and Goodlet, 1990).

The depletion of brain GSH by ethanol during development may make the developing fetus more susceptible to the toxic effects of alcohol or free radicals generated by its metabolism. It may be used in etiology of fetal alcohol syndrome (Rayes et al, 1993).

Presence of potential acetaldehyde producing enzymes, ADH, catalase and cytochrome P-450 IIE1 in the brain could help in understanding potential acetaldehyde mediated CNS effects of ethanol (Gill et al. 1992; Anandatheerthavarada et al.,1993; Simitskii and Kharchenko, 1994). Nonoxidative metabolism of ethanol, with esterification of ethanol with fatty acids via fatty acid ethyl ester synthetase is potentially related to terminal organ damage in the brain (Isenberg et al., 1992; Bora and Lange, 1993).

Ethanol is rich source of nonnutritive calories so, alcoholism is often complicated by malnutrition and vitamin deficiencies (Leevy et al., 1970; Thomson et al, 1983).

Individual differences in ethanol metabolism appear to be genetically controlled. Neurologic complications of alcoholism may be under genetic control. Genetically determined differences in alcohol metabolism as gastric ADH in ethiology of alcoholism in woman, different ADH3 allele frequencies in patients with alcohol related organ damage compared with control subjects may explain differences in susceptibility to alcohol - related disease (Agraval and Goedde, 1992). Possible biologic markers for genetic susceptibility to alcoholism are altered stimulation of cyclic AMP production and increased inhibition of monoamine oxidase by ethanol in platelets (Tabakoff et al., 1988) and increased synthesis of phosphatidylethanol in lymphocytes (Mueller et al..,1988). Genetically determinated abnormalities in the thiamine dependent enzyme transketolase may explain Wernicke-Korsakoff syndrome in one group of alcoholics (Blass and Gibson, 1977; Martin et al., 1993).

Fatty-acid ethyl esters generated by nonoxidative metabolism of ethanol correlates with susceptibility to ethanol-induced lesions (Laposata and Lange, 1986). Acetate, a product of ethanol metabolism also has significant CNS effects (Carmichael et al. 1991).

Alcohol intoxication and central nervous system

According to physiological and psihological actions drugs of abuse were classified into groups as stimulants, sedatives, hallucinogens. Although a single drug may have several actions. Alcohol acts as a stimulant in the early phase of intoxication, but as a brain concentration increase sedation occurs.

The clinical symptoms of acute alcohol intoxication are well known The intoxicating effects of ethanol correlate with blood concentrations which reflect levels present in the brain. Blood ethanol concentration about 5.0 mmol/l are associated with mild intoxication in persons which are not alcoholic. At concentrations higher than 21.0 mmol/l occurs vestibular and cerebelar disfunction. Blood ethanol levels in excess of 100.0 mmol/l are associated with death from respiratory depression. It is known that existing the direct relationship between concentration and depressant effect of the drug.

Alcoholism, as a disease of addiction is a major health problem in the world. Besides to multisistemic derangaments in the body this problem has some specific caracteristics such as tolerance., physical dependence and withdrowal syndrom.

Tolerance to ethanol is defined as a reduction in the behavioral and physiological effects on repeated exposure to the drug. Tolerance, which develops in prolonged and excessive ethanol use, leads to survive in ethanol concentration exceeding 217 mmol/l. There is decrease in sensitivity to the effect of drug. Various factors are involve in developing ethanol tolerance, behavioral, pharmacological and genetic (Le, 1990).

There is functional and dispositional tolerance. Dispositional tolerance include changes in absorption, distribution, metabolism and secretion of the ethanol. Acceleration of metabolism of ethanol in the liver is accomplished by induction of MEOS. In the neuron, adaptive changes in alcohol dehydrogenase may metabolize ethanol more rapidly when tolerance occurs. Functional tolerance (tissue tolerance) produced by chronic ethanol intake consider the possible roles of the biogenec amine neurotransmitters.

Tolerance to ethanol involve developing adaptive changes in neurocellular constituents such as neuromodulators, neurotransmitter receptors, ion channels, G proteins, intercellular second messengers, membrane lipids. (Hoffman et al., 1990; Gandhi and Ross, 1989; Cloninger, 1987; Charness et al., 1989 ; Kalant,1993).

Acute and chronic alcohol intoxication associated with alterations in neuronal activity of the central nervous system. The changes of behavior produced by ethanol arise as a result of the effect of ethanol or its metabolic products on the biochemical or physiological processes occurring in the central nervous system.

The precise molecular mechanisms responsible for ethanol intoxications are unknown. Disordering effects of ethanol on nerve membranes are responsible for many of the abnormalities in function that are mediated by membrane bound proteins as adenylate cyclase and Na+,K+,-ATP-ase and also there are changes in membrane lipid composition as demyelination and lipid loss in brain of chronic alcoholics (Leonard, 1987). Ethanol alters the metabolism of the biogenic amines e.g. serotonin, norepinephrine, dopamine, triptamine. This effect is mediated by acetaldehyde (Majchowich, 1973). Acetaldehyde can react with keto acids , pyruvate and alfa-ketoglutarate and form acetone, with sulfhydril groups that could be involve in active sites of enzymes and with amino groups many enzymes. Interaction of acetaldehyde with amino groups of various biogenic amines leads to form Schiff's base intermediates (Walsh, 1973). Tetrahydoisoquinoline (TIQ) alkaloids can be formed from catecholamines by spontaneous condensation with acetaldehyde. TIQ alkaloids are capable of functioning as false adrenergic neurotransmitters (Cohen, 1973). One of the most likely sites of ethanol intoxication is a complex of membrane proteins containing a receptor for the GABA, the inhibitory neurotransmitter, and an associated chloride-ion channel (Ticku, 1980; Ticku, 1989).

Typical alcohol related CNS effects are functional tolerance, physical dependence, intoxication, withdrawal syndrome and serious clinical manifestation of ethanol withdrawal is delirium tremens, movement disorders, hepatic encephalopathy, and brain damage in the form of various neurological syndromes including dementia, Wernicke-Korsakoff syndrome, corpus callous degeneration, cerebral atrophy, cerebellar degeneration, alcoholic myopathy, peripheral disorders and fetal alcoholic syndrome in the children of alcoholic mothers (Nakada and Kuight, 1984; Chareness et al., 1989; Neiman et al., 1990; Lishman, 1981; Lishman,1986).

Vitamins, ethanol and brain

Pyridoxine, thiamine, nicotinic acid and vitamin B_{12} are essential for normal cerebral functioning so, deficiencies many of them cause neurrological disorders in alcoholics (Dastur et al., 1976; Mezey, 1978 ; Veitch, et al., 1976). Decarboxilases as a cofactor contained pyridoxal-5-phosphate. They are involved in neurotransmitter synthesis. Pyridoxine deficiency can produce behavioral and electrophysiologial changes. Nicotinic acid deficiency may result in pellagra in which diffuse lesions are found.

Thiamin depletion is found in alcoholics (Thomson et al., 1983). Ethanol exerts a marked influence on the tissue levels of the thiamin metabolizing enzymes. Activity of thiamin pyrophosphate synthesizing enzyme is reduced whereas the enzymes dephosphorylating thiamin phosphates is increased (Laforenza et al., 1990). Thiamin. as thiamin pyrophosphate TPP functions as a cofactor for three cerebral enzymes, transketolase, pyruvate decarboxilase and alfa-ketoglutarate dehydrogenase. The metabolic consequence of the changes in these enzymes include decreased glucose utilisation and possibly decreased ATP synthesis.Depression of transketolase activity impairs the generation of NADPH from the pentose phosphate pathway, thus fatty acid synthesis might be reduced and myelinisation of brain components impaired.

Reduced glycolytic rates in the brain in thiamin depleted rats may reduced incorporation of glucose into cerebral amino acid . Thiamin deficiency causes a decrease in net cerebral protein synthesis and synthesis of acetylholine and other neurotransmitter substances.

The effects of alcoholic intoxication on brain metabolism

Ethanol is one of the most commonly used neurotoxic drugs. Besides in well - known pharmacological effects alcohol has a different metabolic actions on the central nervous system.

The brain is the organ with biosynthetic and metabolic pathways of carbohydrates, lipids, nucleic acids and proteins. All of these enzyme systems occur in the brain as well enzymes of tricarboxylic acid cycle.

Glucose is primary source of energy for the brain. The fall in blood glucose correlated with a cerebral dysfunction that ranged from mild behavioral impairment to coma.

Glutamate is compound of great interest for CNS. It was required as an energy source, for protein synthesis, as a precursor of GABA and proline, in ammonia fixation in CNS in form glutamine, for synthesis of tripeptide glutathione and as a excitatory amino acid. Glutathone is tripeptide that consisting glutamic acid,

cysteine and glycine. It acts as an antioxidants as a substrate in y-glutamyl cycle as the substrate for GGT in amino acid transport. The disulfide form of glutathione is rapidly reduced by NADPH -dependent glutathione reductase.
Transamination of glutamate with oxalacetate leads to aspartate. Aspartate is excitatory amino acid . It can serve as a available energy source, for de novo nucleotide synthesis and in the arginino succinic acid synthetase reaction in urea synthesis in the brain.
The brain depends on an adequate supply of glucose both as an energy source an as a donor of carbon fragments for protein and lipid biosynthesis. Glucose utilization is thiamin deficient and it is likely that this is a major factor in the production of tiamine deficient brain damage.
Metabolic changes in high-energy phosphates in rat brain occur after acute and chronic ethanol intoxication. Increased breakdown of membrane phospholipids may be a reason for the development of cerebral atrophy in chronic alcoholism (Denays et al., 1993).

Ethanol, aminoacid and protein metabolism

The adversive effects of ethanol upon body protein metabolism are well known. Ethanol, administrated acutely inhibits protein synthesis in a number of rat tissues in vivo (Thernan and Ward, 1986).
Amino acid and protein metabolism are of essential importance for brain function. Ethanol, or its metabolites, changes many of biochemical processes in the brain , and these which related to amino acid and protein (Goertz, 1982; Noble and Tewari,1973) Ethanol inhibits cerebral protein synthesis caused a decrease in RNA content in the Purkinje cells. The effect of ethanol on brain amino acids included an increase in GABA, glutamic acid, and aspartic acid. Alanine, glycine serine taurine levels were unchanged, whereas keto acids and glutamine were decreased. The effect of ethanol on brain amino acids are of interest because tyrosine and tryptophan serve as a precursors to monoaminoregic neurotransmitters. Dopamine, noradrenaline and adrenaline are synthesized from tyrosine and 5-hydroxitriptamine from tryptophan.
Acetaldehyde reactivity with amino or sulphydryl groups may influence protein synthesis by decreasing amino acid availability. Significant decrease in the

concentrations of most amino acids in plasma following an acute dose of ethanol to the rats have been noted (Siegel et al., 1964; Bekary et al., 1991). Fetal serum glutamate was elevated by ethanol action (Karl et al., 1995). Changes in the relative concentrations of amino acids in plasma may thus influence the availability of various amino acids in the brain. Protein metabolites such as ammonia, indoles, amines, peptides, free amino acids, endotoxins, methionine metabollytes and GABA are assumed to play as important role in the multifactorial pathogenesis of hepatic encephalopaty (Bode and Schifer, 1985).

Normal brain function is dependent upon the nutrient environment. The brain has no storage form of energy, it requires a constant supply of oxygen, its mayor energy source, glucose, as well as amino acid and other trace elements. Circulating levels of these essential brain constituents, in particular amino acids, are dependent upon normal liver function and the supply of glucose to the brain may be affected in alcoholic by recurrent hypoglycemia or impaired hepatic gluconeogenesis.

Chronic ethanol ingestion has many effects on hepatic metabolism which become more pronounced as an increasing number of liver cells are destroyed. Reduction of functional liver mass is associated with a decrease in hepatic storage of nutritients (Mezey, 1978) so, increasing liver damage is associated with structural brain changes (Acker et al. 1982). Changes in the metabolic profiles of ammonia occur as a consequence of loading of high doses of ethanol (Nikoli}, 1987). There is interference of ammonia causing neurotoxicity especially in severe alcoholic liver damage and alterations in hepatic and brain amino acid and protein metabolism.

Ammonium ions play an important role as a cerebral irritant. Ammonia reaching the brain reacts with amino acids such as glutamate, thus reducing cerebral concentrations of excitatory neurotransmitters and resulting in high levels of glutamine.. Ammonia is toxic to brain tissue in many ways, including inhibition of $Na^{+}K$+ATP-ase and chloride pumps on nerve cell membranes. This leads in turn to impaired conduction and reduced transfer of electrolytes across membranes so that water balance may be disturbed.

It is suggested that the toxic effect of ammonia on the oxidative metabolism of mitochondria is related to pyruvate and alfa-ketoglutarate oxidase stages. The content of ammonia in living brain is kept at a relatively low concentration. It is suggested that a detoxification mechanism for ammonia will operate efficiently in brain where ammonium ions show a toxic action. Removal of free ammonia in the brain with glutamine synthatase system is energy consumption process.

Ammonia neutralization mainly occurs in the liver of ureothelic by forming of urea by urea cycle enzymes (Ratner, 1973).

Brain tissue possessed no mayor activity of carbamoil-P synthetase and ornithine-transcarbaylase. Role of urea formation in the brain is in the neutralization of argininosuccinic synthetase which has toxic effect in the brain, or in the synthesis of arginine from citruline in nonhepatic tissue which is important for the replenishment of the arginine deficit, necessary for protein synthesis. There is the possibility that the small amounts of urea in brain have an effect on the structure and properties of brain proteins. The effect of chronic ethanol abuse on hepatic metabolism and concentration of amino acids have been shown to influence brain neurotransmitter synthesis. This may be relevant to states of dependence, depression and withdrawal associated with ethanol abuse. It has been suggested that changes in plasma amino acid balance in alcoholics may lead to disturbances in neurological functioning and possibly damage via the production of false neurotransmitters (Smith and Amit, 1987). Severe liver damage through failure to catabolise aromatic amino acids allows increased plasma levels of phenylalanine, tyrosine and tryptophan in relation to the branched-chain amino acid. Increased amounts of phenylalanine, tyrosine and triptophan are transferred across the blood-brain barrier. Phenylalanine acts as a competitive inhibitor of tyrosine decarboxylase, thus inhibiting the synthesis of dopamine and noradrenaline from tyrosine. Tyrosine may then be converted via tyramine to the false neurotransmitter octopamine. Octopamine, which is a weak neurotransmitter, blocks receptor sites to dopamine and noradrenaline. The decrease in level on nerve cell stimulation which would result from such blockade might conceivably lead to neuronal atrophy.

Nutritional deficits may reflect the toxicity of ethanol and acetaldehyde. Depletion in cellular glutathione level was ameliorated by methionine. Polyenyl-phosphatidylcholine corrects the alcohol-induced hepatic phosphatidylcholine depletion. Ethanol toxicity may be correct and with vitamin supplementation or by some amino acids, as glycine, cisteine, methionine, arginine ((Lieber, 1994; Shaw et al, 1981; Garcia-Ruiz et al, 1995; Kricevskaya et al., 1985).

The importance of arginine in ethanol intoxication

This study was designed to investigate the influence of subchronic ethanol consumption on oxidative damage of brain by measuring malondialdehyde (MDA) and finding the possible association between MDA and activity of glutamine synthetase, arginase and urea content. The second of interest is investigation the supplementation with arginine on those processes.

Experiment was performed on male albino Wistar rats weighing about 200 g. Animals were divided in four groups:

1- control rats (n=6),
2- rats (n=6) treated with ethanol in drinking water as 15% solution;
3- rats (n=6) treated with arginine as 0.5% water solution;
4- rats (n=6) treated simuntanely with ethanol and arginine.

Animals were killed after three weeks.

Some blood constituents as urea, creatinine, glucose, uric acid, triglycerides and cholesterol were determined in the blood plasma obtained from aorta abdominalis using standard biochemical analysis. Brains from experimental animals were rapidly removed, washed and homogenized in ice cold water. As a marker of lipid peroxidation brain malondialdehyde (MDA) was measured by tiobarbituric acid reaction (Stroev and Makarova, 1989). Brain urea and blood plasma citruline were determined diacethyl monoxime reaction (Rahmatulah and Boyde, 1980).

Arginase activity was measured by the method of Porembska and Kedra (1975) on the basis of formed ornithine with ninhydrine color reaction; glutamine synthetase by the method of Levintow (1954). Enzyme activities were expressed in units per mg. protein (U / mg. prot.). The level of protein was determined by the method of Lowry et al. (1951). Statistical analyzes of data was calculated by Student' t-test.

Blood biochemistry was changed after ethanol ingestion compared to control. As showen on Table 1 intake of ethanol for three weeks changes some blood

constituents such as urea, uric acid, triglycerides and cholesterol which are increased. The level of glucose was slowly depressed.
Ethanol-arginine intake lead to elevation of urea content. All parameters, except glucose, show decreased levels compared to ethanol group. These data may be of clinical importance.
Brain MDA was increased ($p < 0.01$) as a result of oxidative stress provocated by ethanol. Supplementation with arginine slowly decreases MDA but depression is not statistically important. (Fig.3.).

Table 1. Blood biochemistry in ethanol and ethanol-arginine treated rats (mean ± SD).

	control	ethanol	ethanol-arginine
urea, mmol / L	8.15 ± 0.26	8.42 ± 0.63	8.58 ±0.48
creatinine, μmol/ L	50.16 ± 2.6	45.16 ± 2.08	52.1 ± 7.7
citruline, mmol / L	1.09 ± 0.05	1.01 ± 0.045	1.35 ± 0.04 •
uric acid, μmol / L	117.8 ± 20.5	174.1 ± 35.7 ••	134.2 ± 35.3
glucose, mmol / L	5.5 ± 0.9	4.8 ± 0.8	6.01 ± 0.7
triglycerides,mmol/l	0.423 ± 0.04	0.884±0.25 •••	0.667±0.14 ••
cholesterol,mmol / l	1.22 ± 0.09	1.42 ± 0.13 •	1.21 ± 0.14

•$p < 0.05$, •• $p < 0.01$, ••• $p < 0.001$

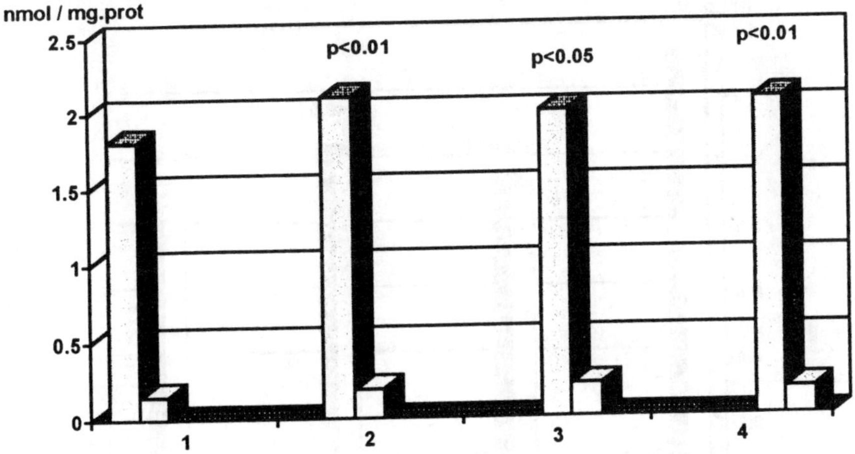

Fig. 3. MDA level (mean ■ ± SD □) in 1-control, 2-ethanol, 3-arginine and 4-ethanol-arginine group.

There is a considerable amount of evidence in regarding the important role of lipid peroxidation in development of ethanol-induced tissue injuries (Lieber, 1988; Comporti, 1985). Ethanol-induced lipid peroxidation is manifested by an increase in MDA content. Lipid peroxide content in the brain was determined by measuring tiobarbituric acid (TBA) reactants and expressed as MDA content in nmol/ mg. protein.

Glutamine synthetase is the enzyme that reflect the presence of excess levels of reactive oxygen species in the brain provocated by ethanol (Bondy and Guo, 1995). Its activity was significantly depressed in the group of animals treated with ethanol ($p < 0.001$) , while, ethanol-arginine treatment did not modify enzyme activity (Fig 4).

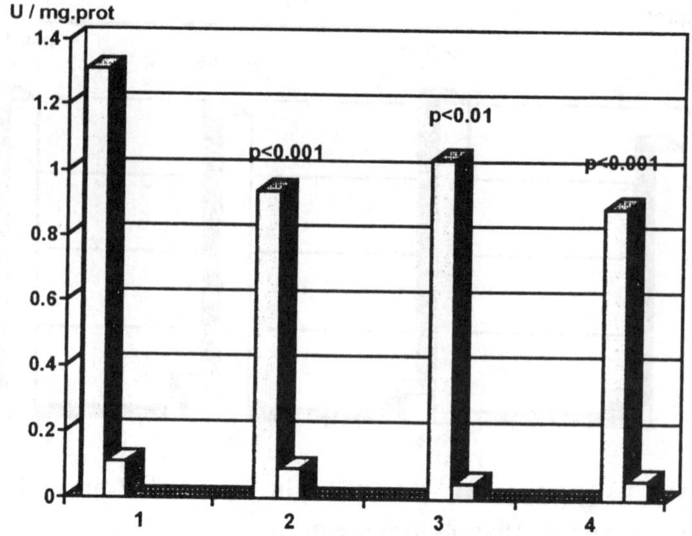

Fig . 4. Glutamine synthetase (mean ■ ± SD □) in 1-control, 2-ethanol, 3-arginine and 4-ethanol-arginine group.

There is association between elevation of MDA and depression of glutamine synthetase activity (Fig.5 and 6).

Arginine is essential amino acid that required for normal growth rate in young animals as a component of proteins, in urea synthesis, in synthesis of creatine, in synthesis of nitric oxide (Fig. 7).

Metabolism of arginine in the brain is very active. Arginase is the enzyme that hydrolysis arginine to ornithine and urea. The high level of arginase is found in the liver of ureothelic, but enzyme activity was detected in many of other tissues as brain, kidney, mammary gland (Porembska, 1983). The importance of non-hepatic arginase is not completely defined. The importance is given to other product of reaction, ornithine, that serve in production of polyamines, proline and glutamic acid or other guanidino compounds.

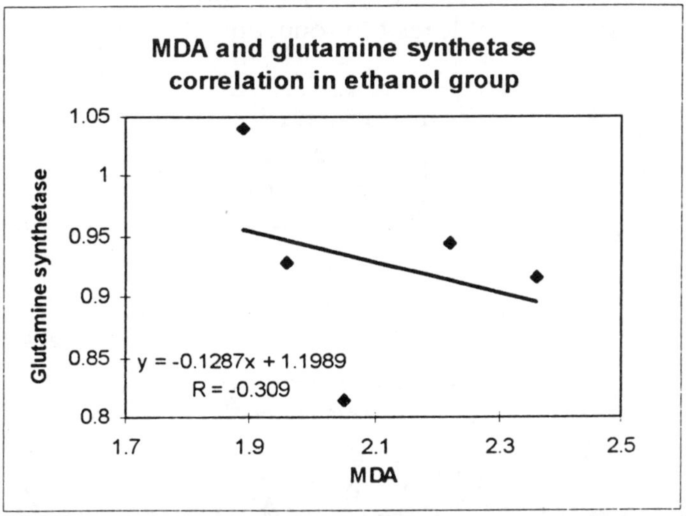

Fig 5. Relationship between MDA and glutamine synthetase in ethanol group of animals.

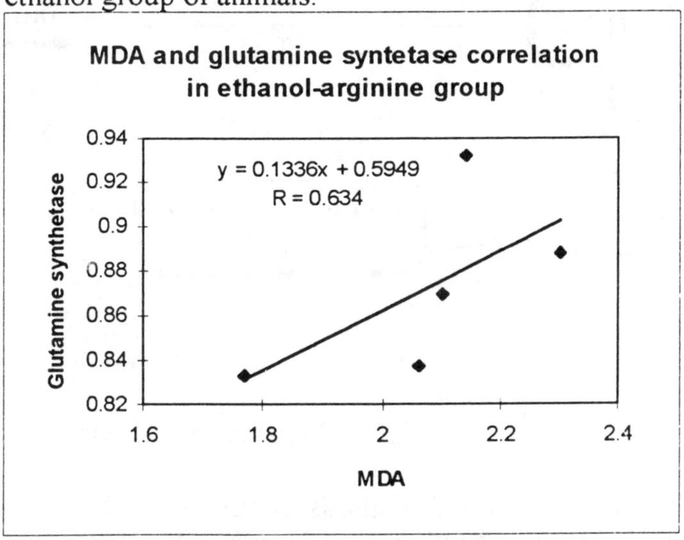

Fig 6. Relationship between MDA and glutamine synthetase in ethanol-arginine group of animals.

The last investigation gives more attention to oxidative pathway in metabolism of arginine that leads to production of nitro oxide (NO).

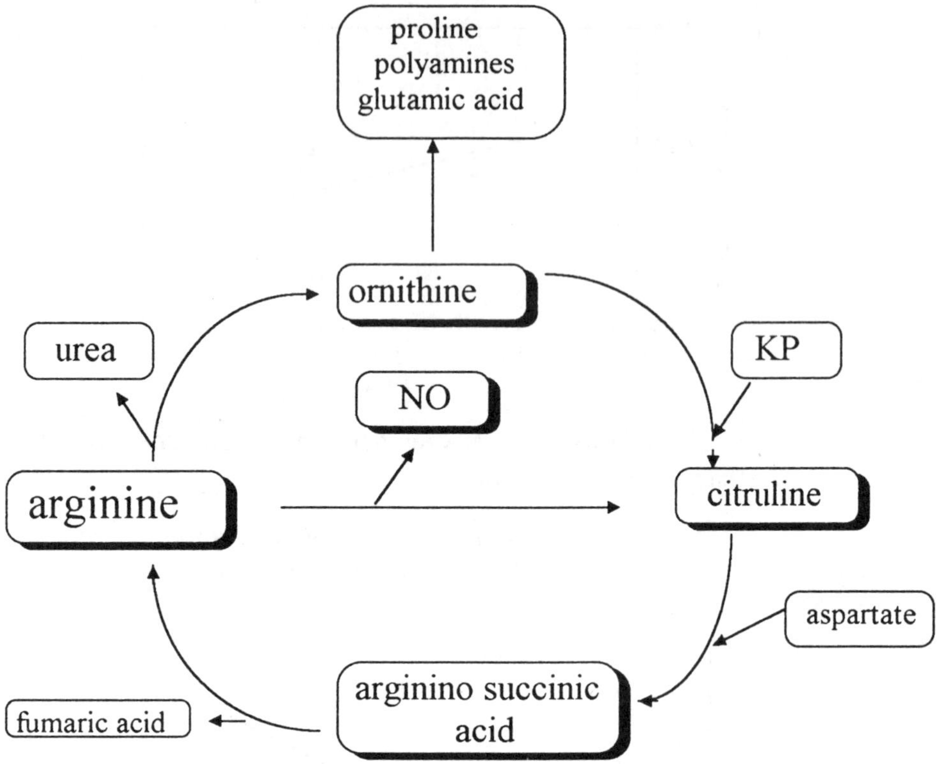

Fig. 7. Metabolism arginine.

Arginine is a precursor in the synthesis of nitric oxide by cells containing nitric oxide synthetase (NOS). NOS enzymaticaly incorporates molecular oxygen into L-arginine and converts it to L-citruline and nitric oxide (NO) which plays a significant role in numerous homeostatic processes. Nitric oxide plays a key role in the regulation of vascular tone, immune system function , neurotransmission, platelet agregation and adhesion (Kosti}, 1993). In the rats treated with ethanol there is increased activity of arginase ($p < 0.01$). Supplementation with arginine leads to elevation of enzyme activity also (Fig..8).

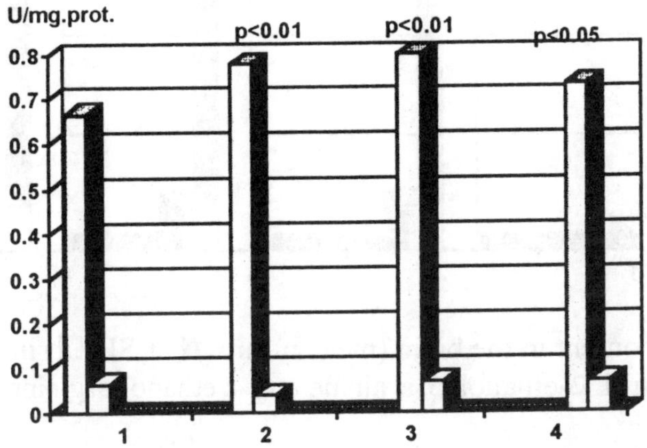

Fig. 8. Arginase activity (mean ■ ± SD □) in 1-control, 2-ethanol, 3-arginine and 4-ethanol-arginine group.

As shown brain urea was increased in ethanol and ethanol-arginine group ($p<0.01$) (Fig. 9).Increase occurs probably as a result of arginase action.

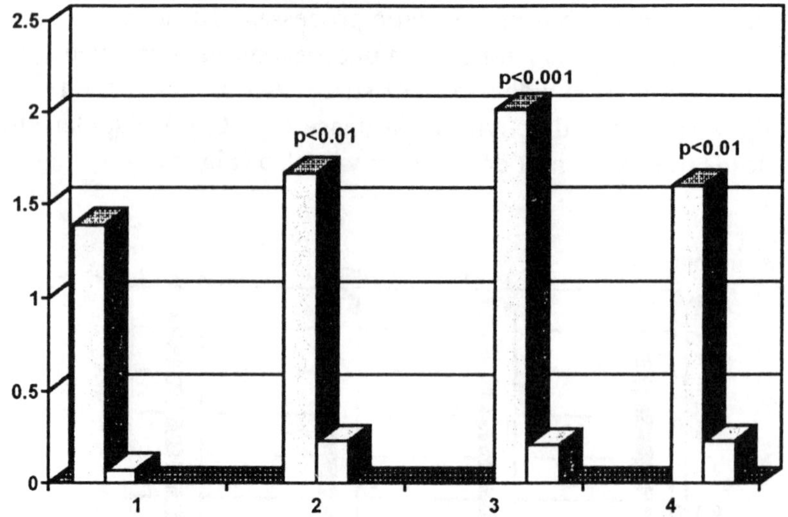

Fig. 9. Urea content in the brain (mean mean ■ ± SD □) in 1-control, 2-ethanol, 3-arginine and 4-ethanol-arginine group.

Association between brain arginase activity and urea content was given on Fig. 9 and 10, and association between urea and MDA on Fig. 11 and 12.Increased level of MDA is folowing by increasing of urea level .

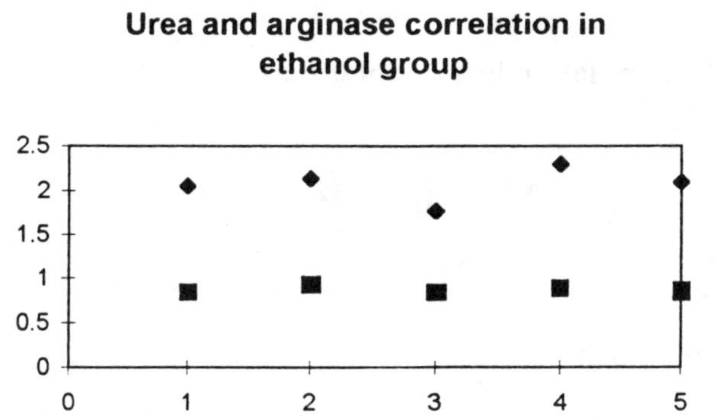

Fig 9. Relationship between urea and arginase in ethanol group of animals (R = 0.634).

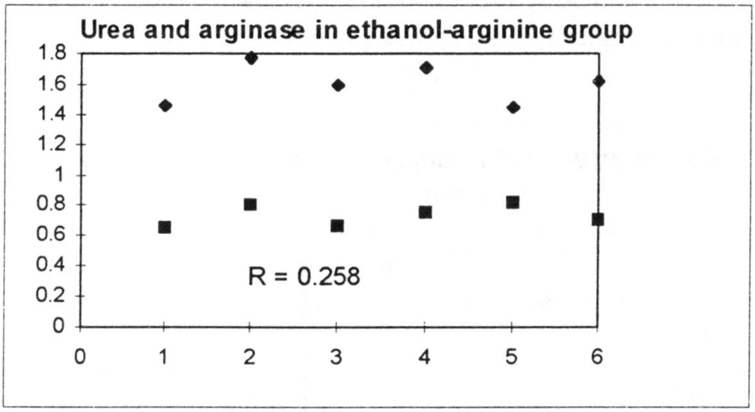

Fig. 10 . Relationship between urea and arginase in ethanol-arginine group of animals (R= 0.258.

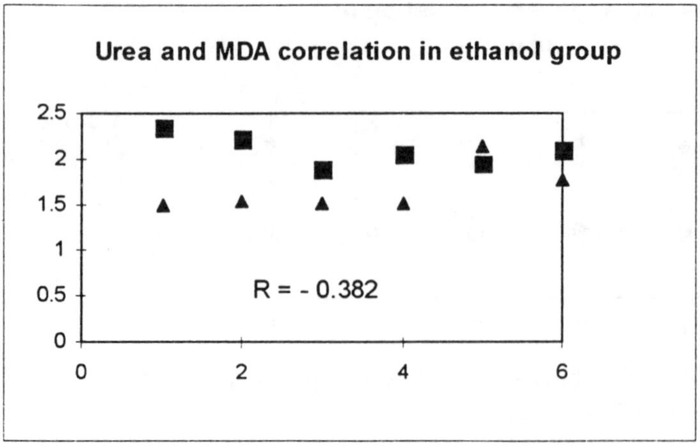

Fig. 11. Relationship between urea ◆ and MDA ■ in the brain of animals treated with ethanol.

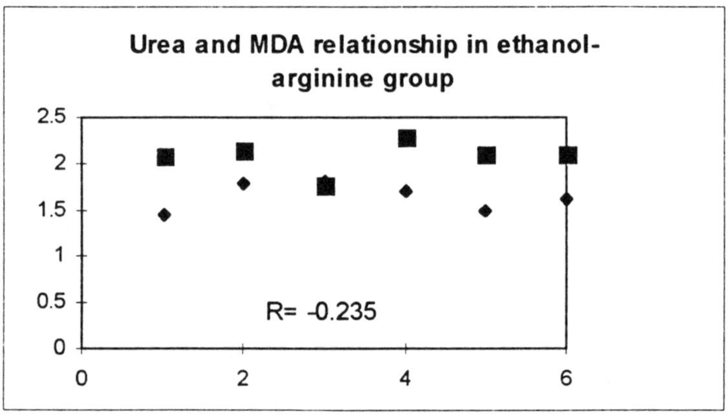

Fig. 12. Urea ◆ and MDA ■ relationship in ethanol-arginine group.

The mechanism by which ethanol produces its effects, and the neuronal components involved in its action have been a matter of debate over the years. It is believed that depressant effects of ethanol are mediates via GABA and N-methyl-D-aspartate (NMDA) receptor systems. According to current views there is a connection between ethanol, nitric oxide and neurotoxicity in chronic ethanol exposure (Lancaster, 1992; Dawson, 1991). There is also evidence that factors facilitating a release and/or synthesis of NO might influenting to the brain function (D`olji},1996.) Nitric oxide is a major messenger molecule in the cardiovascular, immune and central nervous system. In the brain NO is responsible for the glutamate-linked enhancement of 3' 5' cyclic guanosine monophosphate levels (cGMP). Because NO can not be stored in vesicles like other neurotransmitters, its release is regulated by the activity of the enzyme, NO synthetase.

According to Mori and Natsuki, (1996) expression of nitric oxide synthetase is responsible for the induction of neuronal cell disorders attributed to chronic ethanol damage .We have investigated the effect of nitric oxid precursor, L-arginine on some brain metabolic effects of ethanol.

In the CNS the NO stimulates the release of various of various neurotransmitters, including excitatory amino acids L-glutamate and L-aspartate (Lawrence and Jarrott, 1993) Glutamate is the most abundant excitatory neurotransmitter in the brain. Under certain conditions it may become a potent excitotoxin and contribute to neurodegeneration. The role of free radicals involments is shown in alcoholic disorders (Comporti, 1985; Bonddy and Guo, 1994) such as reduction of glutamine synthetase activity which is particularly vulnerable to oxidative stress (Bondy and Guo, 1995). One of free radicals, nitric oxide has been associated with some disorders. Excessive production of NO has neurotoxic effect. Alcohol provocates the stimulation of excitatory amino acids, glutamate, aspartate, and excitatory amino acid receptor N-methyl-D-aspartate (NMDA). Excitotoxicity is one of consequence in neurodegenerative disorders. The role of free radicals, as well as nitro oxide in cell death is induced by activation of excitatory amino acid receptors. According to Mori and Natsuki (1996) NO synthetase activity was higher in ethanol group in cerebral culture cells.

The possible mechanism in explanation of increasing brain arginase activity upon the ethanol may be secondary action of hormone released as growth hormone, prolactine, cortisol (Stokes, 1973; Badger et al., 1993). Increased urea production has compensatory protective effect in the brain. Arginine has similar action on hormone release specifically catecholamines, corticosterone, glucagon, growth hormone, insulin, prolactin and somatostatin. (Merimee et al., 1965, Palmer et al. 1975; Walter et al., 1980), so, arginine supplementation does not ameliorates the effects of ethanol on arginase activity and urea level.

References

Acker,W., Aps, E.J., Majumdar, S.K., Shaw, G.K., Thomson, A.D. (1982): The relationship between brain and liver damage in chronic alcoholic patients. J. Neurol. Neurosurg., and Psych., 45, 984-987.

Agarval, D.P., Goedde, H.W. (1992) Mediobiological abd genetic studies on alcoholism. Clin. Investig., 70, 465-477.

Anandatheerthavarada, H.K., Shankar, S.K., Bhamre, S., Boyd, M.R., Song, B.J., Ravindranath, V. (1993): Induction of brain cytochrome P 450 IIE1 by chronic ethanol treatment. Brain Res, 601 (1-2), 279 - 285.

Badger, T.M., Ronis, M.J., Lumpkin, C.K., Valentine, C.R., et al. (1993): Effects of chronic ethanol on growth hormone secretion and hepatic cytochrome P-450 isozymes of the rat. J. Pharm. Exp. Ther., 264/1, 438-447.

Bekairi, A.M., Abulaban, F.S., Ahmed, M.M., Qureshi, S. (1991): Plasma amino acid levels in rats following concurrent administration with alcohol, nicotine and caffeine. Res. Common. Subst. Abuse, 12/1-2, 27 - 39.

Bing, R.J. (1978): Cardiac metabolism; its contributions to alcoholic heart disease and myocardial failure.Circulation 58, 6, 965-970.

Blass, J.P., Gibson, G.E. (1977): Abnormality of a thiamine -requiring enzyme in patients with the Wernicke-Korsakoff syndrome. New England Journal of Medicine,297,25, 1367-1370.

Bode, J.Ch. Schafer, K (1985).: Pathophysiology of chronic hepatic encephalophaty. Hepato-gastroeneterol., 32, 259-266.

Bondy, S.C., Guo, S.X. (1994): Effect of ethanol treatment on indices of cumulative oxidative stresss. Eur. J. Pharmacol., 270 (4), 349-355.

Bondy, S.C., Guo, S.X. (1995): Regional selectivity in ethanol-induced pro-oxidant events within the brain . Biochem. Pharacol. 49 (1), 69-72.

Bora, P.S., Lange, L.G. (1993): Molecular mechanism of ethanol metabolism by human brain to fatty acid ethyl esters. Alcohol Clin. Exp. Res., 17/1, 28 - 30.

Caballeria, J.; Baraona,E., Deuleofeu R. (1991): Effect of H2 -receptor antagonists on gastric alcohol dehydrogenase activity.. Dig. Dis. Sci., 36, 1673-1697.

Carmichael, F.J., Israel, Y., Crawford, M., Minhas, K, Saldivia, V., Sandrin, S., Canpisi, P., Orrego, H. (1991): Central nervous system effects of acetate. J. Pharm. Exp. Ther., 259/1, 403 - 408.

Charness, M.E. (1989): Ethanol and opioid receptor signalling. Experientia, 45, 418-428..

Charness, M.E., Simon, R.P., Greenberg, D.A. (1989): Ethanol and the nervous system. New Engl. J. Med. 321, no 7, 442-454.

Cloninger, C.R. (1987): Neurogenetic adaptive mechanisms in alcoholism. Science 236, 410-416.

Cohen, G. (1983): Tetrahydroisoquinoline alkaloids: uptake, storage, and secretion by the adrenal medula and by adrenergic nerves. Ann. N.Y.Acad Sci.,215, 116-119.

Comporti, M..:(1985) Biology of disease. Lipid peroxidation and cellular damage in toxic liver injury. Lab. Invest., 53, 599-623.

Dancy, M., Maxwell, J.D. (1986): Alcohol and dilated cardiomyipathy. Alcohol and Alcoholism, 21, No.2, 198.

Dastur, D.K., Santhadevi, N., Quadros, E.V., Avari, F. C.R. (1979): The B-vitamins in malnutrition with alcoholism. Br.J. Nutr., 36, 143 - 159.

Dawson, V.L., Dawson, T.M., London, E.D., Bredt, D.S., Snyder, S.H. (1991): Nitric oxide mediates glutamate neurotoxicity in primary cortical cultures. Proc. Natl. Acad. Sci, USA, 88/14, 5368-6371.

Denays, R., Chao, S.L., Mathur-Davre, R., Jeghers, O., Fruhling, J., Noel, P., Ham, H.R. (1993): Metabolic changes in the rat brain after acute and chronic ethanol intoxication: A (31)P NMR spectroscopy study. Magn. Reson. Med., 29/6, 719 - 723.

DiPadova, C., Worner, T.M., Julkunen, R.J.K, Lieber, C.S. (1989): Effects of fasting and chronic alcohol consumption on the first pass metabolism of ethanol. Gastroenterology, 92, 1169 - 1173.

D`olji}, M.R (1996): Nitric oxide- a novel modulator of brain function. Acta Biol. Med. Exper., 212, 3-5.

Frezza, M., Di Padova, C., Pozzato,G., Terpin, M., Baraona, E., Lieber, C.S. (1990): High blood alcohol levels in women. The role of decreased gastric alcohol dehydrogenase activity and first-pass metabolism. N. Engl. J. Med., 322, 2, 95-99.

Gandhi, C.R., Ross, D.E. (1989): Influence of ethanol on calcium, inositol phospholipids and intracellular signalling mechanisms. Experientia, 45, 407-413.

Garcia-Ruiz, C, Morales, A., Colell, A., Balldesta A., Rodes, A., Kaplowitz, N. Fernandez, C/J.C. (1995): Feeding S-adenosyl—L-methionine attenuates both ethanol-induced depletion of mitochondrial glutathione and mitochobdrial dysfunction in periportal and perivenous rat hepatocytes. Hepatology, 21 (1),

207-214.

Gill, K., Menezm J,F., Lucas, D., Deitrich, R.A. (1992): Enzymatic production of acetaldehyde from ethanol in rat brain tissue. Alcohol Clin. Exp. Res. 16/5, 910 - 915.

Goertz, B. (1982): Effect of ethanol on rat brain protein synthesis. Exp. Brain Res., 48, 438-442.

Hawkins, R.D., Kalant, H. (1972): The metabolism of ethanol and its metabolic effects. Pharm. Rev., 24, no 1., 67-157.

Hillbom, M., Kaste, M. (1990):Alcohol abuse and brain infarction. Ann. Med., 22, 347-352.

Hoffman, P.L., Ishizawa, H., Girri,P.R., Dave,J.R., Grant,K.A., Liu,l., Gulya,K.:(1990) 22, 269-274.

Isenberg, K.E.,,Bora, P.S., Zhon, X., Wu, X., Moore, B.W., Lange, L.G. (1992): Nonoxidative ethanol metabolism.: Expression of fatty acid ethyl ester synthetase III in cultured neural cells. Biochem. Biophus. Res. commun., 185/3, 938 - 943.

Isselbacher, K.J. (1977): Metabolic and hepatic effects of alcohol. N.Engl. J. Med., 296, no 11, 612-616.

Kalant, H. (1993): Problems in the search for mechanisms of tolerance. Alcohol. Alcohol, Suppl., 2. 1-8.

Karl, P.I., Kwun, R., Slonim, A., Fisher, S.E. (1995): Ethanol elevates fetal serum glutamate levels in the rat. Alcohol.Clin. Exp. Res., 19/1, 177-181.

Kostic, M.M. (1993): A new bioregulatory system: nitric oxide from L-arginine. (review). Iug. Physiol. Pharmacol. Acta, 29/1, 3-34.

Krichevskaya, A.A., Shugaley, V.S., Ananyan, A.A., Zigova, J.G. (1985): Protective effect of arginine in cold stress. Vopr. Med. Himm., 6, 50-53.

Laforenza, U., Patrini, C., Gastaldi, G., Rindi, G. (1990): Effects of acute and chronic ethanol administration on thiamine metabolizing enzymes in some brain areas and in other organs of the rat. Alcohol. Alcohol., 25/6, 591 - 603.

Lanchester, F.E. (1992): Alcohol, nitric oxide, and neurotoxicity. Is there is connection/., (review). Alcohol, Clin. Exp. Res, 16, 539-541.

Laposata, E.A., Lange, L.G. (1986): Presence of nonoxidative ethanol metabolism in human organs commonly damaged by ethanol abuse. Science, 231, 497-449.

Lê A.D.: Factors regulating ethanol tolerance. Ann. Med., 22, 265-268, 1990.

Leevy, C.M., Tamburm, C.H., Smith, F. (1970): Alcoholism, drug addiction and nutrition. Med. Clin. North. Am., 54, 1567-1573.

Leonard, B.E. (1987): Ethanol as a neurotoxin.(comment). Biochem. Pharmacol., 36,2055-2059.

Levintow, l., Meister, A. (1954): Reversibility of the enzymatic synthesis of glutamine. J. Biol. Chem., 209, 265 - 280.

Lieber, C.S., Lasker, J.M., Alderman, J., Leo, M.A. (1987): The microsomal ethanol oxidising system and its interaction with other drugs, carcinogens and vitamins. Ann. N.Y. acad. Sci, 492, 11-24.

Lieber, C.S. (1988): Biochemical and molecular basis of alcohol-induced injury to liver and other tissues. N. Engl. J. Med., 319, No 25, 1639-1650.

Lieber, C.S. (1984): Metabolism and metabolic effects of alcohol. Med. Clin. North. Am., 68, 3-31.

Lieber, CS. (1994): Hepatic and metabolic effects of ethanol: pathogenesis and prevention. Ann Med, 26 (5) 325-30.

Lishman, W.A. (1981): Cerebral disorders in alcoholism. Brain, 104, 1-20.

Lishman, W.A. (1986): Alcoholic dementia: a hypothesis. Lancet, 1, 1184-1186.

Li, T-K., Bosron, W.F. (1987): Distribution and properties of human alcohol dehydrogenase isoenzymes. Ann.N.Y.Acad. Sci, 492, 1-10.

Lovinger, D.M., White, G., Weight, F.F. (1990): Ethanol inhibition of neuronal glutamate receptor function. Ann. Med., 22, 247-252.

Lowrence, A.J., Jarrott, B. (1993): Nitric oxide increases interstitial excitatory amino acids release in rat dorsomedial medulla oblongata. Neurosci Lett., 151, 126-129.

Lowry, O.H. Rosenbrought, N.J., Farr,A.I., Randall, R.J. (1951): Protein measurement with the Folin phenol reagent. J. Biol. Chem., 193, 265 - 273.

Majchrowicz, E. (1973): Alcohol, aldehydes and biogenic amines. Ann.N.Y.Acad Sci, 215, 84-88.

Martin,P.R., McCool, B.A., Singleton,C.K. (1993):Genetic sensitivity to thiamine deficiency and development of alcoholic organic brain disease.. Alcohol. Clin. Exp. Res., 17/1, 31- 37.

Merimee, T.J., Lillierap, D.A., Rabinowitz, D. (1965): Effect of arginine on serum levels of human growth hormone. Lancet, 2, 668-670.

Mezey, E. (1978): Liver disease and nutrition. Gastroenterology, 74/4, 770 - 783.

Mori, C., Natsuki, R. (1996): Effect of ethanol on expression of nitric oxide synthrtases in the cerebral culture cells from chick embrio. Nippon, Yakurigako, Zasshi, 107 (4),197-203.

Morton, S., Mitchell, M.C. (1985): Effect of chronic ethanol feeding on glutathione turnover in the rat. Biochem. Pharmacol. 34, 1559-1563.

Mueller,G.C., Fleming, M.F., LeMahieu, M.A., Lybrand,G.S., Barry, K.J. (1988): Synthesis of phosphatidylethanol - potential marker for adult males at risk for alcoholism. Proc.Natl. Acad Sci. USA, 85, 9778- 9782.

Nakada, T., Kuight, R.T. (1984): Alcohol and central nervous system. Med. Clin North Am. 68, 121-131.

Neiman, J., Lang, A.E., Fornazzari, L., Carlen, P.L.(1990): Movement disorders in alcoholism. (review). Neurology, 40, 741-746.

Nikolic, J. (1987): The influence of ethanol on the level of ammonia, urea and glutamine in rats tissue. Acta Med. Medianae, 7 (1), 42-47.

Noble, E. P., Tewari, S. (1973): Protein and ribonucleic acid metabolism in brains of mice folowing chronic alcohol consumption. Ann. N.Y.Acad Sci., 215, 333- 345.

Palmer, J.P., Walter, J.M., Ensinck, J.W. (1975): Arginine stimulated acute phase of insulin and glucagon secretion. In normal man. Diabetes, 24, 735-740.

Porembska, Z. (1973): Different speciens of arginase in animal tissues. Enzyme, 15, 198-209.

Porembska, Z., Kedra, M. (1975): Early diagnosis of myocardial infarction by arginase activity determination. Clin. Chim. Acta, 60, 355 - 361.

Quintanilla, M.E., Tampier, L.(1992): Ethanol intake: effect on liver and brain mitochondrial function and acetaldehyde oxidation. Alcohol, 9/5, 375 - 380.

Rahmatullah, M., Boyde, T.C. (1980): Improvements in the determination of urea using diacetyl monoxime; method with and without deproteinisation. Clin. Chim. Acta, 107, 3-9.

Raskin, N. H. (1973): Alcohol dehydrogenase in brain. A toxicologic role. Ann. N.Y.Acad. Sci, 215, 49-53.

Ratner, S.(1973): Enzymes of arginine and urea synthesis. Advenc. Enzymol., 39, 1-90.

Rawat, A.K., Kuriyama,K. (1972): Ethanol oxidation: Effect on the redox state of brain n mouse. Science, 176, 1133-1135.

Roine, R., Gentry,R.Y., Hernandez-Munoz, R., Baraona, e., Lieber, CS. (1990):Aspirin increases blood alcohol concentrations in human after ingestion of ethanol. JAMA, 264, 2406-2408.

Simitskii, V.N., Kharchenko, N.K. (1994): the role of acetaldehyde in the pathogenetic mechanisms of the development of chronic alcoholism. Fiziol. Zh., 40, (3-4), 94-100.

Shaw, S., Jayatilleke, E., Ross, W.A., Gordon, E.R., Lieber, C.S. (1981): Ethanol-induced lipid peroxidation potentiation by long-term alcohol feeding and attenuation by methionine. J. Lab. Clin. Med., 98, 417-424.

Siegel, F.L., Roach, M.K., Pomerou,, I.R. (1964): Plasma amino acid patternt in alcoholism: the effect of ethanol loading. Proc.. Natl. Acad. Sci.USA, 51,605-609.

Smith, B.R., Amit, Z. (1987): False neurotransmitters and the effects of ethanol on the brain. Ann. N.Y. Acad. Sci., 492, 384-389.

Stewart, J.A., Caron, H. (1977): Arginases of mouse brain and liver. J. Neurochem, 29, 657-663.

Stokes, P. E. (1973): Adrenocortical activation in alcoholics during chronic drinking. Ann. N.Y. Acad. Sci., 215, 77-83

Stroev, E.A., Makarova, V.G. (1989): Laboratory manual in biochemistry, Mir, Moscow,251-255.

Takahashi, T., Lasker, J.M., Rosman, A.S., Lieber, C.S. (1993): Induction of cytochrome P-4502E1, in the human liver by ethanol is caused by a corresponding increase in encoding messenger RNA. Hepatology, 17, 236-245.

Tabakof, B., Hoffman, P.L., Lee, J.M., Saito,T., Willard, B., De Leon-Jones, F. (1988): Differences in platelet enzyme activity between alcoholics and nonalcoholics. N.Engl. J. Med,318, 134-139.

Ticku, M.K. (1980): The effects of acute and chronic ethanol administration and its withdrawal on γ- aminobutyric acid receptor binding in rat brain. Br. J. Pharmac., 79, 403 - 410.

Ticku, M.K. (1980): Ethanol and the benzodiazepine-GABA receptor-ionophore complex. Experientia, 45, 413-418.

Thomson,A.D, Ryle, P. R, Shaw, G. K. (1983): Ethanol, thiamine and brain damage. Alcohol and Alcoholism, 18/1, 27-43.

Tiernan, J.M., Ward, L.C. (1986): Acute effects of ethanol on protein synthesis in the rat. Alcohol and alcoholism, 21, No. 2., 171-179.

Veitch, R.L., Lumeng, L., Li, T.K. (1976): Vitamin B6 metabolism in chronic alcohol abuse. Gastroenterology, 70 / 2, 293.

Walsh, M.J. (1973): Biogenesis of biologically active alkaloids from amines by alcohol and acetaldehyde. Ann. N.Y.Acad. Sci., 215, 98-110.

Walter, R.M., Gold, E.M., Michas, C.A., Ensinck, J.W (1980): Portal and peripheral vein concentrations of insulin and glucagon after arginine infusion in morbidly obese subjects. Metabolism, 29, 11, 1037-1040.

West, J.R., Goodlett, C.R. (1990): Teratogenic effects of alcohol on brain development. Ann. Med., 22, 319-325.

Zimatkin, S.M., Rout, U.K., Koivusalo, M., Buhler, R., Lindros, K.o. (1992): Regional distribution of low-K(m) mitochondrial aldehyde dehydrogenase in the rat central nervous system. Alcohol Clin. Exp. Res., 16/6, 1162 - 1167.

Progress in HPLC-HPCE, Vol. 7, pp. 223-244
G. Ali Qureshi *et al.* (Eds)

AMPHETAMINE NEUROTOXICITY: ROLES FOR DOPAMINE, GLUTAMATE AND OXIDATIVE STRESS

Bryan K. Yamamoto, Gary A. Gudelsky[1], and Stacy E. Stephans*

Program in Basic and Clinical Neuroscience,Department of Psychiatry,Case Western Reserve University School of Medicine,Cleveland, Ohio 44106,U.S.A.,[1]College of Pharmacy

University of Cincinnati Medical Center,Cincinnati, Ohio 45267, U.S.A.

Introduction

Methamphetamine (METH) and its derivative 3,4-methylenedioxymethamphetamine (MDMA) are sympathomimetic amines. METH has potent effects on the CNS resulting in psychomotor activation, mood elevation, anorexia, increased mental alertness, enhanced physical endurance, and hyperthermia. The mood-elevating and positive reinforcing effects most likely contribute to the high abuse liability of METH. The abuse of METH by humans that continuously administer high doses results in a behavioral pattern resembling paranoid schizophrenia that is characterized by symptoms such as paranoid delusions, aggressive behavior, hallucinations and disordered thought processes (Ellinwood, 1967). MDMA has become increasingly popular in Europe and North America among young people over the last decade because of its mild stimulant properties and euphoria-inducing effects. (Steele et al, 1994; Cuomo et al., 1994; McKenna and Peroutka, 1990; Green et al., 1995). The ability of the drug to enhance oneís sensitivity to emotions and ability to interact with others also has contributed to the abuse of this psychostimulant. MDMA also can produce acute toxic reactions including malignant hyperthermia, convulsions, hepatitis and death (Randall, 1992, Green et al., 1995; Dykhuizen et al., 1995). These behavioral side effects and abuse potential have prompted a concerted effort to determine the long-lasting effects of METH and MDMA on the brain.

Characteristics of Amphetamine ToxicityMethamphetamine

High doses of METH are neurotoxic to rodents and subhuman primates. This neurotoxicity is manifested as a loss of dopamine (DA) terminals and DA uptake sites (Wagner et al., 1980), decreased tyrosine hydroxylase-immunoreactive fibers (Ryan et al., 1988; Pu et al., 1994), decreases in tyrosine hydroxylase and tryptophan hydroxylase activity (Hotchkiss and Gibb, 1980), an increase in fluorescent swollen axons (Ellison, et al, 1978), and a depletion of DA and serotonin tissue concentration (Ricaurte et al., 1980; 1982). Other evidence of neurotoxicity comes from studies employing the Fink-Heimer silver degeneration method in which degenerating fibers have been observed within the striatum following METH exposure (Ricaurte et al., 1982; 1984). Furthermore, these toxic effects have been reported to endure for months (Bittner et al., 1981; Seiden et al., 1975/76).

The neurotoxic effects of METH also exhibit brain region specificity. Striatal DA neurons are affected most severely by METH while the DA terminals in nucleus accumbens, olfactory area, and frontal cortex are affected only minimally (Ricaurte et al., 1980). The substantia nigra and ventral tegmental area, the regions containing the cell bodies of these DA projection neurons are relatively spared from METH-induced damage although there is evidence of a loss of TH immunoreactivity in the substantia nigra of mice exposed to high doses of METH (Trulson et al, 1985 ; Sonsalla et al., 1996). The brain regions that are the most prone to METH-induced serotonergic toxicity are the amygdala, frontal cortex and striatum. The hypothalamus is resistant to both tissue DA and serotonin depletions (Ricaurte et al., 1982). Moreover, depletions of DA produced by METH also exhibit subregional selectivity such that the ventral and medial areas of the striatum appear to be the most vulnerable to the METH-induced DA depletions (Eisch et al., 1992) . Despite the fact that METH is neurotoxic to both DA and 5HT neurons in most brain regions, it does not appear to damage norepinephrine (Wagner et al., 1980), acetylcholine (Hotchkiss et al., 1979), or GABA (Hotchkiss et al., 1979) systems. It is not known why METH uniquely affects particular neuron types or what might dictate the brain region specific response to the toxic effects of METH.

MDMA

It is well documented that the single or repeated administration of MDMA and structurally related compounds produces evidence of selective neurotoxicity to 5-HT axon terminals. In this regard, treatment of rats or nonhuman primates with MDMA has been shown to result in a decrease in a) the concentration of 5-HT in many forebrain regions (Ricaurte et al., 1985, 1988; Stone et al, 1986; Mokler et al., 1987; Schmidt, 1987) b) the activity of tryptophan hydroxylase (Stone et al., 1986), c) the number of 5-HT uptake sites labeled by [3H]-paroxetine (Battaglia et al., 1987) and d) the density of fine axon terminals of 5-HT neurons (O'Hearn et al., 1988).

MDMA has a biphasic effect on brain 5-HT concentrations; an early, reversible phase of 5-HT depletion occurs within 3-6 hours after its administration, after which 5-HT concentrations return to normal values (Schmidt, 1987; Gibb et al., 1990). A delayed and long-lasting depletion of brain 5-HT occur 2-3 days after drug treatment. There appears to be only partial recovery of the long-lasting depletion of brain 5-HT; 5-HT concentrations and uptake sites remain reduced in some brain regions for up to one year, whereas there is some recovery after 4 months in other brain regions (e.g., striatum, hippocampus) (Scanzello et al., 1993; Lew et al., 1996; Sabol et al., 1996).

The toxicity of MDMA differs from that produced by METH in several aspects. First, whereas METH toxicity in the rat affects both 5-HT and DA terminals, MDMA produces toxicity that is predominately selective for 5-HT terminals. However, it is noteworthy that MDMA produces depletions of DA in the mouse (Logan et al., 1988; Miller and OíCallaghan, 1994). Secondly, the neurotoxicity produced by MDMA is much less brain region specific than that produced by METH. Long-term depletions of 5-HT have been documented in most brain regions containing 5-HT axon terminals (Battaglia et al, 1991; Sabol et al., 1996).

Mechanisms of Methamphetamine and MDMA-induced neurotoxicity

Role for Dopamine

There is considerable evidence in support of the hypothesis that long-term neurotoxicity produced by METH and MDMA is due, in part, to the acute and sustained release of DA. Inhibition of DA transmission through the inhibition of tyrosine hydroxylase (Schmidt et al., 1985), the blockade of transporter mediated DA release with uptake blockers (Marek et al., 1990; Pu et al., 1994; Schmidt et al., 1985) and DA antagonists (Buening and Gibb, 1974; Hotchkiss and BIll, 1980; Sonsalla et al., 1986) all attenuate the long-term depletion of DA produced by METH. Disruption of dopaminergic function by inhibition of synthesis, storage or by lesion also attenuates the MDMA-induced depletion of brain 5-HT (Stone et al., 1988; Schmidt et al., 1990; Brodkin et al., 1993). In addition, restoration of dopaminergic function with the administration of L-Dopa results in reinstitution of MDMA-induced 5-HT neurotoxicity in rats treated with a 5-HT antagonist (Schmidt et al., 1991). Furthermore, a positive correlation exists between the magnitude of the acute release of DA and the extent of long-term 5-HT depletion induced by MDMA and structurally related compounds (Nash and Nichols, 1991).

Role for Glutamate

Despite the well-documented evidence that DA may play a role in mediating METH toxicity, much of the evidence is indirect and is based on pharmacological manipulations of the DA system and the subsequent observed effects on the long-term METH-induced DA depletions. However, many of these manipulations of DA transmission also modify METH-induced changes in the extracellular concentrations of glutamate (Nash and Yamamoto, 1992; 1993, Stephans and Yamamoto, 1994; 1996a; 1996b).

Glutamate and other excitatory amino acids (EAAs) have been linked to a number of neurodegenerative disorders including Huntington's disease, brain hypoxia/ischemia, and epilepsy (reviewed by Lipton and Rosenberg, 1994; Olney, 1990). Despite the wealth of studies on EAAs and neurotoxicity, little is known with regard to their potential role in psychostimulant-induced neurodegeneration. Sonsalla et al., (1989) were the first to implicate EAAs in METH-induced neurotoxicity by demonstrating that a N-methyl-d-aspartate (NMDA) antagonist, MK801, blocks the decreases in tyrosine hydroxylase activity and DA tissue content after METH. Others have since extended their original findings by demonstrating that both noncompetitive and competitive NMDA antagonists also prevent the neurotoxicity associated with METH (Baldwin et al, 1993; Weihmuller et al., 1992; Sonsalla et al., 1991). Similarly, Fuller et al.(1992) reported that NMDA antagonists also block neurotoxicity toward striatal DA neurons produced by d-amphetamine administered to iprindole-treated rats. Our laboratory was the first to demonstrate that METH itself or d-amphetamine administered to iprindole-treated rats, increases the extracellular concentration of striatal glutamate measured in vivo (Nash and Yamamoto, 1992; 1993). These results have been confirmed subsequently by others (Abekawa et al., 1994; Bowyer, et al., 1993; Mora and Porras, 1993). Furthermore, the increase in glutamate along with the subsequent long-term depletion of DA tissue content, were blocked by agents that decrease DAergic function (e.g. "-methyl-p-tyrosine and haloperidol) (Nash and Yamamoto, 1992; Stephans and Yamamoto, 1994). It appears that the increase in glutamate occurs only in brain regions that are susceptible to the long-term DA depletions produced by METH since there is no significant increase in glutamate efflux in brain regions such as the medial prefrontal cortex (Stephans and Yamamoto, 1996) and nucleus accumbens (Abekawa et al., 1994) that are relatively resistant to the DA-depleting effects of METH.

Although METH increases the extracellular concentrations of glutamate, to our knowledge, there is no direct "glutamate-releasing" property of METH. Therefore, METH may increase glutamate through a polysynaptic pathway. Glutamate is a primary neurotransmitter of intercortical projections and the widespread cortical efferent projections to the striatum (Hassler et al., 1982). There is evidence that NMDA stimulates DA release via NMDA receptors located on striatal DA terminals (Krebs et al., 1991). It is possible that elevated glutamate may stimulate NMDA heteroreceptors on DA terminals and augment the METH-induced efflux of DA to concentrations that are toxic to nerve terminals. Alternatively, it is possible that a toxic regimen of METH indirectly increases corticostriatal glutamate transmission by producing sustained increases in DA release, stimulating striatal efferent activity through enhanced corticostriatal glutamate release, and thereby increasing glutamate to concentrations that are excitotoxic to DA terminals. In fact, striatal cell body lesions with the excitotoxin, quinolinic acid, decrease METH-induced DA depletions in the striatum (O'Dell et al., 1994). This is suggestive of the possibility that the local interneuronal striatal circuitry or postsynaptic elements on striatal efferents are critical to METH-induced DA terminal injury.

The two main efferent projections of the striatum are the pallidum and the substantia nigra pars reticulata (Alexander and Crutcher, 1990). Both GABAergic projections eventually converge in the ventral thalamic nuclei (Albin et al., 1989; Graybiel, 1990) and then diverge to cortical areas before returning to the striatum. This thalamocortical-corticostriatal path is a positive feedback loop that utilizes glutamate as a transmitter (Araki and Endo, 1976). Along these lines, we found that a neurotoxic regimen of METH increases GABA efflux in the substantia nigra, decreases extracellular GABA in the ventral thalamus, and increases extracellular glutamate in the motor cortex (Table 1).

TABLE 1.

METHAMPHETAMINE AND MICRODIALYSIS OF GABA AND GLUTAMATE

	Baseline	Post-Methamphetamine# 1 hour	6 hours
GABA (pg/20 :l)			
Substantia Nigra	20 + 2	33 + 8*	28 + 3*
Ventral Thalamus	28 + 5	19 + 2*	13 + 7*
Glutamate (ng/20 :l)			
Motor Cortex	1.4 + 0.2	1.3 + 0.2	3.0 + 0.9*

#Methamphetamine (10 mg/kg, ip) was administered once every 2 hrs over an 8 hr period. The times of 1 hr and 6 hrs indicate the periods after the first administration.
*Indicates significantly different from baseline ($p<0.05$)

Although the evidence in support of a role for glutamate in METH-induced toxicity to DA terminals is substantial, the role of excitotoxic processes in MDMA-induced depletion of brain 5-HT is controversial. Although MDMA, in contrast to METH, does not produce an increase in the extracellular concentration of glutamate (Nash and Yamamoto, 1992), blockade of NMDA receptors by MK-801 or dextrorphan has been reported to prevent the MDMA-induced depletion of brain 5-HT (Laverty and Logan, 1990; Finnegan et al., 1990; Farfel et al., 1992; Colado and Green, 1993). However, the neuroprotective effect of MK-801 is not observed in all brain regions (Colado and Green, 1993), and the pharmacological specificity of dextrorphan has been questioned (Henderson and Fuller, 1992). Moreover, there is evidence that the neuroprotective effect of MK-801 against MDMA toxicity is due to drug-induced hypothermia (Farfel and Seiden, 1995). Thus, the role of glutamate and NMDA receptors in MDMA toxicity remains unresolved.

Role for 5-HT

There is evidence that the acute serotonergic effects of MDMA also contribute, in part, through the modulation of DA release, to the long-term 5-HT toxicity. Blockade of 5-HT2 receptors has been reported to attenuate the long-term depletion of 5-HT produced by MDMA (Nash et al., 1990; Schmidt et al., 1991). Conversely, it has been reported that 5-HT2 agonists enhance MDMA induced 5-HT toxicity (Gudelsky et al., 1994). These effects may be due, in part, to the facilitation and suppression by 5-HT2 agonists and 5-HT2 antagonists, respectively, of the MDMA-induced release of DA (Gudelsky et al., 1994; Yamamoto et al., 1995; Schmidt et al., 1992; 1994). Thus, the acute release of 5-HT by MDMA appears to facilitate stimulated DA release, thereby contributing to the DA-dependent processes that ultimately result in 5-HT toxicity.

Although excessive extracellular DA appears to be necessary for MDMA-induced 5-HT toxicity, it alone, or even in combination with 5-HT2 receptor activation, is not sufficient to produce selective toxicity to 5-HT axon terminals (Huang and Nichols, 1993). It is proposed that the acute activation of the 5-HT transporter by MDMA in the presence of increased extracellular DA is required for MDMA-induced neurotoxicity (Henderson et al., 1993; Johnson and Nichols, 1991). This contention is supported by the findings that 1) inactivation of the 5-HT transporter prevents MDMA-induced 5-HT release and long-term 5-HT depletion (Gudelsky and Nash, 1996; Schmidt, 1987; Malberg et al., 1996) and 2) 5-HT toxicity can be produced by the co-administration of agents that enhance, selectively, transporter-mediated DA and 5-HT release (Johnson and Nichols, 1991). Thus, it can be envisioned that activation of the 5-HT transporter by MDMA may render 5-HT terminals vulnerable to DA-mediated toxicity.

Oxidative and Energetic Stress

Oxidative stress can be defined as the cytotoxic consequences of reactive oxygen species (e.g., ï O2-, ïOH) generated as byproducts of oxidative metabolism. The targets of free-radical mediated damage are protein, lipids, and DNA. The protein, lipids and DNA of the brain are particularly vulnerable to oxidative stress for several reasons: (1) the brain receives a high percentage of the total oxygen in the body, (2) the brain is relatively deficient in protective mechanisms in that it has very little catalase; and the glutathione system in the brain, although the most important defense mechanism of the brain against oxygen free radical damage, is significantly less active and easily saturated compared to other organ systems, (3) Iron, which catalyzes the conversion of H2O2, to ïOH via the Fenton reaction, is present in high concentrations in specific brain regions, and (4) the brain contains a high concentration of polyunsaturated lipids which can readily react with free radicals and undergo peroxidation. .

There is some evidence that free radicals may mediate toxicity to amphetamine analogues. The administration of antioxidants (e.g., ascorbic acid) acting as free radical scavengers attenuate the depletion of striatal DA and 5HT in response to METH and MDMA (DeVito and Wagner, 1989; Gudelsky, 1996). In addition, the spin trap phenyl-t-butylnitrone, which is used to trap free radicals, has been shown to prevent the neurotoxicity produced by METH (Cappon et al., 1995; Schmidt and Taylor, 1995) and MDMA (Colado and Green, 1995). Furthermore, sulfhydryl reducing conditions reverse the decrease in tryptophan hydroxylase activity produced by MDMA (Stone et al., 1988). Cadet et al. (1994) showed that transgenic mice expressing the human CuZn-superoxide dismutase (SOD) gene were relatively resistant to the DA-depleting effects of METH. Since SOD is responsible for the degradation of ï O2-, this is evidence supportive of the role for the formation of reactive oxygen species in METH neurotoxicity. Along similar lines, METH increases the tissue content of the ïOH-salicylate adduct, 2,3-dihydroxybenzoic acid measured ex vivo (Giovanni et al., 1992; Fleckenstein et al., 1996) and degenerates neurite outgrowth in DA neuron cultures through a DA-dependent increase in intracellular oxidation as indicated by dichlorohydrofluorescin fluorescence (Cubells et al., 1994). MDMA also as been reported to increase the concentration of 2,3 DHBA in dialysis samples of the striatum (Gudelsky and Yamamoto, 1994).

It can be hypothesized that the mechanism of psychostimulant-induced oxidative stress involves processes initiated by the acute and sustained increase in the extracellular concentration of DA produced by METH and MDMA. Indeed, DA has been shown to be neurotoxic to cells when applied either in vitro or in vivo (Filloux and Townsend, 1993; Michel and Hefti, 1990; Rosenberg, 1988). Conversely, DA depletions attenuate the neurotoxic effects of EAAs (Chapman et al., 1989). The mechanism involved in DA-induced cytotoxicity may involve the generation of free radical species and quinones. Both the enzymatic and nonenzymatic oxidation (autoxidation) of DA result in the formation of ï O2 - and H2O2. DA is enzymatically metabolized to form H2O2 which is then catalyzed non-enzymatically by iron to form ïOH (Olanow, 1992). In addition, DA autoxidation also results in the formation of cytotoxic quinones, which presumably attack thiol containing proteins and exist as 5-cysteinyl adducts of DA (Fornstedt et al., 1989). Consistent with this contention are the findings that the intrastriatal injection of DA results in the formation of protein bound, cysteinyl adducts of DA and DA terminal neurotoxicity; both of these effects are prevented by the co-administration of antioxidants (Hastings et al., 1996). Furthermore, there is evidence that free radicals, and possibly DA quinones, can modify DA transporter function (Berman et al., 1996). Finally, MDMA has been shown to produce cellular changes, e.g., formation of malonyldialdehyde-related substances, that are consistent with free radical-induced lipid peroxidation (Sprague and Nichols, 1995). Therefore, the massive increase in the extracellular concentrations of DA, such as that produced by METH and MDMA, could result in oxidative stress and damage due to the production of hydroxyl free radicals.

Glutamate and glutamate receptor activation can also lead to the production of free radicals and may involve a number of different pathways that cause oxidative stress and the generation of oxygen radicals. Glutamate-induced increases in intracellular calcium can activate calcium-dependent peptidases such as calpain I (Siman et al., 1989). This enzyme can catalyze the conversion of xanthine dehydrogenase to xanthine oxidase which promotes the catabolism of the purine bases xanthine and hypoxanthine to uric acid yielding oxygen free radicals in the process (Dykens et al., 1987). Subsequent to NMDA receptor stimulation, phospholipase A2 activates the arachidonic acid cascade. The conversion of arachidonic acid by lipoxygenases and cyclo-oxygenases to eicosanoids results in the liberation of damaging oxygen free radicals (Dumuis et al., 1988; Lazararewiz et al., 1988). In addition, arachidonic acid and free radical species that are generated enhance glutamate release and inhibit reuptake (Pellegrini-Giampietro et al., 1988; Williams, 1989; Volterra et al.,1994) thus promoting a feed-forward cycle. An additional mechanism by which free radicals are generated by glutamate involves iron and the Fenton reaction. The acidic environment caused by increased synaptic glutamate leads to liberation of cellular stores of reduced iron which then promote the decomposition of hydrogen peroxide to yield hydroxyl radicals. NMDA receptor stimulation also has been linked to the activation of nitric oxide synthase which is involved in the synthesis of nitric oxide from arginine (Garthwaite et al, 1988; Dawson et al., 1991). Once nitric oxide accumulates, it combines with superoxide radicals and forms a peroxynitrite anion which decomposes eventually into a hydroxyl radical (Beckman et al., 1990). Along these lines, it has been reported that inhibition of nitric oxide synthase, protects against METH -induced neurotoxicity (Itzhak and Ali, 1996; Abekawa et al., 1996).

An impairment in energy metabolism also may contribute to METH-induced neurotoxicity. Alterations in brain energy utilization by low doses of amphetamine and related analogues were reported originally in the early 1960's. The results of these early experiments show that low doses of amphetamine and METH increase metabolism in cerebral cortex or whole brain as measured by formation of lactate and changes in high energy substrates such as ATP and phosphocreatine (Lewis and Van Petten, 1962; Nahorski and Rogers, 1973; Sylvia et al., 1977). However, it was difficult to correlate specifically the changes in brain metabolism with the stimulant effects of these drugs as the fluctuation in neuronal energy metabolites precedes the characteristic stimulant-induced behaviors (Lewis and Van Petten, 1962). The pattern of glucose utilization following systemic administration of amphetamine or METH reveals that brain energy utilization increases in a region specific manner. It appears overall, that the increase is greatest in those brain regions that are most susceptible to the toxic effects of METH. Therefore, the brain region selective effect of METH-induced energy consumption may be related directly to the mechanism by which METH produces neuronal damage. METH produces a massive increase in DA efflux via reversal of the transporter protein (Fisher and Cho, 1979; Liang and Rutledge, 1982). This mechanism is highly dependent upon an outward sodium gradient (Raiteri et al, 1979; Liang and Rutledge, 1982). Therefore, energy utilization may increase in response to the disrupted ion gradient that facilitates the reversal of the DA transporter protein. Since it has been estimated that under steady state conditions neurons expend more than half of their ATP supply solely to maintain ion homeostasis (Seisjo, 1978), the effects of METH on ionic gradients would increase substantially the ATP requirement of the cell and thereby affect tissue ATP stores. In this manner, METH would enhance indirectly energy consumption through the activation of sodium/potassium ATPase protein (Na+/K+ ATPase) which restores the inward sodium gradient to maintain ion homeostasis. However, there also is evidence that DA reuptake via the DA transporter protein directly involves ATP hydrolysis (Cao et al., 1990). It is possible that ATP also is consumed by METH-induced reverse transport of DA. Regardless, the massive increase in extracellular DA that results after administration of high doses of METH and transporter-mediated uptake of METH itself into the neuron eventually will activate the uptake protein in the inward direction. This direct interaction with the transporter protein might enhance ATP consumption to a point where the demand for ATP exceeds the available stores.

Compounds which block energy production can cause neuronal damage. L-methyl-4-phenyl-1,2,3,6-tetrahydropyridine (MPTP) results in a decrease in striatal ATP and damage to DA neurons (Chan et al., 1991). This effect is thought to be mediated by the inactivation of NADH dehydrogenase at complex I of the mitochondria by MPP+. Malonate, a reversible inhibitor of succinate dehydrogenase in complex II of the mitochondria, causes a depletion of ATP and an accumulation of lactate (Beal et al., 1993a; 1993b). Moreover, malonate also potentiates the depletion of DA produced by either MPTP or methamphetamine suggestive of the hypothesis that perturbations in energy metabolism may contribute to the neuronal damage produced by MPP+ and methamphetamine (Albers et al., 1996).

To date, there have not been any investigations of a direct inhibition of metabolic enzymes by METH as occurs with the DA neurotoxin MPP+. However, a recent report shows that 6-OHDA inhibits select complexes within the electron transport chain (Glinka and Youdim, 1995; Glinka et al., 1996). The putative increase in 6-OHDA by METH (Seiden and Vosmer, 1984) may therefore create an energy impairment. This effect of METH on energy stores could be due to a prolonged and sustained increase in brain energy utilization and/or a direct metabolic impairment caused by 6-OHDA. The combined effects of metabolic compromise and oxidative stress induced by DA may enhance the excitotoxic action of glutamate (Hoyt et al., 1995; Bridges et al., 1991). Cortical cultures exposed to DA are more susceptible to glutamate-induced cell death which can be blocked by the antioxidant n-acetylcysteine (Hoyt et al., 1995). Glutamate can form reactive oxygen species within the mitochondria (Reynolds and Hastings, 1995, Dugan et al., 1995) which are then followed by changes in mitochondrial membrane potential. Along parallel lines, DA has been shown to inhibit mitochondrial respiration (Ben-Shachar et al., 1995). Therefore, DA and glutamate may have additive or synergistic effects on the inhibition of mitochondrial function to produce eventual nerve terminal degeneration.

Considerable attention has been given to the potential role of environmental and core body temperature in METH- and MDMA-induced neurotoxicity. For example, it has been shown that cold ambient temperature attenuates the neurotoxicity produced by either METH or MDMA (Ali et al., 1994; Broening et al., 1995; Miller and OíCallaghan, 1994). In addition, Broening et al. (1995) have demonstrated that the magnitude of thermal response to MDMA is correlated with the magnitude of MDMA toxicity. Importantly, many drugs (e.g., dizolcipine, ketanserin, haloperidol) that provide neuroprotection against METH or MDMA toxicity also lower core body temperature and/or prevent stimulant-induced hyperthermia (Farfel and Seiden, 1995; Malberg et al., 1996; Miller and OíCallaghan, 1994; Albers and Sonsalla, 1995; Bowyer et al., 1994; Hewitt and Green, 1994). Thus, drugs that influence DA, glutamate, and 5-HT2 receptor mechanisms may alter stimulant-induced neurotoxicity through modulation of body temperature, as well as through modulation of specific neurotransmitter regulation of cellular neurotoxic processes (Albers and Sonsalla, 1995). Although hyperthermia may contribute to the extent of amphetamine-induced neurotoxicity, hyperthermia does not account for the main neurotoxic processes induced by these drugs. MDMA has been shown to produce 5-HT toxicity in the absence of drug-induced hyperthermia (Farfel and Seiden, 1995) and even under conditions in which MDMA evokes a hypothermic response (Broening et al., 1995). In addition, certain drugs (e.g., reserpine) attenuate METH-induced hyperthermia without providing protection against METH-induced toxicity (Albers and Sonsalla,

1995). It can be envisioned that the neuroprotective effects of hypothermia are related to the proposed cellular mechanisms of METH- and MDMA-induced neurotoxicity. The rate of uptake of METH or MDMA into DA and 5-HT neurons and/or subsequent activation of transporter-mediated DA and 5-HT release could be diminished at a lower ambient and core body temperature. Hypothermia also may retard the free radical-initiated reactions that ultimately result in neurotoxicity.

Models of Amphetamine Neurotoxicity and Conclusions

It has been hypothesized that neither DA nor glutamate alone may be sufficient to account for the neurotoxicity of METH (Nash and Yamamoto, 1992; Stephans and Yamamoto, 1994). However, when DA and glutamate are sequentially increased, they may have an additive or synergistic effect to destroy a metabolically compromised DA axon terminal that has been continually stimulated by METH (Nash and Yamamoto, 1992; 1993; Stephans and Yamamoto, 1994).

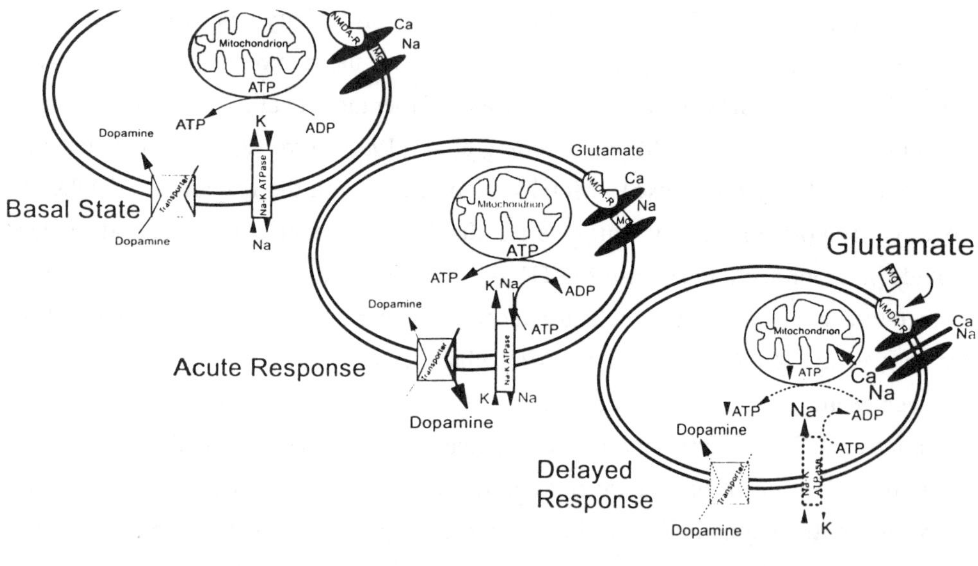

Figure 1. illustrates that the acute response to METH is evidenced by an efflux of DA and the utilization of ATP by Na+/K+ ATPase.

During this time, extracellular glutamate concentrations are normal and the magnesium block of the NMDA channel remains intact. In the delayed response to METH, continued efflux of dopamine and activation of Na+/K+ ATPase are followed by a compromised energy state. This compromised energy state produced by METH results from the continued activation of Na+/K+ ATPase, an eventual ATP deficit, and the subsequent inactivation of Na+/K+ ATPase. These events, in combination with high extracellular concentrations of glutamate, lead to an increase in intracellular sodium, the depolarization of the neuron, and the removal of the magnesium block in the NMDA ion channel. The influx of calcium will activate calcium-dependent proteins which damage the DA neuron or enhance the vulnerability of the DA neuron to the long-term neurotoxic effects of glutamate or oxygen-derived free radicals.

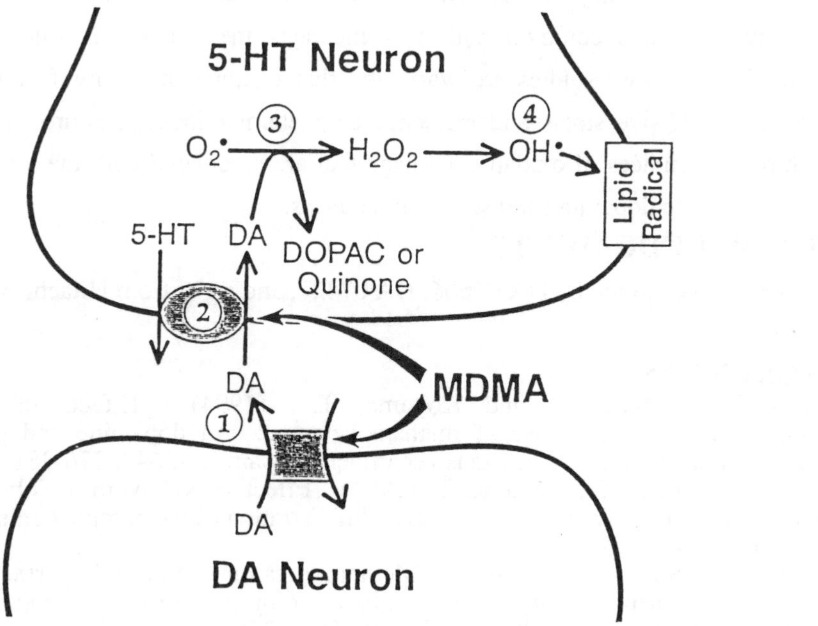

A model depicting a potential mechanism of MDMA toxicity is presented in Figure 2.

The somewhat generalized processes and reactions that broadly define oxidative stress may appear to fall short of explaining the extent of toxicity exhibited by amphetamine analogues and the pattern of selective toxicity to 5-HT terminals produced by MDMA. It is proposed that it is the concomitant increase in DA release and activation of the 5-HT transporter produced by MDMA that are critical for imparting these toxic processes on 5-HT terminals.

It is hypothesized that MDMA-induced 5-HT neurotoxicity is due, in part, to (1) an increased release of DA and activation of the 5-HT transporter, (2) the enzymatic and/or autoxidation of DA within the 5-HT terminal and the subsequent formation of hydroxyl free radicals therein and (3) the cellular damage, e.g., lipid peroxidation, protein oxidation, associated with their formation.

In conclusion, it is clear that multiple factors contribute to amphetamine toxicity and that either one alone may be necessary but not sufficient. Nevertheless, oxidative stress may be a final common path that mediates the damage produced by the amphetamines. Future studies are warranted that examine not only factors which contribute to oxidative stress and influence amphetamine-induced neurotoxicity, but more important, indices of oxidative damage that are associated with the toxic effects of amphetamine to dopamine and serotonin neurons.

ACKNOWLEDGMENTS

This work was supported by DA 07606, DA 07427, and a gift from Hitachi America

REFERENCES

Abekawa, T., Ohmori, T., and Koyama, T. (1994). Effects of repeated administration of high dose of methamphetamine on dopamine and glutamate release in rat striatum and nucleus accumbens. Brain Res. 643, 276-281.

Abekawa, T., Ohmori, T., Koyama, T. (1994). Effect of NO synthase inhibition on behavioral changes induced by a single administration of methamphetamine. Brain Res., 666, 147-150.

Albers, D.S. and Sonsalla, P.K. (1995). Methamphetamine-induced hyperthermia and dopaminergic neurotoxicity in mice: pharmacological profile of protective and nonprotective agents. J. Pharmacol. Exp. Ther. 275, 1104-1114.

Albers, D. S., Zeevalk, G. D., and Sonsalla, P. K. (1996). Damage to dopaminergic nerve terminals in mice by combined treatment of intrastriatal malonate with systemic methamphetamine or MPTP. Brain Res. 718, 217-220.

Albin, R. L., Young, A. B. and Penney, J. B. (1989). The functional anatomy of basal ganglia disorders. Trends in Neuroscience 12, 366-375.

Alexander, G. E. and Crutcher, M. D. (1990). Functional architecture of basal ganglia circuits: Neural substrates of parallel processing. Trends in Neuroscience 13, 266-271.

Ali, S. Y., Newport, G. D., Holson, R., Slikker, W. and Bowyer, J. F. (1994). Low environmental temperatures or pharmacological agents that produce hypothermia decrease methamphetamine neurotoxicity in mice. Brain Res. 658, 33-38.

Araki, T. and Endo, K. (1976). Short latency EPSPs of pyramidal tract cells evoked by stimulation of the centrum medianum-parafascicular complex and the nucleus ventralis anterior of the thalamus. Brain Res. 113, 405-410.

Baldwin, H. A., Colado, M. I., Murray, T. K., De Souza, R., J., and Green, A. R. (1993) Striatal dopamine release in vivo following neurotoxic doses of methamphetamine and effect of the neuroprotective drugs, chlormethiazole and dizocilpine. Br. J. Pharm. 108, 590-596.

Battaglia, G., Yeh, S.Y., O'Hearn, E., Molliver, M.E., Kuhar, M.J., De Souza, E.B. (1987). 3,4-Methylenedioxymethamphetamine and 3,4-methylenedioxyamphetamine destroy serotonin terminals in rat brain: Quantification of neurodegeneration by measurement of [3H]-paroxetine-labeled serotonin uptake sites. J. Pharmacol. Exp. Ther. 242, 911.

Battaglia, G, Sharkey, J, Kuhar, M and De Souza, E (1991). Neuroanatomic specificity and time course of alterations in rat brain serotonergic pathways induced by MDMA (3,4-methylenedioxymethamphetamine): assessment using quantitative autoradiography. Synapse 8, 249-260.

Beal, M. F., Brouillet, E., Jenkins, B. G., Ferrante, R. J., Kowall, N. W., Miller, J. M., Storey, E., Srivastava, R., Rosen, B. R. and Hyman, B. T. (1993a). Neurochemical and histologic characterization of striatal excitotoxic lesions produced by the mitochondrial toxin 3-nitropropironic acid. J. Neurosci. 13, 4181-4192.

Beal, M. F., Brouillet, E., Jenkins, B. G., Henshaw, R., Rosen, B. and Hyman, B. T. (1993b). Age-depenent striatal excitotoxic lesions produced by the endogenous mitochondrial inhibitor malonate. J. Neurochem. 61: 1147-1150.

Beckman, J. S., Beckman, T. W., Chen, J., Marshall, P. A. and Freeman, B. A. (1990). Apparent hydroxyl radical production by peroxynitrite: Implications for endothelial injury form nitric oxide and superoxide. Proc. Natl. Acad. Sci. USA, 87, 1620-1624.

Ben-Shachar, D., Zuk, R., Glinka, Y. (1995). Dopamine neurotoxicity: inhibition of mitochondrial respiration. J. Neurochem. 64, 718-723.

Berman, S. B., Zigmond, M. J. and Hastings, T. G. (1996). Modification of dopamine transporter function: effect of reactive oxygen species and dopamine. J. Neurochem. 67, 593-600.

Bittner, S. E., Wagner, G. C., Aigner, T. G. and Seiden, L. S. (1981). Effects of a high-dose treatment of methamphetamine on caudate dopamine and anorexia in rats. Pharmacology Biochemistry & Behavior 14, 481-486.

Bowyer, J.F., Davies, D., Schmued, L., Broening, H., Newport, G., Slikker, W. and Holson, R. (1994). Further studies on the role of hyperthermia in methamphetamine neurotoxicity. J. Pharmacol. Exp. Ther. 268, 1571-1580.

Bowyer, J. F., Gough, B., Slikker, W., Lipe, G. W., Newport, G. D., and Holson, R. R. (1993). Effects of cold environment or age on methamphetamine-induced dopamine release in the caudate putamen of female rats. Pharm. Biochem. Behav. 44, 87-98.

Bridges, R. J., Koh, J-Y, Hatalski, C. G., Cotman, C. W. (1991). Increased excitotoxic vulnerability of cortical cultures with reduced levels of glutathione. Eur. J. Pharm. 42, 199-200.

Brodkin, J., Malyala, A. and Nash, J. F. (1993). Effect of acute monoamine depletion on 3,4-methylene-dioxymethamphetamine-induced neurotoxicity, Pharmacol. Biochem. & Behav. 45, 647.

Broening, H., Bowyer, J. F. and Slikker, W. (1995). Age-dependent sensitivity of rats to the long-term effects of the serotonergic neurotoxicant (±)-3,4-methylenedioxymethamphetamine (MDMA) correlates with the magnitude of the MDMA-induced thermal response. J. Pharmacol. Exp. Ther. 275, 325-333.

Cadet, J. L., Sheng, P., Ali, S., Rothman, R., Carlson, E. and Epstein, C. (1994). Attenuation of methamphetamine-induced neurotoxicity in copper/zinc superoxide dismutase transgenic mice. J. Neurochem., 62, 380-383.

Cao, C. J., Eldefrawi, A. T. and Eldefrawi, M. E. (1990). ATP-regulated neuronal catecholamine uptake: A new mechanism. Life Sci. 47, 655-667.

Chan, P., Di Monte, D. A., Luo, J-J., DeLanney, L. E., Irwin, I. and Langston, J. W. (1994). Rapid ATP loss caused by methamphetamine in the mouse striatum: Relationship between energy impairment and dopaminergic neurotoxicity. J. Neurochem. 62, 2484-2487.

Chapman, A.G., Durmuller, N., Lees, G. J., and Meldrum, B. S. (1989). Excitotoxicity of NMDA and kainic acid is modulated by nigrostriatal dopaminergic fibres. Neurosci. Letts. 107, 256-260.

Colado, M. and Green, A. R. (1995). The spin trap agent phenyl-N-tert-butylnitrone prevents ecstasy induced neurodegeneration of 5-hydroxytryptamine neurons. Eur. J. Pharmacol. 280, 343-336.

Colado, M. I, Murray, T. K. and Green, A. R. (1993). 5-HT loss in rat brain following 3,4-methylenedioxymethamphetamine (MDMA), p-chloroamphetmamine and fenfluramine administration and effects of chlormethiazole and dizolcipine. Br. J. Pharmacol. 108, 583-589.

Cubells, J. F., Rayport, S., Rajendran, G. and Sulzer, D. (1994). Methamphetamine neurotoxicity involves vacuolation of endocytic organelles and dopamine-dependent intracellular oxidative stress. J. Neurosci., 14, 2260-227.

Cuomo, M. J., Dyment, P. G. and Gammino, V. M. (1994). Increaseing use of ìEcstasyî (MDMA) and other hallucinogens on a college campus. J. Am. Coll. Health 42, 271-274.

Dawson, B. L., Dawson, T. M., London, E. D., Bredt, D. S., and Snyder, S. H. (1991). Nitric oxide mediates glutamate neurotoxicity in primary cortical culture. Proc. Natl. Acad. Sci. USA, 88, 6368-6371.

Dugan, L., L., Sensi, S. L., Canzoniero, L., M., T., Handran, S. D., Rothman, S. M., Lin, T-S., Goldberg, M. P., Choi, D. W. (1995). Mitochondrial production of reactive oxygen species in cortical neruons following exposure to n-methyl-d-aspartate. J. Neurosci. 15, 6377-6388.

Dumius, A., Sebben, M., Haynes, L., Pin, J.-P. And Bockaert, J. (1988) NMDA receptors activate the arachidonic acid cascade system in striatal neurons. Nature 336, 68-70.

Dykens, J. A., Stern, A. and Trenkner, E. (1987). Mechanism of kainate toxicity to cerebellar neurons in vitro is analogous to reperfusion tissue injury. J. Neurochem. 49, 1222-1228.

Dykhuizen, R. S., Brunt, P. W., Atkinson, P., Simpson, J. G. and Smith, C. C. (1995). Ecstasy induced hepatitis mimicking viral hepatitis. Gut 36, 939-941.

Eisch A. J., Gaffney, M., Weihmuller, F. B., OíDell, S. J. and Marshall, J. F. (1992) Striatal subregions are differentially vulnerable to thenerutoxic effects of methamphetamine. Brain Research 598:321-326.

Ellinwood, E. . (1967). Amphetamine psychosis: I. Description of the individuals and process. J. Nervous and Mental Disease, 144, 273-283.

Ellison, G., Eison, M. S., Huberman, H. S. and Daniel, F. (1978) Long-term changes in dopaminergic innervation of caudate nucleus after continuous amphetamine administration. Science 210, 276-278.

Farfel, G. M., Vosmer, G. and Seiden, L. S. (1992). The N-methyl-D-aspartate antagonist MK 801 protects against serotonin depletions induced by methamphetamine, 3,4-methylendioxymethamphetamine and p-chloroamphetamine. Brain Res. 595, 121-127.

Farfel, G. and Seiden, L. (1995). Role of hypothermia in the mechanism of protection against serotonergic toxicity: I. Experiments using 3,4-

methylenedioxymethamphetamine, dizolcipine, CGS 19755 and NBQX. J. Pharmacol. Exp. Ther. 272, 860-867.

Filloux, F. and Townsend, J. J. (1993). Pre-and postsynaptic neurotoxic effects of dopamine demonstrated by intrastriatal injection. Exp. Neurology, 119, 79-88.

Finnegan, K. T., Skratt, J. J., Irwin, I. and Langston, J. W. (1990). The N-methyl-D-aspartate (NMDA) receptor antagonist, dextrorphan, prevents the neurotoxic effects of 3,4-methylenedioxymethamphetamine (MDMA) in rats. Neurosci. Lett. 105, 300-306.

Fisher, J. F. and Cho, A. K. (1976). Chemical release of dopamine from striatal homogenates: Evidence for an exchange diffusion model. J. Pharm. Exptl Ther. 208, 203-209.

Fleckenstein, A, Gibb, J. W. and Hanson, G. T. (1996) Acute effects of methamphetamine on tryptophan hydroxylase activity and oxygen radical formation in rat brain: role of hyperthermia. Soc. Neurosci. Abstract 22, 2077.

Fuller, R. W., Hemrick-Luecke, S. K. and Ornstein, P. L. (1992). Protection against amphetamine-induced neurotoxicity toward striatal dopamine neurons in rodents by LY274614, an excitatory amino acid antagonist. Neuropharmacology 31, 1027-1032.

Garthwaite, J., Charles, S. L. and Chess-Williams, R. (1988). Endothelium-derived relaxing factor release on activation of NMDA receptors suggests role as intercellular messenger in the brain. Nature 336, 385-388.

Gibb, J., Johnson, M. and Hanson, G. (1990). Neurochemical basis of neurotoxicity. Neurotoxicology 11, 317.

Giovanni, A., Liang, L. P., Hastings, T. G., and Zigmond, M. J. (1995). The estimation of hydroxyl radical content in rat brain using salicylate: impact of methamphetamine. J. Neurochem. 64, 1819-1825.

Glinka, Y. Y. and Youdim, M. B. H. (1995). Inhibition of mitochondrial complexes I and IV by 6-hydroxydopamine. Eur. J. Pharm., 292, 329-332.

Glinka, Y. Y., Tipton, K. F. and Youdim, M. B. H. (1996) Nature of inhibition of mitochondrial respiratory complex I by 6-hydroxydopamine. J. Neurochem. 66, 2004-2010.

Graybiel, A. M. (1990). Neurotransmitters and neuromodulators in the basal ganglia. Trends in Neuroscience 13, 244-254,

Green, A. R., Cross, A. J. and Goodwin, G. M. (1995). Review of the pharamcology and clinical pharmacology of 3,4-methylenedioxymethamphetamine (MDMA or ìEcstasyî). Psychopharmacology 119, 247-260.

Gudelsky, G. A. (1996). Effect of ascorbate and cysteine on the 3,4-methylenedioxymethamphetamine-induced depletion of brain serotonin. J. Neural Transm. 103, 1397-1404.

Gudelsky, G. A. and Nash, J. F. (1996). Carrier-mediated release of serotonin by 3,4-methylenedioxymethamphetamine: implications for serotonin-dopamine interactions. J. Neurochem. 66, 243-249.

Gudelsky, G. A., Yamamoto, B. K. and Nash, J. F. (1994). Potentiation of 3,4-methylenedioxymethamphetamine -induced dopamine release and serotonin neurotoxicity by 5-HT2 agonists. Eur. J. Pharmacol. 264, 325-330.

Gudelsky, G. A. and Yamamoto, B. K. (1994). MDMA increases the extracellular concentration of 2,3-dihydroxybenzoic acid in the striatum: evidence for increased hydroxyl radical formation. Soc. Neurosci. Abstr. 20, 343.

Hassler, R., Haug, P., Nitsch, C., Kim, J. S. and Paik, K. (1982). Effect of motor and premotor cortex ablation on concentrations of amino acids, monoamines, and acetylcholine, and on the ultrastructure in rat striatum: A confirmation of glutamate as the specific cortico-striatal transmitter. J. Neurochem. 38, 1087-1098.

Hastings, T. G., Lewis, D. A. and Zigmond, M. J. (1996). Role of oxidation in the neurotoxic effects of intrastriatal dopamine injections. Proc. Natl. Acad. Sci. 93, 1956-1961.

Henderson, M. G. and Fuller, R. W. (1992). Dextromethorphan antagonizes the actue depletion of brain serotonin by p-choloroamphetamine and H75/12 in rats. Brain Res. 594, 323-326.

Henderson, M. G., Perry, K. and Fuller, R. W. (1993). Possible involvement of dopamine in the long-term depletion by p-chloroamphetamine and B,B-difluoro-p-chloroamphetamine in rats. J. Pharmacol. Exp. Ther. 267, 417.

Hewitt, K. E. and Green, A. R. (1994). Chlormethiazole, dizolcipine and haloperidol prevent the degeneration of serotonergic nerve terminals incduced by adminstration of MDMA (Ecstasy) to rats. Neuropharmacology 33, 1589-1595.

Hotchkiss, A. J. and Gibb, J. W. (1980). Long-term effects of multiple doses of methamphetamine on tryptophan hydroxylase and tyrosine hydroxylase activity in rat brain. J. Pharm. Exp. Ther., 214, 257-262.

Hotchkiss, A.J., Morgan, M.E. and Gibb, J.W. (1979). The long-term effects of multiple doses of methamphetamine on neostriatal tryptophan hydroxylase, tyrosine hydroxylase, choline acetyltransferase and glutamate decarboxylase activities. Life Sci. 25, 1373-1378.

Hoyt, K. R., Hastings, T. G., and Reynolds, I. J. (1995). Dopamine toxicity in cultured neurons: effects on neuronal viability, glutathione, and intracellular free calcium. Neurodegenerative Diseases ë95: Molecular and Cellular Mechanisms, Washington, D.C.

Huang, X. and Nichols, D. E. (1993). 5-HT2 receptor-mediated potentiation of dopamine synthesis and central serotonergic deficits. Eur. J. Pharmacol. 238, 291.

Itzhak, Y. and Ali, S. F. (1996) The neuronal nitric oxide inhibitor, 7-nitroindazole, protects against methamphetamine-induced neurotoxicity in vivo. J. Neurochem. 67, 1770-1773.

Johnson, M. P and Nichols, D. E. (1991). Combined administration of a non-neurotoxic 3,4-methylenedioxymethamphetamine analogue with amphetamine produces serotonin neurotoxicity in rats. Neuropharmacology 30, 819-824.

Krebs, M. O., Desce, J. M., Kemel, M. L., Gauchy, C., Godeheu, G., Cheramy, A. and Glowinski, J. (1991). Glutamatergic control of dopamine release in the rat striatum: Evidence for presynaptic N-methyl-D-aspartate receptors on dopaminergic nerve terminals. J. Neurochem. 56, 81-85.

Lasarewicz, J. W., Wroblewski, J. T., Palmer, M.E. and Costa, E. (1988). Activation of n-methyl-D-aspartate-sensitive glutamate receptors stimulates arachidonic acid release in primary cultures of cerebellar granule cells. Neuropharm. 27, 765-769.

Laverty, R. and Logan, B. J. (199) Protection by MK-801 and other drugs of methylenedioxymethamphetamine (MDMA) neurotoxicity in rats and mice. Eur. J. Pharmacol. 183, 451-452.

Lew, R., Sabol, K., Chou, C., Vosmer, G., Richards, J. and Seiden, L. (1996). Methylenedioxymethamphetamine-induced serotonin deficits are followed by partial recovery over a 52-week period. Part II: radioligand binding and autoradiography studies. J. Pharmacol. Exp. Ther. 276, 855-865.

Lewis, J. J. and Van Petten, G. T. (1962). The effect of amphetamine and related compounds on the concentration of adenind nucleotides, inorganic phosphate and creatine phosphate in the rat brain. J. Pharm. Exptl. Ther. 136, 372-377.

Liang, N. Y. and Rutledge, C. O. (1982). Comparison of the release of [3H]-dopamine from isolated corpus striatum by amphetamine, fenfluramine and unlabelled dopamine. Biochem. Pharm. 31, 983-992.

Lipton, S. A. and Rosenberg, P. A. (1994). Excitatory amino acids as a final common pathway for neurologic disorders. Mechanisms of Disease 330, 613-622.

Logan, B. J., Laverty, R., Sanderson, W. D. and Yee, Y. B. (1988). Differences between rats and mice in MDMA (methylenedioxymethamphetamine) neurotoxicity. Eur. J. Pharmacol. 152, 227-234.

Malberg, J., Sabol, K. and Seiden, L. (1996). Co-administration of MDMA with drugs that protect against MDMA neurotoxicity produces different effects on body temperature in the rat. J. Pharmacol. Exp. Ther. 278, 258-267.

McKenna, D. J. and Peroutka, S. J. (1990). Neurochemistry and neurotoxicity of 3,4-methylenedioxymethamphetamine (MDMA, ìEcstasyî). J. Neurochem. 54, 14-22.

Michel, P.P. and Hefti, F. (1990). Toxicity of 6-hydroxydopamine and dopamine for dopaminergic neurons in culture. J. Neurosci. Res. 26, 428-435.

Miller, D. B. and OíCallaghan, J. P. (1994). Environment, drug and stress-induced alterations in body temperature affect the neurotoxicity of substituted amphetamines in the C57Bl/6J mouse. J. Pharmacol. Exp. Ther. 270, 752-760.

Mokler, D. J., Robinson, S. E., and Rosecrans, J. A. (1987). (+)-3,4-Methylenedioxymethamphetamine produces long-term reductions in 5-hydroxytryptamine in rats. Eur. J. Pharmacol. 138, 265-268.

Mora, F. and Porras, A. (1993). Effects of amphetamine in the release of excitatory amino acid neurotransmitters in the basal ganglia of the conscious rat. Canadian J. Pharmacol. 71, 348-351.

Nahorski, S. R. and Rogers, K. J. (1973). In vivo effects of amphetamine on metabolites and metabolic rate in rat brain. J. Neurochem. 21, 679-686.

Nash, J. F., Meltzer, H. Y., and Gudelsky, G. A. (1990). Effect of 3,4-methylenedioxymethamphetamine on 3,4-dihydroxyphenylalanine accumulation in the striatum and nucleus accumbens. J. Neurochem. 54, 1062-1067.

Nash, J. F. and Nichols, D. E. (1991). Microdialysis studies on 3,4-methylenedioxyamphetamine (MDA) and structurally related analogues, Eur. J. Pharmacol. 200, 53-58.

Nash, J. F. and Yamamoto, B. K. (1992). Methamphetamine neurotoxicity and striatal glutamate release: comparison to 3,4-methylenedioxymethamphetamine. Brain Res. 581, 237-243.

Nash, J. F. and Yamamoto, B. K . (1993). Effect of d-amphetamine on the extracellular concentrations of glutamate and dopamine in iprindole-treated rats. Brain Res. 627, 1-8.

O'Dell, S. J., Weihmuller, F. B., McPherson, R. J., and Marshall, J. F. (1994). Excitotoxic striatal lesions protect against subsequent methamphetamine-induced dopamine depletions. J. Pharm. Exp. Ther. 269, 1319-1325.

O'Hearn, E., Battaglia, G., De Souza, E. B., Kuhar, M. J. and Molliver, M. E. - (1988). Methylenedioxyamphetamine (MDA) and methylenedioxymethamphetamine (MDMA) cause selective ablation of serotonergic axon terminals in forebrain: Immunocytochemical evidence for neurotoxicity, J. Neurosci., 8, 2788-27.

Olney, J.W.(1990). Excitotoxic amino acids and neuropsychiatric disorders Annual Rev. Pharmacol. Toxicol. 30, 47-71,

Pellegrini-Giampietro, D. E., Cherici, G., Alesiani, M., Carla, V., and Moroni, F. (1990). Excitatory amino acid release and free radical formation may cooperate in the genesis of ischemia-induced neuronal damage. J. Neurosci., 10, 1035-1041.

Pu, C., Fisher, J. E., Cappon, G. D. and Vorhees, C. V. (1994). The effects of amfonelic acid, a dopamine uptake inhibitor on methamphetamine-induced dopaminergic terminal degeneration and astrocytic response in rat striatum. Brain Res. 649, 217-224.

Raiteri, M., Cerrito, F., Cervoni, A. M. and Levi, G. (1976). Dopamine can be released by two mechanisms differentially affected by the dopamine transport inhibitor nomifensine. J. Pharm. Exptl. Ther. 208, 195-202.

Randall, J. (1992). Ecstasy-fueled ìraveî parties become dances of death for English youths. JAMA 268, 1505-1506.

Ricaurte, G. A., Guillery, R. W., Seiden, L. S., Schuster, C. R. and Moore, R. Y. (1982). Dopamine nerve terminal degeneration produced by high doses of methylamphetamine in the rat brain. Brain Res. 235, 93-103.

Ricaurte, G. A., Schuster, C. R. and Seiden, L. S. (1980). Long-term effects of repeated methylamphetamine administration on dopamine and serotonin neurons in the rat brain: A regional study. Brain Res. 193,153-163

Ricaurte, G. A., Seiden, L. S., and Schuster, C. R. (1984). Further evidence that amphetamines produce long-lasting dopamine neurochemical deficits by destroying dopamine nerve fibers. Brain Res. 303, 359-364.

Ricaurte, G., Bryan, G., Strauss, G., Seiden, L. and Schuster, C. (1985). Hallucinogenic amphetamine selectively destroys brain serotonin nerve terminals. Science 229: 986.

Ricaurte, G. A., Forno, L., Wilson M. A., DeLanney, L., Irwin, I., Molliver, M. E. and Langston, J. W. (1988). (+)-3,4-Methylenedioxymethamphetamine selectively damages central serotonergic neurons in nonhuman primates. J. Am. Med. Assoc. 260, 51-55.

Olanow, C. F. (1992). An introduction to the free radical hypothesis in parkinson's disease. Ann. Neurol. 32, S2-S9.

Reynolds, I. J., Hastings, T. G. (1995). Glutamate induces the production of reactive oxygen species in cultured forebrain neurons following N-methyl-D-aspartate receptor activation. J. Neurosci. 15, 3318-3327.

Rosenberg, P. A. (1988). Catecholamine toxicity in cerebral cortex in dissociated cell culture. J. Neurosci. 8, 2887-2894.

Ryan, L. J. , Martone, M. E., Linder, J. C. and Groves, P. M. (1988). Continuous amphetamine administration induces tyrosine hydroxylase immunoreactive patches in the adult rat neostriatum. Brain Res. Bull., 21, 133-137.

Sabol, K., Lew R., Richards, J., Vosmer, G. and Seiden, L. (1996). Methylenedioxymethamphetamine-induced serotonin deficits are followed by partial recovery over a 52 -week period. Part I: Synaptosomal uptake and tissue concentrations. J. Pharmacol. Exp. Ther. 276, 846-854.

Scanzello, C. R., Hatzidimitriou, G., Martello, A., Katz, J. and Ricaurte, G. A. (1993). Serotonergic recovery after 3,4-methylenedioxymethamphetamine injury: observations in rats. J. Pharmacol. Exp. Ther. 264, 1484-1492.

Schmidt, C. J. (1988). Neurotoxicity of the psychedelic amphetamine, methylenedioxymethamphetamine. J. Pharmacol. Exp. Ther. 240, 1-7.

Schmidt, C. J., Fadayel, G. M., Sullivan, C. K. and Taylor, V. L. (1992). 5-HT2 receptors exert a state-dependent regulation of dopaminergic function: studies with MDL 100,907 and the amphetamine analogue, 3,4-methylenedioxymethamphetamine. Eur. J. Pharmacol. 223, 65-74.

Schmidt, C. J., Sullivan, C. K. and Fadayel, G. M. (1994). Blockade of striatal 5-hydroxytryptamine2 receptors reduces the increase in extracellular concentrations of dopamine produced by the amphetamine analogue 3,4-methylenedioxymethamphetamine. J. Neurochem. 62, 1382-1389.

Schmidt, C. J., Black, C. K. and Taylor, V. L. (1990). Antagonism of the neurotoxicity due to a single administration of methylenedioxymethamphetamine. Eur. J. Pharmacol. 181, 59-70 Schmidt, C. J., Taylor, V. L., Abbate, G. and Nieduzak, T. (1991). 5-HT2 antagonists stereoselectively prevent the neurotoxicity of 3,4-methylenedioxymethamphetamine by blocking the acute stimulation of dopamine synthesis: reversal by L-DOPA. J. Pharmacol. Exp. Ther. 256, 230-235.

Seiden, L. S., Fishman, M. W. and Schuster, C. R. (1975/76). Long-term methamphetamine induced changes in brain catecholamines in tolerant rhesus monkeys. Drug and Alcohol Dependence 1, 215-219.

Seiden, L. S., and Vosmer, G. A. (1984) Formation of 6-hydroxydopamine in caudate nucleus of the rat brain after a single large dose of methylamphetamine. Pharm. Biochem. Behav. 21, 29-31.

Siesjo, B. K. (1978). Brain Energy Metabolism. John Wiley and Sons, New York.

Siman, R., Noszek, J. C. and Kegerise, C. (1989). Calpain activation is specifically related to excitatory amino acid induction of hippocampal damage, J. Neurosci. 9,1579-1590.

Sonsalla, P. K., Riordan, D. E. and Heikkila, R. E. (1991). Competitive and noncompetitive antagonists at N-methyl-D-aspartate receptors protect against methamphetamine-induced dopaminergic damage in mice. J. Pharm. Exp. Ther. 256, 506-512.

Sonsalla , P. K., Jochnowitz, N. D., Zeevalk, G. D., Oostveen, J. A. and Hall, E. D. (1996). Treatment of mice with methamphetamine produces cell loss in the substantia nigra. Brain Res. 738, 172-175.

Sprague, J. E. and Nichols, D. E. (1995). The monoamine oxidase-B inhibitor L-deprenyl protects against 3,4-methylenedioxymethamphetamine-induced lipid peroxidation and long-term serotonergic deficits. J. Pharmacol. Exp. Ther. 273, 667-673.

Steele, T. D., McCann, U. D. and Ricaurte, G. A. (1994). 3,4-methylenedioxymethamphetamine (MDMA, ìEcstasyî): pharmacology and toxicology in animals and humans. Addiction 89, 539-551.

Stephans, S. E. and Yamamoto, B. K. (1994). Methamphetamine-induced neurotoxicity: Roles for glutamate and dopamine efflux. Synapse 17, 203-209.

Stephans, S. E. and Yamamoto, B. K. (1996a). Effect of repeated methamphetamine administrations on dopamine and glutamate efflux in rat prefrontal cortex. Brain Res. 700, 99-106.

Stephans, S. E. and Yamamoto, B. K. (1996b). Methamphetamine sensitization increases the vulnerability of the prefrontal cortex to methamphetamine neurotoxicity. Neuroscience 72, 593-600.

Stone, D. M., Stahl, D., Hanson, G. R. and Gibb J. W. (1986) The effects of 3,4-methylenedioxymethamphetamine (MDMA) on monoaminergic systems in rat brain. Eur. J. Pharmacol. 128, 41-49.

Stone D. M., Johnson M., Hanson G. R., Gibb J. W. (1988). Role of endogenous dopamine in the central serotonergic deficits induced by 3,4-methylenedioxymethamphetamine. J. Pharmacol. Exp. Ther. 247, 79-87.

Sylvia, A. L., LaManna, J. C., Rosenthal, M. and Jobsis, F. F. (1977). Metabolite studies of methamphetamine effects based upon mitochondrial respiratory state in rat brain. J. Pharm. Exptl. Ther. 210, 117-125.

Trulson, M. E., Cannon, M. S., Faegg, T. S., and Raese, J. D. (1985). Effects of chronic methamphetamine on the nigral-striatal dopamine system in rat brain: Tyrosine hydroxylase immunochemistry and quatitative light microscopic studies. Brain Res. Bull., 15, 569-577.

Volterra, A., Trotti, D., Tromba, C., Floridi, S. and Racagni, G. (1994). Glutamate uptake inhibition by oxygen free radicals in rat cortical astrocytes. J. Neurosci. 14, 2924-2932.

Wagner, G. C, Ricaurte, G. A., Seiden, L. S., Schuster, C. R., Miller, R. J. and Westley, J. (1980). Long-lasting depletions of striatal dopamine and loss of dopamine uptake sites following repeated administration of methamphetamine. Brain Res., 181,151-160 .

Weihmuller, F. B., OíDell, S. J. and Marshall, J. F. (1992). MK-801 protection against methamphetamine-induced striatal dopamine terminal injury is associated with attenuated dopamine overflow. Synapse 11, 155-163.

Williams, J.H., Errington, M.L., Lynch, M.A. and Bliss, T.V.P. (1989). Arachidonic acid induces a long-term activity-dependent enhancement of synaptic transmission in the hippocampus. Nature 341, 739-742.

Yamamoto, B. K., Nash, J. F. and Gudelsky, G. A. (1995). Modulation of methylenedioxymethamphetamine-induced striatal dopamine release by the interaction between serotonin and (-aminobutyric acid in the substantia nigra. J. Pharmacol. Exp. Ther. 273, 1063-1070.

Progress in HPLC-HPCE, Vol. 7, pp. 245-263
G. Ali Qureshi *et al.* (Eds)

Identification of emesis induced by centrally acting drugs and anticancer drugs

Minami, M. [1], Caudy, P. [2], Endo, T. [1], Hirafuji, M. [1], Hamaue, N. [1], Akita, K. [1], Saito, H. [3], Yoshioka, M. [3], Parvez, S. [4], Parvez, S.H. [4] and Qureshi, G.A. [5]

[1] *Department of Pharmacology, Faculty of Pharmaceutical Sciences, Health Sciences University of Hokkaido, Ishikari-Tobetsu, Hokkaido 061-02, Japan,* [2] *TSP Corporation, France, Les Ulis, France,* [3] *Department of Pharmacology, Hokkaido University, School of Medicine, Sapporo, Hokkaido, Japan,* [4] *Neuroendocrinologie & Neuropharmacologie du Developpement, Universite de Reims, UFR Sciences, B.P. 1039, 51687 Reims, Cedex, France, and* [5] *Clinical Research Center, Novum, Karolinska Institute, Huddinge University Hospital, S-14157 Huddinge, Sweden.*

INTRODUCTION

Emesis is an instinctive defense reaction caused by the somato-autonomic nerve reflex, which is integrated in the medulla oblongata. Neurochemical and neuronal mechanisms involved in the emetic response and relevant receptors seem to be rather complicated relative to emetic stimuli. It is well known that dopamine receptor agonists and L-DOPA are effective drugs for the treatment of Parkinson's disease. These drugs, however, frequently (approximately 30%) cause nausea and vomiting as unavoidable side effects by stimulating dopamine receptors in peripheral and brain tissues. The emetic action of dopamine agonists has generally been assumed to be mediated by action on dopamine receptors located in the chemoreceptor trigger zone (CTZ), the tract of the nucleus solitarius (NTS), or the stomach vagus neurons, and to be well inhibited by dopamine receptor antagonists such as metoclopramide and domperidone (Pinder *et al.*, 1976; Andrews *et al.*, 1990; Mitchelson, 1992; Andrews and Bhandari, 1993). The emetic effects of talipexole have appeared to be triggered by stimulation of the above dopamine receptors, because haloperidol (Hsu *et al.*, 1986) and metoclopramide dose-dependently suppress the drug induced vomiting in dogs, which were the animals with the most vigorous vomiting among the examined animal species. Talipexole, an azepine-derivative D_2 receptor agonist, showed a greater effectiveness and fewer emetic side effects when compared with bromocriptine, the most widely used dopamine receptor agonist (Nakanishi *et al.*, 1993). Talipexole, in a dose sufficient to produce antiparkinsonian activity, did not induce emesis to the extent found with bromocriptine in cynomolgus monkeys (Kohno *et al.*, 1996a). Recently, Nishio *et al.* (1996) have found that talipexole has a weak serotonin 5-HT_3 receptor blocking effect, while bromocriptine lacks this effect.

A number of studies indicate that 5-HT and 5-HT_3 receptors are involved in the emetic response. Emetic compounds elevate 5-HT release from intestinal enterochromaffin (EC) cells, activate 5-HT_3 receptors on abdominal afferent vagal axons, and evoke an abnormal discharge in the vagal fibers projecting to the area postrema, which in turn cause the vomiting reflex by releasing 5-HT and stimulating 5-HT_3 receptors in the area postrema (Fig. 1) (Andrews *et al.*, 1990; Mitchelson, 1992; Andrews and Bhandari, 1993).

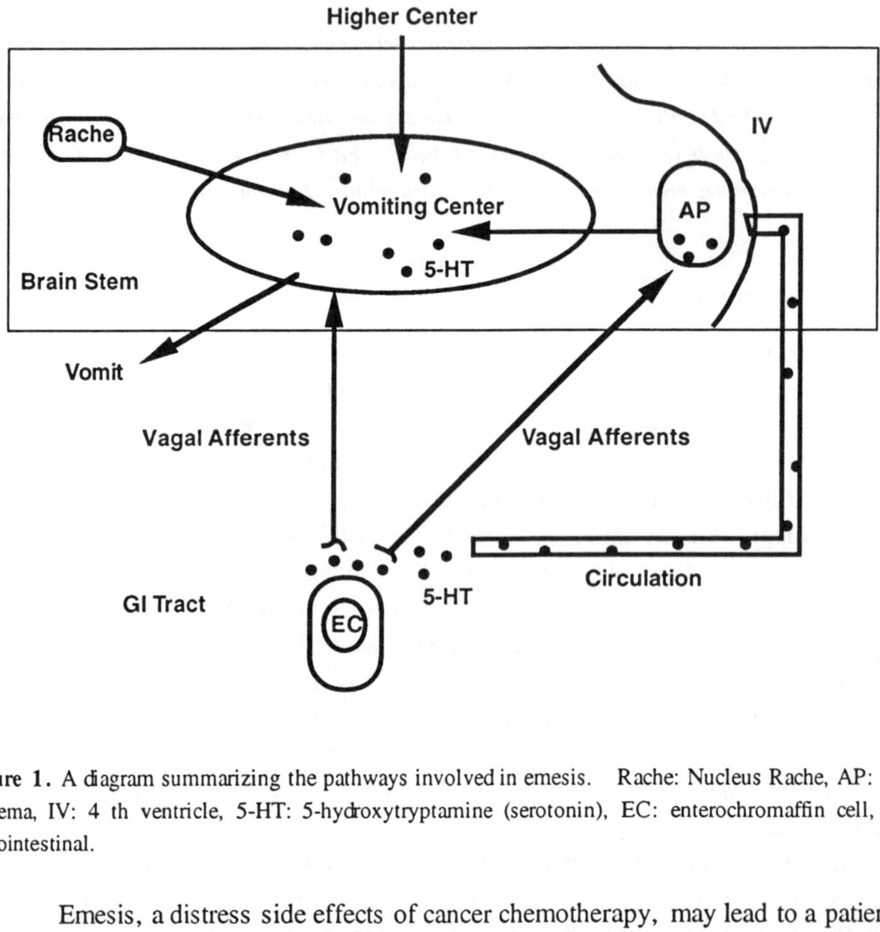

Figure 1. A diagram summarizing the pathways involved in emesis. Rache: Nucleus Rache, AP: area postrema, IV: 4 th ventricle, 5-HT: 5-hydroxytryptamine (serotonin), EC: enterochromaffin cell, GI: gastrointestinal.

Emesis, a distress side effects of cancer chemotherapy, may lead to a patient's refusal of therapy. We have previously reported that cisplatin and copper sulfate increase abdominal afferent vagal activity in a time course parallel to retching and vomiting in ferrets (Endo *et al.*, 1995), and the administration of 5-HT_3 receptor antagonist suppress both vagal activity and emesis.

This chapter will discuss the identification of emesis as an unavoidable side effect induced by centrally acting drugs and anti-cancer drugs. This review may be summarized as follows: 1) Cytotoxic drugs or centrally acting emetic drugs have an initial action within the gut that results in histological changes, activation of the biosynthesis of 5-HT, and an increase in the concentration of 5-HT in the gastrointestinal tract. 2) The 5-HT is released from the EC cells of the intestinal mucosa, which in turn stimulates 5- HT_3 receptors on the vagal afferent nerve fibers. 3) This stimulation of vagal afferent fibers results in an increase in 5-HT in the vomiting center rather than a direct action of emetic drugs at this level and leads to emesis. 4) The 5-HT in circulation trapped by platelets is metabolized to 5-hydroxyindole acetic acid (5-HIAA) in the liver and 5-HIAA is excreted in the urine.

Animal model for emesis and behavioral experiments

The initial studies which identified the emetic and anti-emetic effects using the ferret and dog were rapidly followed by a demonstration of the efficacy in man (Andrews, 1992). Once cats and dogs have vomited as a result of an injection, even anti-emetic drugs may induce emesis due to the animal's learning ability. *Suncus murinus*, the best animal model for motion sickness (Matsuki *et al.*, 1988), is too small to evaluate the chemical data for the vomiting center. The ferret has been scientifically evaluated for use in testing drugs for emetic and anti-emetic activity (Floczyk *et al.*, 1982; Costall *et al.*, 1986; Miner and Sanger, 1986). The toxic effects of anti-cancer drugs in the ferret are qualitatively similar to those observed in other species (London *et al.*, 1979; Miner and Sanger, 1986). The 5- HT_3 receptor antagonists have now been shown to be effective to a varying degree against cisplatin-induced emesis in the cat (Smith *et al.*, 1988), the pigeon (Preziosi *et al.*, 1989), and *Suncus murinus* (Torii *et al.*, 1991). *Suncus murinus* appears to be the least sensitive of the emetic species to cisplatin, requiring doses of 20 mg/kg, i.p. to 40 mg/kg, i.v. to induce emesis (Matsuki *et al.*, 1992) in contrast to 3 mg/kg, i.v. for the dog and 10 mg/kg for the cat, ferret, and pigeon.

According to the methods of Stables *et al.* (1987), ferrets given drugs are observed for the onset of emesis (latency), duration, the number of emetic episodes (retching or vomiting) over a 360 minutes period after injection (Fig. 2). Emesis usually ceases in control animals within this time in acute emesis. Emesis is induced by intraperitoneal injection (a convenient method of drug administration to ferrets) of cisplatin (10 mg/kg) or cyclophosphamide (200 mg/kg). These doses of the cytotoxic drugs are the lowest required to cause repeated and reproducible vomiting (Endo *et al.*, 1990a). Anti-emetic drugs are administered 30 minutes before cisplatin or cyclophosphamide administration. In the study for the delayed phase emesis, their behavior is observed during 72 hours.

Abdominal vagotomy in ferrets (Mackay and Andrews, 1983) prevents the cisplatin-induced vomiting in ferrets (Hawthorn *et al.*, 1988). Abdominal visceral lesions can also abolish cyclophosphamide-induced emesis (Andrews *et al.*, 1988). The ability of abdominal vagotomy to antagonize the initial phase of retching and vomiting caused by whole body irradiation of the ferrets may also be indicative of a peripheral site of action (Andrews and Hawthorn, 1987).

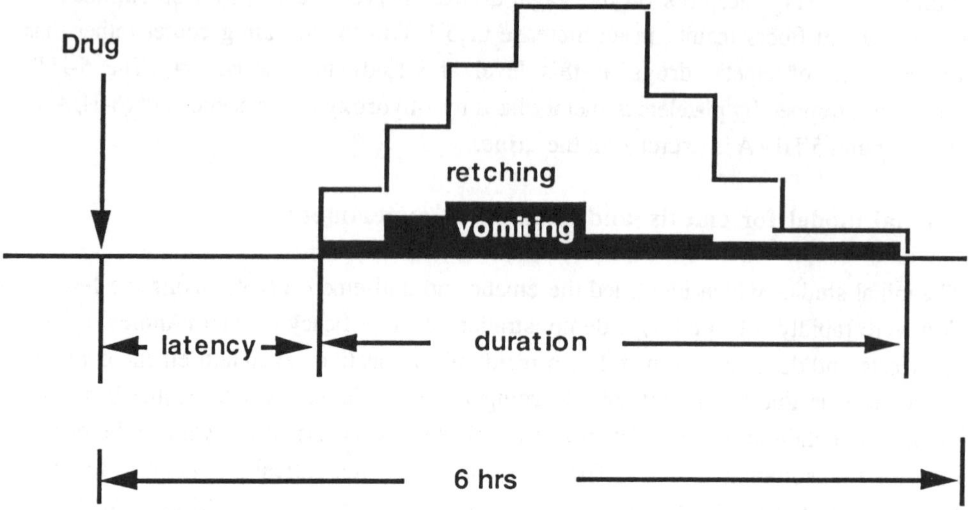

Figure 2. Behavioral experiments for emesis: According to the methods of Stables *et al.* (1987), ferrets given drugs are observed for the onset of emesis (latency), duration, the number of emetic episodes (retching or vomiting) over a 360 minutes period after injection of emetic agent.

Intracerebroventricular administration of cisplatin produces an emesis within 4 minutes; intravenous or intraperitoneal injection of cisplatin produces vomiting 90-120 minutes after drug administration (Minami *et al.*, 1991). The difference in anticancer drugs may be related to different time latencies for vomiting. It is possible that cisplatin may transform emesis-producing substances via drug formation in the intestine and the liver and that these emesis-producing substances may cause the vomiting. With regard to the mechanism of cisplatin-induced emesis, involvement of free radicals has also been proposed. Torii *et al.*(1992 & 1993) have found that the free radical formation of cisplatin produces emesis without a timelag in *Suncus murinus*. They have also reported that both antioxidant and 5-HT_3 receptor antagonists abolished the vomiting induced by a radical generator, pyrogallol. Our group also found that the phosphadidylcholine hydroperoxide levels are reduced after two administrations of cisplatin (Minami *et al.*, unpublished data).

Neurophysiological studies of abdominal afferent vagal nerve activity

Rats and ferrets were anesthetized with urethane (500 mg/kg, i.p.) and α-chloralose (50 mg/kg, i.p.), and immobilized with gallamine triethiodide (10 mg/kg, i.v.) under artificial respiration. Afferent abdominal vagus nerve activity was recorded from the peripheral cut end of the nerve with bipolar platinum-iridium wire electrodes (Fig. 3) (Yoshioka *et al.*, 1992). The nerve was immersed in a pool of warm paraffin oil. The nerve activity was amplified, passed through a filter (time constant of 6.7 msec, and a high cutoff filter of 1000 Hz) and displayed by a thermal array recorder (RTA-1300A, Nihon Kohden Co. Ltd., Japan).

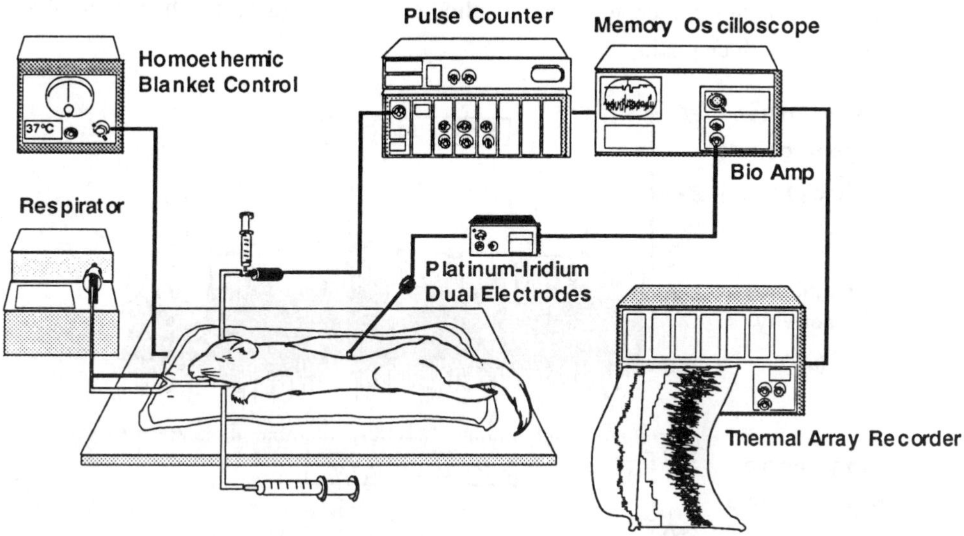

Figure 3. Diagram showing the record of afferent abdominal vagus nerve activity from a ferret according to the method of Yoshioka *et al.* (1992).

Afferent vagal terminals are located in areas adjacent to the EC cells. Therefore, the EC cell-vagal afferent nerve unit appears to be the anatomical mucosal chemoreceptor of the gastrointestinal tract (Newson *et al.*, 1982). In fact, vagal afferent nerves are affected by the luminal environment of the gut (osmolarity, pH and temperature) as well as by particles of food passing through it (Andrews, 1986). The central terminal of the abdominal vagal afferent nerves projects into the area postrema and vomiting center. The vagal afferent nerve itself does not release 5-HT, but activates the 5-hydroxytryptaminenergic neurons. Abdominal vagotomy is effective for emetic stimuli which release 5-HT from the gut and stimulate the vagal afferent nerves. The fact that abdominal vagotomy inhibits 85% of the cisplatin-induced emesis indicates that abdominal vagal afferent nerves seem to be the major pathway for cisplatin-induced emesis (Endo et al., 1992). On the other hand, abdominal vagotomy tends to be in effective for emesis in which humoral agents excite the vomiting center or the CTZ.

The increased firing induced by talipexole has been prevented by pretreatment with granisetron (a selective 5-HT_3 receptor antagonist), but not with metoclopramide or by spinal section. These results indicate that afferent vagal firing is mediated via stimulation of 5-HT_3 receptors on the neurons and is secondarily caused by stimulation of the dopamine receptors. Since talipexole blocks the 5-HT_3 receptors (Nishio *et al.*, 1996), the direct stimulation of 5-HT receptors on the vagus by talipexole is negligible. Both talipexole and bromocriptine increase vagal activity dose-dependently, and the effect of 10 mg/kg of bromocriptine is significantly higher than that noted with 1.0 mg/kg of talipexole (Minami *et al.*, 1997). The emesis-related serotoner-gic mechanisms of dopamine receptor agonists, however, seem to be much less involved than those found with cisplatin.

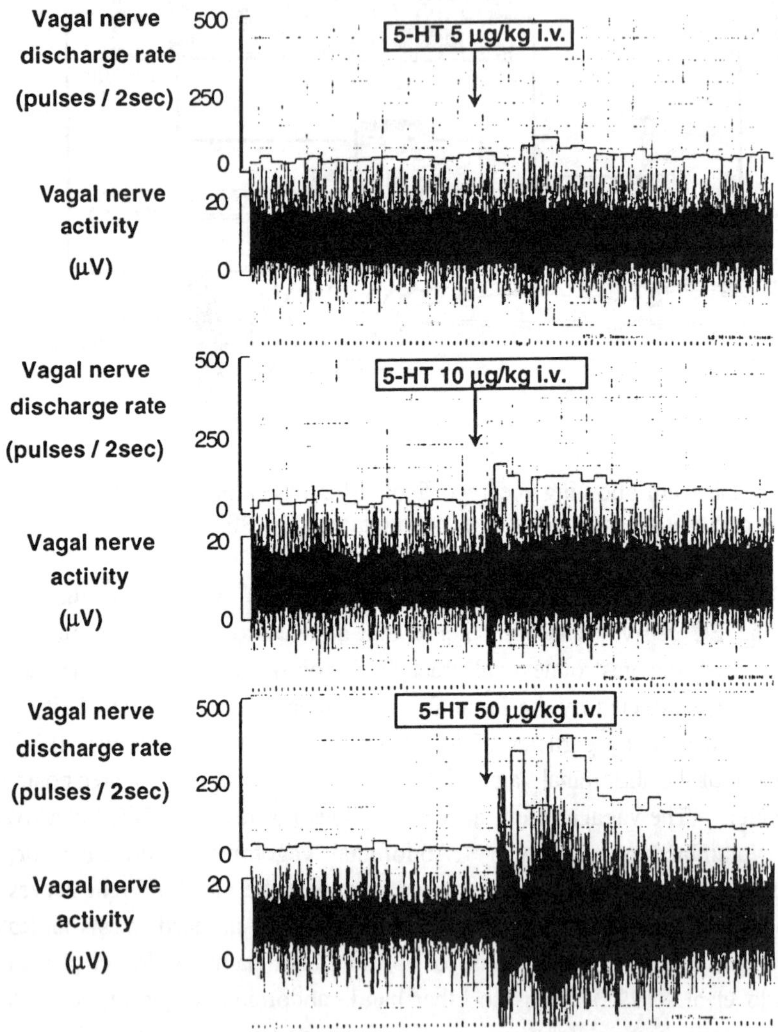

Figure 4. Effects of intravenous injection of 5-HT (serotonin) on abdominal vagal nerve activity in an anesthetized ferret. All records are from the same ferret (from Endo *et al.*, 1995 with permission).

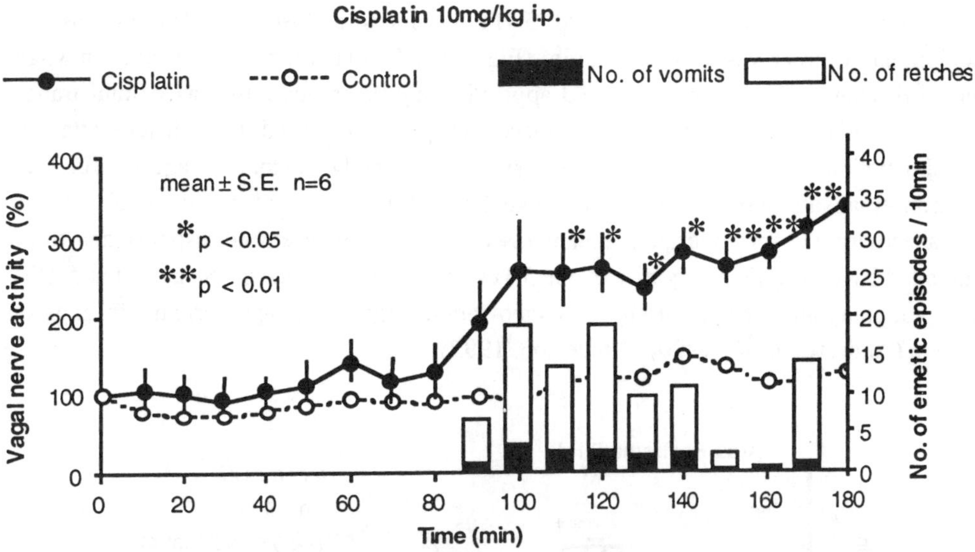

Figure 5. Effects of cisplatin on the abdominal vagal afferent nerve activity in ferrets. Cisplatin (solid line) produced a significant increase in afferent abdominal vagal nerve activity. Vehicle (dotted line) did not affect the afferents. The columns show the mean data of the time course of emetic responses of 6 different ferrets. The time course of cisplatin-induced emesis paralleled the change in the afferents (from Endo *et al.*, 1995 with permission).

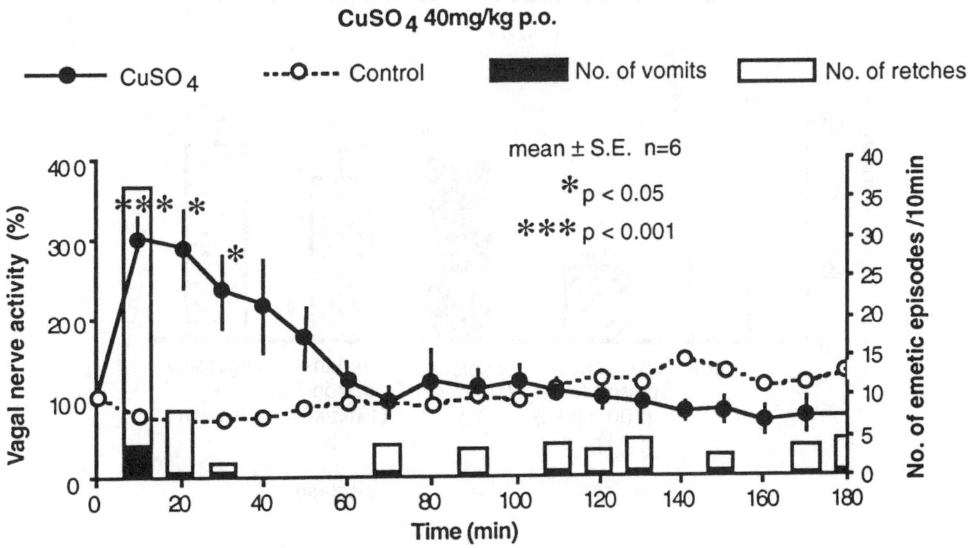

Figure 6. Effects of copper sulfate on the abdominal vagal afferent nerve activity in ferrets. Emesis were evoked immediately after its administration through a stomach tube. The time course of copper sulfate-induced emesis also paralleled the change in the afferents (from Endo *et al.*, 1995 with permission).

Intravenous bolus injection of 5-HT produces a dose-dependent increase in abdominal afferent vagal nerve activity (Fig. 4). Cisplatin-induced increases in vagal nerve discharges have been observed approximately 90 minutes after its administration (Fig. 5), whereas copper sulfate-induced firing has occurred immediately after its administration (Fig. 6)(Endo *et al.*, 1995). Namely, cytotoxic agents such as anticancer agents and copper sulfate have been shown to involve abdominal afferent vagal activity and serotonergic mechanisms in parallel to their emetic response and to be inhibited by a specific antagonist of the 5-HT_3 receptor. On the other hand, the 5-HT_3 receptor antagonists have not shown antiemetic activity against apomorphine in dogs or ferrets (Andrews *et al.*, 1990; Mitchelson, 1992).

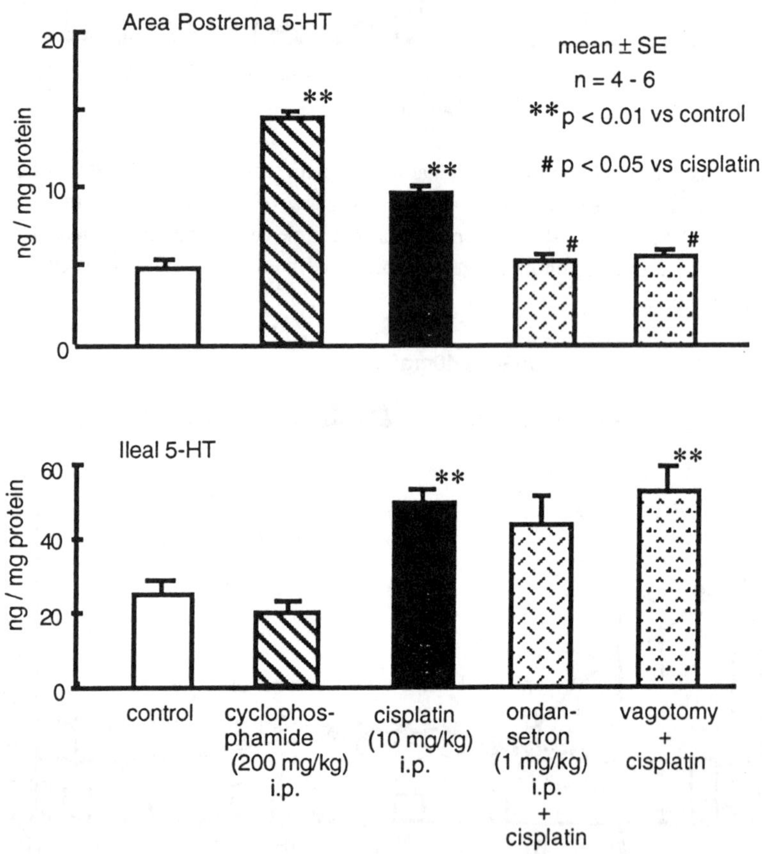

Figure 7. Effects of ondansetron and vagotomy on the increase in 5-HT concentration induced by cisplatin in the area postrema and ileum of ferrets (from Endo *et al.*, 1992 with permission).

Evidence for the involvement of serotonin in emesis

The 5-HT in the mucosal enterochromaffin (EC) cells is the sourse most relevant to emesis. The gut is estimated to contain over 80% of the 5-HT in the body, and of this, 95% is located within the EC cells (Erspamer, 1966). It is regionally distributed with the levels closely related to the distribution and numbers of EC cells. The 5-HT is stored in large electron-dense granules. Many secretory granules, which are predominantly located in the supranuclear region (Golgi area), move towards the base of the EC cells and are released from the basal surface (Minami *et al.*, 1995a).

Dosing with cisplatin (Gunning *et al.*, 1987; Endo *et al.*, 1990a) or copper sulfate (Endo *et al.*, 1991) on two consecutive days increased the level of 5-HT and 5-HIAA and norepinephrine (but not dopamine) in the ileal, but not gastric, mucosa in a ferret killed 3 hours after the second dose (Fig. 7).

Determination of monoamine concentration and relevant enzyme activities

The ileum (taken 20 cm from the pylorus and isolated as a piece 3 cm long) and the area postrema (caudal region of the floor of the fourth ventricle) were dissected out at a relevant time after the drug injection. These tissues were stored frozen (-80°C) until assayed. Tissue concentrations of serotonin (5-HT) and 5-hydroxyindole acetic acid (5-HIAA) were measured using HPLC with an electrochemical detector (ECD) (Minami *et al.*, 1988). Tryptophan hydroxylase (TPH) activity was measured by a modification of the method of Kuhn *et al.* (1978). The enzymatic reaction byproduct, 5-hydroxytryptophan (5-HTP) was determined with HPLC-ECD. Monoamine oxidase (MAO) activity was determined by Kraml's fluorometric method using kynuramine as a substrate. Protein assays were performed by the method of Lowry *et al.* (1951) with bovine serum albumin as the standard.

Mucosal levels of 5-HT

Anticancer drugs congregate at rapid turnover tissues such as the gastrointestinal tract, and therefore, impairment of the epithelium of the mucous membrane often occurs. The release of 5-HT from the EC cells induced by radiation and cytotoxic drugs may be related to a triggering of the secretory mechanism of the EC cells in the mucous membrane of the ileum. The most probable mode of 5-HT release may depend upon calcium ion influx (Racké and Schwörer, 1991), although this mechanism has not been clarified. We have reported that ileal mucosal TPH activity increases in cisplatin- or cyclophosphamide-treated ferrets, as compared to that observed in non-control ferrets. MAO activity, on the contrary, is inhibited in cisplatin- or cyclophosphamide-treated ferrets as compared to that observed in the non-drug controls (Endo *et al.*, 1993). This increased TPH activity appears to be due to the activation of ileal 5-HT biosynthesis induced by cytotoxic drugs. Through this mechanism, the life span of 5-HT might be prolonged. Cisplatin-treated ferrets have shown an increase in ileal 5-HT levels (Gunning *et al.*, 1987; Endo *et al.*, 1990a & 1990b). Our group has also clarified that intragastric copper sulfate administration produces an increase in ileal 5-HT levels (Endo *et al.*, 1991).

Bromocriptine (5 mg/kg) at a dose producing antiparkinsonian activity (Irifune *et al.*, 1993; Kohno *et al.*, 1996b) has caused a significant increase in 5-HT turnover and the relevant enzyme activities in intestinal tissues, in addition to a greater increase in the spontaneous firing of afferent vagal neurons, when compared with talipexole. This result shows that bromocriptine-induced emesis is presumably facilitated by increased 5-HT release from EC cells, which results in an increase in afferent vagal activity. Neither drugs influences monoamine metabolism in the area postrema. Furthermore, the talipexole-induced increase in afferent vagal activity is not prevented by sectioning the spinal cord. These results suggest that the emetic responses produced by the anti-Parkinson drugs, bromocriptine and talipexole, might involve serotonergic mechanisms in the abdominal tissues, but not in the area postrema.

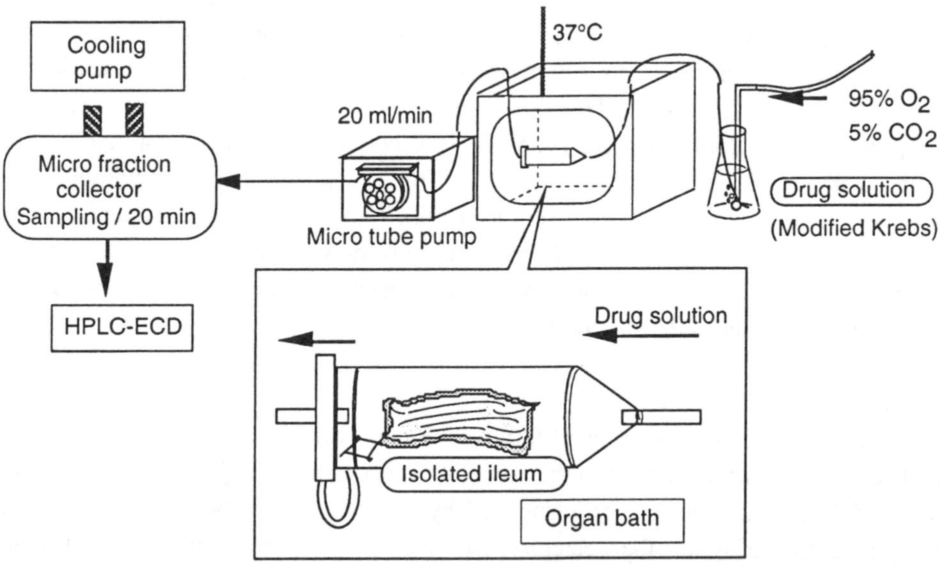

Figure 8. Block diagram showing the modified method detecting 5-HT release from the isolated ileum of ferrets or rats according to Milano *et al.* (1991).

5-HT release from EC cells

In ferrets, the ileum was isolated and a piece of 3 cm length (weight 1.0-1.5 g), taken 20 cm from the pylorus, was dissected according to Milano's method (1991). Namely, ileal segments isolated from ferrets were placed in organ baths, which contained modified Krebs solution and gassed with 95% O_2 and 5% CO_2 (Fig. 8). After a 60 min equilibration period, a tissue-bathing buffer solution was collected in 20 min fractions. 5-HT released from the ileum was measured using HPLC-ECD.

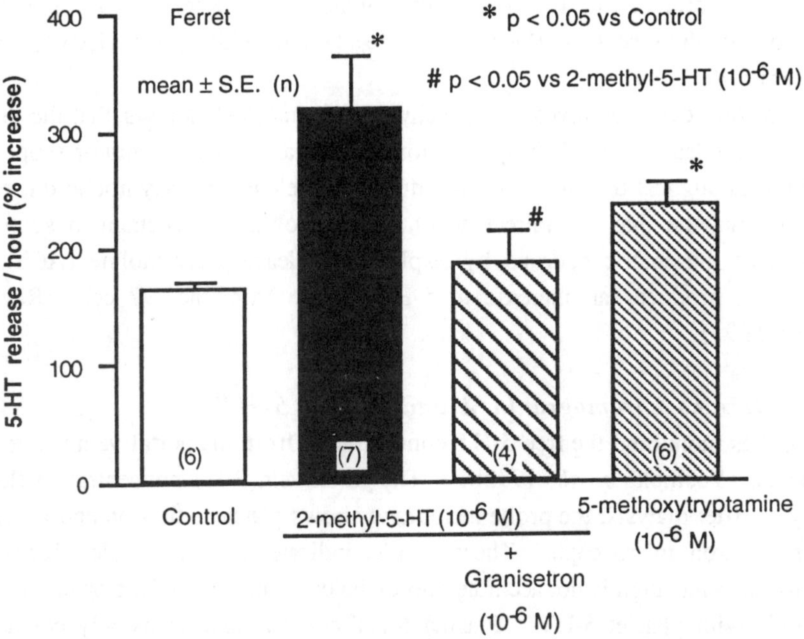

Figure 9. Effect of granisetron on the 5-HT release induced by 2-methyl-5-HT and effect of 5-methoxytryptamine on 5-HT release from ferret isolated ileum.

Cisplatin has increased 5-HT release from both an isolated guinea pig vascularly perfused ileum (Schwörer *et al.*, 1991) and an isolated cat ileum (Milano *et al.*, 1991). Schwörer *et al.* (1991) have found that spontaneous outflow of 5-HT into the portal circulation reflects an exocytotic release from EC cells. The 5-HT release from EC cells is regulated by a polymodal mechanism and is calcium dependent. Gebauer *et al.* (1993) have studied whether 5-HT autoreceptors are involved in the regulation of 5-HT release. A selective 5-HT_3 receptor antagonist, 2-methyl-5-HT, produces a dose-dependent 5-HT release (Minami *et al.*, 1995b). The 5-HT_3 receptor agonist-induced 5-HT release is prevented by selective 5-HT_3 receptor antagonists, but other kinds of 5-HT_3 receptor antagonists do not prevent the 5-HT release induced by 5-HT_3 receptor agonists. Gebauer *et al.* (1993) have reported that stimulation of 5-HT_4 receptors causes inhibition of release. On the other hand, we have found that a selective 5-HT_4 agonist, 5-methoxytryptamine, produces 5-HT release dose-dependently; it has been assumed that this release occurs indirectly via the cholinergic mechanisms (Fig. 9)(Minami *et al.*, 1995b). 5-HT_{1A} agonist possesses a broad spectrum of antiemetic action; one such mechanism appears to be associated with 5-HT release. Not only autoreceptors, but also many heteroreceptors on the EC cells might be associated with 5-HT release from the EC cells(Schwörer *et al.*, 1987 & 1989; Rack *et al.*, 1988; Minami *et al.*, 1995b). 5-HT release is facilitated by the stimulation of nicotinic and muscarinic receptors (Schwörer *et al.*, 1987) and beta-adrenoceptors (Racké *et al.*, 1988) and is inhibited by α_2 adrenoceptors (Racké *et al.*, 1988) and $GABA_A$ and $GABA_B$ receptors (Schwörer *et al.*, 1989).

Interesting evidence revealed by Schwörer *et al.*(1991) shows that the cisplatin-induced 5-HT release is blocked by tetrodotoxin, hexamethonium, and/or scopolamine. These findings suggest that the cisplatin-induced EC cell effects may not be due to direct action, but may be due to indirect action via a cholinergic mechanism such as the stimulation of the enteric neurons by cisplatin to release acetylcholine (ACh). This released ACh produces an increase in 5-HT release from the EC cells (Racké and Schwörer, 1991).

Plasma 5-HIAA, chromogranin A and platelet 5-HT

Our group has monitored the free 5-HT concentration from the portal vein using in vivo microdialysis (Yoshioka *et al.*, 1992b). The portal vein 5-HT concentration fluctuates markedly. After dialysis, the probe membrane shows platelet adhesion and aggregation under an electron microscope. These results indicate that using microdialysis as a method for determination is not accurate due to the contamination of the probe membrane and cannula with platelet 5-HT. Plasma 5-HT concentration is usually contaminated with platelet 5-HT. Cubeddu *et al.* (1992) have reported that the plasma 5-HT levels of cancer patients under cytotoxic drugs increases slightly, although it never achieved significance, but as would be expected the plasma 5-HIAA levels increase significantly. The gastrointestinal tract has the highest body content of 5-HT and chromogranin A (CgA), and the great majority is contained within the EC cells. In EC cells, CgA is co-stored along with 5-HT in the core of the 5-HT storage vesicles (Bargsten and Grube, 1992). Therefore, when a large amount of 5-HT (and hence CgA) is released from the EC cells, this release is reflected by increases in plasma 5-HIAA and CgA, and in the urinary 5-HIAA. When serial samples are collected and processed for 5-HIAA and CgA, the time course for the release of 5-HT from EC cells can be estimated (Cubeddu *et al.*, 1990 & 1992). No difference in platelet 5-HT levels have been observed between healthy subjects and cancer patients. Platelet 5-HT does not increase during cisplatin- or cyclophosphamide-induced emesis.

Urinary level of 5-HIAA

A number of studies in man have examined 5-HT metobolism following cisplatin administration. Cubeddu *et al.* (1992) have measured plasma and urinary 5-HT concentrations in numerous cancer patients. They have proposed that the determination of urinary 5-HIAA concentration, a major metabolite of 5-HT, might be an indicator of gut mucosal 5-HT release. A high dose of cisplatin (> 50 mg/m^2) produces a marked increase in urinary 5-HIAA levels in patients by 350 ± 39% within 2 hours, and the increased urinary 5-HIAA levels paralleled the occurrence of vomiting (Cubeddu *et al.*, 1990). This suggests that urinary 5-HIAA might be a convenient index of 5-HT release from the gastrointestinal tract. Using the urinary 5-HIAA concentration, the severity of acute emesis and the degree of delayed emesis may be predictable. Urinary 5-HIAA levels might be useful for predicting anticipatory emesis in the female and for determining whether anticipatory emesis occurs due to psychological phenomenon, to the disease itself, or to anticancer drugs. The urinary 5-HIAA determination might also be useful in the analysis of the mechanism of pregnancy sickness.

5-HT levels in the vomiting center

An additional source of 5-HT would be from the central terminals of abdominal vagal afferents projecting to the area postrema. In addition, although the vagal afferents may not themselves release 5-HT, they may activate serotonergic neurons (Andrews *et al.*, 1988; Andrews, 1994).

Receptor-binding studies of ferret tissues have shown that the area postrema contains high concentrations of 5-HT$_3$ receptors, and these receptors are likely to be important in the control of emesis (Higgins *et al.*, 1989). One of the 5-HT-containing neuronal groups in the brain stem is the raphe nucleus. The NTS receives input from the raphe nucleus. However, it is uncertain whether the NTS-raphe nuceus pathway is an important component of emetic response (Leslie *et al.*, 1992). Namely, the vomiting center includes NTS-raphe, the area postrema, and adjacent structures.

The neurons within the area postrema are excited not only by 5-HT, but also by gastrin, histamine, neurotensin, substance P, and enkephaline (Carpenter *et al.*, 1983; Borison *et al.*, 1984). All these substances are found in the gastrointestinal tracts. Stimulation of afferent vagal nerve fibers appears to be related to the increased 5-HT levels in the vomiting center. In fact, the electrical stimulation of abdominal vagal nerves produces an increase in the vomiting center and evoked vomiting in anesthetized ferrets. Our group has reported that 5-HT levels of the vomiting center (area postrema, NTS and adjacent structures) increases significantly after cisplatin or cyclophosphamide administration (i.p.)(Endo *et al.*, 1992). We have found that tryptophane hydroxylase activity, the rate-limiting enzyme of 5-HT, increases significantly in cisplatin-treated ferrets (Endo *et al.*, 1993). Pretreatment with 5-HT$_3$ receptor antagonists or abdominal vagotomy significantly inhibit the cytotoxic drug-induced increase in the 5-HT concentration of the area postrema (Fig. 7)(Endo *et al.*, 1993). It has been reported that the area postrema-5-HT release occurs as platelets are passed through the area postrema; an abdominal vagotomy clearly inhibits the area postrema-5-HT increase induced by cisplatin. These findings suggest that an area postrema-5-HT increase induced by cisplatin is due to emetic stimuli originating from the vagal afferent nerves. Area postrema-5-HT levels correlated well with the frequency of emesis (Minami *et al.*, 1995c). Therefore, the area postrema appears to perform a functional role in relaying signals between the visceral afferents and the vomiting center (Fig. 1). Pretreatment with ondansetron diminishes the area postrema-5-HT increase, but does not alter the increased ileum 5-HT levels (Endo *et al.*, 1992). The increased 5-HT levels in the vomiting center which are induced by anticancer drugs are inhibited by the chemical vagotomy caused by 5-HT_3 receptor antagonists. These findings suggest that the site of action of 5-HT_3 receptor antagonists are the 5-HT_3 receptor sites on afferent vagal nerve endings.

Intracellular Ca^{2+} dynamics in EC cells

The secretion of 5-HT from EC cells may occur via exocytosis. This has been supported by many observations that 5-HT release is largely dependent on extracellular Ca^{2+} (Racké and Schwörer, 1991). However, almost all of these studies have been carried out using isolated segments of the small intestine. Direct evidences showing the role of Ca^{2+} in the basic mechanisms of 5-HT release from EC cells is still lacking. Although it has been considered that 5-HT release is regulated by multiple receptor-mediated mechanisms (Racké and Schwörer, 1991), little is known about the intracellular Ca^{2+} dynamics in response to these receptor stimuli in EC cells.

Our group has isolated mouse ileal crypts and investigated the intracellular Ca^{2+} dynamics using a fluorescence dye, fura-2/AM, according to the method of Satoh *et al.* (1995), who for the first time demonstrated the intracellular Ca^{2+} dynamics of crypt cells including epithelial cells, Paneth cells, or EC cells in isolated mouse ileal crypts. The ileum 3-4 cm length was excised, and the muscle layer was peeled off under a binocular microscope. The mucosal-comprising epithelium and the underlying connective tissue were incubated in Ringer solution containing collagenase in the presence of trypsin inhibitor and serine protease inhibitors. Crypts were isolated by moderate pipetting and resuspended in Ringer solution. The isolated crypts were transferred to an incubation chamber attached to a coverslip coated with Cell-Tak, an adhesive protein. The crypts were then loaded with fura-2/AM. The the cytosolic Ca^{2+} concentration was then analyzed using digital video-imaging system (Argus-50/CA, Hamamatsu Photonics, Japan).

Satoh *et al.*, (1995) have reported that norepinephrine, but not isoproterenol, induces a transient elevation of cytosolic Ca^{2+} concentration in some EC cells in the isolated mouse crypts. This effect may be mediated via the a1-adrenoceptor subtype, since we have observed that the norepinephrine-induced Ca^{2+} mobilization is reduced by the pretreatment of a1 antagonists (Hirafuji *et al.*, unpublished data). Carbachol increases the cytosolic Ca^{2+} concentration in Paneth cells via muscarinic receptors, but not in EC cells (Satoh *et al.*, 1995). We have also evaluated the effects of a number of pharmacological substances on the Ca^{2+} dynamics in the isolated mouse crypts in order tofind out what other kinds of transmitters can induce the Ca^{2+} mobilization in EC cells. However, none of the tested substances including 5-HT, 2-methyl-5-HT, substance P, histamine, angiotensin II, L-glutamic acid, morphine, cisplatin, and cyclophosphamide have induced the Ca^{2+} mobilization in the isolated mouse crypts (Hirafuji *et al.*, unpublished data).

Consistent with the results for Ca^{2+} dynamics in EC cells, our *in vitro* experiment has shown that norepinephrine, but not isoproterenol, induces a 5-HT release from isolated mouse ileal tissue (Hirafuji *et al.*, unpublished data). However, catecholamines seem to stimulate 5-HT secretion from the intestines of cats, guinea-pigs, and rats through cAMP-mediated mechanisms via ß-adrenoceptors (Ahlman *et al.*, 1976; Pettersson *et al.*, 1978; Racké *et al.*, 1988). Therefore, there may be species differences in ß-adrenoceptor-mediated 5-HT release from EC cells.

Acknowledgment

We thank Mr. Takashi Ogawa, Kenji Kato, Kazuhiko Kikuchi, Hideya Yokota, Naoki Ohmae, Ms. Emi Ihira, Chiaki Kudo, Tomoko Suzuguchi, and Yumi Teramoto for their valuable technical assistance in this research. The study was partly supported by a grant from SmithKline Beecham Pharmaceuticals.

REFERENCES

Ahlman, H., Lundberg, J.M., Dahlström, A. and Kewenter, J. (1976). A possible vagal adrenergic release of serotonin from enterochromaffin cells in the cat. *Acta physiol. scand.* **98**, 366-375.

Andrews, P.L.R. (1986). Vagal afferent innervation of the gastrointestinal tract. In: *Visceral Sensation Progress in Brain Research*. Cervero, F. and Morrison, J.F.B. (Eds), Elsevier, Amsterdam, The Netherland, pp. 65-86.

Andrews, P.L.R. and Hawthorn, J.(1987). Evidence for an extra-abdominal site of action for the 5-HT_3 receptor antagonist BRL24924 in the inhibition of radiation evoked emesis in the ferret. *Neuropharmacol.* **26**, 1367-1370.

Andrews, P.L.R., Rapeport, W.G. and Sanger, G.L.(1988). Neuropharmacology of emesis induced by anti-cancer therapy. *Trends Pharmacol. Sci.* **9**, 334-341.

Andrews, P.L.R., Davis, C.J., Bingham, S., Davidson, H.I.M., Hawthorn, J. and Maskell, L. (1990). The abdominal visceral innervation and the emetic reflex: pathways, pharmacology and plasticity. *Canad. J. Physiol. Pharmacol.* **68**, 325-345.

Andrews, P.L.R.(1992). Physiology of nausea and vomiting. *Br J. Anest.* **69**, 2s-19s.

Andrews, P.L.R. and Bhandari, P. (1993). The 5-hydroxytryptamine receptor antagonists as antiemetics: preclinical evaluation and mechanism of action. *Eur. J. Cancer* **29**A, S11-S16.

Andrews, P.L.R. and Davis, C.J. (1995). The physiology of emesis induced by anti-cancer therapy. In: *Serotonin and the scientific basis of anti-emetic therapy*. Reynolds, D.J.M., Andrews, P.L.R., Davis, C.J. (Eds), Oxford Clin. Comm., Oxford, pp. 25-49.

Bargsten, G. and Grube, D. (1992). Serotonin storage and chromogranins: an experimental study in rat gastric endocrine cells. *J.Histochem.Cytochem.* **40**, 1147-1155.

Borison, H.L., Borison, R. and McCarthy, L.E. (1984). Role of the area postrema in vomiting and related functions. *Fed. Proc.* **43**, 2955-2958.

Carpenter, D.O., Briggs, D.B. and Strominger, N.L. (1983). Responses of neurons of canine area postrema to neurotransmitters and peptides. *Cell. Mol. Neurol.* **3**, 113-126.

Cubeddu, L.X., Hoffman, I.S., Fleunmayor, N.T. and Finn, A.L.(1990). Efficacy of ondansetron and the role of serotonin in cisplatin-induced nausea and vomiting. *N.Engl.J.Med.* **322**, 810-816.

Cubeddu, L.X., Hoffman, I.S., Fleunmayor, N.T. and Malave, J.J. (1992). Changes in serotonin metabolism in cancer patients: its relationship to nausea and vomiting induced by chemotherapeutic drugs. *Br. J. Cancer* **66**, 198-203.

Endo, T. , Minami, M. , Monma, Y. , Yoshioka, M. , Saito, H., Kinami, J. , Toshimitsu, Y. and Parvez, H.(1990a). Effect of GR38032F on cisplatin- and cyclophosphamide-induced emesis in the ferret. *Biogenic Amines* **7**(6), 525-533.

Endo, T. , Minami, M. , Monma, Y., Saito, H.and Takeuchi, M. (1990b). Emesis-related biochemical and histopathological changes induced by cisplatin in the ferret. *J. Toxicol. Sci.* **15**, 235-244.

Endo, T., Minami, M., Monma, Y., Yoshioka, M., Saito, H., Toshimitsu, Y. and Parvez, H.(1991). Effect of ondansetron, a 5-HT_3 antagonist, on copper sulfate-induced emesis in the ferret. *Biogenic Amines* **8**, 79-86.

Endo, T., Minami, M., Monma, Y., Yoshioka, M., Saito, H., and Parvez, H.(1992). Vagotomy and ondansetron (5-HT_3 antagonist) inhibited the increase of serotonin concentration induced by cytotoxic drugs in the area postrema of ferrets. *Biogenic Amines* **9**, 163-175.

Endo, T., Takahashi, M., Minami, M., Monma, Y., Yoshioka, M., Saito, H., and Parvez, H.(1993). Effects of anticancer drugs on enzyme activities and serotonin release from ileal tissue in ferrets. *Biogenic Amines* **9**, 479-489.

Endo, T., Nemoto, M., Minami, M., Yoshioka, M., Saito, H. and Parvez, S.H. (1995). Changes in the afferent abdorminal vagal nerve activity induced by cisplatin and copper sulfate in the ferret. *Biogenic Amines* **11**, 399 - 407.

Erspamer, V.(1966). Occurrence of indolalkyl amines in nature. In: *Hand Book Exp. Pharmacol.*, Vol 19. Eeichler, O., Farah, A. (Eds), Springer, Berlin, Heidelberg, New York, pp. 132-181.

Florczyk, A. P. , Schurig, J. E. and Bradner, W.T.(1982). Cisplatin-induced emesis in the ferret : a new animal model. *Cancer Treat Rep.* **66**(1), 187-189(1982).

Gebauer, A., Merger, M. and Kilbinger, H. (1993). Modulation by 5-HT_3 and 5-HT_4 receptors of the release of 5-hydroxytryptamine from the guinea-pig small intestine. *Naunyn-Schmiedeberg's Arch Pharmacol.* **347**, 137-140.

Gunning, S.J., Hagan, R.M. and Tyers, M.B. (1987). Cisplatin induced biochemical and histological changes in the small intestine of the ferret (abstract). *Br. J. Pharmacol.* **90**, 135p.

Hawthorn, J., Ostler, K.J. and Andrews, P.L.R. (1988). The role of the abdominal visceral innervation and 5-Hydroxy-tryptamine M-receptors in vomiting induced by the cytotoxic drugs cyclo-phosphamide and cis-platin in the ferret. *J. Exp. Physiol.* **73**, 7-21.

Higgins, G.A., Kilpatrick, G.J., Bunee, K.T., Jones, B.J. and Tyers, M.B. (1989). 5-HT_3 receptor antagonists injected into the area postrema inhibit cisplatin-induced emsis in the ferret. *Br. J. Pharmacol.* **97**, 247-255.

Hsu, W.H., Schaffer, D.D. and Dyer, D.C. (1986). The emetic effect of B-HT 920 and apomorphine in the dog : antagonism by haloperidol. *Life Sci.* **39**, 1021 -1026.

Irifune, M., Nomoto, M. and Fukuda, T.(1993). Effects of talipexole on motor behavior in normal and MPTP-treated common marmosets. *Eur.J.Pharmacol.* **238**, 235 - 240.

Kohno, Y., Fukuzaki, K., Kitahara, K. and Koja, T. (1996a). Anti-tremor activity of talipexole produced by selective dopamine D_2 receptor stimulation in cynomolgus monkeys with unilateral lesions in the ventromedial tegmentum. *Eur. J. Pharmacol.*, in press.

Kohno, Y., Kitamura, Y., Nakazawa, M. and Nomura, Y. (1996b). The effects of talipexole on central monoamine receptors and metabolism, with special reference to its effects on dopaminergic systems. *Pharmacometrics* **52**, 233 - 241.

Kuhn, D.M., Vogel, R.T. and Lovenberg, W. (1978). Calcium-dependent activation of tryptophan hydroxylase by ATP and magnesium. *Biochem.Biophys.Res. Comm.* **82**, 759 - 766.

Leslie, R.A., Reynolds, J.M. and Lawes, I.N.C. (1992). Central connections of the nuclei of the vagus nerve. In: *Neuroanatomy and Physiology of Abdominal Vagal Afferents.* Ritter, S., Ritter, R.C. and Barnes, C.D. (Eds.), CRC Press, Boca Raton, FL, pp. 81-98.

London, S. W. , McCarthy, L. E. and Borison, H. L (1979). Suppression of cancer chemotherapy-induced vomiting in the cat by nabilone, a syntheticcannabinoid (40465). *Proc. Soc. Exp. Biol. Med.* **160**, 437-440.

Lowry, O. H., Rosenbrough, N.J., Farr, A.L. and Randall, R.J.(1951). Protein measurement with the folin phenol reagent. *J. Biol. Chem.* **193**, 265-275.

Mackay, T.W. and Andrews, P.L.R. (1983). A comparative study of the vagal innervation of the stomach in man and the ferret. *I.Anat.* **136**, 449-481.

Matsuki, N., Ueno, S., Kaji, T., Ishihara, A., Wang, C-H. and Saito, H.(1988). Emesis induced by cancer chmotherapeutic agents in the *suncus murinus*: a new experimental model. *Jpn. J. Pharmacol.* **48**,303-306.

Matsuki, N., Torii, Y., Ueno, S. and Saito, H. (1992). *Suncus murinus* as an experimental animal model for emesis and motion sickness. In: *Mechanisms and control of emesis.* Bianchi, A.L., Grelot, L., Miller, A.D. and King, G.L. (Eds.), Colloque INSERM, John Libbey Eurotext, 223, pp. 323-329.

Milano, S,.Simon, C. and Grelot, L.(1991). *In vitro* release and tissue levels off ileal serotonin after cisplatin-induced emesis in the cat. *Clin. Auto. Res.* **1**, 275-280.

Minami, M., Sano, M., Togashi, H., Sakurai, H. and Saito, H. (1988). Stroke-related plasma noradrenaline, angiotensin II, arginine-vasopressin and serotonin concentrations in stroke-prone spontaneously hypertensive rats. In: *Progress in Hypertension,* Vol.1, H. Saito, H. Parvez, S. Parvez and T. Nagatsu (Eds.), VSP, Utrecht, pp. 89-114.

Minami, M., Endo, T., Monma, Y. and Shiroshita, Y. (1991). Pharmacology of emesis induced by anticancer drugs. *J. Toxicol. Sci.* **16** (Supple.II), 35-39.

Minami, M., Endo, T., Nemot, M., Hamau, N., Hirafuji, M., Monma, Y., Yajima, T., Yoshioka, M. and Saito, H. (1995a). How do toxic emetic stimuli cause serotonin release in the gut and the brain? In: *Serotonin and thescientific basis of anti-emetic therapy.* Reynolds, D.J.M., Andrews, P.L.R. and Davis, C.J. (Eds.), Oxford Clinical Comm., Oxford, pp. 68-76.

Minami, M., Tamakai, H., Ogawa, T., Endo, T., Hamaue, N., Hirafuji, M., Yoshioka, M. and Blower, P. (1995b). Chemical modulation of 5-HT3 and 5-HT4 receptors affects the release of 5-hydroxytryptamine from the ferret and rat intestine. *Res. Comm. Mol. Pathol. Pharmacol.* **89**(2), 131-142.

Minami, M., Saito, H. and Yoshioka, M. (1995c). Toxicological aspects of cispatin-induced emesis with emphasis on serotonin release. *J.Toxicol. Sci.*, **20** (Append), 77-85.

Minami, M., Nemoto, M., Endo, T., Hamaue, N. and Kohno, Y. (1997). Effects of talipexole on emesis-related changes in abdominal afferent vagal activity and ileal serotonin metabolism in rats. *Res. Comm. Mol. Pathol. Pharmacol.*, **95**(1), 67-83.

Miner, W.D. and Sanger, G.J.(1986). Inihibition of cisplatin-induced vomiting by selective 5-hydroxytryptamine M-receptor antagonism. *Br. J. Pharmac.* **88**, 497-499.

Mitchelson F. (1992). Pharmacological agents affecting emesis. A review (Part I). *Drugs* **43**, 295-315.

Nakanishi, T., Kowa, H., Mizuno, Y. and Yanagisawa, N. (1993). The clinical evaluation of B-HT 920 (talipexole hydrochloride) in Parkinson's disease. *Clin. Eval.* **21**, 59-110.

Nishio, N., Kohno, Y., Fujii, A., Negishi, Y., Inoue, A. and Nakata, Y. (1996). 5-HT_3 receptor blocking properties of the antiparkinsonian agent, talipexole. *Gen. Pharmacol.* **27**, 779 - 785.

Newson, B., Ahlman, H., Dahlstrom, A. and Nyhus, L.M. (1982). Ultrastructural observations in the rat ileal mucosa of possible epithelial "taste cells" and submucosal sensory neurons. *Acta. Physiol. Scand.* **114**, 161-164.

Pettersson, G., Dahlström, A., Larsson, I., Lundberg, J.M., Ahlman, H. and Kewenter, J. (1978). The release of serotonin from rat duodenal enterochromaffin cells by adrenoceptor agonists studied in vitro. *Acta physiol. scand.* **103**, 219-224.

Pinder,R. M. ,Brogden, R.N. ,Sawyer, P.R., Speight, T.M. and Avery, G.S. (1976). Metoclopramide: a review of its pharmacological prooperties and clinical use. *Drugs* **12**, 81-131.

Preziosi, P., Ripanti, G., Navarra, P., D'Amato, M. and Vacca, M. (1989). Comparison among 5-HT3 antagonists on cisplatin emesis in pigeons. *Arch. Int. Pharmacodyn. Ther.* **298**, 301-302.

Racké, K., Schwörer, H. and Kilbinger, H.(1988). Adrenergic modulation of the release of 5-hydroxytryptamine from the vascularly perfused ileum of the guinea-pig. *Br. J. Pharmacol.* **95**, 923-931.

Racké, K. and Schwörer, H. (1991). Regulation of serotonin release from the intestinal mucosa. *Pharmacol. Res.* **23**, 13-25.

Satoh, Y., Habara, Y., Ono, K. and Kanno, T. (1995). Carbamylcholine- and catecholamine-induced intracellular calcium dynamics of epithelial cells in mouse ileal crypts. *Gastroenterology* **108**, 1345-1356.

Schwörer, H., Racké, K. and Kilbinger, H.(1987). Cholinergic modulation of the release of 5-hydroxytryptamine from the guinea pig ileum. *Naunyn-Schmiedebergs' Arch Pharmacol.* **336**, 127-132.

Schwörer, H., Racké, K, K. and Kilbinger, H.(1989). GABA eceptors are invlolved in the modulation of the release of 5-hydroxytryptamine from the vascularly perfused small intestine of the guinea-pig. Eur. J. Pharmacol. **165**, 29-37.

Schwörer, H., Racké, K, K. and Kilbinger, H.(1991). Cisplatin increases the release of 5-hydroxytryptamine (5-HT) from the isolated vascularly perfused small intestine of the guinea-pig: Involvement of 5-HT_3 receptors. *Naunyn-Schmiede-berg's Arch. Pharmacol.* **344**, 143-149.

Smith, W.L., Callaham, E.M. and Alphin, R.S.(1988). The emetic activity of centrally administered cisplatin in cats and its antagonism by zacopride. *J. Pharm. Pharmacol.* **40**, 142-143..

Stables, R. , Andrews, P. L. R. , Bailey, H. E. , Costall, B. , Gunning, S. J. , Hawthorn, J. , Naylor, R. J. and Tyers, M. B.(1987). Antiemetic properties of the 5-HT3-receptor antagonist, GR38032F. *Cancer Treat. Rev.* **14**, 333-336.

Torii,Y. , Saito, H. and Matsuki, N.(1991). Selective blockade of cytotoxic drug-induced emesis by 5-HT3 receptor antagonists in suncus murinus. *Jpn. J. Pharmacol.* **55**, 107-113.

Torii, Y., Saito, H. and Matsuki, N. (1992). Pyrogallol, a generator of free radicals in emetogenic in Suncus murinus. In: *Mechanisms and Control of Emesis*. Bianchi, A.L., Grelot, L., Miller, A.D. and King, G.L. (Eds.), Colloque INSERM, John Libbey Eurotext, 223, pp. 359-360.

Torii, Y., Mutoh, M., Saito, H. and Matsuki, N. (1993). Involvement of free radicals in cisplatin-induced emesis in suncus murinus. *Eur. J. Pharmacol.* **248**, 131-135.

Yoshioka, M., Kikuchi, A., Matsumoto, M., Ushiki, T., Minami, M. and Saito H. (1992a). Evaluation of 5-hydroxytryptamine concentration in portal vein measured by microdialysis. *Res. Comm. Chem. Pathol. Pharmacol.* **79**(3), 370-376.

Yoshioka, M., Ikeda, T., Abe, M., Togashi, H, Minami, M. and Saito H. (1992b). Pharmacological characterization of 5-hydroxytryptamine-induced excitation of afferent cervical vagus nerve in anaesthetized rats. *Br. J. Pharmac.* **106**, 544-549.

Progress in HPLC-HPCE, Vol. 7, pp. 265-284
G. Ali Qureshi *et al.* (Eds)

Vitamin B_{12} concentration as a marker of dementia

MASURU MINAMI[1], SHIN-ICHI KIMURA [1], KEN-ICHI UENO[1], TORU ENDO[1], MASHAHIKO HIRAFUJI[1], YOSHIO MONMA[1], HIROKO TOGASHI[1], HIDEYA SAITO [2], H.S.PARVEZ [3]AND G.A.QURESHI[4] .

[1]Department of Pharmacology, Faculty of Pharmaceutical Sciences, Health Sciences University of Hokkaido, Ishikari-Tobetsu 061-02, Japan,; [2] The First Department of Pharmacogy, Hokkaido University School of Medicine, Sapporo 060, Japan; [3]Neuropharmacology Unit, University Paris XI, 91405 Orsay Cedex, France ; [4]Clinical Research Center, Novum, Karolinska Institute, Huddinge University Hospital, S-14157 Huddinge, Sweden.

INTRODUCTION

Mental deterioration associated with V-B_{12} deficiency is well documented (Droller and Dossett, 1959). Laboratory tests recommended by the National Institutes of Health Consensus Conference on Differential Diagnosis of Dementing Diseases include a complete blood counts, electrolyte and metabolic screening, thyroid panel, Vitamin B_{12} (V-B_{12}) and folate levels, syphilis serology, urinalysis, chest X-ray and ECG, and head CT scan. These evaluations are sufficient to diagnose the majority of treatable dementia.

V-B_{12} is necessary for nervous system function as well as blood cell formation. V-B_{12} deficiency may sometimes present neurologic disorders in the absence of anemia (Lindenbaum *et al.*, 1988). There is good evidence of a progressive fall in serum V-B_{12} levels with age (Dawson and Donald, 1966). Recent studies have shown that serum V-B_{12} levels decline 18 to 28 pmol/L annually in the normal elderly patients (Yao *et al.*, 1994). It is reported that the serum V-B_{12} levels were low in patients with primary degeneration dementia but in none of the patients with secondary dementia (Karnaze and Carmel, 1987). V-B_{12} deficiency does not always induce megaloblastic anemia, is not always due to malabsorption of free V-B_{12} and can express itself in different ways (Carmel *et al.*, 1995). Although the nature of the association of low serum V-B_{12} and dementia is unclear, it is well known that V-B_{12} could improve early and mild demented patients.

At first, we reviewed the reports concerning with the dementia and human V-B_{12} concentration. Next, cerebral V-B_{12} concentration of stroke-prone spontaneously hypertensive rats (SHRSP), one of the vascular dementia-animal models, was measured as well as cerebral DNA levels. Furthermore, this chapter was attempted to elucidate the role of V-B_{12} on learning ability and acetylcholine (ACh) levels after chronic oral administration of V-B_{12}.

Human serum V-B_{12} levels and dementia

There is increasing incidence for an association between senile dementia of Alzheimer-type (SDAT) and nutritionally independent V-B_{12} deficiency (McCaddon *et al.*, 1995). Low serum V-B_{12} levels are common in conditions such as dementia (Carmel *et al.*, 1995 & 1996; Gomes *et al.*, 1995). Carmel *et al.* (1996) reported that serum V-B_{12} levels in 36 spouses correlated with those of the 36 demented patients related to them (r=0.46, p=0.004). These observations identify a hitherto unsuspected group of people at high risk for V-B_{12} deficiency and suggest that spouses of demented patients with low V-B_{12} levels should have their V-B_{12} levels measured.

Mean serum V-B_{12} levels were significantly lower in SDAT (219 μg/dl, RIA) than in non-dementia patients (367 μg/dl, RIA, $p < 0.01$) (Cole and Prchal, 1984). It is unclear if patients with dementia represent definable deficiency and what the mechanisms are. According to a prospective study of Carmel *et al* (1995), 50% of patients tested had abnormal deoxyuridine suppression and 44% had increased serum methylmalonic acid and/or homocysteine levels; these test results correlated with each other (Carmel *et al.*, 1995). Perhaps most importantly, this is accompanied not only by metabolic changes but by evidence of mild neurologic dysfunction. Cunha *et al.* (1995) reported that 16 of 19 demented patients with low V-B_{12} levels persisted in showing progressive decline during 3 to 24 months despite V-B_{12} replacement. The non-responder to V-B_{12} replacement seems to reflect the presence of associated irreversible dementia or a follow-up of shorter duration.

Thus, screening for V-B_{12} deficiency should be considered in patients with recent onset of mild mental status changes. Although the long-standing dementia does not improve, treating such patients with V-B_{12} has other concrete benefit (Carmel *et al.*, 1995). Current state-of-the-art testing uses serum V-B_{12} levels as a screening test and serum or urine homocysteine and methylmalonic acid (MMA) determinations as confirmatory tests (Swain, 1995).

Stages of vitamin B_{12} deficiency

The recommended daily allowance of V-B_{12} is 2 microgram/day in the average adult, and the average daily intake is between 6 and 7 microgram/day (Food and National Board, 1989). The body stores 2 to 5 mg of the vitamin, 50-% of which is stored in the liver (Kano *et al.*, 1985). A reference interval of 190-1016 ng/l determined by radioimmunoassay (RIA) was established in 332 healthy volunteers by Carmel *et al.* (1995). Serum V-B_{12} should possibly be considered a screening test, with high levels ruling out a V-B_{12} deficiency, levels between 150 and 350 pmol/L requiring confirmation, and levels lower than 150 pmol/L probably not needing confirmation. Some asymptomatic patients may have borderline or even subnormal serum levels.

V-B_{12} deficiency is thought to be a continuum with five stages (Herbert, 1987). These stages include: I, normality; II, negative V-B_{12} balance; III, V-B_{12} depletion with possible clinical signs and symptoms (reversible neuropsychiatric findings); IV, V-B_{12}-deficient erythropoiesis with possible clinical signs or symptoms (potentially reversible neuropsychologic symptoms); V, V-B_{12} deficiency anemia with probable clinical signs and symptoms, including irreversible lateral column involvement.

Diagnostic laboratory testing of vitamin B_{12} deficiency

Serum V-B_{12} assay It was later found that early assays overestimated actual V-B_{12} levels owing to the measurement of nonactive cobalamin analogues (Swain, 1995).

Serum V-B_{12} concentrations were determined with a radioisotope assay with purified intrinsic factor (Amersham Int., Bucks., England); normal 0.95 reference interval from 135-590 pmol/l, mean 358 pmol/l.

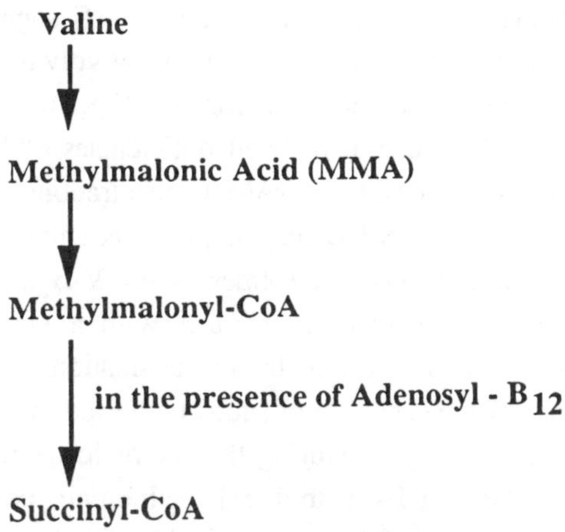

Figure 1 Enzymatic reactions dependent on the presence of vitamin B_{12}

Serum methylmalonic acid (MMA) and serum homocystein concentration Various investigators have proposed defects in activity of one or the all of the two V-B_{12}-dependent enzyme, L-methylmalonyl-CoA mutase or methionine synthetase, to explain the neuropsychiatric abnormalities seen in V-B_{12} deficiency (Scott *et al.*, 1981; Stabler *et al.*, 1988)(Fig. 1). Blocking the reaction promoted by these enzymes would result in the accumulation of methylmalonic acid (MMA), in the case of L-methylmalonyl-CoA synthetase (Fig. 1) and homocysteine in the case of methionine synthetase (Fig. 2)(Scott *et al.*, 1981; Stabler *et al.*, 1988). MMA is a metabolite accumulating early in V-B_{12} deficiency. The V-B_{12} concentration was significantly lower in SDAT (Kristensen *et al.*, 1993). Correspondingly, the mean MMA levels was significantly higher in Alzheimer's disease (AD) ($p=0.009$) (Kristensen *et al.*, 1993). Serum MMA concentrations were measured by a stable-isotope-dilution method with solid-phase extraction described previously (Rasmussen, 1989).

Serum MMA and total homocysteine concentrations could be measured by recently developed techniques using capillary gas chromatography and mass spectrometry (Stabler *et al.*, 1988). Total serum homocysteine can be determined by high performance liquid chromatography with fluorescence detection (Jacobsen *et al.*, 1994). These metabolites may accumulate in the blood stream resulting in elevated levels. Serum levels for homocystein and MMA may be more specific but in most hospitals only a spot urine test for homocystein and a 24-hour urine collection for testing MMA are available. Savage *et al.* (1994) have shown that elevated levels of MMA are 96% to 99% accurate as an indicator for V-B_{12} deficiency. Only 7 of 434 V-B_{12}-deficient patients tested showed normal levels of MMA; 6 of these 7 had elevations in homocystein levels.

Methionine-S-adenosyltransferase (MAT) activity In the group of SDAT patients who had low plasma levels of V-B_{12}, the reduction in methionine-S-adenosyltransferase (MAT) activity was significantly correlated with an increase in the serum levels of homocysteine (Gomes *et al.*, 1995). S-adenosylmethionine (SAM) is required in numerous transmethylation reactions involving nucleic acid, proteins, phospholipids, amines and other neurotransmitters (Fig.2). The synthesis of SAM is intimately linked with V-B_{12} metabolism, and deficiencies of V-B_{12} have been found to reduce central nervous system (CNS) SAM concentrations (Bottiglieri *et al.*, 1994). Preliminary studies indicate that SAM may improve cognitive function in patients with dementia (Bottiglieri *et al.*, 1994). Treatment with V-B_{12}, SAM and folate cause a significant decrease in homocysteine in parallel with a significant increase in MAT (Gomes *et al.*, 1995). It is suggested that the evaluation of erythrocyte MAT activity may be a useful marker for the detection of such an aberration.

MAT activity was assayed by measuring the rate of formation of S-adenosyl-[methyl-^{3}H] methionine from ATP and L-[methyl-^{3}H] methionine, using a modification of the procedure of Oden and Clarke (1983), as described by Gomes *et al.* (1994).

Antibodies to parietal cells and intrinsic factor Tests for antibodies to parietal cells and intrinsic factor may be useful for diagnosis. These are present in the majority (90%) of cases of pernicious anemia but are also present in atrophic gastritis without pernicious anemia; however, the absence of these markers does not rule out possible V-B_{12} deficiency (Nimo and Carmel, 1987; Fairbanks *et al.*, 1983). Serum gastrin levels may also be a helpful marker when interpreting an equivocal V-B_{12} level (Swain, 1995). The criteria for diagnosing pernicious anemia were considered when the patients had a V-B_{12} level below 180 pmol/l (250 pg/ml, RIA) (Carmel *et al.*, 1988).

Shilling test In the first stage of the test, the patient is given radiolabeled (cobalt^{57}) V-B_{12} orally, followed by a flushing dose of intramuscular V-B_{12} 1 to 2 hours later. Radioactivity is then measured in the urine. Stage II consists of repeating the test with intrinsic factor also being administered. Normal results on stage II, but not stage I, indicate that V-B_{12} deficiency is probably due to intrinsic factor deficiency. Problems with this test include a 14% chance of false positives (Carethers, 1988), a 32% chance of equivocal results, and a relatively high cost in nuclear medicine departments. Lindenbaum *et al.* (1988) have found that up to 25% of patients with pernicious anemia will still have low results on stage II for 2 months after vitamin replacement has begun owing to problems in the absorption of the intrinsic factor-V-B_{12} complex. Another limitation of this test is that some patients have problems absorbing the vitamin in its protein-bound state. Therefore, this test has limited value.

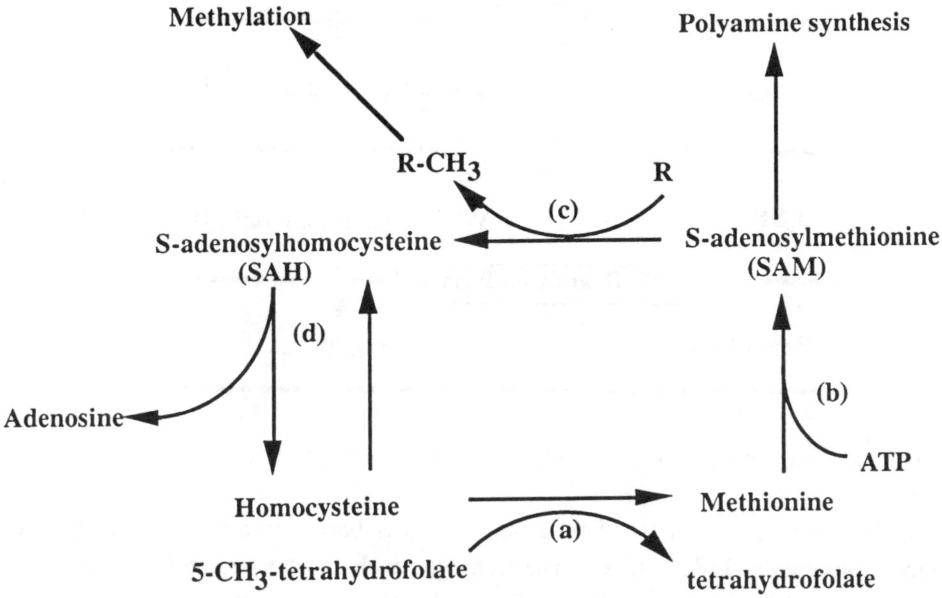

Figure 2 S-adenosylmethionine metabolism: (a) methionine synthetase; (b) methionine adenosyltransferase; (c) R-methyl transferase; (d) SAH hydrolase.

V- B_{12} in Cerebrospinal fluid (CSF) The vitamin is protein bound and is separated from the protein complex by pepsin in the presence of stomach acid. It is then bound by intrinsic factor and transported from the proximal ileum to the distal ileum, where it is then bound to transcobalamin II for transport to the liver and bone marrow (Fig. 3). V-B_{12} in serum and Cerebrospinal fluid (CSF) is bound to specific transport proteins (transcobalamins). The choroid plexus is known to actively regulate concentrations in the CSF and to make the blood CSF barrier selective (Spector and Johansson, 1989). CSF V-B_{12} concentrations correlated significantly with serum V-B_{12} concentrations (Regland *et al.*, 1992). CSF V-B_{12} concentrations and CSF/serum V-B_{12} ratios were significantly lower in the demented patients than in the control group (Regland *et al.*, 1992). CSF and serum V-B_{12} levels appear insufficient as measures of the true brain V-B_{12} status. It may be a more dynamic approach to use the CSF/serum V-B_{12} ratios as an indicator of transport function across the blood brain barrier, and possibly across the CSF brain cell barrier. Thus, assessment of a CSF/serum ration is probably more useful than assessment of CSF V-B_{12} alone, since it also takes into account the possibility of a dysfunctional transport of V-B_{12} across membranes (Regland *et al.*, 1992)(Fig.3).

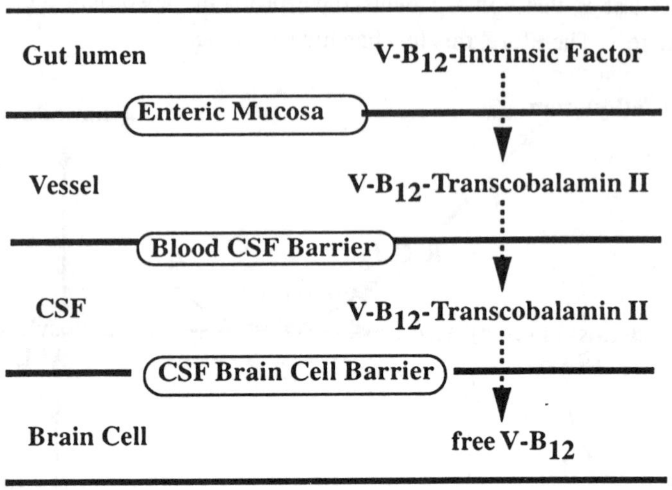

Figure 3 The transportation of V-B_{12} from the gut to the brain cell. CSF: cerebrospinal fluid.

The RIA technique used (Regland *et al.*, 1992) is based on a kit from Diagnostic Products Corporation, L.A., USA. The technique includes conversion by means of a boiling step and the separation of free and bound fractions with dextran-charcoal. By increasing the sample volume to 1.3 ml, retiter the antibody and tracer concentrations, increasing the antibody incubation time and using a very well controlled and reproducible separation step, the coefficient of variation can be kept low. At a concentration of 5 pmol/l, there is a coefficient of variation within-day of 3-4% and a day-to-day variation of 4-5%.

V-B_{12} deficiency is unlikely to be a common cause of mental disturbance in the elderly. For patients with no neurological involvement, CSF V-B_{12} was found to be 17.7 ± 7.8 pg/ml (13.7 ± 6.0 pmol/l, RIA)(Frenkel *et al.*, 1971). They were the first to report on isotopic measurement of V-B_{12} in CSF. They described a technique whereby concentrations as low as 1 pg/ml (<1 pmol/L) can be measured accurately, which is necessary in CSF. Despite normal serum V-B_{12} levels, 13 out of 17 patients with SDAT and alcohol related brain damage have a CSF levels of V-B_{12} lower than 5 pg/ml (radio assay method)(Van Tiggelen, 1984). According to Frenkel *et al.* (1973) the pathological level found in patients with untreated pernicious anemia. In multi-infarct dementia (MID) the normal serum V-B_{12} corresponded with a normal CSF V-B_{12} (above 10 pg/ml, radio assay). By using competitive protein-binding assay method, the mean V-B_{12}

level in serum was 742 ± 359 pg/ml for the SDAT group, and 962 ± 254 pg/ml for the MID group (Ikeda *et al.*, 1990). The mean V-B_{12} level in the CSF for the SDAT (28 ± 7 pg/ml) was significantly lower than that of the MID group ($p < 0.01$) (Ikeda *et al.*, 1990). This indicates abnormal functions of the choroid plexus and possibly of the blood brain barrier in SDAT, not in MID. The V-B_{12} level in CSF might serve as a biological marker for SDAT.

V-B_{12} deficiency and AIDS dementia complex

V-B_{12} deficiency is also associated with acquired immunodeficiency syndrome (AIDS) (McCaddon *et al.*, 1995). Decreased serum V-B_{12} levels occurred in up to 20% of AIDS patients (Herzlich & Schiano, 1993). There was an associated low serum V-B_{12} resulting from malabsorption due to low gastric intrinsic factor secretion. V-B_{12} deficiency may correlate with conditions involving decreased stomach acidity because the protein-bound V-B_{12} must be separated by acid and pepsin. The human immunodeficiency virus (HIV) causes achlorhydria secondary to decreased secretion of acid and pepsin (Herzlich et al., 1992). Patients with HIV also exhibit decreased secretion of intrinsic factor, which would lead to an increased incidence of pernicious anemia. Therefore, patients with AIDS also should be considered for periodic screening for V-B_{12} levels. Protease inhibition is a common factor in Alzheimer and AIDS dementia resulting in protein-bound V-B_{12} malabsorption and disrupted V-B_{12} metabolism (McCaddon *et al.*, 1995). Following treatment with V-B_{12} the symptoms resolved over a 2-months period (Herzlich and Schiano, 1993). AIDS dementia complex represented a reversible adverse synergistic interaction between HIV and V-B_{12} deficiency.

Brain V-B_{12} and the rhythm disturbance in human and SHRSP

Human

Kamgar-Parsi *et al.* (1983) first reported that a patient with a non-24 hour sleep-wake cycle was able to follow a normal 24 hour sleep-wake regimen after V-B_{12} treatment. Sugita *et al.* (1988) and Okawa *et al.* (1990) reported successful treatment with V-B_{12} of the sleep-wake rhythm disorder. The patient's improvement after treatment with V-B_{12} is striking. It appeared to enable the patient to advance his sleep period and accomplish at a normal sleep schedule. V-B_{12} deficiency might be associated with central nervous system deficits, including insomnia.

Stroke-prone spontaneously hypertensive rats (SHRSP)

Using stroke-prone spontaneously hypertensive rats (SHRSP) as one of the vascular dementia-animal models (Minami *et al*, 1985 & 1995; Saito *et al.*, 1995), we attempted to evaluate the role of V-B_{12} on behavioral changes including ambulatory activity and passive avoidance test after chronic oral administration of V-B_{12}. The lower part of Fig.

4 indicates the results of power spectral analysis of the behavioral changes in the SHRSP shown in Fig. 5 which died from cerebral infarction. At an age of 18 weeks (the lower part of Fig. 5, this SHRSP had a 24 hr τ (tau) value for ambulation. However, at the onset of stroke (the lower part of Fig. 4), a much longer periodicity was observed in addition to the 24 hr periodicity.

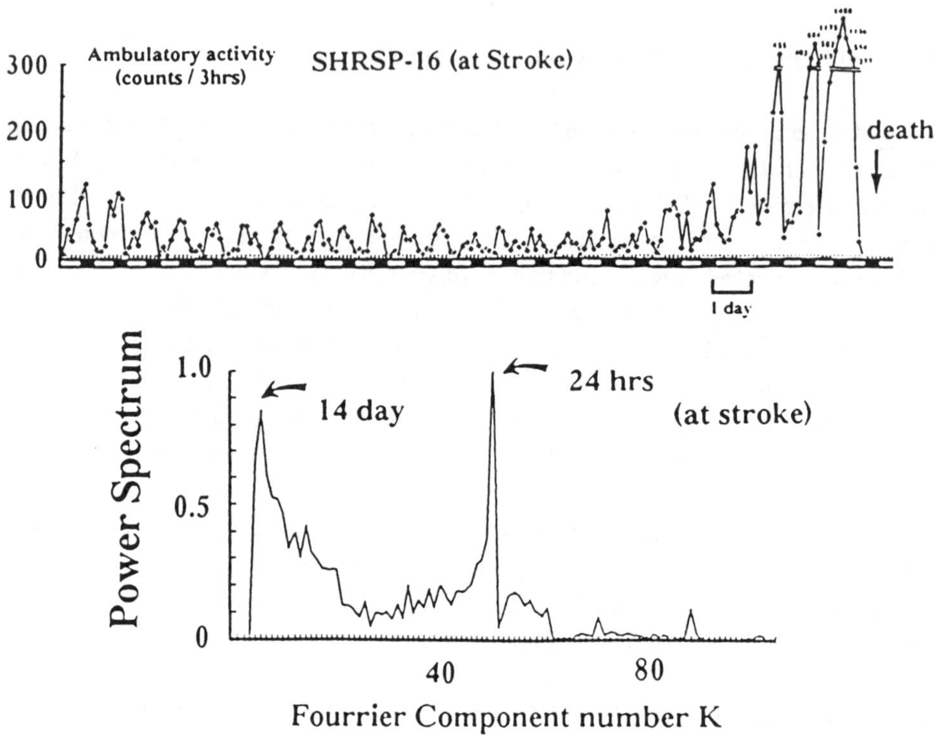

Figure 4 The upper part of the figure indicates the behavioral changes of SHRSP which died from cerebral infarction. The lower part of the figure indicates the power spectral analysis of ambulation in SHRSP at the onset of stroke. A much longer periodicity was observed in addition to the circadian rhythm. This SHRSP is the same rat described in Fig. 5. (from Minami *et al.*, 1995 with permission)

All 27 SHRSP which died from cerebral infarction and/or cerebral bleeding showed a much longer periodicity in addition to the circadian rhythm such as Fig. 4. While, only one of 10 SHRSP treated by methylcobalamin (Eisai Co. Ltd., Japan: 2 mg/kg, p.o.) showed new rhythm (a much longer periodicity in addition to the circadian rhythm). With respect to frequency of new rhythm, there is a significant difference between the control SHRSP and methylcobalamin-treated SHRSP.

V-B_{12} ceased the abrupt increase of ambulatory activity; disturbance of biological rhythms before death in SHRSP might correspond to behavioral changes such as delirium-state observed in patients with dementia caused by cerebrovascular lesions. Brain V-B_{12}/DNA ratios of SHR and SHRSP were significantly lower than that of normotensive Wistar Kyoto rats (WKY) as shown in Fig. 6. After treatment of methylcobalamin, brain V-B_{12}/DNA ratios of SHR and SHRSP were significantly increased. The mechanism of these effects by oral administered methylcobalamin are unknown. The efficacy of methylcobalamin in our cases suggests that V-B_{12} might be beneficial for the treatment in the disturbance of biological rhythms as well as sleep-wake disturbance.

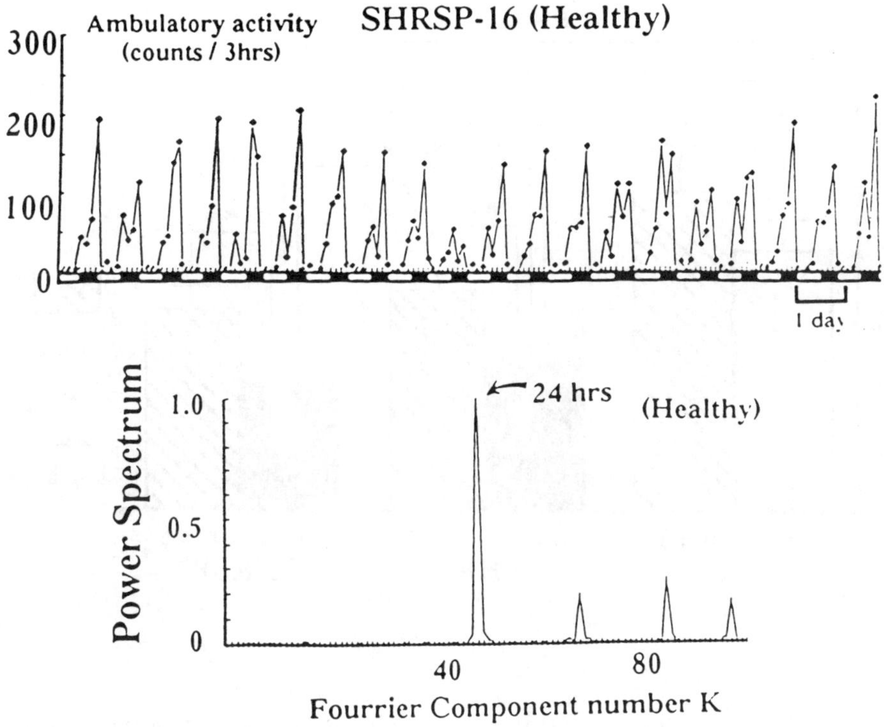

Figure 5 Behavioral changes (the upper part of the figure) and the power spectral analysis of ambulation (the lower part of the figure) in a SHRSP at the younger age (healthy). Synchronization with the light and dark phases was observed and circadian rhythms were noted in healthy SHRSP. (from Minami *et al.*, 1995 with permission)

Characteristics of SHRSP From an early age, systolic blood pressure of SHRSP (186.8 ± 18.8 mmHg, 15 weeks old) was significantly higher than that of WKY (137.3 ± 14.7 mmHg)($p < 0.001$). Death occurred at 35.5 ± 15.2 weeks in 27 male SHRSP: 11 died from cerebral infarction, 10 from cerebral hemorrhage, and 6 from cerebral hemorrhage with cerebral infarction. Senility was the most frequent cause of death in the control male WKY who died at 82.7 ± 26.0 weeks. There was a significant difference between the age at death of SHRSP and WKY. The SHRSP has a specific humoral make-up including high renin activity, a high aldosterone concentration and a high angiotensin II (AII) concentration (Kawashima *et al.*, 1980). In our study, SHRSP-stroke case had significantly higher plasma norepinephrine (NE), AII and arginine vasopressin (AVP) levels (Minami *et al.*, 1988).

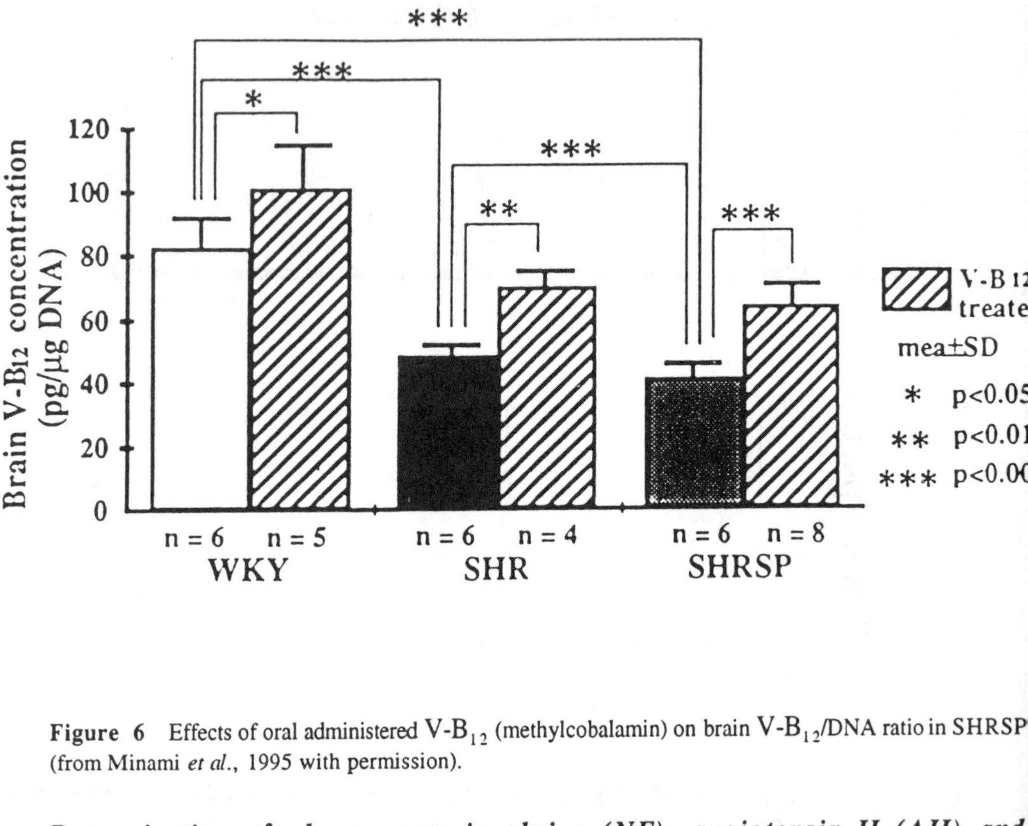

Figure 6 Effects of oral administered V-B_{12} (methylcobalamin) on brain V-B_{12}/DNA ratio in SHRSP (from Minami *et al.*, 1995 with permission).

Determination of plasma norepinephrine (NE), angiotensin II (AII) and arginie vasopressin (AVP) After purification with aluminum oxide (Bioanalytical System Inc., USA), catecholamines were separated by high performance liquid chromatography (HPLC) using a Waters Associates (USA) ODS reverse phase column (4.6 mm x 250 mm). Plasma NE was determined using HPLC with electrochemical detection (HPLC-ECD). We used a HPLC system (EP-10, Eicom, Japan) attached to an ECD (ECD-100, Eicom, Japan) to build a system based on the principle of amperometory. Plasma NE concentration was corrected with protein determined by the method of Lowry *et al.* (1951), using bovine serum albumin as the standard.

Plasma AII and AVP were determined using radioimmunoassay (RIA). The determination of AII and AVP was previously described in detail (Minami *et al.*, 1988).

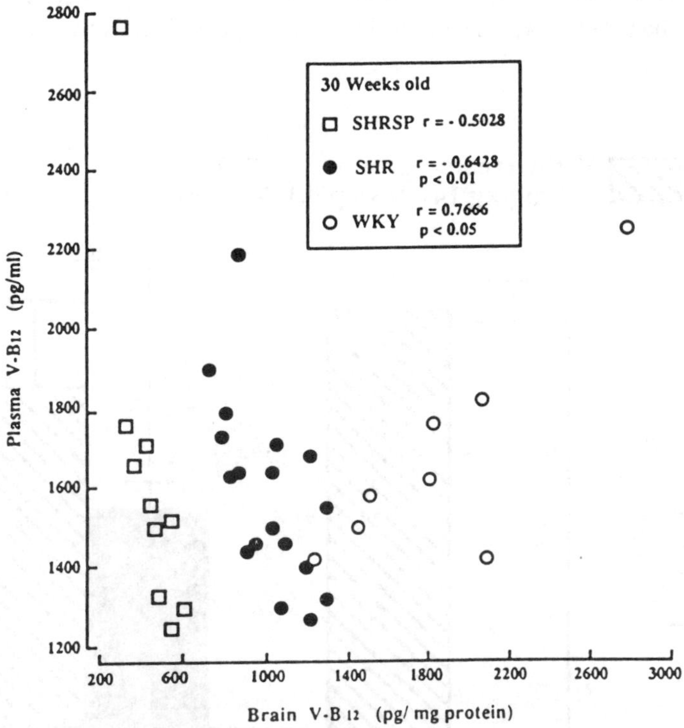

Figure 7 Correlationship between brain V-B_{12} levels and plasma V-B_{12} levels (from Minami *et al.*, 1995 with permission).

Determination of Ambulatory activity in SHRSP Age-matched male SHRSP and WKY were subjected to a 12-hr light (L) and dark (D) alternation cycle (LD). Illumination was provided by fluorescent light (100 lux). Room temperature was maintained at 22 ± 2°C throughout the experiment. Rats were fed a regular diet (Clea CE-2, Clea Japan, Inc., Tokyo). The automatic Ambulo-Drinkometer (O'Hara & Co., Ltd., Tokyo) was used to determine ambulatory activity and water drinking activity simultaneously. This apparatus is composed of ten separated steel cages equipped with micro-switches which are activated by the tilting of the floor as the rats move around. A drinking spout is connected to a water tank via an infusion machine made of acrylfiber and stainless tubes which delivers 0.05-ml drops. The water falls drop by drop as the rat drinks. There is a weak electrical current in the upper part of the infusion machine. Short circuits generated by individual drops are amplified to activate an electromagnetic counter. Using this apparatus, we could determine exactly amount and rate of water drinking. Water drinking activity was recorded once every hour. This apparatus has been described in detail previously (Togashi *et al.*, 1982; Minami *et al.*, 1985). Ambulatory and drinking activity data were obtained at hourly intervals and were subjected to autocorrelation and power spectral analysis. The average period of the rhythm (τ_{LD})(Stephan, 1983) was determined by power spectral analysis.

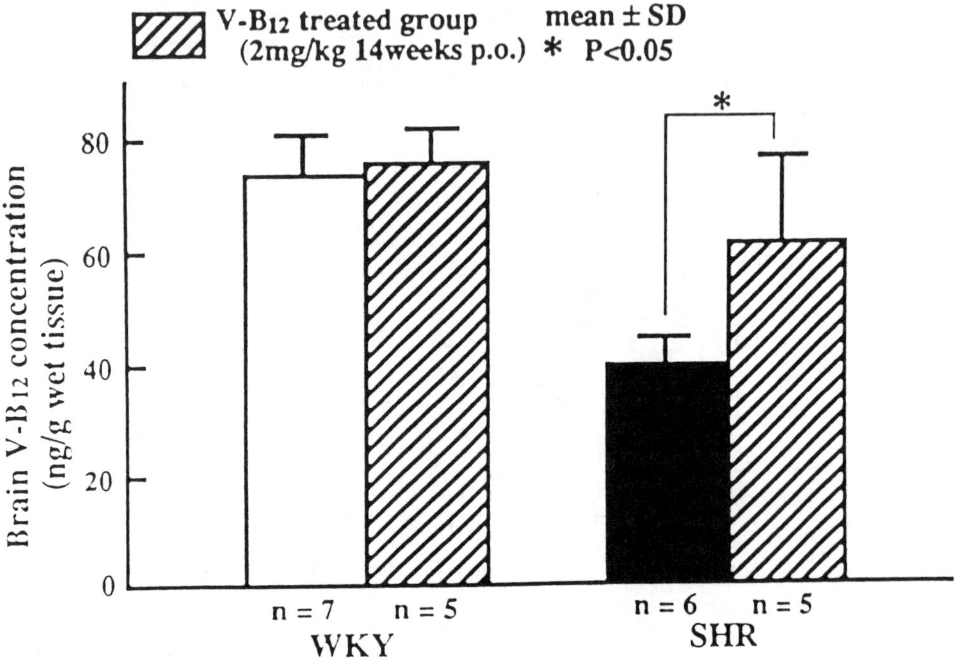

Figure 8 Effect of oral V-B_{12} (methylcobalamin) on the brain V-B_{12} concentration of WKY and SHR (from Minami *et al.*, 1995 with permission).

Cerebral V-B_{12} concentration and cerebral DNA concentration Plasma and cerebral V-B_{12} levels were determined using competitive protein binding assay (Baxter, USA). Cerebral DNA contents were assayed by fluorescence method.

After absorption, V-B_{12} is transported in the serum, bound primarily to an $\alpha 2$ globulin called transcobalamin II. Another V-B_{12} binding protein called transcobalamin I also exists and is probably derived from leucocytes. Transcobalamin II is a key protein for transport of V-B_{12} across cell membrane. As shown in Fig. 7, the plasma V-B_{12} (transcobalamin I and II bound forms) was negatively correlated to the cerebral V-B_{12} in SHR and SHRSP. While, in normotensive WKY, there were significant positive correlations between the plasma V-B_{12} and cerebral V-B_{12}. Low levels in cerebral V-B_{12}

were thought to be caused by such a deficiency in the transportation of V-B_{12} across the blood brain barrier. However, the cerebral V-B_{12} level increased significantly after oral administration of methylcobalamin (2 mg/kg/day, p.o.) in SHR (Fig. 8). These results suggest that methyl-cobalamin was absorbed sufficiently in SHR.

V-B_{12} and DNA levels of same sample in the rat brain were simultaneously determined (Fig. 9). The brain V-B_{12} concentration correlated significantly with the brain DNA concentration in WKY, SHR and SHRSP. Linear regression lines of SHR and SHRSP shift to the left from that of WKY.

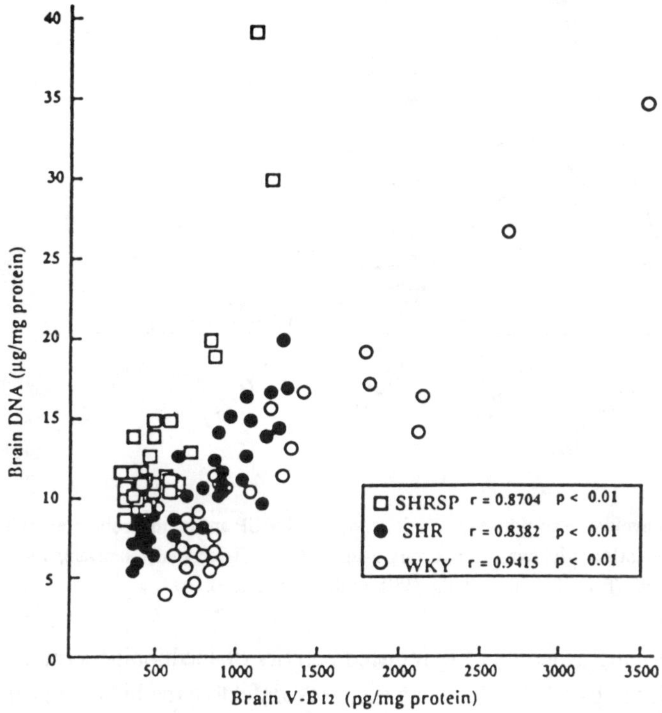

Figure 9 Correlation between V-B12 levels and DNA levels in rat brain (from Minami *et al.*, 1995 with permission).

Cerebral acetylcholine (ACh) level and dementia in V-B_{12} deficiency

Human

The involvement of acetylcholine (ACh) as the main neurotransmitter in SDAT is now well documented. V-B_{12} involved as coenzyme in biochemical reactions related to the liberation of methyl groups in the brain which might directly but selectively influence the ACh synthesis through the bioavailability of choline (Ch), the enzyme substrate of Ch acetyl-transferase (Nadeau and Roberge, 1988).

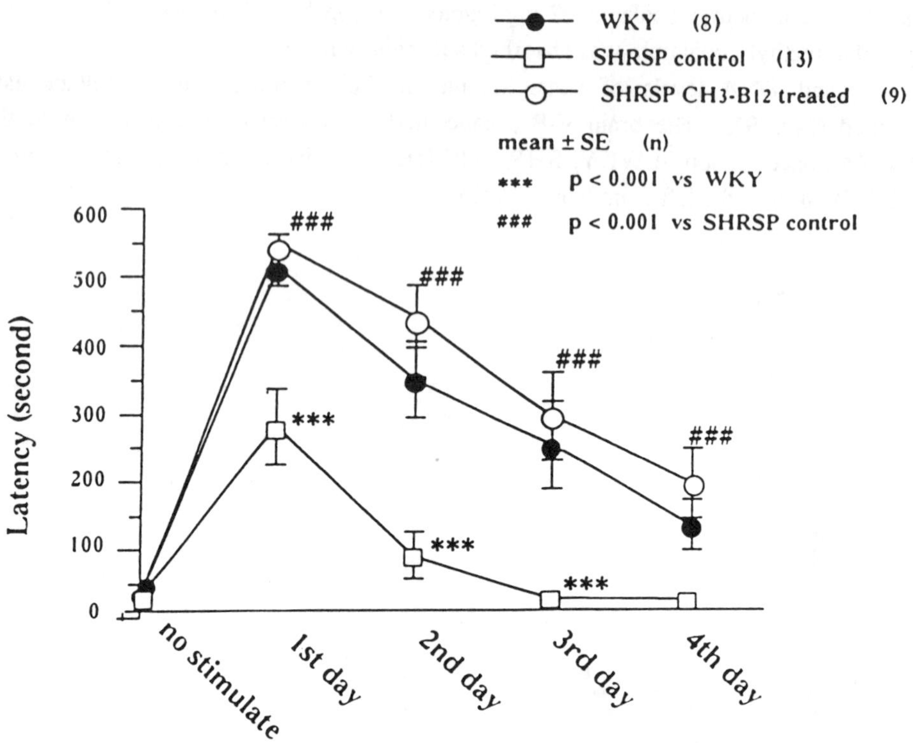

Figure 10 Passive avoidance response in WKY, control-SHRSP and methylcobalamin (CH_3-B_{12}) treated SHRSP. The first retention trial was carried out 24-h. After the acquisition trials, four further trials followed at 24-h intervals (from Minami *et al.*, 1995 with permission).

Hakim *et al.* (1983) suggested that the reduced activity of methionine synthetase, shown to be the result induced by V-B_{12} deficiency, causes defective methylation in the liver and leads to insufficient formation of Ch, the necessary precursor of acetylcholine in the brain. It is presumed that prolonged low serum V-B_{12} levels might produce irreversible changes in the basal nucleus of Meynert such as a disorder of cholinergic innervation (Coyle *et al.*, 1983). CSF ACh levels were reported to correlate negatively with degree of dementia evaluated by Memory and Information Test (Davis *et al.*, 1983). The aging process of the central cholinergic neuron is known to involve major changes in ACh synthesis, storage and release mechanism (Allen *et al.*, 1983; Perry, 1980). A disturbance in cholinergic systems was also reported in patients with vascular dementia (Wallin *et al.*, 1989). A report about ACh determination in human CSF demonstrated that CSF ACh concentrations in middle-aged and elderly normal healthy subjects were markedly lower than that in younger group (Duvoisin and Dettbarn, 1967). A disturbance in the central cholinergic systems is reported in patients with vascular dementia (Wallin *et al.*, 1989) as well as in the normal aging human and in patients with SDAT (Allen *et al.*, 1983; Perry, 1980).

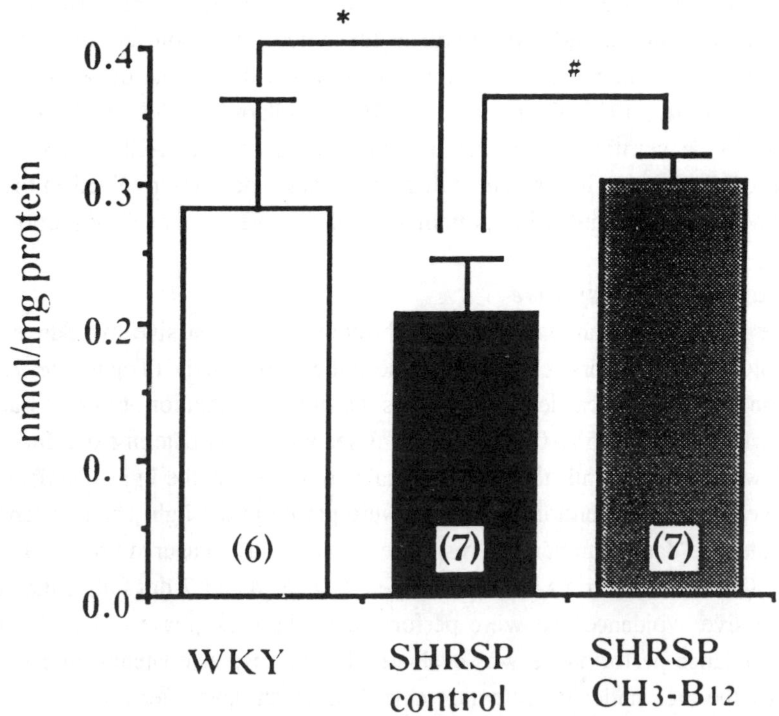

Figure 11 Effect of methylcobalamin (CH3-B12) on the hypothalamus acetylcholine concentration of WKY and SHRSP. *$p < 0.05$ vs WKY, #$p < 0.05$ vs SHRSP-control (from Minami *et al.*, 1995 with permission).

SHRSP

The levels of ACh in the CSF of rats did not differ from those of humans (Duvoisin and Dettbarn, 1967). CSF ACh concentration was significantly lower in SHRSP than that in WKY (Togashi *et al.*, 1990 & 1994). In our another study, the significant impairment in passive avoidance response observed in SHRSP may reflect a disturbance of memory (Togashi *et al.*, 1990). In the passive avoidance task, the response latency of nondrug control SHRSP was significantly shortened as compared with that in WKY. The response latency in V-B_{12}-treated SHRSP was significantly restored (Fig. 10). ACh levels of brain cortex and hypothalamus in nondrug control SHRSP were significantly lower than those in WKY. While, brain ACh levels of V-B_{12}-treated SHRSP were increased significantly (Fig. 11). We lately reported that the decrease in the choline passive avoidance task was observed in SHRSP with positive correlations between the hippocampal ACh levels and the latency of the passive avoidance response (Togashi *et al.*, 1996). Chronic administration of V-B_{12} ameliorated cerebral V-B_{12} deficiency as well as behavioral abnormalities in SHRSP. It is known that biosynthesis of Ch occurs in a series of reactions requiring V-B_{12} (Nadeau and Roberge, 1988). Brain cholinergic systems are generally thought to be critical for memory function and disturbance of the central cholinergic system has been shown in patients with vascular dementia as well as in SDAT (Coyle *et al.*, 1983; Gattfreis *et al.*, 1983; Collerton, 1986; Wallin *et al.*, 1989). It remains to be clarified whether the change in the central ACh levels are a direct consequence of V-B_{12}. It is possible that neurotransmitters are involved in the observed effects of V-B_{12} on learning ability, although no defenitive data have been reported.

Passive avoidance response

Rats were tested using an one-way step-through type of passive avoidance apparatus divided into light and dark chambers as described previously (Togashi *et al.*, 1996). Acquisition trials were carried out 24 hrs before the retention trials. Each animal received a foot shock (75V, 0.2 msec) for 3 seconds upon entering the dark chamber. This trial was repeated until the rats eventually remained in the light chamber for more than 300 sec. In the retention trials, rats were placed in the light chamber and the time taken to enter the dark chamber, the response latency, was measured up to a maximum of 600 sec. The retention trials were performed 1, 2, 3, 4 and 7 days after the acquisition trials. Passive avoidance test were performed in the dark phase (13:00-16:00). The passive avoidance performance was evaluated by the response latency in each retention trial and by the area under the retention curve for seven days after acquisition trials. On the day of the final session, the rat tissue ACh concentration was determined by HPLC-ECD after the preparation of microwave irradiation (Matsumoto *et al.*, 1990 & 1991).

Acethylcholine (ACh) and choline (Ch) determination

After the passive avoidance test, rats were killed by microwave irradiation (5kW for 1.5 sec), brains were removed and right hemispheres were dissected into 7 regions (cortex, cerebellum, midbrain, hippocampus, hypothalamus, medulla oblongata and striatum) according to the method of Glowinski and Iversen (1966) to measure ACh and Ch contents. The tissues were stored at -80°C until they were assayed. Extraction of the ACh and Ch tissue was carried out with aliquots of 0.2 N perchloric acid including ethylhomocholine as an internal standard. The homogenates were centrifuged for 10 min (10,000 r.p.m., 2°C). The supernatant, which was neutralized with 0.2 N potassium hydrogen carbonate, was filtrated (0.22 mm Millipore filter, Bedford, MA, USA) and injected into a HPLC (Eicom, Japan) system connected with an immobilized enzyme reactor and an ECD (Eicom, Japan) as reported previously (Matsumoto *et al.*, 1990 & 1991).

CONCLUSION

Vitamin B_{12} (V-B_{12}) deficiency may sometimes present as a neurological disorder. Mental deterioration in elderly and Alzheimer-type dementia have also been associated with cerebral V-B_{12} deficiency. Behavioral impairment in the passive avoidance task was observed in SHRSP with positive correlations between the hippocampal ACh levels and the latency of the passive avoidance response. These studies suggest that both V-B_{12} and ACh deficiencies might be involved in the mechanisms underlying the pathogenesis of dementia. Methylcobalamin might be beneficial in the treatment of biological rhythm disturbance and impairment of learning and memory in the patients with dementia. V-B_{12} deficiency may be more widespread than initially thought. General screening with V-B_{12} levels should be considered for all elderly. The impairment might be improved by oral administration of V-B_{12}.

REFERENCES

Allen, S.J., Benton, J.S., Goodhardt, M.J., Haan, E.A., Sims, N.R., Smith, C.C.T., Spillane, J.A., Bowen, D.M. and Davidson, A.N. (1983). Biochemical evidence of selective nerve cell changes in the normal aging human and rat brain. *J. Neurochem.* **41**, 256-265.

Bottiglieri, T., Hyland, K. and Reynolds, E.H. (1994). The clinical potential of ademetionine (S-adenosylmethionine) in neurological disorders. *Drugs* **48**, 137-152.

Carethers, M. (1988). Diagnosing vitamin B12 deficiency: a common geriatric disorder. *Geriatrics* **43**, 89-94; 105-107; 111-112.

Carmel, R., Sinow, R.W., Siefel, M.E. and Samloff, I.M. (1988). Food cobalamin malabsorption occurs frequently in patients with unexplained low serum cobalamin levels. *Arch. Intern. Med.* **148**, 1715-1719.

Carmel., R., Gott, P.S., Waters, C.H., Cairo, K., Green, R., Bondareff, W., DeGiorgio, C.M., Cummings, J.L., Jacobsen, D.W. and Buckwalter, G. (1995). The frequently low cobalamin levels in dementia usually signify treatable metabolic, neurologic and electrophysiologic abnormalities. *Eur. J. Haematol.* **54**, 245-253.

Carmel, R., Cairo, K., Bondareff, W., Gott, P.S., Cummings, J.L. and Henderson, V.W. (1996). Spouses of demented patients with low cobalamin levels: a new risk group for cobalamin deficiency. *Eur. J. Haematol.* **57**, 62-67.

Cole, M.G. and Prchal, J.F. (1984). Low serum vitamin B_{12} in Alzheimer-type dementia. *Age and Ageing* **13**, 101-105.

Collerton, D. (1986). Cholinergic function and intellectual decline in Alzheimer's disease. *Neuroscience* **19**, 1-28.

Coyle, J., Price, D. and DeLong, M. (1983). Alzheimer's disease: a disorder of cortical cholinergic innervation. *Science* **219**, 1184-1189.

Cunha, U.G., Rocha, F.L., Peixoto, J.M., Motta, M.F. and Barbosa, M.T. (1995). Vitamin B_{12} deficiency and dementia. *Int. Psychogeriatr.* **7**, 85-88.

Davis, K.L., Hsieh, JY-K., Levy, M.I., Horvath, T.B., Davis, B.M. and Mohs, R.C. (1983).

Cerebrospinal acetylcholine, choline and senile dementia of the Alzheimer type. *Psychophamacol. Bull.* **18**, 193-19 .

Dawson, A.A. and Donald, D. (1966). The serum Vitamin B_{12} in the elderly. *Geront. Clin.* **8**, 220-225.

Droller, H. and Dossett, J.A. (1959). Vitamin B_{12} levels in senile dementia and confusional states. *Geront. Clin.* **1**, 96-106.

Duvoisin, R.C. and Dettbarn, W.D. (1967). Cerebrrospinal fluid acetylcholine in man. *Neurology* **17**, 1077-1081.

Fairbanks, V.F., Lennon, V.A., Kokmem, E. and Howard, F.M.Jr. (1983). Tests for pernicious anemia: serum intrinsic factor antibody. *Mayo Clin. Proc.* **58**, 203-204.

Food and Nutritional Board, National Research Council (1989). Recommended daily allowances. 1oth ed. Washington, DC: National Academy Press.

Frenkel, E.P., McCall, M.S. and White, J.D. (1971). An isotope measurement of vitamin- B_{12} in cerebrospinal fluid. *Am. J. Clin. Pathol.* **55**, 58-64.

Frenkel, E.P. and McCall, M.S. (1973). Cerebrospinal fluid folate and vitamin B_{12} in anti-convulsant induced megaloblastosis. *J. Lab. Clin. Med.* **81**, 105.

Gattfreis, C-G, Adorfsson, R., Aquilonius, S-M, Carlsson, A., Eckernäs, S.Å, Nordberg, A., Oreland, L., Svennerholm, L., Wiberg, Å. and Winblad, B. (1983). Biochemical changes in dementia disorders of Alzheimer type (AD/SDAT). *Neurobiol. Aging* **4**, 261-271.

Glowinski, J. and Iversen, L.L. (1966). Regional studies of catecholamines in the rat brain. I. The disposition of [^{3}H]norepinephrine, [^{3}H]dopamine and [^{3}H]dopa in the various regions of the brain. *J. Neurochem.* **13**, 655-669.

Gomes Trolin, C., Löftberg, C., Trolin, G. and Oreland (1994). Brain ATP: L-methionine S-adenosyltransferase (MAT), S-adenosylmethionine (SAM) and S-adenosylhomocysteine (SAH): Regional distribution and age-related changes. *Eur. Neuropsychopharmacol.* **4**, 469-478.

Gomes, T.C., Regland, B. and Oreland, LK. (1995). Decreased methionine adenosyltransferase activity in erythrocytes of patients with dementia disorders. *Eur. Neuropsycopharmacol.* **5**, 107-114.

Hakim, A.M., Cooper, B.A., Rosinblatt, D.S. and Pappius, M. (1983). Local cerebral glucose utilization in two models of B_{12} deficiency. *J. Neurochem.* **40**, 1155-1160

Herbert, V. (1987). Nutrition science as a continually unfolding story: the folate and vitamin B12 paradigm. The Herman Award Lecture. *Am. J. Clin. Nutr.* **46**, 387-402.

Herzlich, B.C., Schiano, T.D., Moussa, Z., Zimbalist, E. and Panagopolis, G. (1992). Decreased intrinsic factor secretion in AIDS: relation to pariatal cell acid secretory capacity and vitamin B12 malabsorption. *Am. J. Gastroenterol.* **87**, 1781-1788.

Herzlich, B.C. and Schiano, T.D. (1993). Reversal of apparent AIDS dementia complex following treatment with vitamin B_{12}. *J. Intern. Med.* 233: 495-497.

Ikeda, T., Furukawa, I., Mashimoto, S., Takahashi, K. and Yamada, M. (1990). Vitmain B12 levelss in serum and cerebrospinal fluid of people with Alzheimer's disease. *Acta Psychiatr. Scand.* **82**, 327-329.

Jacobsen, D.W., Gatautis, V.J. and Green, N. (1994). Rapid HPLC determination of total homocysteine and other thiols in serum and plasma: sex differences and correlation with cobalamin and folate concentraions in healthy subjects. *Clin. Chem.* **40**, 873-881.

Kamgar-Parsi, B., Wehr, T.A. and Gillin, J.C. (1983). Successful treatment of human non-24-hour sleep-wake syndrome. *Sleep* **6**, 257-264.

Kano, Y., Sakamoto, S., Miura, Y. and Takaku, F. (1985). Disorders of cobalamin metqabolism. *Crit. Rev. Oncol. Hematol.* **3**, 1-34.

Karnaze, D.S. and Carmel, R. (1987). Low serum cobalamin levels in primary degenerative dementia. *Arch. Intern. Med.* 429-431.

Kawashima, K., Shiono, K. and Sokabe, H. (1980). Variation of plasma and kidney renin activities among substrains of spontaneously hypertensive rats. *Clin. Exp. Hypertens.* **2**, 229-245.

Kristensen, M.O., Gulmann, N.C., Christensen, J.E., Ostergaard, K. and Rasmussen, K. (1993). Serum cobalamin and methylmalonic acid in Alzheimer dementia. *Acta Neurol Scand.* **87**, 475-481.

Lindenbaum, J., Healton, E.B., Savage, D.G., Brust, J.C. M. Garret, T.J., Podell, E.R., Marcell, P.D., Stabler, S.P. and Allen, R.H. (1988). Neuropsychiatric disorders caused by cobalamin deficiency in the absence of anemia or macrocytosis. *N. Engl. J. Med.* **318**, 1720-1728.

Lowry, O.H., Rosenbrough, N.J., Farr, A.L. and Randall, R.J. (1951). Protein measurement with folin phenol reagent. J. Biol. Chem. 193: 265-275.

Matsumoto, M., Togashi, H., Yoshioka, M., Hirokami, M., Morii, K. and Saito, H. (1990) Simultaneous high-performance liquid chromatographic determination of norepinephrine, serotonin, acethylcholine and their metabolites in cerebrospinal fluid of anaesthetized normotensive rats. *J. Chromatography.* **526**, 1-10.

Matsumoto, M., Togashi, H., Yoshioka, M., Morii, K., Hirokami, M., Toshihara, M., Ikeda, T., Saito, Y. and Saito, H. (1991). Significant correlation between cerebrospinal fluid and brain levels of norepinephrine, serotonin and acetylcholine in anaesthetized rats. *Life Sci.* **48**, 823-829.

McCaddon, A., Regland, B. and Fear, C.F. (1995). Trypsin inhibition: a potential cause of cobalamin deficiency common to the pathologenesis of Alzheimer-type dementia and AIDS dementia complex? *Med. Hypotheses* **45**, 200-204.

Minami, M., Togashi, H., Koike, Y., Saito, H., Nakamura, N. and Yasuda, H. (1985). Changes in ambulation and drinking behavior related to stroke in stroke-prone spontaneously hypertensive rats. *Stroke* **16**, 44-48.

Minami, M., Sano, M., Togashi, H., Sakurai, H. and Saito, H. (1988). Stroke-related plasma noradrenaline, angiotensin II, arginine-vasopressin and serotonin concentrations in stroke-prone spontaneously hypertensive rats. In: *Neurotransmitters as modulators of blood pressure.* Progress in Hypertension vol. 1., Saito, H., Parvez, H., Parvez, S. and Nagatsu, T. (Eds), VSP, The Netherland, pp.89-114.

Minami, M., Kimura, S., Ueno, K., Endo, T., Monma, Y., Togashi, H., Matsumoto, M., Yoshioka, M., Saito, H. and Parvez, H.S. (1995). Role of vitamin B_{12} on behavioral cahnges in SHRSP. In: *New Advances in SHR Research-Pathophysiology and Pharmacology.* Progress in Hypertension vol.3., Saito, H., Yamori, Y., Minami, M. and Parvez, H.S. (Eds), VSP, The Netherland, pp.173-188.

Nadeau, A. and Roberge, A.G. (1988). Effects of vitamin B_{12} supplementation on choline acetyltransferase activity in cat brain. *Internat. J. Vit. Nutr. Res.* **58**, 402-406.

Nimo, R.E. and Carmel, R. (1987). Increased sensitivity of the detection of the blocking anti-intrinsic factor antibody. *Am. J. Cardiol.* **88**, 729-733.

Oden, K.L. and Clarke, S. (1983). S-Adenosyl-L-methionine synthetase from human erythrocytes: role in the regulation of cellular S-adenosyl methionine levels. *Biochemistry* **22**, 2978-2986.

Okamoto, K., Yamori, Y. and Nagaoka, A. (1974). Establishment of the stroke-prone SHR. *Circ. Res.* **34/35** (Suppl I), 143-153.

Okawa, M., Mishima, K., Nanami, T., Shimizu, T., Iijima, S., Hishikawa, Y. and Takahashi, K. (1990). Vitamin B_{12} treatment for sleep-wake rhythm disorders. *Sleep* **13**, 15-23.

Perry, E.K. (1980). The cholinergic system in old age and Alzheimer's disease. *Age Aging* 9, 1-8.

Rasmussen, K. (1989). Solid-phase sample extraction for rapid determination of methylmalonic acid in serum and urine by a stable-isotope-dilution method. *Clin. Chem.* **35**, 260-264.

Regland, B., Abrahamsson, L., Blennow, K., Gottfries, C.G. and Wallin, A. (1992). Vitamin B12 in CSF: reduced CSF/serum B_{12} ratio in demented men. *Acta Neurol. Scand.* **85**, 276-281.

Saito, H., Togashi, H., Yoshioka, M., Nakamura, N., Minami, M. and Parvez, H.S. Animal models of vascular dementia with emphasis of stroke-prone spontaneously hypertenisive rats. In: *New advances in SHR research- Pathophysiology and Pharmacology.* Progress in Hypertenision, vol. 3, H. Saito, Y.

Yamori, M. Minami and S.H. Parvez (eds), VSP, Netherlands, 1995, pp.159-171.

Savage, D.G., Lindenbaum, J., Stabler, S.P. and Allen, R.H. (1994). Sensitivity of methylmalonic acid and total homocysteine determinations for diagnosing cobalamin and folate deficiencies. *Am.J.Med.* **96**, 239-246.

Scott, J. M., Dinn, J. J., Wilson, P. and Weir, D. G. (1981). Pathogenesis of subacute combined degeneration : a result of methyl group deficiency. *Lancet* **2**, 334-337.

Spector, R. and Johansson, C.E. (1989). The mammalian choroid plexus. *Scintific American*, November, 48-53.

Stabler, S.P., Marcell, P.D., Podell, E.R., Allen, R.H., Savage, D.G. and Lindenbaum, J. (1988). Elevation of total homocysteine in the serum of patients with cobalamin or folate deficiency detected by capillary gas chromatography mass spectrometry. *J. Clin. Invest.* **81**, 466-476.

Stephane, F.K. (1983). Circadian rhythms in the rat: constant darkness, entrainment to T cycles and to skeleton photoperiods. *Physiol. Behav.* **30**, 451-462.

Sugita, Y., Mikami, A., Teshima, Y., Egawa, I. and Tsutsumi, T. (1988). Successful treatment with vitamin B_{12} and taking sunlight for a case of hypernychthermeral syndrome. *Jpn. J. Psychiatry Neurol.* **42**, 177-179.

Swain, R. (1995). An update of vitamin B_{12} metabolism and deficiency states. *J. Fam. Pract.* **41**, 595-600.

Togashi, H., Minami, M., Bando, Y., Koike, Y., Shimamura, K. and Saito, H. (1982). Effects of clonidine and guanfacine on drinking and ambulation in spontaneously hypertensive rats. *Pharmacol. Biochem. Behav.* **17**, 519-522.

Togashi, H., Matsumoto, M., Yoshioka, M., Minami, M. and Saito, H. Behavioral and Neurochemical evaluation of stroke-prone spontaneously hypertensive rats for vascular dementia-animal model. In: *Basic, Clinical and therapeutic Aspects of Alzheimer's and Parkinson's disease*, T. Nagatsu *et al.* (Eds), Vol. 1, Plenum Press, New York, 1990, pp. 473-476.

Togashi, H., Matsumoto, M., Yoshioka, M., Hirokami, M., Minami, M. and Saito, H. (1994). Neurochemical profiles in cerebrospinal fluid of stroke-prone spontaneously hypertensive rats. *Neurosci. Lett.* **166**, 117-120.

Togashi, H., Kimura, S., Matsumoto, M., Yoshioka, M., Minami, M. and Saito, H. (1996). Cholinergic changes in the hippocampus of stroke-prone spontaneously hypertensive rats. *Stroke* **27**(3), 520-525.

Van Tiggelen C.J.M. (1984). Alzheimers Disease/ Alcohol dementia : Association with zinc deficiency and cerebral vitamin B_{12} deficiency. *J. Orthomolecul. Psychiat.* **13**, 97-104.

Wallin, A., Alafuzoff, I., Carlsson, A., Eckernäa, S.-A., Gottfries, C.-G., Karlsson, I., Svennerfolm, L. and Winblad, B. (1989). Neurotransmitter deficits in a non-multi-infarct category of vascular dementia. *Acta. Neurol. Scand.* **79**, 397-406.

Yao, Y., Yao, G., Mesches, D.N. and Lou, W. (1994). Decline of serum cobalamin levels with increasing age among geriatiric outpatients. *Arch. Fam. Med.* **3**, 918-922.

Progress in HPLC-HPCE, Vol. 7, pp. 285-349
G. Ali Qureshi *et al.* (Eds)

Histamine

Kzutaka Maeyama and Takehiko Watanabe*

Department of Pharmacology, Ehime University School of Medicine, Ehime, 791-02, and
** Department of Pharmacology I, Tohoku University School of Medicine, Sendai, 980-77*
Japan.

I. Introductuion

Histamine plays important roles in allergy, gastric secretion and neuronal functions. Therefore, the regulation of histamine functions is important to treat pathological conditions and has been dealt in several monographs or reviews. Since the last two books on histamine were published (Watanabe and Wada, 1991; Uvnas 1991), advances in molecular and biochemical researches on histamine have disclosed its roles which had been uncertain. Namely, the clonings of enzymes responsible for histamine synthesis and metabolism and histamine H1 and H2 receptors are introducing the new molecular approaches to clarify the interaction between histamine agonists or antagonists and receptors using site directed mutagenesis to achieve rational drug design, and the functions of histamine using knockout mice in which genes for histamine receptors or histamine related enzymes are targeted to disclose its physiological roles. With regard to histamine H3 receptors, its characteristics has been clarified using its newly developed antagonists and agonists, though its identification at the molecular level has not been completed yet. Clinically, new generations of histamine H1 and H2 receptor antagonists have higher selectivity and potency and less adverse effects than classical or first generation drugs. Thus, nonsedative H1 receptor antagonists which are less penetrable to blood brain barrier due to their hydrophilicity have been prescribed for allergic diseases in place of old drugs. New experimental techniques have been applied to histamine research. For example, *in vivo* brain microdialysis methods gave us real time changes in extracellular histamine concentrations in brains, and positron emission tomography (PET) studies using [^{11}C]-mepyramine or [^{11}C]-doxepin as a radiotracer developed a new field to visualize histamine H1 receptors in a living human brain under the physiological and pathological conditions.

In this chapter, we shall describe histamine analysis, characterizations of histamine-metabolizing enzymes and receptors, and finally its neuropathological roles from the view of recent approaches.

II. Assay of histamine

The analysis of histamine in biomaterials such as tissues, blood and urine is important to elucidate the etiology of pathophysiological status. Historically, histamine was first measured by bioassay such as the decrease of blood pressure in anesthetized and atropinized cats and the contraction of guinea pig ileum. Even though these bioassay methods are sensitive to histamine, they are not enough to be applied to accurate determinations of histamine in biosamples because of their less selectivity, less sensitivity and more variability. Next, the colorimetry using the Pauly's diazo reaction was established but the sensitivity was not sufficient to determine brain histamine levels. Shore et al. (1959) introduced a sensitive fluorometric method by using the conjugation of histamine with *o*-phthalaldehyde (OPA) which has been sensitive enough to measure histamine levels in mast cells and gastric glands containing high amounts of histamine. But ammonium, histidine and polyamines in crude biomaterial samples interfere with the measurement of histamine, because this method produces strong fluorescences to these materials. To solve this problem, the fluorometry coupled with high performance liquid chromatography (HPLC) was developed.

Using the principle of methylation of histamine by histamine *N*-methyltransferase (see p. 9), the radioenzymatic assay for histamine was developed by Snyder et al. (1966). Especially, using histamine *N*-methyltransferase purifiied from rat kidney, which contains the highest activity among all tissues, the radioenzymatic method was improved to show higher sensitivity and has been used widely (Shaff and Beaven, 1979). The detail of the radioenzymatic assay was described by Beaven (1991). Gas chromatography mass spectrometry (GCMS) has many advantages to measure both histamine and N^{τ}-methylhistamine with the highest sensitivity and selectivity (Hough et al., 1981).

Features of various methods for asaay of histamine are compared in **Table 1**. In this section, HPLC assay system, radioimmunoassay and newly developed luminometry of histamine will be discussed.

1. HPLC-fluorometry

The post column derivatization method in which histamine is separated with a cation ion-exchanger column and reacted with OPA automatically has advantages such as high sensitivity and easiness for manipulating numbers of samples (Yamatodani et al., 1985). Briefly, brain tissues are homogenized by sonication in 5-10 volumes of 3 % of perchloric acid (PCA) containing 5 mM

Table 1 Comparison of methods for assay of histamine and N^{τ}-methylhistamine.

Methods	Reagents	Usable range(amount or concentration)	Reference
I.Assay for histamine			
1)HPLC-fluorometry	OPA (Post)	0.05 pmol	Yamatodani et all.(1985)
2)HPLC-electrochemistry	OPA, ME (Pre)	0.45pmol	Harsing et al. (1986)
3)Radioenzymatic assay			
a. Double isotope	HMT, ^{14}C-SAMe ^{3}H-Histamine	50-5000 pmol/ml	Beaven et al (1972)
b. Single isotope			
Single step	HMT, ^{14}C-SAMe	25-250 pmol/ml	Beaven et al.(1972)
Multiple step	HMT, ^{3}H-SAMe, Bovine serum albumin	0.25-100 pmol/ml	Beaven et al.(1982)
4)Immunoassay			
a. Radioimmunoassay	Acylation regent	1- 50 pmol/ml	Morel & Delaage(1988)
b. Enzyme immunoassay	1,4-Benzoquinone	3-200 pmol/ml	Chevrier et al.(1986)
II. Assay for both histamine and N^{τ}-methylhistamine			
1)HPLC-fluorometry	OPA , ME (Pre)	0.2 pmol/ml	Tsuruta et al.(1981)
2)HPLC-electrochemistry	Bolton-Hunter reagent (Pre)	0.1-1000 pmol	Mine et al.(1986)
3)HPLC-luminometry	Luminol, immobilzed DAO column (Post)	5 pmol	Alam, et al.(1995)
4)GC-MS	Heptafluorobutyric anhydride	0.1 pmol	Hough et al.(1981)
5)Radioimmunoassay	1,4-Benzoquinone	0.05 pmol	Garbarg et al.(1989)

Abbreviations: Pre, precolumn derivatization; Post, postcolumn derivatization; OPA, *o*-phthalaldehyde; HMT, histamine *N*-methyltransferase; SAMe, *S*-adenosyl-*L*-methionine; ME, 2-mercaptoethanol; DAO, diamine oxidase.

EDTA-Na_2 and the homogenate is centrifuged at 10, 000 x g for 15 min at 4 C. Fifty μ l of the supernatant is directly applied into a cation exchanger column (TSK-gel SP-2SW; 6 mm i.d. x 150 mm) and eluted with 0.25 M KH_2PO_4 at a flow rate of 0.6 ml/min. The eluate from the column is mixed with 0.1 % OPA solution and 2 M NaOH in a reaction coil made of polytetrafluoroethylene tubing (0.5 mm i.d. x 5 m) at 45° C at pH 12.0 and finally 25 % of phosphoric acid is added to reduce pH to 3.0. Each solution of OPA, NaOH and phosphoric acid is pumped out at 0.2 ml/min. The fluorescence is monitored in a fluoromonitor (F1080, Hitachi,Japan) using excitation and emission wavelengths of 360 and 450nm, respectively. Plasma sample (0.5 ml) is mixed with 0.25 ml of 7.5% PCA and centrifuged at 10,000 x g at 4° C, and 50 μ l of the supernatant is applied to HPLC. Samples from brain microdialysis experiments are directly injected into an HPLC column without treatment with PCA.

In studies using (R)-α-methylhistamine, a histamine H3 agonist, it interfered with histamine determination in usual HPLC-fluorometry since it also causes fluorescence by reaction with OPA and the retention time is the same as that of histamine on separation with a strong cation exchanger column (TSK-gel SP-2SW, Tosoh, Japan). Instead of this column, a weak cation exchanger (TSK-gel CM-2SW, 6 mm i.d. x 150 mm, Tosoh, Japan) is used for separation of (R)-α-methylhistamine and histamine by elution with a mixture of 15 % (v/v) methanol, 0.18 M imidazole and 37.5 mM citric acid, pH 6.8, as a mobile phase at a flow rate of 0.6 ml/min. (R)-α-Methylhistamine and histamine are postlabeled sequentially with OPA and their retention times are 24.0 and 28.8 min, respectively (Sakai et al., 1993). Intraperitoneal (i.p.) treatment of mast cell-deficient mice (*W/W^v*) with (R)-α-methylhistamine (50 mg/kg) increased the histamine level in whole brain to 1.93-fold of the control but caused no significant changes in locomotor activity.

The precolumn derivatization method in which histamine is first reacted with OPA in the presence of 2-mercaptoethanol and the derivatized histamine is applied to a reversed-phase ODS column also has big advantages. It causes high sensitivity and selectivity, and N $^{\tau}$-methylhistamine, a major metabolite of histamine in brain, can be also mesaured (Turuta et al. 1981). For this measurement, the prepurification step is necessary because primary amines react easily with OPA and their peaks of fluorecence on column chromatography overlap with those of histamine and N $^{\tau}$-methylhistamine. Oishi et al. (1988) showed the changes in histamine dynamics after treatment of (R)-α-methylhistamine using the precolumn deriatization method. At the dose of 6.3 mg/kg of (R)-α-methylhistamine, the histamine level in the mouse brain did not change significantly, but the N^{τ}-methylhistamine level decreased 1hr after i.p. treatment.

2. Radioimmunoassay

Radioimmunoassay is one of other attempts to determine histamine or N^{τ}-methylhistamine with high sensitivity and selectivity. Panula et al. (1984) showed the localization of histamine neuron using the immunohistochemical technique with an antibody against histamine. It is important to get specific and potent antibodies against histamine or N^{τ}-methylhistamine coupled to serum albumin (Garbarg et al., 1989). Histamine or N^{τ}-methylhistamine in tissue extracts are prederivatized and determined from the competition between them and the respective [^{125}I] iodinated tracers. The derivatized histamine or N^{τ}-methylhistamine is 1000 times more specific for each antibody than nonderivatized ones, histidine and other histamine derivatives. This radioimmunoassay is so sensitive as to be applied to the quantification of the two amine immunoreactivities in less than 1 mg of hypothalamus samples. The radioimmunoassay kit is now commercially available and widely used for measurements of histamine in clinical samples of blood and urine and that released from basophils. In the radioimmunoassay system from Immunotech (Marseille, France), histamine is first converted to its succinylglycinamide derivative and subsequently subjected to a competitive binding assay wi.h an [^{125}I]-conjugated tracer. The linear range is between 1-150 nM (Morel and Delaage, 1988).

In addition to radioimmunoaasay, the enzyme immunoassay method of histamine was developed by Chevrier et al. (1986). In this assay, the histamine treated with 1,4-benzoquinone is allowed to compete with a histamine-peroxidase conjugate for binding to an anti-histamine monoclonal antibody that was coated on a microtitre plate. Histamine concentrations from 3 to 200 pmol/ml can be determined by this method. It has an advantage to be able to keep this measurement set stably for a longer time than the radioimmunoassay kit using an [^{125}I] iodinated tracer.

3. HPLC-luminometry

Diamine oxidase (DAO, see p.10 for details), which has been originally known as histaminase, oxidizes not only histamine but also N^{τ}-methylhistamine to imidazole acetoaldehyde and N^{τ}-methylimidazole acetaldehye, respectively, in addition to various polyamines. One of the reaction

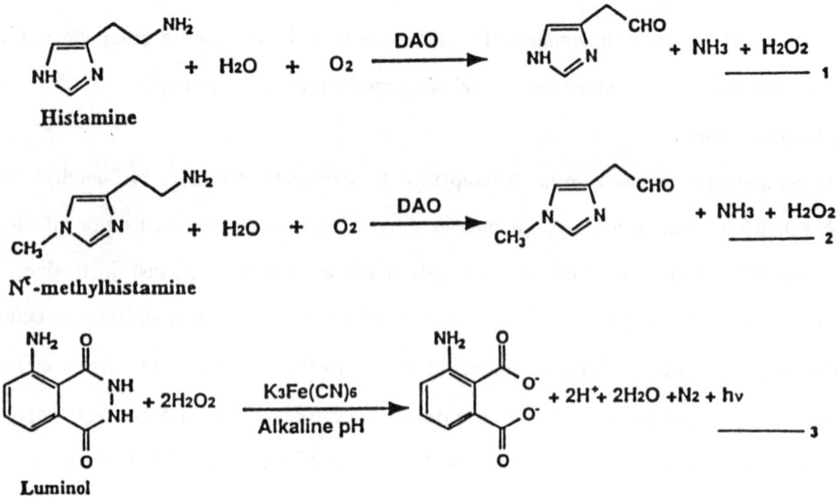

Fig. 1. The principle of the chemiluminescent detection of histamine and N^{τ}-methylhistamine. Histamine and N^{τ}-methylhistamine are oxidized by DAO to give rise to H_2O_2 in Reaction (1) and (2), respectively. H_2O_2 reacts with luminol in the presence of ferricyanide in alkaline solution to emit the luminescence in Reaction (3) (Alam et al., 1995).

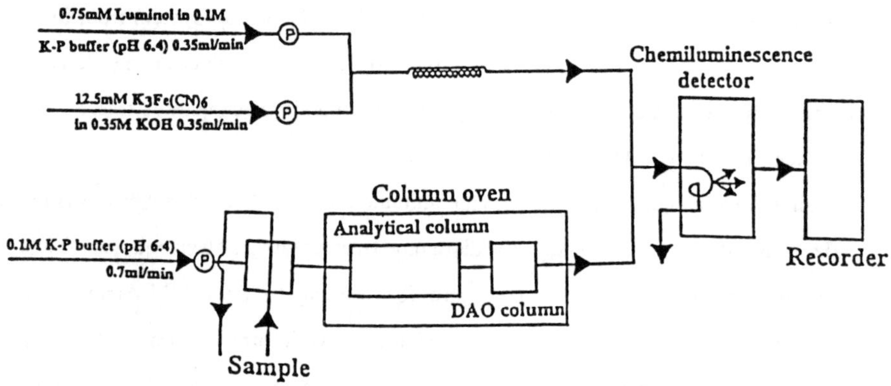

Fig. 2. Flow diagram of the HPLC-chemiluminescent method for assay of histamine and N^{τ}-methylhistamine. Analytical column: TSK-gel ODS90Ts(150 mm x 4.6 mm i.d.). DAO column: DAO conjugated with TSK-gel Tresyl-5PW (30 mm x 4.6 mm i.d.). Reaction solution consists of 0.75 mM luminol in 0.1 M potassium phosphate buffer (pH 6.4) and 12.5 mM potassium ferricyanide in 0.35 M potassium hydroxide at the flow rate of 0.35 ml/min. Elution buffer: 0.1 M potassium phosphate buffer (pH 6.4) at the flow rate of 0.7 ml/min. P : a pump.

products, hydrogen peroxide, reacts with luminol to produce chemiluminescence (**Fig. 1**). Using immobilized DAO, a new HPLC-chemiluminescent method for determination of histamine and N^{τ}-methylhistamine was developed in our laboratory (Alam et al., 1995). Briefly, DAO was partially purified 158-fold from lyophilized powder preparation of porcine kidney by two steps of colume chromatography, *i.e.*, DEAE cellulose and hydroxylapatite column chromatographies. Especially, it is important to remove the catalase activity from the DAO fraction, which was achieved by careful hydroxylapatite column chromatography. DAO was immobilized covalently on TSK-gel Tresyl-5PW (Tosoh, Japan) and packed into a peek column (4.6 mm i.d. x 30 mm length). Histamine and N^{τ}-methylhistamine were eluted in a reversed-phase of an analytical column (TSK-gel ODS Ts; 4.6 mm i.d. x 150 mm length, Tosoh, Japan) with 0.1 M potassium phosphate buffer (pH 6.4) at a flow rate of 0.4 ml/min and passed sequentially through the immobilized DAO column. Hydrogen peroxide that was produced from histamine and N^{τ}-methylhistamine by the immobilzed DAO was reacted with the reaction agent, luminol, in the presence of ferricyanide as a catalyst in alkaline solution to emit luminescence. The luminescence was monitored with a chemiluminescence detector (JASCO 825-CL , Japan Spectroscopic Co. Japan). The diagram of this luminescence method was shown in **Fig. 2**. The retention time of histamine and N^{τ}- methylhistamine were 6.1 and 12.7 min, respectively (**Fig. 3**). The lower limit of detection of histamine and N^{τ}-methylhistamine were 5 and 10 pmol, respectively. With regard to the measurements of histamine and N^{τ}-methylhistamine, there are good correlations between the values obtained with HPLC-fluorometry and HPLC-luminometry (Alam et al., 1995). As an application of HPLC-luminometry method, the turnover of histamine in mouse brain after treatment with immepip, a new H3 receptor agonist developed by Leurs and Timmerman (1992) was measured. To confirm of the effect of H3 agonists or antagonists, it is important to determine histamine and N^{τ}-methylhistamine in brain simultaneously. Immepip increased the histamine level but decreased the N^{τ}-methylhistamine level dose-dependently in the forebrain and midbrain of mice. These changes in the levels of histamine and N^{τ}- methylhistamine were due to the agonistic action of H3 receptors as inhibition of histamine release from the histaminergic neuron, being in good agreement with results obtained with (R)- α -methylhistamine (Qishi et al., 1989).

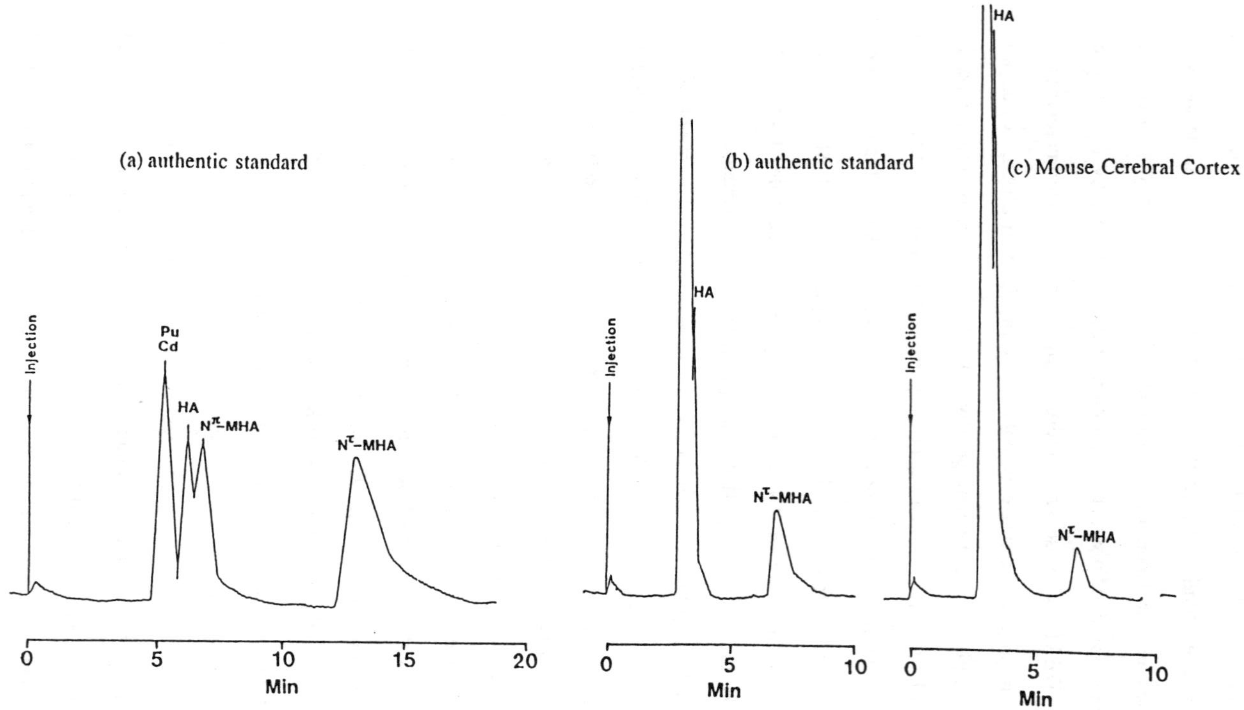

Fig. 3. Typical chromatograms of standard amines and brain sample on HPLC-chemiluminescent method. Standard amines (HA: histamine, N$^{\tau}$-MHA: N$^{\tau}$-methylhistamine, N$^{\pi}$-MHA: N$^{\pi}$-methylhistamine, Pu: putrescine, Cd: cadaverine) were injected into the HPLC-luminometry at a flow rate of the elution buffer of 0.4 ml/min and of the reaction solutions of 0.2 ml/min (**a**). For obtaining a sharp peak of N$^{\tau}$-methylhistamine, standard amines and sample of brain cortex were injected into the HPLC-luminometry at a flow rate of the elution buffer of 0. 7 ml/min and of the reaction solutions of 0.35 ml/min (**b,c**).

II. Biochemistry

1. Enzymes involved in histamine metabolism

a. Metabolism of histamine

Histamine is formed by the decarboxylation of *L*-histidine, which is catalyzed by histidine decarboxylase (*L*-histidine carboxylyase, E.C.4.1.1.22, HDC), and inactivated by one of the two pathways; i.e. the methylation and oxidation (**Fig. 4**). In the methylation pathway, histamine is methylated to N^{τ}-methylhistamine by the action of histamine *N*-methyltransferase (*S*-adenosyl-*L*-methionine: histamine *N*-methyltransferase, E.C.2.1.1.8, HMT) with *S*-adenosyl-*L*-methionine as a methyl donor. N^{τ}-Methylhistamine is oxidized by a B type of monoamine oxidase [amine: oxygen oxidoreductase (deaminating) (flavin-containing). E.C.1.4.3.4, MAO] to N^{τ}-methyl-imidazole acetaldehyde, which is further oxidized by aldehyde dehydrogenase (aldehyde: NADP oxido-reductase, E.C.1.2.1.3, ALDH) to N^{τ}-methylimidazole acetic acid. In the oxidation pathway, histamine is oxidized by diamine oxidase (diamine: oxygen oxidoreductase, E.C.1.4.3.6, DAO, histaminase) to imidazole acetaldehyde, which is further oxidized by ALDH to imidazole acetic acid (ImAA)*. Which is a major metabolic pathway of the two depends on animal species and tissues, but generally speaking, the methylation pathway is dominant in the brain, because the brain has little DAO activity**.

Tissue distributions and enzymatic properties of HDC, HMT and DAO are summarized in **Tables 2** and 3, respectively, for general information.

b. Histidine decarboxylase (HDC)

----

*The further metabolism of the two imidazole acetic acid derivatives, ImAA and N^{τ}-MeImAA, to their corresponding ribosides and ribotides is out of the scope of this review, though, in **Fig. 4**, ImAA is metabolized to ImAA riboside by action of ImAA phosphoribosyltransferase and phsphatase. It should be also mentioned that ImAA is produced from L-histidine by the action of aminotransferase and decarboxylation of imidazole pyruvate (Prell, personal communication).

**Thomas & Prell (1995) recently found that there is a considrable activity of DAO in the brain.

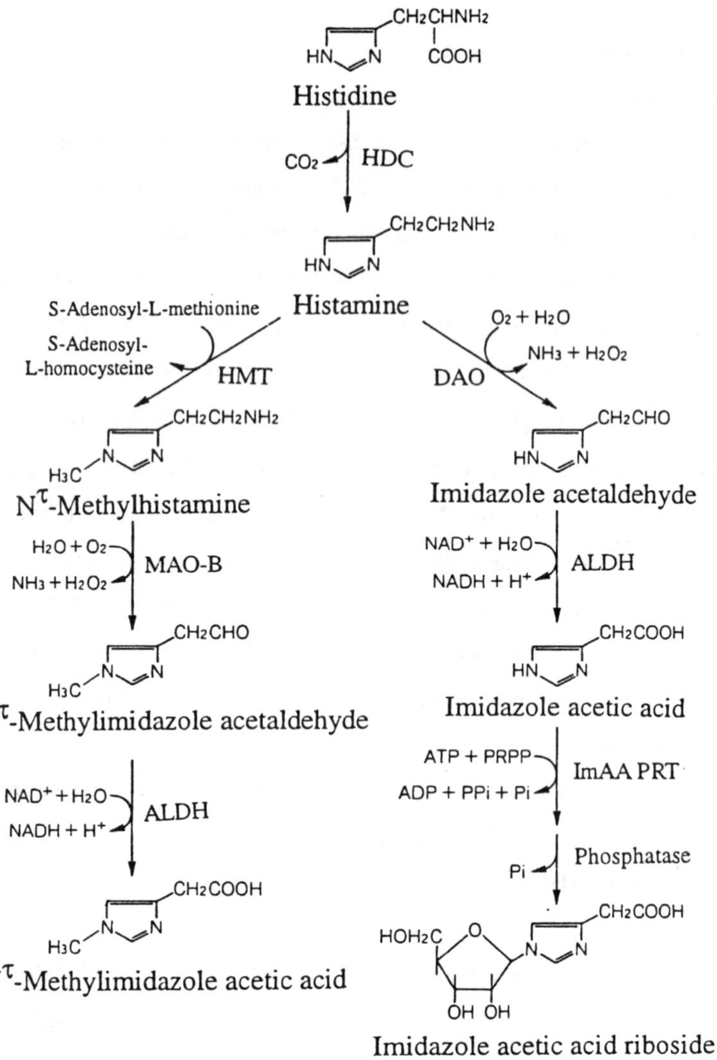

Fig. 4. Metabolic pathways of histamine. For abbreviations, see the text.

Table 2 Distribution of HDC, HMT and DAO in rat tissues (Hayashi and Watanabe, 1993).

	HDC (pmol/min/mg protein)	HMT (nmol/min/g tissue)	DAO (nmol/min/g tissue)
Kidney	N.D.	1370	0.008
Ileum		98	5.38
Brain	0.76	22	0.002
Spleen	0.13	0.060	0.002
Heart	N.D.	0.055	0.006
Lung	0.07	0.051	0.028
Stomach	10.4	0.035	0.25
Liver	0.01	0.033	0.01
Skin	0.50		0.002
Mast cell	118		
Fetus	16.6		
Placenta			13.4

Table 3 Properties of HDC, HMT and DAO

	HDC	HMT	DAO
Km for substrate	0.25 mM	0.035mM (for histamine) 0.38 mM (for SAM)	0.02 mM
Vmax	0.81 μ mol/min/mg	0.01 μ mol/min/mg	1.08 μ mol/min/mg
Coenzyme	Pyridoxal 5'-phosphate	None	Topaquinone
Substrate	L-Histidine N ' -methylhistidine	Histamine Imidazole compounds	Histamine Polyamines
Inhibitor	α-Fluoromethylhistidine	Metoprine	Aminoguanidine
Molecular weight (amino acid)	11 kDa (5.5 kDa x2)* ( )	3.3 kDa	18.5 kDa (9.3 kDa x2) (751)
Gene (exon)	23 kb (12)		9 kb (4)

*A nascent HDC has a molecular weight of 74 kb. which is posttranslationally processed to give 5.5 kb monomer (see text for details).

1) Distribution

The HDC activities in most mammalian tissues are very low (Compare the HDC activities with those of HMT and DAO in **Table 1**). The highest activities are found in the mast cell and stomach. The HDC activity is increased by treatments with various agents such as phorbol ester, endotoxin and staphylococcal enterotoxin, and under some physiological conditions such as gestation or hormones such as gastrin and estrogen (Watanabe and Wada, 1983; Watanabe et al., 1991; Hayashi and Watanabe, 1993), indicating that HDC is an inducible enzyme. For example, very high ativities are found in the fetal rat liver (18-20 days of gestation), fetal mouse skin, hamster placenta, mouse mastocytoma P-815, and kidneys of mice treated with estradiol, all of which are good sources for purification of HDC (Watanabe and Wada, 1983; Watanabe et al., 1991; Hayashi and Watanabe, 1993).

In the brain, the HDC activity is the highest in the hypothalamus, followed by the cortex, striatum and midbrain, and lowest in the cerebellum and spinal cord (**Table 4**). On subcellular fractionation, the HDC activity is found in the supernatant fraction of synaptosomes, but some activity is bound to membrane fractions (**Table 4**).

2) Purification and characterization.

The enzymes have been purified from fetal rat liver (Taguchi et al., 1984), kidneys of mice treated with estrogen (Martin & Bishop, 1986), and mouse mastocytoma P-815 (Ohmori et al., 1990). The summary of purification protocol of HDC from mastocytoma P-815 cells is shown in **Table 5** (Ohmori et al., 1990).

The properties of HDC are summarized in **Table 3**. The coenzyme is pyridoxal 5'-phosphate, the substrate specificity is very strict to L-histidine with only N^{π}-methylhistidine 15 % as active as L-histidine (Watanabe et al., 1991), and the specific inhibitor is (S)-α-fluoromethylhistidine (FMH) (Kollonitsch et al., 1978; Watanabe et al., 1990). The molecular weight of purified enzyme is 110 kDa composing of two 55 kDa subunits, whereas the molecular weight calculated from amino acid sequence of the cloned enzyme subunit is 74 kDa, suggesting the posttranslational processing of the nascent enzyme to 55 kDa subunit. The processing enzyme is not clear, but Ichikawa et al. (1995) showed that elastase-like protease(s) is responsible for the conversion. Yatsunami et al.

Table 4 Distribution of histamine content and HDC and HMT activities in brain regions and their subcellular distributions.

Region	Histamine (ng/g)	HDC (pmol/h/mg protein)	HMT (nmol/h/mg protein)
Hypothalamus	209 ± 14	28. 8 ± 2.4	1.05 ± 0.09
Pons-meddula	25 ± 4	4.7 ± 0.4	0.47 ± 0.03
Cerebellum	26 ± 5	3.3 ± 0.2	0.37 ± 0.02
Hippocampus	50 ± 6	6.8± 0.4	0.83± 0.05
Striatum	56 ± 4	10.4 ± 0.8	0.92 ± 0.05
Thalamus-Midbrain	75 ± 8	12.6 ± 0.9	0.91 ± 0.06
Cerebral cortex	39 ± 5	6.3 ± 0.4	0.82 ± 0.05
Subfraction	**Histamine (%)**	**HDC (%)**	**HMT (%)**
P_1(Nucleus)	20.2	10.4	5.55
P_2(Mitochondria)	46.4	24.7	13.7
P_3(Miclosomes)	15.6	5.7	0.85
S(Supernatant)	17.7	61.1	79.9
M_1*	64.1	37.3	25.3
M_2	31.2	10.3	5.0
M_3	4.5	52.4	69.7

*P_2 fractions were subjected to hypotonic shock and centrifuged at 10,000 x g for 20 min. The precipitate is M_1 and the suprenatant was further centrifuged at 100,000 x g for 60 min. The precipitate and supernatant were M_2 and M_3.

Table 5 Purification of HDC from mastocytoma P815 cells (Ohmori et al., 1990)

Steps	Protein (mg)	Activity (nmol/min)	Specific activity (nmol/min/mg)	Yield (%)
1. Cytosol	2,700	174	0.064	100
2. Ammonium sulfate (25-55%) and pH 7.5 dialysis	1,350	164	0.122	94
3. pH 6.0 dialysis	1,190	147	0.124	84
4. DEAE-Sepharose CL-6B	260	91.1	0.350	52
5. Phenyl-Sepharose CL-4B	19.5	52.4	2.69	30
6. Hydroxylapatite	2.57	34.0	13.2	20
7. Phenyl-Superose (HPLC)	0.158	26.9	170	15
8. Mono Q (HPLC)	0.035	18.2	520	10
9. Diol-200 gel filtration (HPLC)	0.015	12.2	800	7

(1995) showed that 74- and 55 kDa forms of human HDC have equivalent enzyme activity and are present in the particulate and supernatant fractions, respectively, and the 74 kDa form of human HDC exists as a monomer***.

The HDC cDNA was cloned from rat, human and mouse (Joseph et al., 1990; Yamauchi et al., 1990; Yamamoto et al., 1990), and the amino acid sequences deduced form the nucleotide sequences are very similar among the 3 enzymes, sharing 85 % homology in 2/3 of the N-terminal portion (**Fig. 5**). The HDC gene was elucidated from humans (Yatsunami et al., 1994), and then from mice (Ohtsu et al., unpublished), being 24 kb in length composing of 12 exons with the last 12th exon encoding 2/5 of the whole cDNA (**Fig. 6**). The uptream region from the initiation site of the human HDC gene contains various consensus sequences of regulatory sites such as 1 TATA-like, 1 GC box, 4 GATA, and 6 LTB-binding domains (Yatsunami et al., 1994). The functions and tissue-specific gene expression are under current studies.

c. Histamine *N*-methyltransferase (HMT)

The content of HMT in tissues are higher than that of HDC (**Table 2**), and HMT was purified from various sources such as rat kidney, guinea pig brain and bovine brain (Watanabe and Wada, 1983; Verburg and Henry, 1986; Hayashi and Watanabe, 1993). In brain, it is distributed in all regions with highest in the hypothalamus, but the unevenness of the activities in various regions is not as remarkable as that of HDC (**Table 4**), suggesting the wide distribution of the histaminergic neuron system in brain. On cell fractionation, most activity is found in the supernatant (**Table 4**). The HMT activity is not influenced by various conditions, except that the activity is higher in females than in males (Snyder & Axelrod, 1965), suggesting that HMT is a house-keeping enzyme.

The properties of HMT are summarized in **Table 3**. The enzyme is very specific to histamine as a natural substrate, which is the basis of radioenzymatic assay of histamine by HMT reaction with *S*-adenosyl-*L*-methionine as a methyl donor as described previously (p. 5), though HMT methylates many imidazole compounds. Specific HMT inhibitors are metoprine and SKF 91488 with the respective Ki values of 0.1 and 0.9 mM (Watanabe and Wada., 1983; Verburg and Henry.

***This is not true for mouse HDC (Yatsunami, personal communication).

```
          1        10              20         30         40         50          60          70
Rat   HDC MMEPSEYHEY-Q---ARGKEMVDYICQYLSTVRERQVTPNVKPGYLRAQIPSSAPEEPDSWDSIFGDIEQIIMP
Mouse HDC M   C  R  REYYR           S               Q       L A                 RV
Human HDC M   E  R ------- R                R   D Q        L E     D            R
Morg. HDC                      MTLSINDQNK-LDAFWAYC-VKNQYFNIGYPESA--DFDYTNLERFLRFSINNC
Rat   DDC     MDSREF-R---RRGKEMVDYIADYLDGIEGRPVYPDVEPGYLRALIPTTAPQEPETYEDIIRDIEKIIMP
                               +   +               +   +     +  +    +

                   80          90        100         110       120        130        140
Rat   HDC GVVHWQSPHMHAYYPALTSWPSLLGDMLADASTAWGSRGLPA-ACTELEMNIMDWLAKMLGLPDFFLHHHP
Mouse HDC                                            -                   EY
Human HDC                                            -       V           EH
Morg. HDC GDWGEYCNYLLNSFDFEKEVMEYFAD-LFK-IPFEQSWGYVTNGGTEGNMFGC-YLGREI-FPDGTL----
Rat   DDC GVTHWHSPYFFAYFPTASSYPAMLADMLCGAIGCIGFSWAASPACTELETVMMDWLGKMLELPEAFLAGRA
          +                         + +         ● ●      ++   ●      +      +   +

                 150        160        170        180        190       200        210
Rat   HDC SSQGGGVLQRTVSESTLIALLAARKNKILEMKAHEPNADESSLNARLVAYASDQAHSSVEKAGLISLVKI
Mouse HDC   R      S                    A T C  D N          T                  M
Human HDC          Q                      TS   D    C                          M
Morg. HDC ---------------------------------------------------YYSKDTHYSVAKIVKLLRIKS
Rat   DDC GE-GGGVIQGSASEATLVALLAARTKMIRQLQAASPELTQAALMEKLVAYTSDQAHSSVERAGLIGGVKI
                                                              + +   + ++ ●

                 220        230        240        250        260            270        280
Rat   HDC KFLPVDDNFSLRGEALQKAIEEDKQQGLVPVFVCATLGTTGVCAFDKLSEL-----GPICAREGLWLHVDAAYAG
Mouse HDC R                                             R   -----     S
Human HDC                           R                   C   -----              I
Morg. HDC QVVESQPNGEIDYDDLMKKIADDKEAH--PIIF-ANIGTTVRGAIDDIAEIQKRLKAAGIKREDYYLHADAALSG
Rat   DDC KAIPSDGNYSMRAAALREALERDKAAGLIPFFVVVTLGTTSCCSFDNLLEV-----GPICNQEGVWLHIDAAYAG
                +       + ● ●  ++     +     ●  +++   ● +   +              +    + +++  +

                 290        300        310        320        330        340         350
Rat   HDC TAFLRPELRGFLKGIEYADSFTFNPSKWMMVHFDCTGFWVKDKYKLQQTFSVNPIYLRHAN--SGVATDFMH
Mouse HDC     C       E                                                   --  A
Human HDC     C  F                                                        --
Morg. HDC MILPFVDDAQPFTFADGIDSIGVSGHKMIGSPIPCGIVVAKKENVDRI--SVEIDYI--------SAHD---
Rat   DDC SAFICPEFRYLLNGVEFADSFNFNPHKWLLVNFDCSAMWVKKRTDLTEAFNMDPVYLRHSHQDSGLITDYRH
                             ●+        +         +      +         ●●    +         ● +

                 360        370        380        390        400        410            420
Rat   HDC WQIPLSRRFRS-IKLWFVIRSFGVKNLQAHVRHGTDMAKYFESLVRSDPVFEIPAERHLGLVVFRLKG-----PNC
Mouse HDC            -                       E            S     K                -----
Human HDC            -V                      E          N  S    K                -----
Morg. HDC KTITGSRNGHTPLMLWEAIRSHSTEEWKRRITRSLDMAQYAVDRMQKAGINAWRNKNSIT-VVFPCPSERVWREHC
Rat   DDC WQIPLGRRFRS-LKMWFVFRMYGVKGLQAYIRKHVKLSHEFESLVRQDPRFEICTEVILGLVCFRLKG-----SNQ
             +  ●+         ●+   ●+●                ●● ●                   +●+              ●

                 430        440        450        460        470        480        490
Rat   HDC LTESVLKEIAKTGQVFLIPATIQ-DKLIIRFTVTSQFTTKDDILRDWNLIREAANLVLSQHCTSQPSPR
Mouse HDC           A  L                             E      H  Q               R
Human HDC    N      A RL                           RD          D  T  I         R
Morg. HDC LATSGDVAHLITTAHHLD--TVQIDKLI------------DDVIADFNL--HAA
Rat   DDC LNETLLQRINSAKKIHLVPCRLR-DKFVLRFAVCSRTVESAHVQLAWEHIRDLASSVLRAEKE
          +                +    ●  ● +++●                ●     ●     ●     ●+
```

Fig. 5. Amino acid sequences of HDC's from rat, mouse, human and *Morganella morganii* and rat DDC (DOPA decarboxylase).

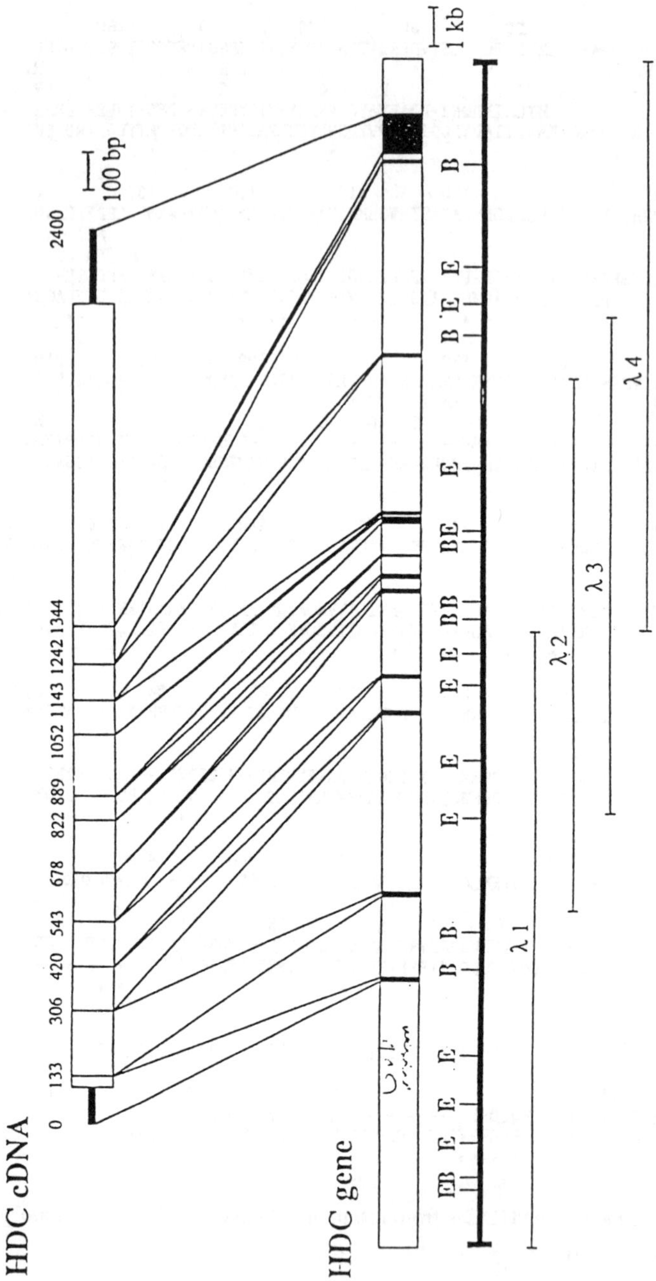

Fig. 6. Structure of human HDC gene.

1986). HMT cDNA's of 3.3 kb were cloned from rat and human kidneys (Takemura et al., 1992; Girard et al., 1994; Yamauchi et al., 1994), and their primary sequences composing of 293 and 295, respectively, amino acids are shown in **Fig. 7**. Both enzymes share 84 % homology in the primary sequences composing of 281 amino acids. The structure of HMT gene was determined by Aksoy et al. (1996)*[4] (**Fig. 8**). The HMT gene is found to be located on chromosome 2 with DNA as a template and two different sites of the sequence as primers (Aksoy et al., 1996), being in contrast in 1p32 in the first paper determined by FISH with cDNA as a probe (Yamauchi et al., 1994).

d. Diamine oxidase (DAO)

The enzyme has been purified from various sources and the properties are summarized in **Table 3** (Mondovi, 1985; Hayashi and Watanabe, 1993). The substrate specificity is broad, and so polyamines like putrescine, spermidine, spermine are oxidized better than histamine and N^{τ}-methylhistamine. The identification of the coenzyme has been controversial, but recently it turned out that the coenzyme is topaquinone (Steinbache et al., 1995). The DAO activity is changeable by many compounds and conditions, such as tumor bearing and gestation, suggesting the possible role of DAO in situations of excess hisatamine production. Its inhibitors are carbonyl reagents like aminoguanidine. The human DAO cDNA was determined and amino acid sequences the number of which is 751 are shown in **Fig. 9**, together with the gene structure (Chassanda et al., 1994).

e. Monoamine oxidase-B (MAO-B)

It is now clear that there are two types of MAO, A and B. The numbers of amino acids of human MAO-A and B are 527 and 520, respectively, and the corresponding molecular weights are 59.7 and 58.8 kDa (Shin et al., 1994). MAO-A and B share 70 % amino acids and exist as a homodimer, which carries FAD on each subunit. The location on chromosome is X11.23-11.4. Histamine is not a good substrate, but N^{τ}-methylhistamine is oxidized by MAO-B. Immuno-

---

But it seems that the identification of the initiation site is not evidenced well according to Yamauchi et al. (personal communication), who suggested that there is the first exon 2 kb upstream of that reported by Aksoy et al. (1996).

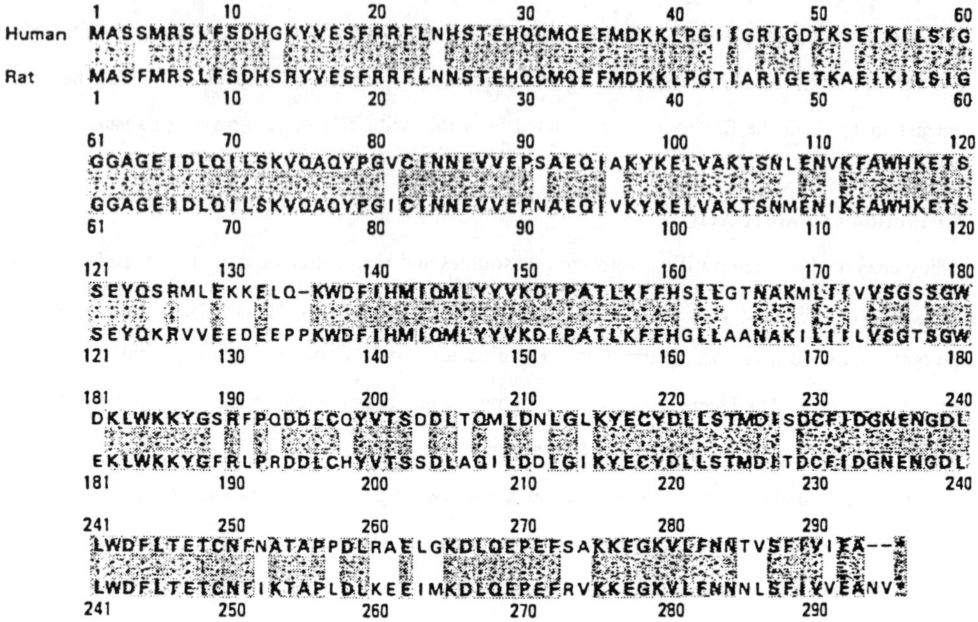

Fig. 7 Amino acid sequences of human and rat HMT's.

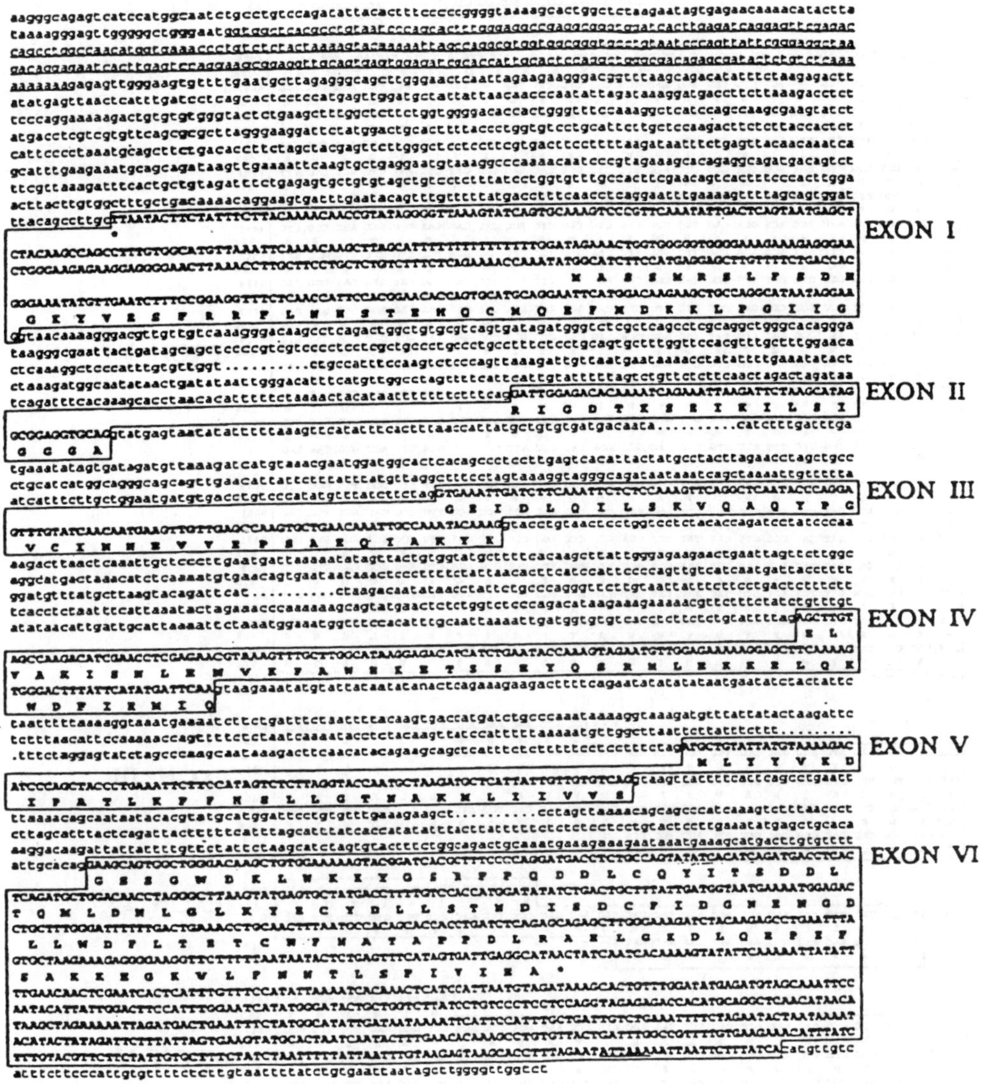

Fig. 8. Nucleotide sequence of human HMT gene.

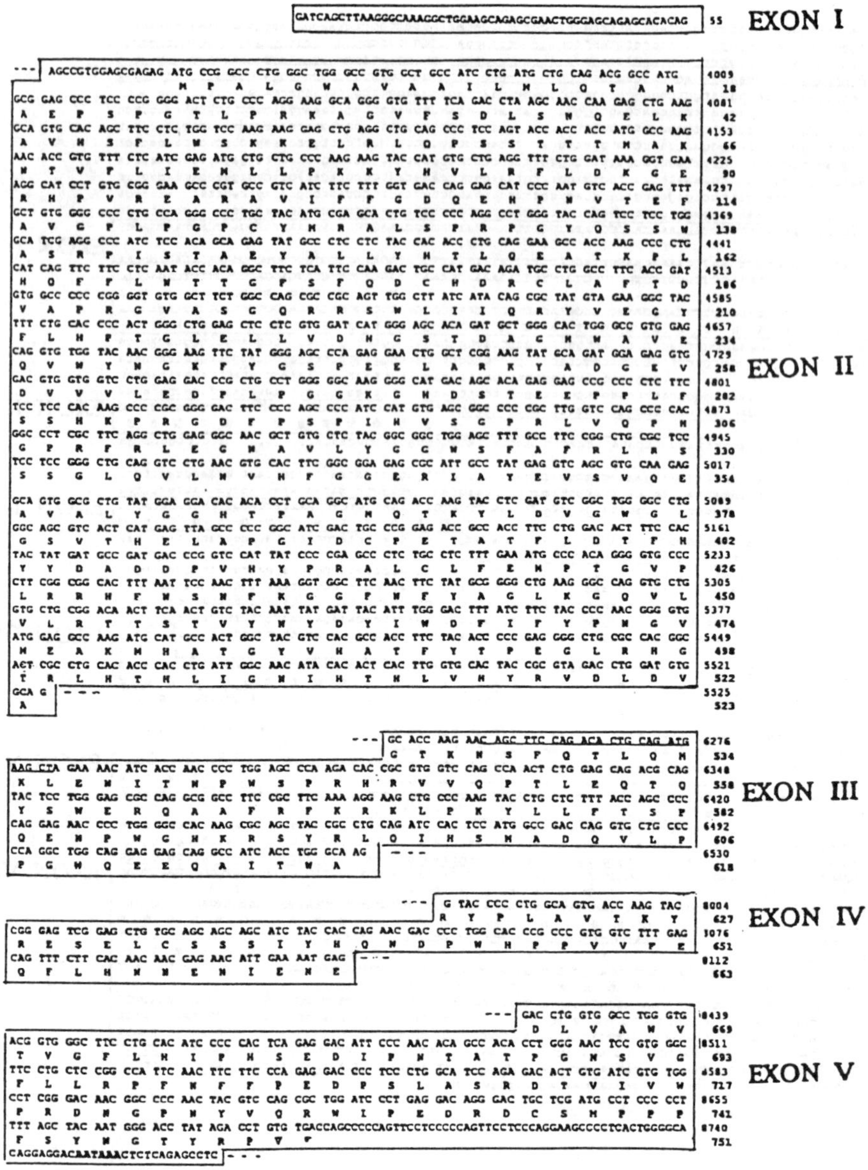

Fig. 9. cDNA and amino acid sequences of DAO.

histochemically, MAO-B is shown to be present in the TM nucleus (Arai et al., 1986), suggesting that histamine is uptaken presynaptically and metabolized, though the presence of presynaptic re-uptake system has not been unequivocally demonstrated.

2. Histamine receptors

There are 3 types of histamine receptors, H1, H2 and H3 (**Table 6**). The H1 and H2 receptors have been shown to cause various functions in the central (postsynaptically) and peripheral tissues, while the H3 receptors are located not only on the presynaptic terminals of histaminergic neurons that had been shown to be autoreceptors initially (Arrang et al., 1983, 1987) but also on other neurons as heteroreceptors (Schlicker et al., 1988, 1992). Moreover, H3 receptors are also located on the postsynaptic structures in the CNS as well as in the periphery (Fukui and Yanai, 1993; Leurs et al., 1995).

a. H1 receptors

1) Second messengers

When histamine binds to H1 receptors on the postsynaptic membranes, phospholipase C $_{\gamma}$ (PL C $_{\gamma}$) is activated through Gq protein, and inositol 1,4,5-trisphosphate (IP_3) and diacylglycerol (DG) are produced from phosphatidylinositol bisphosphate (PIP_2) (**Fig. 10**). IP_3 binds to IP_3 receptors on the intracellular Ca store (endoplasmic reticulum) and releases Ca^{2+} from it giving rise to the increase of the intracellular concentration of Ca^{2+}, $[Ca^{2+}]_i$. The increase in $[Ca^{2+}]_i$ activates calmodulin kinase (CaMK) or protein kinase C (PKC) (in the presense of DG), both of which phosphorylate various kinds of protein giving rise to cellular responses, or activates cAMP, cGMP, arachidonate cascade or NO systems. For details, see Fukui and Yanai (1993), Leurs et al. (1995) and Fukui and Horio (1996).

2) cDNA and amino acid sequences

cDNA of H1 receptors was cloned by the expression cloning strategy from cDNA library of bovine adrenal medulla taking advantage of increase in Cl^- conductance by the activation of Ca^{2+}-dependent Cl^- channel by histamine (Yamashita et al., 1991). The full length of cDNA clone is about 2.9 kb, and the number of the deduced amino acid sequence of the coding region is 487, which has 7 transmembrane domains, a long 3rd intracellular loop and consensus glycosylation sites in the

Table 6 Histamine H1, H2 and H3 receptors.

Property	H1 Receptor	H2 Receptor	H3 Receptor
Localization	Blood vessels, Smooth muscle CNS	Stomach, Heart CNS. Uterus(rat)	Presynapsis Heteroreceptors
Chromosome	3	5	?
Molecular weight (Amino acids)	56 kDa (487)	40 kDa (359)	?
Second messenger	IP_3/Ca	cAMP	? (Ca?)
Effector	Gq/11	Gs	Gi/o (?)
Agonist	2-Methylhistamine 2-Thizolylethylamine	4-Methylhistamine Impromidine	(R) α Methylhistamine Immepip, immetit
Antagonist	Mepyramine(Pyrilamine) Doxepin	Cimetidine Famotidine	Thioperamide Clobenpropit

N-terminal region (**Fig. 11**), all of which are characteristic of G-protein-coupled receptors. Subsequently, cDNAs of rat, guinea pig, bovine, mouse and human H1 receptors were isolated and these six H1 receptors possess 90 % homology of amino acid sequences in transmembrane domains and their vicinity. The H1 receptor genes were isolated from several species including humans, in which the chromosome of H1 receptors is locted on 3p25 (Fukui et al., 1994). The H1 receptor gene knocked-out (H1KO) mice were produced by Inoue et al. (1996) and their behavioral assessments were carried out: In short, H1KO mice are more anxious than wild type mice, particularly when they are introduced to a novel environment (Yanai et al., unpublished).

b. H2 receptors

1) Second messengers

The H2 receptors are coupled to the adenylate cyclase system through Gs protein, and cAMP is produced from ATP (**Fig. 10**). cAMP activates protein kinase A (PKA) which phosphorylates various proteins. Recently, other signal transduction systems are also activated by H2 receptors, i.e. release of arachidonic acid (Traiffort et al., 1992) and increase in $[Ca^{2+}]_i$ (see Leurs et al., 1995).

2) cDNA and amino acid sequences

cDNA of H2 receptors was cloned by Gantz al. (1991) by an elegant strategy of analogy cloning. Namely, cDNA library of canine gastric parietal cells, a richest source of H2 receptors, was screened by analogy to other amine receptors with the oligonucleotides of the third and sixth transmembrane domains, which are most homologous in G-protein-coupled receptors, as primers for PCR. The clones were expressed in L cells and assayed for the capacity of increase in cAMP and [^{3}H]tiotidine binding. The amino acid sequences deduced from nucleotide sequences of H2 receptor cDNA are shown in **Fig. 11** and the number of amino acids are 359. Later, H2 receptor cDNAs were cloned from rats, guinea-pigs, mice and humans. H2 receptors are also G-protein coupled ones with 7 transmembrane domains but with a shorter 3rd intracellular loop than H1 receptors (For details see Gantz et al., 1992, 1993; Leurs et al., 1995). The genes were isolated from various sources including humans, and are all intronless, single exons as well as the H1 receptor gene.

c. H3 receptors

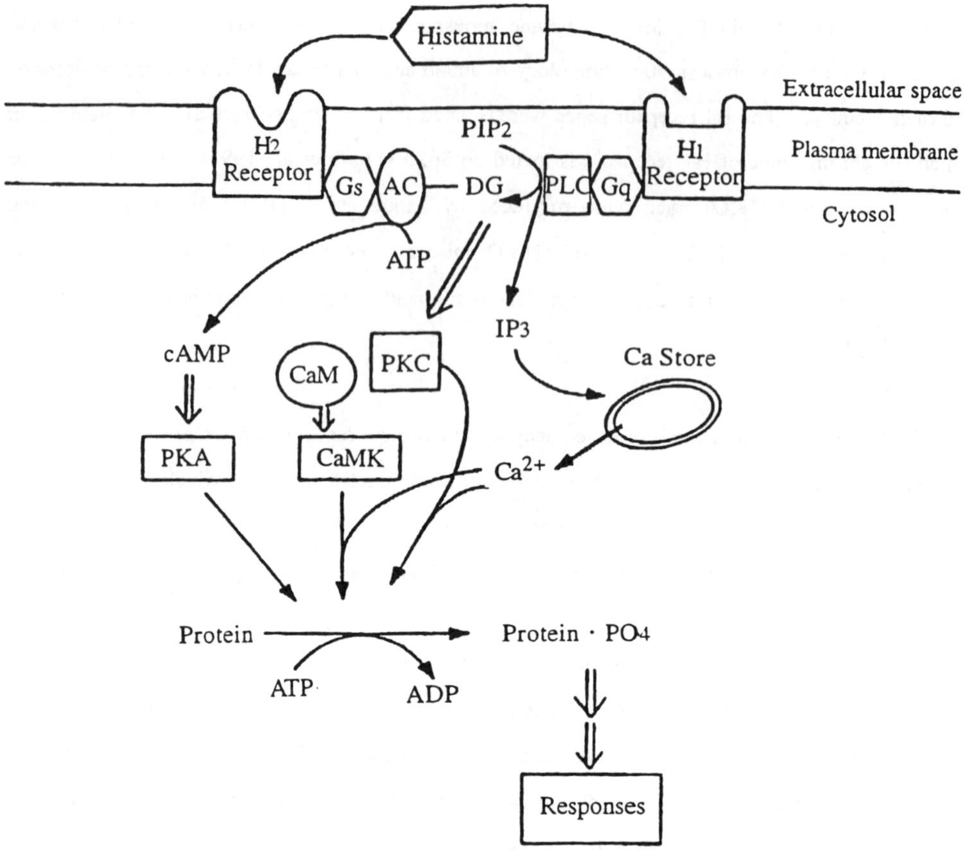

Fig. 10. Signal transductions via H1 and H2 receptors. For abbreviations see text.

(A) H1 Receptor

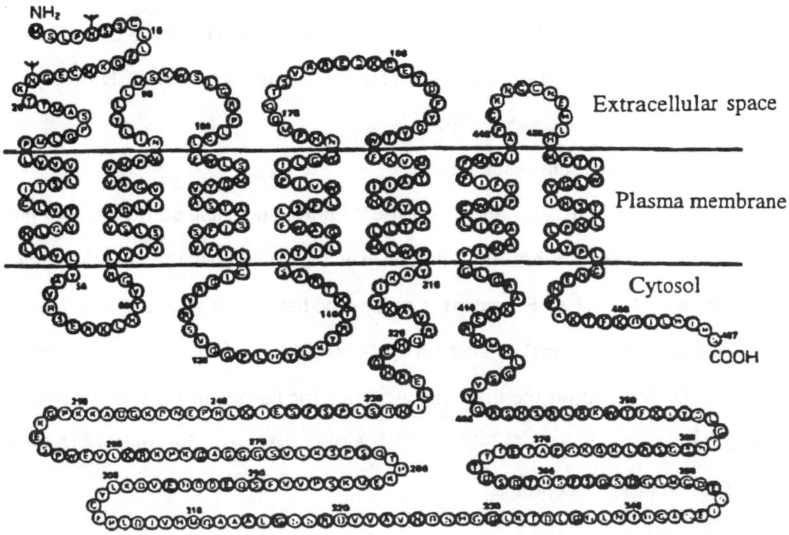

(B) H2 Receptor

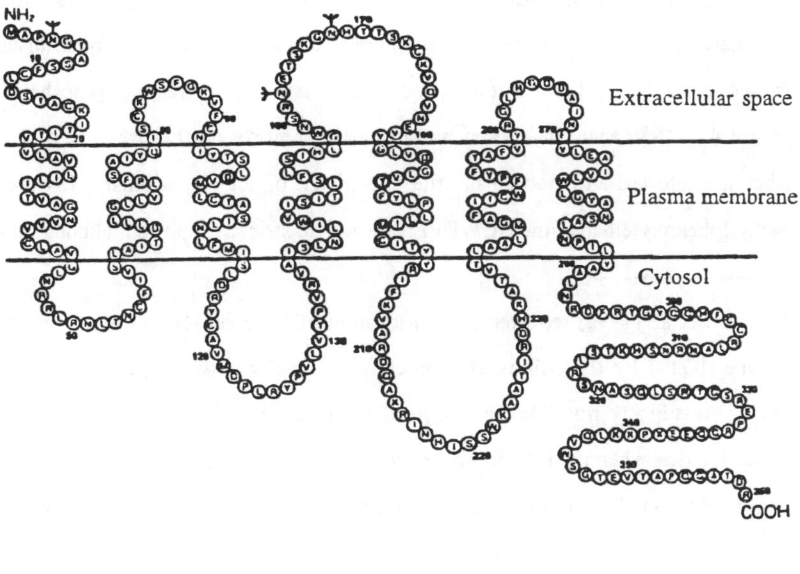

Fig. 11. Amino acid sequences of H1 and H2 receptors.

1) Second messengers

The detail of H3 receptor signal transduction is not known, but G protein may be involved (Clark and Hill, 1996). It is likely that L-type Ca channel is coupled to H3 receptors (Oike et al., 1992; Endou et al., 1993) and $[Ca^{2+}]_i$ is responsible for the action of H3 receptors similarly to other presynaptic receptors (Langer, 1997).

2) cDNA and amino acid sequences

H3 receptors have been neither purified*5 nor cloned, and so the detailed nature of H3 receptors at the molecular level is not known. It may be stated that the homology cloning strategy of H3 receptors with the H1 or H2 receptor as a probe has been so far unsuccessful in several laboratories (personal communications). There may be subtypes of H3 receptors: West et al. (1990) named H3A and H3B based on the different affinities for ligands in binding assays. Recently developed newer H3 ligands such as GT-2016, UCL6199, iodoproxyfan and GR168320 may be useful for classification of H3 receptors.

d. H_{IC} receptors

Brandes and his coworkers have presented evidence that histamine is an intracellular messenger, mediating the platelet aggregation caused by PMA and murine lymphocyte mitogenesis caused by concanavallin A, and designated H_{IC} receptors to distinguish it from the other histamine receptors (Saxena et al., 1989; Brandes et al., 1990). This is a very interesting hypothesis, but since its first proposal in 1989 there have been no supporting reports on the presence of H_{IC} receptors from other laboratories. In particulalr, the specificity of an H_{IC} agonist. N,N-diethyl-2-(4-[phenyl-methyl]phenoxy)ethanamine HCl (DPPE), is not so strict as reported. although this agent and its

----

*5Cherifi et al. (1992) reported the purification of H3 receptor protein from human gastric tumor cell line HGT-1 by the affinity column chromatography with thioperamide. an H3 antagonist. as a ligand. It is highly possible that the purified protein may be a type of cytochrome P-450. because we remind that when the H1 receptor protein was purified by affinity column chromatography with mepyramine. an H1 antagonist, as a ligand by Fukui et al. (1990. 1995), the purified protein was turned out to be a cytochrome $P\text{-}450_{2D}$.

analogs have been successfully used as anticancer drugs (Brandes et al., 1993). Meanwhile it turned out that the H_{IC} recptor is a cytochrome P-450 (LaBella et al., 1992), and thus serious questions were raised on its identity as an independent histamine receptor.

III. Pathophysiology

1. Overview of the histaminergic neuron system

a. Location and distribution: Input to and output from the neurons.

Cell bodies of the histaminergic neuron system are located exclusively in the tuberomammillary (TM) nucleus of the posterior hypothalamus (Watanabe et al., 1984, Panula et al., 1984, Steinbusch et al., 1984) in five cell clusters, E1-E5 (Inagaki et al., 1990) (**Fig. 12a**).

Efferent fibers of the histaminergic neuron system possessing varicosities are distributed in almost all regions of the brain from the olfactory bulb to the spinal cord (Inagaki et al.,1988; Panula et al., 1989) (**Fig. 12a**). The density of nerve fibers in rat brain is the highest in the hypothalamic nuclei, medial septum and diagonal band, moderate in the cerebral cortex, basal ganglia and amygdaloid complex, and low in the olfactory bulb, hippocampus, caudate-putamen, brain stem, cerebellum, and spinal cord. This pattern of distribution is well conserved phylogenetically (Airaksinen et al., 1992). Most histaminergic fibers of varicosities do not show typical synaptic configurations with cell bodies or dendrites of other neurons, but are diffusely scattered in the brain (Inagaki et al.,1988; Panula et al., 1989). Thus, histamine seems to be released non-synaptically from the varicosities, resulting in the formation of a gradient of histamine concentrations in the neighborhood of its release (Hayashi et al., 1984; Takagi et al., 1986). Only a few of their varicosities form synaptic contacts with soma of the other neuron systems such as the mesencephalic trigminal nucleus (Inagaki et al., 1987) and nucleus accumbens (Airaksinen et al., 1992). The projections from the TM nucleus to various brain regions are bilateral with ispsilateral predominance, and the 5 clusters of cell bodies named groups E1-E5 (Inagaki et al., 1988) have similar efferent projection patterns, suggesting no functional differences bewteen the five groups (Takeda et al., 1984; Steinbusch et al., 1986; Ericson et al., 1989; Inagaki et al., 1990).

The histaminergic neurons receive afferent fibers from various regions of the brain such as the

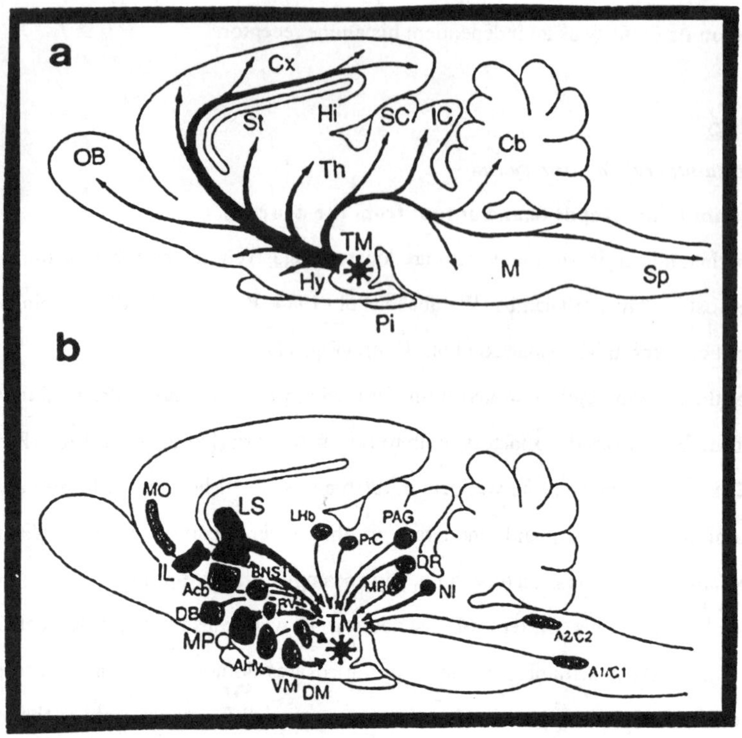

Fig. 12. Schematic representation of histaminergic neuron system. **a**. Input; **b**. Output.

prefrontal cortex, medial preoptic nucleus, and septum-diagonal band complex (Wouterlood et al., 1987, 1988). Ericson et al. (1991) determined details of the origin of the neuronal inputs to the TM nuceleus using comprehensive anterograde tracing techniques with *Phaseolus vulgaris*-leucoagglutinin as a tracer. Their results indicated that the main afferent fibers originated from some limbic forebrain regions such as the infralimbic cortex, septum and preoptic region (**Fig. 12b**). The histaminergic neurons also receive synaptic contacts from varicose fibers containing neuropeptide Y (Tamiya et al., 1989), substance P (Tamiya et al., 1990), GABA (Ericson et al., 1991), adrenaline (C1-C3), noradrenaline (A1-A2) and serotonin (B5-B9) (Ericson et al., 1989).

Like many other neuron systems, the histaminergic neurons contain other neuroactive substances, such as GABA, adenosine, Met-enkephalin-Arg-Phe, substance P and galanin (Watanabe & Wada, 1991; Yamatodani et al., 1991).

b. Functions

Brain histamine has been suggested to be involved in a variety of physiological functions such as sleep-wakefulness cycle, circadian rhythm, locomotor activity, thermoregulation, drinking, food intake, hormonal regulation, convulsion, stress and so on, as summarized in **Table 7** (Prell and Green, 1986; Schwartz et al., 1991, Watanabe and Wada, 1991; Yamatodani et al., 1991; Onodera et al., 1994; Ghi et al., 1995). From the organization of the central histaminergic neuron system as described above, it is likely that histamine plays a fundamental role in supporting the brain activity as a whole (Wada et al., 1991). Therefore, various neuropharmacological effects of histamine could have been anticipated from the morphological features described above; namely a very broad distribution of varicose nerve fibers from a limited numbers of cell bodies located in the center of the brain (1,500 per each side of rat brain) (Prell and Green, 1986; Schwartz et al., 1991, Watanabe and Wada, 1991; Yamatodani et al., 1991; Onodera et al., 1994).

There are also many indications of interactions of histamine with various neurotransmitter sytems. For example, histamine modifies the release of other neuroactive substances, such as noradrenaline, serotonin, acetylcholine (Schlicker et al., 1988, 1992, Clapham et al., 1992; Mochizuki et al., 1994), and dopamine (Schlicker et al., 1993), whereas glutamate, aspartate, acetylcholine and Met-enkephalin stimulate histamine release, but GABA, substance P and

Table 7 Possible functions of histaminergic neuron system (Onodera et al., 1994)

Response to stimulation	Receptors
ACTH secretion	
increase	H1
decrease	H2
Analgesia	H1 < H2
Antidiuresis vasopressin secretion	H1
Arousal	H1
Avoidance response	H1
Catalepsy	H1
Convulsions	H1 > H2
Drinking behavior	H1 > H2
Emesis	H1 > H2
Food intake inhibition	H1
Foot shock aggression	
inhibition	H1
facilitation	H2
Glycogenolysis	H1
Growth hormone secretion decrease	H1
Hyperglycemia	H1
Hypothermia	
lowering set point	H1
increasing heat loss	H2
Locomotor activity	
hyperactivity	H1
hypoactivity	H2
Prolactin secretion	
increase	H1
decrease	H2
Pina reflex	H1
Ptosis	H1
Tachycardia and hypertension	H1

Table 8 Effects of histaminergic drugs on the duration of electroconvulsion.

Drug	Mechanism	Histamine level	Activity of histamine neurons	Duration of convulsion
L-Histidine	Precursor	Increase	Activation	Shortening
α FMH	Inhibition of synthesis	Decrease	Inactivation	Prolongation
Metoprine	Inhibition of breakdown	Increase	Activation	Shortening
Thioperamide	H3 antagonism	Decrease	Activation	Shortening
(R) α Methylhistamine	H3 agonists	Increase	Inactivation	Prolongation*

*Only in young mice.

adenosine inhibit it (Philippu et al., 1991; Okakura et al., 1992; Okakura-Mochizuki et al., 1996).

In the following sections, the roles of histamine in some pathophysiological aspects will be discussed. Attempts will be made to correlate clinical findings with basic studies as far as possible. Drugs used for studies on the histaminergic neuron system are shown in **Fig. 13**.

2. Aging/dementia and learning-memory

During normal aging of humans, levels of both brain histamine and N^{τ}-methylhistamine in CSF increased (Mazurkiewicz-Kwilecki & Prell, 1984; Prell et al., 1988, 1991), suggesting that the metabolism of brain histamine increases with aging. The histamine levels in the CSF and brain of Alzheimer's patients have been found to be higher than in controls (Cacabelos et al., 1986, 1987,1989; Perez et al., 1988), but Mazurkiewicz-Kwilecki & Nososwah (1989) observed the lower histamine level in this disease. It should be noted that in Alzheimer's disease high densities of neurofibrillary tangles are also found in the visinity of cortically-projecting histamine neurons (Saper et al., 1987) and are concentrated in the TM nucleus (Airaksinen et al., 1991). It should be also noted that the mammillary body where the histaminergic neuron is located is a focus responsible for Wernicke-Korsakoff syndrome caused by thiamine deficiency (Onodera et al., 1994). In this respect, histamine may contribute significantly to NMDA receptor-mediated neuronal cell death in thiamine deficiency-induced Wernicke's encephalopathy (Langlais et al., 1994).

A PET study with [^{11}C]pyrilamine or doxepin as a probe showed age-dependent decreases in the density of H1-receptors in human brain, although a clear age-dependent decrease in [^{3}H]pyrilamine binding capacity was not observed in *in vitro* studies using membrane fractions of human brains (Yanai et al., 1992). The desitiy of H1 receptors in the brain is found to be lower in patients of Alzheimer's disease or brain infarction than that of sex- and age-matched healthy volunteers (Yanai et al., unpublished). The increases of histamine level in the brain may down regulate H1 receptors.

The senescence accelerated mice (SAM) are a good model system for studies on abnormal aging (Takeda et al., 1991). The brain histamine level was higher in SAM/P8, abnormal aging mice, than that in SAM/R1, normal aging mice (Meguro et al., 1992). This increase in histamine level may be derived from mast cells, because the number of mast cells in the brain increase with aging. De

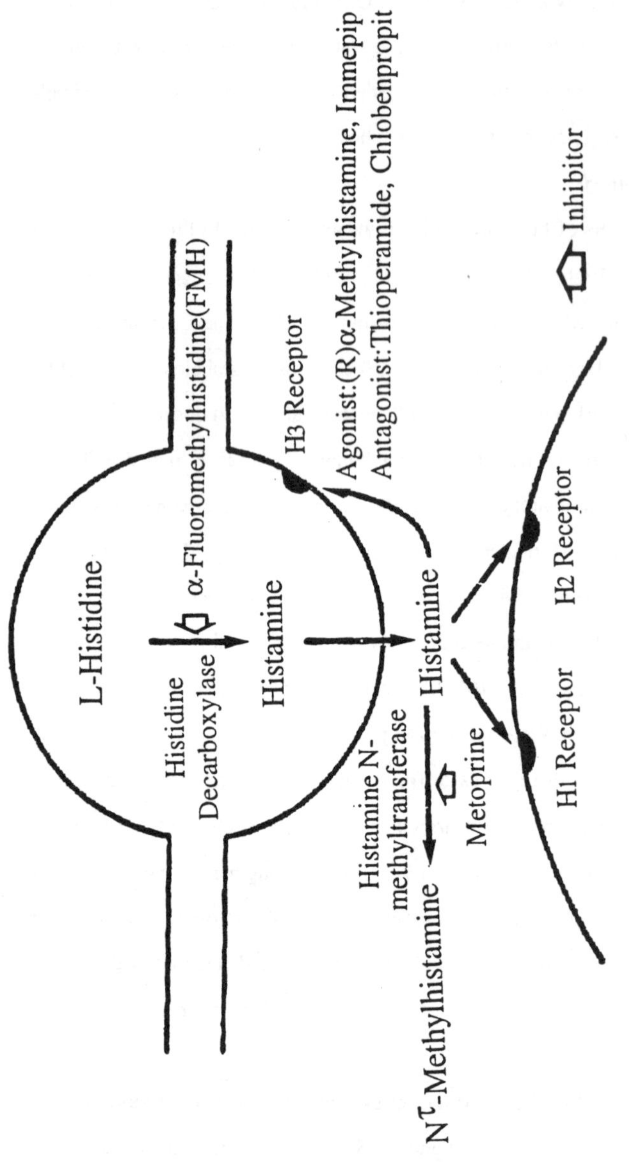

Fig. 13. Histaminergic terminal and sites of action of various drugs.

Almeida et al. (1986, 1988) examined the effect of intracerebroventricular treatment of rats with histamine on the step-through passive avoidance response and observed improvement of learing and memory retention. Similarly, the step-through and conditioned active avoidances of rats were inhibited by H1-antagonists and these inhibitions were antagonized by icv administration of histamine (Tasaka et al., 1986; Kamei et al., 1990, 1991). Accordingly, FMH also inibited step-through active avoidance (Kamei et al., 1993). In this respect, Meguro et al. (1994) showed that thioperamide, an H3 antagonist, improved the passive avoidance test in SAM-P/8 mice. Prast et al. (1996) reported that histamine facilitated short-term memory. On the other hand, Cacabelos et al. (1990) and Sakai et al. (1997) reported opposite results; increase in learning task ability was found by depletion of histamine in brain by administration by FMH. This apparent discrepancy may be derived from differences in the methods of evaluation of learning/memory, but the idea that the TM nucleus is an inhibitory center of self-potentiation (Huston et al., 1997) is very suggestive in the interpretation of these studies. These authors advocate that the TM nucleus is involved in neural plasticity and functional recovery following damage to the CNS and may function as an inibitory neural system in the control of reinforcement and mnemonic processes on the basis of their extensive behavioral studies. In addition, THA (9-amino-1,2,3,4-tetrahydroacridine), which is an inhibitor of acetylcholine esterase and has been proposed to be effective for treatment of Alzheimer's disease (Summers et al., 1986), was found to be a potent inhibitor of histamine *N*-methyltransferase (Cumming et al., 1990). This was confirmed by Nishibori et al. (1991) and Morisset et al. (1996). Thus, activations of both the cholinergic and histaminergic neurons by THA may improve cognitive performance. It is noteworthy that in advanced Alzheimer disease the TM neurons are severely affected by neurofibrillary degeneration, senile plaques and amyloid deposition (Saleh et al., 1995). Interestingly, onset of Alzheimer disease is unexpectedly delayed in those reporting extended use of H2 blockers (Breitner et al., 1995).

Gerbils are good models for brain infarction, because they have few anastomoses between the anterior and posterior branches of the Willis arterial circle of the brain. Thus, the occlusion of the two carotic acteries evokes brain ischemia in gerbils more reproducibly than in rats. Sugimoto et al. (1994) showed that the delayed type of neuronal cell death of hippocampal CA2 region of the gerbil subjected to the 3 min occlusion of carotid arteries on both sides is enhanced by the pretreatment of FMH, suggesting that histamine is protective against neuronal cell death.

3. Epilepsy/ Electroconvulsion

It has been known for a long time that high doses of H1-antagonsits occasionally produce convulsions in children of pre-school age (Wyngaarden et al., 1951, Schwartz & Patterson, 1978; Mueller, 1983; Yokoyama et al., 1993a,b,c, 1996). Moreover, the combination of an H1-antagonist (tripelennamine) with a narcotics (pentazocine) as a drug abuse (see p. 23) causes major motor seizures (Lahmeyer et al., 1980; Polkis & Whyatt, 1980). Cimetidine is also known to evoke convulsions, though very rarely (Itil et al., 1980).

In EEG studies, Shimoda et al. (1960) observed that diphendydramine increased epileptic discharges in epilpetic patients, whereas Churchill & Gammon (1949) found that diphenhydramine decreased them. King & Weeks (1965) showed that tripelennamine increased epileptic discharges in 562 patients. Recently, Yokoyama et al. (1993c) also found that *d*-chlorpheniramine increased epileptic spikes in an epileptic child by EEG activation study. A PET study with [^{11}C]doxepin as a probe showed that the levels of H1 receptors are increased in the epileptic foci in the brains of patients with complex partial seizure (Iinuma et al., 1993). It is likely that H1-receptors are up-regulated as one mechanism of self-defense against spreading of abnormal discharges from the focus to the surrounding regions like opioid μ-receptors (Frost et al., 1988).

Animal experiments using various epileptic models have shown that activation of the central histaminergic system reduced convulsions and its suppression had the opposite effect (Tuomisto et al., 1986, 1987; Scherkl et al., 1991a,b; Yokoyama et al., 1992). The histamine levels in several regions of the brain (striatum, hippocampus, amygdala, midbrain, thalamus and hypothalamus) of audiogenic seizure-sensitive Krushinsky-Molodkina rats were significantly lower than those of the epilepsy-resistant, parental Wistar strain (Onodera et al., 1992b). The duration of electrically-induced clonic convulsions in mice was increased by administration of high and moderate doses of H1-antagonists in adult and young mice, respectively (Freeman & Sturman, 1990; Yokoyama et al., 1993b). The protective effect of L-histidine, a precursor of histamine, against electrically-induced clonic convulsions in mice was abolished by H1-antagonists, but not by H2-antagonists (Yokoyama et al., 1992). H2-Receptors seem to have no appreciable effect in seizures, because H2 antagonists such as zolantidine have been shown either to have no modulating effect on electrically- and pentylenetetrazole-induced seizures (Freeman & Sturman, 1990; Scherkl et al., 1991a,b; Yokoyama et al., 1992) or to induce only slight sensitization (Gerald & Richter, 1976).

Concerning H3-receptor ligands and seizures, Scherkl et al. (1991a) reported that neither an agonist nor an antagonist influenced the seizure threshold for electrically-induced convulsions. On the other hand, thioperamide was found to decrease the durations of the tonic, clonic and postconvulsive coma phases of electrically-induced seizures in mice (Yokoyama et al., 1993a). This was confirmed by a stronger H3 antagonist, clobenpropit (Yokoyama et al., 1994) (**Table 8**). Recently, Mochizuki et al. (1997) studied the penetration of H3 antagonists by *ex vivo* experiments, showing that clobenpropit is worse in penetration through the blood-brain-barrier than thioperamide. Sakurai et al. (1994) carried out pharmacokinetic studies of thioperamide in details, showing that thioperamide does not penetrate the brain easily but not so poorly as Mochizuki et al. (1997).

4. Schizophrenia/Reverse tolerance of methamphetamine

Neuroleptics such as chlorpromazine, perphenazine and mesoridazine competitively antagonize the binding of [^{3}H]doxepin to H1 receptors (Richelson & Nelson, 1984; Kanba & Richelson, 1991). This is no wonder when we remind of the history of initial development of chlorpromazine as an antihistaminics. Nakai et al. (1989) reported that the [^{3}H]pyrilamine binding capacity to brain membrane fractions was lower in schizophrenics than in control. On the other hand, phenothiazine derivatives are reported to be more potent than haloperidol in antagonizing H2 receptors (Kanba & Richelson, 1991). Kaminsky et al. (1990) reported a case in which administration of famotidine, an H2 antagonist, that had been prescribed for treatment of peptic ulcer was associated with improvement of the deficit symptoms of schizophrenia. There are several reports on the occurrence of visual or auditory hallucination and mania in patients taking cimetidine or ranitidine (Brunton, 1996), particularly in elderly people. However, these drugs are less selective to H2-receptors than famotidine. It is interesting that clozapine has an affinity to H3 receptors (Kathman et al., 1994; Rodrigues et al., 1995). Recently, Prell et al.(1995) found that the levels histmamine metabolites in CSF is increased in schizophrenic patients compared to normals.

Ito et al. studied in details on the role of the histaminergic neuron system in methamphetamine (MAP)-reverse tolerance or behavioral sensitization, a supersensitivity that is caused by the repeated administration of MAP (Ito et al., 1996a,b.c; 1997a). Pretreatment with L-histidine inhibited MAP-induced stereotyped behavior in rats, whereas FMH, pyrilamine and zolantidine enhanced it. The inhibitory effect of L-histidine administration on behavioral sensitization was antagonized by pyrilamine and zolantidine, suggesting that the inhibitory effects are mediated through both H1- and H2-receptors in brain (**Fig. 14**). When stereotyped behavior was observed in MAP-treated rats, histamine release was increased in the striatum measured by brain microdialysis (**Fig. 15**). Sulpiride and haloperidol (D2-antagonists) blocked the MAP-induced increases in histamine release, but SCH23390 (D1 antagonist) did not. Haloperidol reduced the intensity of stereotyped behavior induced by MAP. Terefore, the MAP-induced histamine release seems to be mediated through D2-receptors in the striatum. Thus, MAP stimulates the histaminergic neurons under a control of the dopamine D2-receptor-mediated transmission, which in turn ultimately inhibits stereotyped behavior. Cocaine also facilitates histamine release from the striatum (Ito et al., 1997b). In this respect, it is interesting that the discriminative stimulus effect of cocaine and methamphetamine were potentiated by H1 antagonists (Suzuki et al., 1996) and a similar effect was observed in humans (Johanson et al., 1996). Furthermore, there is a cross-generalization between a cocaine cue and H1 blockers (Gauvin et al., 1995).

5. Food intake

Clinically it is known that patients taking H1 antagonists sometimes suffer from obesity. The histamine levels in cerebrospinal fluids of patients of obesity and anorexia nervosa are lower and higher, respectively, than those of normal controls. These clinical observations suggest that histamine suppresses the appetite. In fact, icv injection of histamine deppressed the feeding behavior of rats (Clineschmidt & Lotti, 1973; Itowi et al., 1988), whereas microinjection of H1-antagonists into the ventromedial hypothalamus or icv infusion of FMH induced transient increase in feeding in the early light period (Sakata et al., 1988; Fukagawa et al.. 1989). Histamine depletion by chronic administration of FMH increased feeding and body weight (Orthen-Cambill and Salmon. 1992; Tuomisto et al., 1993). These results indicate the histaminergic tonic control of the feeding

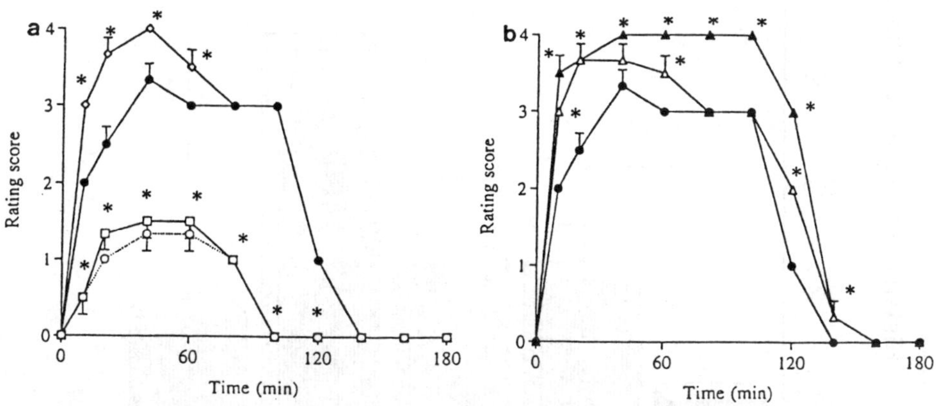

Fig. 14. Effects of histaminergic agents on behavioral sensitization induced by MAP (Ito et al., 1997a). Rats were injected with saline (○), MAP (3 mg/kg) (●), MAP plus L-histidine (750 mg /kg) (□), MAP plus FMH (100 mg/kg) (◇), MAP plus pyrilamine (5 mg/kg) (△), and MAP plus zolantidine (5 mg/kg) (▲) once daily i.p. for 14 consecutive days. After drug-free interval of 7 days, MAP (1 mg/kg) was recahllenged at Time 0, and the degree of sterotyped behavior was estimated for 3 hrs according to the scores from 0 to 4 described by Creese and Iversen.

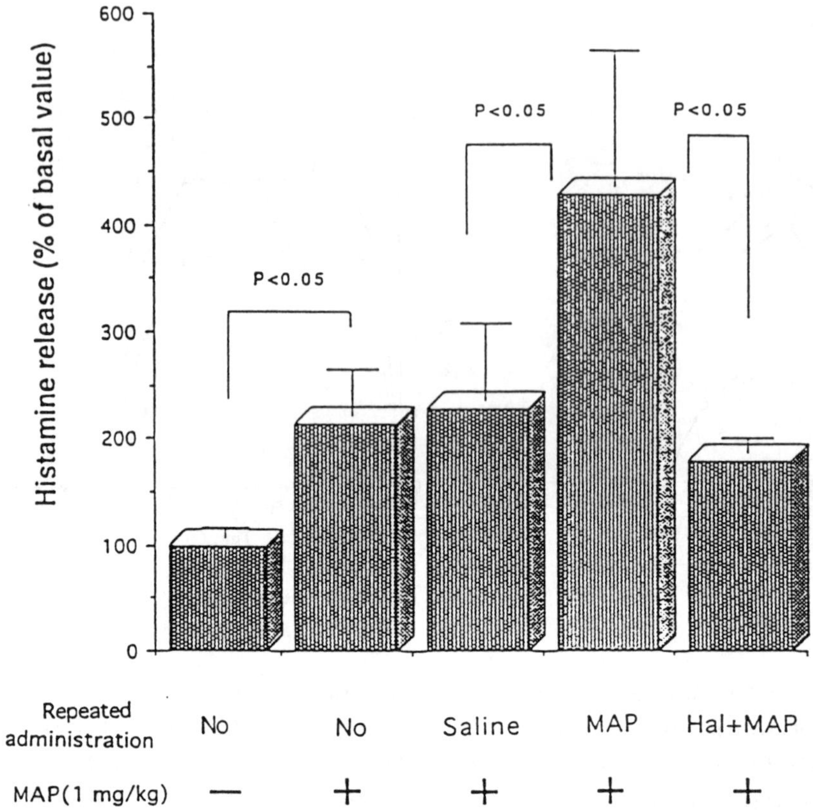

Fig. 15. Effects of histaminergic agents on histamine release in the striatum of rats subjected to behavioral sensitization caused by repeated aministration of MAP. Brain histamine release was measured by *in vivo* microdialysis. Hal; haloperidol.

behavior through H1 receptors and brain histamine is suggested to be a physiological anorectic substance. For details, see a review article by Sakata (1991).

7. Sedation-sleep/Circadian rhythms

It has been well known that H1 antagonists cause sedation and drowsiness and sometimes impair psychomotor performacne (Nicholson et al., 1985; White & Rumbold, 1988). In clinical studies, H1 antagonists had little effect on wakefulness during nocturnal sleep, but increased the tendency to sleep during daytime (Nicholson et al., 1985, White & Rumbold, 1988). This is a basis of the development of H1 antagonists with little sedation, non-sedative or second generation of H1 antagonists.

Intracerebroventricular (icv) administration of histamine or H1 agonists induced behavioral arousal in conscious rats and decreased the duration of narcosis of pentobarbital-treated rats (Kalivas, 1982). This H1-mediated effect of inducing wakefulness was also demonstrated by EEG studies (Monti et al., 1986; Tasaka et al., 1989). FMH increased the duration of slow-wave sleep and decreased that of wakefulness during the dark period in rats (Kiyono et al., 1985; Monti et al., 1988; Itowi et al., 1991) and in cats (Lin et al., 1988). Furthermore, increased release of endogenous histamine induced by thioperamide induced wakefulness, and the effect was diminished by pretreatment with an H1-antagonist (Lin et al., 1990). Recently, Lin et al. (1996) proposed the importance of the interactions between histaminergic and cholinergic neurons in the circuit in cortical activation during wakefulness. Electrophysiologically, histamine excites nucleus basalis cholinergic neurons by a concomitant activation of H1 and H2 receptors, suggesting that the TM neurons accordingly facilitate tonic firing of cholinergic nuerons during wakefulness (Khateb et al., 1995). Thus, histamine seems to maintain wakefulness as a "wake amine", confirming the old paradigm proposed by Monnier et al. (1971).

Itowi et al. (1990, 1991) examined the possible histaminergic modulation of the oscillation or synchronization of free running circadian phases of the spontaneous locomotor and drinking rhythms in rats. Their results showed that icv injections of histamine had two different types of effect on the spontaneous activity phase: it caused transient phase-advance when it was injected during the subjective light period (advance-sensitive period) and a permanent phase delay when injected during the subjective dark period (delay-sensitive period). A phase-response curve plotted for the permanent phase shifting effect of histamine resembled that induced by brief light pulses (Turek, 1987), suggesting that histamine is involved in the system relaying light inofrmation within the suprachiasmatic nucleus itself (Yamatodani et al., 1991). For details, see Yamatodnai et al. (1991) or Onodera et al. (1994).

In fact, histamine release from the hypothalamus shows a clear ciradian rhythm, being high in the dark period and low in the light period in rats (Mochizuki et al., 1992) as demonstrated using an *in vivo* brain microdialysis method (Mochizuki et al., 1991). This was confirmed in a diurnal animal, the Rhesus monkey (Onoe et al., 1992); In this monkey, the hypothalamic histamine output was higher in the daytime than at night, and was well correlated with the sleep-wakefulness state. Thus, both in nocturnal and diurnal animals, the activity of the central histaminergic system increases in the active phase. These observations are supported by the recent finding that photic phase shifts in wheel-running rhythms are significantly attenuated by FMH administration, suggesting that histamine modulates photic input to the circadian clock (Eaton et al., 1995).

8. Catalepsy, Extrapyramidal system/Parkinson's disease

Catalepsy is an extrapyramidal symptom observed in advanced Parkinson's disease and as a side effect of neuroleptics (Klemm, 1989). Some histamine H1 antagonists are clinically useful to reduce the neuroleptics-induced extrapyramidal symptoms (Peggs et al., 1995) as well as H2 antagonist, famotidine (Malinai et al., 1995).

The central administration of histamine has been shown to induce catalepsy dose-dependently in mice (Onodera and Ogura, 1983; Malec and Langwinski, 1993) and rats (Maslinski et al., 1973; Nowak and Pilk, 1977). In mice, histamine-induced catalepsy was antagonized by H1 antagonists, but not by H2 antagonists and no histamine metabolite induced catalepsy in mice (Onodera, 1991). These findings suggest that histamine *per se* is a cataleptogen, and the histamine-induced catalepsy seems to be mediated by H1 receptors (Klem, 1989; Onodera, 1991). Nakamura and Vincent (1986) suggested that the conversion of MPTP (1-methyl-4-phenyl-1,2,3,6-tetradydropyridine), a causative agent of experimental Parkinson disease (Bradbury et al., 1986), to MPP^+ (1-methyl-4-phenylpyridinium) occurs via MAO-B within histaminergic neurons in the hypothalamus that innervates the substantia nigra where the toxic MPP^+ could be released and then taken up into the dopaminergic neurons. Cumming et al. (1989) showed that MPTP did not affect the histamine levels in the brain. In this respect, Garbarg et al. (1983) had reported that the HDC activities in postmortem brain specimens of patients with Parkinson's disease were not significantly different from those of normals. Recently, Nakamura et al. (1996) reported that the number of large size neurons in the TM nucleus in patients of Parkinson's desease was not altered when compared with controls, and Lewy bodies were rarely present in the TM nucleus. These data suggest that the degeneration of histaminergic neurons may not occur and histaminergic neurons may not correlate with the onset of Parkinson's disease. However, Coelho et al. (1991) suggested that there may occur the neuronal destruction of histaminergic sytem in Parkinson's disease, because blood histamine reflects the brain damages showing the decrease in blood histamine in drug-treated Parkinson patients.

Recently, Ryu et al. (1994a,b, 1996) suggested that histamine H3 receptors were closely related to the dopaminergic neuron system, especially D1 receptors, demonstating the heterogenous distribution of H3-, D1- and D2-receptors in the extrapyramidal system of rats. The changes of the H3, D1, and D2-receptor binding sites were reported after treatment with quinolinic acid and 6-hydroxydopamine. The up-regulation of H3 receptors occurred in the substantia nigra and striatum after 6-hydroxydopamine injection into the rat barin (Ryu et al., 1994a). The treatment with quinolinic acid resulted in similar increases in H3- and D1-receptor binding sites in the ipsilateral striatum and substantia nigra. D2 receptor binding sites were relatively well conserved, whereas H3- and D1- receptors increased considerably (Ryu et al., 1994b). These data suggest that H3- and D1-receptor binding sites are locallized on the striatonigral projection neurons which are together sensitive to quinolinic acid, and that the distributional compartment of D2 receptor bindind sites is quite different from those fo H3 and D1 recptors. H3 receptors are also up-regulated by physical denervation, i.e. enucleation of eye ball (Nakagawa et al., 1994).

9. Morphine-induced rewarding effect/dependence

In USA, tripelennamine, an H1 antagonist, has been used illicitly in combination with paragoric "blue velvet" or pentazocine "T's" blues by heroin addicts (Burton, 1965; Lahmeyer &

Steingold, 1980; Senay, 1985). In Japan, the abuse of over-counter cough syrups, which usually contains active ingredients such as dihydrocodeine, methylephedrine, caffeine and *d*-chlorpheniramine, has been spreading among adolescents and is a serious social issue of petty crime (Ueyama et al., 1984; Watanabe et al., 1984). The rewarding effects of the opioids pentazocine and dihydrocodeine are potentiated by combination with antihistaminics such as tripelennamine and *d*-chlorpheniramine. Furthermore, the refinforcing effect of methamphetamine was enhanced in combination with morphine and *d*-chlorpheniramine, which produced additive and potentiative effects, respectively (Masukawa et al., 1993). These results suggest that antihistaminics have potentiative effects on reinforcing effects of opioids and methamphetamine, as similarly as cocaine, a dopamine reuptake inhibitor (Masukawa et al., 1993; Suzuki et al., 1995).

Moreover, zolantidine, a histamine H2-antagonist and FMH, a histamine synthesis inhibitor, potentiated morphine-induced place preference (Suzuki et al., 1995). These potentiations by two agents were well antagonized by SCH 23390, a dopamine D1-angtagonist. Conversely, administration of L-histidine (400 and 800 mg/kg, s.c.) attenuated morphine-induced place preference (Suzuki et al., 1995). Thus, there seems to be a close relationship between the histaminergic and dopaminergic systems in morphine-induced conditioned place preference and the histaminergic neurons appear to play inhibitory roles in this paradigm of behavior test.

V. Concluding remarks

History of histamine research began in early 1900's and many epoch-making breakthroughes have clarified the functions of histamine. As far as histamine in the brain is concerned, it has been difficult to specify its roles until the histaminergic neuron system was visualized by immunohistochemistry in mid 1980's. In fact, studies on the functions of histaminergic neurons have been reinvestiged after the elucidation of their localization and distribution were achieved. In retrospect, a number of efforts to develop the methods of assay for histamine and its metabolites propelled the accurate analysis of the activity in histaminergic neurons. Each measurement system has both advantages and disadvantages, but now we can choose a good and convenient one from a list of these methods depending on our experimental protocol and equipments. With regard to histamine receptors, the characterization and clonings were almost completed except the H3 receptor, and the gene technology including knock-out mice targetting the enzymes related with the metabolism of histamine or the receptors has just begun to elucidate the roles of histamine in brain.

In this chapter, we summarized pathophysiological conditions under which histamine affected directly or indirectly according to recent pharmacological data obtained by animal models (aging, learning-memory, epilepsy and so on) and by new techniques such as *in vivo* brain microdialysis and PET. Though the increase in new information on histamine and its functions gave us some clues for treatment of neuronal diseases, further advance is highly expected.

Acknowledgements

We acknowledge critical readings of the manuscript and useful discussions with Drs. K.Yanai, T. Mochizuki and C.Ito. Studies carried out in our laboratories are supported by grants-in-aid from the Department of Education, Science, Culture and Sports of the Japanese Governments (#08407705).

References

Airaksinen, M.S., Paetau, A., Paljaervi, L., Reinkainen, K., Riekinen, P., Soumalainen, P. and Panula, P.(1991). Histamine neurons in human hypothalamus: Anatomy in normal and Alzheimer disease brains. Neuroscience 44, 465-481.

Airaksinen, M.S., Alanen, S., Szabat, E., Visser, T.J. and Panula, P.(1992). Multiple neurotransmitters in the tuberomammillary nucleus: Comparison of rat, mouse and guinea pig. J. Comp. Neurol. 323, 103-116.

Aksoy, S., Raftogianis, R. and Weinshilboum, R.(1996). Human histamine N-methyltransferse gene: Structural characterization and chromosomal localization. Biochem. Biophys. Res. Commun. 219, 548-554.

Alam, M.K., Sasaki, M., Watanabe, T. and Maeyama, K. (1995). Simultaneous determinations of histamine and N^{τ}-methylhistamine by high-performance liquid chromatography-chemiluminescence coupled with immobilized diamine oxidase. Anal. Biochem. 229, 26-34.

Arai, R., Kimura, H. and Maeda, T.(1986). Topographic atlas of monoamine-oxidase containing neurons in the rat brain studied by an improved histochemical method. Neuroscience 19, 905-926.

Arrang, J.M., Garbarg, M. and Schwartz, J.C.(1983). Autoregulation of histamine release mediated by a novel class (H3) of histamine receptor. Nature 302, 832-837.

Arrang, J.M., Garbarg, M., Lancelot, J.C., Lacomte, J.M., Pollard, H., Robba, M., Schunach, W. and Schwartz, J.C.(1987). Highly potent selective ligands for histamine H3 receptors. Nature 327, 117-123.

Arrang, J.M., Defontain, N. and Schwartz, J.C.(1988). Phencyclidine blocks histamine H3 receptors in rat brain. Eur. J. Pharmacol. 157, 31-35.

Beaven, M.A.(1991). Radioenzymatic assays in Biological fluids. In: Uvnas, B. (ed) Histamine and histamine antagonists. (Handbook of Experimental Pharmacology, vol 97) pp 39-48. Splinger-Verlag, Berlin Heidelberg New York.

Beaven, M.A., Jacobsen, S. and Horakova, Z. (1972). Modification of the enzymatic isotopic assay of histamine and its application to measurement of histamine in tissues, serum and urine. Clin. Chim. Acta 37, 91-103.

Beaven, M.A., Robinson-White, A., Roderick, N.B. and Kaufman, G.L.(1982). The demonstration of histamine release in clinical conditions: A review of past and present assay procedures. Klin. Wochenchr. 60, 873-881.

Bradbury, A.J., Costall, B., Domeney, A.M., Jenner, P., Kelly, M.E., Marsden, C.D. and Naylor, R.J.(1986). 1-Methyl-4-phenylpyridine is neurotoxic to the nigrostriatal dopamine pathway. Nature 319, 56-57.

Brandes, L.J., LaBella, F.S., Glavin, G.B., Paraskeva, F., Sxena, S.P., McNicol, A. and Gerrard, J.M.(1990). Histamine as an intracellular messenger. Biochem. Pharmacol. 40, 1677-1681.

Brandes, L.J. and LaBella, F.S.(1993). Identification of intracellular histamine receptors (H_{IC}) that regulate cell proliferation. Adv. Biosci. 89, 31-41.

Breitner, J.S., Welsh, K.A., Helms, K.J., Gaskell, P.C., Gau, B.A., Roses, A.D., Pericak-Vance, M.A. and Saunders, A.M.(1995). Delayed onset of Alzheimer's disease with nonsteroidal anti-inflammtory and histamine H2 blocking drugs. Neurobiol. Aging, 16, 523-530.

Brunton, L.L.(1996). Agents for control of gastric acidity and treatment of peptic ulcers. In: Pharmacological Basis of Therapeutics. McGraw-Hill, pp.901-916.

Burton, J.F.(1965). Mainliners and blue velvet. J. Forensic Sci. 10, 466-472.

Cacabelos, R., Niigawa, H. and Ikemura, Y.(1986). Neuroendocrie correlates in senile dementia of the Alzheimer type. Prog. Clin. Neurosci. 2, 231-247.

Cacabelos, R., Niigawa, H. and Nishimura, T.(1987). Neuroendocrine function in psychiatric disroders and dementia. Ja. J. Neuropsychopharmacol. 9, 13-19.

Cacabelos, R., Yamatodani, A., Niigawa, H., Hariguchi, S., Tada, K., Nishimura, T. and Wada, H.(1989). Brain histamine in Alzheimer's disease. Meth. Find. Exp. Clin. Pharmacol. 11, 353-360.

Cacabelos, R. and Alvarez, X.A.(1990). Histidine decarboxylase (HDC) inhibition induced by FMH provokes learning-related hypokinetic activity. Agents Actions. 33, 131-134.

Chassande ,O., Renard, S., Barbry ,P. and Lazdunski, M.(1994). The human gene for diamine oxidase, an amiloride binding protein. Molecular cloning, sequencing, and characteization of the promoter. J. Biol. Chem. 269, 14484-14489.

Cherifi, Y., Pigeon, C., Leromacner, M., Bado, A., Reyldesmars, S. and Lewin, M.J.M.(1992). Purification of a histamine H3-receptor negatively coupled to phoshpoinositide turnover in the human gastric cell line Hgt-1. J. Biol. Chem., 267, 25315-25320.

Chevrier, D., Guesdon, J.-L., Mazie, J.-C. and Avrameas, S. (1986) Enzyme immunoassay for the measurement of histamine. J. Immunol. Methods 94, 119-125.

Churchill, J.A. and Gammon, G.D.(1949). The effect of antihistaminic drugs on convulsive seizures. J. Am. Med. Assoc. 141, 18-21.

Clapham, J. and Kilpatrick, G.J. (1992). Histamine H3 receptors modulate the release of [^{3}H]acetylcholine from slices of rat entrhinal cortex: Evidence for the possible existence of H3 receptor subtypes. Br. J. Pharmacol. 107, 919-923.

Clark, E.A. and Hill, S.J.(1996). Sensitivity of histamine H3 receptor agonist-stimulated [^{35}S]GTP γ S binding to pertusis toxin. Eur. J. Pharmacol. 296, 223-225.

Clineschmidt, B.V. and Lotti, V.J.(1973). Histamine intraventricular injection suppresses ingestive behavior of cat. Arch. Int. Pharmacoldyn. Ther. 206, 228-298. 3

Coelho, M.H., Silva, I.J., Azevedo, M.S. and manso, C.F.(1991). Decrease in blood histamine in drug-treate parkinson patients. Mol. Chem.Neurophathol. 14, 77-85.

Cumming, P., Jakubovic, A. and Vincent, S.R.(1989). Cerebral histamine levels are unaffected by MPTP aministration in mouse. Eur. J. Pharmacol. 166, 299-301.

Cumming, P., Reiner, P.B. and Vincent, S.R.(1990). Inhibition of rat brain histamine N-methyltransferase by 9-amino-1,2,3,4-tetrahydroacridine (THA). Biochem. Pharmacol., 40, 1345-1350.

DeAlmeida, M.A.M.R. and Izquierdo, I.(1986). Memory facilitation by histamine. Arch. Int. Pharmacoldyn. Ther. 283, 193-198.

DeAlmeida, M.A.M.R. and Izquierdo, I.(1988). Intracerebroventricular histamine, but not 48/80, causes potentiating memory facilitaion in rats. Arch. Int. Pharmacoldyn. Ther. 283, 193-198.

Eaton, S.J., Cote, N.K. and Harrington, M.E.)1995). Histamine synthesis inhibition reduces light-induced phase shifts of circadian rhythms. Brain Res. 695, 227-230.

Endou, M., Poli, E. and Levi, R.(1993). Histamine H3 receptor signalling in the heart: possible involvement of Gi/Go proteins and N-type Ca channel. J. Pharmacol. Exp. Ther. 269, 221-???.

Ericson, H., Blomqvist, A. and Koehler, C. (1989). Brainstem afferents to the tuberomammillary nucleus in the rat brain with special reference to monoaminergic innervation. J. Comp. Neurol. 281, 169-192.

Ericson, H., Blomqvist, A. and Koehler, C. (1991). Origin of neuronal inputs to the region of the tuberomammillary nucleus in the rat brain. J. Comp. Neurol. 311, 45-64.

Freeman, P. and Sturman, G.(1990). Histaminergic H1 antagonists modulate seizure pattern induced by electroshock in mice. Br. J. Psychopharmac. 100, 389P.

Frost, J.J., Mayberg, H.S., Fisher, R.S., Dannals, R.F., Wilson, A.A., Ravert, H.T. and Wagner, H.N.(1988). Mu-opiate recptors measured by positron emission tomography re increased in temporal lobe epeilepsy. Ann. Neurol. 23, 231-237.

Fukagawa, K., Sakata,T., Shiraishi, T., Yoshimatsu ,H., Tujimoto, K. Ookuma, K. and Wada, H. (1989). Neuronal histamine modulates feeding behavior through H1-receptor in rat hypothalamus. Am. J. Physiol. 256, R605-R611.

Fukui, H., Mizuguchi, H., Liu, Y.Q., Leurs, R., Kangawa, K., Matsuo, H. and Wada, H. (1990). Purification of [^{3}H]mepyramine binding protein, a possible histamine H1 receptor from rat liver and its amino acid sequence homo logy with debrisoquine 4-hydroxylase cytochrone P-450. Eur J Pharmacol 183, 1727-1728.

Fukui, H and Yanai, K.(1993). Histamine receptors. In: Methods in Neurotransmitter and Neuropeptide Research. (S.H. Parves, M. Naoi, T. Nagatsu, S. Pravez, Eds.), Elsevier, Amsterdam, pp.141-188.

Fukui, H., Fujimoto, K., Mizuguchi, K.Sakamoto, K., Horio, Y., Takai, S., Yamada, K. and Ito, S.(1994). Molecular cloning of the human histamine H1 receptor gene.Biochem. Biophys. Res. Commun., 201, 894-901.

Fukui, H., Mizuguchi, H., Liu, Y.Q., Wang, N.P., Hayashi, H., Kangawa, K., Wakamiya, T., Leurs, R., Shiba, T. and Matsuo, H.(1995). Purification and characterization of [^{3}H]mepyramine (histamine H1 antagonist)-binding protein from rat liver: A highly homologous protein with cytochrome P-450 2D. J. Biochem., 117, 993-998.

Fukui, H. and Horio, Y.(1996). Mechanism of histamine H1 receptor desensitization. In: Receptor Desensitization and Ca^{2+}-Signaling (M.K. Uchida, Ed.), Japan Scientific Societies Press, Tokyo, pp.107-118.

Gantz, I., Schafffer, M., DelValle, J., Logdson, G., Campbell, V., Uhler, M. and T.Yamada (1991). Molecular cloning of a gene encoding the histamine H2 receptor. Proc. Natl. Acad. Sci. USA 88, 429-433.

Gantz, I., DelValle, J., Wang, L., Tashiro, T., Munzert, G., Guo, Y., Konda, Y. and Yamada,T. (1992). Molecular basis for the interaction of histamine with the histamine H2 receptor. J. Biol. Chem. 267, 20840-20843.

Gantz, I., DelValle, J. and Yamada, T.(1993). Molecular physiology of the histamine receptor. J. Tokyo Women's Medical College, 63, 29-39.

Garbarg, M., Javoy-Agid, F., Schwartz, J.C. and Agid, Y.(1983). Brain histidine decarboxylase activity in Parkinson's disease. Lancet 8, 74-75.

Garbarg, M., Pollard, H., Trung Tuong, M.D., Schwartz, J.C. and Gros, C. (1989). Sensitive radioimmunoassays for histamine and tele-methylhistamine in the brain. J. Neurochem. 53, 1724-1730.

Gauvin, D.V., Carl, K.L., Briscoe, R.J., Vallet, M. and Holloway, F.A.(1995). Cross-generalization between a cocaine cue and two antihistamines. Eur. J. Pharmacol. 294, 281-288.

Gerald, M.C. and Richter,N.A.(1976). Studies on the effects of histaminergic agents on seizure susceptibility in mice. Psychopharmacologia 46, 277-282.

Ghi, P., Ferretti, C. and Blengio, M.(1995). Effects of different types of stress on histamine H3 receptors in the rat cortex. Brain Res. 690, 104-107.

Girard, B., Otterness, D.M., Wood, T.C., Honchel, R., Wieben, E.D. and Weinshilboum, R.M.(1994). Human histamine N-methyltransferase pharmacogenetics: Cloning and expression of kidney cDNA. Mol. Pharmacol., 45, 461-468.

Harsing, L.G. Jr., Nagashima, H., Duncalf, D., Vizi, E.S. and Goldiner, P.L. (1986). Determination of histamine concentration in plasma by liquid chromatography/electrochemistry. Clin. Chem. 32, 1823-1827.

Hayashi, H., Takagi, H., Takeda, N., Kubota, Y., Tohyama, M., Watanabe, T. and Wada, H.(1984). Fine structure of histaminergic neurons in the caudal magnocellular nucleus of the rat as demonstrated by immunocytochmistry using histidine decarboxylase as a marker. J. Comp. Neurol. 229, 233-241.

Hayashi. H. and Watanabe, T.(1993). Histamine and related enzymes. In: Methods in Neurotransmitter and Neuropeptide Research(Parvez, S.H., Naoi, M, Nagatsu, T. and Parvez, S, Eds.), Elsevier, Amsterdam, pp.111-139.

Hough, L.B., Khandelwal, J.K., Morrishow, A.M. and Green, J.P. (1981). An improved GC-MS method to measure tele-methylhistamine. J. Pharmacol. Methods, 5, 143-148.

Huston. J.P., Wagner, U. and Hasenoehrl ,R.U.(1997). The tuberomanmillary nucleus projections in the control of learning, memory and reinforcement processes: evidence for an inhibitory role. Behav. Brain Res. 83, 97-105.

Ichikawa, A., Fukui, T., Yamamoto ,J., Ohgoh, M., Tanaka, S. & Funakoshi, S. (1995). *De novo* synthesis and posttranslational processing of L-histidine decarboxylase in mice. Meth. Find. Exp. Clin. Pharmacol. 17(C), 5-9.

Iinuma, K., Yokoyama, H., Otsuki, T., Yanai, K., Watanabe,T., Ido,T. and Ito, M.(1993). Histamine H1 receptors are increased in parietal epilepsy: A PET study. Lancet 341, 238.

Inagaki, N., Yamatodani, A., Shinoda,K., Shiotani, Y., Tohyama, M., Watanabe, T. and Wada, H. (1987). The histaminergic innervation of the mesencephalic nucleus of the trigeminal nerve in the rat brain. Brain Res. 418, 388-391.

Inagaki, N., Yamatodani, A., Ando-Yamamoto, M., Tohyama, M., Watanabe, T. and Wada, H. (1988). Organization of histaminergic fibers in the rat brain. J. Comp. Neurol. 273, 283-300.

Inagaki ,N., Tada, K., Taniuchi, I., Yamatodani, A., Tohyama,M., Watanabe, T. and Wada, H. (1990). An analysis of histaminergic efferents of the tuberomammillary nucleus to the medial preoptic area and inferior colliculus of the rat. Exp. Brain Res. 80, 374-390.

Inoue, I., Yanai, K., Kitamura, D., Taniguchi, I, Kobayashi, T., Watanabe, T. and Watanabe, T.

(1996). Impaired locomotor activity and exploratory behavior in mice lacking in histamine H1 receptors. Proc. Nat. Acad. Sci. USA 93, 13316-13320.

Itil, T.M. and Soldatos, S.(1980). Epileptogenic side effect of psychotropic drugs. J. Am. Med. Assoc. 244, 2460-1464.

Ito, C., Onodera, K., Sakurai, E., Sato,M. and Watanabe,T.(1996a). Effects of dopamine antagonists on neuronal histamine release in the striatum of rats subjected to acute and chronic treatments with methamphetamine. J. Pharmacol. Exp. Ther. 279, 271-277.

Ito, C., Sato, M., Onodera, K. and Watanabe,T.(1996b). The role of brain histaminergic neuron system on methamphetamine-induced behavioral sensitization in rats. New York Acad. Sci. 801, 353-360.

Ito, C., Onodera. K., Sakurai, E., Sato, M. and Watanabe, T.(1996c). The effect of methamphetamine on histamine level and histidine decarboxylase activity in rats. Brain Res. 734, 98-102.

Ito, C., Onodera, K., Watnabe, T. and Sato, M.(1997a). Effects of histamine agents on methampheramine-induced stereotyped behavior and behavioral sensitization in rats. Psychopharmacology 130, 363-367.

Ito, C., Onodera, K., Sakurai, E., Sato, M. and Watanabe, T.(1997b). Cocaine and histaminergic neuron. J. Neurochem. in press.

Itowi, N., Nagai, K., Nakagawa, H., Watanabe T. and Wada, H.(1988). Changes in feeding of rats elicited by histamine infusion. Physiol. Behav. 44, 221-226.

Itowi, N., Yamatodani, A., Nagai ,K., Nakagawa, H. and Wada, H.(1990). Effects of histamine and α-fluoromethylhistidine injections on circadian phase of free-running rhythm.. Physiol. Behav. 47, 549-554.

Itowi, N., Yamatodani, A., Kiyono, S., Hiraiwa, M.L. and Wada. H.(1991). Effect of histamine depletion on the circadian amplitude of the sleep-wakefulness cycle. Physiol. Behav. 49, 643-646.

Johanson, C.-E., Evans, S. and Henningfield, J.(1996). The discriminative stimulus effects of tripelennamine in humans. Psychopharmacology 126, 140-146.

Joseph, D.R., Sullivan, P.M., Vang, Y.-M., Kozak, C., Fenstermacher, D.A., Behrendsen, M.E. and Zahnow, C.A.(1990). Characterization and expression of the complementary DNA encoding rat histiidne decarboxylase. Proc. Nat. Acad. Sci. USA 87, 733-737.

Kalivas, P.W.(1982). Histamine-induced arousal in the conscious and pentobarbital-pretreated rat. J. Pharmacol. Exp. Ther. 222, 37-42.

Kamei, C., Chung, Y.H. and Tasaka, K.(1990). Influence of certain H1 blockers on the step-through active avoidance response in rats. Psychopharmacology 102, 312-318.

Kamei, C. and Tasaka, K.(1991). Participation of histamine in step-through active avoidance response and its inhibition by H1-blockers. Jap. J. Pharmacol. 57, 473-482.

Kamei, C., Okumura, Y. and Tasaka, K.(1993). Influence of histamine depletion on learning and memory recollection in rat. Psychopharmacology 111, 376-382.

Kaminsky, R., Moriarty, T.M., Bodine, J., Wolfe,D.E. & Davidson,M.(1990). Effect of famotidine on deficit symptoms of schzophrinia.Lancet 335, 1351-1352.

Kanba, S. and Richelson, E.(1991). Interractions with psychotropic drugs. In: Histminergic Neurons: Morphology and Function (T.Watanbe and H.Wada, Eds.). CRC Press, Boca Raton. pp, 271-282.

Khateb, A., Fort, P., Pegna, A., Jones, B.E. and Muehlethaler, M.(1995). Cholinergic nucleus basalis neurons are excited by histamine in vitro. Neuroscience 69, 495-506.

Kathmann, M., Schlicker, E. and Gothert, M.: Intermediate affinity and potency of clozapine and low affinity of other neuroleptics and of antidepressants at H3 receptors. Psychopharmacology, 116: 464-468, 1994.

Klemm, W.R.(1989). Drug effects on acive immobility responses: What they tell us about neurotransmitter systems and motor functions. Prog. Neurol. 32, 4-3-422.

King, G. and Weeks, S.D. (1965). Pyribenzamine activation of the electroencephalogram. Electroenceph. Clin. Neurophysiol. 18, 503-507.

Kiyono, S., Seo, M.L., Shibagki, m., Watanabe, T., Maeyama, K. and Wada, H.(1985). Effect of α -fluoromethylhistidine on sleep^waking parameters in rats. Physiol. Behav. 34, 615-617.

Kollonitsch, J., Perkin, L.M., Patchett, A.A., Doldouras, G.A., Marburg, S., Duggan, D.E.,

Maycock, A.L. and Aster, S.D.(1978). Selective inhibitor of aminergic transmitter. Nature 274, 906-908.

LaBella, F.S., Queen, G., Glavin, G., Durant, C., Stein, D. and Brandes, L.J.(1992). H3 antagonist, thioperamide, inhibits adrenal steroidogenesis and histamine binding to adrenocortical microsomes and binds to cytochrome P450. Br. J. Pharmacol. 107, 161-164.

Lahmeyer, H.W. and Steingold, R.G. (1980). Pentazocine and tripelennamine: A drug abuse epidemic? Int. J. Addict. 15, 1219-1232.

Langer, S.(1997). 25 years since the discovery of presyanptic receptors: Present knowledge and future perspectives. Trends Pharmacol. Sci. 18, 95-99.

Langlais, P.J., Zhang, S.Z., Weilersbacher, G., Hough, L.B. and Barke, K.E.(1994). Histamine-dediated neuornal death in a rat model of Wernicke's encephalopathy. J. Neurosci. Res. 38, 565-574.

Leurs, R. and Timmerman, H.(1992). The histamine H3 receptor: a target for developing new drugs. Prog. Drug. Res. 39, 127-165.

Leurs, R., Smit, M. and Timmerman, H.(1995). Molecular pharmacological aspects of histamine receptors. Pharmacol. Ther. 66 413-426.

Lin,J.S., Sakai,K. and Jouvet,M.(1988). Evidence for histaminergic arousal mechanism in the hypothalamus of cat. Neuroendocrinology 27, 111-122.

Lin, J.S., Sakai, K., Vanni-Mercier,G., Arrang,J.M., Garbarg,M., Schwartz,J.-C. and Jouvet,M. (1990). Involvement of histaminergic neurons in arousal mechanisms demonstrated with H3 receptor ligands in the cat. Brain Res. 523, 325-330.

Lin, J.-S., Hou, Y., Sakai, K. and Jouvet, M.(1996). Histaminergic descending inputs to the mesopontine tegmentum and their role in the comtrol of cortical activation and wakefulness in the cat. J. Seurosci. 16, 1523-1537.

Malec, D. and Langwinski, R.(1993). Is the brain histamine involved in cataleptogenic aciton of analgesics and haloperidorl? Life Sci., 33, 623-625.

Malinari, S.P., Kaminski, R., Di-Rocco, A. and Yahr, M.D.(1995). The use of famotidine in the treatment of parkinson' s disease: a pilot study. J. Neural Transm., Parkinson Disease and

Dementia Sec. 9, 243-247.

Martin, S.A.M. and Bishop, J.O.(1986). Purification and characterization of histidine decarboxylase from mouse kidney. Biochem. J. 234, 349-354.

Maslinski, C.Z., lebrecht, U., Nowak, J.Z., Pilc, A. and Wieczorek-Fila, Z.(1973). Catalepsia-like symptoms produced by histidine in rats. Agents Actions, 3, 185-186.

Masukawa, Y., Suzuki, T. and Misawa, Y.(1993). Differential modification of the rewarding effects of methamphetamine and cocaine by opioids and antihistaminics. Psycho-pharmacology 111, 139-143.

Mazurkiewicz-Kwilecki, .M. and Prell, G.(1984). Age-related changes in brain histamine. Agents Actions 14, 554-557.

Mazurkiewicz-Kwilecki, I.M. and Nososwah, S.(1989). Changes in the regional barin histamine and histidine levels in postmortem brains of Alzheimer's patients. Can. J. Physiol. Pharmacol. 67, 75-78.

Meguro, K., Yanai, K., Yokoyama, H., Sakurai, E., Maeyama, K., Watanabe, T. and Matsuzawa, T. (1992). Neurochemical studies on central histaminergic neuron system of senescence accelerated mice. Biog. Amines 8, 129-132.

Meguro, K., Sakai, N., Yanai, K., Sakurai, E., Maeyama, K., Watanabe, T. and Sasaki, H.(1995). Effects of thioperamide, a histamine H3-antagonist, on the step-through passive avoidance response in senescence accelerated mice. Pharmacol. Biochem. Behav. 50, 321-325.

Mine, K., Jacobson, K.A., Kirk, K.L., Kitajima, Y. and Linnoila, M. (1986). Simultaneous determination of histamine and Nt-methylhistamine with high-performance liquid chromatography using electrochemical detection. Anal. Biochem. 152, 127-135.

Mochizuki, T., Yamatodani, A., Okakura, K., Takemura, M., Inagaki, N. and Wada, H.(1991). In vivo release of neuronal histamine in the hypothalamus of rats measured by microdialysis. Naunyn-Schmiedeberg Arch. Pharmacol. 343, 190-195.

Mochizuki, T., Yamatodani, A., Okakura, K., Horii, A., Inagaki, N. and Wada, H. (1992). Circadian rhythm of histamine release from the hypothalamus of freely-moving rats. Physiol. Behav. 51, 391-394.

Mochizuki, T., Okakura-Mochizuki,K., Horii, A., Yamamoto, Y. and Yamatodani, A.(1994). Histaminergic modualtion of hippocampal acetylcholine release in vivo. J. Neurochem. 62, 2275-2282.

Mochizuki, T., Jansen, F.P., Leurs, R., Windhorst, A.D., Yamatodani, A., Maeyama, K. and Timmerman, H.(1997). Brain penetration of the histamine H3 rreceptor antagonists thiopermide and clobenpropit in rat and mouse, determined with ex vivo [^{125}I]iodophenpropit binding. Brain Res. 743, 178-183.

Mondovi, B.(1985). Structure and Functions of Amine Oxidases, CRC Press, Boca Raton.

Monnier, M., Sauer, R. and Hatt, A.M. (1971). The activating effect of histamine on the central nervous system. Int. Rev. Neurobiol. 12, 265-305.

Monti, J.M., Pellejero, T. and Jantos, H. (1986). Effects of H1- and H2-histamine receptor agonists and antagonists on sleep and wakefulness in the rat. J. Neural Transm. 66, 1-11.

Monti, J.M., D'Angelo, L., Jantos, H. and Pazos, S.(1988). Effects of α-fluoromethyl-histidine on sleep and wakefulness in the rat. J. Neural Transm. 72, 141-145.

Morel, A.M. and Delaage, M.A. (1988). Immunoanalysis of histamine through a novel chemical derivatization. J. Allergy Clin. Immunol. 82, 646-654.

Morisset, S., Taiffort, E. and Schwartz, J.-C.(1996). Inhibition of histamine versus acetylcholine metabolism as a mechanism of tacrine activity. Europ. J. Pharmacol. 315, R1-R2.

Mueller, M.S.(1983). Phenylpropanolamine, a non-prescription drug with potentially fatal side effects. New Eng. J. Med. 308, 653.

Nakagawa, Y., Yanai, K., Ryu, J.H., Kiyosawa, M., Tamai, M., Watanabe, T.(1994). Marked increase in [^{3}H](R)α-methylhistamine binding in the superior colliculus of visually deprived rats after unilateral enucleation. Brain Res. 643,74-80.

Nakai, T., Kitamura, N., Hashimoto, T., Kajimoto, Y., Nishino, N., Shirakawa, O. and Mita,T. (1989). Study on histamine H1 receptor in brains from schizophrenics and Parkinsonians. Rep. Pharmacopsychiat. Res. Found. 20, 267-272.

Nakamura, S. and Vincent, S.R.(1986). histaochemistry of MPTP oxidation in thr rat brain: sites of synthesis of the parkinsoniam-inducing toxin MPP+. Neurosci. Lett. 65, 321-325.

Nakamura, S., Ohnishi, K., Nishimura, M., Suenaga, T., Akiguchi, K., Kimura, J. and Kimura, T.(1996). Large neurons in the tuberlmammillary nucleus in patients with Parkinson's disease and multiple system atrophy. Neurology 46, 1693-1696.

Nicholson, A.N., Pascoe, P.A. and Stone, B.M.(1985). Histaminergic systems and sleep: Studies in man with H1 and H2 antagonists. Neuropharmacology 24, 245-250.

Nishibori, M., Oishi, R., Itoh, Y. and Saeki, K.(1991). 9-Amino-1,2,3,4-tertrahydroacridine is a potent inhibitor fof histamine N-methyltransferase. Jpn. J. Pharmacol. 55, 539-

Nowak, J.Z. and Pilk, A.(1977). Direct and indirect action of histamine on dopamine metabolism in the rat striatum. Eur. J. Pharmacol. 46, 171-175.

Ohmori, E, Fukui, T., Imanishi, N., Yatsunami, K. and Ichikawa, A.(1990). Purification and characterization of L-histidine decarboxylase from mouse mastocytome P-815 cells. J. Biochem. 107, 834-839.

Oike M, Kitamura, K., and Kuriyama, H. (1992). Histamine H3 receptor activation auguments voltage-dependent Ca^{2+} current via GTP hydrolysis in rabit saphenous artery. J. Physiol. (London) 448, 133-152.

Oishi, R., Nishibori, M., Itoh, Y., Saeki, K., Fukuda, T. and Araki, Y.(1988). Histamine turnover in the brain of morphine-dependent mice. Naunyn-Schmiedebergs Arch. Pharmacol. 337, 58-63.

Oishi, R., Itoh, Y., Nishibori, M. and Saeki, K.(1989). Effects of the H3 agonist, (R)-α-methylhistamine and the antagonist thioperamide on histamine metabolism in the mouse and rat brain. J. Neurochem. 52, 1388-1392.

Okakura, K., Yamatodani, A., Mochizuki, T., Horii, A. and Wada, H.(1992). Glutamatergic regulation of histamine release from rat hypothalamus. Eur. J. Pharmacol. 213, 189-192.

Okakura-MOchizuki, K., Mochizuki, T., Yamamoto, Y., Horii, A. and Yamatodani. A. (1996) Endogenous GABA modulates histamine release from the anterior hypothalamus of the rat. J. Neurochem. 67, 171-176.

Onodera, K. and Ogura, Y.(1983). Opiate-like actions by intracerebraoventricular histamine in mice and its antagonism by naloxone. Jpn. J. Oral Biol. 25, 1144-1147.

Onodera, K.(1991). Effects of antidepressants and antihistaminics on catalepsy induced by intracerebraoventricular administration of hisatamine in mice. Meth. Find. Exp. Clin. Pharmacol 13, 397-403.

Onodera, K.(1992). Aggressive behavior in thiamine deficient rats and histaminergic system. Anuario Psiguiat. Basc Clin. neurosci. 3, 31-40.

Onodera, K., Tuomisto, L., Tacke, U. and Airaksinen, M.(1992b). Strain differences in regional brain histamine levels between genetically epilepsy-prone and resistant rats. Meth. Find. Exp. Clin. Pharmacol. 14, 13-16.

Onodera, K., Yamatodani, A., Watanabe, T. and Wada, H. (1994). Neuropharmacology of the histaminergic neuron system in the brain and its relationship with behavioral disorders. Progress Neurobiol. 42, 685-702.

Onoe, H., Yamatodani, A., Watanabe, Y., Ono, K., Mochizuki, T., Wada, H. and Hayaishi, O. (1992). Prostaglandin E2 and histamine in the posterior hypothalamus.

Orthen-Gambill, N. and Salomon, M.(1992). FMH-induced decrease in central histamine levels produces increased feeding and body weight in rats. Physiol. Behav. 51, 891-893.

Panula, P., Yang, H-Y.T. and Costa, E.(1984). Histamine-containig neurons in the rat hypothalamus. Proc. Nat. Acad. Sci. USA 81, 2572-2576.

Panula, P., Pirvola, U., Auvinen, S. and Airaksinen, M.S.(1989). Histamine-immunoreactive neurons in the rat hypothalamus. Neuroscience 28, 585-610.

Peggs, J.F., Shimp, L.A. and Opdycke, R.A.(1995). Antihistamines: the old and the news. Am. Fam. Physician 52, 593-600.

Perez, A., Albarraran, M.A. and Cacabelos, R.(1988). Biochemical studies of body fluids in senile dementia of Alzheimer type (SDAT) and multiinfarct demntia (MID). Acta Neurol. Scand. 77, 129.

Philippu, A.(1991). Interactions with other neuron systems. In: Histaminergic Neurons: Morphology and Function, (T.Watanabe and H.Wada, Eds), CRC Press, Boca Raton, pp.323-343.

Polkis, A. and Whyatt, P.L.(1980). Current trends in the abuse of pentazocin and tripelennamine: The metropolitan St.Louis experience. J. Forens. Sci., 25, 72-78.

Prast, H., Argyriou, A. and Philippu, A.(1996). Histaminergic neurons facilitates social memory in rats. Brain Res. 734, 316-318.

Prell, G.D. and Green, J.P.(1986). Histamine as a neuroregulator. Ann. Rec. Neurosci. 9, 209-254.

Prell, G.D., Khandelwal, J.K., Burns, R.S., Lewitt, P.A. and Green, J.P.(1988). Elevated levels of histamine metabolites in cerebrospinal fluid of aging, healthy humans. Comp. Gerontol. 2, 114-119.

Prell, G.D., Khandelwal, J.K., Burns, R.S. and Lewitt, P.A.(1991). Effects of age and gender on the levels of histamine metabolites and pros-methylimidazole acetic acid in human cerebroventricular fluid. Arch. Gerontol. Geriatr. 12, 1-12.

Prell, G.D., Green, J.P., Kaufmann, C.A., Khandelwal, J.K., Morrishow, A.M., Kirch, D.G., Linnoila, M. & Wyatt, R.J. (1995). Histamine metabolites in cerebrospinal fluid of patients with chronic schizophrenia: their relationship to levels of other aminergic transmitters and ratings of symptoms. Schizophr. Res. 14, 93-104.

Richelson, E. and Nelson, A.(1984). Antagonism by neuroleptics of neurotransmitter receptors of normal human brain in vitro. Eur. J. Pharmacol. 103, 197-204.

Rodrigues, A.A., Jansen, F.P., Leurs, R., Timmerman, H. and Prell, G.(1995). Interaction of clozapine with the histamine H3 receptor in rat brain. Br. J. Pharmacol. 114, 1523-1524.

Ryu, J.H., yanai, K. and Watanabe, T.(1994a). Marked increase in histamine H3 receptors in the striatum and substantia nigra fter 6-hydroxydopamine-induced denervation of dopamine neurons: An autoradiographic study. Neurosci. Lett. 178, 19-22.

Ryu, J.H., Yanai, K., Iwata, R., Ido, T. and Watanabe, T.(1994b). Heterogenous distributions of histamine H3, dopamine D1 and D2 receptors in rat brain. NeuroReport 5, 621-624.

Ryu, J.H. Yanai, K., Zhao, X.L. Watanabe T. (1996). The effect of dopamine D1 receptor stimulation of histamine H3-receptors following destruction of the ascending dopaminergic neurons. Brit. J. Pharmacol. 118, 585-592.

Sakai, N., Yamasaki. S., Onodera K., Yanai, K., Maeyama, K. and Watanabe, T.(1991). Effects of (S)-α-fluoromethylhistidine and (R)-α-methylhistamine on locomotion of *W/Wv* mice. Pharmacol. Biochem. Behav. 46, 95-99.

Sakai, N., Yanai, K., Sakurai, E., Sakurai, E., Miura, Y. and Watanabe, T. (1997). Depletion of brain hisatmine induced by α-fluoromethylhistidine enhances radial maze performance in rats with a modualtion of brain amino acid levels. Life Sci. submitted.

Sakata ,T., Fukagawa, K., Ookuma, K., Fujimoto, K., Yashimatsu, H., Yamatodani, A. and Wada, H.(1988). Modulation of neuronal hisamine in control of food intake. Physiol. Behav. 44, 539-547.

Sakata, T.(1991). Feeding behavior. In:Histaminergic Neurons: Morphology and Function. CRC Press, Boca Raton, pp.297-313.

Sakurai, E., Gunji, E., Iizuka, Y., Hikichi,N., Maeyama, K. and Watanabe, T.(1994). The disopsition of thioperamide, a histamine H3 antagonsit, in rats. J. Pharmacol. Pharmac. 46, 209-212.

Saleh, A., Heyn, S., Conatas, N.K. and Swaab, D.F.(1995). Decreased protein synthetic activity of the hypothalamic tuberomammillary nucleus in Alzheimer's disease as suggested by smaller Gongi apparatus.

Saper, C,B. and German, D.C.(1987). Hypothalamic pathology of Alzheimer's disease. Neurosci. Lett. 74, 364-370.

Saxena, S.P., Brandes, L.J., Becker, A.B., Simons, K.J., LaBella, F.S. and Gerrard, J.M.(1989). Histamine is an intracellular messenger mediating platelet aggregation. Science 243, 1596-1599.

Scherkl, R., Hashem, A. and Frey, H.H.(1991a). Importance of histamine in seizure susceptibility. In: New Perspectives in Histamine Research, H.Timmerman and H.VanDerGroot, Eds., Birkhauser, Basel, pp.85-89.

Scherkl, R., Hashem, A. and Frey, H.H.(1991b). Histamine in brain---its role in regulation of seizure susceptibility. Epilepsy Res. 10, 111-118.

Schlicker, E., Betz, R. and Goethert, M.(1988). Histamine H3 receptor-mediated inhibition of serotonin release in the rat brain cortex. Naunyn-Schmiedebergs Arch. Pharmacol. 337, 588-590.

Schlicker, E., Behling, A., Lummen, G. and Goethert, M.(1992). Histamine $H3_A$ receptor-mediated inhibition of noradrenaline release in the mouse brain cortex. Naunyn-Schmiedebergs Arch. Pharmacol. 345, 489-493.

Schlicker, E., Fink, K., Detzner, M. Goethert, M.(1993) Histamine inhibits dopamine release in the mouse striatum via presynaptic H3-receptors. J. Neural Transm. 93, 1-10.

Schwartz, J.C., Arrang, J.M., Garbarg, M. M., Pollard, H., Ruat, M. (1991). Histaminergic transmission in the mammalian brain. Physiol. Rev. 71, 1-51.

Schwartz, J.F. and Patterson,J.H.(1978). The toxic effect of antihistaminic drugs. J. Am. Med. Assoc. 145, 2772-283.

Senay, E.(1985). Clinical experience with T's and B's. Drug Alcohol Depend. 14, 305-311.

Shaff, R.E. and Beaven, M.A. (1979) Increased sensitivity of the enzymatic isotopic assay of histamine:measurement of histamine in plasma and serum. Anal. Biochem. 94, 425-430.

Shimoda, Y., Koizumi, A. and Tanaka ,K.(1960). Electroencephalographic acitivation of "diencephalic epilepsy" with diphenhydramine. Yonago Acta Med. 4, 99-102.

Shin, J.C., Chen, K., Grimsby, J.S., Zhu, Q.-S. & Gqllaher, T.K.(1994). Structure, gene and promoter organiztion of monoamine oxidases A and B. In: Monoamine Oxidase Inhibitors in Neurological Diseases (Liberman,A, Olanow,W, Youdim,M.B.H. & Tipton,K.F.), Wiley, New York, pp. 77-94.

Shore, P.A., Burkhalter, A. and Cohn, V.H.Jr. (1959) A method for the fluorometric assay of histamine in tissues. J Pharmacol. Exp. Ther. 127, 182-186.

Snyder, S. and Axelrod, J. (1965). Sex differences and hormonal control of histamine methyltransferase acrtivity. biochim. Biophys. Acta 111, 416-421.

Snyder, S.H., Baldessarini, R.J. and Axelrod, J. (1966) A sensitive and specific enzymatic isotopic assay for tissue histamine. J. Pharmacol. Exp. Ther. 153, 544-549.

Steinbache, V., Groen, B.W., Wijmenga, S.S., Niesenn, W.M.A., Jongejan, J.A. and Duine, J.A. (1995). Identification of topaquinone as illustrated for pig kidney diamine oxidase and Escherichia coli amine oxidase. Anal. Biochem. 230, 159-166.

Steinbusch, H.W.M. and Mulder, A.H.(1984). Immunohistochemical localization of histamine in neurons and mst cells in the rat brain. In: Classical Transmitters and Transmitter Receptors in the CNS, Part II, Handbook of Chemical Neuroanatomy, A.Bjoerklund, T.Hoeckfelt and M.J.K uhar, Eds., Elsevier, Amsterdam, pp.126-140.

Steinbusch,H.W.M., Sauren, Y., Groenwegen, H., Watanabe, T. and Mulder, A.H. (1986). Histaminergic projections form the premammillary and posteiror hypothalamus region to the caudate putamen complex in the rat. Brain Res. 368, 389-393.

Sugimoto, K., Abe, K., Lee, T.-H., Sakurai, E., Yanai, K., Kogure, K., Itoyama, Y. and Watanabe, T. (1994). Histamine depletion in brain caused by treatment with (S)-α-fluoromethylhistidine enhances ischemic damage of gerbil hippocampal CA2 neurons. Brain Res. 666, 279-283.

Summers, W.P., Majovski, L.V., Marsh, G.M., Tachiki, K. and Kling, A.(1986). Oral tetrahydro-aminoacridine in long-term treatment of senile dementia, Alzheimer type. New Eng. J. Med. 315, 1241-1245.

Suzuki, T., Takamori, T., Misawa, M. and Onodera, K.(1995). Effects of histaminergic system on the morphine-induced conditioned place preference in mice. Brain Res. 675, 195-202.

Suzuki, T., Mori, T., Takamori, K., Onodera, K. and Misawa, M.(1996). Effects of H1-antagonists on discriminative stimulus effects of cocaine and methamphetamine in rats. Behav. Pharmacol. 7, 111-118.

Taguchi, Y., Watanabe, T., Kubota, H., Hayashi, H. and Wada, H.(1984). Purification of histidine decarboxylase from the liver of fetal rats and its immunochemical and immunohistochemical characterization. J. Biol. Chem. 259, 5214-5221.

Takagi, H., Mosrishima, Y., Matsuyama, T., Hayashi, H., Watanabe, T. and Wada, H.(1986). Histaminergic axons in the neostriatum and cerebral cortex of the rat. Brain Res. 364, 114-123.

Takeda, N., Inagaki, S., Taguchi, Y., Tohyama, M., Watanabe, T. and Wada, H.(1984). Origins of histamine containing fibers in the cerebral cortex of rats studied by immunohistochemistry with histidine decarboxylase as a marker and transection. Brain Res. 323, 55-63.

Takeda, T., Hosokawa, M. and Higuchi, K.(1991). Senescence-accelerated mouse (SAM): A novel murine model of accelerated senescnece. J. Aging Geriat. Sci. 39, 911-919.

Tekemura, M., Tanaka, T., Taguchi, Y, Imamura, I., Mizuguchi, H., Kubota, H., Fukui, H., Yamatodani, A. and Wada, H.(1992). Histamine N-methyltransferase from rat kidney-----

Cloning. nucleotide sequence, and expression in E. coli cells. J. Biol. Chem. 267, 15687-15691.

Tamiya, R., Hanada, M., Narita, N., Kawai, Y., Tohyama, M., and Takagi, H.(1989). Neuropeptide Y afferents have synaptic interactions with histaminergic neurons in the rat brain. Neurosci. Lett. 99, 241-245.

Tamiya, R., Hanada, M., Kawai, Y., Narita ,N., Inagaki, S., Tohyama, M., and Takagi, H.(1990). Histaminergic neurons receives substance P-ergic inputs in the posterior hypothalamus. Exp. Brain Res. 79, 261-265.

Tasaka, K., Kamei, C., Kitayama, S., Kitaizumi, K., Akahori H. and Hokonohara, T.(1986). Comparative study of various H1-blockers on neuropharmacological and behavioral effects including KB-2414, a new antiallergic agent. Archs. Int. pharmacodyna. Ther. 280, 275-291.

Tasaka, K., Chung, Y.H., Sawada, K. and Mio, M. (1989). Excitatory effect of histamine on the arousal asystem and its inhibition by H1 blockers. Brain Res. Bull. 22, 271-275.

Thomas ,B. & Prell, G.D.(1995). Imidazoleacetic acid, a gamma-aminobutyric acid receptor agonist, can be formed in rat brain by oxidation of histamine. J. Neurochem. 65, 818-826.

Traiffort, E., Ruat, M., Arrang, J.M., Leurs, R., Piomelli, D. and Schwartz, J.-C. (1992). Expression of a cloned rat histamine H2 receptor mediating inhibition of arachidonate release and activation of cAMP accumulation. Proc. Nat. Acad. Sci. USA 89, 2649-2653.

Tsuruta, Y., Kohashi, K. and Ohkura, Y. (1981). Simultaneous determination of histamine and Nt-methylhistamine in human urine and rat brain by high-performance liquid chromatography with fluorescence detection. J .Chromatogr. 224, 105-110.

Tuomisto, L. and Tacke, U.(1986). Is histamine an anticonvulsive inhibitory transmitter? Neuropharmacology 25, 955-958.

Tuomisto, L., Tacke, U. and Willman, A.(1987). Inhibition of sound-induced convulsions by metoprine in the audiogenic seizure susceptible rat. Agents Actions 14, 558-560.

Tuomisto, L., Yamatodani, A., Jolkkonen, J., Sainiok, E.L. and Airaksinen, K.M. (1993). Inhibition of brain histamine synthesis increases food intake and inhibits vasopressin response to salt loading in rats. Meth. Find. Exp. Clin. Pharmacol. 12, 100-105(?).

Turek,F.W.(1987). Pharmacological probes of the mammalian circadian clock: Use of phase response curve approach. Trends Phamacol. Sci. 8, 212-217.

Uvnaes, B.(1991). Handbook of Experimental Pharmacology, Vol. 97, Springer, Berlin.

Ueyama, K., Tominaga, H., Mstsumoto, K. and Nakae, T.(1984). A case of hallucinatory-paranoid state induced by chronic methylephedrine addiction. Kyushu N-Psychol. 30, 412-415.

Verburg, K.M. and Henry, D.P.(1986). Histamine N-methyltransferase. Enzymology and functional aspects. In: Neurotransmitte Enzymes. Neuromethods, Series 1, Vol. 5 P, Humana Press, Clifton, pp.147-204.

Wada, H., Inagaki, N., Yamatodani, A. and Watanabe, T.(1991). Is the histaminergic neuron system a regulatory center for whole brain activity? Trends Neurosci. 14, 415-418.

Watanabe, N., Sakai, T. and Tada, K.(1984). a case withmanic-depressive state induced by brain ephedrine preparation. Clin.Psychiatry 26, 1120-1122.

Watanabe, T. and Wada, H.(1983). Histidine decarboxylase and histamine N-methyltransferase. In: Methods in Biogenic Amine Research (Parves, S., Nagatsu, T., Nagatsu, I. and parves, H., Eds.). Elsevier, Amsterdam, pp.689-720.

Watanabe, T., Taguchi, Y., Shiosaka, S., Tanaka, J., Kubota, H., Terano, Y., Tohyama, M. and Wada, H.(1984). Distribution of the histaminergic neuron system in the cntral nercvous system of rat: a fluorescnt immmunohistochemical analysis with histidine decrboxylase as a marker. Brain Res. 295, 13-25.

Watanabe ,T., Yamatodani, A., Maeyama, K. and Wada, H.(1990). Pharmacology of α-fluoromethylhistidine, a specific inhbitor of histidine decarboxylase. Trends Pharmacol. Sci. 11, 363-367.

Watanabe, T., Taguchi, Y., Maeyama, Y. and Wada, H.(1991). Histidine decarboxylase. In: Handbook of Experimental Pharmacology, (Uvnas, B., Ed.), Vol. 97, Springer, Berlin, pp.145-163.

Watanabe, T. and Wada, H. (Eds.) (1991). Histminergic neurons: Morphology and Function. CRC Press, Boca Raton.

West, R.E.J., Zweig, A;, Shih, N.-Y., Siegel, M.I., Egan, R.W. and Clarck, M.A.(1990). identification of two H3-histamine receptor subtypes. Mol. Pharmacol. 38, 610-613.

White, J.M. and Rumbold, G.R.(1988). Behavioral effect of histamine and its antagonists. A review. Psychopharmcology 95, 1014.

Wouterlood, F.G., Steinbusch, H.W.M.,Lutiten, P.G.M. and Bol, J.G.J.M.(1987). Projection from the prefrontal cortex to histminergic cell groups in the posterior hypothalamic region of the rat. Anterogrde tracing with Phseolus vulgaris-leucoagglutinin combined with immunocyto-chemistry of histidine decarbozylase. Brain Res. 406, 330-336.

Wouterlood, F.G., Gaykema, R.P.A., Steinbusch, H.W.M. and Bol, J.G.J.M.(1988). The connections between septum-diagonal band complex and histaminergiic neurons in the posterior hypothalamus of the rat. Anterogrde tracing with Phseolus vulgaris-leucoagglutinin combined with immunocyotchemistry of histidine decarbozylase. Neuroscience 26, 827-845.

Wyngaarden, J.B. and Seevers, M.H.(1951). The toxic effects of antihistaminic drugs. J. Am. Med. Assoc. 145, 277-283.

Yamamoto, J., Yatsunami, K., Ohmori, E, Sugimoto, Y., Fukui, T., Ktayama, T. and Ichikawa, A. (1990). cDNA-derived amino acid sequence of L-histidine decaboxylase form mouse mastocytoma P-815 cells. FEBS Lett. 188, 215-218.

Yamashita, M., Fukui, H., Sugama, K., Horio, Y., Ito, S., Mizuguchi, H. and Wada, H.(1991). Expression cloning of a cDNA encoding the bovine histamine H1 receptor. Proc. Nat. Acad. Sci. USA 88, 11515-11519.

Yamatodani, A., Fukuda, H., Wada, H., Iwaeda, T. and Watanabe, T.(1985). High-performance liquid chromatographic determination of plasma and brain histamine without previous purification of biological samples:cation-exchange chromatography coupled with post-column derivatization fluorometry. J. Chromatogr., 344, 115-123.

Yamatodani, A., Inagaki, N., Panula, P., Itowi, N., Watanabe, T. and Wada, H. (1991). Structure and functions of histaminergic neuron system. In: Handbook of Experimental Pharmacology (Uvnaes, B., Ed.), Vol. 97, Springer, Berlin, pp.243-283.

Yamauchi, K., Sato, R., Tanno, Y., Ohkawara, Y., Maeyama, K., Watanabe, T., Satoh, K., Yoshizawa, M., Shibahara, S. and Takishima, T.(1990). Nucleotide sequence of the cDNA encoding L-histidine decaboxylase derived from human basophilic neukemia cell line, KU-812-F.

Nucl. Acid. Res. 18, 5891.

Yamauchi, K., Sekizawa, K., Suzuki, H., Nakazawa, H., Ohkawara,Y., Katayose, D., Ohtsu, H., Tamura, G., Shibahara, S., Takemura, M., Maeyama, K., Watanabe ,T., Sasaki, H., Shirato, K. and Takishima, T.(1994). Structure and function of human histamine N-methylransferase: Critical enzyme in histmine metabolism in air way. Am. J. Physiol., 267, L-342-L349.

Yanai, K., Watanabe, T., Meguro, K., Yokoyama, H., Sato, I., Sasano, K., Itoh, M., Iwata, R., Ishiwata, K., Takahashi, T. and Ido, T. (1992). Age-dependent increase in histamine H1 receptor in human brains revealed by PET. NeuroReport 3, 433-436.

Yatsunami, K., Ohtsu,H., Tsuchikawa,M., Higuchi,T., Ishibashi,K., Shida,A., Shima,Y., Nakagawa, S.,Yamauchi,k., Yamamoto,M.,Hayashi,N., Watanabe,T. and Ichikawa, A.(1994). Structure of the L-histidine decarbozylase gene. J. Biol. Chem. 269, 1554-1559.

Yatsunami. K., Tsuchikawa, M., Kamada, M., Hori, K. and Higuchi, T.(1995). Comparative studies of human recombinant 74- and 54kDa L-histidine decarboxylases. J. Biol. Chem. 270, 3081330817.

Yokoyama, H., Onodera, K., Maeyama, K., Yanai, K., Iinuma, K., Tuomisto, L., and Watanabe. T. (1992), Histamine levels and clonic convulsions of electrically-induced seizure in mice: the effects of α-fluoromethylhistidine and metoprine. Naunyn-Smiedbergs Arch. Pharmacol. 346, 40-45.

Yokoyama, H., Onodera. K., Iinuma, K. and Watanabe, T.(1993a). Effect of thioperamide, a histamine H3 antagonist on electrically-induced convulsions in mice. Eur. J. Pharmacol. 234, 129-133.

Yokoyama, H., Onodera, K., Iinuma, K. and Watanabe, T.(1993b). The proconvulsant effect of histamine H1 antagonists in developing mice. Psychopharmacology 112, 199-203.

Yokoyama, H., Iinuma, K., Yanai, K., Watanabe, T., Sakurai, E. and Onodera, K.(1993c). A histamine H1 antagonist, ketotifen produced seizures in an epileptic child: activation study of d-chlorphenilamine on EEG. Meth. Find. Exp. Clin. Pharmacol. 15, 183-188.

Yokoyama, H., Onodera. K., Maeyama, K., Sakurai, E., Iinuma, K., Timmerman, H. and Watanabe, T.(1994). Clobenpropit, a new histamine H3 receptor angagonist, and electrically-induced convulsions. Eur. J. Pharmacol. 260, 23-28.

Yokoyama, H. and Iinuma, K.(1996). Histamine seizures. Implications for the treatment of epilepsy. CNS Drugs 5, 321-330.

Progress in HPLC-HPCE, Vol. 7, pp. 351-382
G. Ali Qureshi *et al.* (Eds)

Abnormal tryptophan metabolism, neurologic/ psychiatric disturbances and its relationship to immune activation

Thomas Gasse[1], Bernhard Widner[2], Gabriele Baier-Bitterlich[2], Barbara Sperner-Unterweger[3], Friedrich Leblhuber[4], Helmut Wachter[2], Dietmar Fuchs[2]

[1]Department of Neurology, University of Innsbruck, [2]Institute of Medical Cemistry and Biochemistry, University of Innsbruck, and Ludwig Boltzmann Institute for AIDS-Research, Innsbruck; [3]Department of Psychiatry, University of Innsbruck, and [4]Department of Gerontology, Landesnervenklinik Wagner-Jauregg, Linz, Austria

INTRODUCTION

Tryptophan is an essential amino acid. It is essential for protein synthesis, and inadequate dietary intake of tryptophan leads to a negative nitrogen balance and loss of muscle mass finally causing weight loss (Brown 1996). Further, tryptophan is substrate for two enzymic pathways (Fig.1): (1) tryptophan-5-hydroxylase converts tryptophan to 5-hydroxy-tryptophan to finally form 5-hydroxy-tryptamine (serotonin). Tryptophan-5-hydroxylase is mainly located in neuroendocrine tissues which are the primary site of serotonin production. (2) Tryptophan (2,3)-oxygenases, tryptophan pyrrolase and indoleamine (2,3)-dioxygenase (IDO) respectively, catalyze the degradation of tryptophan via the kynurenine pathway. Tryptophan pyrrolase is active in the liver and in contrast, IDO is present in a variety of cells like fibroblasts human fibroblasts, monocytes/macrophages or tumour cell lines like He La cells (Werner *et al.* 1987b). Importantly, IDO is inducible by cytokines like interferon-g, and thus, in situations of enhanced endogenous formation of cytokines increased degradation of tryptophan takes place (see chapters below). Degradation of tryptophan by pyrrolase and IDO leads in a first step to the formation of N-formyl-kynurenine and kynurenine. Further conversion of which leads among other products to production of quinolinic acid and finally niacinamide cofactors NAD and NADP (Fernstrom and Wurtman, 1972, Byrne *et al.* 1986, Freese *et al.* 1990).

The 5,6,7,8-tetrahydrobiopterin-dependent mono-oxygenase tryptophan-5-hydroxylase converts tryptophan to 5-hydroxy-tryptophan, which after decarboxylation forms 5-hydroxytryptamine (serotonin). Serotonin is widely distributed throughout the central nervous system (Fernstrom and Wurtman_1972, McEntee and Crook 1991), and it acts as a neurotransmitter. Among other biochemical activities, serotonin stimulates peristalsis of the intestine, and causes a dose-dependent constriction of smooth muscle. Serotonin is implicated in a variety of neural functions such as pain, feeding and sleep, and it also causes dose-dependent constriction of smooth muscle (Litman and Correia 1983, McEntee and Crook 1991). Consequently, impaired availability of serotonin may manifest manifold consequences to the central nervous system.

Brain serotonin levels are controlled by the plasma concentrations of tryptophan which is transported into the brain by an active process, and a tryptophan-free diet was found to impair the synthesis and release of brain serotonin in vivo in rodents (Biggio *et al.* 1974, Gartside *et al.* 1992), which is also reflected by decrease of cerebrospinal fluid (CSF) concentrations of the serotonin metabolite, 5-hydroxyindoleacetic acid, in non-human primates (Young *et al.* 1989). Furthermore, there is data indicating that a tryptophan-free amino acid mixture decreases brain serotonin synthesis also in humans (Young *et al.* 1989, Benkelfast *et al.* 1995).

In the paraventricular nucleus of the hypothalamus serotonin acts to adjust the relationship between hunger and eating (Leibowitz *et al.* 1989), and higher levels of brain serotonin also seem to direct the desire towards protein-rich foods away from carbohydrate-rich foods. Protein-rich food will increase the plasma concentrations of tyrosine, phenylalanine, leucine, isoleucine and valine which compete with the transport of tryptophan into the brain and lower the rate of serotonin production (Fernstrom and Wurtman 1972, Leibowitz and Shor-Posner 1986, Curzon 1996, Wurtman and Wurtman 1996). Following this line of arguments an increase of serotonin in the brain is like an equivalent of a carbohydrate-rich meal. Therefore, serotonin agonists can be used as an adjunct to treatment of obesity.

IDO catalyzes the oxidation of tryptophan to form N-formyl-kynurenine which is the first product in the biosynthesis of nicotineamide dinucleotides (Fig.1). Several years ago, it was demonstrated that virus infections are associated with enhanced activity of IDO, and it was shown in vitro and in vivo, that IDO activation is induced by interferon-g, a T-cell-derived cytokine (Byrne *et al.* 1986). Currently it is well established that in a variety of cells and in various species IDO activity is inducible by interferon-g (Werner *et al.* 1987a and 1987b).

Interferon-g is produced preferentially from activated T-lymphocytes during the so-called cell-mediated immune response which is directed against modified self surface structures of cells. Cell-mediated immune response plays a key role during virus infections, allograft rejection, autoimmune disorders and certain types of malignancy, and it centrally involves a subset of T-lymphocytes, namely the Th1-subtype which is prone to release cytokines interleukin-2 and interferon-g (Romagnani *et al.* 1994). These cytokines are crucial for the interaction of T-cells and monocytes/macrophages to mount a successful cytotoxic response. From the background of an enhanced endogenous formation of interferon-g during cell-mediated immune response one has to anticipate that in such situations tryptophan degradation should be enhanced, and vice versa, enhanced tryptophan degradation in body fluids of patients - estimated from tryptophan and kynurenine measurements - indirectly indicate endogenous formation of interferon-g and TH1-type immune response (Fig.2).

Depletion of tryptophan upon stimulation with interferon-g may be part of the cytotoxic repertoire of immunocompetent cells, since intracellular growth of Toxoplasma gondii and Chlamydia psittaci in human fibroblasts in vitro is inhibited by interferon-g (Pfefferkorn 1984). This inhibition cannot be abrogated by adding back tryptophan to the medium which indicates that the formation of toxic products from tryptophan is more important than the reduced tryptophan concentration of the medium to stop growth of bacteria. In contrast, also the growth of tumor cells can be arrested in vitro when tryptophan is depleted, but in this case adding back tryptophan restores cell growth indicating that the depletion of tryptophan itself is limiting the growth rate of the tumor rather than the toxic metabolites formed.

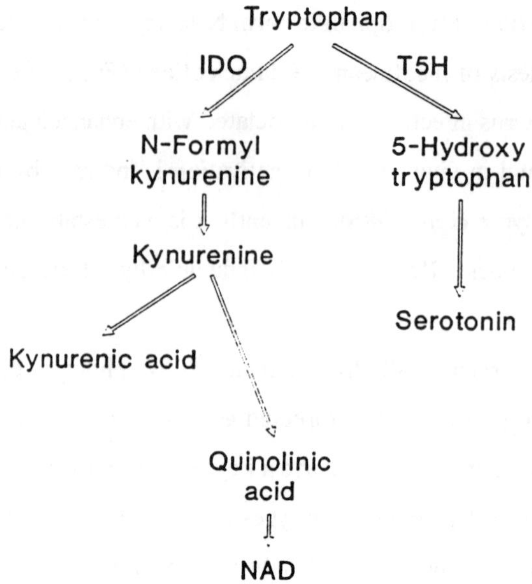

Fig.1: Tryptophan is converted by two enzymic pathways: (1) tryptophan-5-hydroxylase (T5H) converts tryptophan to 5-hydroxy-tryptophan, (2) tryptophan pyrrolase and indoleamine (2,3)-dioxygenase (IDO) converts tryptophan to N-formyl kynurenine in a first step.
NAD...Niacinamide adenine dinucleotide

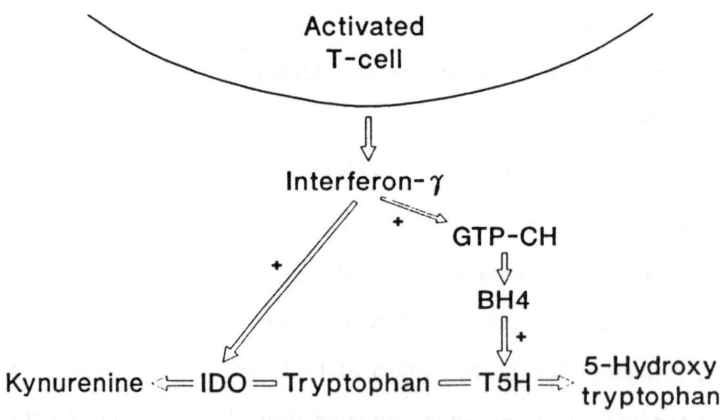

Fig.2: Activation of T-cells during the cell-mediated (TH1-type) immune response leads to cytokine-inducible catabolism of tryptophan: interferon-g induces (1) indoleamine (2,3)-dioxygenase (IDO) and (2) GTP-cyclohydrolase I (GTP-CH), the key enzyme for 5,6,7,8-tetrahydrobiopterin (BH4), the cofactor of tryptophan-5-hydroxylase.

Interferon-g may indirectly influence the biosynthesis of tryptophan-derived products including serotonin (Fig.3) because their production rate depends on the concentration of tryptophan in the blood and brain (Fernstrom and Wurtman 1972). Interferon-g may have additional impact on serotonin availability because it also interferes with its release from storage sites (Coleman *et al.* 1991). In addition to the disturbed bioavailability of serotonin which can be induced by interferon-g, the formation of tryptophan degradation products could further contribute to biochemical and neurological deterioration (Fig.3). Some of these products like kynurenic acid and quinolinic acid are potent neurotoxins (Freese *et al.* 1990, Heyes *et al.* 1992b), and increased amounts of 3-hydroxy-anthranilic acid may interfere with protein biosynthesis (Benavente and Truscott 1991).

Two metabolic pathways competing for serotonin availability

The availability of tryptophan as well as the activity of tryptophan-5-hydroxylase are crucial for the production of serotonin. On the one hand, low tryptophan concentrations are rate-limiting for tryptophan-5-hydroxylase, on the other hand, the activity of the enzyme depends on the presence of the cofactor pteridine, 5,6,7,8-tetrahydrobiopterin. Low tryptophan concentrations may result from reduced dietary intake but also be a consequence of enhanced degradation.

Interestingly, the activity of tryptophan-5-hydroxylase as well as degradation of tryptophan by IDO are both directly and indirectly associated with immune activation (Fig.2). Interferon-g, released during states of immune activation, will activate GTP-cyclohydrolase I in various cells, the key enzyme for the biosynthesis of pteridines. In most cells this will lead to increased formation of 5,6,7,8-tetrahydrobiopterin and initiate the formation of serotonin if tryptophan-5-hydroxylase is present (Fuchs *et al.* 1991c). In human monocytes/macrophages large amounts of neopterin are formed instead of 5,6,7,8-tetrahydrobiopterin (Fuchs *et al.* 1988). In parallel, interferon-g also induces IDO degrading tryptophan and therefore reducing the availability of tryptophan for the biosynthesis of serotonin. Thus, immune activation and production of interferon-g will have a divergent effect on serotonin production, on the one hand increasing the activity of tryptophan-5-hydroxylase, on the other hand reducing the availability of tryptophan

(Fig.2). The net effect is unclear from the biochemical background and may differ between specific cells, it may depend on the distribution of inducible GTP cyclohdrolase as compared with tryptophan-5-hydroxylase. Finally, one has to keep in mind that IDO is not specific for free tryptophan, it also cleaves serotonin as it cleaves protein-bound tryptophan.

Data obtained in patients with cancer and with HIV infection (Launay *et al.* 1988) demonstrate that serotonin concentrations in the blood of these patients are usually low. These clinical situations are associated with immune activation and enhanced circulating levels of interferon-g (Fuchs *et al.* 1989/90). Thus, the data may indicate that at least in situations of chronic immune activation the effect on tryptophan degradation in the peripheral blood is more expressed than the potential upregulation of tryptophan-5-hydroxylase by overproduction of 5,6,7,8-tetrahydrobiopterin.

Quinolinic acid

IDO converts tryptophan to form N-formyl kynurenine. Depending on the availability of the necessary enzymes in cells this product is further metabolized. In human monocytes/macrophages it was demonstrated that kynurenine formed from tryptophan due to activation of IDO is further catabolized to anthranilic acid and 3-hydroxyanthranilic acid (Werner *et al.* 1987) and later on also quinolinic acid was shown to accumulate (Heyes *et al.* 1992d). However, in most other cells and cell lines only kynurenine is produced and further conversion does not take place (Werner *et al.* 1987a and 1987b).

The tryptophan metabolite quinolinic acid behaves as an excitotoxin and it has been detected in the central nervous system. In the mammalian brain quinolinic acid has been localized in glial cells, and the concentrations increase with age. The neurotoxicity of quinolinic acid is mediated via the Mg(2+)-sensitive N-methyl-D-aspartate (NMDA) receptor (Freese *et al.* 1990). The toxicity of quinolinic acid depends on the presence of an intact glutamate-aspartate afferent input to the target area (Jhamandas *et al.* 1994). Injections of quinolinic acid induce loss of cholinergic neuron markers in the neocortex and amygdala. The neurotoxic action of quinolinic acid impairs the performance on memory-related tasks. In the striatum, focal quinolinic acid injections produce neurotransmitter deficits, and quinolinic acid produces a profound loss of the NADPH diaphorase staining neurons in the area of injection.

Excitotoxicity may contribute to the neuronal degeneration associated with a number of neurological diseases, Alzheimer's disease, Huntington's disease, Parkinson's disease. However, the metabolic tryptophan degradation pathway yielding quinolinic acid produces a number of intermediates that may act as excitotoxin antagonists (Freese *et al.* 1990), e.g., kynurenic acid is an antagonist of NMDA receptors and can modulate the neurotoxic effects of quinolinic acid as well as disrupt excitatory amino acid neurotransmission. Consequently, quinolinic acid and kynurenic acid may represent important mediators of neuronal dysfunction and nerve cell death (Heyes *et al.* 1992b).

Association between tryptophan degradation and neopterin formation

Large amounts of neopterin are produced from human monocytes/macrophages on stimulation with interferon-g (Wachter *et al.* 1991). Consequently, in diseases which are associated with activation of T-cells and endogenous formation of interferon-g enhanced concentrations of neopterin can be observed in body fluids such as serum, plasma, cerebrospinal fluid and urine (Fuchs *et al.* 1988). Monitoring of neopterin concentrations turned out to be useful for follow-up and to predict the further clinical course in patients with infections including human immunodeficiency virus (HIV) infection, in autoimmune diseases, malignant disorders and in various types of malignancies (Fuchs *et al.* 1988, 1992b and 1993). Monitoring of neopterin concentrations in patients allows insight into the biology of interferon-g in patients and the assay is more sensitive than that generally applied for direct measurement of the cytokine concentrations. The sensitivity of these tests is limited by the short biological halflife of interferon-g in vivo and therefore the assays appear to be not sensitive enough for routine studies in patients (Fuchs *et al.* 1989b). Since interferon-g stimulates both enzyme pathways in parallel, namely the degradation of tryptophan and the formation of neopterin, in patients with various diseases usually a close association exists between increased neopterin and kynurenine concentrations and decreased tryptophan levels.

Tryptophan catabolism induced by exogenous cytokines

In vivo administration of cytokines like interferons is applied for treatment of certain malignant disorders and also persistent infections. The antiproliferative effects of these cytokines are important for beneficial therapy and immune system activation is induced. In agreement with in vitro observations, exogenous interferons are also able to induce IDO in the patients. Therefore, the administration of cytokines like interferons, interleukin-2 and tumor necrosis factor-a causes significant degradation of tryptophan (Brown *et al.* 1991) which can be detected by a decrease of tryptophan levels in, e.g. serum or plasma, and an increased formation of tryptophan degradation products like kynurenine (Brown *et al.* 1989). In cancer patients treated not only with interferon-g but also with interferon-a or -ß or other cytokines IDO is induced which most probably takes place via an induction of endogenous formation of interferon-g. One may speculate that in vivo antineoplastic effects of these cytokines and clinical side effects are mediated, at least in part, by a general or localized depletion of tryptophan (Brown *et al.* 1991). Interestingly, growth of a number of human cell lines in vitro is inhibited by interferon-g and addition of extra tryptophan restores growth. Thus, in some in vitro systems, antiproliferative effects of interferon-g may be even mediated by degradation of tryptophan.

It is well established that exogenously administered cytokines, such as interferons, sometimes induce side-effects like fatigue, malaise, and neurologic deterioration. Certainly, as has been observed, the severity of neurologic/psychiatry symptoms is usually much less expressed than in patients with, e.g., HIV-1 infection (see below). However, if side-effects turn out to be severe in patients, cytokine therapy is interrupted. It appears that the severity of neuropsychiatric symptoms in patients with endogenous formation or exogenous administration of interferons just depends on the strength and duration of the immunologic stimuli as well as on the intensity of the biochemical changes induced.

Tryptophan catabolism in patients and diseases

Because interferon-g is critical for the induction of IDO, clinical situations which are associated with activation of cellular immune response and endogenous formation of interferon-g should be associated with stimulated tryptophan catabolism (Table 1). Cell-mediated immune reaction or TH-1 type immune response is known to take place during allograft rejection and infectious diseases, among them particularly virus infection as well as infections with intracellular parasites and bacteria, e.g. malaria and M.tuberculosis infection. In addition, several autoimmune diseases like rheumatoid arthritis and systemic lupus erythematosus and also certain malignant diseases and neurodegenerative disorders are linked with disturbances of the cell-mediated immunity. As expected, these disorders are also characterized by increased production of neopterin in patients (Fuchs *et al.* 1988 and 1993, Wachter *et al.* 1991), concentrations reflecting the extent and activity of the diseases as well as being of some predictive value for the further course of the disease. From this background a variety of diseases can be expected to be associated with significant disturbances of tryptophan metabolism. As a consequence also concentrations of tryptophan metabolites are expected to be influenced, e.g. tryptophan degradation products like kynurenine and quinolinic acid can be expected to be increased whereas, e.g., serotonin might be decreased.

Abnormal tryptophan metabolism has been reported in a wide spectrum of diseases. Some of which are clearly involving the central nervous system and are characterized by neuropsychiatric symptoms, endogenous depression and neurodegenerative disorders like Huntington's disease and Alzheimer's disease belonging to this group of diseases. Since longe time in these disorders abnormalities of neurotransmitter metabolism were considered to play a role, and recent data imply that the immune system might be critical to interfere with tryptophan metabolism at least in some of these diseases. In addition, several other diseases are found to be associated with tryptophan metabolic abnormalities which at the first glance fall into the category of immunopathologies. This is particularly true in infections and in chronic inflammatory diseases comprising HIV-1 infection, autoimmune diseases and certain types of malignancy (Table 1). In this group of disorders abnormalities of tryptophan metabolism do not seem to be involved directly in the pathogenesis of the disease, but they may well contribute to neuropsychiatric

Table 1: Diseases which are associated with cellular immune activation, formation of interferon-g and/or degradation of tryptophan

	Immune activation	Tryptophan degradation
Infections		
systemic viruses, intracellular bacteria	Blood	Blood
intrathecal meningitis	CSF (Blood)	CSF (Blood)
Autoimmune diseases E.g., rheumatoid arthritis	Blood	Blood
Multiple sclerosis	CSF (Blood ?)	CSF (Blood ?)
Malignant diseases E.g., haematological neoplasias	Blood	Blood
Neurodegenerative disorders Alzheimer's and Huntington's Disease	Blood (CSF ?)	Blood (CSF ?)
Depression	Blood (CSF ?)	Blood (CSF ?)
Pregnancy	Blood	Blood

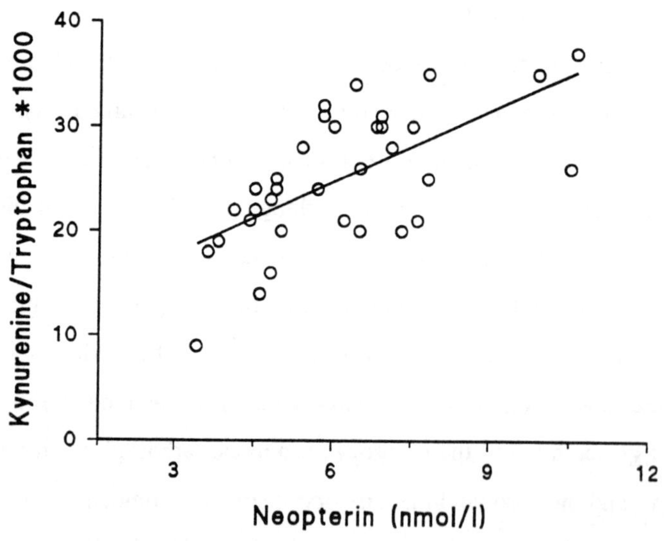

Fig.5: Correlation between plasma neopterin concentrations and the kynurenine per tryptophan ratio during pregnancy.

symptoms which are often seen in patients especially in later stages. Particularly in cerebral infections it is known that immunological alterations in the brain can be dramatic and there is ongoing overproduction of cytokines, which is reflected by, e.g. enourmosly increased neopterin concentrations in the CSF (Griffin *et al.* 1990, Hagberg *et al.* 1993).

Depression

Decreased tryptophan appears to be involved in depression. Despite much research, the pathophysiology underlying lower tryptophan availability in major depression has remained elusive. Lower tryptophan levels may be related to immune activation which may occur in that illness and is known to modulate tryptophan metabolism. From this background, especially a role of chronic inflamatory process, e.g. caused by virus infections, must be taken into consideration.

Indeed, some relationships between lower plasma tryptophan levels and immune-inflammatory variables were found in depression (Maes *et al.* 1993). Both, plasma levels of tryptophan and the ratio between tryptophan and competing amino acids, valine, leucine, isoleucine, tyrosine and phenylalanine, were significantly lower in major depressed subjects as compared with healthy controls. There were significant inverse correlations between plasma tryptophan levels and interleukin-6 production. Thus, also in depression lower plasma tryptophan availability may be related to the immune response.

Further evidence for an immune response in major depression was found when increased neopterin and interferon-g secretion were found to correlate with the lower availability of tryptophan (Maes *et al.* 1994). Plasma neopterin levels were significantly higher in depressed subjects than in normal controls; 61% of melancholic patients had increased neopterin levels. Patients with major depression had significantly lower tryptophan values which were significantly and negatively correlated with plasma neopterin levels. The results further support the hypothesis that major depression is accompanied by an immune response and that the lower tryptophan availability in that illness may be a result of immune activation.

Decreased plasma tryptophan levels will lower tryptophan availability to the brain and will decrease brain and serotonin synthesis and release (Biggio *et al.* 1974, Gartside *et al.* 1992, Young *et al.* 1989). Lowering of brain serotonin function can precipitate clinical depressive symptoms in individuals who are vulnerable to major depressive disorders (Smith *et al.* 1997). This observation supports a role for the neurotransmitter serotonin in depressed patients. Depression is associated with altered brain serotonin activity (Cowen 1996), in addition, drugs that support serotonin neurotransmission are effective antipressants (Kasper *et al.* 1992). It is not clear, however, whether impaired serotonin function is responsible for the clinical signs of major depressive disorders (Biggio *et al.* 1974, Young *et al.* 1985, Cowen 1996).

Impaired brain serotonin function may relate to symptoms such as anxiety and irritability (van Praag 1991) and tryptophan-free died taken by recovered depressed patients induced symptoms typical of major depression (Moreno *et al.* 1995). In parallel, psychomotor disturbance was observed, a key symptom in the diagnosis of biological or melancholic depression (Parker *et al.* 1994). Patients also presented with congnitive disortions which seem to play a causal part in depressive episodes (Teasdale 1988). Decreased serotonin function was also found to be associated with suicidal thoughts and suicidal behaviour (Linnola and Virkkunen 1992).

On the opposite, tryptophan-free amino acid mixtures can reverse the therapeutic effects of some classes of antipressant drugs (Delgado *et al.* 1990). This observation suggests that brain serotonin pathways are involved in the mode of actions of certain antidepressants. Impaired serotonin neurotransmission may also play a pathophysiologic role in the development of depression in vulnerable individuals (Smith *et al.* 1997), although tryptophan depletion in patients experiencing a current acute episode of depression does not cause a worsening of symptoms (Delgado *et al.* 1994).

Pregnancy

There is a signficant degradation of tryptophan going on during pregnancy which correlates to the gestational week. Overall, plasma tryptophan concentrations in healthy pregnant women are similar to the nonpregnant controls (Schröcksnadel *et al*, 1996).

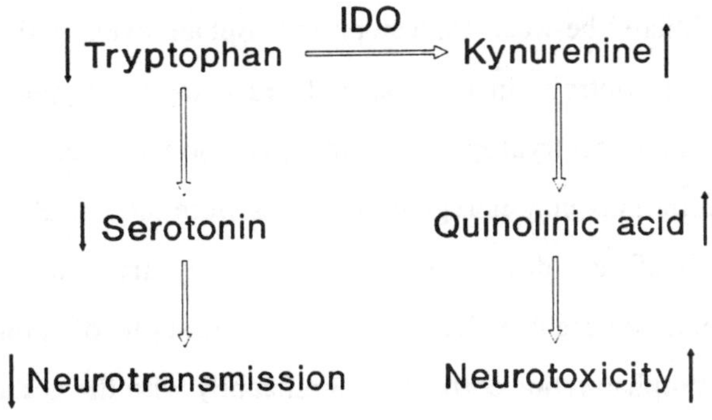

Fig.3: Endogenously formed interferon-g may have a bidirected effect to trigger neuropsychiatric disturbances in patients. Interferon-g stimulates indoleamine (2,3)-dioxygenase (IDO), consequently tryptophan is depleted and kynurenine is formed. In turn (1) production of serotonin is reduced due to reduced bioavailability of tryptophan and (2) quinolinic acid, a potent neurotoxin, accumulates upon further conversion of kynurenine.

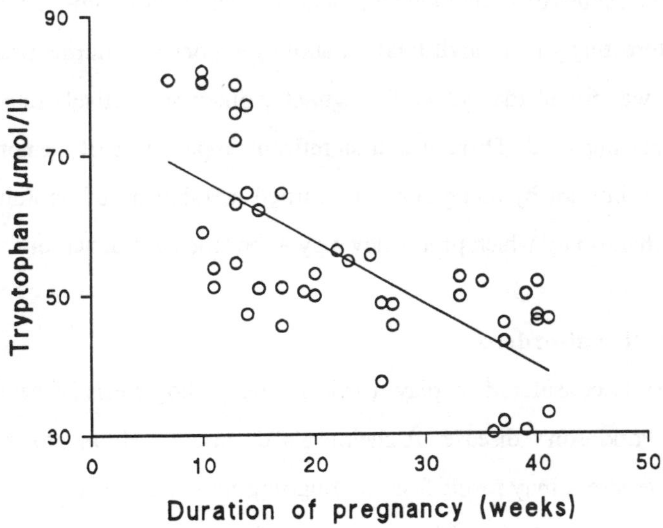

Fig.4: Decline of plasma tryptophan concentrations during pregnancy.

However, values differed between trimesters: tryptophan levels in the 1st trimester were higher than those of controls, in the 2nd and 3rd trimester tryptophan medians were lower. The correlation of tryptophan to the gestational week was inverse (Fig.4). Kynurenine concentrations in nonpregnant controls were indentical to those of pregnant women and no significant differences existed to any trimester and there was no correlation between kynurenine levels and the duration of pregnancy. However, kynurenine per tryptophan ratios correlated significantly with the gestational week.

Tryptophan concentrations correlated inversely to serum neopterin, and a positive correlation was found between the kynurenine per tryptophan ratio and neopterin levels (Fig.5). The increase of the kynurenine per tryptophan ratio indicates that kynurenine is increased relatively to the decreased concentrations of tryptophan (Fuchs *et al.* 1996c). Tryptophan degradation seems to decline rapidly during the puerperium. In an independent study (Gard *et al.* 1986) post partum depression and blues were found to be associated with reduced plasma tryptophan.

Decreased serum tryptophan during the later time of pregnancy may incresae the risk for neurotransmitter disturbances and development of related symptoms like memory loss or depression. Interestingly in an investigation about memory loss during pregnancy (Sharp *et al.* 1993) it was found that 80% of pregnant women subjectively rated their current memory as being impaired. There was a significant impairment of memory as tested by recall or priming but not by recognition. Disturbed metabolism of serotonin may be one underlying mechanism by which pregnancy may affect the central nervous system.

Neurodegenerative disorders

Oxidative stress is considered to play a role in the pathogenesis of neurodegenerative disorders like Parkinson's disease, Alzheimers's disease or Huntington's disease (Beal 1992). Oxidative stress may result from an ongoing immune reaction, however, signs of inflammationare only rarely found in patients. But if immune reaction would play a role, abnormalities of tryptophan metabolism could be anticipated.

Huntington's disease

Huntington's disease is an autosomal dominantly inherited, chronic neurodegenerative disorder with progressive neuronal loss in the striatum and cortex, characterized by choreiform movements and dementia (Martin *et al.* 1984). The defective gene has been identified recently (The Huntington's Disease collaborative Research Group 1993).

In patients with definitive Huntington's disease some serum parameters of immune activation, e.g., neopterin, soluble cytokine receptors of interleukin-2, sIL-2Rs, and tumour necrosis factor, sTNF-R (Diez-Ruiz *et al.* 1995), were found increased compared to controls (Leblhuber *et al.* 1995a). Serum tryptophan was decreased in these patients, and there were associations between cognitive impairment as reflected by mini mental test scores and serum tryptophan and neopterin levels. A significant correlation was found between neopterin and tryptophan and the kynurenine per tryptophan ratio, and there were also asssociations between tryptophan, sTNF-R and neopterin.

Importantly, changes of soluble immune parameters were detected in sera of patients, in the CSF the abnormalities of tryptophan metabolism and also of the immune system parameters were much less expressed. This observation agrees with earlier findings made in chronic neurodegenerative disorders, depression and myoclonic seizure disorders (Heyes *et al.* 1992b) when no increases of quinolinic acid was found in CSF, while CSF kynurenic acid concentrations were even lower in Huntington's disease and Alzheimer's disease than in controls. This data implies that systemic effects of the immune system could be important in the pathogenesis of Huntington's disease. Likewise, the decreased serum tryptophan correlated inversely to the concentrations of neopterin. The data suggest that immune activation could be involved in degradation of tryptophan, and indeed a strong correlation between neopterin and the kynurenine to tryptophan ratio further indicates increased activity of cytokine-inducible IDO. Thus, the significant correlation found between neopterin concentrations and the tryptophan catabolism would support the concept for its possible role in the formation and accumulation of neurotoxins and in the pathogenesis of the disease.

As was mentioned above, earlier studies of quinolinic acid concentrations in patients with Huntington's disease lead to some controversy. There were no clearcut findings about increased concentrations of quinolinic acid in the CSF of patients (Schwarcz *et al.* 1988,

Heyes *et al.* 1992b). The more recent data about tryptophan metabolic changes in Huntington's disease (Leblhuber *et al.* 1995) demonstrate that systemically there exist more significant changes but not intrathecally as was investigated in the studies mentioned before. It needs to be demonstrated whether the abnormalities of tryptophan metabolism in the blood of patients are involved in the pathogenesis of the disease or whether this may simply represent an epiphenomenon.

Similar relationships seem to exist in patients with Alzheimer's dementia. Decreased tryptophan and increased kynurenine concentrations were demonstrated in the serum of patients but were less drastic in the CSF (Leblhuber *et al.* 1995b). A chronic type of cell-mediated immune response seems to go on in the patients which however appears to be established systemically rather than intrathecally.

Infections

From an immunological standpoint especially systemic infections with viruses and parasites will be most relevant for tryptophan metabolic changes induced by interferon-g. However, from a neurological point of view, more attention might be drawn to intrathecal infections, since especially the potential neutotoxicity of quinolinic acid would be most relevant but also the impact of cytokines to disturb the biosynthesis of serotonin.

Meniningitis

Meningitis in children is associated with intrathecal formation of cytokines, particularly in bacterial and viral infections significant formation of cytokines like tumor necrosis factor-a and interleukin-6 but also interferon-g seems to be critically involved in the pathogenesis of symptoms (Griffin *et al.* 1990). Compared to children with febrile convulsions, children with meningitis have elevated CSF concentrations of neopterin and interferon-g (Zaknun *et al.* 1994). In parallel, significant degradation of tryptophan could be detected: CSF kynurenine was increased and tryptophan was decreased in a subgroup of patients representing those with the highest concentrations of neopterin. In addition, a significant correlation between neopterin concentrations and the kynurenine per tryptophan ratio was observed indicating a role of IDO to be involved in the production of kynurenine from tryptophan.

Also levels of quinolinic acid were found increased in the CSF of children with infections of the central nervous system (Heyes *et al.* 1995), extremely high quinolinic acid concentrations were found in patients with bacterial infections even despite treatment with antimicrobial agents. CSF kynurenine levels were found increased in parallel.

Experiments in an animal model for intrathecal immune response to infection agrees well with these clinical findings. As a model of localized inflammatory neurologic disease, macaques with intrathecal poliovirus infection showed enhanced IDO activation and quinolinic acid concentrations in CSF correlating to the degree of inflammatory responses and neurologic damage (Heyes *et al.* 1992c). Local synthesis of quinolinic acid from tryptophan within the central nervous system seemed to be related to activated glia cells.

Lyme neuroborreliosis and late Lyme encephalopathy

Decreased tryptophan concentrations and increased kynurenine and neopterin were found in the CSF of patients with Lyme neuroborreliosis. The correlation found between tryptophan metabolic changes and increased neopterin production suggests that the decrease of tryptophan is due to degradation rather than a reduced dietary intake of this essential amino acid (Fuchs *et al.* 1996a). About two thirds of patients with Lyme neuroborreliosis had detectable CSF kynurenine, and there was a positive correlation between neopterin and kynurenine levels (Fuchs *et al.* 1991b, Gasse *et al.* 1994): patients who presented with undetectable kynurenine levels, had relatively low CSF neopterin concentrations in parallel. CSF tryptophan was low in only a few patients, this was the case again preferentially in those patients who showed the highest neopterin and kynurenine levels (Fuchs *et al.* 1991, Gasse *et al.* 1994). When kynurenine and neopterin concentrations in the CSF were highest, tryptophan was no longer detectable in the CSF. Abnormalities were related to activity of the disease, and during antibiotic treatment concentrations tended to normalize (Dotevall *et al.* 1990).

The increased formation of neopterin and degradation of tryptophan was located inside the blood-brain-barrier. There were only minor changes of the analytes' concentrations seen in the serum specimens, and only a low percentage of patients had increased neopterin compared to healthy controls (Dotevall *et al.* 1990, Fuchs *et al.* 1991, Gasse *et al.* 1994). Serum tryptophan tended to lower levels and kynurenine to higher concentrations compared to controls (Fuchs *et al.* 1991, Gasse *et al.* 1994), however, the absolute changes were much lower than those observed intrathecally.

In patients with late Lyme encephalopathy the deviations from normal of the analytes' concentrations in CSF were only marginal compared to patients with acute Lyme neuroborreliosis (Gasse *et al.* 1994). However, the direction of changes was in the same way as it was observed in patients with Lyme neuroborreliosis. There was the impression that serum concentrations of tryptophan become lower with longer disease duration. Serum tryptophan levels were decreased compared to the healthy controls (Gasse *et al.* 1994).

Intrathecal degradation of tryptophan and formation of neopterin suggests a pathogenetic role of intrathecally released interferon-g and activated monocytic cells in patients with Lyme neuroborreliosis. It seems plausible that interferon-g is formed in response to certain spirochetes' surface proteins which are known to stimulate release of the cytokine in vitro (Fuchs *et al.* 1996). An immune response established locally and monocytic cells like microglia or astrocytes or monocytes/macrophages, invaded through the blood-brain-barrier, may be involved in the degradation of tryptophan and the formation of increased neopterin. Interestingly a correlation was found between CSF cell counts and neopterin levels (Dotevall *et al.* 1990).

Human immunodeficiency virus type 1 (HIV-1) infection

HIV-1 infection starts almost immediately with detectable immune activation which can be extremely long-lasting (Fuchs *et al.* 1988). Usually, HIV-1 infection is extreme with regard to the length and intensity of chronic immune activation; patients with HIV-1 infection and, specifically, patients with, e.g., AIDS may reach extremely high neopterin concentrations which can hardly be seen in other patients groups (Fuchs *et al.* 1988,

Fahey *et al.* 1990, Wachter *et al.* 1991). In HIV-1 infection, significant immune activation usually takes places inside the blood/brain barrier and in the peripheral blood. Reduced tryptophan and increased kynurenine concentrations have been reported in patients with HIV-1 infection (Werner *et al.* 1988), and various significant correlations were found between tryptophan, kynurenine, interferon-g and neopterin concentrations (Fuchs *et al.* 1990b and 1991a). The highest degree of correlation was found between neopterin, interferon-g and the kynurenine per tryptophan ratio which implies that activated IDO is involved in this metabolic changes. It is likely that immune stimulation, which takes place inside and outside the blood-brain barrier, is caused by HIV-1 and/or by opportunistic pathogens, and activated immunocompetent cells may be involved in brain tissue damage observed in patients with HIV infection.

Higher neopterin concentrations were observed in the CSF of patients with HIV-1 associated polyneuropathy and dementia as compared to patients with HIV-1 infection without such symptoms (Fuchs *et al.* 1989a). The data indicated an association between neuropsychiatric abnormalities and immune activation. Also a close correlation between advanced stages of AIDS dementia complex and higher neopterin concentrations in CSF was observed (Brew *et al.* 1990; Griffin *et al. 1991*). Higher serum neopterin concentrations and, thus, higher degree of immune stimulation were associated with more severe neuropsychiatric abnormalities (Fuchs *et al.* 1989a and 1990a). Employing a scoring system, which was developed to classify patients with dementia of Alzheimer type and multi-infarct dementia (Zaudig *et al.* 1991), to detect early cognitive abnormalities in patients with HIV-1 infection, neopterin and tryptophan concentrations correlated well with the degree of cognitive impairment, and lower tryptophan concentrations in serum and CSF were associated with dementia in the patients (Fuchs *et al.* 1990a, Möller *et al.* 1990). In addition, a significant correlation existed between the extent of brain lesions, measured by computed tomography (expressed as ventricle/brain ratio), and the degree of immune activation (Fuchs *et al.* 1990a). A greater sizer of brain-tissue damage correlated with lower tryptophan and higher neopterin concentrations.

In patients with HIV infection low tryptophan concentrations were also found to be associated with impairment of normal sleep (Wiegand *et al.* 1991). Because of the known influence of serotonin on sleep behaviour, the data may suggest that reduced biosynthesis of serotonin is the underlying mechanism how tryptophan changes result in disturbances of the central nervous system in patients with HIV infection. This conclusion is further supported by data showing that whole-blood serotonin content is significantly decreased in AIDS-patients (Launay *et al.* 1988).

Decreased tryptophan, and neurologic/psychiatric symptoms

Patients with HIV-1 infection frequently suffer from neurological and/or psychiatric symptoms. Sometimes AIDS results in severe cognitive and motor disturbances, formerly termed AIDS dementia complex, myelopathy, peripheral neuropathy, and visual dysfunction (Price *et al.* 1988). Long before the appearance of such clinical signs of central nervous system involvement, HIV-1 may enter the brain, and the virus can frequently be isolated from cerebrospinal fluid (CSF) of patients (Price *et al.* 1988; Sharer *et al.* 1992).

The pathogenesis of dementia in HIV-1 infection is still unclear, but the process is probably multifactorial. Because of its neurotoxicity, the tryptophan degradation products kynureninic acid and quinolinic acid have been suggested as possible mediators for development of AIDS dementia complex (Lipton *et al.* 1992). It still needs to be demonstrated if the changes are persistent or whether the development of clinical signs of HIV dementia may be linked to any eventual reduction in tryptophan concentration.

Brain damage is considered to be, at least to some part, directly mediated by HIV-1. However, cognitive impairment sometimes develops also in the case when neurons cannot be demonstrated to be infected with HIV-1. In HIV-1 infection lowest tryptophan concentrations and highest neopterin levels were specifically found in patients with neurologic and psychiatric symptoms, those patients with AIDS-dementia complex having the most abnormal values (Brew *et al.* 1990, Fuchs *et al.* 1990a and 1992a). The data suggest that immunologically-induced degradation of tryptophan is involved in the pathogenesis of HIV-1 related neuropsychiatric symptoms. Also in HIV-1 infection tryptophan degradation may disturb biosynthesis of serotonin, and may lead to the accumulation of neurotoxic degradation products like kynurenic acid and quinolinic acid (Fuchs *et al.* 1994).

Biochemical abnormalities, either directly or indirectly induced by the infection with HIV-1 or with associated opportunistic pathogens, appear to contribute to the spectrum of neuropsychiatric symptoms, because immune system activation substantially interferes with tryptophan metabolic pathways and, as a consequence, with neurobiochemistry. Tryptophan is not only depleted in the CSF of these patients but also in the blood, which may accelerate the development of symptoms related to tryptophan abnormalities. The decrease of plasma tryptophan levels reduces the transport of the amino acid through the blood-brain barrier, which additionally should decrease its availability for serotonin biosynthesis also in brain tissues. However, the situation in the brain is still less clear because serotonin concentrations in the CSF of patients with HIV-1 infection and neurological symptoms were only marginally lower than in asymptomatic controls (Larsson *et al.* 1989).

In HIV-1 infected patients reductions of tryptophan and concomitant increases of kynurenine in serum and CSF are accompanied by increases of quinolinic acid (Heyes *et al.* 1991 and 1992a). Further, close correlations exist between quinolinic acid, kynurenic acid and kynurenine with neopterin in CSF and serum. While CSF kynurenic acid levels were increased in HIV-1-infected patients, the magnitude of the increases were smaller than those of quinolinic acid and the molar concentrations of kynurenic acid were consistently lower than quinolinic acid by at least one order of magnitude.

Studies in macaques infected with retrovirus confirm the increase of both, IDO activity and quinolinic acid concentrations, in the brain particularly in those animals with local inflammatory lesions (Heyes *et al.* 1992). Correlations between CSF quinolinic acid, kynurenic acid and kynurenine with markers of immune stimulation, neopterin, white blood cell counts and IgG levels, indicate a relationship between accelerated kynurenine pathway metabolism and the degree of intracerebral immune stimulation.

Tryptophan in patients with HIV-1 infection during antiretroviral therapy

When comapring concentrations of tryptophan, kynurenine and neopterin in blood and CSF from HIV-1 infected individuals before and after antiretroviral treatment with zidovudine, pretreatment concentrations of tryptophan in CSF and blood were low, and an increase was observed after treatment. In parallel, the neopterin concentrations in CSF decreased (Gisslén *et al.* 1994). Serum neopterin concentrations behaved similarly. The pretreatment serum concentrations of neopterin and of tryptophan correlated inversely. Thus, tryptophan deficiency is reversible by zidovudine treatment. Again, the parallelism between the trends towards normalization of tryptophan and of neopterin, points to the view that zidovudine therapy also achieves some sort of down-regulation of the stimulated immune system in patients with HIV infection (Gisslén *et al.* 1994, Fuchs *et al.* 1996b).

Malignant diseases and weight loss

In patients with haematological neoplasias increased serum neopterin concentrations were found to be associated with significantly decreased tryptophan concentrations and increased kynurenine levels (Denz *et al.* 1993). When comparing loss of body weight in these patients with concentrations of serum tryptophan, kynurenine and neopterin, it was found that weight loss was seen particularly in patients with lower tryptophan and higher neopterin values. An association between higher neopterin levels and greater weight loss was apparent at study entry and during the follow-up of patients. In a similar study, decreased serum tryptophan and increased neopterin concentrations were found in patients with gastrointestinal tumors representing cancer cachexia (Iwagaki *et al.* 1995). This data support the concept that weight loss is closely linked to endogenous interferon-g activity and to the degradation of tryptophan.

Weight loss is the main symptom of so-called cachexia which is associated with the later stages of chronic disorders like malignant diseases or virus infections such as HIV-1 infection. The pathogenetic mechanisms underlying cachexia are poorly understood; however, it appears that enhanced formation of cytokines such as interferon-g and tumor necrosis factor-a are involved.

An inverse correlation of tryptophan and neopterin concentrations was found in patients with cancer. Because tryptophan is essential for protein synthesis, inadequate availability of tryptophan due to insufficient diet or enhanced degradation may lead to negative nitrogen balance and loss of muscle mass consequently causing weight loss and cachexia (Brown 1996). The relationship between low tryptophan levels and development of cachexia could be of particular relevance in patients with HIV infection. IDO is induced chronically by HIV infection, and it is further increased by opportunistic infections. The chronic loss of tryptophan may not only initiate mechanisms responsible for the dementia and diarrhea but also for cachexia which is seen preferentially in AIDS patients. In these symptoms, AIDS resembles classical pellagra due to dietary deficiency of tryptophan and niacin (Brown *et al.* 1991).

Inflammatory disorders

In autoimmune and chronic inflammatory diseases elevated urinary tryptophan metabolites were found, and it seems likely that systemic or local depletion of tryptophan occurs in autoimmune diseases and may relate to degeneration, wasting and other symptoms in such diseases (Brown *et al.* 1991). High levels of IDO were found, e.g., in cells isolated from synovia of arthritic joints.

Decreased serum tryptophan levels were found in patients with systemic sclerosis (Csipo *et al.* 1995) and similarly, serum kynurenine and neopterin concentrations were found increased in patients with cardiomyopathy (Rudzite *et al.* 1992). in the latter study serum kynurenine and neopterin concentrations were determined in patients with hypertrophic and dilated cardiomyopathy. Increased kynurenine levels have been observed in cases of dilated cardiomyopathy and a positive correlation existed between serum kynurenine and neopterin levels. There were several associations between kynurenine levels and clinical findings, such as left ventricle systolic volume and circumferential fiber shortening velocity in patients with dilated cardiomyopathy. Thus increased kynurenine accumulation in serum was associated with the severity of dilated cardiomyopathy.

Multiple sclerosis

Tryptophan disturbance could be also of relevance in patients suffering from multiple sclerosis (Fuchs *et al.* 1990). In fact, increased degradation of tryptophan was reported in patients with relapsing-remitting multiple sclerosis (Ott *et al.* 1993). Decreased tryptophan concentrations in CSF correlated with increased neopterin concentrations indicating a role of IDO activated by interferon-g .

Tryptophan therapy, eosinophilia-myalgia syndrome (EMS)

Because of the suggested association between decreased tryptophan avaialability and mood disturbances, sleep disorders and depression, exogenous administration of tryptophan was frequently used in patients suffering from these or related symptoms. In past several years tryptophan was well tolerated and no gross side effects were apparent. However, at the end of the eighties tryptophan produced by a specific manufacturer has been implicated in eosinophilia-myalgia syndrome (EMS) which is a multisystem illness of uncertain pathogenesis (McKinley *et al.* 1993). Contaminations of tryptophan preparations were suspected to play a role, but also formation of tryptophan catabolites could be critical. The possibility is remaining that tryptophan degradation products like kynurenic acid and quinolinic acid are involved in the development of EMS in tryptophan treated patients. On the one hand, the possible contamination of tryptophan preparations with bacterial endotoxins like lipopolysaccharide may directly stimulate the degradation of tryptophan in treated patients. On the other hand, circulating cytokines in patients induced by the underlying disease may initiate degradation of tryptophan and accumulation of toxic products. In fact, serum concentrations of sIL-2R, were found increased in sera of EMS which is indicative for immune activation. Patients suffering from chronic symptoms of myalgia, arthralgia, muscle cramps, fatigue, or subjective memory impairment had persistently elevated sIL-2R levels, whereas patients with near-resolution of EMS had normal sIL-2R levels.

Even if the association between the accumulation of tryptophan degradation products and development of EMS will turn out to be false, one has to keep in mind that administration of tryptophan may represent a special risk in patients who suffer from diseases which are associated with features of immune activation. In situations when endogenous interfereon-g is increased in patients there is the fear of adverse effects. It seems reasonable to test patients to for excluding states of immune activation when they are considered for tryptophan administration.

Conclusion

Abnormal metabolism of tryptophan is found in a wide range of diseases which at first glance seem to be rather unrelated. A decrease of tryptophan with a concomitant increase of tryptophan catabolites downstream the kynurenine degradation pathway indicates that the reduced tryptophan levels are not due to reduced dietary intake of the amino acid, rather enhanced degradation of tryptophan by activation of IDO is the most likely explanation. This assumption is confirmed by additional results showing that tryptophan degradation correlates with enhanced immune activation as reflected by increased neopterin levels and concentrations of cytokines or their serum soluble receptors.

Because of its role in the biosynthesis of the neurotransmitter serotonin and also of the formation of toxic metabolites, cytokine-induced degradation of tryptophan might be of relevance in the pathogenesis of neuropsychiatric symptoms during disorders which are associated with immune activation. With this respect, immunologically induced tryptophan metabolic changes could play a role in neurodegenerative disorders. However, the most consistent abnormalities of tryptophan are found in the peripheral blood, and abnormalities of tryptophan and serotonin concentrations in the brain are rarely being found. Significantly decreased concentrations of tryptophan in the CSF can be detected only in situations when immune activation is dramatic and long-lasting as in, e.g., HIV infection or Lyme neuroboreliosis. However, from the presently available data there is evidence that immune activation has significant impact on tryptophan availability and correlates well with neuropsychiatric symptoms in patients.

The investigation of tryptophan metabolic changes induced by interferon-g in patients with an activated cellular immune system clearly discovers a link between immune activation and the neuroendocrine system. The immunologically induced degradation of tryptophan may be part of the cytotoxic repertoir of immunocompetent cells, however it may cause significant depletion of tryptophan pools. Especially in long-lasting situations of immune deterioration the impact of the immune system to diminish tryptophan availability may be strong enough to precipitate neuropsychiatric symptoms. This might even be true when immunologically induced loss of tryptophan by IDO activation is not induced intrathecally but is initiated systemically or even at local sites of immune response, as it is the case in, e.g. malignant diseases. With longer duration the decrease of tryptophan will become significant also in the blood, and it will reduce the transport of tryptophan into the brain and thus lower, e.g., the biosynthesis of serotonin.

In conclusion, tryptophan metabolic changes induced by the immune system may represent a possibility for communication between the immunoendocrine and the neuroendocrine systems.

Acknowledgements

This work was financially supported by the Austrian Ministry of Science and Reserach, "Sektion Forschung".

References

Beal, M.F. (1992). Does impairmant of energy metabolism result in excitotoxic neuronal death in neurodegenerative illnesses? *Ann. Neurol.* **31,** 119-130.

Benavente, M.G. and Truscott, R.J.W. (1991). Modification of proteins by 3-hydroxyanthranilic acid: the role of lysine residues. *Arch. Biochem. Biophys.* **290,** 451-457.

Benkelfat, C., Nishizawa, S., Young, S.N., et al. (1995). Tryptophan depletion markedlyreduces the rate of serotinin synthesis in normal healthy volunteers. *Am. Coll. Neuropsychopharmacol.* **34,** 124.

Biggio, G., Fadda, F., Fanni, P., et al. (1974). Rapid depletion of serum tryptophan, brain tryptophan, serotonin and 5-hydroxyindoleacetic acid by a tryptophan, serotonin and 5-hydroxyindoleacetic acid by a tryptophan free diet. *Life Sci.* **14,** 1321-1329.

Brew, B.J., Bhalla, R.B., Paul, M., Sidtis, S.S., Keilp, J.J., Sadler, A.E., Gallardo, H., McArthur, J.C., Schwartz, M.K. and Price, R.W. (1990). Cerebrospinal fluid neopterin in human immunodeficiency virus type 1 infection. *Ann. Neurol.* **28**, 556-560.

Brown, R.R., Lee, C.M., Kohler, Lopez, K.D., Kohler, S., Oette, D.H., Bruno, E., Kozakewich, H., Nathan, D.G. and Hoffman, R. (1989). Altered tryptophan and neopterin metabolism in cancer patients treated with recombinant interleukin 2. *Cancer Res.* **49,** 4941-4944.

Brown, R.R., Ozaki, Y., Datta, S.P., Borden, E.C., Sondel, P.M., and Malone, D.G. (1991). Implications of interferon-induced tryptophan catabolism in cancer, autoimmune diseases and AIDS. *Adv. Exp. Med. Biol.* **294,** 425-435.

Brown, R.R. (1996). Metabolism and biology of tryptophan. Some clinical implications. *Adv. Exp. Med. Biol.* **398,** 15-25.

Byrne, G.I., Lehmann, L.K., Kirschbaum, J.G., Borden E.C., Lee C.M., and Brown R.R. (1986). Induction of tryptophan degradation in vitro and in vivo: A gamma interferon stimulated activity. *J. Interferon Res.* **6,** 389-396.

Coleman, J.W., Buckely, M.G., Holliday, M.R. and Morris, A.G. (1991). Interferon gamma inhibits serotonin release from mouse peritoneal mast cells. *Eur. J. Immunol.* **21,** 2559-2564.

Cowen, P.J. (1996). The serotonin hypothesis: necessary but not sufficient. In: *Selective Serotonin Reuptake Inhibitors.* J Feighner and W. Boyer (Eds). 2nd edn. John Wiley, Chichester pp. 63-86.

Csipo, I., Czirjak, L., Szanto, S., Szerafin, L., Sipka, S. and Szegedi, G. (1995). Decreased serum tryptophan and elevated neopterin levels in systemic sclerosis, *Clin. Exp. Rheumatol.* **13,** 269-270.

Curzon, G. (1996). Brain tryptophan. *Adv. Exp. Med. Biol.* **294,** 27-34.

Delgado, P.L., Charney, D.S., Price, L.H., et al. (1990). Serotonin function and the mechanism of antidepressant action: reversal of antidepressant induced remission by apid depletion of plasma tryptopan. *Arch. Gen. Psych.* **47,** 411-418.

Delgado, P.L., Price, L.H., Miller, H.L., Salomon, R.M., Aghajanian, G.K., Heninger, G.R., and Charney, D.S. (1994). Serotonin and the neurobiology of depression: effects of tryptophan depletion in drug-free depressed patients. *Arch. Gen. Psych.* **51,** 865-874.

Denz, H., Orth, B., Weiss, G., Herrmann, R., Huber, P., Wachter, H. and Fuchs, D. (1993). Weight loss in patients with hematological neoplasias is associated with immune system stimulation. *Clin. Investig.* **71,** 37-41.

Diez Ruiz, A., Tilz, G.P., Zangerle, R., Baier Bitterlich, G., Wachter, H. and Fuchs, D. (1995). Soluble receptors for tumor necrosis factor in clinical laboratory diagnosis. *Eur. J. Haematol.* **54,** 1-8.

Dotevall, L., Fuchs, D., Reibnegger, G., Wachter, H. and Hagberg, L. (1990). Cerebrospinal fluid and serum neopterin levels in patients with Lyme neuroborreliosis. *Infection* **18,** 210-213.

Fahey, J.L., Taylor, J.M.G., Detels, R., Hofmann, B., Melmed, R., Nishanian, P. and Giorgi, J.V. (1990). The prognostic value of cellular and serologic markers in infection with human immunodeficiency virus typre 1. *N. Engl. J. Med.* **322,** 166-172.

Fernstrom, J.D. and Wurtman, R.J. (1972). Brain serotonin content: physiological regulation by plasma neutral amino acids. *Science* **178,** 414-416.

Freese, K.J., Swartz, M.J., During, and Martin, J.B. (1990). Kynurenine metabolites of tryptophan: Implications for neurologic diseases. *Neurology* **40,** 691-695.

Fuchs, D., Hausen, A., Reibnegger, G., Werner, E.R., Dierich, M.P. and Wachter, H. (1988). Neopterin as a marker for acrtivated cell-mediated immunity: application in HIV infection. *Immunol. Today* **9,** 150-155.

Fuchs, D., Chiodi, F., Albert, J., Hagberg, L., Hausen, A., Norkrans, G., Reibnegger, G., Werner, E.R. and Wachter, H. (1989a). Neopterin concentrations in cerebrospinal fluid and serum of individuals infected with HIV-1. *AIDS* **3**, 285-288.

Fuchs, D., Hausen, A., Reibnegger, G., Werner, E.R., Werner-Felmayer, G., Dierich, M.P. and Wachter, H. (1989b). Interferon-gamma concentrations are increased in sera from individuals with human immunodeficinecy virus type 1. *J. Acquir. Immune Defic. Syndr.* **2,** 158-162.

Fuchs, D., Malkowsky, M., Reibnegger, G., Werner, E.R., Forni, G. and Wachter, H. (1989/90). Endogenous release of interferon gamma and diminished response of peripheral blood mononuclear cells to antigenic stimulation. *Immunol. Lett.* **23**, 103-108.

Fuchs, D., Möller, A.A., Reibnegger, G., Stöckle, E., Werner, E.R. and Wachter, H. (1990a). Decreased serum tryptophan in patients with HIV-1 infection correlates with increased serum neopterin and with neurologic/ psychiatric symptoms, *J. Acquir. Immune Defic. Syndr.* **3**, 873-876.

Fuchs, D., Forsman, A., Hagberg, L., Larsson, M., Norkrans, G., Reibnegger, G., Werner, E.R. and Wachter, H. (1990b). Immune activation and decreased tryptophan in patientswith HIV-1 infection, *J. Interferon Res.* **10**, 599-603.

Fuchs, D., Reibnegger, G., Werner, E.R., and Wachter, H. (1990c). Immune-med mechanisms in multiple sclerosis. J. Neurol. **237,** 125

Fuchs, D., Möller, A.A., Reibnegger, G., Werner, E.R., Werner-Felmayer, G., Dierich, M.P. and Wachter, H. (1991a). Increased endogenous interferon-gamma and neopterin correlate with increased degradation of tryptophan in human immunodeficiency virus type 1 infection. *Immunol. Lett.* **28**, 207-211.

Fuchs, D., Dotevall, L., Hagberg, L., Werner, E.R. and Wachter, H. (1991b). Kynurenine in cerebrospinal fluid of patients with Lyme neuroborreliosis. *Immunol. Infect. Dis.* **1**, 271-274.

Fuchs, D., Weiss, G., Werner-Felmayer, G. and Wachter, H. (1991c). Cytokine-induced increase in liver serotonin. *Immunol. Lett.* **28,** 259.

Fuchs, D., Möller, A.A., Weiss, G., Dierich, M.P. and Wachter, H. (1992a). Immune activation, impaired tryptophan metabolism, and neurologic/ psychiatric symptoms in HIV infection. In: HIV-1 infection of the central nervous system (Weiss S, Hippius H, eds). *Hogrefe & Nagel Publishers*, Seattle, Toronto, Göttingen, Bern, pp103-119.

Fuchs, D., Weiss, G., Reibnegger, G. and Wachter, H. (1992b). The role of neopterin as a monitor of cellular immune activation in transplantation, inflammatory, infectious and malignant diseases. *Crit. Rev. Clin. Lab. Sci.* **29**, 307-341.

Fuchs, D., Weiss, G. and Wachter, H. (1993). Neopterin, biochemistry and clinical use as a marker for cellular immune reactions. *Int. Arch. Allergy Immunol.* **101**, 1-6.

Fuchs, D., Dierich, M.P. and Wachter, H. (1994). Human immunodeficiency virus infection of the nervous system. *Ann. Neurol.* **34**:125.

Fuchs, D., Dotevall, L., Gasse, T., Hagberg, L., Tilz, G.P., Baier-Bitterlich, G., Demel, U., Schmutzhard, E. and Wachter, H. (1996a). Enhanced intrathecal neopterin production and tryptophan degradation in patients with Lyme neuroborreliosis. *Acta Dermatol.* **5**,137-141.

Fuchs, D., Gisslen, M., Larsson, M., Norkrans, G., Hagberg, L. and Wachter, H. (1996b). Increase of tryptophan in serum and in cerebrospinal fluid of patients with HIV infection during zidovudine therapy. *Adv. Exp. Med. Biol.* **398**, 131-134.

Fuchs, D., Schröcksnadel, H., Baier-Bitterlich, G., Dapunt, O. and Wachter, H. (1996c). Activated cellular immunity and decreased serum tryptophan in healthy pregnancy. *Adv. Exp. Med. Biol.* **398**, 149-153.

Gard, P.R., Handley, S.L., Parsons, A.D. and Waldron, G. (1986). A multivariate investigation of postpartum mood disturbance. *Br. J. Psychiatry* **148**, 567-575.

Gasse, T., Murr, C., Meyersbach, P., Schmutzhard, E., Wachter, H. and Fuchs, D. (1994). Neopterin production and tryptophan degradation in acute Lyme neuroborreliosis versus late Lyme encephalopathy. *Eur. J. Clin. Chem. Clin. Biochem.* **32**, 685-689.

Gisslén, M., Larsson, M., Norkrans G., Fuchs, D., Wachter H. and Hagberg H.L. (1994). Tryptophan concentrations increase in cerebrospinal fluid and blood after zidovudine treatment in patients with HIV type 1 infection. *AIDS Res. Human Retroviruses* **10**, 947-951.

Griffin, D.E., Ward, B.J., Jauregi, E., Johnson, R.T. and Vaisberg, A. (1990). Immune activation during measles: interferon-g and neopterin in plasma and cerebrospinal fluid in complicated and uncomplicated disease. *J. Infect. Dis.* **161**, 449-453.

Griffin, D.E., McArthur, J.C. and Cornblath, D.R. (1991). Neopterin and interferon gamma in serum and cerebrospinal fluid of patients with HIV-associated neurological disease. *Neurology* **41**, 69-74.

Hagberg, L., Dotevall, L., Norkrans, G., Larsson, M., Wachter, H. and Fuchs, D. (1993). Cerebrospinal fluid neopterin concentrations in central nervous system infection. *J Infect. Dis.* **168**,1285-1288.

Heyes, M.P., Brew, B.J., Matin, A., Price, R.W., Salazar, A.M., Sidtis, J.J., Yergey, J.A., Mouradian, M.M., Sadler, A.E. and Keilp, J. (1991). Quinolinic acid in cerebrospinal fluid and serum in HIV-1 infeection: Relationship to clinical and neurological status. *Ann. Neurol.* **29**, 202-209.

Heyes, M.P., Brew, B.J., Saito, K., Quearry, B.J., Price, R.W., Lee, K., Bhalla, R.B., Der, M., Markey, S.P. (1992a). Inter-relationships between quinolinic acid, neuroactive kynurenines, neopterin and beta 2-microglobulin in cerebrospinal fluid and serum of HIV-1-infected patients. *J. Neuroimmunol.* **40**, 71-80.

Heyes, M.P., Saito, K., Crowley, J.S., Davis, L.E., Demitrack, M.A., Dez, M., Dilling, L.A., Elia, J., Krnesi, M.J., Lackner, A. (1992b). Quinolinic acid and kynurenine pathway metabolism in inflammatory and non-inflammatory neurological disease. *Brain* **115**, 1249-1273.

Heyes, M.P., Saito, K., Jacobowitz, D., Markey, S.P., Takikawa, O. and Vickers, J.H.(1992c). Poliovirus induces indoleamine-2,3-dioxygenase and quinolinic acid synthesis in macaque brain. *FASEB J.* **6**, 2977-2989.

Heyes, M.P., Saito, K. and Markey, S.P. (1992d). Human macrophages convert L-tryptophan into the neurotoxin quinolinic acid. *Biochem. J.* **283**, 633-635

Heyes, M.P., Saito, K., Milstien, S. and Schiff, S.J. (1995). Quinolinic acid in tumors, hemorrhage and bacterial infections of the central nervous system in children. *J. Neurol. Sci.* **133**,112-118.

Iwagaki, H., Hizuta, A., Tanaka, N. and Orita, K. (1995). Decreased serum tryptophan in patients with cancer cachexia correlates with increased serum neopterin. *Immunol. Invest.* **24**, 467-478.

Jhamandas, K.H., Boegman, R.J. and Beninger, R.J. (1994). Quinolinic acid induced brain neurotransmitter deficits: modulation by endogenous excitotoxin antagonists. *Can. J. Physiol. Pharmacol.* **72**,1473-1482.

Kasper, S., Fuger, J. and Moller, H.J. (1992). Comparative efficacy of antidepressants. *Drugs* **43,** 11-23.

Larrson, M., Hagberg, G., Norkrans, G. and Forsman, A. (1989). Indole amine deficiency in blood and cerebrospinal fluid from patients with human immunodeficiency virus infection. *J. Neurosci. Res.* **23**, 441-446.

Launay, J.M., Copel, L. and Callebert, J. et al. (1988). Decreased whole blood 5-hydroxytryptamine(serotonin). in AIDS patients. *J. Acquir. Immune Defic. Syndr.* **1**, 324-325.

Leblhuber, F., Tilz, G.P., Walli, J., Wachter, H. and Fuchs, D. (1995a). Activated immune system in patients with Huntington's disease. *Pteridines* **6,** 26-27.

Leblhuber, F., Tilz, G.P., Walli, J., Zachhuber, C., Wachter, H. and Fuchs, D. (1995b). Immune activation in patients with Alzheimer's disease. *Pteridines* **6,** 26.

Leibowitz, S.F. and Shor-Posner, G. (1986). Brain serotonin and eating behaviour. *Appetite* **7 (Suppl.),** 1-14.

Leibowitz, S.F., Weiss, G.F., Walsh, U.A. and Viswanath, D. (1989). Medical hypothalamic serotonin: role in circadian patterns of feeding and macronutrient selection. *Brain Res.* **503,** 132-140.

Linnola, M. and Virkkunen, M. (1992). aggression, suicidality and serotonin. *J Clin. Psychiatry* **53,** 46-51.

Lipton, S.A. (1992). Models of neuronal injury in AIDS: another role for the NMDA receptor? *Trends Neurosci.* **15**, 75-79.

Maes, M., Meltzer, H.Y., Scharpe, S., Bosmans, E., Suy, E., De Meester, I., Calabrese, J.and Cosyns, P. (1993). Relationships between lower plasma L-tryptophan levels and immune-inflammatory variables in depression. *Psychiatry Res.* **49**,151-165.

Maes, M., Scharpe, S., Meltzer, H.Y., Okayli, G., Bosmans, E., D'Hondt, P., Vanden-Bossche, B.V. and Cosyns, P. (1994). Increased neopterin and interferon-gamma secretion and lower availability of L-tryptophan in major depression: further evidence for an immune response. *Psychiatry Res.* **54**,143-60.

Martin, J.B. (1984). Huntington's disease: new approaches to an old problem. *Neurology* **34**,1059-1072.

McEntee, W.J. and Crook, T.H. (1991). Serotonin, memory, and the aging brain. *Psychopharmacol.* **103**, 143-149.

McKinley, K.L., Harati, Y. and Schneider, L.W. (1993). Chronic immune activation in the eosinophilia-myalgia syndrome. *Muscle Nerve* **16**, 947-951.

Möller, A.A., Fuchs, D., Reibnegger, G., Zaudig, M., Werner, E.R. and Wachter, H. (1990). Changes of neopterin and tryptophan in serum and cerebrospinal fluid of patients with different stages of HIV-1 infection. *Pteridines* **2**, 1-93.

Moreno, F.L., Straver, G., Heninger, G., Delgado, P. (1995). Mood response to tryptophan and vulnerability to depression. *Soc. Neurosci.* **21A**.

Ott, M., Demisch, L., Engelhardt, W. and Fischer, P.A. (1993). Interleukin-2, soluble interleukin-2-receptor, neopterin, L-tryptophan and beta 2-microglobulin levels in CSF and serum of patients with relapsing-remitting or chronic-progressive multiple sclerosis. *J. Neurol.* **241**; 108-14.

Parker, G., Hadzi-Pavlovic, D., Wilhelm, K., Hickie, I., Bradaty, H., Boyce, P., Mitchell, P. and Eyers, K. (1994). defining melancholia; properties of a refined signed-based measure. *Br. J. Psych.* **164**, 316-326.

Pfefferkorn, E.R. (1984). Interferon-gamma blocks the growth of Toxoplasma gondii in human fibroblasts by inducing the host cells to degrade tryptopühan. *Proc. Natl. Acad. Sci.* **81,** 908-912.

Price, R.W., Brew, B., Sidti, J., Rosenblum, M., Scheck, A.C. and Cleary, P. (1988). The brain in AIDS: central nervous system HIV-1infection and AIDS dementia complex. *Science* **239,** 586-592.

Romagnani, S., Del Prete, G., Manetti, R., Ravina, A., Annunziato F., Delarli, M., Mazzetti, M., Piccinni, M.P., D'Elios, M.M. and Parronchi, P. (1994). Role of TH1/TH2 cytokines in HIV infection. *Immunol. Rev.* **140,** 73-92.

Rudzite, V., Skards, J.I., Fuchs, D., Reibnegger, G. and Wachter, H. (1992). Serum kynurenine and neopterin concentrations in patients with cardiomyopathy. *Immun Lett.* **32,**125-129.

Schröcksnadel, H., Baier-Bitterlich, G., Dapunt, O., Wachter, H. and Fuchs D. (1996). Decreased plasma tryptophan in pregnancy. *Obstet. Gynecol.* **88,** 47-50.

Schwarcz, R., Tamminga, C.A., Kurlan, R. and Shoulson, I. (1988). Cerebrospinal fluid levels of quinolinic acid in Huntington's disease and Schizophrenia. *Ann. Neurol.* **24**, 580-582.

Sharer, L.R. (1992). Pathology of HIV-1 infection of the central nervous system A review. *J. Neurophatol. Exp. Neurol.* **51,** 3-11.

Sharp, K., Brindle, P.M., Brown, M.W. and Turner, G.M. (1993). Memory loss during pregnancy. *Br. J. Obstet. Gynaecol.* **100**, 209-215.

Smith, K.A., Fairburn, C.G. and Cowen, P.J. (1997). Relapse of depression after rapid depletion of tryptophan. Lancet **349,** 915-919.

Teasdale, J.D. (1988). Cognitive vulnerability to persistent depression. *Cognit. Emot.* **2**, 247-274.

The Huntington's Disease Collaborative Research Group. (1993). A novel gene containing a trinucleotide repeat that is expanded and unstable on Huntington's disease chromosomes. *Cell* **72**, 971-983.

Van Praag, H.M. (1991). Serotonin dysfunction and aggression control. *Psychol. Med.* **21,**15-19.

Wachter, H., Fuchs, D., Hausen, A., Reibnegger, G., Weiss, G., Werner, E.R., and Werner-Felmayer, G. (1991). In: *Neopterin: Biochemistry-Methods-Clinical Application.* de Gruyter Berlin, New York.

Werner, E.R., Bitterlich, G., Fuchs, D., Hausen, A., Reibnegger, G., Szabo, G., Dierich, M.P. and Wachter, H. (1987a). Human macrophages degrade tryptophan upon induction by interferon-gamma. *Life Sci.* **41**, 273-280.

Werner, E.R., Hirsch-Kaufmann, M., Fuchs, D., Hausen, A., Reibnegger, G., Schweiger, M. and Wachter, H. (1987b) Interferon-g-induced degradation of tryptophan by human cells in vitro. *Biol. Chem. Hoppe Seyler.* **368**, 1407-1412.

Werner, E.R., Fuchs, D., Hausen, A., Jäger, H., Reibnegger, G., Werner-Felmayer, G., Dierich, M.P and Wachter, H. (1987a). Tryptophan degradation in patients infected by human immunodeficiency virus. *Biol. Chem. Hoppe Seyler.* **369**, 337-40.

Wiegand, M., Möller, A.A., Schreiber, W., Krieg, J.C., Fuchs, D., Wachter, H. and Holsboer, F. (1991). Nocturnal sleep in patients with HIV infection. *Psych. Clin. Neurosci.* **240**, 153-158.

Wurtman, R.J., and Wurtman J.J. (1996). Brain serotonin, carbohydrate-craving, obesity and depression. *Adv. Exp. Med. Biol.* **398,** 35-41.

Young, S.N., Smith, S.E., Phil, R.O. and Ervin, F.R. (1985). Tryptophan depletion causes a rapid lowering of mood in normal males. *Psychopharmacology* **87**,173-177.

Young, S.N., Ervin, F.R., Phil, R.O. and Finn, P. (1989). Biochemical aspects of tryptophan,depletion in primates. *Psychopharmacology* **98**, 508-511.

Zaudig, M., Mittelhammer, J. and Hiller, W. Pauls, A., Thora, C., Moringo, A. and Mombour, W. (1991). Description and test-retest reliability of the SIDAM - a structured interview for the diagnosis of dementia of the Alzheimer type, multiinfarct demential of other etiology according to ICD-10 and DSM-III-R. *Psychol. Med.* **21**, 225-236.

Zaknun, D., Zaknun, J., Unsinn, K., Wachter, H., and Fuchs, D. (1994). Interferon gamma-induced formation of neopterin and degradation of tryptophan in cerebrospinal fluid o children with meningitis but not with febrile convulsions. *Pteridines* **5**, 102-106.

Progress in HPLC-HPCE, Vol. 7, pp. 383-411
G. Ali Qureshi *et al.* (Eds)

Acetylcholine, Catecholamine, Indolamine, and GABA levels in the Brain of Rats submitted to a Convulsant

HEVOR, T.K., RICHARD, O., HELARY, K., PICHON, J., AND CLOIX J.-F.

UPR 9074. Université d'Orléans. B.P. 6759 F-45067 Orléans CEDEX (France).

INTRODUCTION

1) Involvement of neurotransmitters in epileptogenesis

Epilepsy is one of the major neurologic disorders relatively frequent among human population. Until now, the actual mechanisms that lead to electroencephalographic abnormalities and seizure genesis are not yet understood. The researchers have focused their attention in many directions comprizing neurotransmission, ions, amino acids, energy metabolism, etc. The obvious sign of epilepsy onset is the appearance of an abnormal electroencephalogram either in a given small region of the brain or in a large or very large region of the brain. In these regions, there is an abnormal synchronisation of neuronal discharges, thus leading to the formation of characteristic waves on electroencephalogram. This triggers a large panel of various disorders. Because of this variety, this pathology is often termed epilepsies in the plural, but, there may be a basic mechanism underlying the neuronal synchronisation. Among many possibilities of synchronisation, one can be neurotransmission since the latter is the ground of neuronal communication in neuronal circuits (Hevor and Gayet, 1979). That is why many investigators focuse their studies on neurotransmitters (Jobe and Laird, 1987). In these investigations, many models have been used. There is no worldwide model and, more or less, each model mimics one aspect of the pathology. Thus, the results obtained from different models are often conflicting. Some years ago, Engel *et al.* (1982) have shown that there was an obvious hypometabolism during interictal period in man. This finding was performed using 2-deoxyfluoroglucose. There was a model of "grand mal" seizures generated by the administration of a derivative of methionine, methionine sulfoximine. Many years ago, this model has been described as follows : "seizures induced by methionine sulfoximine are of particular interest to the neurochemist and neurophysiologist because of their resemblance to human epilepsy and their readily reproducibility in a variety of experimental animals» (Wolfe and Elliot, 1962). In the field of metabolism, this model also mimics human epilepsy since, animals submitted to methionine sulfoximine display an important increase in brain glycogen which has been considered as a sign of hypometabolism (Hevor *et al.,* 1986). A direct link between the metabolic changes and epileptogenesis induced by methionine sulfoximine has not been established. We postulated that the impairment of glucose metabolism induced by the convulsant may result in an impairment of the neuron metabolism (Hevor *et al.,* 1986). Thus, the neurotransmission of the neuronal cells could be disturbed. In this paper, we present changes in the concentration of common neurotransmitters, i.e., acetylcholine, catecholamines, indolamines, and GABA in the encephalon of rats submitted to the convulsant methionine sulfoximine.

2) ASSESSMENT OF NEUROTRANSMITTER LEVELS IN THE CENTRAL NERVOUS SYSTEM

During a long time, the determination of neurotransmitter concentrations in the nervous system was laborious. In many methods, radioactive compounds were used and then, the different neurotransmitters or their metabolites were analysed. These methods were time consuming and lacked precision and sensitivity. The availability of high performance liquid chromatography (HPLC) allows the development of rapid and sensitive methods for the measurement of neurotransmitter concentrations in tissues.

2.1) Principles

The principle of neurotransmitter determination is the separation of different neurotransmitters by a column and then, the measurement of the separated compounds by a detector (for a revue see Mefford, 1981, figure 1). The suitable column commonly used is a reversed phase column containing nonpolar stationary phase with a polar mobile phase. The stationary phase is silica which is bonded with hydrocarbon chain such as octadecyl silyl chain, commonly termed C-18 columns. The retention of polar compounds by such a column is not enough high so that these compounds are eluted quickly with the polar phase, with the most polar compounds coming off first. This chromatographic separation can be sufficient for some neurotransmitters such as serotonin. For other neurotransmitters, the separation is improved by adding a ion-pair reagent to increase their retention. A common suitable pairing agent is sodium dodecyl sulfate. Taking into account the pKa of the neurotransmitter, the pH and/or the molarity of the mobile phase can be changed for a better separation of two overlapping compounds.

The detection is made by an electrochemical detector. The principle of such a detection is the measurement of electrons exchanged during oxidation or reduction of the compoud to be measured. In the case of neurotransmitters, the oxidation procedure is often used. At the end of the column, the neurotransmitters enter a small space of one-microliter range size. There, the neurotransmitters are oxidized by the application of a direct current of 1 volt range. The current is applied throught a reference Ag/AgCl electrode and a working electrode. The latter is made of carbon paste. A third electrode termed auxiliary electrode is added to the system for a potentiostatic control of the potential. Electrons coming from the oxidation procedure are guided toward a large gain amplifier, the outpout of which is connected to a recorder or to a data analysis system. The data analysis allows to get directly the neurotransmitter concentrations after a calibration curve is made. There are two main designs for electrochemical detectors : the amperometric detector and the coulometric detector. The former oxidized 1 to 5 % of the compound to be measured and the latter oxidized 100 % of the compound. Theoretically, the coulometric detector should be more sensitive. In practice, the two detector have the same sensitivity because of the high electrical noise of the coulometric detector. Amperometric detector are more often used.

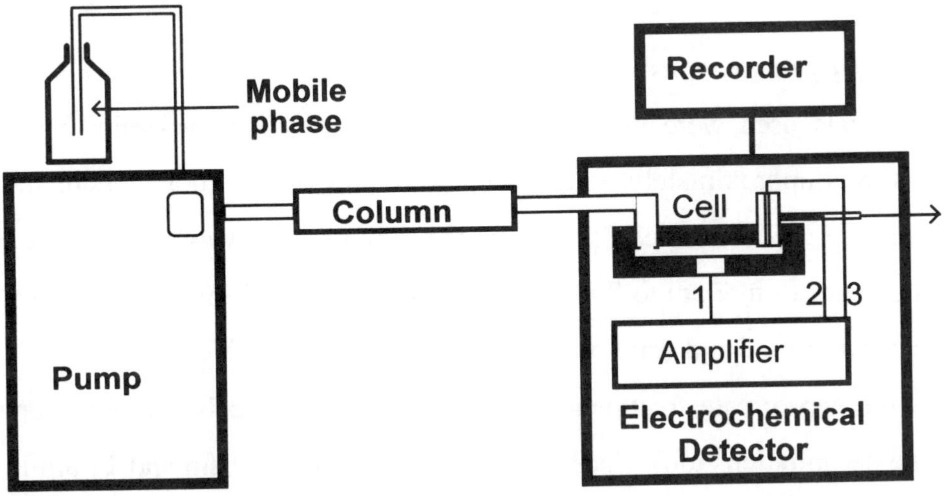

Figure 1 : Principle of neurotransmitter measurement using high performance liquid chromatography with electrochemical detection. 1, carbon paste or platinum electrode; 2, auxiliary electrode; 3, Ag/AgCl reference electrode.

2.2) Equipements and special compounds

All the materials and apparatus for HPLC are now available from many salers (Waters, Beckman, Hitachi, Spectra-Physics, etc.). The electrochemical detectors can be purchased from Bioanalytical System (BAS, West Lafayette, Ind, USA) or from Tacussel Electronique (Villeurbanne, France). The pairing agent sodium octyl sulfate was purchased from Kodak (Eastman-Kodak, Rochester, USA) or from Sigma (Sigma Chemicals, St Louis, Missouri, USA). The decarboxylase inhibitor benserazide was a gift from Hoffman-La Roche (Basel, Switzerland). Because of the high sensitivity of the neurotransmitter determinations, the common products to be used must be of the best grade. However, for common use, the chromatographic grade which is more expensive is not always indispensable.

2.3) Sample preparations

The animals used were males rats of the Sprague Dawley strain. Their weight was approximately 300g. They were caged in a room where the temperature was controlled (22°C) and the light regimen was standard, i.e., 12 H of light from 7 am to 7 pm and 12 H of darkness from 7 pm to 7 am. The access to food and water was free.

After the appropriate treatment, the animals were killed by decapitation. All the decapitations took place in the morning between 9 am and 11 am in order to avoid natural changes in the neurotransmitter concentrations due to circadian rythm. The heads of the animals were immediately immersed in liquid nitrogen. These fast and deep freezing is necessary because the neurotransmitter concentrations largely change with ischemia and anoxia. The better way to avoid post-mortem changes is a microwave fixation. This consists in focusing high energy microwaves on the heads of animals, thus inactivating all the enzymes in few seconds. In this way, neurotransmitter concentrations are close to the *in vivo* concentrations. The difficulty is the high price of the microwave apparatus of high energy equiped with a wave guider which focuses the energy. Because of the price, this apparatus is not commonly available. The immediate freezing in liquid nitrogen is an acceptable alternative. With a good handling, the freezing method gives results close to the microwave method. The heads of the animals remained in the liquid nitrogen for about 10 min in order to freeze the deep parts of the tissue. Then, they were taken off and stored in a freezer at -30°C. The dissection was carried out in the frozen state and several regions were separated (Glowinski and Iversen, 1966). Each frozen region was weighed and then homogenized in 3 volumes of 0.2 M perchloric acid by sonication.

The acidification precipitates all the proteins and no further enzymatic changes takes place. A part of the homogenate was taken off for the determination of protein concentration which was made according to Lowry *et al.,* (1951) or Miller (1959) or Bradford (1976). The remaining homogenate was centrifuged at high speed in order to get a very clean supernatant. Ultracentrifugation is not indispensable. The gravity of 50.000 *g* during 15 min is sufficient. In our hands, centrifigation at 30.000 *g* during 30 min gives suitable results. Anyway, the centrifugation has to be performed at low temperature, by 4°C, in order to reduce chemical degradation. Sometimes, the supernatant was purified and concentrated using alumina adsorption (Mefford, 1981). This might be necessary for some samples from blood and urine. In the case of brain extracts, the supernatant can be directly injected into the HPLC system.

2.4) Neurotransmitter measurements

The pump of the HPLC apparatus was an isocratic system from Beckman (model 112 solvent delivery module, Berkeley, California, USA). In our hands, gradient pump was not necessary provided a suitable column is chosen. A reverse phase of 5-µm-ultrasphere ODS column was used. The column internal length and width were respectively 15 cm and 0.46 cm. The separation performance of a column depends on its number of separation units termed plates. Now, small columns with a large number of plates are available. The use of such columns can reduce the time necessary for the analysis. The detector that we used was an electrochemical one equipped with a glassy carbon working electrode from Bioanalytical System. The detector was coupled with an amperometric amplifier from Tacussel (Villeurbanne, France). This amplifier allows to apply in the detector a chosen voltage to the compounds already separated by the column. Many electrochemical detectors need a long time (several hours) before electrical stabilisation. The amplifier is connected to a data analysis system from Waters (Waters 740 data module integrator, Milford, Ma, USA). This system allowed to get directly the neurotransmitter concentrations after a previous calibration.

The statistical analysis of the results depend of the kind of experiment. In our case, the comparison of treated animals with controls were made using the common Student's *t* test with a level of at least 0.05.

3) ACETYLCHOLINE

The possible involvement of acetylcholine in epileptogenesis has been less studied than that of the other neurotransmitters. One reason of this apparent lack of interest was the difficulty in obtaining reliable results using hard-working methods. By 1980, a series of investigations showed that acetylcholine can be measured using HPLC with electrochemical detection with high reproducibility and sensitivity (Potter *et al.*, 1983, Damsa and al. 1985). Evidences show that the disturbances of cholinergic system can lead to epileptogenesis.

3.1) Measurement

The principle of neurotransmitter determination described above can be used for acetylcholine determination. However, the equipment has to be completed. The principle follows. Acetylcholine is separated from other compounds by the HPLC column. The reverse phase column also separates choline. The outlet of the separation column is connected to a bioreator column containing two enzymes, i.e., acetylcholinesterase and choline oxidase. These enzymes are immobilized onto the spheres of the column (Biometra, Göttingen, Germany). So, acetylcholine is cleaved into acetate and choline by acetylcholinesterase. Choline is transformed into betaïne and hydrogen peroxide by choline oxidase. The outlet of the bioreactor is connected to the electrochemical detector cell. Hydrogen peroxide is easily detected by this cell, thus offering a high sensitivity to the method. Note that the separated choline will be measured together with acetylcholine in the same run. It is important to equip the detector cell with a platinum working electrode instead of glassy carbon electrode. This kind of cell is commonly available (Bioanalytical System, Tacussel, Waters). Apart from the preceding precisions, the apparatus is the same as that described in the general method.

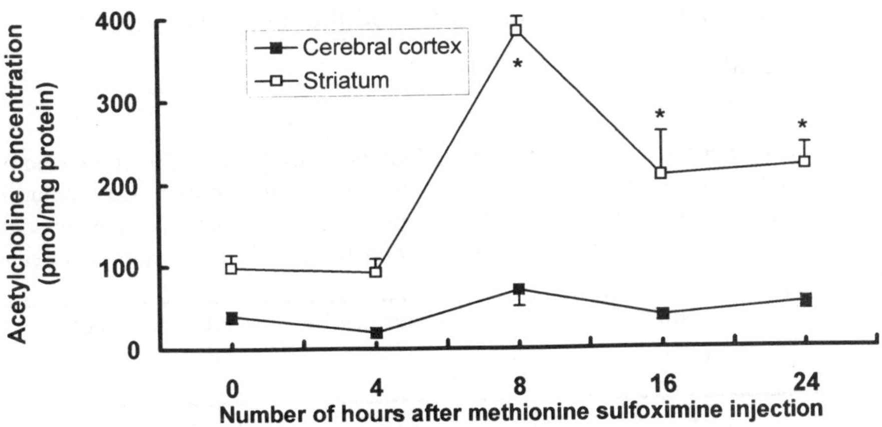

Figure 2 : Effect of the intraperitoneal injection of methionine sulfoximine (200 mg/kg body weight) on acetylcholine concentration in the rat cerebral cortex and striatum. Each point is the mean ± standard deviation of at least 5 independent determinations. Comparisons were made using Student's *t* test with $p<0.05$ and the asterisks indicate significant differences.

The mobile phase that we used contained 0.05 M potassium phosphate, 1 mM sodium octyl sulfate, and 5 mM tetramethylammonium chloride. The pH of the mixture was adjusted to 7.4 with KOH and the flow rate was set to 1 ml/min. A potential of +0.5 volt was applied to the electrodes. The brain samples were prepared as described above.

3.2) Methionine sulfoximine-induced changes in brain acetylcholine concentration

The concentration of acetylcholine greatly changed from one region to another in the encephalon of control rats. The value in the cerebellum was within the limit of the detection range. It was 39 ± 7 pmol/mg protein in the cerebral cortex and 97 ± 15 pmol/mg protein in the striatum. The distribution of choline was different from that of acetylcholine since the choline concentration was more constant from a region to another. The lowest concentration was recorded in cerebral cortex (581 ± 82 pmol/mg protein) and the highest in pons plus medulla oblongata (1275 ± 204 pmol/mg protein).

Table 1.
Effect of the intraperitoneal injection of methionine sulfoximine (200 mg/kg body weight) on choline concentration in the rat cerebral cortex and striatum. Each point is the mean ± standard deviation of at least 5 independent determinations. Comparisons were made using Student's t test with $p<0.05$ and the asterisks indicate significant differences.

	Number of hours after methionine sulfoximine injection				
	0	4	8	16	24
Cerebral cortex (pmol/ mg protein)	581 ± 82	682 ± 128	276 ± 72 *	480 ± 81	492 ± 165
Striatum (pmol /mg protein)	1120 ± 130	1375 ± 181	515 ± 86 *	1125 ± 167	849 ± 108

When the rats were given methionine sulfoximine by intraperitoneal injection (100 mg/kg body weight), they displayed tonic convulsions by 4-6 hours after dosing. This was followed by violent tonic and clonic convulsions that began by 7-8 hours after dosing and lasted for several hours. Then, recovery was observed and this recovery was complete the next day. The period separating the injection from the appearance of the first sign of convulsions was termed preconvulsive period; the period where the seizures were obvious was convulsive period; the recovery was postconvulsive period. After the drug administration (100 mg/kg or 200 mg/kg), a general increase in acetylcholine level was observed in many regions of the encephalon, in agreement with other observations (Robinson and Hambretch, 1986; Lallement *et al.,* 1992; Baptista *et al.,* 1994). However, this increase was significant only in hypothalamus, thalamus, and principally in striatum (figure 2). The changes observed in choline concentration were opposite to those in acetylcholine concentration. There were decreases in choline concentration which were sometime significant in some regions. (table 1). In order to know whether the cholinergic system is involved in the epileptogenesis induced by methionine sulfoximine, atropine (5 mg/kg body weight) was injected to the animals cumulatively with methionine sulfoximine. The modifications in animal behaviour, brain acetylcholine and choline concentrations were almost the same as those following the injection of methionine sulfoximine alone. An experiment opposite to the preceding one was displayed by reinforcing the cholinergic system in order to know whether the seizures will be more important. For this, physostigmine (eserine) was intraperitoneally administered at the dose of 0.5 mg/kg body weight simultaneously with methionine sulfoximine. In these experiments, the increase in acetylcholine concentration was more important. The control values of 36 ± 6 pmol/mg protein in the hypothalamus, 98 ± 15 pmol/mg protein in the striatum, and 40 ± 6 in the thalamus increased to respectively 151 ± 32, 319 ± 18, and 103 ± 8 pmol/mg protein. Despite these important increases in acetylcholine concentration, the behaviour of the animals was the same as when methionine sulfoximine was injected alone apart from the tremor that was observed early after physostigmine administration. Note that physostigmine alone significantly increased acetylcholine concentration in the three regions mentioned above.

3.3) Acetylcholine is not involved in methionine sulfoximine epileptogenesis

The heterogeneous distribution of acetylcholine in the encephalon can be explained by the topography of cholinergic neuronal circuit. This explains the high value encountered in the striatum which is a region rich in cholinergic pathways. The absence of such pathways in the cerebellum results in an almost undetectable level of acetylcholine in this brain area. Methionine sulfoximine induced an increase in specific regions of the brain. Even when the convulsant dose is increased, there was no significant increase in cerebral cortex, hippocampus, and pons and medulla oblongata. One can assume that the changes in the specific regions could be involved in the epileptogenesis induced by methionine sulfoximine. This possibility is unlikely since the antagonization or the reinforcement of cholinergic system did not modify the profile of the seizures generated by methionine sulfoximine. In other models, atropine was able to inhibit seizures (Jope *et al.,* 1986). The mechanism by which methionine sulfoximine increases acetylcholine concentration is currently under investigation.

4) CATECHOLAMINES

Biogenic amines have been involved in the genesis of epilepsy in some models (Suñol *et al.,* 1988). The two main catecholaminergic neurotransmitters in the nervous system are norepinephrine and dopamine.

4.1) Catecholamine measurement

The general methodology described above can be used to measure norepinephrine and dopamine in one run. The mobile phase contained 0.1 M sodium acetate, 0.2 M citric acid, 0.15 m M EDTA, 0.43 m M sodium octyl sulfate, and 8% methanol (v/v). The percentage of methanol highly influences the time and the quality of the separation. When the mobile phase contains a low level of methanol, the retention time will be long, but the peaks will be better separated although they will be wide. With a high level of methanol in the phase, a quick separation will be performed, the peaks will be sharp, but these peaks can overlap. For each kind of measurement, a suitable percentage of methanol has to be found for a good separation of catecholamines. In general, about 10 % methanol is a convenient value. In our hand, the retention times was 4.5 min for norepinephrine and 18.9 min for dopamine using 8% methanol in the mobile phase. So the peaks were well separated. However, if the aim of the study is to measure only dopamine content, the rate of methanol can be increased. Norepinephrine peak will be mixed with front peak and the retention time of dopamine will be much smaller. This allows to reduce the time necessary for the analyses. For all the catecholamine determinations, the potential applied to the electrodes was +0.6 volt.

4.2) Norepinephrine

4.3) Norepinephrine concentration was measured in seven regions of the rat encephalon : cerebral cortex, hypothalamus, hippocampus, striatum, thalamus, cerebelum, and pons and medulla oblongata. In control animals, the values ranged from 108 ± 15 pmol/mg protein in the cerebelum to 581 ± 39 pmol/mg protein in the hypothalamus. When the animals were submitted to methionine sulfoximine (100 mg/kg body weight), a little decrease in norepinephrine concentrations was observed before the onset of the seizures (figure 3). This decrease was rarely significant at the level of $p<0.05$ (Student's *t* test). When the animals displayed seizures during the convulsive period (8 H after dosing), the norepinephrine concentration was already statistically the same as in the controls except in the cerebral cortex where a little decrease was observed. Then, no further changes were observed up to 48 H after dosing. In order to know if the decrease in the norepinephrine content is associated to the convulsions, a series of experiments were done with the smallest dose of methionine sulfoximine that induces convulsions, i.e., 80 mg/kg. In this case, no change was observed anywhere during the studied periods. The preceding results show that there is no clear link between the genesis of seizures induced by methionine sulfoximine and norepinephrine content although a little decrease was observed. A putative mechanism could be an increase in the rate of the neurotransmitter turnover. In this case, the stimulation of the norepinephrine receptors would increase and lead to convulsive state. Our experimental design does not allow to get informations on the turnover. However, since we used different doses of methionine sulfoximine and different times for norepinephrine determination, there would be a discrepancy at a moment between the synthesis and degradation of the neurotransmitter, leading to visible changes in its concentration (Hevor *et al.,* 1990). No important change was observed any time, especially with the dose of 80 mg/kg body weight.

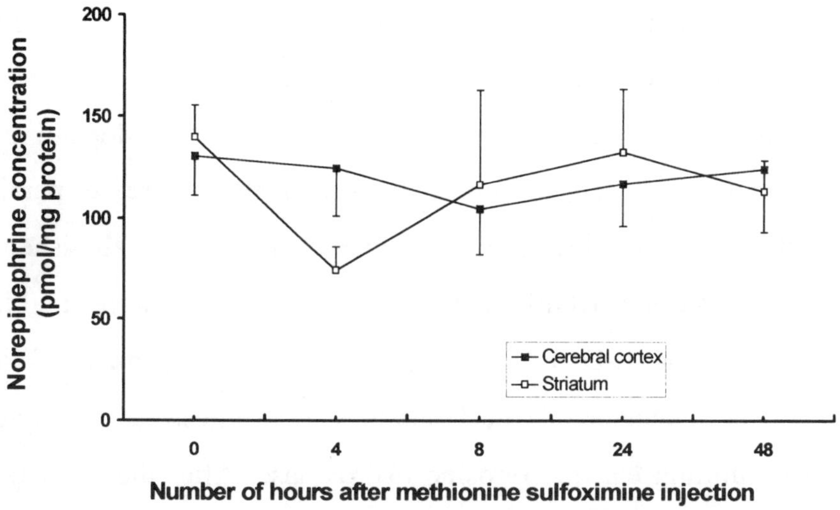

Figure 3 : Effect of the intraperitoneal injection of methionine sulfoximine (100 mg/kg body weight) on norepinephrine concentration in the rat cerebral cortex and striatum. Each point is the mean ± standard deviation of at least 5 independent determinations. Comparisons were made using Student's *t* test with $p<0.05$ and the asterisks indicate significant differences.

Anyway, we can conclude that no change in norepinephrine concentration parallels epileptogenesis induced by methionine sulfoximine.

4.3) Dopamine

Methionine sulfoximine induced a marked decrease in the dopamine concentration in the rat encephalon (figure 4). This decrease was already observed 4 H after dosing in the cerebral cortex, hypothalamus, and striatum. The dopamine level was significantly lower for a long time and the control values were reached 48 H after dosing. Animals were injected with L-DOPA (100 mg/kg) together with benserazide (10 mg/kg) and methionine sulfoximine (100 mg/kg) in order to counteract the decrease of dopamine. In these experiments, no change was observed in the dopamine level 8 H after dosing. Also, the small decreases observed during the preconvulsive period in norepinephrine content in the hypothalamus, striatum, and pons and medulla oblongata no longer existed. During these experiments, no convulsion was observed at any time.

Methionine sulfoximine is able to greatly decrease the dopamine level in many regions of the rat encephalon and this decrease parallels the convulsive state. When the dopamine decrease was abolished by the administration of L-DOPA, the epileptiform seizures were inhibited. The rationale of the L-DOPA utilization in these experiments follows. L-DOPA, the precursor of catecholamines, is synthesized from tyrosine by the key limiting enzyme tyrosine hydroxylase. After the formation of L-DOPA, there is no limiting step for the synthesis of dopamine, norepinephrine, or epinephrine. This means that the supply of the nervous cells with more L-DOPA will result in the synthesis of more catecholamines. The difficulty is that L-DOPA can be destroyed by peripheral decarboxylases before reaching the central nervous system. To reduce this possibility, benserazide, an inhibitor of decarboxylases must be injected concomitantly with L-DOPA. Thus, there will be a limited peripheral degradation of L-DOPA. Since benserazide cannot cross the blood brain barrier, L-DOPA that enters the central nervous system could be decarboxylated with the local decarboxylase and the catecholamine syntheses will take place in the nervous tissue. This is the strategy which is used in the treatment of some diseases such as Parkinson disease when the central dopamine level has to be increased. In our experiments, the decrease in dopamine level is concomittant with seizure genesis.

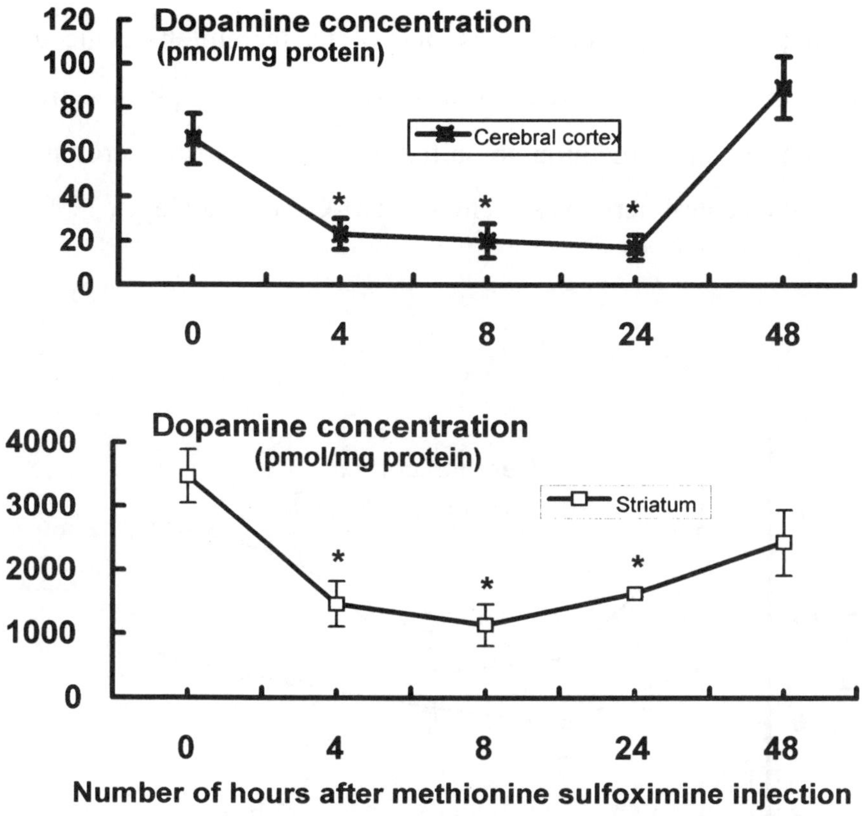

Figure 4 : Changes in the dopamine concentration of cerebral cortex and striatum of rats given methionine sulfoximine (100 mg/kg body weight) by intraperitoneal injection. Each point is the mean ± standard deviation of at least 5 independent determinations. Comparisons were made using Student's *t* test with $p<0.05$ and the asterisks indicate significant differences.

Dopamine is often involved in some inhitory pathways in the central nervous system principally in the striatum. It is possible that the impairment of such circuits provokes locomotor disorders that lead to convulsions. The selective destruction of some nuclei that utilize dopamine will perhaps explain more clearly the possible role of dopamine in epileptogenesis induced by methionine sulfoximine.

5) SEROTONIN

Serotonin (5-hydroxytryptamine) is one of the central biogenic amines involved in many fonctions. Predictively, it can play a role in epileptogenesis since methionine sulfoximine disturbs the serotoninergic system in the central nervous system (Ginefri-Gayet and Gayet, 1992 a and b).

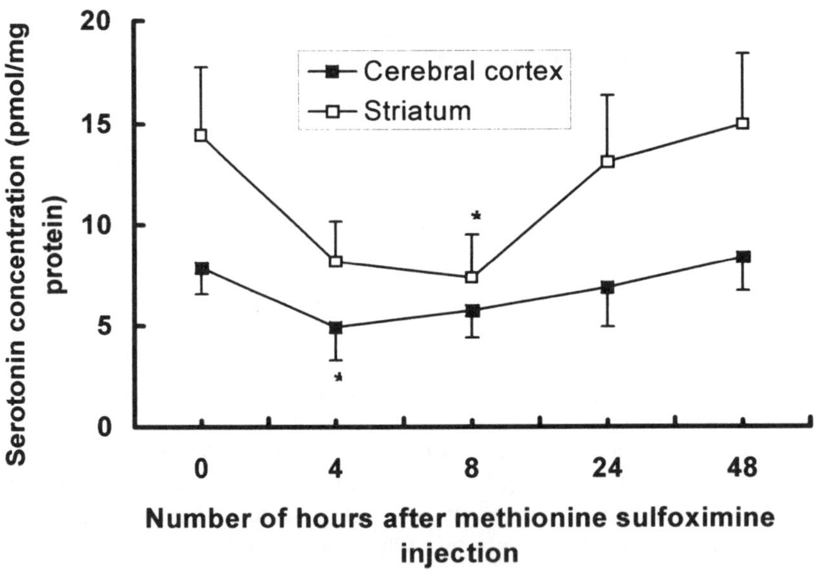

Figure 5 : Changes in the serotonin concentration of cerebral cortex and striatum of rats given methionine sulfoximine (100 mg/kg body weight) by intraperitoneal injection. Each point is the mean ± standard deviation of at least 5 independent determinations. Comparisons were made using Student's t test with $p<0.05$ and the asterisks indicate significant differences.

5.1) Serotonin measurement

The general methodology for neurotransmitter determination can be used for serotonin measurement. The particularity is only the mobile phase composition which was 0.1 M disodium phosphate, 0.1 M citric acid, and 8 % methanol. No ion-pairing agent is needed here. The flow rate was 1 ml/min and the polarisation applied to the detector electrodes was +0.85 volt. This method allows to measure the concentration of serotonin and of its precursor tryptophan in the same run. In control animals the serotonin concentration ranged from 7.9 ± 1.3 pmol/mg protein in the cerebral cortex to 19.6 ± 4.4 pmol/mg protein in the hypothalamus, apart from the cerebellum where the concentration was very low, within the limit of the detection range. Tryptophan concentration was far higher than that of serotonin since it was about 100 pmol/mg protein in all the areas including cerebellum.

After the administration of 100 mg/kg body weight of methionine sulfoximine, there was a deep and significant decrease in serotonin concentration in all the studied regions during the convulsive period. The recovery was complete 24 H after dosing (figure 5). Tryptophan concentration did not change any time. The strategy of the neurotransmitter precursor injection was also developed here, as in the case of catecholamine study.

For this, 5-hydroxytryptophan (100 mg/kg) together with benserazide (10 mg/kg) and methionine sulfoximine (100 mg/kg) were silmultaneously administered to the animals. Under these conditions, no change was observed in serotonin concentration as well as in tryptophan concentration. The convulsions were also inhibited.

Table 2.
Effect of the intraperitoneal injection of methionine sulfoximine (100 mg/kg body weight) on tryptophan concentration in the rat cerebral cortex and striatum. Each point is the mean ± standard deviation of at least 5 independent determinations. Comparisons were made using Student's *t* test with $p<0.05$. No difference was observed.

	Number of hours after methionine sulfoximine injection				
	0	4	8	24	48
Cerebral cortex (pmol/ mg protein)	104 ± 16	88 ± 21	83 ± 16	95 ± 20	98 ± 15
Striatum (pmol / mg protein)	107 ± 15	92 ± 22	98 ± 20	98 ± 23	107 ± 19

5.2) Possible involvement of serotonin in methionine sulfoximine epileptogenesis

The preceding results show a striking parallel between the decrease in serotonin level and the seizure onset. When this decrease was counteracted, the seizures were inhibited. It is tempting to conclude that the serotonergic system may play a role in the epileptogenesis induced by methionine sulfoximine (Hevor and Delorme, 1990). Now, our results do not allow to know how a decrease in serotonin concentration can lead to epileptogenesis. Future studies will help to gain insight into this question. In previous investigations, Blizard and Balkoski (1982) have shown that a restriction in tryptophan availability explained the decrease in serotonin level induced by methionine sulfoximine. Such a possibility seems improbable in our experiments since the tryptophan concentration remained unchanged all along the study.

6) GABA

During the present time, gamma-aminobutyric acid (GABA) is often involved in epileptogenesis (Fine *et al.,* 1990; Stringer and Lothman, 1991; Lallement *et al.,* 1993). Many antiepileptic drugs target the GABAergic system (Chapman *et al.*, 1982; Gonsalves *et al.,* 1989; Collins *et al.,* 1994). One of the first property of methionine sulfoximine which was well established is its powerful inhibition of glutamine synthetase (Lamar, 1968; Meiter, 1978). Since glutamate, the precursor of GABA, is involved in many brain pathways, it is possible that the inhibition of glutamine synthetase play a role in methionine sulfoximine epileptogenesis (Toussi *et al.,* 1987). The first explored possibility was a probable accumulation of the excitatory amino acid glutamate. Engelsen and Fonum (1985), Fonnum and Paulsen (1990) and Somers and Beckstead (1990) have shown that methionine sulfoximine does not increase glutamate level in rat *globus pallidus* and neostriatum. Another possibility is the impairment of GABA synthesis. This possibility is investigated in the following experiments.

6.1) GABA measurement

The HPLC system is the same as that described above in the general methodology. GABA is well separated by the column but, it cannot be directely detected by the electrochemical detector. So it was derivatized according to the method of Kehr and Ungerstedt (1988). The extraction of GABA from the frozen tissue was made using methanol instead of perchloric acid. The frozen tissue was sonicated in 5 volumes of 50 % (v/v) cold methanol. The homogenate was centrifuged at 30 000 g during 15 min at 4 °C. Because of the high concentration of GABA in brain tissue, the supernatant was diluted in 50 % methanol at the ratio of 1 : 200. The mixture for derivatization contained 0.1 M carbonate buffer pH 9.6, 20 mM orthophthaldialdehyde, and 0.45 % (v/v) t-buthylthiol. One hundred microliters of the derivatization solution were added to 10 µl of the diluted supernatant and the mixture was thoroughly shaken during 2 min. Then, 20 µl of the latter mixture were immediately injected into HPLC. The HPLC mobile phase contained 0.15 M sodium acetate buffer pH 5.4, 1 mM EDTA, and 50 % (v/v) acetonitrile. The flow rate was 0.8 ml/min and the electrode polarisation was set to 0.75 volt.

6.2) GABA level in the brain of rats submitted to methionine sulfoximine

The concentration of GABA in the controls was far higher than thoses of the other neurotransmitters described above. It ranges from 44 ± 12 nmol/mg protein in the cerebral cortex to 112 ± 18 nmol/mg protein in the hypothalamus. When methionine sulfoximine was administered to the animals, no change was observed in the GABA concentration in the brain regions apart from the hippocampus where the concentration decreased from 89 ± 19 nmol/mg protein in controls to 38 ± 7 nmol/mg protein in treated animals during the convulsive period (figure 6).

To know whether the decrease in hippocampus could account for seizure genesis, the minimal dose of methionine sulfoximine which is able to induce convulsions (80 mg/kg body weight) was injected. In this case, GABA concentration was unchanged in the encephalon although the animals displayed typical convulsions.

6.3) GABA does not play any role in methionine sulfoximine epileptogenesis

Many drugs that impair the GABAergic system are potent convulsants. It is the case of compounds such as bicuculline, an antagonist of GABA receptors (Meldrum and Horton, 1971) or mercaptopropionic acid, an inhibitor of glutamate decarboxylase, the enzyme that synthesizes GABA, (Lamar, 1970). Since methionine sulfoximine disturbs glutamate pathways by inhibiting glutamine synthetase, it was expected that the convulsions induced by this convulsant would be mediated by GABA metabolism impairment (Janjua *et al.*, 1991).

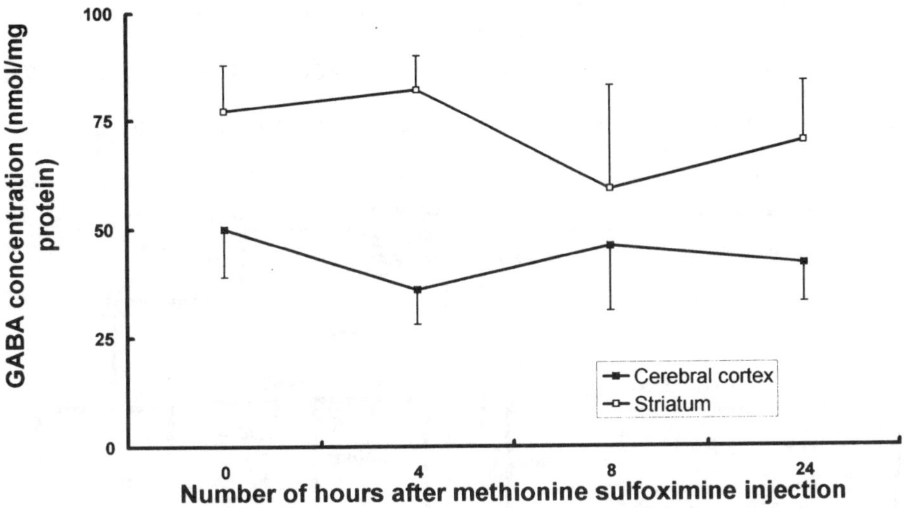

Figure 6 : Effect of the intraperitoneal injection of methionine sulfoximine (100 mg/kg body weight) on GABA concentration in the rat cerebral cortex and striatum. Each point is the mean ± standard deviation of at least 5 independent determinations. Comparisons were made using Student's *t* test with $p<0.05$. No difference was observed.

The preceding results show that methionine sulfoximine did not alter GABA concentration. The possibility of an increased turnover without change in actual GABA concentration seems unlikely as shown by the dose studies performed by Hevor (1996). So, it can be concluded that GABA probably is not involved in epileptogenesis induced by methionine sulfoximine. This lack of involvement has been observed in other models of epilepsy (Maggio *et al.*, 1991).

7) POSSIBLE MECHANISM OF METHIONINE SULFOXIMINE EPILEPTOGENESIS

The study on the concentrations of several neurotransmitters in the encephalon of rats submitted to the convulsant methionine sulfoximine can be summarized as follows. The convulsion onset is parallel to an increase in acetylcholine content, a decrease in dopamine and serotonin contents, and no important change in norepinephrine and GABA contents. The experiments that were performed allow to consider that norepinephrine and GABA may not be involved in methionine sulfoximine epileptogenesis.

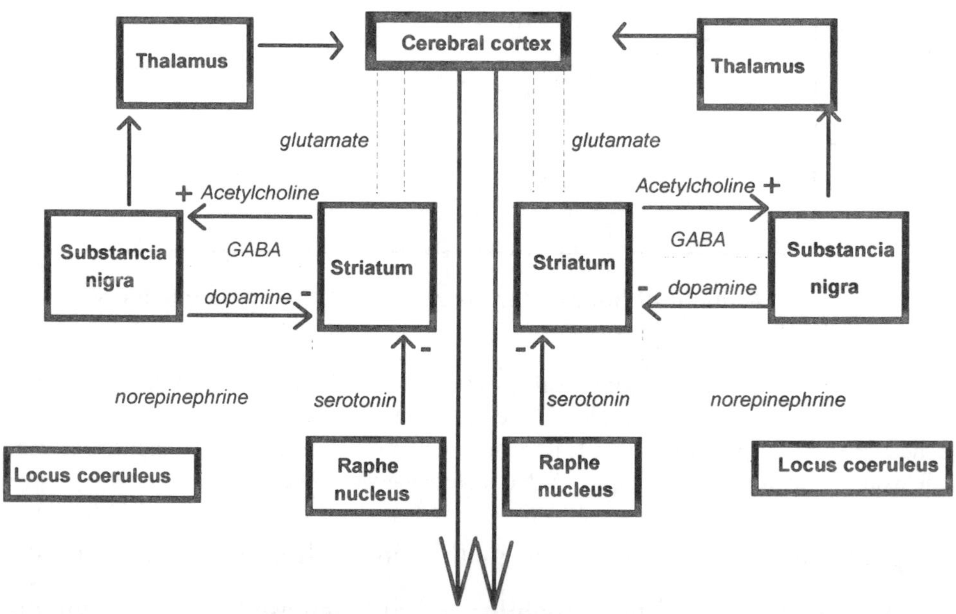

Figure 7 : A hypothetical diagram of methionine sulfoximine-induced impairments in neuronal pathways of the rat brain. The convulsant decreases the contents in dopamine and serotonin of fibers from respectively *locus coeruleus* and *raphe nuclei* to striatum. Conversely, the convulsant does not modify glutamate content and thus, the cortical stimulation of striatum via glutamatergic fibers does not change. Consequently, the regulatory role of striatum on locomotor actvity will be imbalanced and result in motor disorder. Moreover, the impairment of striatum will imbalance the inputs that the latter sends to *substancia nigra*. These inputs are not probably cholinergic or GABAergic fibers, as shown in our results. Changes in *substancia nigra* will modify some cortical inputs, via thalamus. Then, cerebral cortex may change some of its output states via glutamatergic pathways and induce convulsions. (Dot lines, unchanged pathways).

Now, what can be the meaning of changes observed in the concentration of the other neurotransmitters ? The ratio of intracellular neurotransmitter versus released neurotransmitter is very high and it can be estimated to be 1000 : 1. On this ground, it can be considered that the concentrations measured in the whole tissue essentially reflects the concentration of intracellular neurotransmitter and this is the case in our experiments. Nowadays, the actual effect of methionine sulfoximine on the neurotransmitter synthesizing enzymes such as choline acetyltransferase, tyrosine hydroxylase, and tryptophan hydroxylase is not yet known. It is also the case of the degrading enzymes, i.e., acetylcholinesterase, monoamine oxidases, and catechol-O-methyltransferase. Assuming that methionine sulfoximine does not disturb the activities of the preceding enzymes, its targets would be limited to the existing neurotransmitters. On the ground of this supposition, an increase in intracellular acetylcholine content would reflect a reduced liberation of the neurotransmitter. So, in the cholinergic system where acetylcholine concentration increased, the postsynaptic neurons would have an hypoactivity. In the dopaminergic system and serotonergic system where the corresponding neurotransmitters decreased, the postsynaptic neurons would have an hyperactivity if their receptors are not altered. Whatever the validity of the preceding hypotheses, methionine sufoximine impairs cholinergic, dopaminergic, and serotonergic neurotransmissions. Inside the encephalon, striatum is the region which is particularly concerned with the methionine sulfoximine action. There is a high probability that the alterations of the striatal neurotransmission pathways impairs the regulation of locomotor activity (Turski *et al.,* 1989; Richard and Hevor, 1995). The cortical glutamatergic neurons send motor informations to the striatum which is an important center for locomotor activity regulation. The disturbances of such a center may lead to seizure genesis (figure 7). The participation of cholinergic system to the motor disorder is not obvious in the case of methionine sulfoximine, according to our results. This is not the case of other convulsants such as soman or pilocarpine which induce convulsions via cholinergic system (Lallement *et al.,* 1992, Millan *et al.,* 1993). Many points remain unclear in the actual mechanism of methionine sulfoximine epileptogenesis. Future experiments will clarify the many possible hypotheses.

Acknowledgement

This work was supported by grants from CNRS (UPR 9074), from the Ministère de l'Education Nationale, from the Mairie d'Orléans, and from Région Centre. Many thanks to M. P. Moreau for his technical help.

REFERENCES

Baptista, T., Weiss, S.R.B., Zocchi, A., Sitcoske, M., and Post R. (1994) Electrical kindling is associated with increases in amygdala acetylcholine levels : an *in vivo* microdialysis study. *Neurosci. Lett.,* 167:133-136.

Blizard, D.A. and Balkoski, V. (1982) Tryptophan availability, ventral serotonergic function and methionine sulphoximine-induced convulsions. *Neuropharmacol.*, 21:27-30.

Bradford, M.M. (1976) A rapid and sensitive method for the quantitation of microgram quantities of protein utilizing the principle of protein-dye binding. *Anal. Biochem.,* 72:248-254.

Chapman, A.G., Riley, K., Evans, M.C., and Meldrum, B.S. (1982). Acute effects of sodium valproate and γ-vinyl GABA on regional amino acid metabolism in the rat brain. *Neurochem Res.* 7:1089-1105.

Collins, R.M.Jr., Zielke, H.R., and Woody, R.C. (1994). Valproate increases glutaminase and decreases glutamine synthetase activities in primary cultures of rat brain astrocytes. *J. Neurochem.* 62:1137-1143.

Damsma G., Westerink, B.H.C. and Horn, A.S. A simple, sensitive and economic assay for choline and acetylcholine using HPLC, an enzyme reactor, and an elctrochemical detector (1985). *J. Neurochem.,* 45:1649-1652.

Engel, J., Brown, W.J., Kuhl, D.E., Phelps, M.E., Mazziota, J.C., and Crandal P.H. (1982) Pathological findings underlying focal temporal lobe hypometabolism in partial epilepsy. *Ann. Neurol.*, 12:518-528.

Engelsen, B. and Fonnum, F. (1985). The effect of methionine sulfoximine, an inhibitor of glutamine synthetase, on the levels of amino acids in intact an decorticated rat *neostriatum. Brain Res.* 338:165-168.

Fine, A., Meldrum, B.S., and Patel, S. (1990). Modulation of experimentally induced epilepsy by intracerebral grafts of fetal GABAergic neurons. *Neuropsychologia* 28:627-632.

Fonnum, F. and Paulsen, R.E. (1990). Comparison of transmitter amino acid levels in rat *globus pallidus* and *neostriatum* during hypoglycemia or after treatment with methionine sulfoximine or gamma-vinyl gamma-aminobutyric acid. *J. Neurochem.* 54:1253-1257.

Ginefri-Gayet M. and Gayet J. (1992) Possible link between brain serotonin metabolism and methionine sulfoximine-induced hypothermia and associated behavior in the rat. *Pharmacol. Biochem. Behav.*, 43: 173-179.

Ginefri-Gayet, M. and Gayet, J. (1992) Involvement of serotonin receptors in methionine sulfoximine-induced hypothermia in the rat. *Eur. J. Pharmacol.*, 217:85-91.

Glowinski, J. and Iversen, L.L. (1966). Regional studies of catecholamines in the rat brain. I - The disposition of [^{3}H]norepinephrine, [^{3}H]dopamine, and [^{3}H]dopa in various regions of the brain. *J. Neurochem.* 13:665-669.

Gonsalves, S.F., Twitchell, B., Harbaugh, R.E., Krogsgaard-Larsen, P., and Schousboe, A. (1989). Anticonvulsant activity of the glial GABA uptake inhibitor, THAO, in chemical seizures. Eur. J. *Pharmacol.* 168:265-268.

Hevor T.K. (1996) Methionine sulfoximine has no major effect on gamma-aminobutyric acid concentration in the rat brain. *Biog. Amines, 12:445-461.*

Hevor, T.K., Aissi, E. and Delorme, P. (1990) Correlation between carbohydrate and catecholamine level impairments in methionine sulfoximine epileptogenetic rat brain. *Neurochem. Res.*, 15:861-868.

Hevor T.K. and Delorme P. (1990) Possible involvement of indolamines in the glycogenic effect of the convulsant methionine sulfoximine in rat brain. *Experientia,* 46:710-713.

Hevor, T.K., Delorme, P., and Beauvillain, J.C. (1986) Glycogen synthesis and immunocytochemical study of fructose-1,6-diphosphatase in methionine sulfoximine epileptogenic rodent brain. *J. Cereb. Blood Flow Metab.*, 6:292-297.

Hevor, T.K. and Gayet, J. (1979). Cyclic nucleotides in the brain of mice and rats submitted to the convulsant methionine sulfoximine. *Biochem. Pharmacol.* 28:3507-3512.

Janjua, N.A., Mori, A., and Hiramatsu, M. (1991). Gamma-aminobutyric acid uptake is decreased in the hippocampus in a genetic model of human temporal lobe epilepsy. *Epilepsy Res.* 8:71-75.

Jobe C. and Laird H.E. (1987) Neurotransmitters and epilepsy, Humana Press, N.Y, pp400.

Jope, R.S., Morrisett, R.A., and Carter-Snead, O. (1986) Characterization of lithium potentiation of pilocarpine-induced status epilepticus in rats. *Exp. Neurol.*, 91:471-480.

Kehr, J. and Ungerstedt, U. (1988). Fast HPLC estimation of gamma-aminobutyric acid in microdialysis perfusates : effect of nipecotic and mercaptopropionic acids. *J. Neurochem.* 51:1308-1310.

Lallement G., Carpentier P., Collet A., Baubichon D., Pernot-Marino I. and Blanchet G. (1992). Extracellular acetylcholine changes in rat limbic structures during soman-induced seizures. *Neurotoxicology.* 13:557-568.

Lallement, G., Carpentier, P., Pernot-Marino, I., Baubichon, D., Collet, A., and Blanchet, G. (1993). The transient impairment of the gabaergic function during initiation of soman-induced seizures. *Brain Res.*, 629:239-244.

Lallement, G., Denoyer, M., Collet A., Pernot-Marino, I., Baubichon, D., Monmaur, P., and Blanchet G. (1992) Changes in hippocampal acetylcholine and glutamate extracellular levels during soman-induced seizures : influence of septal cholinoceptive cells. *Neurosci. Lett.*, 139:104-107.

Lamar C. Jr (1968) The duration of the inhibition of glutamine synthetase by methionine sulfoximine. *Biochem. Pharmacol.*, 17:636-640.

Lamar, C. Jr (1970). Mercaptopropionic acid : a convulsant that inhibits glutamate decarboxylase. *J. Neurochem.* 17:165-170.

Lowry, O.H., Rosebrough, N.J., Farr, A.L., and Randall, R.J. (1951). Protein measurement with Folin phenol reagent. *J. Biol. Chem.*, 193:265-275.

Maggio, R., Sohn, E., and Gale, K. (1991). Lack of proconvulsant action of GABA depletion in *substantia nigra* in several seizures models. *Brain Res.* 547:1-7.

Mefford, I.N. (1981) Application of high performance liquid chromatography with electrochemical detection to neurochemical analysis : measurement of catecholamines, serotonin and metabolites in rat brain *J. Neurosci. Methods*, 3:207-224.

Meister, A. (1978). Inhibition of glutamine synthetase and gamma-glutamylcysteine synthetase by methionine sulfoximine and related compounds. *In:* Enzyme activated irreversible inhibitors, N. Seiller, M.J. Jung, and J. Kosh-Weser, (Eds.). Elsevier, North Holland, pp 187-210.

Meldrum, B.S. and Horton, R.W. (1971). Convulsive effects of 4-deoxypyridoxine and bicuculline in photosensitive baboon (Papio-papio) and rhesus monkey (Macaca mulatta). *Brain Res.* 35:419-436.

Millan, M., Chapman, A.G. and Meldrum, B.S. (1993) Extracellular amino acid levels in hippocampus during pilocarpine-induced seizures. *Epilepsy Res.*, 14:139-148.

Miller, G.L. (1959) Protein determination for large numbers of samples. *Analyt. Chem.*, 31:964.

Potter, P.E., Meek, J.L., and Neff, N.H. (1983). Acetylcholine and choline in neuronal tissue measured by HPLC with electrochemical detection. *J. Neurochem.,* 41:188-194.

Richard, O. and Hevor, T.K. (1995). Methionine sulfoximine increases acetylcholine level in the rat brain: no relation with epileptogenesis. *NeuroReport* 6:2027-2032.

Robinson, S.E. and Hambrecht, K.L. (1986) Effect of intrastriatal injection of soman on acetylcholine, dopamine and serotonin metabolism. *Life Sci.*, 42:2331-2339.

Somers, D.L. and Beckstead, R. (1990). Chronic methionine sulfoximine administration reduces synaptosomal aspartate and glutamate in rat striatum. *Neurochem Lett.* 115:335-340.

Stringer, J.L. and Lothman, E.W. (1991). Pharmacological evidence indicating a role of GABAergic systems in termination of limbic seizures. *Epilepsy Res.* 7:197-205.

Suñol, C., Tussel, J.M., Gelpi, E., and Rodriguez-Farre, E. (1988). Regional concentrations of GABA, serotonin and noradrenaline in brain at onset of seizures induced by lindane (gamma-hexachlorocyclohexane). *Neuropharmacol.* 27:677-681.

Toussi, H.R., Schatz, R.A., and Waszczak, B.L. (1987). Suppression of methionine sulfoximine seizures by intranigral gamma-vinyl GABA injection. *Eur. J. Pharmacol.* 137:261-264.

Turski, L., Ikonomidou, C., Turski, W.A., Bortolotto, Z.A., and Cavalheiro E.A. (1989) Cholinergic mechanisms and epileptogenesis. The seizures induced by pilocarpine : a novel experimental model of intractable epilepsy. *Synapse*, 3:154-171.

Wolfe, L.S. and Elliot, K.A.C. (1962). Chemical studies in relation to convulsive conditions. *In:* Neurochemistry, K.A.C. Elliot, I.H. Page, and J.H. Quastel, (Eds.). Thomas, Springfield, pp 694-727.

Progress in HPLC-HPCE, Vol. 7, pp. 413-460
G. Ali Qureshi *et al.* (Eds)

ANALYSIS OF SALSOLINOLS, ENDOGENOUS NEUROTOXINS, IN HUMAN MATERIALS

MAKOTO NAOI[1], WAKAKO MARUYAMA[2], KAZUO MATSUBARA[3], KEATH TIPTON[4], MARGHERITA STROLIN BENEDETTI[5], and HASAN PARVEZ[6]

[1]*Department of Biosciences, Nagoya Institute of Technology, Nagoya 466, Japan,*
[2]*Laboratory of Biochemistry and Metabolism, Department of Basic Gerontology, National Institute for Longevity Sciences, Obu 474, Aichi, Japan,*
[3]*Department of Pharmacy, Asahikawa Medical College, Asahikawa 078, Japan,*
[4]*Department of Biochemistry, Trinity College, Dublin, Ireland,*
[5]*Preclinical Development and Human Pharmacology, Zambon Group, Milan, Italy,*
[6]*CNRS, Reim, France*

I. INTRODUCTION

The pathogenesis of major neurodegenerative diseases, such as Parkinson's disease (PD), Alzheimer's and Huntington's disease, has been intensively studied and in the familiar cases responsible genes have been partially established. However, in the so-called sporadic cases the endogenous and environmental factors are considered to interfere each other to induce the pathological process. PD is now one of the most common neurodegenerative disorders in the aged and the characteristic pathological finding is the selective cell death of dopamine neurons in the pars compacta of the substantia nigra. After the discovery of 1-methyl-4-phenyl-1,2,3,6-tetrahydropyridine (MPTP), which induces parkinsonism in human (Davis *et al.*, 1979; Langston *et al.*, 1983), endogenous or xenobiotic neurotoxins have been proposed to be involved in the pathogenesis of PD. Three groups of endogenous alkaloids are now proposed as candidates of neurotoxins; dopamine-derived salsolinol, 1-methyl-6,7-dihydroxy-1,2,3,4-tetrahydroisoquinoline (Naoi *et al.*, 1997a), β-phenethylamine-derived 1,2,3,4-tetrahydroisoquinolines (Nagatsu and Yoshida, 1988) and indoleamine-derived β-carbolines (Collins *et al.*, 1992). 1,2,3,4-Tetrahydroisoquinoline (TIQ) was reported to induce parkinsonism in monkey, but the selective cell death of dopamine neurons in the nigrostriatal system has never been confirmed with this type of isoquinolines without catechol structure.

Salsolinol, 1-methyl-6,7-dihydroxy-1,2,3,4-tetrahydroisoquinoline, is an alkaloid synthesized from dopamine and acetaldehyde. The origin of its synthesis suggests that it may be implicated in the mechanism underlying the additive drinking of alcoholism (Cohen and Collins, 1970; Hamilton and Hirst, 1980: Clow *et al.*, 1985; Faraj *et al.*, 1989). Salsolinol and tetrahydropapaveroline [THP, 1-(3,4-dihydroxyphenyl)-methyl-6,7-dihydro-1,2,3,4-tetrahydroisoquinoline], were shown to induce alcohol preference in the rats (Meyers, 1989), but the following results were inconsistent, and it has never been clarified whether it can elicit withdrawal-like symptoms (Duncan and Deitrich, 1980).

However, recently salsolinol derivatives are gathering attention as neurotoxin candidates to induce PD in humans. Out of salsolinol derivatives, *N*-methyl(*R*)salsolinol was proved to be a most probable candidate of dopaminergic neurotoxins. By injection of *N*-methyl(*R*)salsolinol in rat striatum, an animal model of PD was prepared and the biochemical, histopathological and behavioral changes were quite similar to those observed in parkinsonian patients (Naoi *et al.*, 1996a). In the substantia nigra dopamine neurons were found to be depleted selectively after one week continuous infusion of *N*-methyl(*R*)salsolinol in the striatum. The cell death was proved to be apoptotic using dopaminergic neuroblastoma SH-SY5Y cells (Maruyama *et al.*, 1997b). The neurotoxicity was found to be enantio-selective and only the (*R*)enantiomer of *N*-methylsalsolinol induced parkinsonism in rats and apoptosis in SH-SY5Y cells.

Salsolinol was detected in human urine for the first time by gas chromatography (GC) (Sandler *et al.*, 1973), and then in cerebrospinal fluid and brain tissue (Collins *et al.*, 1982; Sjöquist *et al.*, 1981a and b). Salsolinol had been considered to be synthesized by the non-enzymatic Pictet-Spengler reaction as a racemic form from dopamine and acetaldehyde (Robbins, 1968). However, it has an asymmetric carbon at C1 position and predominant occurrence of the (*R*)-enantiomer in human urine suggests that (*R*)salsolinol may be synthesized enzymatically (Dostert *et al.*, 1987a and b; Strolin Benedetti *et al.*, 1989a). The proposed biosynthesis pathway of (*R*)salsolinol and *N*-methyl(*R*)salsolinol is shown in Figure 1.

(*R*)Salsolinol was considered to be synthesized through (*R*)salsolinol 1-carboxylic acid, followed by decarboxylation into 1,2-dehydrosalsolinol and subsequent reduction to (*R*)salsolinol (Brossi 1982; Dostert *et al.*, 1990a). Salsolinol-1-carboxylic acid was confirmed to occur in human brain and urine (Sjöquist and Ljungquist, 1985) and 1,2-dehydrosalsolinol in human urine (Dostert *et al.*, 1990a). However, in human brains we could not detect the enzyme(s) catalyzing the enzymatic reduction of 1,2-dehydrosalsolinol (Reaction 4) or direct oxidative decarboxylation of salsolinol-1-carboxylic acid (Reaction 5) into (*R*)salsolinol. Recently, we purified a novel (*R*)salsolinol synthase from human brain (Naoi *et al.*, 1996b). This enzyme condenses dopamine with acetaldehyde into (*R*)salsolinol. At present the biosynthesis by Reaction 1 is the only confirmed pathway of (*R*)salsolinol in the human brain. Neither *N*-methyldopamine (epinine), adrenaline nor noradrenaline are the substrat, suggesting that *N*-methyl(R)saloslinol is not synthesized directly from *N*-methyldopamine. The *N*-methylation of (*R*)salsolinol into *N*-methyl(*R*)salsolinol was proved in rat striatum by *in vivo* microdialysis method (Maruyama *et al.*, 1992). The oxidation of *N*-methyl(*R*)salsolinol into 1,2-dimethyl-6,7-dihydroxyisoquinolinium ion (DMDHIQ+), a compound with chemical structure similar to an oxidation product of MPTP, 1-methyl-4-phenylpyridinium ion (MPP+), is catalyzed by an oxidase sensitive to semicarbazide (Naoi *et al.*, 1995a) (Reaction 7).

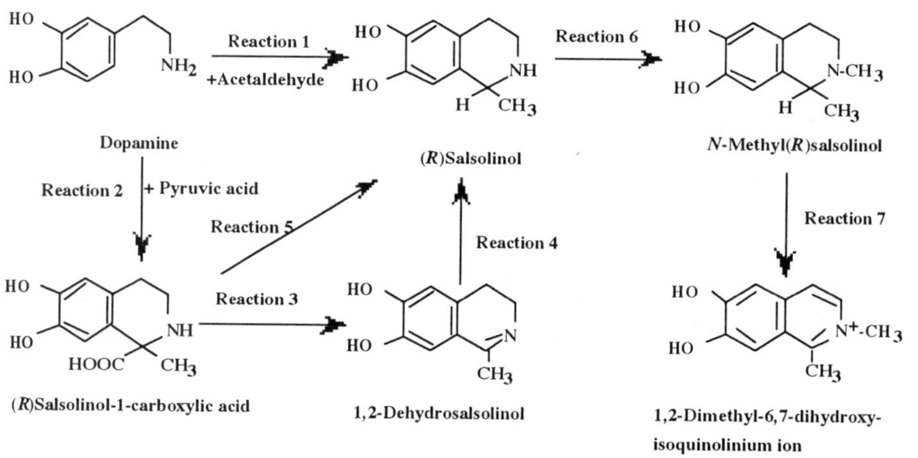

Figure 1. Biosynthesis pathway of (*R*)salsolinol and the metabolism into *N*-methylsalsolinol and 1,2-dimethyl-6,7-dihydroxyisoquinolinium ion. Reaction 1 and 2 are catalyzed by a (*R*)salsolinol synthase, Reaction 6 by a (*R*)salsolinol *N*-methyl-transferase and Reaction 7 by a *N*-methyl(*R*)salsolinol oxidase or by auto-oxidation. Reaction 3 is non-enzymatic and Reaction 4 and 5 cannot be confirmed in human brain.

These results suggest that the analysis of salsolinol derivatives in human materials, such as the brain tissue, cerebrospinal fluid (CSF) and intraventricular fluid (IVF), is essential for the study of their possible involvement to the pathogenesis of neurodegenerative diseases. This review summarizes the analytical methods to detect salsolinol derivatives by gas chromatography-mass spectrometry (GC-MS) and high performance liquid chromatography (HPLC)-electrochemical detection (ECD) or fluorometric detection (FD). The chirality of the catechol isoquinolines was confirmed to account for the neurotoxicity and other biological activity of the isoquinolines, and recent advantages to measure the enantiomers of salsolinol derivatives in the human samples are also reviewed. The principle of chiral separation (Deng *et al.*, 1997a) and application to clinical samples (Maruyama *et al.*, 1997c) was reviewed in the Volume 6 of this Progress in HPLC-HPCE. The methods to measure the activity of the enzymes related to the metabolism of *N*-methyl(*R*)salsolinol are also presented.

II. GC/MS ANALYSIS OF SALSOLINOLS

For the precise identification of novel compounds, gas chromatography/mass spectrometry (GC/MS)is still the *sine qua non* of the methodology. Selected ion monitoring (SIM) is a quantitative technique with GC/MS, by which molecular and/or fragment ions specific to the target compound are recorded. The first application of GC/MS for the analysis of salsolinol derivatives in mammalian tissues was reported by Sjöquist and Magnuson (1980), although their results appear to be different from the recent data. The recent accurate and sensitive analytical methods for salsolinol derivatives are ascribed to great improvement of GC/MS instruments and capillary columns.

II-1 Materials and methods

Sample preparation

Active alumina (Sjöquist *et al.*, 1981a; Matsubara *et al.*, 1987; Collins *et al.*, 1990), a phenylboronic acid (PBA) cartridge (Pianezzola *et al.*, 1989; Allievi *et al.*, 1991; Haber *et al.*, 1996) and a C_{18} cartridge (Collins *et al.*, 1990; Musshoff *et al.*, 1993) are commonly used for the extraction of salsolinol from biological samples for the CC/MS analysis. In general, the sample and the buffer used in the extraction procedure should be treated with an aldehyde trapping reagent, semicarbazide hydrochloride, to prevent artifact formation of salsolinol during sample preparation. Active alumina and PBA methods are briefly described as follows:

Alumina method. For extraction of salsolinol from biological samples, the most conventional cleanup procedure is the activated alumina method, which has been originally developed for catecholamine analysis. Salsolinol is adsorbed on the alumina at pH 8.0 - 8.5, followed by washing with 0.01 M $NaHSO_3$. Then it is eluted under an acidic condition from the alumina. Caution should be taken to avoid the extensive interaction with alumina, which leads the oxidation of catechol hydroxyl groups.

PBA method. The sample is buffered at pH 8.5 - 9.0, and then passed through a PBA cartridge, which has been pre-washed with the buffer, under low-pressure aspiration. After washing twice with 1 ml water and 2 ml methanol, the adsorbed salsolinol is eluted form the cartridge with 1 ml of 5 M HCOOH (or 0.1 M HCl)/methanol. However, as the solvent acetonitrile would be used instead of methanol, because the latter contains aldehydes which might produce salsolinol as an artifact during the extraction procedure.

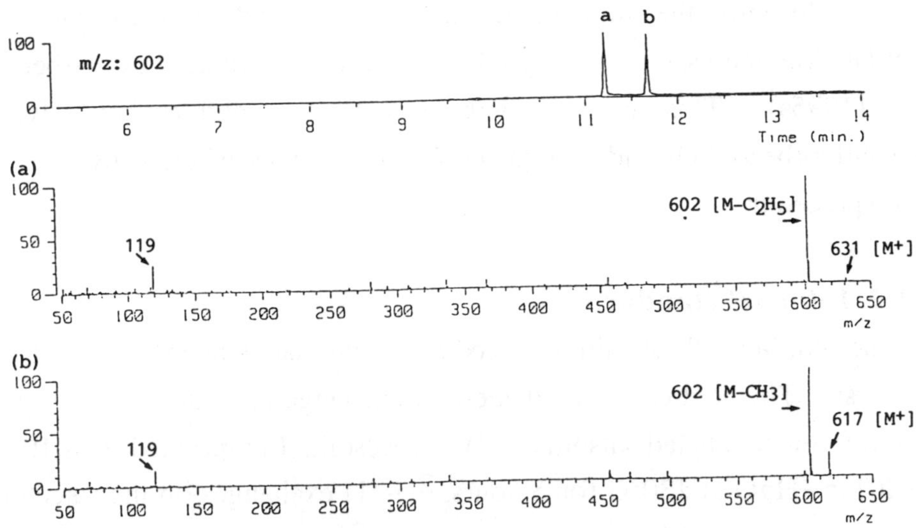

Figure 2. GC/MS analysis of PFP derivatives of salsolinol. Selective-ion chromatogram at m/g 60z (upper), and mass spectra of salsolinol (a) and 1-ethyl-6,7-dihydroxy-1,2,3,4,-tetrahydroisoquinoline (b).

In the GC/MS analysis, salsolinol should be derivatized to its volatile derivative. After the cleanup step, the sample is dried by freeze-drying or under nitrogen gas. Then salsolinol is derivatized with an adequate derivatizing reagent prior to the injection into an GC/MS apparatus. Salsolinol is well reacted with acylation reagents, that is, trifluoroacetic anhydride (Ung-Chhun *et al.*, 1985), pentafluoropropionic anhydride (PFPA) (Sjöquist and Magnuson, 1980; Matsubara *et al.*, 1985; Allievi *et al.*, 1991; Musshoff *et al.*, 1993) and heptafluorobutyric acid (Origitano *et al*, 1981). Among these acylation regents, PFPA seems to be the most suitable because of the mass range and chemical stability of the derivative. The reaction with salsolinol and PFPA is easily achieved by 30-minute heating at 60°C, and the derivative can be washed with toluene and water (Matsubara *et al.*, 1985). The mass spectrum of this derivative is shown in Figure 2.

Recently, the simultaneous derivatization and extraction method for GC/MS analysis for urinary salsolinol has been reported by Harber *et al* (1995a). This method combines extraction and acylation with pentafluorbezoyl chloride for catechol isoquinolines in urine by a one-step procedure.

II-2 Internal standard

Deuterium labeled salsolinol is used as an internal standard for GC/MS analysis. The acid-catalyzed deuterium exchange procedure can be utilized to make labeled salsolinol. When salsolinol is refluxed with 10% ^2HCl in ^2H_2O at 130°C for 5 hours, ^2H-^1H exchanges on the catechol ring at 5 and 8 positions, occur to yield 5,8-$^2H_2$-salsolinol (Matsubara *et al.*, 1985). From 2,5,6-$^2H_3$-dopamine with acetaldehyde, 5,8-$^2H_2$-salsolinol also can be easily synthesized by the Pictet-Spengler reaction (Sjöquist and Magnuson, 1980; Haber *et al.*, 1995b). Deuterium on catechol structure is slowly re-exchanged with hydrogen in the aqueous solution. The reaction of deuterated acetaldehyde ($C^2H_3C^2HO$) with dopamine (Allievi *et al.*, 1991) followed by acid catalyzed exchange gives 1-C^2H_3-1,5,8-$^2H_3$-salsolinol (Matsubara *et al.*, 1985). However, the methyl group at position 1 is cleaved in the electron ionization chamber (Fig. 2). Researchers should be aware that commercially available deuterated compounds, such as deuterated dopamine and acetaldehyde, contain trace amount of non-deuterated form. In general, the 3 to 5 mass shifts in selected ions of the internal standard from those of the target compound are suitable for accurate GC/MS assay.

The compounds chemically and structurally similar to salsolinol, such as 1-ethyl-6,7-dihydroxytetrahydroisoquinoline (the synthesis method is described in Matsubara *et al.*, 1994), are also a suitable internal standard (Fig. 2). The use of these compounds having structure similar to salsolinol allows contaminant-free analysis.

For further precise quantitation, an addition of deuterated dopamine to the sample prior to the cleanup step makes possible to ascertain any artifactual formation of salsolinol during sample preparation (Collins *et al.*, 1990).

II-3 Enantiomer analysis

For the diasteroisomers analysis, two technically different methods can be utilized for CC/MS analysis. One is based on the derivatization with a chiral derivatizing reagent. Salsolinol is derivatized with *N*-methyl-*N*-trimethyl-silyltrifluoracetamide, which reacts only on hydroxyl groups, for 30 min at 70°C, then further reacted with the chiral reagent, (*R*)-(-)-2-phenylbutyryl chloride, for 1 min (Haber *et al.*, 1995c). After removal of excess reagents, the (*R*)- and (*S*)-salsolinol can be separated on a conventional column, such as BGM-silaren capillary column. However, the chiral derivatizing reagent is contaminated small amount of the opposite diasteroisomer. Thus researchers should carefully evaluate the ratio of enantiomers in the sample. The other method is based on recently developed chiral active capillary columns, such as a Chirasil-Dex CB column, which are easily applied for diasteroisomer analysis without use of any specific reagents.

II-4 Sensitivity

The sensitivity for salsolinol determination is primary influenced by the utilized GC/MS system. Following 4 systems can be employed for the quantitative analysis of salsolinol and its analogs.

Electron ionization (EI)/SIM. The most established and utilized means of ionization is EI, where sample is volatilized into the gas phase and passed through a beam of energetic electrons. The multiple ion detection for salsolinol increases the accuracy of the assay. For tissue samples, the lower detection limit for salsolinol is reported to be ca. 0.5 - 2 pmol/sample (Matsubara *et al.*, 1987; Collins *et al.*, 1990). The sensitivity also depends on the clean-up method and the instrument employed. Harber *et al.* (1995a) reported that the detection limits for salsolinol is 10 fmol/ml of urine using one step extraction and derivatization method for sample preparation described above.

Negative chemical ionization (NCI)/SIM. NCI/SIM is a very sensitive technique for halogen-containing compounds, but is rather less informative than EI mode because the NCI spectrum is too simple. The reported lower detection limit for the PFPA derivative, 0.55 pmol/ml of urine (Allievi *et al.*, 1991), has not greatly exceeded that reported by EI/SIM mode.

High resolution SIM. Measurement of exact (mill) mass of selected ions leads to allow simultaneous identification, when the full mass spectrum cannot be obtained due to the low concentration. This method cannot be undertaken by a quadrupole-type mass detector. Salsolinol analysis by this mode has not been reported.

Tandem SIM (MS/MS). MS/MS can analyze secondary (daughter) ions, so that it extremely improves the resolution and sensitivity. This system for salsolinol analysis has not been reported, however, the detection limits for tetrahydroisoquinoline derivatives without a catechol structure have been reported to be ca. 25 fmol (Kajita *et al.*, 1995).

Table 1
Salsolinol distribution in tissues and body fluid of alcoholic and nonalcoholic humans.

Tissue		Amount	*Reference*
Urine		0.8 - 17.3 (pmol/mg creatinine)	Kajita *et al.*, 1995
		1.0 - 6.1 (pmol/ml)	Allievi *et al.*, 1991
Plasma			
	control	93 (0 - 232)* (pg/ml)	Faraj *et al.*, 1989
	alcoholic	497 (50 - 1331)*	
CSF			
		11 ± 6 (pmol/ml)	Sjöquist *et al.*, 1981
	control	3.1 ± 0.8	Dordain *et al.*, 1984
	parkinsonian	2.2 ± 0.4	
	alcoholic	6.1 ± 3.4	
		2.3 ± 2.5*	Faraj *et al.*, 1991
Hypothalamus			
		73 ± 61 (pmol/g)	Sjöquist *et al.*, 1982b
Cortex			
	frontalis	83 ± 63 (pmol/g)	Sjöquist *et al.*, 1982b
	gyrus precentralis	14 ± 5	

* Data by radio-immuno assay. Other values were obtained by GC/MS method.

II-5 Distribution of salsolinol in human tissues detected by GC/MS

Originally, pharmacological interest in salsolinol, as well as related biogenic alkaloid condensations, in mammals came from their postulated roles in the development of alcoholism. Thus earlier works on salsolinol in mammals mainly concerned its formation and level after alcohol intake. The reported salsolinol concentrations in the brain and body fluid from alcoholic and nonalcoholic humans are summarized in Table 1. The quantitative data reported by Sjöquist *et al.* (1981a and 1982) are extremely higher than those reported others.

III DETECTION OF SALSOLINOL DERIVATIVES BY HPLC-ECD

High performance liquid chromatography (HPLC) is a convenient method for analysis of salsolinol derivatives in human materials, and electrochemical detection (ECD) is commonly used because of the high electrochemical reactivity of the catechol structure of salsolinols. Fluorometric detection (FD) is also used for the oxidized isoquinolinium ion. The sensitivity of HPLC-ECD or FD is quite higher than GC/MS, and the amounts of the sample required for the analysis can be dramatically reduced. Especially by use of multi-electrode detection system [Coulochem Electrode Array System (CEAS), ESA, Chelmsford, MA, USA], less than 100 μl of human CSF is enough for analysis of salsolinols, catecholamines and their precursor amino acids (Maruyama *et al.*, 1997c). This section will describe the recent advances in the HPLC analyses of human samples.

Salsolinol derivatives in serum and urine should be purified and concentrated by use of a bonded column or activated alumina, as in the case of GC/MS. A phenylboric acid (PBA) column (Varian, Harbo City, CA, USA) is not adequate for analysis of *N*-methylated salsolinol, because charged isoquinolines are scarcely held on the column. The combined column system with a primary-secondary amine column (PSA) and a PBA column connected in a series is commonly used for *N*-methylated salsolinols (Sällström Baum and Rommelspacher, 1994). The cartridges are conditioned prior to application of samples; washing first only the PSA column with 1 ml methanol and 1 ml of 0.1 M HCl, then both the columns are connected and washed with 1 ml of methanol, 2.5 ml of 0.3% ammonium hydroxide and 5 ml of 5 mM phosphate buffer, pH 8.5. Then, the sample buffered around pH 5.0 is loaded on the top of the PSA column and passed through the cartridges. After 4 washings with 2.5 ml of 5 mM phosphate buffer, pH 8.5 and subsequently with 1 ml methanol, the PBA is removed from the PSA column and salsolinols are eluted with 1 ml of 0.1 M HCl. The sample is injected into an HPLC apparatus directly, or after lyophilized and dissolved in 0.1 M HCl.

For conjugated salsolinol derivatives, urine sample should be hydrolyzed. Urine sample (10 ml) is brought to pH 4.5 by addition of 5 ml of 0.2 M acetate buffer and submitted to hydrolysis with 200 μl of β-glucuronidase-arylsulphatase (*Helix Pomatia*, Merck, Darmstadt, Germany) at 37°C for 16 hours (Dostert *et al.*, 1991). For analysis of sulfo-conjugated 1,2-dehydrosalsolinol, the urine sample is hydrolysis with arylsulphatase (Sigma, S-1629) at pH 7.5 (Dostert *et al.*, 1991). Salsolinol derivatives were extracted with $CHCl_3$ and purified using a PBA column. As another hydrolysis method, the plasma is mixed with ascorbic acid (3 μg/ml plasma) and perchloric acid (75 μl/ml) and centrifuged. The supernatant is hydrolyzed by boiling for 20 min. The sample was brought to pH 5 and subjected to the solid phase extraction.

III-1 Materials and methods

Standards

The *(R)*- and (*S*)-enantiomers of salsolinol and *N*-methylsalsolinol, and 2(*N*)-methyl-6,7-dihydroxy-1,2,3,4-tetrahydroisoquinoline (*N*-methylnorsalsolinol) were synthesized according to Teitel *et al.* (1972), and 1,2-dimethyl-6,7-dihydroxyisoquinolinium ion$^+$ to Bembenek *et al.* (1990). Racemic salsolinol and 1-(3',4'-dihydroxybenzyl)-6,7-

dihydroxy-1,2,3,4-tetrahydroisoquinoline (tetrahydropapaveroline) are available from Sigma (St .Louis, MO, USA) and 6,7-dihydroxy-1,2,3,4-tetrahydroisoquinoline (norsalsolinol) from Janssen (Beerse, Belgium).

Sample preparation

Human brain tissues, cerebrospinal fluid (CSF) and intraventricular fluid (IVF) can be applied to HPLC directly after deproteinization by mixing with perchloric acid containing EDTA and sodium metabisulfite, centrifugation and filtration through a Millex filter (pore size; 0.45 μm). For analysis of salsolinols in brain regions, the brain samples prepared by punching out a block (ca. 0.1 mg wet weight) is suspended in 0.1 M perchloric acid solution containing 100 μM EDTA and 100 μM sodium metabisulfite (10 volume/ weight) and sonicated for 30 s at a power level of 20. The sample is centrifuged at 22 000 g for 10 min and filtered through a Millipore filter.

Electrochemical detectors and other HPLC apparatus

ECD is sensitive enough to detect salsolinol derivatives in human tissue samples without purification procedure. The most common detector reported in the literature is a Coulochem or Coulochem-II detector (ESA, Chelmsford, MA, USA) or a multi-electrode detection system [Coulochem Electrode Array System (CEAS), ESA]. Salsolinol derivatives can be detected with oxidation voltage lower than 400 mV and in CEAS the highest response is obtained at 240 mV (Maruyama *et al.*, 1996b). Also salsolinols can be detected in a redox mode; the voltage of the first electrode is set at +500 mV and that of the second at -200 mV. With other types of ECD, such as a Bioanalytical Systems amperometric detector (Bioanalytical Systems, West Lafayette, IN, USA), much higher voltage of oxidation, 0.7 - 0.9 V, is required (Riggin and Kissinger, 1977).

A reversed-phase column is commonly used for the analysis. The mobile phase is citric-phosphate buffer, pH 2.4 to 4.3 containing EDTA, sodium 1-octanesulfonate and 10-16% methanol.

The oxidation product of *N*-methylsalsolinol, 1,2-dimethyl-6,7-dihydroxyisoquinolinium ion is detected by HPLC-FD using a reversed-phase column, Inertosil-3 (4.6 mm i.d. x 250 mm, GL Sciences) and the mobile phase iss composed of 90 mM sodium acetate-35 mM citric acid buffer, pH 4.35, containing 130 μM disodium EDTA and 460 μM sodium octanesulfonate and 11% methanol. The fluorescence at 500 nm is measured with excitation at 355 nm.

III-2 Occurrence of salsolinol derivatives in human tissues

As shown in Figure 1, the biosynthesis intermediates and the metabolites of salsolinol are detected in human tissues and fluid. Table 2 summarizes the data on the amounts of salsolinol derivatives reported in literature. The occurrence of salsolinol derivatives has been studied in the human brain, CSF, plasma and urine, but systematic study in other tissues has never been reported.

The presence of salsolinol in foods and beverages, such as banana and wine, suggests that some parts of salsolinol found in human peripheral tissues and plasma may be derived from them. However, salsolinol was reported not to be transported though the blood-brain barrier (Origitano *et al.*, 1981), indicating that salsolinol in the brain should be synthesized *in situ*. The presence of salsolinol-1-carboxylic acid and 1,2-dehydrosalsolinol in the brain tissue suggests that salsolinol is synthesized from dopamine and pyruvic acid through salsolinol 1-carboxylic acid into 1,2-dehydrosalsolinol followed by reduction (Collins *et al.*,

1979; Brossi, 1982). In rat kidney an enzyme was detected to catalyze the decarboxylation of salsolinol 1-carboxylic acid into 1,2-dehydrosalsolinol (Collins and Cheng, 1988). However, this enzyme was not found in the rat brain (Collins and Cheng, 1988) and human brain (Naoi *et al.*, 1996b). The enantiomeric determination of salsolinols and assay for the activity of the enzymes related to metabolism of salsolinols in the brain will be reviewed in the following section of this review.

The physiological function of salsolinol derivatives in the human brain remains to be elucidated. However, salsolinol was found to have scavenging capacity of hydroxyl radical, as shown by *in vivo* microdialysis at rat striatum (Maruyama *et al.*, 1995a). By perfusion of (*R*)salsolinol the basal level of hydroxyl radical was reduced significantly, whereas *N*-methyl(*R*)salsolinol increased the level. On the other hand, the generation of hydroxyl radical by oxidation of *N*-methyl(*R*)salsolinol was confirmed by *in vitro* experiment (Maruyama *et al.*, 1995b). These results suggest that "oxidative stress" may be involved in the neurotoxicity of *N*-methyl(*R*)salsolinol, which was proved further by the protection of dopaminergic neuroblastoma SH-SY5Y cells by antioxidants against apoptosis induced by *N*-methyl(*R*)salsolinol (Maruyama *et al.*, 1997b). Apoptosis can be prevented with reduced glutathione and catalase, indicating that oxidative stress initiates the chain reactions to the cell death. Salsolinol may be neuroprotective as a scavenger of hydroxyl radicals, not only as a precursor of a neurotoxin, *N*-methyl(*R*)salsolinol.

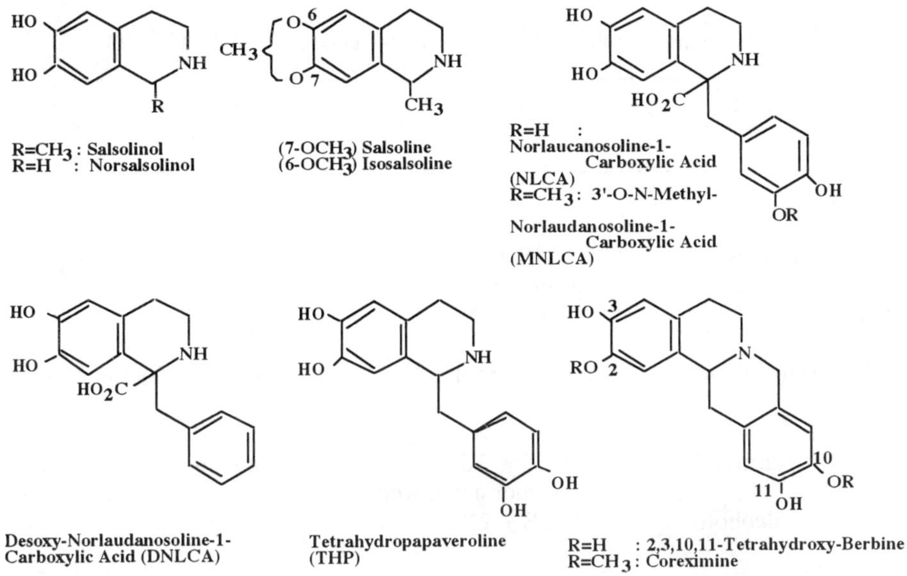

Figure 3. Complex tetrahydroisoquinoline detected in human tissues.

Another group of dopamine-derived isoquinolines were also detected in human tissues, which are called as "complex tetrahydroisoquinolines or "morphine-like" alkaloids (Cohen, 1978) (Figure 3). In human brain dopamine is oxidized by monoamine oxidase into 3,4-dihydroxy-phenylacetaldehyde, which can condense with dopamine into 1-(3,4-dihydroxyphenyl])-methyl-6,7-dihydro-1,2,3,4-tetrahydroisoquinoline, tetrahydropapaveroline (THP). THP was once considered to be a precursor of "morphine-like" alkaloids and to be involved in the molecular mechanism for addiction to ethanol. At present THP is not recognized as a product of dopamine in mammalian tissues under basal conditions, whereas it was recognized in the urine of parkinsonian patients receiving both L-DOPA therapy and alcohol (Sandler *et al.*, 1973).

Table 2
Analysis of salsolinol and the derivatives in human materials by HPLC-ECD

Salsolinol derivatives	Tissue	Content	Reference
Salsolinol			
	Caudate nucleus	218 ± 95 (pmol/g wet weight)	Sasaoka *et al.*, 1988
	Putamen	207 ± 86	
	CSF parkinsonian	< 5 pmol/ml	Moser and Kämpf, 1992
	Striatum		
	control	19 ± 10 (pmol/g wet weight)	Ung-Chhun *et al.*, 1985
	alcoholic	58 ± 21	
	Hypothalamus		
	control	43 ± 13 (pmol/g wet weight)	Ung-Chhun *et al.*, 1985
	alcoholic	87 ± 49	
	Urine		
	control	1.1 ± 0.3 (μg/day)	Collins *et al.*, 1979
	alcoholic	28.8 ± 8.7	
	control	197 ± 54* (nmol/day)	Dordain *et al.*, 1984
	parkinsonian	39 ± 7	
	control	2.8 ±2.6 (mg/mg creatinine)	Adachi *et al.*, 1986
		18.1 ± 9.9 ng/ml (free + conjugated)	Strolin Benedetti *et al* 1989a
		2.9 - 3.4 (ng/ml)	Dostert *et al.*, 1989
Salsolinol-1-carboxylic acid			
	Striatum	161 ± 69 (pmol/g wet weight)	Ung-Chhun *et al.*, 1985
	Hypothalamus	83 ± 42	

Table 2-continued
Analysis of salsolinol and the derivatives in human materials by HPLC

Salsolinol derivatives	Tissue	Content	Reference
1,2-Dehydrosalsolinol			
Urine		21.5 - 55.3 (ng/ml)	Dostert *et al.*, 1989
	alcoholic	15.3 - 56.5 nmol/day (sulfoconjugate + free)	Dostert *et al.*, 1991
6 or 7-Methyl-salsolinol			
CSF	control	1.3 ± 0.3 (pmol/ml)	Dordain *et al.*, 1984
	parkinsonian	0.6 ± 0.2	
	alcoholic	1.1 ± 0.3	
Striatum			
	control	9 ± 6 (pmol/g wet weight)	Ung-Chhun *et al.*, 1985
	alcoholic	11 ± 6	
Hypothalamus			
	control	8 ± 3 (pmol/g wet weight)	Ung-Chhun *et al.*, 1985
	alcoholic	2 ± 1	
2(*N*)-Methyl-norsalsolinol			
CSF	parkinsonian	15, 26, 60 (pmol/ml)	Moser and Kämpf, 1992
1,2-Dimethyl-6,7-dihydroxyisoquinolinium ion*			
Substantia nigra		254 ± 59.0 (pmol/g wet weight)	Maruyama *et al.*, 1997a
IVF		16.0 ± 16.5 nM	Maruyama *et al.*, 1996b

* Measured by HPLC-FD.

Data on enantiomers of salsolinol derivatives in human materials are summarized in Table 4.

As Cashaw *et al* . (1974) proposed, the failure to detect tetrahydropapaveroline in the brain may be a result of its rapid conversion to tetrahydroberiberines, apomorphines and their O-methylated derivatives. The major metabolites were identified by GS-MS as 2,3,10,11-tetrahydroberiberine, 2,3,9,10-tetrahydro-beriberine and a 2- or 3-monomethylated derivative of tetrahydroberiberIne (Davis *et al.*, 1975).

Table 3.
Other "complex" salsolinol derivatives detected in human tissues

Derivatives	Tissue	Content	References
Tetahydropapaveroline (THP, Norlaudanosaline)			
	Urine (L-DOPA treated)	140 μg/24 h	Coscia *et al.*, 1977
	Urine (L-DOPA/ethanol)	380 μg/24 hr	Sandler *et al.*, 1973
3',4'-Deoxynorlaudanosoline carboxylic acid			
	Urine		
	phenylketonuric	45 - 100 ng/ml	Lasala and Costa, 1979
	control	35 ng/ml	
Norlaudanosaline carboxylic acid			
	Urine		
	control	8.4 μg/24 hr	Coscia *et al.*, 1977
	parkinsonian	127.7	
3'-O-Methylnorlaudanosoline carboxylic acid			
	Urine		
	control	not detected	Coscia *et al.*, 1977
	parkinsonian	2.2 μg/24 hr	

Table 3 summarizes the amounts of "complex" isoquinolines detected in human tissues by GC/MS method. THP and 6-methyl-2,3,10, 11-tetrahydroberiberine were determined also by HPLC-ECD (Cashaw *et al.*, 1987). For determination a reversed phase column and ECD with a LC-4B carbon paste electrode (Bioanalytical Systems, West Lafayette, IN) at +0.65 V vs Ag/AgCl, or a Coulochem analytical cell at +0.30 V. As a condensation product of dopamine with phenylpyruvic acid, 3',4'-deoxynorlaudanosoline-1-carboxylic acid was detected in rat brain, and it increased in the urine of phenyketonurea (Lasala and Costa, 1979). These "complex" isoquinolines are now re-examined for their possible involvement to the neurotoxicity (Goto *et al.*, 1997), in addition to the previous proposal as the alkaloids related to alcoholism and L-DOPA therapy.

III-3 Chiral separation

As discussed above, the chirality of salsolinol were found to be essentially involved in the biological activity. Only the (*R*)enantiomer of *N*-methylsalsolinol could induce parkinsonism in rats (Naoi *et al.*, 1996a), and also induce apoptosis to dopaminergic neuroblastoma SH-SY5Y cells (Maruyama *et al.*, 1997b). The dopamine transport system of the cells can recognize *N*-methyl(*R*)-salsolinol (Takahashi *et al.*, 1994). The results suggest that enantio-selective biosynthesis of (*R*)-enantiomer should account for the specific biological function in the brain. The enzymatic biosynthesis of (*R*)salsolinol has been proposed by the group of Strolin Benedetti from analyses of urine. In healthy control only (*R*)salsolinol was detected (Dostert *et al.*, 1987; Strolin Benedetti *et al.*, 1989a) and the (*S*)enantiomer was detected after alcohol drinking (Strolin Benedetti *et al.*, 1989a) or L-DOPA therapy (Dostert *et al.*, 1989). The presence of racemic salsolinol in foods and beverages indicates that (*S*)-enantiomer is derived from the racemic exogenous salsolinol (Strolin Benedetti *et al.*, 1989b).

For enantiomeric analysis of salsolinol derivatives, three HPLC methods have been applied (see reviews, Deng *et al.*, 1997a: Maruyama *et al.*, 1997c). One method is based on the derivatization of salsolinols to the diastereoisomers and analyzed on an achiral column. The second method uses a column of pure chiral polymer, such as bovine serum albumin (BSA), or silica gel coated or grafted with chiral polymers. The third uses cyclodextrin as an additive in the mobile phase.

Chiral separation by derivatization into the diastereoisomers

The first separation of salsolinol enantiomers by HPLC was reported by Strolin Benedetti *et al.* (1989a). This method was based on derivatization with N-trifluoroacetyl-L-prolyl chloride, selective hydrolysis of the phenolic esters and detection of the resulting amide diastereoisomers by HPLC-ECD. By this method the predominant occurrence of (*R*)salsolinol in urine from healthy subjects and racemic salsolinol in dried banana was detected. This method was improved by derivatization with S-1-(1-naphthal)ethyl isothiocyanate (Pianezzola *et al.*, 1989).

Enantiomeric separation by a chiral column

The chiral column used for separation of salsolinol enantiomer reported for the first time was a Resolvosil^{R}-BSA column (Dostert *et al.*, 1987b), and the derivatized enantiomers in urine and foods were separated by HPLC-ECD. However, the results with this column were hardly reproducible. Another chiral column, LichroCART 250-4 Chira Dex column (4.0 i.d. x 244 mm, Merck) was also used for enantiomeric separation of salsolinol in human urine (Strolin Benedetti *et al.*, 1989a) and brain sample (Deng *et al.*, 1995). This column is fragile to be used for crude biological samples and cannot tolerate the repeated applications. At present the best chiral column for the enantiomeric separation of salsolinol in human samples is a β-cyclodextrin-bonded Nucleodex β-OH (4.0 i.d. x 200 mm: Macherey-Nagel, Düren, Germany) (Sällström Baum and Rommelspacher, 1994). This column was used for the analysis of human plasma and the presence of (*R*)- and (*S*)salsolinol was reported.

This Nucleodex β-OH column was successfully used in combination with a CEAS system for analysis of human brain samples (Maruyama *et al.*, 1997a), CSF (Maruyama *et al.*, 1996a) and IVF (Maruyama *et al.*, 1996b). For analysis of the samples containing dopamine a reversed-phase column, Inertosil-3 (4.6 mm i.d. x 150 mm, GL Sciences, Tokyo, Japan) is connected to the chiral column in series. The mobile phase is 100 mM sodium phosphate buffer, pH 3.6, containing 5% methanol. The oxidation voltage used to detect salsolinol and *N*-methylsalsolinol was 160 mV, 240 mV and 300 mV.

Table 4 summarizes the data on the enantiomeric analysis of salsolinol derivatives in human samples using chiral columns.

As another method for chiral isolation, β–cyclodextrin is used as an additive in the mobile phase, and enantiomeric separation of salsolinol and *N*-methylsalsolinol was achieved in the mode of ion-pair chromatography (Deng *et al.*, 1997a and b). The reversed-phase column was used in the presence of ion-pair agents, such as sodium heptanesulfonate. The theoretical studies on the mechanism of the enantio-separation by this system was reported (Deng *et al.*, 1996). For good separation very high concentrations, usually 10 mM, of β-cyclodextrin are required, but this method may be applicable for preparative separation of the enantiomers (Minami *et al.*, 1997).

Table 4.
Chiral separation of (*R*)- and (*S*)-enantiomer of salsolinol derivatives

Salsolinol derivatives	Tissue	Content	Reference
(*R*)Salsolinol			
	Urine	1.95 - 11 ng/ml	Dostert *et al.*, 1989
		10.2 - 51.75 μg/day	StrolinBenedetti *et al.*, 1989b
		16.2 - 103.3 pmol/ml (14.5 - 124 nmol/day)	Dostert *et al.*, 1990b
	alcoholic	47.5 - 231.1 nmol/day	Dostert *et al.*, 1991
	Plasma	0.397 ng/ml (0 - 0.95) (*R* = *S*)	Sällström Baum and Rommelspacher, 1994
	(S-isomer)*	0.4 - 3.1 pmol/ml	Haber *et al.*, 1996
	(R-isomer)	0.4 - 4.2	
	Frontal cortex	134 ± 125 (pmol/g wet weight)	Maruyama *et al.*, 1997a
	Caudate	73.3 ± 79.9	
	Putamen	37.8 ± 23.0	
	Substantia nigra	94.5 ± 78.7	
	IVF	0.39 ± 0.21 nM	Maruyama *et al.*, 1996b
(*S*)Salsolinol			
	Urine	0 - 23.3 pmol/ml (0 - 12.6 nmol/day)	Dostert *et al.*, 1990b
	alcoholic	56.1 - 139.5 nmol/day	Dostert *et al.*, 1991
N-Methyl(*R*)salsolinol			
	Caudate	65.7 ± 88.3 pmol/g wet weight	Maruyama *et al.*, 1997a
	Putamen	110 ± 126	
	Substantia nigra	76.6 ± 23.0	
	CSF		
	parkinsonian	8.32 ± 2.89 nM	Maruyama *et al.*, 1996a
	control	4.53 ± 2.08 nM	
	IVF	9.15 ± 9.08 nM	Maruyama *et al.*, 1996b

*Measured by GC-MS.

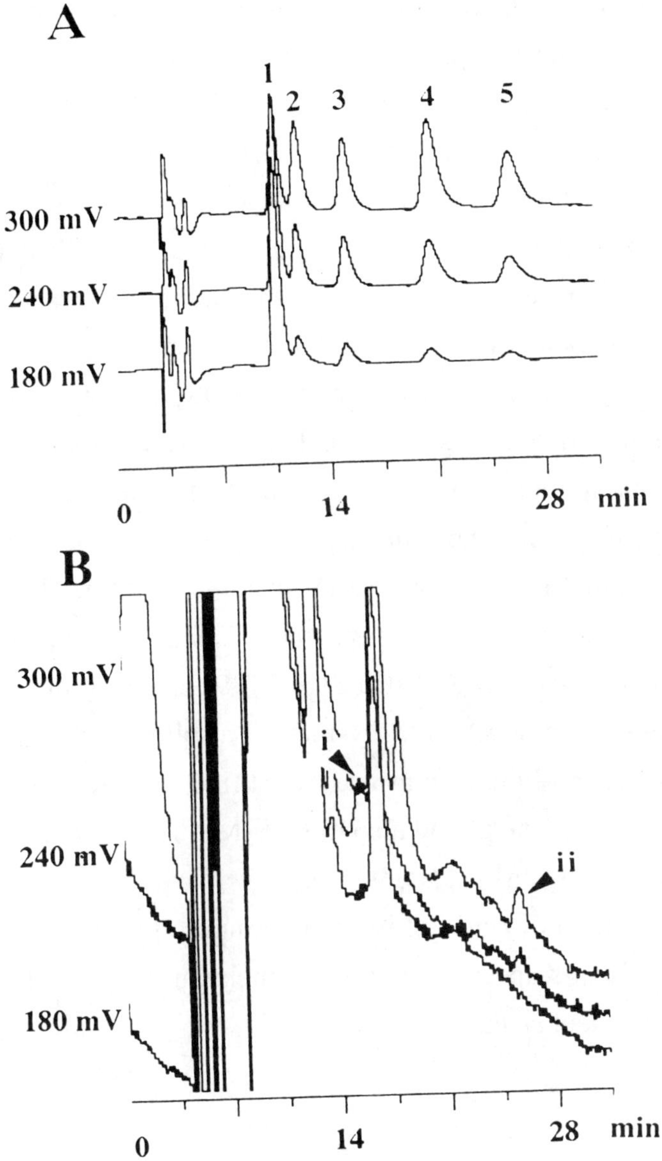

Figure 4. HPLC patterns of salsolinol derivatives in human brain samples. **A**; Standard dopamine (1), (*S*)salsolinol (2), (*R*)salsolinol (3), *N*-methyl(*S*)salsolinol and *N*-methyl(*R*)salsolinol (full scale; 500 μA). **B**; sample prepared from the Putamen, peak (i) and (ii) corresponded to the (*R*)enantiomer of salsolinol and *N*-methylsalsolinol and (iii) to dopamine (**B**, full scale; 50 μA).

Distribution of salsolinol derivatives in human brain regions

In human brain, only (*R*)enantiomer of salsolinol and *N*-methylsalsolinol were detected (Maruyama *et al.*, 1997a). The distribution of salsolinol derivatives was examined and only the (*R*)enantiomer of salsolinol and *N*-methylsalsolinol were detected and the (*S*)enantiomers were under the detection limit, 10 nM. The results are summarized in Table 5, and the typical HPLC patterns by CEAS of the brain samples are shown in Figure 4. (*R*)Salsolinol was found to be distributed non-selectively in four brain regions, the frontal cortex, caudate, putamen and substantia nigra (Table 5), whereas *N*-methyl(*R*)-salsolinol accumulates in the nigro-striatal system. It may be ascribed to the high activity of an *N*-methyltransferase in the substantia nigra, as shown by *in vivo* microdialysis in the rat brain (Maruyama *et al.*, 1992). The occurrence of *N*-methylsalsolinol and *N*-methylnorsalsolinol in human brain was confirmed also by GC/MS (Niwa *et al.*, 1991). As shown in Figure 5, the oxidation product of *N*-methylsalsolinol, 1,2-dimethyl-6,7-dihydroxyisoquinolinium ion was identified in the substantia nigra (Maruyama *et al.*, 1997a) and also in IVF (Maruyama *et al.*, 1996b), but not in CSF (Maruyama *et al.*, 1996a).

The accumulation of catechol isoquinolinium ion in the substantia nigra may be due to its specific binding to melanin, as probed by *in vitro* experiments (Naoi *et al.*, 1994a).

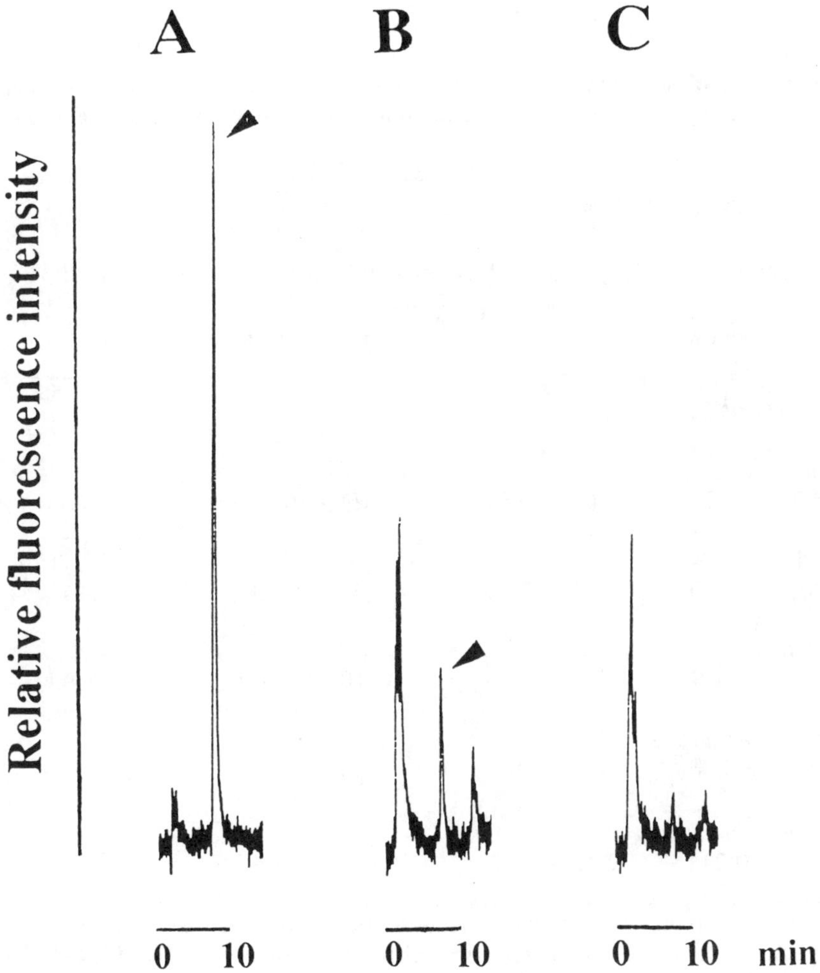

Figure 5. Fluorometric HPLC patterns of 1,2-dimethyl-6,7-dihydroxyisoquinolinium ion in the brain sample. **A**; Standard ion (50 pmol). **B**; A sample prepared from the substantia nigra. **C**; that from the frontal cortex. The fluorescence at 500 nm was measured with the excitation at 365 nm.

These results suggest that *N*-methyl(*R*)-salsolinol increases its selectivity to dopamine neurons by its *N*-methylation, and its toxicity by its oxidation. Such "metabolic bioactivation" may account for the neurotoxicity of *N*-methyl(*R*)-salsolinol specific to dopamine neurons in the nigro-striatal system (Naoi *et al.*, 1994b, 1995b).

Table 5.
Distribution of dopamine, (*R*)salsolinol, *N*-methyl(*R*)salsolinol and 1,2-dimethyl-6,7-dihydroxyisoquinolinium ion in normal human brain regions and human intraventricular fluid

Concentration in the brain regions				
	Dopamine	(*R*)Salsolinol	*N*-Methyl(*R*)salsolinol	1,2-DMDHIQ+
	(nmol/g wet weight)		(pmol/g wet weight[1] or nM[2])	
Brain regions[1]				
Frontal cortex	0.73 ± 0.78	134 ± 125	ND (<10)	ND (<100)
Caudate	26.4 ± 19.0*	73.3 ± 79.9	65.7 ± 88.3	ND
Putamen	20.7 ± 11.9*	37.8 ± 23.0	110 ± 126*	ND
Substantia nigra	3.5 ± 2.7	94.5 ± 78.7	76.6 = 23.0	254 ± 59.0*
Intraventricular fluid[2]	0.39 ± 0.21	9.15 ± 9.08	16.0 ± 16.5	

The brain sample was prepared as described before (Maruyama *et al.*, 1997a). Dopamine, salsolinol and *N*-methylsalsolinol were quantified by HPLC-ECD with a chiral column, and 1,2-dimethyl-6,7-dihydroxy-isoquinolinium ion by HPLC-FD. Value represents mean ± SD of duplicate measurements of 10 samples.
* < 0.01 compared with the vales of the frontal cortex by ANOVA. ND: not detected. IVF from 13 PD patients was analyzed [Maruyama *et al.*, 1996b].

Increase of N-methyl(R)salsolinol in parkinsonian CSF

The lumbar CSF samples from 16 patients with newly diagnosed and untreated PD and from 29 control subjects without neurological disorders were used for the analysis. As a disease control, CSF from 5 patients with multiple system atrophy (MSA) was analyzed. The enantiomers of salsolinol and *N*-methylsalsolinol, dopamine and homovanillic acid (HVA) were analyzed by HPLC-CEAS.

N-Methyl(*R*)salsolinol was detected in CSF from control, patients with PD and MSA. Another enantiomer, *N*-methyl(*S*)salsolinol, was under detection limit (< 0.01 nM). *N*-Methyl(*R*)salsolinol concentration in the control group was not affected by the age from 22 to 76 years ($r = 0.141$) as shown in Figure 6 or by the sex [male; 4.39 ± 1.73 nM, female; 4.89 ± 2.79 nM (mean ± SD)]. In all patients with MSA and almost all control, except 2 out of 29, *N*-methyl(*R*)salsolinol level was lower than 6 nM. On the other hand, in 12 PD patients out of 16 the level was higher than 6 nM. *N*-Methyl(*R*)salsolinol level in the PD patients was significantly higher than that either in control ($p < 0.0001$) or patients with MSA ($p < 0.0022$), as summarized in Table 6. The concentration of (*R*)salsolinol and free dopamine, the metabolic precursors of *N*-methyl(*R*)salsolinol, were under 0.1 nM in CSF. To estimate the biosynthesis rate of this isoquinoline from dopamine, the concentration of *N*-methyl(*R*)-salsolinol was compared with that of HVA, a major metabolite of dopamine. The ratio also significantly increased in PD patients compared with control ($p < 0.0002$) or MSA patients ($p < 0.0455$). The HVA level was reduced in the CSF of PD and MSA, but the difference from control was not statistically significant.

The amounts of *N*-methyl(*R*)salsolinol in CSF do not depend on dopamine turnover and age, suggesting that other factors, such as enzyme activity of the synthesis and metabolism, regulates the level of *N*-methyl(*R*)salsolinol. This hypothesis was proved recently by the increase of a (*R*)salsolinol *N*-methyltransferase in lymphocytes of parkinsonian patients (Naoi *et al.*, 997b).

Table 6.
Analysis of *N*-methyl(*R*)salsolinol in parkinsonian cerebrospinal fluid

	Concentration in CSF		
	N-Methyl(*R*)salsolinol (nM)	HVA (nM)	*N*-Methyl(*R*)salsolinol /HVA
Parkinsonian patients (n = 16)			
	8.32 ± 2.89* (4.54 - 15.7)	141 ± 75 (17.4 - 259.8)	0.086 ± 0.067** (0.028 - 0.261)
Control (n = 29)			
	4.53 ± 2.08 (0.62 - 11.90)	80 ± 92 (57.6 - 422.0)	0.034 ± 0.030 (0.012 - 0.162)
MSA patients (n = 5)			
	3.59 ± 1.52 (1.33 5.25)	156 ± 75 (74.6 - 259.0)	0.034 ± 0.029 (0.017 - 0.070)

The values represent mean and SD. Range is shown in parentheses. n = number of subjects.
* $p < 0.001$ compared with control and MSA patients.
** $p < 0.001$ compared with control and $p < 0.05$ compared with MSA patients.

IV ASSAY FOR THE ACTIVITY OF ENZYMES RELATED IN THE SYNTHESIS OF *N*-METHYL(*R*)SALSOLINOL

As described above, *N*-methyl(*R*)salsolinol is synthesized in the human brain by two enzyme reactions, enantiomeric condensation of dopamine with acetaldehyde and *N*-methylation. The isoquinoline is oxidized by an oxidase present in the cytosol. This section described the assay method of these enzymes.

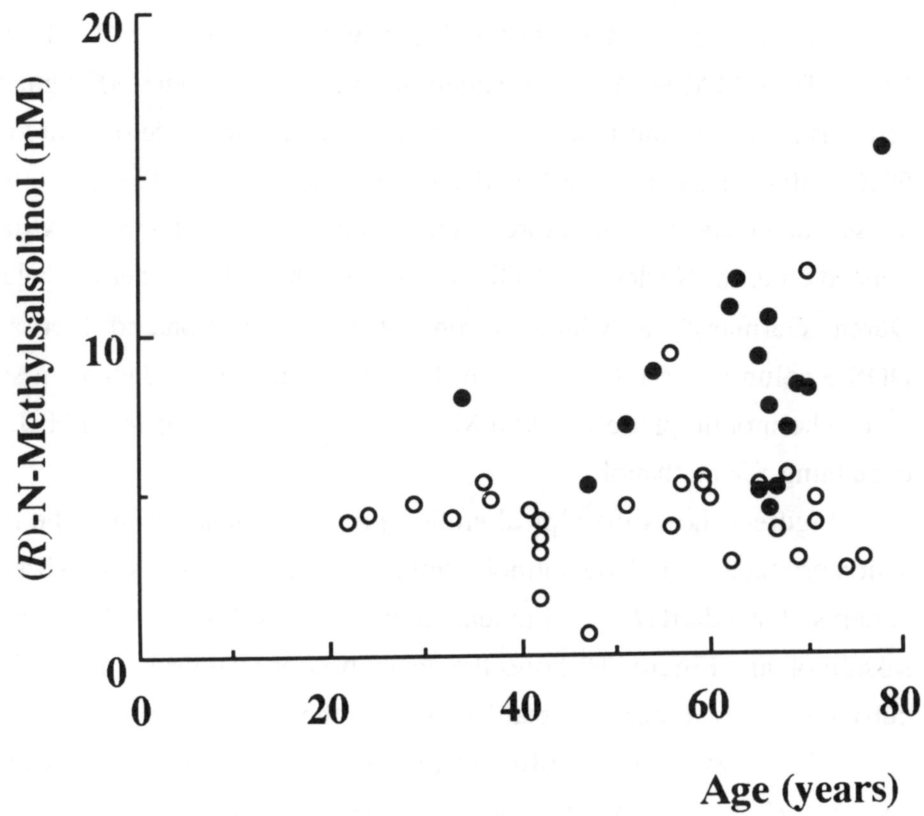

Figure 6. The concentration of *N*-methyl(*R*)salsolinol in parkinsonian and control CSF. The concentration was plotted against the age. Filled circle; parkinsonian CSF. Open circle; control.

IV-1. Determination of the activity of a (*R*)salsolinol synthase

This enzyme activity is detected in human brain and liver, but not in lymphocytes, and in the brain tissues it occurs in the cytosol fraction. The enzyme sample is incubated with 1 mM dopamine and acetaldehyde in 100 μl of 100 mM Tris-HCl buffer, pH 7.4, at 37°C for 20 min. Then, 25 μl of 1 M perchloric acid containing 1 mM EDTA and 1 mM sodium metabisulfite is added to the reaction mixture. After mixed and centrifuged at 22 000 g for 10 min, the sample is filtered though a Millipore HV filter and analyzed by HPLC-ECD, with a Coulochem-II (ESA, Chelmsford, MA, USA). The oxidation voltage of a model 5020 guard cell was 350 mV, and that of the first and second electrode of a model 5010 analytical cell are 50 mV and 300 mV, respectively. The output of the second electrode is monitored. The column used is a β-cyclodextrin-bonded column, Nucleodex β-OH (4.0 i.d. x 200 mm, Macherey-Nagel, Düren, Germany), to which is connected a reverse-phased Inertosil ODS-3 column (4.6 i.d. x 150 mm, GL Sciences, Tokyo, Japan) in series. The mobile phase is 100 mM sodium phosphate buffer, pH 3.6, containing 5% methanol.

Figure 7 shows the typical elution profile of standards and the reaction products of a (*R*)salsolinol synthase. Fig. 8 A shows the HPLC patterns of standard (*R*)- and (*S*)enantiomer of salsolinol and *N*-methylsalsolinol, and Figure 7B shows the generation of (*R*)salsolinol after incubation of dopamine and acetaldehyde with the enzyme sample.

The enzyme was purified from human brain cytosol by successive HPLC using two types of DEAE columns and a gel filtration column (Naoi *et al.*, 1996b). The enzyme was purified by 1730-fold with about 7.1% yield from the activity in the homogenate. The molecular weight was estimated to be 34.3 ± 8.3 KD (mean ± SD).

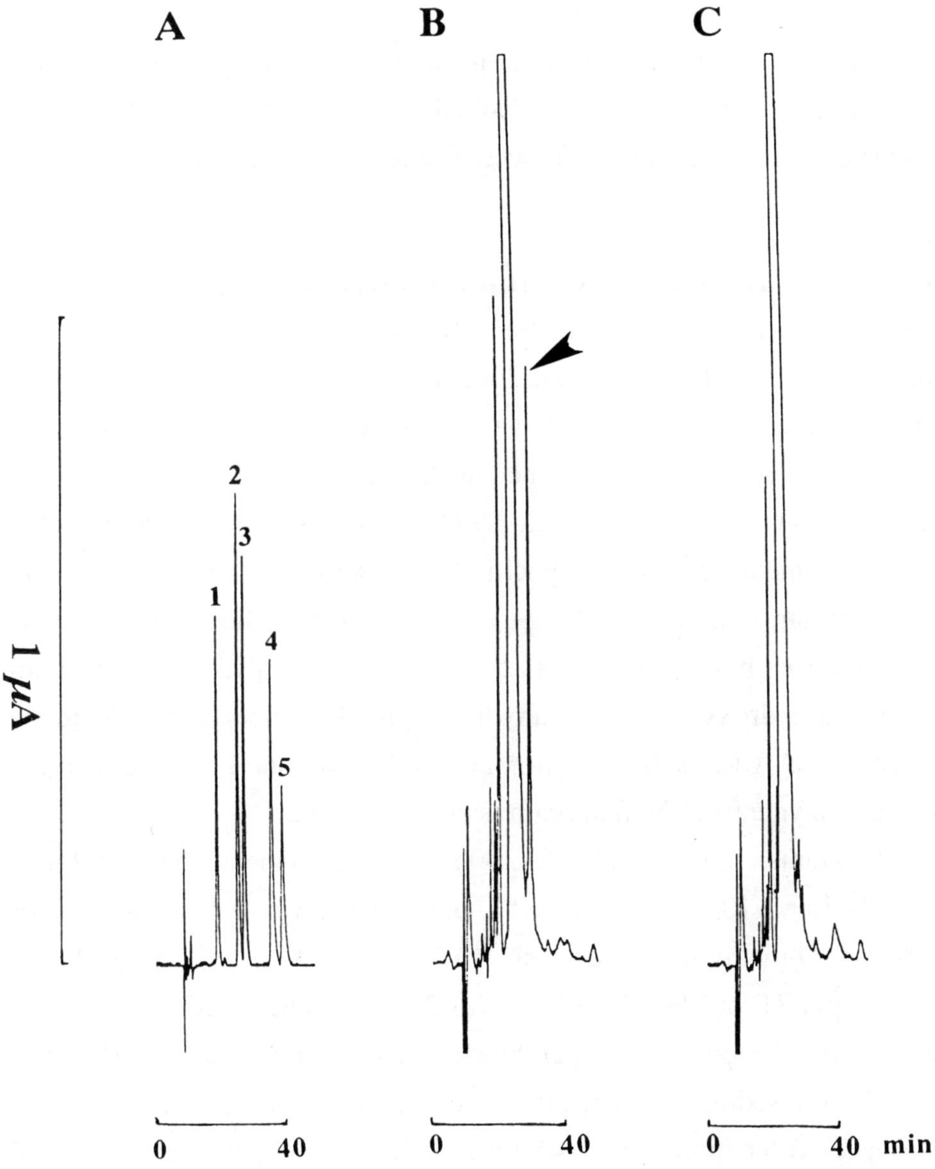

Figure 7. HPLC patterns of standards and the reaction mixture of the purified enzyme. **A**; Standard of (*R*)salsolinol-1-carboxylic acid (peak 1), (*S*)- and (*R*)-salsolinol (2 and 3), *N*-methyl(S)- and *N*-methyl-(*R*)salsolinol. **B**; Reaction mixture of dopamine and acetaldehyde with the purified enzyme. **C**; without enzyme sample. The peak corresponding to (*R*)salsolinol is indicated by an arrow.

Dopamine is the only amine substrate, and *N*-methyldopamine (epinine), adrenaline, noradrenaline or L-DOPA were not. Acetaldehyde, formaldehyde and pyruvic acid were found to be another substrate.

IV-2. Assay of a (*R*)salsolinol *N*-methyltransferase

An *N*-methyltransferase specific (*R*)salsolinol is detected in human brain, liver, lymphocytes and a dopaminergic neuroblastoma SH-SY5Y cells. *In vivo* microdialysis study in rat brain revealed that the activity of this *N*-methyltransferase is higher in the substantia nigra than in other brain regions (Maruyama *et al.*, 1992). Increase of the activity of a neutral *N*-methyltransferase specific for (*R*)salsolinol was confirmed in parkinsonian lymphocytes (Naoi *et al.*, 1997b). The enzyme in the crude sample has optimum pH at 7.0 and 8.0 and neither (*S*)salsolinol nor 6,7-dihydroxy-1,2,3,4-tetrahydroisoquinoline (norsalsolinol) are the substrate. S-Adenosyl-L-methionine is a methyl donor. The purification of this enzyme from human brain is now in progress.

Human brain sample (40-200 μg protein) is incubated with 2 mM (*R*)salsolinol, 10 μM S-adenosyl-L-methionine, 10 μM an inhibitor of catechol-O-methyltransferase, RO 40-7592, in 200 μl of 100 mM Tris buffer, pH 7.0 or pH 8.5, at 37°C for 30 min. The reaction mixture is mixed with 100 μl of 0.1 M perchloric acid containing 1 mM EDTA 2Na and 1 mm sodium metabisulfite, centrifuged and filtered through a Millipore filter (pore size, 0.45 μm). Produced *N*-methyl(*R*)-salsolinol is quantified using HPLC-ECD and a reversed-phase Inertosil ODS-3 column (5 μm, 4.0 mm i.d. x 100 mm, GL Sciences, Tokyo, Japan). A Coulochem-II electrochemical detector is used with a model 5020 Guard cell set at +350 mV and a model 5010 Analytical cell, the first and second electrode of which are set at +50 mV and +300 mV, respectively. The output of the second electrode is monitored.

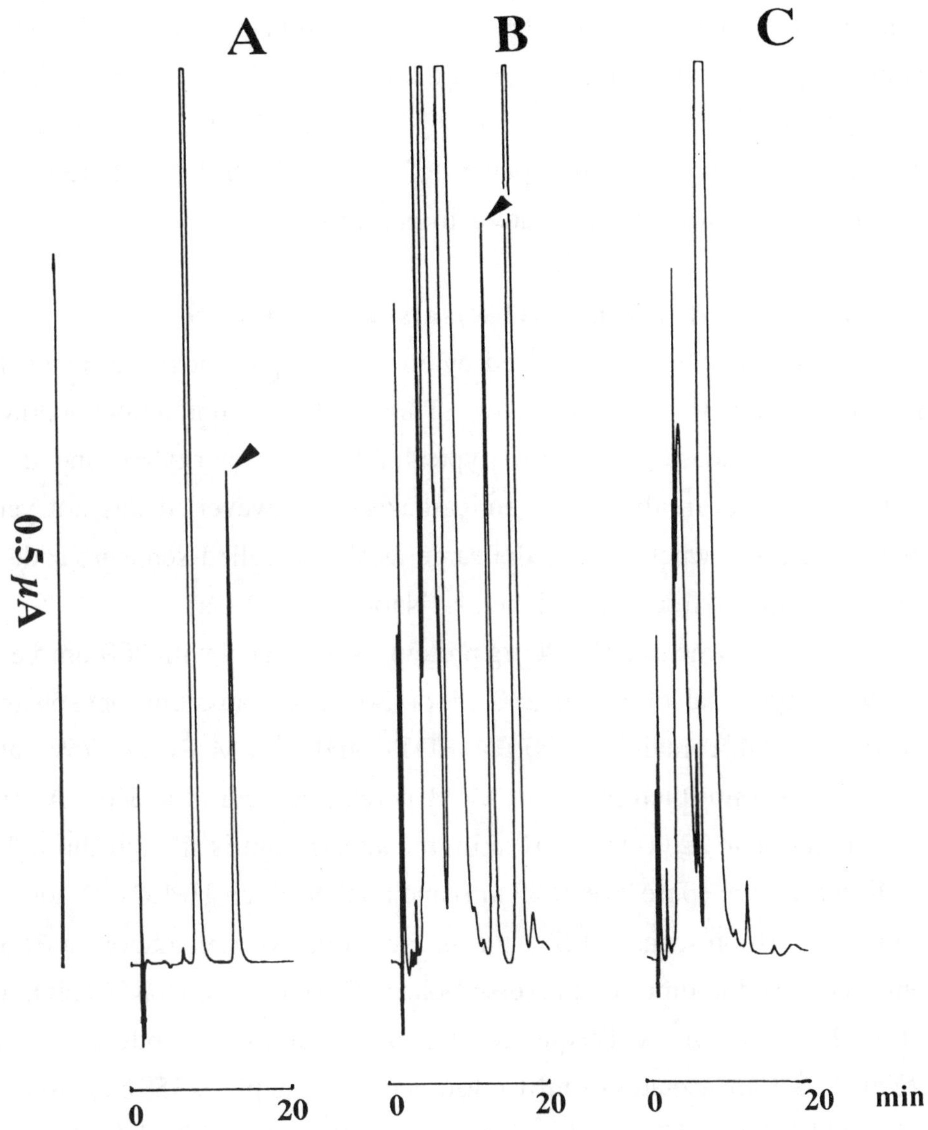

Figure 8. HPLC pattern of standard *N*-methyl(*R*)salsolinol and the reaction product by a (*R*)salsolinol *N*-methyltransferase.

The mobile phase is 25 mM sodium phosphate buffer, pH 3.0, containing 0.1 mM EDTA 2Na, and the flow rate is 1.2 ml/min. The flow rate is 0.6 ml/min for an Inertosil-3 column (3 μm, 4.0 mm i.d. x 100 mm). The typical HPLC pattern of *N*-methyl(*R*)salsolinol after the reaction with brain sample is shown in Figure 8.

IV-3. Assay of a *N*-methyl(*R*)salsolinol oxidase

N-Methyl(*R*)salsolinol is oxidized by an oxidase present in the cytosol fraction of human brain and liver. This oxidase is not mitochondrial monoamine oxidase sensitive to typical inhibitors, clorgyline and deprenyl, but it is inhibited by semicarbazide. However, it has not yet been confirmed whether it is the same as the so-called semicarbazide-sensitive amine oxidase in peripherals (Naoi *et al.*, 1995a).

The enzyme sample (40 μg protein) is incubated with 200 nmoles of *N*-methyl(*R*)salsolinol in 200 μl of 30 mM potassium phosphate buffer, pH 7.4, containing 100 μM EDTA and 10 μM RO 40-7592 at 37°C for 20 min, then 50 μl of 0.1 M perchloric acid is added. After centrifugation at 22 000 g for 10 min, the supernatant is filtered through a Millipore filter (pore size; 0.45 μm) and applied to an HPLC-FD apparatus. The fluorescence at 500 nm is measured with excitation at 365 nm. The used column is a reversed phase Cosmosil 5C18-MS column (4.6 x 150 nm, Nacalai Tesque, Kyoto, Japan), and the mobile phase is 90 mM sodium acetate-35 mM citric acid buffer, pH 4.35, containing 130 μM EDTA and 230 μM sodium octanesulfonate, and 23% methanol.

V. ISOQUINOLINES WITHOUT CATECHOL STRUCTURE IN HUMAN BRAIN

In human brain, 1,2,3,4-tetrahydroisoquinoline (TIQ) without catechol structure was detected by GC/MS (Niwa *et al.*, 1987; Ohta *et al.*, 1987).

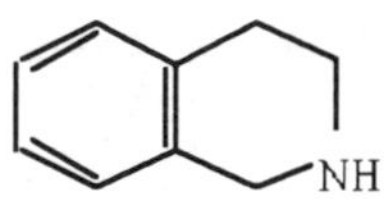

1,2,3,4-Tetrahydro-isoquinoline (TIQ)

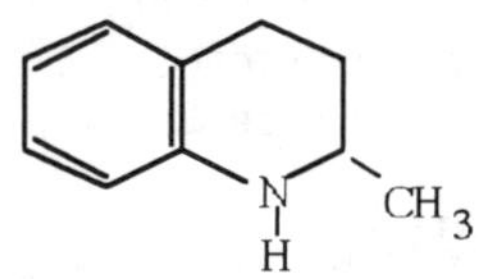

2-Methyl-tetrahydroquinoline

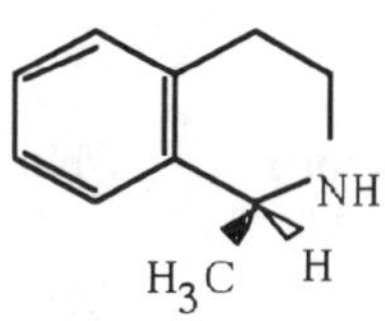

(*S*)-1-Methyl-TIQ

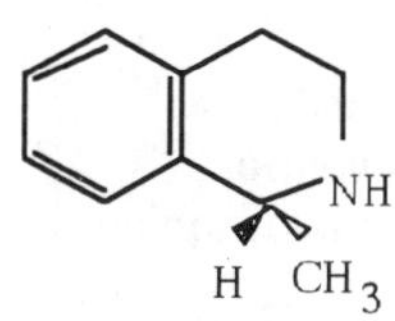

(*R*)-1-Methyl-TIQ

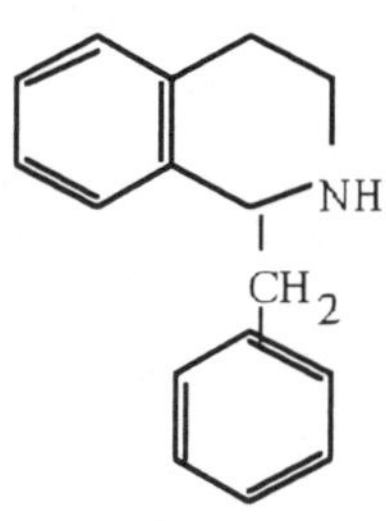

1-Benzyl-TIQ

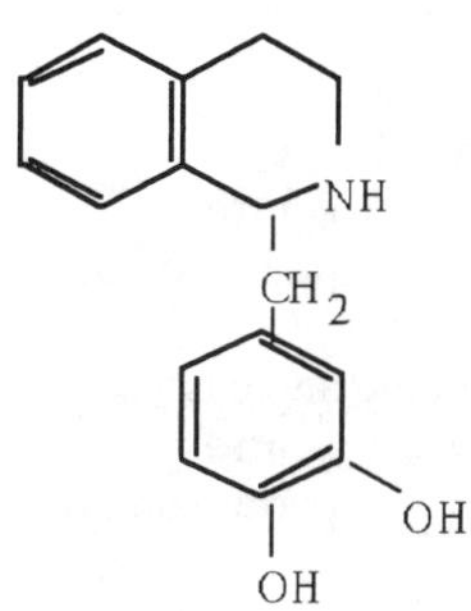

1-(3',4'-Dihydroxybenzyl)-TIQ

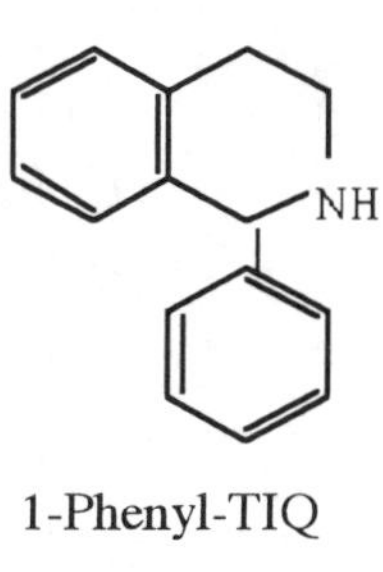

1-Phenyl-TIQ

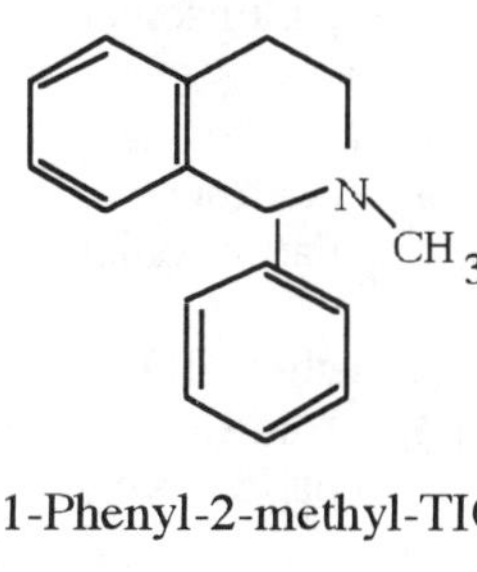

1-Phenyl-2-methyl-TIQ

Figure 9. Isoquinoline derivatives detected in human materials.

Table 5.

1,2,3,4-Tetrahydroisoquinoline derivatives detected in human brain tissues

Table 5.
1,2,3,4-Tetrahydroisoquinoline derivatives detected in human brain tissues

Isoquinoline	Tissue		Concentration	References
1,2,3,4-Tetrahydroisoquinoline (TIQ)				
	Brain	control	1 ng/g	Niwa *et al.*, 1987
		parkinsonian	10 ng/g	
2(*N*)-Methyl-TIQ				
	Brain	control	3 ng/g	
		parkinsonian	2 ng/g	
TIQ				
	Frontal lobe			
		control	0.86 ± 0.23 nmol/g	Ohta *et al.*, 1987
		parkinsonian	0.58 ± 0.20	
	Caudate nucleus			
		control	0.64 ± 0.74	
		parkinsonian	0.25 ± 0.08	
1-Methyl-TIQ				
	Frontal lobe			
		control	0.75 ± 0.25	
		parkinsonian	0.05 ± 0.02	
	Caudate nucleus			
		control	0.52 ± 0.15	
		parkinsonian	0.12 ± 0.03	
1-Benzyl-TIQ				
	CSF	control	0.40 ± 0.23 ng/ml	Kotake *et al.*, 1995
		parkinsonian	1.17 ± 0.74	
1-Phenyl-TIQ				
	Brain	control	19 pg/g	Kajita *et al.*, 1995
		parkinsonian	34, 12	
1-Phenyl-2(*N*)-methyl-TIQ				
	Brain	control	not detected	
		parkinsonian	14, 43	

The chemical structure of the isoquinolines detected in human materials are shown in Figure 9, and the concentrations are presented in Table 5, which were obtained by GC/MS. After repeated injection behavioral changes, mainly bradykinesia, were detected in primates with reduction of dopamine content and tyrosine hydroxylase activity (Nagatsu and Yoshida, 1988). However, the selective death of dopamine neurons in the substantia nigra was not confirmed, and the relation to the pathogenesis to neurodegenerative diseases remains to be proved.

TIQs derived from foods and transported into the brain through the blood-brain barrier (Niwa *et al.*, 1988), or non-enzymatically synthesized by condensation of β-phenethylamine with formaldehyde or acetaldehyde into TIQ or 1-methyl-TIQ. 1-Methyl-TIQ has an asymmetric center at C-1 and the (S)enantiomer is more predominant in wine, cocoa and mouse brains (Makino *et al.*, 1990). At present, the enzyme synthesizing this isoquinoline has never been isolated. In the brain, TIQ is metabolized into *N*(2)-methyl-TIQ (Naoi *et al.*, 1989a). 1-Benzyl-TIQ was proposed to be synthesized by condensation of β-phenethylamine with phenylactaldehyde, an oxidation product of β-phenethylamine itself by monoamine oxidase B, The increase of 1-benzyl-TIQ in parkinsonian CSF was reported (Kotake *et al.*, 1995), and after chronic administration this isoquinoline induced akinesia in a primate, but it was found in the brain ubiquitously and the accumulation in the nigro-striatum was not confirmed (Kotake *et al.*, 1996). Phenyl-TIQ, which is produced non-enzymatically from β-phenethylamine and benzaldehyde, and 1-phenyl-2(*N*)-methyl-TIQ were also detected in human brains by GC/tandem MS (Kajita *et al.*, 1995). 1-Phenyl-2-methyl-TIQ was proved to be cytotoxic to dopaminergic neuroblastoma SH-SY5Y cells (Takahashi *et al.*, 1997).

TIQs without catechol structure require high oxidation voltage for detection with ECD, and most of previous works used GC/MS, which requires large amounts of brain samples and preparation to concentrate isoquinolines. Human brain sample (25 g) is homogenized in 0.4 M perchloric acid (20 ml) containing 0.1% EDTA and 0.1% ascorbic acid, and centrifuged (Niwa *et al.*, 1987). The supernatant is extracted with diethyl ether (20 ml). The aqueous phase is adjusted to pH 11.0 with 6 N NaOH and extracted twice with dichloromethane (20 ml). The isoquinolines are extracted into 0.1 M HCl solution (20 ml), the pH of which is adjusted to 11.0. The organic phase was dehydrated and evaporated. The residue is derivatized with ethyl acetate /heptafluoro-butyric anhydride (HPBA) or pentafluoropropionic anhydride (PFPA) and incubated at 70'C for 30 min. The sample is analyzed by GC/MS. An oxidation product of *N*-methyl-1,2,,3,4-tetrahydroisoquinoline, *N*-methyl-isoquinolinium ion, is detected with HPLC-FD (Naoi *et al.*, 1989). The column is reversed-phase Shimadzu ODS-H (4 mm i.d, x 150 mm, Shimadzu, Kyoto, Japan) and the mobile phase is 90 mM sodium acetate--35 mM citric acid buffer, pH 4.35, containing 130 μM disodium EDTA and 115 μM sodium octanesulfonate, and 10% methanol. The fluorescence at 380 nm is measured with the excitation at 335 nm (Naoi *et al.*, 1989b). By use of a selective derivatization reagent, 3,4-dihydro-6,7-dimethoxy-4-methyl-3-oxoquinoxaline-2-carbonyl chloride, TIQ in the rat brain was determined by HPLC-FD, and the concentration was estimated 42.8 ± 9.4 pmol/g brain (mean ± SD, n = 9) (Ishida *et al.*, 1991). TIQ is extracted from rat brain as described above and the dried sample is reacted with the fluorescence derivatization reagent and applied into an HPLC-FD apparatus. A reversed-phase column, TSK gel ODS-120T (250 x 4.6 mm, Tosoh, Tokyo, Japan) is used and the mobile phase is 50 mM ammonium acetate-methanol mixture (47 ; 53, v/v) and the fluorescence at 476 nm was measured with the excitation at 388 nm.

VI. CONCLUSION

By analyses of human brain samples, the occurrence of salsolinol derivatives has been established by GC-MS and HPLC-ECD. Recent development of a chiral column has made it possible to analyze the enantiomer in small amounts of samples and to assay the activity of a (*R*)salsolinol synthase and an *N*-methyltransferase in clinical samples. The studies of salsolinols has now been developed into a second generation, where the biosynthesis and metabolism can be examined in molecular levels. Our recent results suggest that *N*-methyl(*R*)salsolinol may be a neurotoxin to induce Parkinson's disease in humans, and that the enzymes related to its biosynthesis may be endogenous factors of the pathogenesis of this neurodegeneration. As a most possible candidate, a neutral *N*-methyltransferase has been found to increase in parkinsonian lymphocytes and the activity in the human striatum is well correlated with the level of an endogenous MPP^+-like 1,2-dimethyl-6,7-dihydroxyisoquinolinium ion in the substantia nigra (Naoi *et al.*, in preparation). The molecular studies on the enzymes are now on the way and will bring out the mechanism of the dopamine neurons in Parkinson's disease.

Acknowledgment

This work was supported by a Grant-in-Aid for Scientific Research on Priority Area (M. N. and K. M.), and for Exploratory Research (M. N.) from the Ministry of Education, Science and Education, Japan, and the Grant for Longevity Sciences (9C-03) from the Ministry of health and Welfare, Foundation for Comprehensive Research on Aging and Health (W. M.) of Japan.

REFERENCES

Adachi J., Mizoi Y., Fukunaga T., Ueno Y., Imamichi H., Ninomiya I. and Naito T. (1986) Individual difference in urinary excretion of salsolinol in alcoholic patients. *Alcohol* **3**: 371-375.

Allievi C., Dostert P. and Strolin Benedetti M. (1991) Determination of free salsolinol concentrations in human urine using gas chromatography-mass spectrometry. *J. Chromatogr.* **568**: 271-279.

Bembenek M. E., Abell C. W., Chriset L. A., Rozwadowska M. D., Gresner W. and Brossi A. (1990) Inhibition of monoamine oxidase A and B by simple isoquinoline alkaloids: racemic and optically active 1,2,3,4-tetrahydro-, 3,4-dihydro, and fully aromatic isoquinolines. *J. Med. Chem.* **33**: 147-152.

Brossi A. (1982) Mammalian TIQ's: Products of condensation with aldehydes or pyruvic acid? *Prog. Clin. Biol. Res.* **90**, 123-133.

Cashaw J. L., McMurtrey K. D., Brown H. and Davis V. E. (1974) Identification of catecholamine-derived alkaloids in mammals by gas chromatography and mass spectrometry. *J. Chromatogr.* **99**: 567-573.

Cashaw J. L, Geraghty C. A, McLaughlin B. R and Davis V. (1987) A method for determination of subpicomole concentrations of tetrahydropapaveroline in rat brain by high-performance liquid chromatography with electrochemical detection. *Anal Biochem* **162**: 274-282.

Clow A., Topham A., Saunders J. B., Murray R. and Sandler, M. (1985) The role of salsolinol in alcohol intake and withdrawal. *In*: Aldehyde Adducts in Alcoholism, M. Collins (Ed.) Alan R. Liss. pp. 101-113.

Cohen G. (1978) Alkaloid products in the metabolism of alcohol and biogenic amines. *Biochem. Pharmacol.* **25**: 1123-1128.

Cohen G, and Collins M. (1970) Alkaloids from catecholamines in adrenal tissue: possible role in alcoholics. *Science* **167**: 1749-1751.

Collins M. A., Nijm W. P., Borge C., Teas G. and Goldfarb C. (1979) Dopamine-derived tetrahydroisoquinolines: Increased urinary excretion by alcoholics following alcohol consumption. *Science* **206**: 1184-1186.

Collins M. A., Hannigan J. J., Origitano T., Moura D. and Osswald W. (1982) On the occurrence, assay and metabolism of simple tetrahydroisoquinolines in mammalian tissues. *Prog. Clin. Biol. Res.* **90**: 155-162.

Collins M. A. and Cheng B. Y. (1988) Oxidative decarboxylation of salsolinol-1-car boxylic acid to 1,2-dehydrosalsolinol: Evidence for exclusive catalysis by par ticulate factors in rat kidney. *Arch. Biochem. Biophys.* **264**: 86-96.

Collins M. A., Ung-Chhun N., Cheng B. Y. and Pronger D. (1990) Brain and plasma tetrahydroisoquinolines in rats: effects of chronic ethanol intake and diet. *J. Chromatogr.* **55**: 1507-1514.

Collins MA., Neafsey E. J, Matsubara K, Cobuzzi R. Jr. and Rollema. (1992) Indole-N-methylated β–carbolinium ions as a potential brain-bioactivated neu rotoxins. *Brain Res.* **570**: 154-160.

Coscia C. J., Burke W., Jamrot G., Lasala J. M., McFarlane I., Mitchel I., O'Toole M. M. and Wilson M. L. (1977) Occurrence of a new class of tetrahydroisoquinoline alkaloids in L-dopa-treated parkinsonian patients. *Nature* **269**: 617-619.

Davis V. E., Cashaw J. L. and McMurtrey K. D. (1975) Disposition of cate cholamine-derived alkaloids in mammalian system. In: Alcohol Intoxication and Withdrawal: Experimental Studies. M. M. Gross (Ed) Plenum, New York, pp. 65-78.

Davis G. C. B., Williams A. C., Markey S. P., Ebert M. H, Caine E. D., Beichert C. M. and Kopin I. J. (1979) Chronic parkinsonism secondary to intravenous injection of meperidine analogues. *Psychiatry Res.* **1**: 249-254.

Deng Y., Maruyama W., Dostert P., Takahashi T., Kawai M. and Naoi M. (1995) determination of the (*R*)- and (*S*)-enantiomers of salsolinol and N-methylsalsolinol by use of a chiral high-performance liquid chromatography. *J. Chromatogr. B.* **670**: 47-54.

Deng Y., Maruyama W., Yamamura H., Kawai M., Dostert P., Naoi M. (1996) Mechanism of enantioseparation of salsolinols, endogenous neurotoxins in human brain, with ion-pair chromatography using β-cyclodextrin as mobile phase additive. *Anal. Chem.* **68**: 2826-2931.

Deng Y., Maruyama W., Kawai M., Dostert P. and Naoi M. (1997a) Enantioseparation of biologically active compounds by HPLC. In: Progress in HPLC-HPCE. Vol. 6, Coulometric electrode array detectors for HPLC. I. N. Acworth, M. Naoi, H. Parvez and S. Parvez (Eds.) VSP, Utrecht, pp. 301-338.

Deng Y., Maruyama W., Kawai M., Dostert P., Yamamura H., Takahashi T. and Naoi M. (1997b) Assay for the (*R*)- and (*S*)-enantiomers of salsolinols in biological samples and foods with ion-pair high-performance liquid chromatography using β-cyclodextrin as a chiral mobile phase additive. *J. Chromatogr. B.* **689**: 313-320.

Dordain G., Dostert P., Strolin Benedetti and Rovei V. (1984) Tetrahydroisoquinoline derivatives and Parkinsonism. *In*: Monoamine oxidase and diseases. Prospects for therapy with reversible inhibitors. K. F. Tipton, P. Dostert, M. Strolin Benedetti (Eds.) Academic Press, London, pp. 417-426.

Dostert P., Strolin Benedetti M. and Dedieu M. (1987a) Is salsolinol as racemic mix ture in human urine or does an enantiomer predominate? *Pharmacol. Toxicol.* **60** [Suppl 1]: 12.

Dostert P., Strolin Benedetti M. and Dedieu M. (1987b) Ratio of enantiomers of salsolinol in human urine. *Pharmacol. Toxicol.* **60** [Suppl 1]: 13.

Dostert P., Strolin Benedetti M., Dordain G. and Vernay D. (1989) Enantiomeric composition of urinary salsolinol in Parkinsonian patients after Madopar. *J. Neural Transm. [P-D Sect]* **1**: 269-278.

Dostert S., Strolin Benedetti M., Bellotti V., Allievi C. and Dordain G. (1990a) Biosynthesis of salsolinol, a tetrahydroisoquinoline alkaloid, in healthy subjects. *J. Neural Transm.* **81**: 215-223.

Dostert P., Strolin Benedetti M. and Dordain G. (1990b) Salsolinol and an early detec tion of Parkinson's disease. *In*: Early Markers in Parkinson's and Alzheimer's Diseases, P. Dostert, P. Riederer , M. Strolin Benedetti and R. Roncucci (Eds.) Springer-Verlag, Wien, pp. 93-97.

Dostert P., Strolin Benedetti M., Dordain G. and Vernay D. (1991) Urinary elimination of salsolinol enantiomers in alcoholics. *J. Neural Transm. [GenSect]* **85**: 51-59.

Duncan M. W., and Deitrich R. A. (1980) A critical evaluation of tetrahydroisoquinoline induced ethanol preference in rats. *Pharmacol. Biochem. Behav.* **13**: 265-281.

Faraj B. A., Camp V. M., Davis D. C., Lenton J. D. and Kutner M. (1989) Elevation of plasma salsolinol sulfate in chronic alcoholics as compared to nonalcoholics. *Alcohol. Clin. Exp. Res.* **13**: 155-163.

Goto H., Mochizuki H., Hattori T., Nakamura N. and Mizuno Y, (1997) Neurotoxic effects of papaverine, tetrahydropapaverine and dimethoxyphenylethylamine on dopaminergic neurons in ventral mesencephalic-striatal co-culture. *Brain Res.* **754**: 260-268.

Haber H., Haber H. M. and Melzig M. F. (1995a) A new rapid method for the analysis of catecholic tetrahydroisoquinolines from biological samples by gas chromatography/mass spectrometry. *Anal. Biochem.* **224**: 256-262.

Haber H., Putscher I., Georgi M. and Melzig M. F. (1995b) Influence of ethanol on the salsolinol excretion in healthy subjects. *Alcohol* **12**: 299-303.

Haber H., Henklein P., Georgi M. and Melzig M. F. (1995c) Resolution of catecholic tetrahydroisoquinoline enantiomers and the determination of R- and S-salsolinol in biological samples by gas chromatography-mass spectrometry. *J. Chromatogr.* **672**: 179-187.

Haber H., Winkler A., Putscher I., Henklein P., Baeger I., Georgi M. and Melzig M. F. (1996) Plasma and urine salsolinol in humans: effect of acute ethanol intake on the enantiomeric composition of salsolinol. *Alcohol. Clin. Exp. Res.* **20**: 87-92.

Hamilton M. G. and Hirst M. (1980) Alcohol-related tetrahydroisoquinolines: Pharmacology and identification. *Subst. Alcohol Act./Misuse* **1**: 121-144.

Ishida J., Yamaguchi M. and Nakamura M. (1991) High-performance liquid chromato graphic determination of 1,2,3,4-tetrahydroisoquinoline in rat brain with fluorescence detection. *Anal. Biochem.* **195**: 158-171.

Kajita M., Niwa T., Fujisaki M., Ueki M., Niimura K., Sato M., Egami K., Naoi M., Yoshida M. and Nagatsu T. (1995) Detection of 1-phenyl-N-methyl-1,2,3,4-tetrahydroisoquinoline and 1-phenyl-1,2,3,4-tetrahydroisoquinoline in human brain by gas chromatography-tandem mass spectrometry. *J. Chromatogr.* **669**: 345-351.

Kotake Y., Tasaki Y., Makino Y., Ohta S. and Hirobe M. (1995) 1-Benzyl-1,2,3,4-tetrahydroisoquinoline as a parkinsonism-inducing agent: A novel endogenous amine in mouse brain and parkinsonian CSF. *J. Neurochem.* **65**: 2633-2638

Kotake Y., Yoshida M., Ogawa M., Tasaki Y., Hirobe M. and Ohta S. (1996) Chronic administration of 1-benzyl-1,2,3,4-tetrahydroisoquinoline, an endoge nous amine in the brain, induces parkinsonism in a primate. *Neurosci. Lett.* **217:** 69-71.

Langston J. W., Ballard P., Tetrud J. W. and Irwin I. (1983) Chronic parkinsonism in humans due to a product of meperidine-analog synthesis. *Science* **219**: 979-980.

Lasala J. M. and Costa C. J. (1979) Accumulation of a tetrahydroisoquinoline in phenylketonuria. *Science* **203**: 283-284.

Makino Y., Tasaki Y., Ohta S. and Hirobe M. (1990) Confirmation of the enantiomers of 1-methyl-1,2,3,4- tetrahydroisoquinoline in the mouse brain and foods applying gas chromatography/mass spectrometry with negative ion chemical ionization. *Biomed. Environ. Mass Spect.* **19**: 415-419.

Maruyama W., Nakahara D., Ota M., Takahashi T., Takahashi A., Nagatsu T. and Naoi M. (1992) *N*--Methylation of dopamine-derived 6,7-dihydroxy-1,2,3,4-tetrahydroisoquinoline, (*R*)salsolinol, in rat brains: In vivo microdialysis study. *J. Neurochem.* **59**: 395-400.

Maruyama W., Dostert P. and Naoi M. (1995a) Dopamine-derived 1-methyl-6,7-dihy droxyisoquinolines as hydroxyl radical promoters and scavengers in the rat brain: In vivo and in vitro studies. *J. Neurochem.* **64**: 2635-2643.

Maruyama W., Dostert P., Matsubara K. and Naoi M. (1995b) N-Methyl(R)salsolinol produced hydroxyl radicals : Involvement to neurotoxicity. *Free Rad. Biol. Med.* **19**: 67-75.

Maruyama W., Abe T., Tohgi H., Dostert P. and Naoi M. (1996a) A dopaminergic neurotoxin, (*R*)-N-methylsalsolinol, increases in parkinsonian cerebrospinal fluid. *Ann. Neurol.* **40**: 119-122.

Maruyama W., Narabayashi H., Dostert P. and Naoi M. (1996b) Stereospecific occur rence of a parkinsonism- inducing catechol isoquinoline, *N*-methyl (*R*) salsolinol, in the human intraventricular fluid. *J. Neural Transm.* **103**: 1069-1076.

Maruyama W., Sobue G., Matsubara K., Hashizume Y., Dostert P. and Naoi M. (1997a) A dopaminergic neurotoxin,1(*R*),2(*N*)-dimethyl-6,7-dihydroxy-1,2,3,4-tetrahydroisoquinoline, *N*--methyl(*R*)salsolinol, and its oxidation product, 1,2(*N*)-dimethyl-6,7-dihydroxyisoquinolinium ion, accumulate in the nigro-striatal system of the human brain. *Neurosci. Lett.* **223**: 61-64.

Maruyama W., Naoi M., Kasamatsu T., Hashizume Y., Takahashi T., Kohda K. and Dostert P. (1997b) An endogenous dopaminergic neurotoxin, N-methyl-(*R*)-salsolinol, induces DNA damage in human dopaminergic neuroblastoma SH-SY5Y cells. *J. Neurochem.* **69**, 322-329.

Maruyama W., Deng Y., Dostert P. and Naoi M. (1997c) Analysis of endogenous enantiomers of neurotoxins in clinical samples by a multi-electrode detection system with a chiral column. In: Progress in HPLC-HPCE. Vol. 6, Coulometric electrode array detectors for HPLC. I. N. Acworth, M. Naoi, H. Parvez and S. Parvez (Eds.) VSP, Utrecht, pp. 339-350.

Matsubara K., Akane A., Masuda C., Takahashi S. and Fukui Y. (1985) Salsolinol in the urine of nonalcoholic individuals after long-term moderate drinking. *Alcohol Drug Res.* **6**: 281-288.

Matsubara K., Fukushima S., and Fukui Y. (1987) A systematic regional study of brain salsolinol levels during and immediately following chronic ethanol ingestion in rats. *Brain Res.* **413**: 336-343.

Matsubara K., Ota M., Takahashi T., Maruyama W. and Naoi M. (1994) Structural studies of condensation products of biogenic amines as inhibitors of trypto phan hydroxylase. *Brain Res.* **655**, 121-127.

Meyers R. D. (1989) Isoquinolines, beta-carbolines and alcohol drinking; involvement of opioid and dopaminergic mechanisms. *Experimentia* **45**: 436-443.

Minami C., Deng Y., Urano F., Kaiya T., Kohda K., Kawai M. and Naoi M. (1997) Preparation of (R)-N-methylsalsolinol *via* reversed-phase high-performance liquid chromatography using β-cyclodextrin as a mobile-phase additive. *J. Chromatogr. B.* in press.

Moser A. and Kämpf D. (1992) Presence of methyl-6,7-dihydroxy-1,2,3,4-tetrahy droisoquinoline, derivatives of the neurotoxin isoquinoline, in parkinsonian lumbar CSF. *Life Sci.* **50**: 1885-1891.

Moser A., Scholz J., Nobbe F. Vieregge P., Böhme V. and Bamberg H. (1995) Presence of N-methyl-norsalsolinol in the CSF: correlation with dopamine metabolites of patients with Parkinson's disease. *J. Neurol. Sci.* **131**: 183-189.

Musshoff F., Daldrup T., and Bonte W. (1993) Gas chromatographic-mass spectro metric screening procedure for the identification of formaldehyde-derived tetrahydroisoquinolines in human urine. *J. Chromatogr.* **619**: 235- 242.

Nagatsu T. and Yoshida M. (1988) An endogenous substance of the brain, tetrahy droisoquinoline, produces parkinsonism in primates with decreased dopamine, tyrosine hydroxylase and biopterin in the nigrostriatal regions. *Neurosci. Lett.* **87**: 178-182.

Naoi M., Matsuura S., Takahashi T and Nagatsu T. (1989a) A N-methyltransferase in human brain catalyzes N-methylation of 1,2,3,4-tetrahydroisoquinoline into N-methyl-1,2,3,4-tetrahydroisoquinoline, a precursor of a dopaminergic neurotoxin, N-methylisoquinolinium ion. *Biochem. Biophys. Res. Commun.* **161**: 1213-1219.

Naoi M., Matsuura S., Parvez H, Takahashi T., Hirata Y., Minami M. and Nagatsu T. (1989b) Oxidation of N-methyl-1,2,3,4-tetrahydroisoquinoline into the N-methylisoquinolinium ion by monoamine oxidase. *J. Neurochem.* **52**: 653-655.

Naoi M., Maruyama W. and Dostert P. (1994a) Binding of 1,2(*N*)-dimethyl-6,7-dihy droxyisoquinolinium ion to melanin: effects of ferrous and ferric ion on the binding. *Neurosci. Lett.* **171**: 9-12.

Naoi M., Maruyama W., Niwa T. and Nagatsu T. (1994b) Novel toxins and Parkinson's disease: N-methylation and oxidation as metabolic bioactivation of neurotoxin. *J. Neural Transm.[Supp]* **41**: 197-205.

Naoi M., Maruyama W., Zhang J. H., Takahashi T. Deng Y. and Dostert P. (1995a) Enzymatic oxidation of the dopaminergic neurotoxin, 1(*R*), 2(*N*)-dimethyl-6,7-dihydroxy-1,2,3,4-tetrahydroisoquinoline, into 1,2(*N*)-dimethyl-6,7-dihydroxyisoquinolinium ion. *Life Sci.* **57**: 1061-1066.

Naoi M., Maruyama W. and Dostert P. (1995b) Dopamine-derived 6,7-dihydroxy-1,2,3,4-tetrahydroisoquinolines; oxidation and neurotoxicity. *Progr. Brain Res.* **106**: 227-239.

Naoi M., Maruyama W., Dostert P., Hashizume Y., Nakahara D., Takahashi T. and Ota M. (1996a) Dopamine-derived endogenous 1(*R*),2(*N*)-dimethyl-6,7-dihy droxy-1,2,3,4-tetrahydroisoquinoline, *N*-methyl-(*R*)- salsolinol,induced parkinsonism in rat: biochemical, pathological and behavioral studies. *Brain Res.* **709**: 285-295.

Naoi M., Maruyama W., Dostert P., Kohda K. and Kaiya T. (1996b) A novel enzyme enantioselectively synthesizes (*R*)salsolinol, a precursor of a dopaminergic neurotoxin, *N*-methyl(*R*)salsolinol. *Neurosci. Lett.* **212**: 183-186.

Naoi M., Maruyama W., Dostert P. and Hashizume Y. (1997a) *N*-Methyl-(*R*)salsolinol as a dopaminergic neurotoxin: From an animal model to an early marker of Parkinson's disease. *J. Neural Transm. [Suppl]* **50**: 89-105.

Naoi M. and Maruyama W. (1997b) N-Methyl(R)salsolinol, an endogenous MPTP-like dopamine neurotoxin and its relevance to Parkinson's disease. In: Advances in Neurology, Parkinson's disease G. Stern (Ed.) Lippincott-Rave, Philadelphia, in press.

Niwa T., Takeda N., Kaneda N., Hashizume Y. and Nagatsu T. (1987) Presence of tetrahydroisoquinoline and 2-methyl-tetrahydroisoquinoline in parkinsonian and normal human brains. *Biochem. Biophys. Res. Commun.* **144**: 1084-1089.

Niwa T., Takeda N., Sasaoka T., Kaneda N., Hashizume Y., Yoshizumi H, Tatematsu A. and Nagatsu T. (1988) Migration of tetrahydroisoquinoline, a possible parkinsonian neurotoxin, into monkey brain from blood as proved by gas chromatography-mass spectrometry. *J. Chromatogr.* 452: 85-91.

Niwa T., Takeda N., Yoshizumi H., Tatematsu A., Yoshida M., Dostert P., Naoi M. and Nagatsu T. (1991) Presence of 2-methyl-6,7-dihydroxy-1,2,3,4-tetrahydroisoquinoline and 1,2-dimethyl-6,7-dihydroxy-1,2,3,4-tetrahydroisoquinoline, novel endogenous amines, in parkinsonian and normal human brains. *Biochem. Biophys. Res. Commun.* **177**: 603-609.

Ohta S., Kohno M., Makino Y., Tachikawa O. and Hirobe M. (1987) Tetrahydroisoquinoline and 1-methyl-tetrahydroisoquinoline are present in the human brain: Relation to Parkinson's disease. *Biomed. Res.*, **8**: 453- 456.

Origitano T., Hannigan J. and Collins M. A. (1981) Rat brain salsolinol and blood-brain barrier. *Brain Res.* **224**: 446-451.

Pianezzola E., Bellotti V., Fontana E., Moro E., Gal J., and Desai D. M. (1989) Determination of the enantiomeric composition of salsolinol in biological samples by high-performance liquid chromatography with electrochemical detection. *J. Chromatogr.* **495**: 205-214.

Riggin R. M. and Kissinger P. T. (1977) Determination of tetrahydroisoquinoline alkaloids in biological materials with high performance liquid chromatography. *Anal. Chem.* **49**: 530-533.

Robbins J. H. (1968) Alkaloid formation by condensation of biogenic amines with acetaldehyde. *Clin Res.* **16**: 350.

Sällström Baum S. and Rommelspacher H. (1994) Determination of total dopamine, R- and S-salsolinol in human plasma by cyclodextrin bonded-phase liquid chromatography with electrochemical detection. *J. Chromatogr.* **660**: 235-241.

Sandler M., Bonham Carter S., Hunter K. R. and Stern G. M. (1973) Tetrahydroisoquinoline alkaloids: in vivo metabolites of L-DOPA in man. *Nature* **241**: 439-443.

Sasaoka T., Kaneda N., Niwa T., Hashizume Y. and Nagatsu T. (1988) Analysis of salsolinol in human brain using high-performance liquid chromatography with electrochemical detection. *J. Chromatogr.* **428**: 152-155.

Sjöquist B. and Magnuson E. (1980) Analysis of salsolinol and salsoline in biological samples using deuterium-labeled internal standards and gas chromatography-- mass spectrometry. *J. Chromatogr.* **183**: 17-24.

Sjöquist B., Borg S. and Kvande H. (1981a) Catecholamine derived compounds in urine and cerebrospinal fluid from alcoholics during and after long-standing intoxication. *Subst. Alc. Act/Misuse* **2**: 63-72.

Sjöquist B., Borg S. and Kvande H. (1981b) Salsolinol and methylated salsolinol in urine and cerebrospinal fluid from healthy volunteers. *Subst. Alc. Act/Misuse* **2**: 73-77.

Sjöquist B, Eriksson A. and Windbad B (1982) Salsolinol and catecholamines in hu man brain and their relation to alcoholism. *Prog. Clin. Biol. Res.* **90**; 57-67.

Sjöquist B. and Ljungquist C. (1985) Identification and quantification of 1-car boxysalsolinol and salsolinol in biological samples by gas chromatography-mass spectrometry. *J. Chromatogr. Biomed. Appl.* **343**: 1-8.

Strolin Benedetti M., Bellotti V., Pianezzola E. Moro Carminati P. and Dostert P. (1989a) Ratio of the R and S enantiomers of salsolinol in food and human urine. *J. Neural Transm.* **77**: 47-53.

Strolin Benedetti M., Dostert P. and Carminati P. (1989b) Influence of food intake on the enantiomeric composition of urinary salsolinol in man. *J. Neural Transm.* [GenSect] **78**: 43-51.

Takahashi T., Deng Y., Maruyama W., Dostert P., Kawai M. and Naoi M. (1994) Uptake of a neurotoxin-candidate, (R)-1,2-dimethyl-6,7-dihydroxy-1,2,3,4-tetrahydroisoquinoline into human dopaminergic neuroblastoma SH- SY5Y cells by dopamine transport system. *J. Neural Transm. [GenSect]* **98**: 107-118.

Takahashi T., Maruyama W., Deng Y., Dostert P., Nakahara D., Niwa T., Ohta S. and Naoi M. (1997) Cytotoxicity of endogenous isoquinolines to human dopaminergic neuroblastoma SH-SY5Y cells. *J. Neural Transm.* **104**: 59-66.

Teitel S., O'Brien J. and Brossi A. (1972) Alkaloids in mammalian tissue. II. Synthesis of (+) and (-) substituted 6,7- dihydroxy-1,2,3,4-tetrahydroisoquinolines. *J. Med. Chem.* **15**: 845-846.

Ung-Chhun N., Cheng B. Y., Pronger D. A., Serrano P., Chavez C., Perez R. F., Morales J. and Collins M. A. (1985) Alkaloid adducts in human brain: coexis tence of 1-carboxylated and noncarboxylated isoquinolines and ß-carbolines in alcoholics and nonalcoholics, in Aldehyde Adducts in Alcoholism M. A. Collins (ed), Alan R. Liss, New York, pp. 125-136.

Progress in HPLC-HPCE, Vol. 7, pp. 461-494
G. Ali Qureshi *et al.* (Eds)

POLYAMINES AND RELATED COMPOUNDS AS MARKERS OF PHYSIOPATHOLOGY AND CELL DIFFERENTIATION : ANALYTICAL METHODS

Victor Raul VILLANUEVA
Institut de Chimie des Substances Naturelles, C.N.R.S.,

I - INTRODUCTION

Polyamines are a group of organic polycations represented by spermidine, spermine and their diamine precursor putrescine. These aliphatic amine compounds are ubiquitous in all living systems. Prokaryotic organisms normally contain putrescine and spermidine, eukaryotic cells containing spermine in addition.

Owing to their polycationic character polyamines are able to interact with nucleic acids and other polyanionic cellular components. This could explain their involvement in processes related to cell division, cell growth and cell differentiation. Polyamines have been reported to play an important role in processes such as DNA synthesis, chromosomal condensation, cell-free protein synthesis, early embryonic gene activation, cytokinesis and a considerable number of other cell metabolism processes.

They have been shown to be agonist of the NMDA receptor at the post synaptic level (Trout et al., 1993; Rock and Mcdonal, 1992). In addition, spermine favors the long term potentiation of synaptic activity (Stevens, 1988; Chida et al., 1992) increases the affinity of quisqualate for the NMDA receptor (Alexander et al., 1992) and interferes with the binding of MK-801 to the NMDA receptor (William et al., 1992).

Polyamine contents vary with both cell type and physiological state. Generally, rise of polyamine synthesis precede organ regeneration and elevation of hormone levels in animal following growth stimuli. Polyamines are also increased in rapidly growing normal and neoplastic tissue.

The biochemical and biological interest on polyamines are illustrated by the large number of publications devoted to their study. Readers interested in this field are referred to the reviews or books by Tabor and Tabor (1964), Cohen (1971), Bachrach (1973), Russel (1978), and Smith (1977), for example.

Research on polyamines was relatively paceful prior to 1960 compared to many other fields. In the following period until 1970, there was a dramatic burst of interest on the polyamines in eukaryotic and prokaryotic cells, as well as in many *in vitro* systems. During the last decade and probably after the report of Russel (1971) describing the elevated levels of polyamines in the urine of cancer patients, polyamines have attracted the attention of scientists and clinicians owing to their implication in growth process and their association with cancer. It is also during this period that the most sophisticated analytical methods are developed leaving out far behind the earlier methods like crystallization previously used by biologists. Besides paper and thin layer chromatography, several analytical methods have been developed based on analytical instrumentation like gas chromatographs, mass spectrometers, amino acids analyzers, electrophoresis apparatus, etc...

In general most of these methods concern only the separations of putrescine, cadaverine, spermidine, spermine and a few other related compounds; the analysis of their naturally occuring acetylated derivatives has received some attention. In addition studies concerning the polyamines in plants and microorganisms have led to the identification of a number of unusual polyamines e.g. *sym*-norspermidine, *sym*-norspermine, *sym*-homospermidine, aminopropyl cadaverine, caldopentamine, homo-caldopentamine, etc. (Table 1). This latter group of polyamines, initially believed to be peculiar to certain organisms, has there after been detected in other organisms (Stilway and Walle, 1977; Zappia *et al.,* 1978; Villanueva *et al.*, 1980); on the other hand, the possibility of their occurrence in most living organisms, as minor components cannot be ruled out. They might easily be lost through analytical steps (i.e. extraction procedure, fractionation, derivatisation, etc...) or alternatively they may escape detection because of overlapping with the more common polyamines due to low resolution efficiency of the analytical method used. Moreover, *sym*-norspermidine(N-3-amino propylheptane-1,7 diamine) has often been used as an internal standard in polyamine analysis. Thus, when starting an investigation with a new biological cell system a preliminary analysis on its polyamine composition utilizing a high resolution analytical method its extremely desirable. At present, the analysis of polyamine from biological sources requires methods for the determination of : a) free polyamine composition; b) polyamine conjugates (e.g. acetyl, cinnamyl or peptidyl derivatives) and c) macromolecular polyamine containing compounds.

TABLE 1

AMINE	STRUCTURE	REF*
1,3-DIAMINOPROPANE	$NH_2(CH_2)_3NH_2$	DEROSA *et al.* (1976)
PUTRESCINE	$NH_2(CH_2)_4NH_2$	BACHRACH (1973)
2-OHPUTRESCINE	$NH_2CH_2\text{-}CH_2\text{-}CHOH\text{-}CH_2NH_2$	KULLLNIG *et al.* (1970)
CADAVERINE	$NH_2(CH_2)_5NH_2$	BACHRACH (1973)
AGMATINE	$NH_2(CH_2)_4NH{-}C({=}NH){-}NH_2$	BACHRACH (1973)
3,3'-DIAMINODIPROPYLAMINE (*Sym*-nor-SPERMIDINE or CALDINE)	$NH_2(CH_2)_3NH(CH_2)_3NH_2$	OSHIMA (1983)
SPERMIDINE	$NH_2(CH_2)_3NH(CH_2)_4NH_2$	ROSENHEIM (1924)
Sym-HOMOSPERMIDINE	$NH_2(CH_2)_4NH(CH_2)_4NH_2$	KUTAN *et al.* (1971)
AMINOPROPYL CADAVERINE	$NH_2(CH_2)_3NH(CH_2)_5NH_2$	DION *et al.* (1972)
N,N'-bis-AMINOPROPYL-1,3-PROPANEDIAMINE (*Sym*-nor-SPERMINE or THERMINE)	$NH_2(CH_2)_3NH(CH_2)_3NH(CH_2)_3NH_2$	OSHIMA (1983)
THERMOSPERMINE	$NH_2(CH_2)_3NH(CH_2)_3NH(CH_2)_4NH_2$	OSHIMA (1983)
SPERMINE	$NH_2(CH_2)_3NH(CH_2)_4NH(CH_2)_3NH_2$	ROSENHEIM (1924)
CALDOPENTAMINE	$NH_2(CH_2)_3NH(CH_2)_3)NH(CH_2)_3\text{-}NH(CH_2)_3NH_2$	OSHIMA (1983)
HOMOCALDOPENTAMINE	$NH_2(CH_2)_3NH(CH_2)_3NH(CH_2)_3NH\text{-}(CH_2)_4NH_2$	OSHIMA (1983)

* The complete bibliographic reference list is in alphabetic order in the reference chapter at the end of this article.

TABLE 2

ANALYTICAL METHODS USED FOR POLYAMINE ASSAY

Method	Details	Relative sensitivity
Paper chromatography	Ninhydrin	0.1-0.5 μmoles
Thin layer chromatography (TLC)	Ninhydrin	5-100 nmoles
	Dansyl derivatives	10-100 pmoles
Overpressure thin layer chromatography	Ninhydrin + Cd^{++}	2-60 nmoles
High performance liquid chromatography	Tosyl derivatives	1-10 nmoles
(HPLC)	Fluorescamine derivatives	50-200 pmoles
	Enzymatic + chemiluminescent	5-500 pmoles
	Dansyl derivatives	10-100 pmoles
Ion-exchange chromatography (amino analyzer)	Ninhydrin	1-50 pmoles
	Fluorescamine	5-200 nmoles
	O-Phtalaldehyde	5-200 pmoles
Gas chromatography	*N*-trifluoroacetyl (N-TFA) derivatives	100-600 pmoles
Electrophoresis	Ninhydrin	5-200 nmoles
Mass spectrometry	N-TFA or Dansyl derivatives	10-100 pmoles
Gas chromatography-mass spectrum	N-TFA derivatives + deuterated internal standards	1-10 pmoles
Enzymatic methods	*Serratia marcescens*	5-10 nmoles
Immunonephelometry	Latex beads	1-10 pmoles
Radioimmunoassay (RIA)	Purified antibodies	50-200 fmoles

In this chapter we will discuss the methods currently in use for the determination of polyamine and related compounds (see Table 2). These methods will be subdivised into physico-chemical and biological methods. Particular emphasis will be given to the anlysis by ion exchange chromatography with the use of an aminoacid analyzer which is actually the most favored method for routine analysis and where several contributions from our laboratory have been made.

II - PHYSICO-CHEMICAL METHODS

1. Chromatographic methods

a) Paper chromatography

Paper chromatography is one of the simplest and oldest methods used for the analysis of polyamines. The most commonly used solvents and the Rf values of various polyamines have been summarized by Bachrach (1973).

Ninhydrin (0.1 to 0.5% in either n-butanol, ethanol, acetone or chloroform) is usually employed to detect the amine compounds. After spraying the paper, the chromatogram is heated at 80°-90°C for 5 to 15 minutes. Quantitative assay can be carried out by extraction of the ninhydrin stained spots by 75% ethanol. The intensity of the color is determined photocolorimetrically at 570 nm and the concentration calculated by using a calibration curve obtained by the same chromatographic and extraction methods of known quantities of authentic polyamine compounds.

Although simple, this method is very long and lacks resolution efficiency. This method is at present rarely used and has been replaced by faster methods, such as thin layer chromatography.

b) Thin layer chromatography

This is one of the most popular analytical methods because of its simplicity, rapidity and high sensitivity. Cellulose and Silicic Acid plates are used to analyse free polyamines, or their colored or fluorescent derivatives.

Bachrach (1973) has summarized the most commonly used solvents and the Rf values obtained. Free polyamines are detected after drying the plates, by spraying with ninhydrin solutions and heating at 90-100° for 5 to 15 minutes. The stained spots can be scrapped off the plate and extracted with ethanol as described for paper chromatography. Instead of ninhydrin Shimizu *et al.* (1964) have used 2,4-dinitrofluorobenzene solution and more recently Abe and Samejina (1975) have used fluorescamine which gives fluorescent derivatives and increases sensitivity of detection.

Holder and Bremer (1966) could separate putrescine (Rf 0.39) cadaverine (Rf 0.45) spermidine (Rf 0.28) and spermine (Rf 0.13) using cellulose plates and a solvent system containing isopropanol-37% HCl-H_2O (8:3:2). Blankenship and Walle (1977) have separated the two isomeric monoacetylated (N^1 and N^8)spermidine on silica gel G plates using a solvent system of chloroform-methanol-ammonium hydroxide (2:2:1). N^1-acetyl spermidine migrates with an Rf of 0.37 and N^8-acetyl spermidine with an Rf of 0.47.

The use of the dansylation procedure for polyamine derivatization, introduced by Seiler and Wiechmann in 1965 has considerably improved the performances and posibilities of TLC for amine analysis. TLC has become one of the simplest, least expensive and most sensitive (1-100 picomoles) methods. Moreover, the commercialisation of the so-called high performance thin layer plates HPTLP (Merck, Whatman, Schlicher and Schüll for example) in the course of the last few years has also contributed towards excellent separation of amines by this method specially when the bidimensional technique is employed (Abdel-Monem et al. 1978).

Quantitative determination can be made by *in situ* fluorimetry with the help of a scanner after spraying the chromatograms with triethylamine solution in propanol-2, which enhances fluorescence intensity and stabilizes the Dns derivatives (Seiler, 1970).

Procedures for dansylation of samples has been described by Seiler (1970). The following procedure can be used : 0.1 to 0.5 ml of 0.2 N perchloric acid extracts (or even homogenized material) are placed in glass-stoppered centrifuge tubes. To 0.1 ml of extract 0.3 ml of a solution of dansyl chloride (30 mg/ml) in acetone is added and the reaction mixture saturated by the addition of 200 mg of $Na_2CO_3 \cdot 10\ H_2O$. Reaction is completed by storing the centrifuge tubes at room temperature overnight. 0.1 ml of a proline solution (150 mg/ml of water) is added the next morning to destroy the excess of dansyl chloride and 30 minutes later the samples are mixed with 5 ml of toluene, thoroughly shaken, and the layers separated by centrifugation. Aliquots (20 µl) of the organic layer are then applied to thin layer plates. Seiler and Knödgen (1977) have determined putrescine, spermidine and spermine as dansyl derivatives, from mouse liver tissue, using 5 x 5 cm high-performance thin layer plates (Merck, Darmstadt, G.F.R.) developed with a solvent mixture of cyclohexane-ethyl acetate (1:1) (two runs).

Abdel-Monem *et al.* (1978) have used two dimensional TLC (silica gel GF 250 µm) to analyse the presence of free and acetylated polyamines in human pathological urine. Cyclohexane-diethyl ether (1:9) was used as solvent for the first dimension and chloroform-triethylamine (10:1) in the second dimension to chromatograph the dansylated derivatives. This method has been successfully applied to the analysis of polyamines from different biological sources. For example Cohen *et al.* (1969) have analysed the polyamine content of the t-RNA of *E. coli*. Dion *et al.* (1970) measured the polyamine changes during development of *Drosophila melanogaster*; Fleisher and Russell (1975) measured polyamines from urine, while Seiler and Schmidt-Glenewinker (1975) measured amines in rat nervous system. Seiler *et al.* (1970) have applied mass-spectrometric analysis to the dansyl polyamine derivatives extracted from the various spots of the thin layer chromatogram.

c) Overpressured thin layer chromatography (OPTLC)

An appareil which allows to practice TLC over pressure has also been developed (Tyihak *et al.* 1979) and commercialised by LABOR Mim, and hungarian manufacturer, under the name of "CHROMPRES 25". The OPTLC technique requires a special chromatoplate that is sealed at the edges. This prevent the eluent from flowing off the chromatoplate in an unwanted direction. Linear migration, of the eluent front in the OPLC chamber using the linear developing mode, can be achieved by placing a narrow channel before the eluent inlet. The appareil for OPTLC is an unclosed, horizontal, pressurized chromatographic chamber, where no vapour space exists above the sorbent layer, since a flexible membrane covers the layer under an externally applied pressure of 25 bars. Overpressure is provided by a cushion system filled with water by a pump. The structural parts of the chromatographic chamber and the layout of the system are shown in Fig. 1. Witkiewicz et al. (1993) have recently developed a new appareil, OPLC BS 50 (OPLC-NIT Engineering Company Ltd., Budapest, Hungary) which presents several improvements over the Chrompres-25.

Basdócz *et al.* (1985) have used the Chrompres 25 to separate ornithine, arginine, putrescine, spermidine and spermine in 30 min. using 20 x 20 cm. FIXION 50 X 8 thin layer cation exchange chromatosheets (Reanal, Budapest, Hungary) and 0.3 M potassium phosphate buffer, pH 7.5 containg 2 M sodium chloride.

d) High performance liquid chromatography (HPLC)

HPLC is a fully automated fast and high sensitive column chromatographic method. Polyamine analysis on reversed-phase and adsorption colums has been described using its fluorescamine derivatives (Samejina, 1974), tosyl derivatives (Sugiura *et al.*, 1975), and dansyl derivatives (Abdel-Monem and Ohno, 1975 and Seiler *et al.*, 1977).

HPLC of the dansylated polyamine derivatives has gained in popularity. Currently determination of 10-100 picomoles quantities of amine by fluorimetry can be achieved by this method. Sample preparation is carried out as described in TLC section.

Abdel-Monen *et al.* (1978) have investigated a number of chromatographic systems including adsorption and partition modes and using isocratic solvent elution or programmed gradient elution. Diaminopropane, putrescine, spermidine and spermine have been separated in less than 20 minutes using a micropak CN-10 column (23 cm x 0.25 cm ID) and cyclohexane-methylene chloride-isopropanol (88.5:7.5:40) at a flow-rate of 60 ml/h (1600-1700 psi) at room temperature. Abdel-Monen and Ohno (1977) have also separated polyamines and their natural acetylated derivatives using isocratic

(A)

(B)

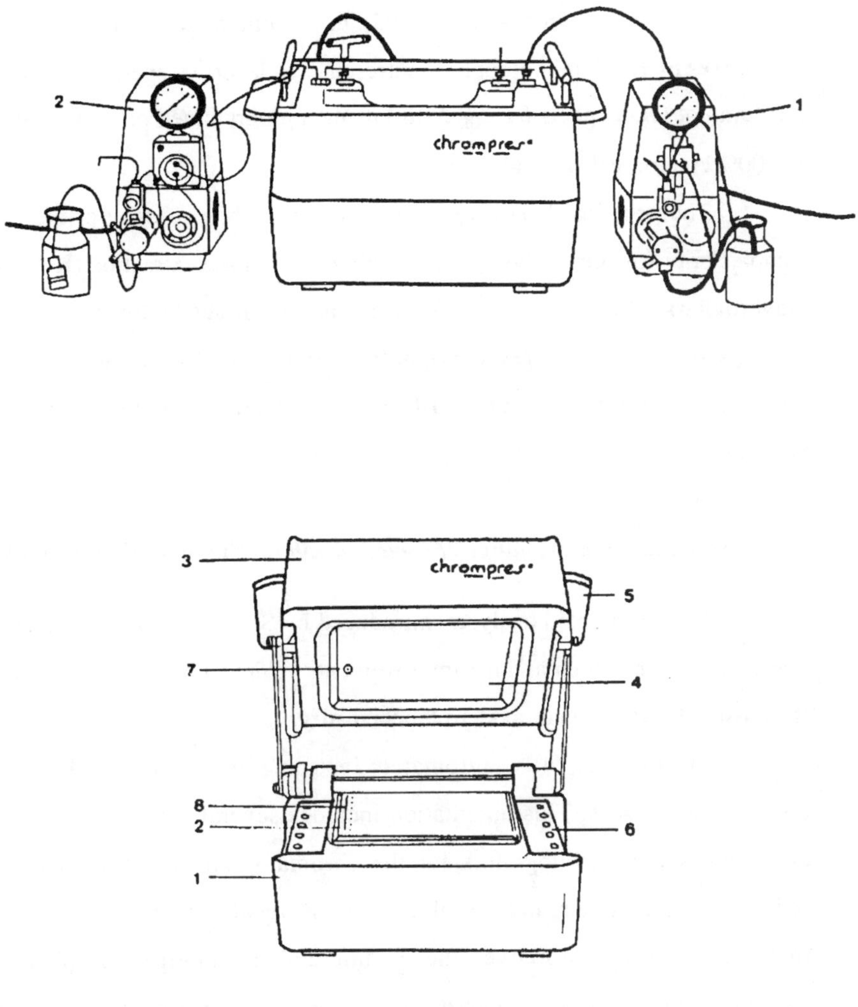

Fig. 1. (A) Chrompres 25 OPLC system. 1 : Water pump for pressure production. 2 : Solvent delivery pump. (B) Chrompres 25 open position. 1 : Bottom support block. 2 : Working plate. 3 : Cover plate. 4 : Cushion system. 5 : Cover plate handle. 6 : Locking device. 7 and 8 : eluant orifices.

elution with chloroform-isopropanol (50:3) and a column (50 x 0.22 cm i.d.) packed either with microparticles of silica gel (Micropak Si-10, Varian Aerograph) or with a column (100 cm x 0.2 mm i.d.) of Corasil II.

Recently Marcé *et al.* (1995) have separated in only 9 min. the dansyl derivatives of 1,3-diaminopropane, putrescine, cadaverine, spermidine and spermine on a column of ODS spheri-5 (C_{18}, 5 μm) using a programmed gradient elution from 70% to 100% Acetonitrile as mobil phase.

The principal handicap of this method is the necessity of the purification of samples prior to their analysis, in general by solid-phase extraction (Kabra *et al.*, 1986). Nevertheless, Kamei *et al.*, 1989 have reported direct hydrolysed urine polyamine analysis by ion-paired, reversed-phase chromatography on an ODS column using enzymatic-chemiluminescent post- column detection without prior purification samples.

e) Automatic ion exchange chromatography using an amino-acid analyzer

The amino-acid analyser, introduced by Spackman *et al.* in 1958, has evolved considerably since then and the improvements achieved have made this method one of the most valuable tools in modern biochemistry.

At present, fully automated instruments are available from different commercial sources. This automation includes several accessories (like refrigerated automatic samples, integrators for the chromatogram quantification, etc...) which reduces manual labor and variability in the results, thereby contributing toward increased output. Moreover the commercial availability of spherical beads of sulfonated crosslinked styrene-divinylbenzene copolymer with a narrow range of diameters of 10 μm and then to 5 μm or less, has improved the resolution and sensitivity of this method.

Sensitivity of the detection method depends also on the reagents used. Ninhydrin allows nanomolar determination by photocolorimetry, while fluorimetry using fluorescamine or o-phtalaldehyde allows picomolar determinations.

Morris *et al.* (1969) reported the use of an aminoacid analyzer for polyamine analysis. Since then a number of methods for the determination of polyamines in biological material has been reported (Abdel-Monem, 1977).

In our laboratory, various automated ion exchange chromatographic methods have been developed and used for the study of polyamines in crude extracts from different biological sources (Villanueva *et al.*, 1977; Villanueva *et al.*, 1978; Villanueva *et al.*, 1980; Adlakha *et al.*, 1980; Bagni *et al.*,1978; Lenfant *et al.*, 1979). Two alternative methods currently used by us will be described here : the first one uses sodium citrate buffers and separated polyamines and phenolicamines which can overlap (for example, dopamine and spermidine; phenylethyl amine, tyramine and spermine). This method also includes some indole-amines and biosynthetically related aminoacids. The second method, which uses lithium citrate buffers separates most of the natural (usual and unusual) polyamines, some related diamines and their biosynthetically related aminoacids.

Both methods use fluorimetric detection employing o-phthalaldehyde as a fluorogenic reagent. The methods described here can be carried out with any commercially available aminoacid analyser equiped at least with a change valve to deliver sequentially three buffers and with a thermostatic system allowing the use of two temperatures during the chromatographic process.

- Preparation of samples for automated ion-exchange chromatography

In our laboratory the routine general procedure used for extration of polyamines consists of suspending the biological material (cell cultures, pieces of animal organs, plants, microorganisms, etc...) into 2-5 volumes of a solution of 5% TCA in 0.01 N HCl. The suspension is homogenized using an ultramixer or sonication, followed by centrifugation to eliminate cell debris. This operation is repeated once or twice and the supernatants pooled. Elimination of TCA from the extracts is not necessary prior to analysis. For the analysis of physiological fluids (e.g. plasma, serum, urine) TCA solution is used for deproteinisation in a final concentration of 8%.

After homogenization and centrifugation 50-100 μl of supernatant aliquots are analyzed directly following the addition of 300 picomoles of 1,7-diaminoheptane which is used as an internal standard. If necessary, samples are hydrolysed by addition of an equal volume of 37% HCl (Merck) in sealed tubed for 24 h at 110°C under nitrogen. After hydrolysis the samples are filtered and vacuum dried in a dessicator containing KOH pellets. HCl is eliminated by repeated aqueous washes and drying. The samples are then dissolved in 0.1 M HCl and aliquots (10-100 μl) are analysed after addition of the internal standard.

Commercial sources of amines are Sigma Chemical Company, St. Louis, Missouri; Merck, Darmstadt, Germany and Eastman Organic Chemical, Rochester, New-York. Salts, acids and alcohols can be obtained from Merck.

Method N° 1

In this method the separation of individual amines is achieved on a 0.45 x 9.5 cm column of DC-4A, Na^+ cation-exchange resin (Durrum Chemical Corp., Palo Alto, CA). The composition of the eluting buffers and the elution program are as indicated in Table 3. Buffers are prepared from a freshly double-distilled deionized water, brought to the required pH with concentrated HCl and filtered through a millipore filter (0.22 μm pore size). After filtration the alcohols are added. This raises the pH in 0.05 units.

The fluorogenic reagent is prepared according to Roth and Hampai (1973) by dissolving 800 mg of *o*-phtalaldehyde in 10 ml of ethanol to which 2 ml of β-mercapto ethanol has been added. This mixture is then added to 1 L of 2.5% boric acid solution (brought to pH 10.40 with 45% KOH) containing 3 ml of an aqueous solution (30%) of Brij 35. Buffers are driven at a flow rate of 20 ml/h. Column back pressure generated during analysis is approximately 25 kg/cm^2. Two temperatures are used : 66°C during the first 64 min. and thereafter 76°C till the end of the chromatogram. The column is regenerated with 0.2 N NaOH (containing 250 mg/L of EDTA) for 8 minutes and reequilibrated by passing the first buffer for 25 minutes. Fig. 2 shows the chromatographic resolution of a standard commercial mixture containing 500

TABLE 3

COMPOSITION AND CONDITIONS OF ELUTING BUFFERS

	Buffers		
	First	**Second**	**Third**
Sodium citrate dyhydrate	0.20 N (19.60 g/L)	0.20 N (19.60 g/L)	0.20 N (19.60 g/L)
Sodium chloride	0.20 N (11.69 g/L)	2.30 N (134.40 g/L)	2.60 N (151.94 g/L)
pH before addition of alcohols	5.50 ± 0.01	5.58 ± 0.01	5.63 ± 0.01
Alcohol added	Ethanol 4.0%	Ethanol 5.0%	Isopropanol 10%
Final pH	5.55 ± 0.01	5.63 ± 0.01	5.68 ± 0.01
Time (min)	39	42	64

picomoles of each aminoacid, mono-di, and polyamines, phenolic amines and indole amines. 4-azaheptamethylene-1,7-diamine is used as an internal standard. The components of the mixture applied to the column, the retention time and the relative fluorescence peak areas are shown in Table 4. The column and the elution program employed give good overall resolution. With the first buffer, acidic and neutral aminoacids (aspartic acid through phenyl alanine included) are eluted in a block. Hydroxytryptophan, tryptophan, histidine, lysine, ethanolamine, ammonia and arginine are subsequently eluted, in that order, with good resolution. The second buffer elutes and separates the mono-, di-, and polyamines and phenolamines. The elution of the indoleamines requires the use of a third buffer of increased ionic strength and containing 10% isopropanol; the column temperature must also be raised from 66°C to 76°C. Complete analysis occurs in 150 min. The presence of alcohols in the buffers is necessary to obtain a good resolution and sharp peaks. The influence and optimal concentration of four different alcohols (methanol, ethanol, propanol and isopropanol)

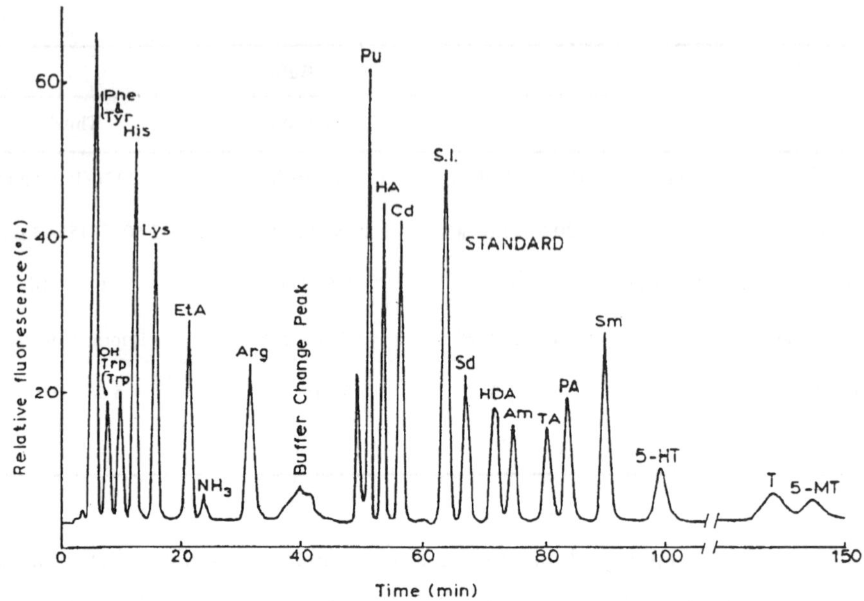

Fig. 2 - Chromatogram of a standard mixture of aminoacids, mono-, di- and polyamines, phenolic- and indoleamines, all at 500 picomoles except for the internal standard (375 picomoles). See Table 4 for the abbreviations and their chromatographic behavior.

in the buffers have been studied. Those which gave better resolution, sharpness of peaks and eluting force were finally adopted. Isopropanol was chosen for the third buffer as it accelerated well the elution of indoleamines without shifting the baseline. The elution capacity of the different alcohols for the indoleamines is as follows : methanol < ethanol < propanol < isopropanol. The percentage recovery of the different compounds after column chromatography has been studied (Villanueva and Adlakha, 1978), and is included in Table 4. Recoveries varied between 70 and 90% for all compounds except agmatine (48%), tryptamine (57%), phenylethylamine (61%) and histamine (67%). Quantification is done with the aid of an integrator coupled to the fluorimeter. The integrator is previously calibrated with a standard mixture and by

TABLE 4

PRECISION AND ACCURACY OF THE ANALYSIS OF MONO-, DI- AND POLYAMINES, PHENOLIC AND INDOLEAMINES*

Compound	Abbreviation	Retention time (min)	Surface area of peaks	% recovery after column chromatogr.
5-Hydroxytryptophane	OHTrp	7.6 ± 0.1	45 661	90.2
Tryptophane	Trp	9.9 ± 0.1	49 974	81.8
Histidine	His	11.7 ± 0.2	87 314	94.6
Lysine	Lys	15.7 ± 0.2	74 514	97.3
Ethanolamine	EtA	22.4 ± 0.1	58 331	91.9
Ammonia	NH_3	24.7 ± 0.2		
Arginine	Arg	32.4 ± 0.2	55 632	77.9
1,3-Diaminopropane	DAP	50.1 ± 0.1	18 790	76.6
Putrescine	Pu	51.6 ± 0.2	68 405	90.1
Histamine	HA	54.4 ± 0.3	54 683	67.2
Cadaverine	Cd	56.6 ± 0.2	58 278	79.7
4-Azaheptamethylenediamine	I.S.	64.0 ± 0.2	89 498	81.2
Spermidine	Sd	67.2 ± 0.3	40 226	80.5
Hexamethylenediamine	HDA	72.5 ± 0.2	35 364	72.7
Agmatine	Am	75.3 ± 0.2	31 325	48.1
Tyramine	TA	80.8 ± 0.2	31 099	85.7
Phenylethylamine	PA	84.1 ± 0.2	39 967	61.6
Spermine	Sm	90.5 ± 0.3	56 743	70.3
Serotonine	5-HT	99.4 ± 0.2	28 434	71.1
Tryptamine	T	140.7 ± 0.2	27 919	57.8
5-Methoxytryptamine	5-MT	147.4 ± 0.2	25 677	77.5

*Average of ten determinations of a mixture containing 500 picomole of each of the products. 4-Aza heptamethylene diamine (I.S.) = 375 picomoles. Quantification was effected with an integrator coupled to the fluorimeter.

employing the method of the internal standard. The method described here analyses in a relatively short time the common basic aminoacids, mono-, di-, poly-, phenolic-, and indole amines in the picomole range in a single crude sample. A three-step change of buffers and two temperatures are employed. Incorporation of alcohols into the buffers dramatically improved the separation and sharpness of peaks. Strongly bound compounds such as tryptamine and 5-methoxy-tryptamine can also be eluted by using isopropanol without much increase in the ionic strength.

If one is not interested in analyzing indoleamines, the third buffer is not necessary and the analysis including spermine takes only 110 minutes.

Method N° 2

A method (Villanueva and Adlakha, 1980) which separates most of the usual and unusual natural polyamines, related basic aminoacids but not the most important of them, namely, ornithine which coelutes with lysine was modified to allow the determination of polyamines and their biogenetically related precursors in the same sample. It uses two lithium citrate buffers, only one column temperature and a resin which is one of the least expensive commercially available.

This method resolves well ornithine, lysine, ammonia, ethanolamine, arginine, carbamylputrescine, histamine, diaminopropane, putrescine, cadaverine, *nor*-spermidine, spermidine, homospermidine, agmatine, *nor*-spermine, spermine and 1,7-diaminoheptane (which serves as an internal standard) in a single sample in 110 minutes. A column (0.4 x 13.0 cm) of HC-X7 cation-exchange resine, Li^+ (Benson Co., Reno, Nevada) is used. Buffers are prepared as for the method N°1 using lithium citrate buffers (see Table 5). Buffers are driven at a flow rate of 23 ml/h and the fluorogenic reagent at 20 ml/h. Column back pressure generated during analysis is less than 5 kg/cm^2 and a unique temperature, 74°C, is used during the chromatography. The column is regenerated with 0.3 N LiOH (containing 250 mg/L of EDTA) for 8 minutes and reequilibrated by pasing the first buffer for 20 minutes. For automated

TABLE 5

COMPOSITION AND CONDITIONS OF ELUTING BUFFERS

	Buffers	
	First	**Second**
Lithium citrate	0.40 N (37.60 g/l)	0.60 N (56.40 g/l)
Potassium chloride	0.12 N (9.00 g/l)	2.00 N (150.00 g/l)
Boric acid	0.30 N (18.60 g/l)	-
pH before addition of ethanol	8.40 (by addition of 3 M LiOH)	5.35 (by addition of conc. HCl)
Ethanol added	-	5.5%
Final pH	-	5.40
Time (min)	44	55

analysis of a series of samples, 10 min. intervals after each analysis is advised. This allows the compressed resin column to come back to its original volume and will prevent column back pressure from increasing. Otherwise resolution efficiency will be lost progressively. Fig. 3 shows the chromatographic resolution of a standard commercial mixture of 300 picomoles of each of aminoacids, mono-, di-, and polyamines. The components of the mixture applied to the column, the retention time and the relative fluorescence peak areas are shown in Table 6.

The column and the elution program employed give good overall resolution. With the first buffer, the acidic and neutral aminoacids (aspartic through phenylalanine) and histidine included are eluted in a block. Ornithine, lysine, ammonia, ethanolamine, arginine, carbamoyl-putrescine and histamine are subsequently eluted in that order with good resolution. The second buffer elutes and separates the mono-, di- and polyamines. The presence of alcohol (5.5% ethanol) in the second buffer is necessary for the sharpness of peaks and to accelerate the elution of certain amines particularly the internal standard 1,7-diaminoheptane.

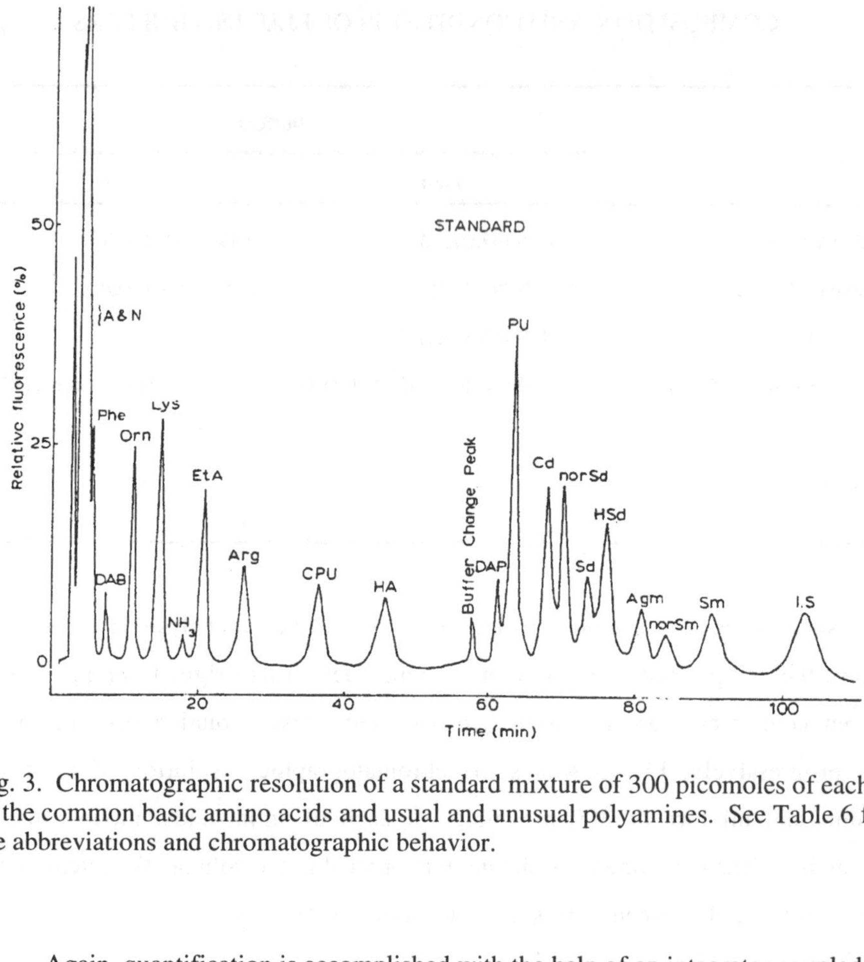

Fig. 3. Chromatographic resolution of a standard mixture of 300 picomoles of each of the common basic amino acids and usual and unusual polyamines. See Table 6 for the abbreviations and chromatographic behavior.

Again, quantification is accomplished with the help of an integrator coupled to the fluorimeter and by employing the method of the internal standard as indicated for the first chromatographic method.

Fig. 4 shows the chromatogram obtained from a crude sample of human erythrocytes. This method is actually routinely used in our laboratory for the analysis of crude samples from many biological sources (plantes, microorganisms, physiological fluids, animal organs extracts, etc...).

TABLE 6

PRECISION AND ACCURACY OF THE ANALYSIS OF MONO-, DI- AND POLYAMINES AND RELATED COMPOUNDS*

Compound	Abbreviation	Retention time (min)	Fluorescence peak areas		
			Mean	SD	RSD %
2,4-Diaminobutyric acid	DAB	8.1 ± 0.1	35 192	423	1.2
Ornithine	Orn	11.3 ± 0.1	145 285	861	0.6
Lysine	Lys	14.8 ± 0.1	207 698	6543	3.2
Ammonia	NH_3	17.8 ± 0.1	25 865	539	2.1
Ethanolamine	EtA	20.8 ± 0.1	163 464	2134	1.3
Arginine	Arg	25.0 ± 0.2	154 943	1387	0.9
Carbamyl-putrescine	CPU	37.1 ± 0.1	158 848	2789	1.8
Histamine	HA	44.5 ± 0.1	171 646	2313	1.4
1,3-Diaminopropane	DAP	61.5 ± 0.1	77 766	1622	2.1
Putrescine	PU	63.6 ± 0.1	354 453	2365	0.8
Cadaverine	Cd	68.8 ± 0.1	227 642	1938	0.9
nor-Spermidine	norSd	70.6 ± 0.2	244 089	2754	1.1
Spermidine	Sd	74.3 ± 0.1	137 811	2293	1.6
Homospermidine	HSd	77.0 ± 0.2	262 455	2455	0.9
Agmatine	Agm	81.5 ± 0.1	119 315	1817	1.5
nor-Spermine	norSm	84.6 ± 0.1	88 389	623	0.7
Spermine	Sm	91.3 ± 0.2	172 568	1537	0.9
Internal standard	IS	103.0 ± 0.1	230 633	2926	1.3

*Average of 10 determinations of the mixture containing 300 picomoles of each of the products. RSD (%) = SM mean x 100.

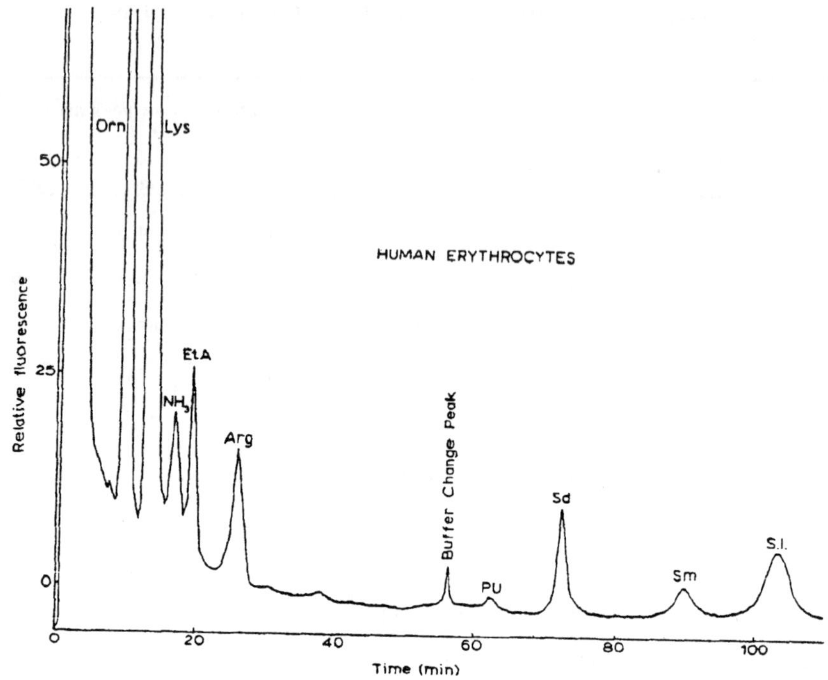

Fig. 4. Chromatogram of a 50 µl sample of a TCA crude extract of erythrocytes. See Table 6 for the identities of the products.

The principal advantage of the two methods presented here is that they can be used for the direct analysis of biological material. The prior purification of samples, which is often time-consuming and tedious are not necessary and thereby the risk of losing amino compounds, which could be present in trace amounts, is avoided. The retention times of different amino compounds are not altered by our procedure. This has been verified by cochromatographing the crude samples with standard mixtures of amino compounds. The methods are sensitive and fast enough to conduct clinical investigation on physiological fluids; their reproductibility is similar to that of automatic analysis of amino acids (see Tables 4 and 6).

Automatic ion-exchange colum chromatography, utilizing an amino acid analyzer, has become the most favored method for the analysis of polyamines in most laboratories. The major disadvantage of this method is the high cost of the instrument and especially its maintenance.

f) Gas chromatography

The use of gas chromatography for the analysis of polyamines in biological samples has never gained much popularity since the analysis of free polyamines result in peak tailing and partial irreversible absorption on gas-chromatographic column packing material due to their high basicity. Results can however be improved with appropriate derivatization. Using 3% SE-30 column Brook and Moore (1969) separated the trifluoroacetylated polyamines of *Clostridium* with a temperature program from 80 to 225°C over 25 minutes. Denton *et al.* (1973) also used trifluoroacetyl derivatives to analyze the polyamine content in human urine on a column packed with 5% OV-17 and a temperature program from 75 to 280°C (15°C/min.). Makita *et al.* (1975) used pentafluorobenzoyl derivatives and Giumanini *et al.* (1976) used the N-permethylated polyamines. Smith *et al.* (1978) have developed a gas chromatographic-mass spectrometric assay for trifluoroacetylated polyamines using deuterium labelled polyamines as internal standards and were capable of measuring serum polyamines at the picomole level. Thus this method offers one of the best combination of high sensitivity, selectivity and accuracy for the assay of polyamines but has several drawbacks. It requires, (i) prior cleanup of the polyamine extracts, (ii) synthesis of deuterated polyamines to be used as internal standards, (iii) two trifluoroacetylation and (iv) two expensive instruments : the gas chromatograph and the mass spectrometer. Despite these limitations this method is a powerful tool for the identification of unkown products (Walle, 1973).

2. Electrophoresis

Fisher and Bohn (1957) and Herbst *et al.* (1958) described the use of paper elecctrophoresis for polyamine analysis. It was adapted by Raina and Cohen (1966) for quantitative assays.

Bachrach (1973) has reviewed some of the buffers used in paper electrophoresis and the migration characteristics of putrescine, cadaverine, spermidine and spermine. For detection and quantitation, ninhydrin spraying can be used as described for paper chromatography. Owing to its relative simplicity this method has been widely used. Results obtained by this method are comparable with those obtained by paper chromatography but paper electrophoresis is also progressively being replaced by other methods such as TLC, HPLC, etc... but particularly by capillary electrophoresis.

Capillary electrophoresis is a family of related techniques emerged in recent years as a new automatic analytical tool for the separation of charged molecules. The separations are based on several mechanisms, such as molecular size, isoelectric focusing, hydrophobicity... (Lukacs *et al.*, 1985; Hjerten, 1990). Mostly, detection is achieved by monitoring UV absorbance directly on-line through a window in the capillary. Sensitive detection methods, such as fluorescence, electrochemical, mass spectroscopy and conductance have been shown to be compatible (Gordon *et al.*, 1988). Other detection options include laser-induced fluorescence and diode assay spectrometry (Ewing *et al.*, 1989).

Zhou *et al.* 1995 have applied this technique to the analysis of polyamines from serum using UV detection (detection limit 10^{-12} mol/L). Oguri *et al.* (1996) have also analysed polyamines and some related compounds using 1-metoxycarbonyl indolizine-3,5-dicarbaldehyde (IDA) as a derivatization reagent and detection at 409 nm.

3. Mass spectrometry

Mass spectrometry is a very sensitive method which can be used for molecular weight and structure determination of polyamines. Tobari and Tchen (1971) have reported the spectra of the hydrochlorides of diamino-propane, putrescine and 2-hydroxyputrescine. Benzoyl derivatives have been used by Kullnig *et al.* (1970) and dansylated derivatives by Seiler *et al.* (1970). As this method needs extremely purified samples its combination with TLC and GLC is actually gaining in popularity (Seiler, 1977; Smith *et al.*, 1978).

Since ionization under electron impact usually requires volatile derivatives of polyamines for their analysis by mas spectrometry, the use of this technique has had limited application so far. However the recently developed procedures such as chemical ionization, field desorption, etc... should allow direct analysis of polyamines by mass spectrometry. Fig. 5 shows the chemical ionization, mass spectra of the two homologous natural polyamine series known at present : (i) diamino-propane, nor-spermidine and nor-spermine and (ii) putrescine, spermidine and spermine and also including homospermidine and aminopropyl cadaverine. As can be seen from the spectra, this technique produces intense protonated molecular ion (MH^+) peaks with virtually no evidence of fragmentation peaks. With this technique however, it is not possible to differentiate isomeric compounds such as aminopropyl cadaverine and homospermidine (see Fig. 5). Thus mass spectrometry remains a very promising analytical tool with the main disadvantage being the high cost of the instrument.

III - BIOLOGICAL METHODS

I. Enzymatic methods

Enzymatic assays for putrescine, spermidine and spermidine/spermine have been described although they never gained much popularity probably due to the

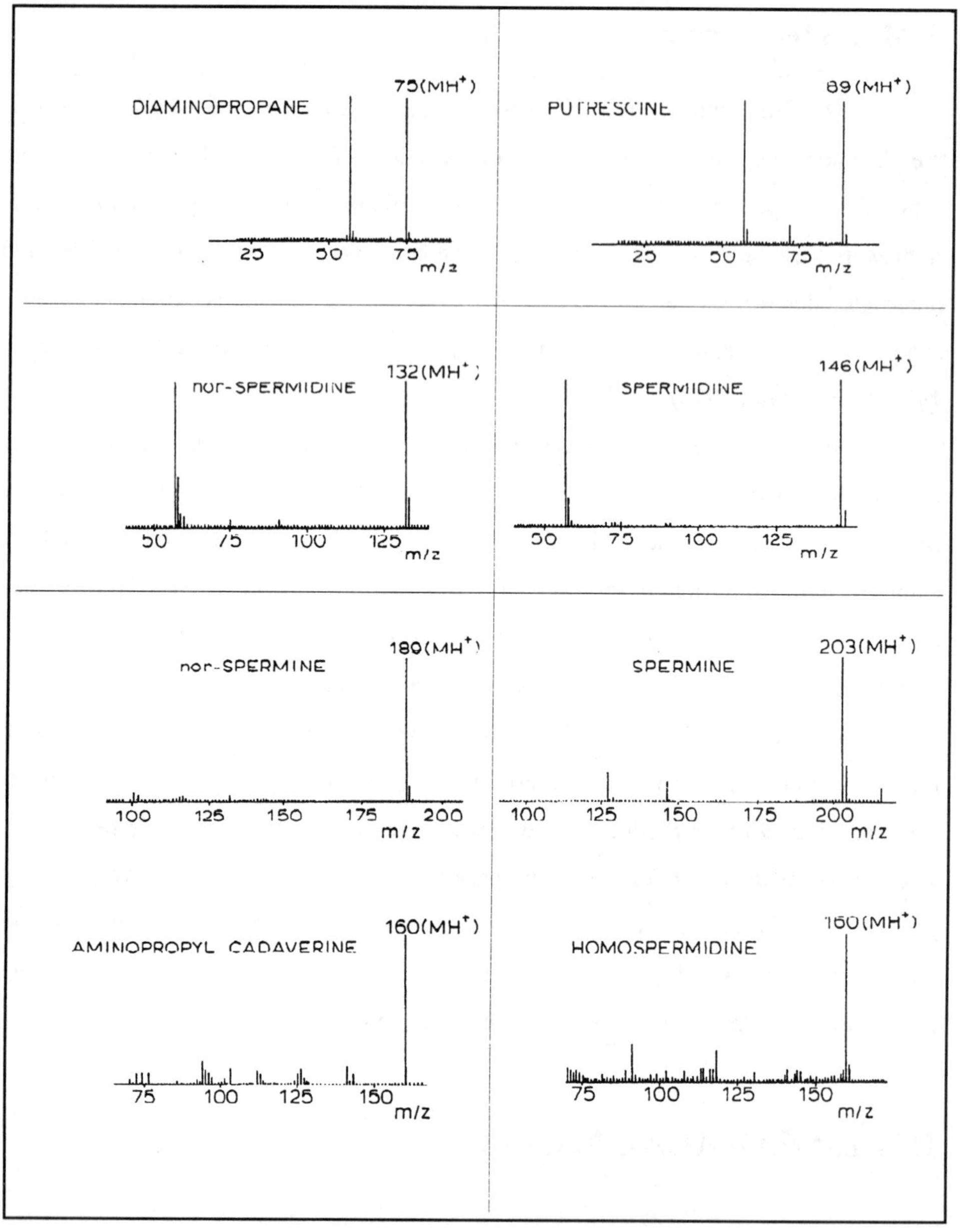

Fig. 5. The chemical ionization mass spectra of the polyamines (hydrochloride form) were obtained by using isobutane as the reactant gas in a modified AEI-MS9 mass-spectrometer through the courtesy of Dr. B.C. Das (Head of the Mass Spectrometry Department of our Institute). The peak at *m/z* 57 is due to C_4H_9 ions from isobutane.

commercial unavailability of the necessary enzymes. Harik *et al.* (1973) described a specific enzymatic assay for putrescine while Bachrach and Oser (1963) used a highly specific oxydase from *Serratia marcescens* to assay spermidine. Bachrach and Reches (1966) also developed an enzymatic method based on the oxidation of both spermine and spermidine by serum amine oxydase followed by colorimetric determination of the oxidation products.

A very interesting work has also been described by Kamei *et al.* (1989) for the detection of polyamines, from hydrolysed urine after separation by reversed phase HPLC ans post-column enzymatic-chemiluminescent reaction.

2. Immunological methods

These methods allow direct determination of both free and bound polyamines thereby obviating the use for hydrolysis. Quash was the first to prepare and characterize antipolyamine antibodies and has contributed considerably to the further development of these methods (Quash and Jonard, 1967; Jonard *et al.*, 1973). Antibodies initially prepared were cross reactive with different polyamines. But with the use of affinity chromatography highly specific antibodies can now be obtained. Using these purified antibodies, Bartos *et al.* (1975) have developed a radioimmuno-assay (RIA) with femtomole sensitivity while Roch and Quash (1978) have developed an immunonephelometric assay based on the covalent binding of antipolyamine antibodies to latex spheres which has picomole sensitivity.

IV - CONCLUSION

The increasing interest in polyamine research has continued to stimulate the development of analytical methods needed for the study of newer aspects of polyamine biochemistry. Basic research and clinical studies have made it imperative to analyze polyamines by methods which are simple, rapid, sensitive and specific; that is to say, adaptable for routine assays. It would seem that TLC of dansylated polyamines, ion exchange chromatography with the help of aminoacid analyzers and immunological methods offer the most promising alternatives to the laboratories engaged in research needing continously a large number of analyses. All these methods are sensitive enough to permit quantitative analysis at the picomol level. Mass spectrometry, which is a potentially useful tool, can help for the structure and molecular weight determination of polyamine compounds, especially when used in combination with TLC or gas-liquid chromatography. Our laboratory, engaged in problems concerning cell metabolism, has adopted the method of ion-exchange chromatography with the help of an aminoacid analyser. Procedures for the analysis of polyamines developed by us are referenced in this paper and methods used at present routinaly in our laboratory are described. Besides sensitivity and rapidity, the aim of all our procedures has been to make them amenable to the determination of functionally and metabolically related amines in a single crude sample. Prior sample purification, which is often time-consuming, is not necessary and thus the risk of losing trace amino compounds is avoided.

Immunological methods are most promising and may become the assay of choice for routine analysis of polyamines in the future.

V - SUMMARY

Sample preparations and two different automatic methods for polyamine analyses are described. Ion-exchange chromatography with the help of an aminoacid analyzer and fluorometric detection is capable of rapid analysis and picomole quantification of natural polyamines; Other methods such as paper chromatography, TLC, OPTLC, GLC, electrophoresis, capillary electrophoresis, mass spectrometry and biological assays are reviewed.

REFERENCES

Abdel-Monem, M.M. (1977) The determination of di- and polyamines by high-pressure liquid and gas chromatography. Advances in Chromatography, M. Dekker, New York.

Abdel-Monem, M.M., Ohno, K., Newton, N.E. and Week, S.C.E. (1978) Thin-layer chromatography and high pressure liquid chromatography of the dansyl polyamines, in R.A. Campbell, D.R. Morris, D. Bartos, G.D. Daves and F. Bartos (Eds) Advances in Polyamine Research, Vol. 2, Raven Press, New York, p. 37-49.

Abe, F. and Samejina, K. (1975) A new fluorimetric method for the determination of spermidine and spermine in tissues by thin layer chromatography. *Anal. Biochem.* **67**, 298-308.

Adbel-Monem, M.M. and Ohno, K. (1977) Polyamine metabolism 1. Synthesis of the 5-(dimethylamino)-1-naphtalene-sulfonyl (danzyl) derivatives of the monoacetyl diamines and polyamines and the elucidation of the monoacetyl diamines and polyamines in the urine of a cancer patient. *J. Pharm. Sci.* **66**, 1089-1094.

Adlakha, R.C. and Villanueva, V.R., Calvayrac, R. and Edmunds Jr. L.N. (1980) Arrest of cell division blocks the utilisation of polyamines in the syncronized cultures of photoautotrophically grown *Euglena*. *Arch. Biochem. Biophys.* **201**, 660-668.

Alexander, S.P.H., Kendall, D.A. and Hill S.J. (1992) Excitatory amino acid-induced phosphoinositide turnover in guinea pig cerebral cortical slices: selective enhancement by spermine of the response to DL-1-aminocyclopentane-trans-1,3-dicarboxylate. *J. Neurochem.* **59**, 610-615.

Bachrach, U. and Oser, I.S. (1963) Enzymatic assay for spermidine. *J. Biol. Chem.* **238**, 2098-2101.

Bachrach, U. and Reches, B. (1966) Enzymic assay for spermine and spermidine. *Anal. Biochem.* **17**, 38-48.

Bachrach, U.`(1973) Function of naturally occuring polyamines. Academic Press, New York.

Bagni, N., Serafini-Fracassini, D., Villanueva, V.R. and Adlakha, R.C. (1978) Contenuto in poliamine, etanolammina e arginina nel polline di melo. *Riv. Ortoflorofrutti Coltura Ital.* **5**, 470-476.

Bardócz, S., Karsai, T. and Elödi, P. (1985) A new technique for the determination of polyamines overpressured ion-exchange thin layer chromatography. *Chromatographia* **20**, 23-24.

Bartos, D;, Campbell, R., Bartos, F. and Grettie, D. (1975) Direct determination of polyamines in human serum by radioimmunoassay. *Cancer Res.* **35**, 2056-2060.

Blankenship, J. and Walle, T. (1977) Acetylation of spermidine and spermine by rat liver and kidney chromatin. *Arch. Biochim. Biophys.* **179**, 235-242.

Brooks, J.B. and Moore, W.E;C. (1969) Gas chromatographic analysis of amines and other compounds produced by several species of *Clostridium. Can. J. Microbiol.* **15**, 1433-1447.

Chida, N., Saito, H. and Abe, K. (1992) Spermine facilitates the generation of long term potentiation of evoked potential in the dentate gyrus of anesthetized rats. *Brain Res*. **593**, 57-62.

Cohen, S.S. (1977) Introduction to the polyamines. PrentinceHal, Inc., Englewood Cliffs, New Jersey.

Cohen, S.S., Morgan, S. and Streibel, E. (1969) The polyamine content of the t-RNA of *E. coli. Proc. Natl. Acad. Sci., USA*. **64**, 669-676.

De Rosa, M., De Rosa, S. and Gambacorta, A. (1976) Occurence and characterization of new-polyamines in the extreme thermophile *Caldariella acidophila. Biochem. Biophys. Res. Commun.* **63**, 1093-1097.

Denton, M.D., Glazer, H.S., Walle, T., Zellner, D.C. and Smith, F.G. (1973) Clinical application of new methods of polyamine analysis. In D.H. Russell (Ed.). Polyamines in normal and neoplastic growth, Raven Press, New York, p. 373-380.

Dion, A.S. and Cohen; S.S. (1972) Polyamine stimulation of nucleic acid synthesis in an uninfected and phage-infected polyamine auxotroph of *Escherichia coli* K-12. *Proc. Natl. Acad. Sci. USA*, **69**, 213-217.

Dion, A.S. and Herbst, E.J. (1970) Polyamines changes during development of *Drosophila melanogaster*, *Ann. N.Y. Acad. Sci.* **171**, 723-734.

Ewing, A.G., Wallingford, R.A. and Olefirowicz T.M. (1989) Capillary electrophoresis. *Anal. Chem.* **61**, 292A-303A.

Fieisher, J.H. and Russell, D.H. (1975) Estimation of urinary diamines and polyamines by thin-layer chromatography. *J. Chromatogr.* **110**, 335-340.

Fisher, F.G. and Bohn, H. (1975) Uber die Bestimmung von Spermin, Spermidin und anderen biogenen Aminen nach papier elektrophoretischer Abrrennung und ihre Mengenverhältnisse in tierischen Organen. Hoppe Seylers *Z. Physiol. Chem.* **308**, 108-115.

Giumanini, A.G., Chiavari, F. and Scarponi, F.L. (1976) N-Permethylation of polyamines for gas chromatographic and mass spectrometric analyses. *Anal. Chem.***48**, 484-489.

Gordon, M.J., Huang, W., Pentoney, S.L. and Zare, R.N. (1988) Capillary electrophoresis. *Science* **242**, 224-228.

Harik, S.I., Pasternak, G.W. and Snyder, S.H. (1973) Putrescine : a sensitive assay and blockade of its synthesis by α-hydrazino ornithine, in D.H. Russell (Ed). Polyamines in normal and neoplastic growth. Raven Press, New York, pp. 307-321.

Herbst, E.J., Keister, D.L. and Weaver, R.H. (1958) The separation of aliphatic amines by paper chromatography or paper electrophoresis. *Arch. Biochem. Biophys.* **75**, 187-185.

Hjerten, S. (1990) Zone broadening in electrophoresis with special reference to high performance electrophoresis in capillaries : an interplay between theory and practice. *Electrophoresis* **11**, 665-690.

Holder, S. and Bremer, H.J. (1966) Bestimmung der Di- und Polyamine in urine. Chromatographische methoden zur anreicherung, trennung und identifizierung. *J. Chromatogr.* **25**, 48-57.

Jonard, J., Quash, G. and Wahl, R. (1967) La spécificité des anticorps antipolyamines. *C.R. Acad. Sci. Paris*, **265 D**, 1099-1102.

Kabra, P.M., Lee, H.K., Lubich, W.P. and Marton, L.J. (1986) Solid-phase extraction and determination of dansyl derivatives of unconjugated and acetylated polyamines by reversed-phase liquid chromatography : improved separation systems for polyamines in cerebrospinal fluid, urine and tissue. *J. Chromatogr.* **380**, 19-32.

Kamai, S. and Ohkubo, A., Saito, S. and Tagaki, S. (1989) Polyamine detection system for high performance liquid chromatography involving enzymatic and chemiluminescent reactions. *Anal. Chem.* **61**, 1921-1924.

Kullnig, R.K., Rosano, C.L. and Hurwitz, C. (1970) Identification of 2-hydroxyputrescine in a pseudomonal lacking spermidine. *Biochem. Biophys. Res. Commun.* **39**, 1145-1148.

Kuttan, R., Padhakrisnan, A.N., Spande, T. and Witkop, B. (1971) *Sym*-Homospermidine, a naturally occurring polyamine. *Biochemistry*, **10**, 361-365.

Lenfant, M., Digiusto, L., Garcia-Giralt, E., Mayadoux, E., Villanueva, V.R. and Adlakha, R.C. (1979) Immunological response assay in mice to purified spleen extract (chalone) and the polyamines spermine and spermidine. *Biomedicine* **31**, 110-113.

Lukacs, K.D. and Jorgenson (1985) Capillary zone electrophoresis : effect of physical parameters on separation efficiency and quantitation. *J. High Res. Chromatogr.* **8**, 407-411.

Makita, M., Yamamoto, S. and Kono, M. (1975) Rapid determination of di- and polyamines in human urine by electron capture gas chromatography. *Clin. Chim. Acta* **61**, 403-405.

Marcé, M., Brown, D.S., Capell, T., Figueras, X. and Tiburcio, A.F. (1995) Rapid high-performance liquid chromatographic method for the quantitation of polyamines as their dansyl derivatives : application to plant and animal tissues. *J. Chromatogr.* **666**, 329-335.

Oguri, S., Fujiyoshi, T. and Miki, Y. (1996) In-capillary derivatization with 1-methoxycarbonylindolizine-3,5-dicarbaldehyde for high-performance cappilary electrophoresis. *Analyst.* **121**, 1683-1688.

Oshima, T. (1983) Unusual polyamines in an extreme thermophile, *Thermus thermophilus* *i*n U. Bachrach, A. Kaye and R. Chayen (Eds) Advances in Polyamine Research, Vol. 4, Raven Press, New York, p. 479-487.

Quash, G., Gresland, L., Delain, E. and Huppert, J. (1973) Antipolyamine antibodies and growth. In D.H. Russel (Ed) Polyamines in Normal and Neoplastic Growth. Raven Press, New York, pp. 157-165.

Quigley, G.J., Teeter, M.M. and Rich, A. (1978) Structural analysis of spermine and magnesium ion binding to yeast phenylalanine transfer RNA. *Proc. Natl. Acad. Sci., USA* **75**, 64-68.

Raina, A. and Cohen, S.S. (1966) Polyamines and RNA synthesis in a polyauxotrophic strain of *E. coli*. *Proc. Natl. Acad. Sci.USA* **55**, 1587-1593.

Roch, A.M. and Quash, G. (1978) Development of immunolatex procedures for measuring bound polyamines. In R.A. Campbell, D.R. Morris, D. Bartos, G., Daves and Bartos (Eds.) in Advances in Polyamin Research, Vol. 2, Raven Press, New York, pp. 55-63.

Rock, D.M. and Mcdonald, R. (1992) Spermine and related polyamines produce a voltage dependent reduction of NMDA receptor single channel conductance. *Mol. Pharm.* 42, 157-164.

Rosenheim, O. (1924) The isolation of spermine phosphate from semen and testis. *Biochem. J.* **18**, 1253-1263.

Roth, M. and Hampai, A. (1973) Column chromatography of aminoacids with fluorescence detection. *J. Chromatogr.* **83**, 353-356.

Russell, D.H. and Durie, B.G.M. (1978) Polyamines as biochemical markers of normal and malignant growth. Raven Pres, New York.

Samejima, K. (1974) Separation of fluorescamine derivatives of aliphatic diamines and poyamines by high speed liquid chromatography. *J. Chromatogr.* **96**, 250-254.

Seiler, N. (1970) Use of the dansyl reaction in biochemical analysis methods. *Biochem. Anal.* **18**, 259-337.

Seiler, N. (1977) Chromatography of biogenic amines. I. Generally applicable separation and detection methods. *J. Chromatogr.* **143**, 221-246.

Seiler, N. and Knodgen, B. (1977) Chromatography of Dns-derivatives on pre-coated high performance thin-layer chromatographic plates. *J. Chromatogr.* **31**, 109-119.

Seiler, N. and Schmidt-Glenewinkel, T. (1975) Regional distribution of putrescine, spermidine and spermine in relation to the distribution of RNA and DNA in the rat nervous system. *J. Neurochem.* **24**, 791-795.

Seiler, N. and Wiechmann, M. (1965) Zum nachweis von aminen in 10^{-10} Mol Masstab : Trennung von 1-Dimethyl aminonaphthalin-5-Sulfonsäure-amiden auf Dunschichtchromatogrammen. *Experiencia* **21**, 203-204.

Seiler, N., Knödgen, B. and Eisenbeiss, F. (1978) Determination of di- and polyamines in tisues by HPLC separation of their Dns-derivatives. *J. Chromatogr.* **145**, 29-39.

Seiler, N., Schneider, H. and Sonnenberg, K.D. (1970) Massenspektrometrische identifizierung von Biogenen Aminen in Form ihrer 1-Dimethyl-aminonaphtalin-5-Sulfonyl-Derivate. *Z. Anal. Chem.* **252**, 127-136.

Shimizu, H., Kakimoto, Y. and Sano, I. (1964) The determination and distribution of polyamines in mammalian nervous system. *J. Pharmacol. Exp. Ther.* **143**, 199-204.

Smith, R.G., Daves, Jr. G.D. and Grettie, D.P. (1978) Gas chromatography mass spectrometry analysis of polyamine in serum using deuterium labelled analogs as internal standards. In R;A. Campbell, D.R. Morris, D. Bartos, G.D. Daves and F. Bartos (Eds), Advances in Polyamines Research, Vol. 2, Raven Press, New York, pp. 23-35.

Smith, T.A. (1977) Recent advances in the biochemistry of plant amines. In L. Reinhold, J.B. Harborne and T. Swain (Eds) Progress in Phytochemistry, Vol. 4, Pergamon Press, Oxford, New York, Toronto, Sydney, Paris, Frankfurt, pp. 27-81.

Spackman, D.H., Stein, H. and Moore, S. (1958) Automatic recording apparatus for use in the chromatography of aminoacids. *Anal. Chem.* **30**, 1190-1206.

Stevens, C.F. (1988) A relaxing factor in brain. *Nature* **336**, 308-309.

Sugiura, T., Hayashi, T., Kawai, S. and Ohno, T. (1975) High-speed liquid chromatographic determination of putrescine, spermidine and spermine. *J. Chromatogr.* **110**, 385-388.

Tabor, H. and Tabor, C;W. (1964) Spermidine, spermine and related amines. *Pharmacol. Rev.* **16**, 245-300.

Tobari, J. and Tchenn, T.T. (1971) Identification of (+) hydroxyputrescine (1-4-diamino butan-2-ol) from a *Pseudomonas* species. *J. Biol. Chem.* **246**, 1262-1265.

Trout, J.J., Koenig, H., Goldstone, A.D., Iqbal, Z. Lu, C.Y. and Siddiqui, F. (1993) NMDA receptor excito-toxicity involves activation of polyamine synthesis: Protection by alpha-difluoro methyl ornithine. *J. Neurochem.* **60**, 352-355.

Varenne, P., Bardey, B., Longevialle, P., Das, B.C. (1977) Modifications d'un spectromètre de mass AEI MS9 pour son utilisation en ionisation chimique. *Bull. Soc. Chim. Fr.* 886-892.

Villanueva, V.R. (1981) Polyamines in *Euglena Gracilis*. In : C.M. Caldarera, V. Zappia and U. Bachrach (Eds), Advances in Polyamine Research, Vol. 3, Raven Press, New York, pp. 389-395.

Villanueva, V.R. and Adlakha, R.C. (1978) Automated analysis of common basic aminoacids mono, di-, and polyamines, phenolicamines and indoleamines in crude biological samples. *Anal. Biochem.* **91**, 264-275.

Villanueva, V.R., Adlakha, R.C. and Cantera Soler, A.M. (1977) Determination rapide de polyamines et de quelques mono- et di-amines dans des extraits végétaux. *J. Chromatogr.* **139**, 381-385.

Villanueva, V.R., Adlakha, R;C. and Calvayrac, R. (1980) Cell cycle related changes in polyamines content in *Euglena. Phytochemistry* **19**, 962-964.

Villanueva, V.R., Adlakha, R.C. and Giret, M. (1979) Analyses automatiques de mono-, di- et polyamines dans les plaquettes, le plasma et les urines. *Feuillets de Biologie* **106**, 141--146.

Walle, T. (1973) Gas chromatography-mass spectrometry of di- and polyamines in human urine : identification of monoacetyl-spermidine as a major metabolic product of spermidine in a patient with acute myelocytic leukemia. In D.H. Russell (Ed.) Polyamines in Normal and Neoplastic Growth. Raven Press, New York, pp. 355-365.

Williams, K. Pullan, L.M., Romano, C., Powel, R.G., Salama, A.I. and Molinoff, P.B. (1992) An antagonist/partial agonist at the polyamine recognition site of the NMDA receptor that alters the properties of the glutamate recognition site. *J. Pharmacol. Exp. Ther.* **262**, 539-543.

Witkiewicz, Z., Mazvrek, M. and Bladek, J. (1993) A new instrument for overpressured layer chromatography. *J. Planar Chromatogr.* **6**, 407-412.

Zhou, G., Yu, Q., Ma, Y., Xue, J., Zhang, Y. and Lin, B. (1995) Determination of polyamines in serum by high-performance capillary zone electrophoresis with indirect ultraviolet detection. *J. Chromatogr.* **717**, 345-349.

Key words : Polyamines; HPLC; Ion-exchange chromatography (Amino acids analyser); Mass spectrometry; TLC; OPTLC; Capillary electrophoresis; Immunological methods; Enzymatic methods.

Progress in HPLC-HPCE, Vol. 7, pp. 495-522
G. Ali Qureshi *et al.* (Eds)

MONOAMINE OXIDASE NEUROTOXICITY AND METHODOLOGY

NICOTRA,[1] A. and PARVEZ,[2] S. H.
[1]*Dipartimento di Biologia Animale e dell'Uomo, Università di Roma I, Viale dell'Università 32, 00185 Roma, Italia and Neuroendocrinologie et Neuropharmacologi du Développement-CNRS, Université de Reims-UFR Sciences, Moulin de la Housse, 51062 Reims Cedex, France*

INTRODUCTION

Monoamine oxidase [MAO; amine: oxygen oxidoreductase (deaminating) flavin-containing; EC 1.4.3.4], is an integral protein of the outer mitochondrial membrane responsible for the degradation of a variety of biogenic and xenobiotic monoamines. MAO may deaminate aliphatic and aromatic primary, secondary and tertiary amines and its most important substrates include neurotransmitters such as serotonin, dopamine, epinephrine and norepinephrine, dietary sympathomimetic amines such as tryptamine, tyramine and β-phenylethylamine (Waldmeier,1987; Youdim *et al.*,1988). Consequently it is important in neurophysiology, regulating the neurotransmitters released by neurons in the central nervous system. In addition it is involved in the regulation of the blood pressure, and in the detoxification of endogenous and dietary amines, preventing their accumulation up to toxic levels.

In mammals MAO exists in two distinct isoenzymes known as MAO A and MAO B, each having different functions and different tissue expression (Weyler *et al.*,1990; Berry *et al.*,1994). As early as the late '60s and early '70s the study of the affinity of the enzyme for the various substrates and of the effect of specific inhibitors such as the two acetylenic amines clorgyline and deprenyl, had revealed the existence of the two forms of the enzyme. MAO A preferentially deaminates serotonin and epinephrine and its activity is inhibited by low concentrations of clorgyline (Johnston,1968) to which the B form is relatively insensitive. In contrast MAO B is inhibited by lower deprenyl concentrations than MAO A and is much more active against β-phenylethylamine and benzylamine (Knoll and Magyar,1972). Some substrates, including tyramine and tryptamine are deaminated by both MAO forms (Sullivan *et al.*,1986;Youdim *et al.*,1988). According to Kinemuchi *et al.(*1984) and Tipton *et al.* (1987) both forms of MAO may deaminate most substrates and a substrate is preferentially deaminated by one form depending on its concentration, affinity and turnover rate.

Cloning of the cDNA for MAO A and MAO B in mammals carried out in the late'80s showed that the two forms are encoded by separate genes (Weyler *et al.*,1990). The human genes for MAO A and MAO B show a high degree of sequence similarity and are both located on the short arm of the X cromosome (Lan *et al.*,1989; Shih,1993) where they are arranged tail-to tail separated by about 40-45 kb (Chen *et al.*,1991).

The properties of the two isoenzymes have been extensively investigated. Much of the research was stimulated, among other things, by the observation that abnormal levels of MAO activity in the nervous system are associated with several diseases, for example, depression, alcohol abuse and schizophrenia (Oreland,1993). Furthermore MAO A inhibitors have been used clinically in the treatment of depression, while MAO B inhibitors may be beneficial in Parkinson's and Alzheimer's diseases (Parkinson Study Group,1989; Cesura and Pletscher,1993; Knoll,1993 a,b). Interestingly in recent studies (Markey *et al.*, 1986) it has been shown that MAO may be involved in transforming the Parkinson-inducing dopaminergic neurotoxin MPTP (N-methyl-4-phenyl-1,2,3,6-tetrahydroxypyridine) (Heikkila *et al*.,1984) to the neuroactive metabolite MPP^+ (N-methyl-4-phenyl-1,2-dihydropyridinium ion).

An evaluation of the levels of activity and of absolute quantities of MAO A o di MAO B in neural tissues can be used as a potential marker for specific psychiatric disorders. We shall now describe several of the methods used to measure both the activity and the amount of MAO A or MAO B in tissue preparations. In some cases it may also prove necessary or interesting to evaluate the amount of MAO A or MAO B directly from tissue sections. Quantitative enzyme radioautography performed using labeled MAO inihibitors may be used for this purpose (Saura *et al*., 1992; Cesura *et al*.,1993; Ekblom *et al*., 1993; Kondoh *et al*.,1994).

ASSAY METHODS FOR MAO A and MAO B ACTIVITIES

As we have reported MAO A and MAO B activities may be distinguished according to their affinity for different substrates and their sensitivity to specific inhibitors such as clorgyline (specific for MAO A) and *l*-deprenyl (specific for MAO B). The activity of either of the forms in a tissue may be tested by using a preferential substrate. When the specificity of a substrate for one MAO form in a tissue is unknown, it must be ascertained by testing the effect of various concentrations of clorgyline and *l*-deprenyl (range 10^{-10}-10^{-4} M). For example, if 5-HT, at a given concentration, is used to measure MAO A activity, the resulting activity must be 100% inhibited by nanomolar concentrations of clorgyline while, in order to obtain the same effect, micromolar concentrations of *l*-deprenyl are required and the inhibition curves obtained must be simple sigmoids. Should the inhibition curves become double sigmoids this would mean that the substrate is deaminated by both forms of MAO (Fowler *et al*.,1978). In that case, the activity must be measured in the presence of deprenyl concentrations that inhibit MAO B activity 100% and do

not significantly affect MAO A. When the effects of irreversible inhibitors such as clorgyline and *l*-deprenyl are to be tested, the enzyme source must be pre-incubated (30-60 min) with inhibitor before adding the substrate and the inhibitor must be present in the reaction mixture also during incubation.
Clorgyline and *l*-deprenyl are the most commonly used inhibitors. Nevertheless numerous other MAO A and MAO B specific inhibitors are available (Cesura and Pletscher,1993).

Several continuous and discontinuous assays for MAO activity have been developed based on the measurement of the quantity of compounds produced by the reaction or of the disappearance of reagents taking part in the reaction. The reaction catalysed by the MAO is generally as follows:

$$\text{Monoamine} + H_2O + O_2 \rightarrow \text{Aldehyde} + NH_3 + H_2\,O_2$$

The immediate product of MAO catalysed deamination is thus an aldehyde. This aldehyde may be oxidized in the presence of oxygen to form the corresponding acid, a process that may be speeded up in tissue homogenates in the presence of aldehyde dehydrogenase and NAD+. Furthermore, if aldehyde reductase and NADH or NADPH are present, the aldehydes may be reduced to the corresponding alcohol. Evaluation of MAO activity may thus be performed by measuring the amount of aldehyde (or of the corresponding acid and alcohol) or of the ammonia or of hydrogen peroxide produced. Another alternative is to measure the quantity of substrate or oxygen consumed.
However it must be take into account that secondary and tertiary monoamines do not form ammonia, and amine derivatives like MPTP do not form either ammonia or aldehyde. Only oxygen consumption and hydrogen peroxide produced are independent of the substrate used.

Generally speaking no particularly complex equipment is needed to carry out these procedures, which include radiometric, fluorometric, spectrophotometric polarographic and luminometric techniques.

It should be emphasized that in many tissues the same monoamines that are deaminated by MAO may be deaminated also by semicarbazide-sensitive amine oxidases [SSAOs, amine:oxygen oxidoreductases (deaminating), copper containing; EC 1.4.3.6]. The reaction catalysed by the SSAOs is the same (Callingham *et al.*,1991), and SSAO activity may therefore be evaluated using the same technique as for the MAO activity. If no literature data are available it is advisable to carry out some preliminary tests to ascertain the presence of any SSAO activity in a tissue by means of inhibition using semicarbazide (10^{-4} M) and then perform the determinations in the presence of semicarbazide if necessary.

Radiochemical assay - Let us first examine the radiochemical method as this is certainly the one most frequently used. There are several reasons why this method is so popular. In the first instance, its high sensitivity allows it to be used also with crude tissue homogenates and it is possible to use many different substrates. Also fluorometric determinations, as described below, have a high sensitivity only that, in this type of assay, the presence of particulate materials in homogenates is the cause of high background. One further advantage of the radiochemical method is its high specificity and simplicity of execution plus the possibility of simultaneously testing a large number of samples. The limitations of the method include its discontinuity and that of being a safety hazard, particularly because it involves the use of radioactive compounds.

The method is based on the observation that, at low pH values in a solution, the substrate monoamines become positively charged while their metabolites remain neutral. This means that the latter may be extracted by means of an organic solvent in which the protonated substrates are insoluble, or else by a negatively charged ion exchange resin. Using this principle, three similar radiochemical methods were developed by Wurtman and Axelrod (1963), Otsuka and Kobayashi (1964) and McCaman *et al.* (1965). In order to evaluate enzymatic activity a suitably labeled substrate, usually with ^{14}C, is used. After the substrate has been deaminated, labeled degradation products such as aldehydes are separated from the non-deaminated substrate by means of acidification and extraction with an organic solvent. From the measurement of the cpm given by the sample, corrected for quenching, it is possible to calculate the amount of substrate metabolized. Enzymatic activity is thus expressed in nmoles (or μmoles) of degraded substrate\mg protein\time. Several modifications were subsequently made to the original procedures mentioned above, mainly for the purpose of enhancing the method of extracting labeled metabolites (Tipton and Youdim,1976; Fowler *et al.*,1980a; Jains *et al.*,1973). A standard assay mixture contained appropriate concentrations of enzyme and labeled substrate in a total volume of 300μl of 0.1M phosphate buffer (pH 7.4). The desired substrate concentration was obtained with or without dilution with non radioactive compound. First the buffer-enzyme mixture was allowed to equilibrate in a shaking waterbath at 37°C for 10 min, then the reaction was started by adding the substrate. After a fixed time the reaction was stopped by acidifying with 300μl of ice cold 2N HCl (or 300μl of 2M citric acid). The deaminated metabolites were extracted by vigorous shaking with a vortex mixer for 1 minute with 5 ml of water-saturated toluene:ethyl acetate (1:1, v/v) containing 0.6% w/v 2,5 diphenyloxazole (PPO). The two layers were separated by centrifugation (1200 x g for 2 min) and the water-phase was frozen in dry ice or in a freezer at -20°C for at least 3 hrs so that the organic layer can be transferred into scintillation vials for counting. Alternatively PPO could be omitted from the extraction mixture, and aliquots of the organic phase after extraction were mixed with 10 ml of scintillation cocktail (Instagel, Packard Instrument Co.; Ready Safe, Beckman Instruments; Omnifluor, NEN Research Products, Co.). Finally, the radioactivity was determined in a scintillation spectrometer.

The extraction method described above is the one used most commonly. Furthermore, also other organic solvents such as anisole, toluene, benzene-ethyl acetate (1:1 v/v) and ether were used. The various monoamines were thus found to display a different extraction efficiency depending on the organic solvent used (Tipton and Youdim,1983). As we have seen, as well as by means of an organic solvent it is possible to separate the substrate from its metabolites by passage through a negatively charged ion exchange resin which adsorbs only the positively charged amines (Jain *et al.*,1973; Tipton and Youdim,1976). Blank values were obtained by adding HCl before the substrate or by using enzyme that had been inactivated by boiling, or omitting the enzyme from the reaction mixture. Blank sample extraction was performed as above.

One critical aspect of this method is that of the efficiency of extraction of metabolites produced in the organic solvent. The percentage recovery of the products is influenced above all by the fact that the aldehydes can be strongly bound to the proteins present in the enzyme source (Alivisatos and Ungar,1968, Southgate and Collins,1969; Jain *et al.*,1973). This effect is greater when crude homogenates are used and depends on a number of experimental conditions such as type of substrate and tissue, quantity of enzyme source for testing, the composition of the reaction mixture and the possibility that the carbonyl products of the substrate deamination are converted to compounds not extractable from the aqueous medium. An accurate calculation of the extraction efficiency is essential above all when

MAO activity values obtained using different tissues and substrates are compared. Fowler and Oreland (1980) and Fowler *et al.* (1980a) used two different methods to calculate extraction efficiency. In the first method a concentrated preparation of enzyme is incubated with the substrate until no further substrate deamination is observed. The efficiency of extraction is calculated as the net radioactivity recovered after the appropriate incubation time / the net radioactivity originally added and expressed as a percentage. This percentage represents a lower limit of the extraction efficiency and does not take into account other possible time dependent factors such as substrate stability. In the second method, extraction of the sample with organic solvent is repeated several times and the percentage radioactivity obtained in the first extraction with respect to the sum of radioactivity values obtained in all the extractions, is calculated. This method provides an upper limit to the extraction efficiency as it does not take into account products that are very poorly extracted. It is advisable to follow both methods and take the mean. Specific MAO activities will thus be corrected for mean efficiency of extraction. It should be borne in mind that when the lower and upper limits for a given substrate differ widely, the mean extraction efficiencies used to correct the data must be regarded as approximate. The data obtained using the radiochemical method should be compared with those obtained using other direct methods, such as the polarographic method to measure 0_2 consumption, which gives absolute values.

As already mentioned, in some tests it may be necessary to include semicarbazide in the incubation mixture in order to inhibit activity due to SSAO. It must not be overlooked however that high semicarbazide concentrations can interfere with aldehyde extraction (Jain *et al.*, 1973; Elliot *et al.*,1989). When using semicarbazide controls should be performed by adding the drug to the sample after the activity has been blocked and by comparing the values with those obtained in the absence of the drug.

Tipton and Singer (1993) have pointed out the pitfalls and errors inherent in using this method, which are often the result of failure to take the necessary precautions to obtain quantitatively valid results. Essentially, the errors that may be committed by researchers are due either to failure to evaluate extraction efficiency under changing experimental conditions, or else the lack of attention paid to the effective measurement of the true initial rates of the reaction. Also this second aspect must be thoroughly re-checked each time the reaction conditions change and a discontinuous method is used.

The reliability of the radiochemical assay for MAO A and B has been examined by Krueger and Singer (1993) who compared the values of kinetic parameters such as V_{max} and K_m for various substrates obtained not only by the radiochemical method but also by various continuous procedures such as polarographic and alcohol dehydrogenase-coupled assays. Beef and rat liver mitochondria and human and beef pure liver enzyme were used as a source of enzyme. In these experiments the authors carefully measured true initial rates of the reactions and efforts were made to control all known sources of error likely to be caused by the radiochemical method. The conclusion reached by the authors is that using substrate for MAO B the radiochemical method yields data comparable to those obtained by continuous methods such as the polarographic assay. However, using 5-hydroxytryptamine (5-HT) and tyramine as substrates for MAO A, a different situation was found. In fact the radiochemical method produced significantly lower V_{max} and higher K_m values compared with other methods. This was true whether rat liver mitochondria or highly purified human liver enzyme were used. The authors thus recommend using the polarographic method rather than the radiochemical one whenever kinetic and inhibition studies with MAO A substrates have to be performed.

Fluorometric assays - These methods are sensitive and convenient and are based mainly on the measurement of hydrogen peroxide or ammonia products, or alternatively on the disappearance of a fluorescent substrate or on the formation of a fluorescent derivative from the substrate. However, there are restrictions on the use of the substrates and furthermore they are subject to the limitations imposed by fluorometry such as the necessity of working at low protein concentrations. A spectrophotofluorometer coupled to a recorder is needed if continuous assay is required.

One of the first assays developed was the one based on the formation of 5-hydroxyquinoline from kynuramine as substrate. This method was developed by Kraml (1965) and subsequently modified by Century and Rupp (1968) and by Morinan and Garratt (1985). The aldehyde produced by the oxidative deamination of kynuramine undergoes spontaneous and complete cyclization to 4-hydroxyquinoline (Weissbach *et al.*,1960). Since this compound has no fluorescent properties at the pH used in the reaction mixture (pH 7.2), 1.0 M sodium hydroxide is added for fluorophore development. Fluorescence is measured at 380 nm after activation at 315 nm or recorded by scanning from 320 nm to 450 nm. Kynuramine displays no fluorescence at these wavelengths. Crude enzyme from a variety of tissues was used and the assay mixture contained 1 ml of aqueous homogenate (at the desired amount of wet weight of tissue), 0.5 ml of kynuramine dihydrobromide (100μg), 0.5ml of phosphate buffer (0.5M pH 7.4) and water to 3 ml. Samples were incubated at 37°C for 30 min and the reaction was stopped by adding 2 ml of 10% trichloroacetic acid (TCA). After rapid centrifugation 1 ml of the supernatant was withdrawn and mixed with 2 ml of 1N NaOH in order to measure fluorescence intensity. The precipitating agent, trichloroacetic acid (TCA), causes a significant quenching effect. To overcome this problem TCA was replaced by perchloric acid (Century and Rupp, 1968; Morinan and Garratt, 1985). Moreover Achee *et al.* (1974) noticed a significant quenching effect due to high substrate concentrations such as might be used in the determination of kinetic constants. This finding has been confirmed by Morinan and Garratt (1985).

Two similar coupled fluorometric assays based on H_2O_2 determination using horseradish peroxidase have been described by Snyder and Hendley (1968) and by Tipton (1969). These methods are based on the finding of Guildbault *et al.* (1967) that hydrogen peroxide formed in oxidase reactions can be measured fluorometrically by coupling it to the process of oxidation of homovanillic acid to a highly fluorescent compound in the presence of peroxidase.

Tipton (1969) tested homogenates, mitochondria and MAO solubilized from rat brain as well as MAO purified from pig brain mitochondria. The reaction mixture contained: phosphate buffer, pH 7.4, 100 μmoles, 50 μl of 0.025M tyramine, 2 mg of homovanillic acid, 0.2 mg of horseradish peroxidase and enzyme (final volume 1.2 ml). Hydrogen peroxide release can be continuously monitored in a cuvette by measuring the increase in fluorescence intensity using a spectrophotofluorometer at 425 nm after activation at 315 nm. When assaying homogenates in continuous assay (Tipton, 1969) it is necessary to work at low protein concentrations to avoid the "inner filter" effect of the particles in solutions that absorb a significant proportion of the excitation or fluorescent radiation. In discontinuous measurement (Snyder and Hendley, 1968) the reaction was stopped by chilling the tube to 0°-3°C. In this case, when crude enzyme was used, the samples were centrifuged at 12,000 x g for 20 min and the supernatant used for fluorescence measurements. To relate fluorescence measurements to enzyme activity a calibration curve was constructed for each experiment by adding increasing amounts of standard hydrogen peroxide solution (range 1-15 nmoles, over which the curve was linear) to the reaction mixture devoid of the enzyme and substrate. The greatest activity was observed using tyramine and tryptamine. However, this method cannot be used with physiologically important amines such as epinephine, norepinephine and serotonin. In fact, when these amines were used, complete inhibition of the formation of fluorescent product was observed.

Nakano *et al.* (1984) described another coupled fluorometric method for MAO in which hydrogen peroxide was measured and which used multiple substrates including serotonin and catecholamines. In this method the hydrogen peroxide produced was reduced in the presence of reduced glutathione (GSH) and glutathione peroxidase (GSH-Px). The oxidized glutathione (GSSG) was converted back to GSH in the presence of glutathione reductase (GR) and of NADPH. The NADP+ produced was assayed by fluorescence at 460 nm with excitation at 360 nm. The reaction was carried out at 37°C using rat brain mitochondria. In a total volume of 250μ, the first reaction mixture contained: 0.1M sodium phosphate buffer, pH 7.4, enzyme source, substrate monoamine, 2 mM GSH, a sufficient amount of GSH-Px, 3 mM NaN_3 (as catalase inhibitor). MAO and GSH-Px reactions were stopped by boiling for 1 min. After cooling to room temperature, GSSG was measured by transferring a 1 ml aliquot of the medium to a second incubation mixture containing 15 μM NADPH, 2 μg of GR, 3 mM $MgCl_2$, 5 mM EDTA in 1ml 0.05 M Tris-HCl buffer (pH 7.4). Incubation was carried out at 25°C for 3 min. After addition of 1.25 ml of 1N HCl and short low speed centrifugation, 40 μl of supernatant were mixed with 400 μl of 8 N NaOH, boiled for 5 min and mixed with 1 ml of water. A calibration curve was prepared by adding known amounts of hydrogen peroxide to the first incubation mixture devoid of enzyme source.The highest specific activity was observed with serotonin. The use of mitochondria instead of homogenate as a source of MAO is recommended. Indeed, one disadvantage of this method is that MAO activity is hard to measure in crude homogenate owing to relatively high blank values. A high level of catalase in the crude extract also affects the assay. A non zero blank rate is always a problem in these assays.

Using mitochondria from bovine brain and liver Harada and Nagatsu (1973) have developed a sensitive discontinuous method consisting in ammonia determination based on coupling with a glutamate dehydrogenase (GLDH) assay system. In this system NAD is formed from NADH and subsequently transformed into a highly fluorescent compound by strong alkali treatment (Kaplan *et al.*, 1951). The standard assay mixture contained the following reagents in a final volume of 250μl: 50μl of 0.5 M sodium phosphate buffer, pH 7.4, 50 μl of 5 mM substrate, 100 μl of mitochondrial suspension or homogenate (0.1-1.5 mg protein) and water to final volume. The reaction was stopped with 3.6 N perchloric acid. After brief centrifugation 20 μl of supernatant were neutralized by adding 1μl of NaOH 1N. This sample, which contained produced NH_3, was used for the GLDH assay and was shaken with 100 μl of the second mixture containing: 0.5 mg of NADH; 1.7 ml of 0.15 M Tris-maleate buffer (pH 8.6),; 100μl of 0.08 M α-ketoglutarate and 5μl (50μg) of GLDH. After 40 min at room temperature 20μl of this mixture were withdrawn and the excess NADH in it destroyed with 20μl of 1N HCl. The NAD+ produced was transformed into its fluorescent derivative by adding 9 ml of 8N NaOH and 1 ml of 0.1% hydrogen peroxide. This mixture was boiled for 5 min and then, after adding 1ml of water, fluorescence was measured at 460 nm after activation at 360 nm. The fluorescence of NAD in alkaline solution has been described by Udenfriend (1962). Standard ammonia solutions were measured with GLDH assay to obtain the calibration curves. Tyramine, serotonin and dopamine were the most readily deaminated substrates.

This kind of discontinuous assay cannot be used with secondary amines such as adrenaline in which ammonia is not a product of the reaction and results may be affected by high levels of ammonia present in many tissues.

Recently Zhou *et al.* (1996) developed the first direct and continuous fluorometric assay for MAO B activity which does away with some of the disadvantages inherent in the currently used fluorometric assays and is also rapid and efficient. The authors have designed and synthesized *E*-2,5-dimethoxycinnamylamine hydrochloride from 2,5-dimethoxybenzaldehyde. Both these compounds fluoresce brightly while the corresponding aldehydes formed after oxidation by MAO do not fluoresce. Whereas 2,5-dimethoxybenzaldehyde is a poor substrate for MAO B, *E*-2,5-dimethoxycinnamylamine hydrochloride has proved to be an excellent substrate for MAO B. *E*-2,5-dimethoxycinnamylamine hydrochloride produces an intense purple fluorescence when irradiated at 343 nm in Tris buffer 100 mM, pH 9.0 or 50 mM sodium phosphate buffer, pH 7.2 at 25°C and MAO B activity can be monitored continuously by measuring the decrease in fluorescence at 393 nm. A fluorescence-concentration calibration curve was constructed by measuring the fluorescence of increasing concentrations (0-164μm) of *E*-2,5-dimethoxycinnamylamine hydrochloride. MAO B purified from beef liver was used as substrate and the reaction mixture contained: 2,5-dimethoxycinnamylamine hydrochloride 20-328 μM, 2930 μl of Tris buffer 100 mM, pH 9.0, MAO B 70 μl (final concentration 0.46 μM). This method is rapid and efficient. However it requires relatively high concentrations of purified enzyme and is limited to the use of a single substrate.

Spectrophotometric assays - Numerous spectrophotometric methods have been developed for the determination of MAO activity. Some of these are based on the direct measurement of absorbance at a given wavelength, when the aldehyde is formed by the oxidative deamination of the monoamine substrate. However, this is possible only for a limited number of substrates, such as benzylamine and its derivatives (Tabor *et al.,* 1954; McEwen *et al*., 1968; Houslay and Tipton, 1974) or kynuramine (Weissenbach *et al*., 1960). For the purpose of spectrophotometrically testing MAO activity against a larger number of substrates, several other assays based on coupled indicator reactions have been developed. In one of these assays catalytic oxidation of the aldehyde is obtained using aldehyde dehydrogenase. During this reaction NAD+ is reduced to NADH. The NAD+ reduction is then measured by spectrophotometer at 340 nm (Houslay and Tipton,1973a,1974). Other coupled indicator reactions are used to measure MAO-dependent production of hydrogen peroxide (Köchli and von Wartburg, 1978; Szutowicz *et al*., 1984) or ammonia (Yamada *et al*., 1972). The following is a more detailed description of the direct procedure based on the use of kynuramine as substrate (Weissenbach *et al*.,1960) and a coupled assay based on hydrogen peroxide measurement (Köchli and von Wartburg, 1978).

According to Weissenbach *et al*. (1960), when kynuramine is oxidized to 5-hydroxyquinoline by MAO, its disappearance may be monitored by continuously recording the decrease in absorbance at 360 nm. The amount of enzyme used in the assay should be such as to guarantee a linear decrease in absorbance with time; both crude homogenates and mitochondrial suspensions can be used. The incubation mixture in 3 ml of 50 mM phosphate buffer (pH 7.4) contains enzyme source (from rat tissues) and kynuramine (1.67×10^{-4} M). The reaction is started by adding the substrate and is carried out at 37°C in a thermostatically controlled cell. Blanks are prepared by omitting the substrate. Absorbance may also be recorded discontinuously. In this case the reaction (final volume 1.5 ml) is stopped after 30 min by adding 0.25 ml of 10% $ZnSO_4$ and 0.05 ml of 1 N NaOH. Samples are boiled for 2 min and the absorbance at 360 nm of the supernatant, obtained after centrifugation, is measured. Blanks are prepared by adding the substrate soon after incubation and the amount of metabolized substrate are obtained by subtracting the value of the absorbance of the sample from that of the blank.

The coupled assay of Köchli and von Wartburg (1978) allows continuous monitoring of the reaction with many physiological substrates. Hydrogen peroxide produced by MAO activity is measured by recording at 502 nm the hydrogen peroxide-dependent oxidation of the colorless leuco-2',7'-dichlorofluorescein to the dye 2',7'-dichlorofluorescein, catalyzed by horseradish peroxidase. The assay mixture (3 ml) contains 0.6 ml of leuco-DCF solution (leuco-2',7'-dichlorofluorescein diacetate 2.5×10^{-4} M dissolved before use in 0.01 N NaOH), 1.8 ml of POD-buffer (8.3 mg of horseradish peroxidase grade I dissolved in 100 ml of 0.1 M sodium phosphate pH 7.15), 0.1 ml substrate at the desired concentration, 0.1-0.5 ml of rat brain mitochondrial suspension (final protein amount 0.02-0.2 mg/ml). The increase in absorbance is recorded at 502 nm and at 25°C. A linear increase is observed after a short time lag. Although simple and sensitive this method nevertheless has several limitations. One of these is the fact that dopamine and norepinephrine cannot be used as substrates. Furthermore, crude tissue preparations may not be suitable owing to the presence of large quantities of catalase, or because leuco-DCF oxidation could occur also in the absence of substrates. Lastly, impurities in the POD and in the substratum may cause some initial oxidation of leuco-DCF and thus alter the results.

Very recently Holt *et al.* (1997) have described a continuous peroxidase-linked spectrophotometric assay for MAO in which 4-aminoantipyrine, after having acted as the proton donor in the peroxidase reaction, condenses with vanillic acid (4-hydroxy-3-methoxybenzoic acid) producing a red quinoneimine dye. The amount of hydrogen peroxyde produced is determined measuring the absorbance at 498 nm. The molar absorption coefficient of quinoneimine dye at pH 7.6 is 4654 M^{-1} cm^{-1}. Assays are made in 96-well polystyrene microtiter plates and each well contains 50μl of rat liver homogenate, 50μl of chromogenic solution and 200μl of substrate, prepared in potassium phosphate buffer 0.2M (pH7.6). The chromogenic solution has the following composition: vanillic acid (1mM), 4-aminoantipyrine (500μM), peroxidase (4 U ml^{-1}) in buffer. This solution is prepared each day and stored at 4°C. Activity is monitored in a THERMOmax microplate reader (Molecular Devices, Menlo, Park,CA) at 490nm and at 37°C. Alternatively assays may be prepared in spectrophotometer cuvettes (final volume 1ml) and in this case absorbance (498 nm) is monitored by spectrophotometer at 37°C. When numerous samples must be analyzed by spectrophotometer, reactions may be followed discontinuously. In this case samples are prepared in glass test tubes (final volune 1ml) and stopped in ice bath by addition of 10μl of phenelzine 10mM. Blanks are prepared omitting the substrate.This method is inexpensive and uses reagents having low toxicity. However it may be used only with substrates which have a higher oxidation-reduction potential than does 4-aminoantipyrine. For instance, 5-hydroxytryptamine, noradrenaline and other catecholamines (Nelson and Huggins, 1974) cannot be used. In addition endogenous catalase activity may be a source of error.

Miscellaneous assays - In addition to those described above, there are several other methods for measuring MAO activity. One of these is the measurement of oxygen consumption, which can be done manometrically (Creasy, 1956) or else by using an oxygen electrode (Tipton and Dawson,1968). Polarographic determinations of oxygen consumption are accurate and simple and may be applied to all MAO substrates. Nevertheless, they have the drawback that high rates of endogenous oxygen consumption may occur, particularly when crude homogenates are used. Cyanide is added to the reaction mixture in order to avoid this problem. However, this compound is known to be a weak inhibitor of MAO activity (Houslay and Tipton,1973b). Other methods are based on measuring the amount of ammonia produced. Mention has already been made of the fluorimetric method of Harada and Nagatsu (1973). The ammonia produced by MAO may also be measured colorimetrically by an indophenol reaction (Nagatsu and Yagi, 1966) or by using an ammonia-selective electrode (Meyerson *et al.*,1978). As we have seen, also in the case of ammonia production based assays, the ammonia present in many tissues may affect specificity and the method cannot be used with secondary amines.

Lastly, MAO activity may be monitored also by luminometric assay. A highly sensitive and rapid assay can be used to measure the bioluminescence produced by a çoupled process in which long-chain aliphatic amines (C_8-C_{16}) used as MAO substrates are converted to long-chain aliphatic aldehydes which can then act as substrates for a bacterial luciferase (Tenne *et al.*,1985). A discontinuous luminometric assay developed by O'Brien *et al., (*1993) is instead used to measure hydrogen peroxide produced by MAO activity. What is measured directly by the luminometer is the light produced by the peroxidase-catalysed chemiluminescent oxidation of luminol (5-amino-2,3,-dihydro-1,4-phthalazinedione) by the hydrogen peroxide. According to O'Brien *et al., (*1993) stock solutions of 10 mM luminol in 50% (v/v) dimethyl sulphoxide and 25 μM horseradish peroxidase are prepared before use. Two reaction mixtures are necessary: the first (3ml) contains 93.3 mM potassium phosphate buffer, pH 7.2, 3.1 mM sodium azide, enzyme source (rat liver mitochondria or MAO B purified from ox liver at the desired concentrations), and substrate. After incubation at 37°C for specified times, aliquots of 50-100 μl are transferred to vials containing the second reaction mixture (67 mM Tris-HCl buffer, pH 8.0, 750 nM horseradish peroxidase and 25 μM luminol). The intensity of the produced light is measured instantaneously. A standard curve for hydrogen peroxide evaluation is prepared each time by adding hydrogen peroxide (11-225 nM) to the second reaction mixture, using an air-tight Hamilton microsyringe. The detection limit of the assay is 0.7 nmol of hydrogen peroxide present/ 3ml MAO assay. As the reaction mixture may contain luminogenic contaminants, it is advisable to measure each time the light emission of the reaction mixture devoid of the substrate. In addition, disposable plastic cuvettes should be avoided becouse ligh signals may be caused by static electricity. This simple and sensitive method can be used with several MAO substrates, including MPTP, which forms neither aldehyde nor ammonia, and milacemide (2-*n*-pentylaminoacetamide), which forms pentylamine and glycinamide.When tyramine and milacemide are used , specific activities are similar to those obtained by the radiochemical and the coupled spectrophotometric assays. However, when benzylamine was the substrate, values were lower. In this case azide should be replaced by the catalase inhibitor 3-amino-1,2,4-triazole. Unfortunately, this luminometric assay cannot be used with 5-hydroxytryptamine, adrenaline or noradrenaline because, in the presence of horseradish peroxidase, they are more readily oxidized by hydrogen peroxide than luminol.

DETERMINATION OF MONOAMINE OXIDASE A AND B CONCENTRATION

The determination of the molar amounts of the two forms of MAO present in a tissue preparation may be performed with two different methods based on the use of irreversible inhibitors of MAO activity and on the assumption that these inhibitors bind stoichiometrically and covalently to the FAD moiety, which is bound to a cysteine residue from each MAO subunit (Singer and Salach, 1981)

Titration of MAO A and MAO B by clorgyline and deprenyl - This method has been described by Fowler *et al.* (1980a and 1980b). MAO A concentration for a fixed amount of human brain homogenate is calculated using 5-HT as substrate and measuring the concentration of clorgyline required to produce 100% inhibition. This concentration corresponds to the enzyme concentration. The same criterion is applied to MAO B measurement whenever benzylamine is used as substrate and deprenyl as selective inhibitor. It is also possible to use substrates deaminated by both forms of MAO. In this case the concentration of MAO A is determined in the presence of deprenyl concentrations that, in the experimental conditions used, completely inhibit MAO B activity without producing any significant decrease in MAO A activity. The same principle is applied for MAO B. Data are expressed as pmoles of MAO A or MAO B per mg protein and are extrapolated from the pmoles of clorgyline or deprenyl required for 100% inhibition. It is assumed that one molecule of inhibitor is bound to the FAD moiety in each MAO subunit and that MAO is composed of two subunits (Weyler *et al.*, 1990). In order to test for the existence of clorgyline and deprenyl binding to sites other than the active centre of MAO, the enzyme concentration was determined in different subcellular fractions having different MAO: protein ratios. In the absence of non specific-binding, the ratio of the V_{max} values in the different fractions should be the same as the ratios of the enzyme concentrations determined by enzyme titration. This test showed that, in the experimental system used by Fowler *et al.*, (1980a), even though in the case of clorgyline binding to measure MAO A no appreciable non specific binding occurred, in the case of deprenyl binding for MAO B measurement, it was remarkable and led to an overestimation of the active centre of MAO B. The likelihood of non specific binding thus imposes a serious limitation on the use of this method.

Quantitation of MAO A and MAO B by ^{3}H-pargyline binding - A more accurate estimate of the active site concentrations of MAO A and MAO B may be obtained by using ^{3}H-labelled pargyline (Parkinson and Callingham, 1980; Gomez *et al*, 1986; O'Carroll *et al.*, 1989; Riley and Denney,1991) an irreversible acetylenic inhibitor capable of binding both MAO A and MAO B. In work by Gomez *et al.* (1986) rat liver mitochondria were incubated in 0.1 ml of 0.1 M potassium phosphate buffer pH 7.8 containing 20-400 pmoles of ^{3}H-pargyline (50 mCi/mmol) for 60 min at 37°C. Separation of the unbound pargyline was achieved by centrifugation or by filtration. In the centrifugation assay, after cooling on ice, 1 ml of buffer was added and the sample then centrifuged at 16,000 x g for 1.5 min. The pellet was resuspended, washed as before and centrifuged again. The second pellet was resuspended in 1 ml of buffer and 10 ml of scintillant (toluene-Triton X-100 [2:1,v/v] containing 0.4% [w/v] PPO) were added for the radioactivity count. When the filtration assay was used, samples were incubated in 50 mM potassium phosphate buffer at pH 7.2 containing varying amounts of ^{3}H-pargyline (3.2 Ci/mmol). The reaction was stopped by adding 4 ml of ice-cold buffer and by rapid filtration, under vacuum, over Whatman GF-C filters. The filters were washed with buffer, dried and their ^{3}H-content measured after addition of 4ml of the same scintillant as that used above. If the tissue contains both forms of MAO, for the determination of either of them, the samples must be preincubated using concentrations of clorgyline (for MAO B) or deprenyl (for MAO A) which ensure 100% inhibition of the activity of the undesired form while leaving the other form unaltered. In order to calculate the non specific binding, the samples were preincubated for 60 min with excess (2 mM) cold pargyline. For the same purpose, it is also possible to use clorgyline at a concentration (1 mM) that can block 100% of the activity of both forms of MAO, and thus block all the active sites of the enzyme. Also a mixture of clorgyline and deprenyl can be used. This is possible because clorgyline and deprenyl bind to the same enzyme site as pargyline. The specific binding was expressed as pmoles of pargyline bound/mg protein and was calculated by subtracting the binding obtained in the presence of unlabeled pargyline from the total binding obtained in its absence. The concentrations of labeled pargyline and the incubation times used in this type of method must be sufficient to ensure the complete labeling of the MAO active sites. For this purpose it is advisable to make time-courses of incorporation of label. If the incubation time required for a given concentration of labeled pargyline is very long, loss of enzyme function could occur. Since pargyline binding to the enzyme requires the latter to retain its catalytic function, it is advisable in this case to increase the concentration of labeled pargyline and to reduce the incubation time (Anderson and Tipton,1994).

The separation of unbound ^{3}H-pargyline by centrifugation or by filtration described by Gomez *et al.* (1986) and described above cannot be used with enzyme preparations that have been rendered soluble. Anderson and Tipton (1994) recently described a simple procedure for performing also this type of separation. After incubating tissues with labeled pargyline as described by Gomez *et al.*, (1986), the unbound inhibitor was removed by mixing the incubation mixture with 1 ml of buffer containing 1% (w/v) activated charcoal and 2% (w/v) Bovine Serum Albumin. Insoluble charcoal-bound inhibitor was separated from the soluble enzyme-bound inhibitor by centrifugation. An aliquot of the supernatant was removed and scintillant added to count the radioactivity of the sample. This procedure is unsuitable for crude tissue preparations.

Riley *et al.*(1989) have described an alternative procedure which can be used when samples contain both MAO A and MAO B. After binding with ^{3}H-pargyline the two enzymes are distinguished by immunoprecipitation with MAO A or MAO B monoclonal antibodies.

REFERENCES

Achee, F. M., Togulga, G. and Gabay, S. (1974). Studies on monoamine oxidase: Properties of the enzyme in bovine and rat brain mitochondria. *J.Neurochem.* **22**, 651-661.

Alivisatos, S.G.A. and Ungar, F. (1968). Incorporation of radioactivity from labeled serotonin and tryptamine into acid-insoluble material from subcellular fractions of brain. I. The nature of the substrate. *Biochemistry* **7**, 285-292.

Anderson, M.C. and Tipton, K.F. (1994). Estimation of monoamine oxidase concentrations in soluble and membrane-bound preparations by inhibitor binding. *J.Neural Transm.* [*Suppl.*] **41**,47-53.

Berry, M.D., Juorio, A.V. and Paterson, I.A. (1994). The functional role of monoamine oxidase A and B in the mammalian central nervous system. *Prog. Neurobiol.* **42**, 375-391.

Callingham, B.A., Holt, A. and Elliot, J. (1991). Properties and functions of the semicarbazide-sensitive amine oxidases. *Biochem.Soc.Trans.* **19**, 228-233.

Century, B. and Rupp, K.L. (1968).Comment on microfluorometric determination of monoamine oxidase. *Biochem. Pharmacol.* **17**, 2012-2013.

Cesura, A.M. and Pletscher, A. (1992). The new generation of monoamine oxidase inhibitors. *Prog. Drug Res.* **38**, 171-297.

Cesura, A.M., Richards, J.G., Picotti, G.B., Kettler, R., Imhof, R. and Da Prada, M. (1993). Reversible monoamine oxidase inhibitors: from laboratory to clinic. In: "Monoamine Oxidase: basic and clinical aspects". Yasuhara, H., Parvez, S.H., Oguchi, K., Sandler, M. and Nagatsu, T., eds.,Utrecht, The Netherlands, pp: 159-176.

Chen,Z.Y., Hotamisligil,G.S., Huang,J.K.,Wen,L.,Ezzeddine,D.,Aydin-Muderrisoglu,N., Powell,J.F., Huang,R.H., Breakefield,X.O., Craig,I. and Hsu,Y.-P.P. (1991). Structure of the human gene for monoamine oxidase type A. *Nucl.Acids Res.* **19**, 4537-4541

Creasy, N.H. (1956). Factors which interfere in the manometric assay of monoamine oxidase. *Biochem.J.* **64**, 178-183.

Elliot ,J., Callingham,B.A. and Sharman, D.E. (1989). Semicarbazide-sensitive amine oxidase (SSAO) from rat aorta. Interactions with some naturally occurring amines and their structural analogues. *Biochem.Pharmacol.* **38**, 1507-1515.

Ekblom,J., Jossan,S.S., Bergström,M., Oreland,L., Walum,E. and Aquilonius,S-M .(1994). Monoamine oxidase-B in astrocytes. *Glia* **8**, 122-132.

Fowler, C.J., Oreland, L. (1980). The nature of the substrate-selective interaction between rat liver mitochondrial monoamine oxidase and oxygen. *Biochem.Pharmacol.* **29**, 2225-2233.

Fowler, C.J., Oreland, L., Marcusson, J. and Winblad, B. (1980a). Titration of human brain monoamine oxidase-A and -B by clorgyline and *l*-deprenyl. *Naunyn-Schmiedeberg's Arch. Pharmacol.* **311**, 263-272.

Fowler,C.J. and Callingham,B.A., Mantle,T.J. and Tipton, K.F. (1978). Monoamine oxidase A and B: a useful concept ? *Biochem.Pharmacol.* **27**, 97-101.

Fowler, C.J., Wiberg, A., Oreland, L., and Winblad, B. (1980b). Titration of human brain type B monoamine oxidase. *Neurochem Res.* **5**, 697-708.

Gomez, N.,Unzeta, M., Tipton, K.F., Anderson, M.C. and O'Carroll, A-M. (1986). Determination of monoamine oxidase concentrations in rat liver by inhibitor binding. *Biochem.Pharmacol.* **35**, 4467-4472.

Guilbault, G.G., Kramer, D.N. and Hackley, E. (1967). A new substrate for fluorimetric determination of oxidative enzymes. *Anal. Chem.* **39**, 271

Harada, M. and Nagatsu, T. (1973). A sensitive fluorometric assay for monoamine oxidase activity. *Anal. Biochem.* **56**, 283-288.

Heikkila, R.E., Hess, A. and Duvoism, R.C. (1984). Dopaminergic neurotoxicity of 1-methyl-4-phenyl-1,2,5,6,tetrahydropyridine in mice. *Science* **224**, 1451-1453.

Holt, A., Sharman, D.F., Baker, G.B. and Palcic, M.M. (1997). A continuous spectrophotometric assay for monoamine oxidase and related enzymes in tissue homogenates. *Anal.Biochem.* **244**, 384-392.

Houslay, M.D. and Tipton, K.F. (1973 a). The nature of the electrophoretically separable multiple forms of rat liver monoamine oxidase. *Biochem.J.* **135**, 173-186.

Houslay, M.D. and Tipton, K.F. (1973 b). The reaction pathway of membrane-bound rat liver mitochondrial monoamine oxidase. *Biochem.J.* **135**, 735-750.

Houslay, M.D. and Tipton, K.F. (1974). A kinetic evaluation of monoamine oxidase activity in rat liver mitochondrial outer membranes. *Biochem. J.* **139**, 645-652.

Jain, M., Sands, F. and Von Korff, R,W. (1973). Monoamine oxidase activity measurements using radioactive substrates. *Anal. Biochem.* **52**, 542-554.

Johnston, J.P. (1968). Some observations upon a new inhibitor of monoamine oxidase in human brain. *Biochem.Pharmacol.* **17**, 1285-1297.

Kaplan, N.O., Colowick, S.P. and Barnes, C.C. (1951) Effect of alkali on diphosphopyridine nucleotide. *J.Biol.Chem.* **191**, 461-472.

Kinemuchi, H.,Fowler, C.J. and Tipton, K.F. (1984). Substrate specificities of the two forms of monoamine oxidase. In:" Monoamine Oxidase and Disease" , Tipton,K.F., Dostert, P. and Strolin Benedetti M., eds., Academic Press, New York. pp: 53-62.

Knoll, J. (1993a). The pharmacological basis of the beneficial effects of (-) deprenyl (selegiline) in Parkinson's and Alzeimer's diseases. *J.Neural Transm.* [*Suppl.*]**40**, 69-91

Knoll, J. (1993b). Some clinical implications of MAO-B inhibition. In: "Monoamine Oxidase: basic and clinical aspects". Yasuhara,H., Parvez,S.H., Oguchi,K., Sandler,M. and Nagatsu,T. eds.,Utrecht, The Netherlands , 197-217.

Knoll, J. and Magyar, K. (1972). Some puzzling pharmacological effects of monoamine oxidase inhibitors. *Adv. Biochem.Psychopharmacol.* **5**, 393-408.

Köchli, H. and von Wartburg, J.P. (1978). A sensitive photometric assay for monoamine oxidase. *Anal.Biochem.* **84**, 127-135.

Kondoh, Y., Murakami, M., Yin, W., Mizusawa, S., Nakamiki, H. and Nagata, K. (1994). Quantitave distribution of rat brain monoamine oxidase A by [^{14}C]clorgyline autoradiography. *Exp.Brain Res.* **99**, 375-382.

Kraml, M. (1965) A rapid microfluorimetric determination of monoamine oxidase. *Biochem.Pharmacol.* **14,** 1683-1685.

Krueger, M.J. and Singer, T.P. (1993). An examination of the reliability of the radiochemical assay for monoamine oxidase A and B. *Anal. Biochem.* **214**, 116-123.

Lan,N.C., Heinzmann,C., Gal,A., Klisak,I., Orth,U.,Lai,E., Grimsby,J., Sparkes,R.S.,Mohandas,T.and Shih,J.C. (1989). Human monoamine oxidase A and B genes map to Xp11.23 and are deleted in a patient with Norrie's disease. *Genomics* **4**, 552-559.

Markey, S.P., Trevor, A. ,Castagnoli, I.J.Jr. and Kopin, I. (eds) (1986). MPTP: a Neurotoxin Producing a Parkinsonian Syndrome, Academic Press, New York.

McCaman, R.E., McCaman, M.W., Hunt, J.M. and Smith, M.S. (1965) Microdetermination of monoamine oxidase and 5-hydroxytryptophan decarboxylase activities in nervous tissues. *J.Neurochem.* **12**, 15-23.

McEwen, C.M., Sasaki, G. and Lenz, W.R. (1968). Human liver monoamine oxidase. I. Kinetic studies and model interactions. *J.Biol.Chem.* **243**, 5217-5225.

Meyerson, L.R., McMurtrey, K.D. and Davis, V.E. (1978). A rapid and sensitive potentiometric assay for monoamine oxidase using an ammonia-selective electrode. *Anal.Biochem.* **86**, 287-297.

Morinan, A. and Garratt, H.M. (1985). An improved fluorimetric assay for brain monoamine oxidase. *J.Pharmacol. Methods* **13**, 213-223.

Nagatsu, T. (1973). Biochemistry of Catecholamines, University Park Press, Baltimore, pp.185-207.

Nagatsu, T. and Yagi, K. (1966). A simple assay of monoamine oxidase and D-amino acid oxidase by measuring ammonia. *J.Biochem.* **60**, 219-221.

Nakano, T., Kawai, S. and Nagatsu, T. (1984). A simple assay for monoamine oxidase using glutathione peroxidase and glutathione reductase. *Japan. J. Pharmacol.* **35**, 163-167.

Nelson, D.R. and Huggins, A.K. (1974). Interference of 5-hydroxytryptamine in the assay of glucose by glucose oxidase: peroxidase: chromogen based methods. *Anal.Biochem.* **59**, 46-53.

O'Brien, E.M., Kiely, K.A., Tipton, K.F. (1993). A discontinuous luminometric assay for monoamine oxidase. *Biochem.Pharmacol.***46**, 1301-1306.

O'Carroll, A.-M., Anderson, M.C., Tobbia, I., Phillips, J.P. and Tipton, K.T. (1989). Determination of the absolute concentrations of monoamine oxidase A and B in human tissues. *Biochem. Pharmacol.* **38**, 901-905.

Oreland, L. (1993). Monoamine oxidase in neuro-psychiatric disorders. In: "Monoamine Oxidase: basic and clinical aspects". Yasuhara,H., Parvez,S.H., Oguchi,K., Sandler,M. and Nagatsu,T. eds.,Utrecht, The Netherlands, pp: 219-247.

Otsuka, S. and Kobayashi, Y. (1964). A radioisotopic assay for monoamine oxidase determinations in human plasma. *Biochem. Pharmacol.* **13**, 995-1006.

Parkinson, D. and Callingham, B.A. (1980). The binding of [^{3}H] pargyline to rat liver mitochondrial monoamine oxidase. *J. Pharm.Pharmacol.* **32**, 49-54.

Parkinson Study Group (1989). Effect of (-) deprenyl on the progression of disability in early Parkinson's disease. *New Engl. J.Med.* **321**, 1364-1371.

Riley, L.A. and Denney, R.M. (1991). Problems with the measurement of monoamine oxidase A protein concentration in mitochondrial preparations. *Biochem. Pharmacol.* **42**, 1953-1959.

Riley,L.A., Waguespack,M.A. and Denney,R.M. (1989). Characterization and quantitation of monoamine oxidase A and B in mitochondria from human placenta. *Mol. Pharmacol.* **36**, 54-60.

Saura, J., Kettler R., Da Prada, M. and Richards, J.G.(1992). Quantitative enzyme radioautography with ^{3}H-Ro 41-1049 and ^{3}H-Ro 19-6327 *in vitro*: localization and abundance of MAO A and MAO B in rat CNS, peripheral organs, and human brain. *J.Neurosci.* **12**, 1977-1999.

Shih.J.C. (1993). cDNA cloning of human liver MAO-A and MAO-B. In: "Monoamine Oxidase: basic and clinical aspects". Yasuhara,H., Parvez,S.H., Oguchi,K., Sandler,M. and Nagatsu,T. eds.,Utrecht, The Netherlands, pp: 15-21.

Singer,T.P. and Salach, J.I. (1981). Interaction of suicide inhibitors with the active site of monoamine oxidase. In "Monoamine Oxidase Inhibitors - The State of the Art ".Youdim, M.B.H. and Paykel, E.S. eds., J.Wiley and Sons, pp: 17-29.

Snyder, S.H. and Hendley, E.D. (1968). A simple and sensitive fluorescence assay for monoamine oxidase and diamine oxidase. *J. Pharm. Exp. Ther.* **163**, 386-392.

Southgate ,J. and Collins, G.G.S. (1969). The estimation of monoamine oxidase using ^{14}C-labelled substrates. *Biochem. Pharmacol.* **18** , 2285-2287.

Sullivan J.P.,McDonnell, L., Hardiman, O.M.,Farrel, M.A., Phillips, J.P. and Tipton, K.F. (1986).The oxidation of tryptamine by the two forms of monoamine oxidase in human tissues. *Biochem. Pharmacol.* **35**, 3255-3260.

Szutowicz, A., Kobes, R.D. and Orsulak, P.J. (1984). Colorimetric assay for monoamine oxidase in tissues using peroxidase and 2,2'-azinodi(3-ethylbenzthiazoline-6-sulfonic acid) as chromogen. *Anal.Biochem.* **138**, 86-94.

Tabor, C. W.,Tabor, H and Rosenthal, S.M. (1954). Purification of amine oxidase from beef plasma. *J. Biol.Chem.* **208**, 645-661.

Tenne, M., Finberg, J.P.M., Youdim, M.B.H. and Ulitzur, S. (1985). A new rapid and sensitive bioluminescence assay for monoamine oxidase activity. *J. Neurochem.* **44**, 1378-1384.

Tipton, K.F. (1969). A sensitive fluorometric assay for monoamine oxidase. *Anal. Biochem.* **28**, 318-325.

Tipton, K.F. and Dawson, A.P. (1968). The distribution of monoamine oxidase and α-glycerophosphate dehydrogenase in pig brain. *Biochem.J.* **108**, 95-99.

Tipton, K.F. and Singer, T.P. (1993). The radiochemical assay for monoamine oxidase activity. Problems and pitfalls. *Biochem.Pharmacol.* **46**, 1311-1316.

Tipton, K.F. and Youdim, M.B.H. (1976). Assay of monoamine oxidase. In: "Monoamine oxidase and its inibition ". *Ciba Foundation Symposium 39*, Wolstenholme, G.E.W. and Knight, J., eds., Elsevier,Exerpta Medica, North Holland, Amsterdam, pp: 393-403.

Tipton, K.F. and Youdim, M.B.H. (1983). The assay of monoamine oxidase activity. In: "Methods in Biogenic Amines Research", Parvez, S., Nagatsu,T., Nagatsu, I. and Parvez, H., eds., Elsevier, Amsterdam, pp: 441-465.

Tipton, K.F., O'Carroll, A.-M. and McCrodden, J.M. (1987). The catalytic behaviour of monoamine oxidase. *J.Neural Transm.* [*Suppl.*] **23**, 73-89.

Waldmeier, P.C. (1987). Amine oxidases and their endogenous substrates (with special reference to monoamine oxidase and the brain). *J. Neural Transm.* [*Suppl.*] **23**, 55-72.

Weissbach, H., Smith, T.E., Daly ,J.W., Witkop, B. and Udenfriend, S. (1960). A rapid spectrophotometric assay of monoamine oxidase based on the disappearence of kynuramine. .*J. Biol. Chem.* **235**, 1160-1163.

Weyler, W., Hsu, Y.P. and Breakefield, X.O. (1990). Biochemistry and genetics of monoamine oxidase. *Pharmac. Ther.*, **47**, 391-417.

Wurtman R.J. and Axelrod ,J. (1963). A sensitive and specific assay for the estimation of monoamine oxidase. *Biochem. Pharmacol.* **12**, 1439-1440.

Yamada, H., Suzuki, H. and Ogura, Y. (1972). Amine oxidases from microorganisms: stoichiometry of the reaction catalyzed by amine oxidase of *Aspergillus niger. Adv. Biochem.Psychopharmacol.* **5**, 185-201.

Youdim, M.B.H.,Finberg, J.P.M. and Tipton, K.F. (1988). Monoamine oxidase. In: "Handbook of Experimental Pharmacology " 90/1, Trendelenburg, U. and Weiner,N. eds., Springer-Verlag, Berlin, pp: 119-192

Udenfriend, S. (1962). Fluorescence Assay in Biology and Medicine. Academic Press, New York.

Zhou J.J.P., Zhong, B. and Silverman, R.B. (1996). Direct continuous fluorometric assay for monoamine oxidase B, *Anal.Biochem.* **234**, 9-12.

Progress in HPLC-HPCE, Vol. 7, pp. 523-546
G. Ali Qureshi *et al.* (Eds)

Dynorphin converting enzymes and strategies for their characterization

Jerzy Silberring

Department of Clinical Neuroscience, Experimental Alcohol and Drug Addiction Research Section, Karolinska Institute. S-171 76 Stockholm, Sweden.

Introduction

During the past two decades, proteolytic conversion and degradation of neuropeptides drew increasing attention due to their potential role as a target for the development of new drugs (Turner and Barnes, 1994). Neuropeptides are biosynthesized in the form of larger, inactive precursors which are consecutively processed to the shorter, biologically active sequences, followed by their termination to inactive fragments (Terenius and Nyberg, 1988;1991; Nyberg and Terenius, 1991). A common feature of precursors is that the peptide sequences are flanked by pairs of basic amino acids. These residues constitute proteolytic cleavage sites for excision of the bioactive fragments. Recent progress in the discovery of several mammalian endo-oligopeptidases urges for further studies on their structures and functional significance in the pathology. The metabolism of dynorphins is of particular interest as these peptides are involved in drug dependence mechanisms, pain and mental disorders. Angiotensin-converting enzyme and enkephalinases terminate the action of shorter fragments towards opioid receptors. Thus, maturation, processing and degradation are the key steps in regulation of their biological activity. There is an increasing interest in involvement of oligopeptidases in the degradation of proteins. For instance, antigen presentation involves the generation of oligopeptides that are displayed to T-cells by major histocompatibility complex (MHC) class I and II molecules (Germain and Margulies, 1993), generating nonapeptides (MHC Class I) or slightly larger peptides (MHC Class II). Many peptides are important signal substances in the endocrine and neuronal systems. The degradation of such peptides is critical for termination of their activity. In the nervous system endo-oligopeptidases transform peptides generated by so-called processing enzymes to fragments with similar, different or essentially no biologic activity. This proteolytic conversion is part of a mechanism which releases the biologically active fragments directed towards various opioid receptors. These proteinases are, in turn, controlled by endogenous inhibitors which are an integral part of metabolic pathways (Bode and Huber, 1992). Both proteinases and endogenous inhibitors have been considered and applied as potential markers of various clinical states. Among the CNS disorders which have been associated with alterations in the neuropeptide - enzyme - inhibitor systems are diseases such as Alzheimers disease and schizophrenia (Terenius and Nyberg, 1988; Näslund et al.,

1994). More recent application of biochemistry and molecular biology has greatly contributed to our understanding of the nature and mode of action of these proteases.
This review will be focused on the prodynorphin system, which is of particular interest due to its highly potent antinociceptive effect, involvement in opioid addiction mechanism(s) and distinct localization in CNS compared to other opioid systems.
Prodynorphin, one of the three known endogenous opioid precursors, generates several opioid fragments, the κ-receptor-selective dynorphins and the shorter δ-receptor-selective enkephalins (see Fig. 1). Prodynorphin-derived peptides are involved in pain mechanisms (Friedman and Hooke, 1981; Hooke et al., 1995; Przewlocki et al., 1986) and opioid dependence (Green and Lee, 1988; Hooke et al. 1995). Other findings describe the view that these peptides may take part in regulation of neuroendocrine functions.

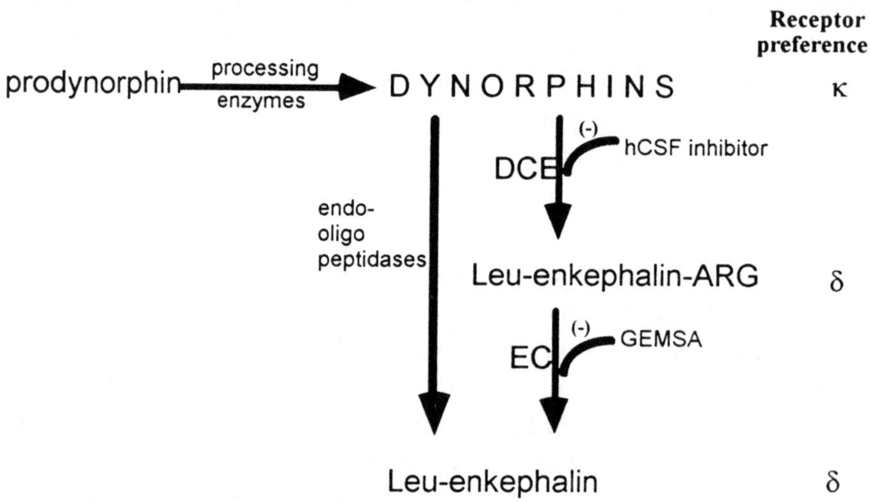

Fig. 1. General scheme for prodynorphin processing to bioactive products and the possible inhibitory mechanisms. (DCE = dynorphin converting enzyme; EC = enkephalin convertase; GEMSA = guanidinoethylmercaptosulfonic acid).

Active products belonging to the dynorphin family of peptides can be cleaved or converted to shorter sequences by several endopeptidases including those recently purified from the CSF (Nyberg et al., 1985), spinal cord (Silberring and Nyberg, 1989; Silberring et al., 1992), human neuroblastoma SH-SY5Y cell lines (Vlaskovska et al., 1994; Vlaskovska et al., 1995) and cortex primary culture (Vlaskovska et al. 1994, 1996). These endopeptidases are operationally called here, dynorphin converting enzymes (DCE:s).
The aim of this paper was to describe present knowledge on endo-oligopeptidases, converting dynorphins (DYN A, DYN B and ANEO) to enkephalins.

Importance of the neuropeptide peptidases

Endo-oligopeptidases have been found to have an important role in endocrinology, immunology and neurobiology. A classic example is angiotensin-converting enzyme (ACE) a known target for important drugs (Terenius and Nyberg, 1988; Turner and Barnes, 1994). A distinct group of oligopeptidases are the converting enzymes that generate peptides with a particular activity from an oligopeptide substrate. ACE (Strittmatter et al., 1985), interleukin-β-converting enzyme (Cerretti et al., 1992) and dynorphin B-29 processing endopeptidase (Berman et al., 1995) are a few examples. The dynorphin peptides are supposed to have a modulatory influence in several CNS areas influencing pain (Nakazawa et al., 1985), motor activity (Herman and Goldstein, 1985) and behavior (Smith and Lee, 1988; Itoch et al., 1994) mainly through action via a κ-receptor (Chavkin et al., 1982). It was early realized that neurons expressing dynorphin peptides also had substantial concentrations of Leu-enkephalin as revealed by immunohistochemistry (McGinty et al., 1983; Zamir et al., 1984). Thus, these neurons must also express enzymes converting dynorphin peptides to enkephalins. This enzymatic conversion is of particular interest, since it conveys a change in biological activity at κ-receptors (dynorphins) to activity at δ–receptors (enkephalins). Searching for such an enzyme, a serine protease has been identified possessing the predicted activity and termed dynorphin converting enzyme (DCE) (Nyberg et al., 1985). Subsequently we have identified a cysteine protease with a similar type of specificity (Silberring and Nyberg, 1989; Silberring et al. 1992). The enzymes primarily generate the hexapeptide, Leu-enkephalin-Arg^6 and not the earlier observed pentapeptide Leu-enkephalin. Using antibodies selective for Leu-enkephalin-Arg^6, we could confirm that high amounts of this peptide is formed in dynorphinergic pathways (Christensson-Nylander et al., 1986). A particular reason for investigating the dynorphins-to-enkephalin enzymatic conversion is the observation that dynorphin peptides are pharmacological antagonists to opiate analgesia (Friedman et al., 1981; Hooke et al., 1995) and restrict the development of opiate tolerance (Green and Lee, 1988; Takemori et al., 1992, 1993).

Neuropeptidases may be pertinent pharmacological targets (as shown for the angiotensin-converting enzyme, "enkephalinase 24.11" and HIV-1 protease) for new drugs acting as inhibitors of these enzymes (Meek et al., 1990). The major research task for neuropeptide peptidases is to reveal their primary structure and physiological relevance and to design specific tools to study their function. For example, as the enzyme converting dynorphin B is upregulated by morphine, inhibition of the proteolytic activity may lead to a novel approach in treating drug addiction in the future (Vlaskovska et al., 1994; 1996).

Conversion of dynorphins

The conversion of dynorphin-derived peptides to enkephalins is associated with changes in receptor specificity. Hence, dynorphins A and B have pronounced affinity for kappa opioid

receptors, while Leu-enkephalin-Arg[6] and Leu-enkephalin show preference for the delta-receptor. Moreover, the Leu-enkephalin-Arg[6] sequence is the shortest one, distinctly deriving from the prodynorphin precursor and, therefore, it may serve as a marker of the metabolic changes within this particular neuropeptide system. As can be concluded from the scheme (Fig. 1), it is likely that the amount (concentration) of a particular peptide fragment strongly depends on the proteolytic activities acting on longer sequences.
The action of enkephalins is terminated by another enzymes, enkephalinases and ACE (Terenius and Nyberg, 1988; Turner and Barnes, 1994) which cleave a number of neuropeptides, including substance P, bradykinin and angiotensin I. They also tend to be more stable in biological fluids or tissues than neuropeptides. For example, chronic arthritis seems to affect both the pro-dynorphin system and dynorphin convertase (Silberring et al., 1992; Persson et al., 1992) and the inhibitors of enkephalinase-24.11 potentiate the actions of natriuretic peptides and affect the metabolism of substance P. It is, therefore, of importance to simultaneously study the entire peptide system including its biosynthesis together with posttranslational modifications and factors regulating these processes. The recent discovery and cloning of several new mammalian endo-proteinases has led to the hypothesis that they are implicated in various phases of prohormone processing. The functional significance of all these enzymes remains to be established.

Table 1. The presently known families of proteolytic enzymes cleaving dynorphins to bioactive fragments (DYN = dynorphin; CSF = cerebrospinal fluid; sp. cord = spinal cord).

Enzyme	origin	cleavage site	substrates
Pitrilysin family (metallo)	testis	dibasic	somatostatin, DYNs (Pierotti, 1994))
Thiol proteases	brain	monobasic	DYN B-29 (Berman, 1995))
	sp.cord	dibasic	DYN A (Silberring, 1992)
	sp.cord	dibasic	DYN B (Silberring, 1989)
	plasma	dibasic	DYN A (Nylander, 1995)
Other enzymes			
Serine/metallo?	CSF	dibasic/monobasic	DYNs (Nyberg, 1985)
Thiol/metallo?	cell lines	dibasic	DYN B (Vlaskovska, 1995)
Metallo	thimet oligopep.	monobasic	DYN B (Chesneau, 1994)
Metallo	choroid plexus	dibasic	DYNs (Nyberg, 1991)

Enzymes from the pitrilysin family (metallo-proteinases) and a thiol proteinase cleaving DYN B-29 at a mono-basic site reveal different specificity towards dynorphin sequences and were not found in the spinal cord. The highly selective metabolism of dynorphins in the spinal cord can be compared only with the pathway found in certain cell lines or in the cortex primary culture (Vlaskovska et al., 1995, 1996), and resembling features related to the DYN B-convertase from the spinal cord tissues. Detailed overview on the DCE derived from the cerebrospinal fluid is given in Chapter 15 of this book (F. Nyberg).

Distribution of dynorphin converting enzymes

Body fluids

Measurement of opioid peptides in CSF has led to the observation, that their levels decrease during prolonged storage of samples at room temperature. Moreover, several dynorphin-related fragments were identified in CSF (Nyberg et al., 1987), namely DYN (7-17) and Leu-enkephalin-Arg^6, which indicated a single cleavage of DYNs between Arg^6-Arg^7 or Arg^6-Lys^7 by an unknown endoproteinase. After extensive studies, it was possible to purifiy and characterize an enzyme capable of releasing Leu-enkephalin-Arg^6. This enzyme was classified as a neutral, serine proteinase with a molecular weight of ca. 40 kDa and with high specificity towards dynorphins (Nyberg et al., 1985). It was therefore named dynorphin converting enzyme, DCE. Further work revealed, that a similar proteinase was present in the rat CSF (Persson et al., 1992a; 1992b), however, this enzyme differed from the human one with regard to its active site (essential thiol group). The rat DCE appeared to be more specific for DYN B than for DYN A or ANEO. Furter details on the CSF enzyme are summarized in Chapter 15 of this book (F. Nyberg).

Direct measurements of dynorphin metabolism in human plasma by mass spectrometry revealed that this body fluid contains high amounts of enzyme, cleaving specifically dynorphin A to Leu-enkephalin-Arg^6 (40 pmol peptide/min/microliter plasma) and which, after prolonged incubation time, was further converted at a lower rate, to Leu-enkephalin (Nylander et al., 1995). This process occurs at neutral pH and was abolished by low concentrations of PHMB, suggesting involvement of the enzyme with a vital thiol residue in the vicinity of its active site. Other dynorphin-related peptides were virtually not converted. Our data, including HPLC and RIA experiments indicate, that DYN A can not be detected in measurable quantities of human plasma due to the rapid degradation of the hectadecapeptide in this body fluid. The more resistant DYN B or ANEO can be used as markers of the prodynorphin system in 2-4 ml of peripheral plasma (Nylander et al., 1995). Enzymatic reactions were performed in diluted plasma which allowed estimation of the primary and secondary cleavage sites. In another report, substantial amounts of DYN A and DYN A(1-13) were incubated with human blood and the fragments were identified by MALDI TOF (Chou et al., 1994). Similar technique was also described by Müller and Hochhaus (1995) where number of protease inhibitors were applied to discriminate between various enzymatic activities. The major metabolite in both studies was DYN A(2-13) formed from DYN A(1-13). Herman et al. (1980) studied in vivo degradation of dynorphin A(1-13) and its analog D-Ala^2-dynorphin A(1-11) after intraventricular administration into the rat. Radiolabeled peptides were extracted from rat brains after in vivo incubation and the mixtures separated by size exclusion chromatography. DYN A(1-13) was substantially degraded after this experiment, whereas its analog was metabolized to a lesser extent. The obtained data may, at least partially, explain discrepancies in analgesic action of opioid peptides injected to various brain structures.

Another important candidate for the processing of dynorphins to enkephalins is endo-oligopeptidase A (Camargo et al., 1973; Camargo et al., 1987). Apparently, this enzyme was found to be identical to the Pz-peptidase or Thimet oligopeptidase (E.C.3.4.22.19) and endopeptidase 24.15 (Orlowski et al., 1983), as a consensus between several research groups (Tisljar, 1993). This enzyme hydrolyzes the Phe^5-Ser^6 bond of bradykinin and releases Leu-enkephalin from DYN B, ANEO and BNEO (Camargo et al. 1987). Its proteolytic activity was effectively inhibited by PHMB and stimulated by thiols. Also, metal chelating compounds EDTA and 1,10-o-phenanthroline reduced enzyme activity by 60 and 80%, respectively. Brain endo-oligopeptidase A was tested towards dynorphin-related peptides containing SH-reactive group (Gomes et al., 1993). Introduction of a Cys-(3-nitro-2-pyridinesulfenyl) group gave rise to several irreversible inhibitors of the enzyme. This elegant study demonstrated clearly that the Cys residue is crucial for the reactivity of endo-oligopeptidase A. The substrate specificity of this enzyme is not determined by the amino acid residues flanking the scissile bonds of the substrates (Camargo et al., 1994). Using competitive studies and various peptide analogs, it has been shown that the length and chain flexibility of the substrates are dominant factors to explain specificity of endo-oligopeptidase A.

CNS tissue

Discovery of the endoproteinase cleaving dynorphins in human CSF raised speculations on the possible tissue source of this enzyme. During this search, two enzymes have been purified on an analytical scale from the spinal cord (human and bovine) to apparent homogeneity (Silberring and Nyberg, 1989; Silberring et al., 1992; Nyberg and Silberring, 1990). They selectively convert dynorphin A (DCE-A) and dynorphin B (DCE-B), thus releasing a major cleavage product, Leu-enkephalin-Arg^6. Other dynorphin/enkephalin-related substrates were not cleaved. Both enzymes were classified as thiol-dependent proteinases and DCE-A had a pH optimum around 5.5, whereas DCE-B was a neutral endopeptidase (Table 2).

Table 2. Recovery of dynorphin convertases from bovine DYN B) and human (DYN A) spinal cords.

Enzyme	substrate	optimum pH	specific activity	purific.factor
DCE-A	Dynorphin A	5.5	100 nmol/min/mg	5,000
DCE-B	Dynorphin B	7.8	17 nmol/min/mg	230

Subcellular fractionation using differential centrifugation and sucrose gradient revealed that DCE-A is not of lysosomal origin (Silberring et al., 1992). Subcellular distribution of DYN A-cleaving enzyme in the rat spinal cord was reported (Fried, 1991). The enzyme activity was highest in a 10,000 g supernatant after differential centrifugation. In the continuous density gradient, the highest activity was found in the top fractions at 0.8-1.05 M sucrose. This distribution indicates localization of the enzyme in a soluble compartment and in synaptosomes. It should be noted that in this experiment the DCE activity was tested at neutral pH which suggests the presence of another, yet uncharacterized, enzyme.

Proteolytic processing of prodynorphin-derived peptides in rat brain was studied with the help of high performance size exclusion chromatography (SEC) linked to electrospray ionization mass spectrometry (Nylander et al., 1995). A new HPLC column was developed for this purpose, capable of separating molecules within the mass range of 0.1-7 kDa. Extracts from rat striatum were incubated with individual dynorphin peptides. Dynorphin A was the most resistant to proteolytic cleavage, converting slowly to Leu-enkephalin (0.3 pmol/min), whereas dynorphin B was processed to this pentapeptide at a 10^4-fold higher rate. A minor fragment was Leu-enkephalin-Arg^6. α-neoendorphin was also rapidly metabolized to Leu-enkephalin (6 nmol/min) without any intermediate step and, to a lesser extent, to Leu-enkephalin-Arg^6. Another experiment revealed that DYN B was differentially cleaved in the striato-nigral pathway. Time-dependent studies, application of inhibitors of carboxypeptidase B-like enzymes, and quantitation by API MS and LC-MS revealed, that only DYN B, but not ANEO, is efficiently converted to a hexapeptide, Leu-enkephalin-Arg^6 in substantia nigra. Physiological significance of this observation remains unknown, but it indicates a distinct enzyme pool(s) within striato-nigral pathway.

Another endoprotease was recovered and purified to apparent homogeneity from soluble extracts of human choroid plexus (Nyberg et al., 1991). The purified enzyme was stained as a single band after sodium dodecyl sulphate polyacrylamide gel electrophoresis with an apparent molecular weight of around 54 kDa. The enzyme effectively cleaved dynorphins generating Leu-enkephalin-Arg^6 and was less active on other neuropeptides containing dibasic stretches. The enzyme was optimally active at neutral pH and was sensitive to EDTA. The serine protease inhibitors DFP and PMSF only slightly affected enzymatic activity. A similar, membrane-bound counterpart was isolated using the same purification strategy and showed almost identical properties with the soluble peptidase.

Degradation of radioactive [^{3}H]DYN A was studied in vivo after injection of the peptide into the periaqueductal gray of rats (Young et al., 1987) and investigated in vitro using rat brain membranes under receptor binding conditions. In both experiments, the hexadecapeptide was N-terminally truncated, forming Des-[Tyr^1]DYN A which was additionally verified by the presence of [3H]-Tyr. Small amounts of [^{3}H]Leu-enkephalin were also observed. In another experiment, DYN A(1-13) was incubated with the enzymes bound to the rat brain membranes (Leslie and Goldstein, 1982). These data are in good agreement with studies performed with mass spectrometry linked to the size-exclusion HPLC, as a detection technique (Nylander et al., 1995; Silberring et al., 1995), where DYN A was slowly converted to its Des[Tyr^1] fragment, in contrast to other prodynorphin-derived peptides: DYN B and ANEO which were efficiently processed to enkephalins, but only in the presence of amastatin- an aminopeptidase inhibitor.

Localization of endo-oligopeptidase, presently known as a Thimet oligopeptidase, was determined by an enzymatic assay and by immunological techniques in the CNS of the rat (Oliveira et al., 1990). The majority of the enzyme was found in solubilized proteins obtained from synaptosomal membranes prepared from hypothalamus, neocortex and striatum, when Western blotting was applied. Immunohistochemical localization was confined to the gray matter in striatum.

Distribution of DYN B-29 convertase, which forms DYN B from DYN B-29, generally followed the distribution of immunoreactive DYN B, with few exceptions (Berman et al., 1994). This pattern does not match the distribution of prohormone-processing enzymes and the enzyme activity was high in the hypothalamus and hippocampus and low in the cerebellum. Complete purification and characterization of the novel metalloendopeptidase (NRD convertase) from rat testis and cortex was recently described by Cohen's group (Chesneau et al., 1994a,b). This enzyme is selective against dibasic stretches and cleaves all three dynorphin-related peptides releasing Leu-enkephalin-Arg^6. NRD convertase has been purified to apparent homogeneity which allowed sequencing and, after screening the rat cDNA library, the full-length cDNA was isolated (Pierotti et al., 1994). This revealed that the enzyme can be classified as belonging to the pitrilysin family of zinc-metalloendopeptidases. No physiological function has been revealed for this enzyme.

Conversion of dynorphins to form the octapeptide DYN A(1-8) was studied in rat brain and spinal cord fractions (Marks et al., 1987). The octapeptide has been reported to exist at high amounts in various brain structures (Weber et al., 1982), but the endogenous substrate and the processing pathway forming this peptide in CNS are not known. By applying various dynorphin fragments the authors provided the evidence that DYN A(1-8) might be formed from DYN (1-9) by the action of a carboxypeptidase B-like enzyme (CP-B) present in the brain membrane preparations, which was active at acidic pH. It is worth noting that neither DYN A(1-17) nor DYN A(1-13) released the octapeptide. This observation is similar to our studies (Silberring et al., 1995) where only trace amounts of the octapeptide were released from Big DYN in the striato-nigral pathway. When the selected substrates were treated by purified preparations containing CP-B and ACE it appeared that DYN A(1-13), but not DYN A(1-17) was a substrate for the octapeptide formation (Marks et al., 1987). The postulated pathway for DYN A(1-13) conversion by particulate brain fractions is presented in the scheme below:

Dynorphin A(1-13) $\Rightarrow$ DYN A(1-12) $\Rightarrow$ DYN A(1-10) $\Rightarrow$ DYN A(1-8)

The process is initiated by the C-terminal truncation of the Lys residue, followed by the consecutive removal of dipeptides by ACE. Membrane fractions obtained from the rat spinal cord were capable of releasing the DYN A(1-8) at a high rate and the enzyme activities were located in the homogenates and particulate fractions of the spinal cord. A purified metalloendopeptidase sensitive to phosphoramidon, and obtained from rat brain membranes, cleaved DYN A(1-17) and DYN A(1-13) at the Gly^3-Phe^4 and Arg^7-Ile^8 bonds (Benuck et al., 1984). However, shorter peptides such as Leu-enkephalin or DYN A(1-8) were much better substrates for this enzyme. Another endopeptidase, phosphoramidon-insensitive, was described by the same group (Benuck et al., 1984), which generates DYN A(1-8) by a single cleavage of dynorphins between the Ile^8-Arg^9 bond.

Cell lines

Dynorphin conversion to Leu-enkephalin-Arg^6 in a cell culture system in vitro was reported (Vlaskovska et al. 1994a,b). Clonal cells, however, can only serve as a complementary model for in vitro studies on the CNS pathology. It has been found that neurons from the cerebral cortex in primary culture contain and release dynorphin converting enzyme particularly specific for DYN B (Vlaskovska et al., 1994a,b). The chemical identity of the DCE is presently unknown but it may be classified as a thiol-dependent neutral endopeptidase, similar to the enzyme recovered from the bovine, human and rat spinal cord tissues (Silberring and Nyberg, 1989; Nyberg and Silberring, 1990). This proteinase exhibited optimal activity at neutral pH and its molecular mass was estimated at approximately 70 kDa by size-exclusion chromatography and SDS PAGE. The enzyme can be distinguished from thimet oligopeptidase (EC 3.4.24.15), which has been classified as a Zn-dependent metallo-peptidase containing an essential Cys residue (Tijslar, 1993), by its enzymatic specificity. The latter enzyme mainly generates Leu-enkephalin, but not Leu-enkephalin-Arg^6, from α-neo-endorphin and dynorphin B (Camargo et al., 1987). Another potential candidate, which may also be excluded, is the recently cloned N-arginine dibasic NRD convertase - a member of the pitrilysin family of zinc-metallo-proteases, which exhibits a different cleavage pattern (Chesneau et al., 1994a) and differs in terms of sensitivity to various proteinase inhibitors. For example, NRD convertase is extremely sensitive to low (micromolar) concentrations of the chelating agents, EDTA and 1,10-o-phenanthroline, while DCE is only partially affected by the millimolar amounts of these agents. Moreover, DCE from the cortical cells in primary culture shows strong preference for dynorphin B and α-neoendorphin, whereas dynorphin A or dynorphin A(1-13) were not cleaved even after prolonged incubation time. This is in contrast to the data obtained for the NRD convertase which preferentially cleaves dynorphin A and dynorphin A(1-13) while dynorphin B and α-neoendorphin were not so effective.

DYN B-29 converting enzyme was localized in the mouse anterior pituitary cell line, AtT-20 and its activity was elevated after stimulation with phorbol ester, a calcium ionophore and forskolin (Devi, 1992). This observation may suggest that the enzyme activity is present in the regulated pathway of secretion.

Two cysteine proteinases involved in cleaving of dynorphins and possessing specificity towards paired basic residues were described in the membrane fraction of neuroblastoma cell (Satoh et al., 1989). The cysteine protease I cleaves DYN A at one position between Arg^6 - Arg^7 whereas cysteine protease II hydrolyzes heptadecapeptide between Lys^{11} - Leu^{12} and Leu^{12} - Lys^{13}. It is interesting that a trypsin-like enzyme having similar specificity was reported earlier by the same group (Satoh et al., 1988).

Role of dynorphin converting enzymes in pathophysiology

Drug dependence

It has been shown that chronic exposure to opiates such as morphine causes tolerance and dependence. The adaptive mechanisms underlying these effects are not yet clearly understood. Recent studies provide evidence that the dynorphinergic system is involved in the development of opiate tolerance (Nylander et al., 1991). In addition, the conversion of dynorphins to enkephalins by DCE was sensitive to opiates both in vivo and in vitro (Vlaskovska et al., 1994). Selective enzyme inhibitors of enkephalin-degrading enzymes reduce the severity of the morphine abstinence symptoms in rats (Maldonaldo et al., 1992).

A particular motive for investigating the effect of opiates on the dynorphin-to-enkephalin enzymatic conversion is the observation that dynorphin peptides are pharmacological antagonists to opiate analgesia (Friedman et al., 1981; Hooke et al., 1995) and restrict the development of opiate tolerance (Green and Lee, 1988; Takemori et al., 1992; 1993; Hooke et al., 1995). Enzymatic conversion of dynorphins to enkephalins could affect the endogenous resistance to the development of opiate dependence. We have also previously observed an increased conversion of dynorphin peptides to Leu-enkephalin-Arg^6 in CNS areas of rats treated chronically with morphine (Nylander et al., 1995 a,b) as well as increased converting enzyme activity in the CSF of these animals (Persson et al., 1989). Recent studies suggest that spontaneous morphine-abstinent animals synthesize and release a putative "withdrawal factor" in the CSF. This factor precipitates the opiate-withdrawal syndrome in morphine-dependent, but not morphine-abstinent, rats (Cappendijk et al., 1996). The component, which is of proteinaceous origin, has been partially purified from the CSF of morphine-abstinent rats by size exclusion and reversed-phase HPLC. It has a molecular mass of approx. 50 kDa and is acid-stable.

Metabolism of dynorphins to Leu-enkephalin-Arg^6 in a cell culture system in vitro was stimulated by chronic exposure to morphine (Vlaskovska et al., 1994a,b). The effect was clearly dependent on opiate receptors, since incorporation of naloxone in the culture medium blocked the effect. Similarly, in a clonal human tumor system (SH-SY5Y neuroblastoma), chronic morphine treatment increases the dynorphin- to- enkephalin processing (Vlaskovska et al., 1995). The mechanism of up-regulation of the enzymatic activity is of considerable interest and needs to be studied. Isolation of the enzyme and molecular biology techniques will be necessary to investigate the possible role of DCE in opiate tolerance.

Rat brain cortical cells in primary culture were used to investigate long term effects of opiates on endopeptidases acting on dynorphin peptides (Vlaskovska et al., 1996). The cultures expressed activity which converted dynorphin B and α-neo-endorphin, but not dynorphin A, to Leu-enkephalin-Arg^6. Disappearance of both peptides used as substrates and the generated products were quantitated by electrospray ionization mass spectrometry. Long term (5 days) treatment with 10 μM morphine increased soluble DCE activity by 370%. This effect was reversed by the presence of naloxone in the culture medium. Since dynorphin peptides influence the development of opiate tolerance, a change in the extent of this transformation may be functionally important.

Enzymatic activity of endo-oligopeptidase was studied in rats, treated with morphine and submitted to naloxone-induced withdrawal (Paik et al., 1994). The enzyme activity was increased, with highest levels detected in dorsal hippocampus, cortex and substantia nigra. Moreover, a significant decrease of endopeptidase 22.19 was observed in peripheral tissues such as pineal gland after morphine treatment. Naloxone-precipitated withdrawal caused increase of the enzymatic activity in lymphoid tissues.

Pain and clinical studies

Only few studies on DCE in CSF samples of clinical origin were performed. Detailed overview is given in a Chapter 15 (F.Nyberg). DCE activity was found to be reduced in CSF collected from pregnant women (Lyrenäs et al., 1988). A negative correlation between DCE activity and prodynorphin-derived opioids was observed in CSF from chronic pain patients (Thörnwall et al. 1994). Moreover, patients with fibromyalgia showed elevated levels of DYN A and decresed DCE activity (Thörnwall et al. 1994). It thus seems that the CSF level of prodynorphin-derived peptides is inversely correlated to DCE activity, hence suggesting a possible involvement of the enzyme in the regulation of peptide level.

DCE activity in the CSF of rats was studied in response to exercise (Persson et al. 1993). Spontaneous running in a wheel for 4-5 weeks with an average running distance of 4 km/24h, caused decrease of the enzymatic activity resulting in the altered processing of dynorphins. As the exercise elevates pain treshold (Shyu et al. 1982), such metabolic changes may influence physiological and behavioral responses at least in rats.

Analysis of DCE in spinal cord tissue of rats with Freund's adjuvant-induced polyarthritis indicated that the enzyme activity was significantly decreased in the dorsal part of spinal cord, whereas its level in the ventral part remained unchanged (Silberring et al. 1992). A similar study was performed with DCE from CSF of poly- and monoarthritic rats. The DCE activity was decreased in the acute, but not chronic, phase of collagen-induced arthritis (Persson et al.,1992).

Inhibitors of dynorphin converting enzymes

Following rapid discoveries in molecular biology, new proteinases have been considered as potential therapeutic targets. Most of these molecules lack detailed information on specificity and selectivity towards various substrate sequences. Such information may also contribute to the design of potent and selective inhibitors. Inhibitors of the proteinases play a key role in the mechanisms regulating protein turnover and metabolism of peptides in the CNS (Turner and Barnes, 1994). Detailde knowledge on their structure and mechanism of binding of endogenous inhibitors is important in development of potential therapeutic agents. A good example may be ACE inhibitors: captopril, enalapril and lisinopril or inhibitors of HIV-protease. Major problem in designing such inhibitory molecules is their specificity towards particular enzyme. An interesting observation illustrating the problem has been described (Turner and Barnes, 1994), where the compound N-1-(R,S)-carboxy-3-phenylpropyl-Ala-Ala-Phe-p-carboxyanilide has been applied as an inhibitor of thimet oligopeptidase. The substance is, however, metabolized by the "endopeptidase 24.11" which converts it to a potent inhibitor of ACE. Hence, it cannot be taken into consideration though its high specificity towards thimet oligopeptidase.

Dynorphin converting enzymes recovered from the spinal cord, peripheral blood plasma, brain structures and cell lines are sensitive to thiol-specific agents, such as PHMB at low concentrations, and are activated by reducing agents like DTT or 2-mercaptoethanol. Therefore, they all were classified as cysteine proteinases. Heavy metal ions (Zn, Hg, Cu) also significantly decrease proteolytic activity. Effective inactivation by PHMB was exploited in an attempt to purify spinal cord DCE by affinity chromatography using this inhibitor in solid phase. The recovery of the enzyme was performed by eluting the column with 2-mercaptoethanol. Unfortunately, it seems that the yield of this technique was extremely low. Brain enzyme, processing DYN B-29 to DYN B was also defined as a cysteine proteinase. Convertases from the cell lines were partially inhibited by EDTA suggesting the presence of the metal binding site necessary for attaining their full activity. Enzyme, purified from the CSF was classified as a metal-dependent serine proteinase, being affected by PMSF and EDTA. N-peptidyl-O-acyl hydroxylamines were applied as possible inhibitors of dynorphin converting enzymes. These substances were developed as effective, irreversible inhibitors of cysteine proteinases. DCEs recovered from the spinal cord were found to be competitively inhibited by this novel class of mechanism-based synthetic inhibitors (Silberring et al., 1993). The most potent (Ki <20 μM) were the N-terminally protected peptides Z-Phe-Phe-NHO-Ma and Boc-Phe-Gly-NHO-Bz, both containing hydrophobic amino acids at the P_2 position. This sequence mirrors position of the Phe^4 residue in enkephalin sequences. No cleavage of the peptide bond within the inhibitor sequence was observed even after prolonged incubation with DCE, as verified by mass spectrometry. Hydrophobicity of these synthetic substances may be changed, thus making them permeable through the cell membrane and therefore, making them useful as potential pharmacological tools for in vitro studies on dynorphin processing.

DCE purified from the human cerebrospinal fluid was only moderately inhibited by enkephalin-analogue diacyl hydroxylamines (Demuth et al., 1991). The unusual specificity of DCEs against very limited number of substrates may indicate that the enzymes, highly specific for cationic amino acid pairs such as Arg-Arg, Arg-Lys, might hydrolyze the peptide bonds by a mechanism of substrate-assisted catalysis (Carter and Wells, 1987). The role of substrate structure in the processing of prohormones has been described in details by Gluschankof et al. (1988).

Several short peptides containing dibasic stretches, similar to that present in dynorphins, were synthesized and tested for their inhibitory properties towards DCE (Silberring and Nyberg, 1989; Silberring et al., 1992). As could be shown in Table 3 , inhibitory potency of these fragments increases with increasing length. Moreover, they were virtually not cleaved by DCE during the reaction time, suggesting distinct mechanism of interaction compared to "classic" proteinases such as trypsin, chymotrypsin and papain. This observation leads to the conclusion, that development of the short chromogenic or fluorogenic substrates will be difficult because the proteinases recognize almost entire sequence of dynorphins.

Sequence	% inhibition
Dynorphin B(5-9)	24
Dynorphin B(4-9)	48
Dynorphin B(1-9)	55

Tab.3: Effect of various peptides on DCE-A activity (20 ng protein). All peptides were applied at a concentration 10^{-4} M.

It is also worth noting that the enzymatic activity, recovered from the cells was efficiently hampered by the structurally-related DYN A, DYN A(1-13) and their des-Tyr^1-analogs, in the following order: DYN A(1-13)>A(2-13)>A(1-17)>A(2-17) (Vlaskovska et al., 1995). It seems possible that dynorphin A fragments may actually act by protecting other dynorphin peptides from degradation. This information might be useful for the rational design of selective inhibitors and, also, for the purification of the enzyme by affinity chromatography using some of these sequences. Interestingly, similar inhibition of DYN A degradation by DYN fragments were reported by Young et al. (1987). DYN A and Des[Tyr^1]DYN A increased stability of [^3H]DYN A while incubated with the rat brain membranes, whereas DYN B, ANEO and DYN A(7-17) showed moderate effect.

The effect of peptidase inhibitors on the antinociception induced by intrathecally administered DYN A and DYN B in the mouse formalin test was examined (Tan-No et al., 1996). As

expected, both peptides produced a dose-dependent reduction of the paw-licking response, which disappeared after 90 and 60 min, respectively. PHMB - a cysteine proteinase inhibitor simultaneously administered with DYN A or DYN B, significantly prolonged antinociception. Neither PMSF (inhibitor of serine proteinases) nor bestatin - a general aminopeptidase inhibitor, were effective in this test. The results indicate the possible involvement of dynorphin converting enzymes, present in the spinal cord, in pain control and suggest an interesting approach comprising design of specific inhibitors influencing activity of these peptidases.
An endogenous, competitive inhibitor of DCE has been purifed from large volumes of human CSF (Silberring, 1995, in preparation). This protein of the molecular weight of ca. 50 kDa is also a potent inhibitor of trypsin, chymotrypsin and plasmin. Its activity is temperature- and pH-dependent and the concentration in human CSF was estimated at about 60 pM. Chemical derivatization revealed that arginine is an essential part of the reactive site. The above finding suggests the co-existence of an inhibitor of DCE in the same compartment and which differs from α_1-proteinase inhibitor, inter-α-trypsin inhibitor, protease nexin-II and α_2-macroglobulin. Its inhibitory properties are similar to those of α_1-antiplasmin, but the estimated molecular weight is about 20 kDa lower and precipitation with specific antibodies did not hamper the inhibitory activity.

Techniques for studies of neuropeptide processing

HPLC and immunological methods

HPLC still remains the most commonly applied technique for identification of fragments, formed after proteolytic cleavage. This method is relatively simple, but requires application of relatively high amounts of substrates and is time-consuming. Another drawback is its poor specificity and, hence, all peaks along UV chromatogram should be identified by means of e.g. mass spectrometry, chemical sequencing or amino acid analysis. Usually, the reversed-phase mode is applied with gradient elution but other techniques such as ion-exchange-, ion-pair- or size exclusion chromatography were successfully applied (Smith and Lew, 1995; Nylander et al. 1995). Time of the analysis and solvent consumption can be significantly reduced by using microbore (1-2 mm i.d.) or capillary column (25-300 μm i.d.) and smaller particle size (3 μm) which also improves sensitivity of the assays. Capillary columns accomplish low flow-rates of around 1-30 μl/min and formation of the gradient under such conditions requires application of flow-splitting (Silberring, 1996). Generally, linear gradient of acetonitrile between 10-60%, supplemented with 0.1% TFA is applied in most cases. The photodiode-array UV detectors have capability of simultaneous scanning over entire spectrum. Therefore, a combination of HPLC separations with multispectral UV detection may significantly improve information obtained from enzymatic experiments. Co-eluting components, impurities and fragments containing aromatic amino acids (Tyr, Phe, Trp) may easily be identified during one run (Silberring, 1991). Another advantage of using photodiode-array detectors is easy detection

of the pirydylethylated cysteine residues along the peptide map obtained after limited proteolytic cleavage of proteins.
Radioimmunoassay and related techniques are applied for the quantitation of neuropeptides and their fragments in tissue extracts and body fluids. All these methods are utilizing antibodies raised, at least theoretically, against one particular peptide. Short sequences, however, are not the good immunogens and often antibodies cross-react with related fragments or with remaining contaminants such as salts. Inspecting formation of only one fragment formed during enzymatic reaction, may lead to confusing results because all other cleavages will be overlooked. For the technical reasons, it would be difficult to predict which peptides will emerge from the experiment and to handle several assays at a time. It is, therefore, recommended to combine HPLC separations with RIA technique, followed by thorough examination of all sequences, as mentioned above.

Substrates for enzymatic reactions

Short chromogenic or fluorogenic substrates offer fast and reliable detection of enzymatic activities. Based on the structure of dynorphin-related peptides, an internally quenched fluorescence substrate Abz-Gly-Gly-Phe-Leu-Arg-Arg-Val-EDDnp was designed for endo-oligopeptidase A (Gomes et al., 1993). This approach was possible to achieve, because the enzyme shows broader specificity and also effectively cleaves shorter peptides such as DYN A(1-8). Dynorphin converting enzymes isolated from the spinal cord do not recognize such sequences and determination of their enzymatic activities needs to be performed using full-lenght peptides which substantially increases complexicity and cost of the assay. Synthetic substrate Dansyl-Phe-Leu-Arg-Arg-Ala-Ser-Leu-Gly-COOH (Dansyl-Phe-Kemptide) was applied for detection and purification of a neutral, divalent cation-sensitive endopeptidase (Supattapone et al., 1988). Using extraction of a cleaved fragment Dansyl-Phe-Leu, the fluorescence was monitored as a measure of the enzyme activity. The enzyme showed high specificity for DYN A and approx. 10 times less cleavage rate was found for BNEO. The drawback of the self-quenched substrates might be an unpredicted truncation at other site than expected by another proteinase present in the incubation mixture.

Mass spectrometry in studies of neuropeptide peptidases

The rapid growth of biotechnology demands development of the sensitive and reliable techniques for analysis of proteins and peptides. ESI MS is designed for the precise estimation of molecular masses but this technique has also been successfully used for the sequencing of a number of peptides, whose structures, for a variety of reasons, e.g. blocked N-termini, purification difficulties, novel amino acids, could not be determined by chemical means (Silberring et al., 1994). Due to the extremely high sensitivity of the method (low femtomolar or even lower) and its structural specificity, it is possible to measure and quantitate components of endogenous origin. Atmospheric pressure ionization (API) interfaces have become powerful and popular tools for sample ionization (so-called soft ionization) and are primarilly used in the analysis of polar and thermolabile compounds (e.g. peptides, proteins). They offer several advantages over other mass spectrometric techniques. Due to their full compatibility with liquid chromatography (Fig. 2) and capillary electrophoresis, they can serve as a sensitive and structure-specific on-line detector for the analysis of complex peptide mixtures. For example, preparation of random peptide libraries consisting of a high number of individual sequences, require rapid detection and verification of bioactive fragments. In such case, mass spectrometry in conjunction with HPLC or capillary electrophoresis, may serve as a convenient tool for the identification of novel peptides with potential inhibitory activity. Isolation and full characterization of compounds of endogenous origin such as neuropeptides, peptides bound to MHC molecules, proteins etc., by a combined LC-MS and LC-MS^n is greatly facilitated with this technique. This technology has been successfully for the verification/identification of peptide fragments and for monitoring the enzymatic reactions (Silberring et al., 1991, Nylander et al., 1995). The API ionization also allows quantitation of the analysed compounds, making possible e.g. kinetic measurement or neuropeptide measurement (Desiderio et al., 1995).

Fig. 2. Schematic representation of the LC-MS setup when (A) reversed-phase in gradient mode and (B) size exclusion (isocratic mode) columns are applied (T - mixing tee).

The extremely high sensitivity of ESI MS has led to development of practical sequencing techniques. A new strategy has been introduced for peptide analysis, based on a nanoliter technology (Mann and Wilm, 1995), where the sample is injected into the instrument at a flow-rate of ca. 25 - 200 nanoliters/min. Thus, as little as 5 μl of analyte can be analyzed for over 1 hour, allowing full structural analysis at a sample consumption of few femtomole/min. This approach was applied for identification of endogenous NPY from the rat brain (Stenfors and Silberring, 1996) and for the sequencing of enkephalins, formed during enzymatic proteolysis of dynorphins, where the first collision experiment was performed in the electrospray source and the selected ion of interest was further fragmented in the Q-sector of the hybrid instrument (Sandin et al., 1996). This MS^3 can provide additional structural information not available by performng simple MS experiments and can be compared with the sequencing potential provided by the ion traps.

Antisense oligodeoxynucleotides

The function of many proteins remains unclear due to the lack of selective tools for their investigation. Oligodeoxynucleotides (ODNs) represent a new approach for drug discovery (Georgieva et al., 1995). The antisense strategy interferes with the information flow from gene to protein in a very specific manner. Thus, it is possible to "knock-down" translation of the mRNA responsible for the biosynthesis of the encoded protein (e.g. peptidases). In this context, antisense ODNs represent useful and powerful tool for studying the consequences of partial inhibiting the expression of individual genes in vitro and in vivo. No data have been presented up to day, concerning alteration of DCEs by this technique.

Combinatorial peptide libraries

Recent trends in the search for novel pharmacological agents have focused on the preparation of combinatorial chemical libraries as potential sources of new leads for drug discovery (Gallop et al., 1994). Such libraries are randomly created sets of molecules which can be screened for their potential biological activity. They should prove particularly useful to the pharmaceutical industry for the rapid identification of novel peptides that bind to receptors, enzymes or antibodies. Techniques for the preparation of random peptide libraries, consisted of millions of individual sequences, have been reported (Meldal et al., 1994). Thus, it is possible to reveal the optimal substrate or inhbitory molecules for proteolytic enzymes and to assign the active site cleft of peptidases. Design of larger libraries, containing more than 100,000 peptide sequences are not commonly used in research due to the increasing costs and technical difficulties in establishing a reliable detection method. Construction of the so-called intelligent libraries containing not more than 100-500 sequences and based on the already existing preliminary data, may often provide similar results. This approach has been successfuly applied for the subsite mapping of subtilisin Carlsberg (Meldal et al., 1994) and for the discovery of various protease substrates (Singh et al., 1995). The results of such studies may contribute to the knowledge on the pathology of the CNS and can also lead to the development and design of specific inhibitors, acting as drugs of potential therapeutic interest. Design of the combinatorial peptide libraries would be an important approach for studying metabolic regulation of neuropeptides in CNS

Final remarks

Functional and clinical studies indicate that the family of dynorphin converting enzymes may play a regulatory role in the prodynorphin system. It is obvious that they act not only as inactivators of the biologically active peptides, but are also capable of releasing a second signal from the parent compound. The enzymatic activity of these proteinases appears to be related to various pathophysiological conditions. Therefore, these enzymes could be considered as highly relevant targets for the development of specific inhibitors, which would not only be applied in further attempts to define functional relevance of endogenous substrates but could also serve as leads for the rational design of potential drugs.

Acknowledgments
I thank Mrs. Delphi Post for the excellent secretarial assistance and Dr. Ingrid Nylander for critical discussions during preparation of the manuscript. This work was supported by the Swedish Alcohol Research Fund and the KBN grant 1147/IA/158/95.

References

Benuck M., Berg M. and Marks N. (1984) Membrane-bound enzymes and their role in processing of the dynorphins and of the proenkephalin octapeptide Met-enkephalin-Arg-Gly-Leu. Neurochem. Res. 9, 733-749.

Berman Y.L., Juliano L. and Devi L.A. (1995) Purification and characterization of a dynorphin-processing endopeptidase. *J. Biol. Chem.* 270, 23845.

Berman Y.L., Rattan A.K., Carr K. and Devi L. (1994) Regional distribution of neuropeptide processing endopeptidases in adult rat brain. Biochimie 76, 245-250.

Bode W. and Huber R. (1992) Natural protein proteinase inhibitors and their interaction with proteinases. Eur. J. Biochem. 204, 433.

Camargo A., Shapanka R. and Greene L. (1973) Preparation, assay, and partial characterization of a neutral endopeptidase from rabbit brain. Biochemistry 12, 1838-1844.

Camargo, A., Oliviera, E., Toffoletto, O., Metters, K., and Rossier, J (1987) Brain endo-oligopeptidase A, a putative enkephalin converting enzyme. *J. Neurochem.*, **48**, 1258-1263.

Camargo A.C., Gomes M., Toffoletto O., Ribeiro M., Ferro E., Fernandes B., Suzuki K., Sasaki Y. and Juliano L. (1994) Structural requirements of bioactive peptides for interaction with endopeptidase 22.19. Neuropeptides 26, 281-287.

Cappendijk S., de Vries R., Silberring J. and Dzoljic M. (1996; personal communication)

Carter P. and Wells J.A. (1987) Engineering enzyme specificity by "substrate-assisted catalysis". Science, 237, 394-399.

Cerreti D.P., Kozlosky C.J., Mosley B., Nelson N., Van Ness K., Greenstreet T.A., March C.J., Kronheim S.R., Druck T., Cannizzaro L.A. and Black R.A. (1992) Molecular cloning of the interleukin-1β converting enzyme. Science 256, 97.

Chavkin, C., James, I.F. and Goldstien, A. (1982) Dynorphin is a specific endogenous ligand of the kappa receptor. *Science,* **215,** 413-415.

Chesneau V., Pierotti A., Barré N., Créminon C., Tougard C. and Cohen P. (1994a) Isolation and characterization of a dibasic selective metalloendopeptidase from rat testes that cleaves at the amino terminus of arginine residues. J. Biol. Chem. 269, 2056-2061.

Chesneau V., Pierotti A., Prat A., Gaudoux F., Foulon T. and Cohen P. (1994b) N-arginine dibasic convertase (NRD convertase): a newcomer to the family of processing endopeptidases. An overview. Biochimie 76, 234-240.

Chou J.Z., Kreek M.J. and Chait B. (1984) Matrix-assisted laser desorption mass spectrometry of biotransformation products of dynorphin A in vitro. J. Am. Chem. Soc. Mass Spectrom. 5, 10-16.

Christensson-Nylander, I., Herrera-Marschitz, M., Stines, W., Hökfelt, T., Terenius, L., Ungerstedt, U., Cuello, C., Oertel, W.H. and Goldstein, M. (1986) Striato-nigral dynorphin and substance P pathways in the rat. I. Biochemical and immunohistochemical studies. *Exp. Brain Res.*, **64**, 169-192.

Demuth H.-U., Silberring J. and Nyberg F. (1991) Inhibition of proteases by enkephalin-analogue diacyl hydroxylamines. J. Enz. Inhibition 4, 289.

Devi L. (1992) Secretion and regulation of a neuropeptide-processing enzyme by AtT-20 cells. Endocrinology 131, 1930-1935.

Fried G., Franck J., Brodin E. and Terenius L. (1991) Subcellular distribution of two enzymes with neuropeptide converting activity in rat spinal cord. Neurochem. Int. 19, 221-226.

Friedman H.J., Jen M.F., Chang J.K., Lee N.M. and Loh H.H. (1981) Dynorphin A, a possible modulatory peptide on morphine or β-endorphin analgesia in mouse. Eur. J. Pharmacol. 69, 357.

Gallop M.A., Barrett R.W., Dower W.J., Fodor S.P. and Gordon E.M. (1994) Applications of combinatorial technologies to drug discovery. Background and peptide combinatorial libraries. J. Mecinal Chem. 37, 1233.

Germain R.N. and Margulies D.H. (1993) The biochemistry and cell biology of antigen processing and presentation. Annu. Rev. Immunol. 11, 403.

Gluschankof P., Gomez S., Lepage A., Créminon C., Nybarg F., Terenius L. and Cohen P. (1988) Role of peptide substrate structure in the selective processing of peptide prohormones at basic amino acid pairs by endoproteases. FEBS Lett. 234, 149-152.

Gomes M., Juliano L., Ferro E., Matsueda R. and Camargo A.M. (1993) Dynorphin-derived peptides reveal the presence of a critical cysteine for the activity of brain endo-oligopeptidase A. Biochem. Biophys. Res. Comm. 197, 501-507.

Green P. and Lee N. (1988) Dynorphin A(1-13) attenuates withdrawal in morphine-dependent rats. Eur. J. Pharmacol. 145, 267.

Herman B.H., Leslie F. and Goldstein A. (1980) Behavioral effects and in vivo degradation of intraventricularly administered dynorphin-(1-13) and D-Ala^2-dynorphin-(1-11) in rats. Life Sci. 27, 883-892.

Herman, B.H. and Goldstein, A. (1985) Antinociception and paralysis induced by intrathecal dynorphin-A. *J. Pharmacol. Exp. Ther.*, **232**, 27-32.

Hooke L.P., He L. and Lee N.M. (1995) Dynorphin A modulates acute and chronic opioid effects. J. Pharmacol. Exp. Ther. 273, 292.

Itoh, J., Ukai, M. and Kameyama, T. (1994) Dynorphin A-(1-13) potently improves the impairment of spontaneous alteration performance induced by the mu-selective opioid receptor agonist DAMGO in mice. *J. Pharmacol. Exp. Ther.* **269**, 15-21.

Leslie F. and Goldstein A. (1982)

Lyrenäs S., Nyberg F., Lindberg B. and Terenius L. (1988) Cerebrospinal fluid activity of dynorphin converting enzyme at term pregnancy. Obstet. Gynecol. 72, 54.

Maldonaldo R., Fournie-Zaluski M.C. and Roques B.P. (1992) Attenuation of the morphine withdrawal syndrome by inhibition of catabolism of endogenous enkephalins in the periaqueductal gray matter. Naunyn-Schmiedebergs Arch. Pharmacol. 345, 466.

Mann M. and Wilm M. (1995) Electrospray mass spectrometry for protein characterization. Trends in Biochem. Sci. 20, 219.

Marks N., Berg M., Benuck M., Lo E.-S., Novachenko H. and Seyfried C. (1987) Prodynorphin processing by rat CNS fractions and purified enzymes: formation of dynorphin A 1-8 by sulfhydryl-activated carboxypeptidase and peptidyl dipeptidase. Neurochem. Int. 10, 413-422.

McGinty, J.F., Henriksen, S.J., Goldstein, A., Terenius, L. and Bloom, F.E. (1983) Dynorphin is contained within hippocampal mossy fibers: immunochemical alterations after kainic acid administration and colchicine-induced neurotoxicity. *Proc. Natl. Acad. Sci. USA*, **80,** 589-593.

Meek T., Lambert D., Dreyer G., Carr T., Tomaszek T., Moore M., Strickler J., Debouck C., Hyland L., Matthews T., Metcalf B. and Petteway S. (1990) Nature 343, 90-92.

Meldal M., Svendsen I., Breddam K. and Auzanneau F.-I. (1994) Portion-mixing peptide libraries of quenched fluorogenic substrates for complete subsite mapping of endopeotease specificity.

Müller S. and Hochhaus G. (1995) Metabolism of dynorphin A 1-13 in human blood and plasma. Pharmaceut. Res. 12, 1165-1170.

Nakazawa, T., Ikeda, M., Kaneko, T., and Yamatsu, K. (1985) Analgesic effects of dynorphin-A and morphine in mice. *Peptides*, **6,** 75-78.

Nyberg F., Nordström F. and Terenius L. (1985) Endopeptidase in human cerebrospinal fluid which cleaves proenkephalin B opioid peptides at consecutive basic amino acids. Biochem. Biophys. Res. Commun. 131, 1069.

Nyberg F., Christensson-Nylander I. and Terenius L. (1987) Measurement of opioid peptides in biologic fluids by radioimmunoassay. In: Handbook of Experimental Pharmacology (C. Patrono and B.A. Peskar eds.). Springer-Verlag, Berlin, pp. 227-253.

Nyberg F. and Silberring J. (1990) Conversion of dynorphins to Leu-enkephalin in human spinal cord. Progr. Clin. Biol. Res. 328, 261.

Nyberg F., Kankaanranta S., Brostedt P. and Silberring J. (1991) Purification and characterization of dynorphin converting enzyme from human choroid plexus. Brain Res. 552, 129.

Nyberg F. and Terenius L. (1991) Enzymatic inactivation of neuropeptides. In: Degradation of bioactive substances: physiology and pathophysiology (J.H.Henriksen ed.), pp. 189-200. CRC Press, New York.

Nylander I., Sakurada T., LeGrevés P. and Terenius L. (1991) Levels of dynorphin peptides, substance P and CGRP in the spinal cord after subchronic administration of morphine in the rat. Neuropharmacol. 26, 1295.

Nylander I., Li Y.-M., Silberring J. and Terenius L. (1995) Dynorphins in plasma; measurable or degradable? Analgesia, 1, 615.

Nylander I., Tan-No K., Winter A. and Silberring J. (1995) Processing of prodynorphin-derived peptides in striatal extracts. Identification by electrospray ionization mass spectrometry linked to size-exclusion chromatography. Life Sci. 57, 123.

Nylander, I., Vlaskovska, M. and Terenius, L. (1995 a) The effects of morphine treatment and morphine withdrawal on the dynorphin and enkephalin systems in Sprague-Dawley rats. *Psychopharmacology*, **118**, 391-400.

Näslund J., Scherhorn A., Hellman U., Lannfelt L., Roses A., Tjernberg L., Silberring J., Gandy S., Winblad B., Greengard P., Nordstedt Ch. and Terenius L. (1994) Relative abundance of Alzheimer Ab amyloid peptide variants in Alzheimer disease and normal aging. Proc. Natl. Acad. Sci. USA. 91, 8378.

Oliveira E.S., Leite P.E., Spillantini M.G., Camargo A. and Hunt S. (1990) Localization of endo-oligopeptidase (EC 3.4.22.19) in the rat nervous tissue. J. Neurochem. 55, 1114-1121.

Paik S., Camarao G., Bauer J., Leite P., Oliveira E., Camargo A.C. and Farges R. (1994) Effects of morphine withdrawal syndrome on endooligopeptidase (EC 3.4.22.19) activity. Eur. J. Pharmacol. 253, 101-106.

Pierotti A.R., Prat A., Chesneau V., Gaudoux F., Leseney A., Foulon T. and Cohen P. (1994) N-arginine dibasic convertase, a metalloendopeptidase as a prototype of a class of processing enzymes. Proc. Natl. Acad. Sci. USA. 91, 6078.

Persson S., Post C., Weil-Fugazza J., Butler S.H. and Nyberg F. (1992a) Decreased cerebrospinal fluid neuropeptide-converting enzyme activity in monoarthritic rats. Neurosci. Lett. 143, 247.

Persson S., Post C., Holmdahl R. and Nyberg F. (1992b) Decreased neuropeptide-converting enzyme activities in cerebrospinal fluid during acute but not chronic phases of collagen induced arthritis in rats. Brain Res. 581, 273.

Persson S., Jönsdöttir I., Thorén P., Post C., Nyberg F. and Hoffmann P. (1993) Cerebrospinal fluid dynorphin converting enzyme activity is increased by voluntary exercise in the spontaneously hypertensive rat. Life Sci. 53, 643-652.

Przewlocki R., Lason W., Silberring J., Herz A. and Przewlocka B. (1986) Release of opioid peptides from the spinal cord of rats subjected to chronic pain. NIDA/INRC Res. Monograph 75, 422.

Sandin J., Nylander I., Silberring J. and Terenius L. (1996) Differential processing of dynorphins in the striato-nigral pathway. Int. J. Biochem. Cell Biol. (submitted).

Satoh M., Yokosava H. and Ishii S.-I. (1988) Membrane-bound trypsin-like enzyme functioning in degradation of dynorphin in neuroblastoma cells. Purification and characterization. J. Biochem. 103, 493-498.

Satoh M., Yokosava H. and Ishii S.-i. (1989) Characterization of cysteine proteases functioning in degradation of dynorphin in neuroblastoma cells: evidence for the presence of a novel enzyme with strict specificity toward paired basic residues. J. Neurochem. 52, 61-68.

Shyu B.-C., Andersson S. and Thorén P. (1982) Endorphin mediated increase in pain treshold induced by long-lasting exercise in rats. Life Sci. 30, 833-840.

Silberring, J., and Nyberg, F. (1989) A novel bovine spinal cord endoprotease with high specificity for dynorphin B. *J. Biol. Chem.*, **264**, 11082-11086.

Silberring J., Brostedt P., Thörnwall M. and Nyberg F. (1991) An approach to study the proteinase specificity by FAB MS combined with photodiode-array UV detection. J. Chromatogr. 554, 83.

Silberring J., Castello M. and Nyberg F. (1992) Characterization of dynorphin A-converting enzyme in human spinal cord. An endoprotease related to a distinct conversion pathway for the opioid heptadecapeptide? *J. Biol. Chem.* **267**, 21324-21328.

Silberring J., Sakurada T. and Nyberg F. (1992) Dynorphin converting enzyme in the rat spinal cord. Decreased activities during acute phase of adjuvant-induced arthritis. Life Sci. 50, 839.

Silberring J., Demuth H.-U. and Nyberg F. (1993) Inhibition of dynorphin converting enzymes from human spinal cord by N-peptidyl-O-acyl hydroxylamines. J. Biochem. 114, 648.

Silberring J., Li Y.-M. and Hjertén S. (1994) Strategies in studies on neuropeptide processing by mass spectrometry. Biochem. Soc. Trans. 22, 136.

Silberring J., Sandin J., Tan-No K., Kasakov L., Winter A. and Nylander I. (1995) Processing of dynorphins in the striato-nigral pathway. Analgesia 1, 734.

Silberring J. (1996) Analysis of neuropeptides by size-exclusion HPLC linked to electrospray mass spectrometry. In:Neuropeptide protocols. Methods in Molecular Biology series (B. Irvine and C. Williams eds.) Humana press, Totowa, NJ. (in press).

Singh J., Allen J., Ator M., Gainor J., Whipple D., Solowiej J., Treasurywala A., Morgan B., Gordon T. and Upson D. (1995) Validation of screening immobilized peptide libraries for discovery of protease substrates. J. Medcinal Chem. 38, 217-219.

Smith, A.P. and Lee, N.M. (1988) Pharmacology of dynorphin. *Ann. Rev. Pharmacol. Toxicol.* **28,** 123-140.

Smith A.I. and Lew R.A. (1995) Combination of high-performance liquid chromatography and radioimmunoassay for characterization of peptide-processing pathways. In: Peptidases and neuropeptide processing. Methods in neuroscience series (Ian Smith ed.) pp. 125-139. Academic press, N.Y.

Stenfors C., Hellman U. and Silberring J. (1996) Identification of neuropeptide Y and its metabolic pathway in the rat brain by nano-spray mass spectrometry. J. Biol. Chem. (submitted).

Strittmatter, S.M., Thiele E.A., Kapiloff M.S. and Snyder S.H. (1985) A rat brain isozyme of angiotensin converting enzyme. J. Biol. Chem. 260, 9825.

Supattapone S., Strittmatter S., Fricker L. and Snyder S. (1988) Characterization of a neutral, divalent cation-sensitive endopeptidase: a possible role in neuropeptide processing. Mol. Brain Res. 3, 173-182.

Takemori, A. E., Loh, H. H., and Lee, N. M. (1993) Suppression by dynorphin A and [des-Tyr1]dynorphin A peptides of the expression of opiate withdrawal and tolerance in morphine-dependent mice. *J. Pharmacol. Exp. Ther.* **266,** 121-124.

Takemori, A. E., Loh, H.H. and Lee, N.M. (1992) Suppression by dynorphin A(1-13) of the expression of opiate withdrawal and tolerance in mice. *Eur. J. Pharmacol.* **221,** 223-226.

Tan-no K., Taira A., Sakurada T., Inoue M., Sakurada S., Tadano T., Sato T., Sakurada C., Nylander I., Silberring J., Terenius L. and Kisara K. (1996) Inhibition of dynorphin converting enzymes prolongs the antinociceptive effect of intrathecally administered dynorphin in the mouse formalin test. Eur. J. Pharmacology (submitted).

Terenius L. and Nyberg F. (1988) Neuropeptide-processing, -converting, and -inactivating enzymes in human cerebrospinal fluid. Int. Rev. Neurobiol. 30, 101.

Terenius L. and Nyberg F. (1991) Peptidases and proteases in cerebrospinal fluid, significance in volume transmission. In: Volume transmission in the brain: novel mechanisms for neural transmission (K.Fuxe and L.F.Agnati, eds.), vol. 1, pp.415-424. Raven Press, New York.

Thörnwall M., Schwarzmeyr S., Almay B., Post C., VaerØy H. and Nyberg F. (1994) Dynorphin converting enzyme in human cerebrospinal fluid. A negative correlation between activity and levels of prodynorphin derived opioid peptides. Regul. Peptides 1, 235.

Tisljar U. (1993) Thimet oligopeptidase--a review of a thiol dependent metallo-endopeptidase also known as Pz-peptidase endopeptidase 24.15 and endo-oligopeptidase. Biol. Chem. Hoppe-Seyler, 374 91-100.

Turner A.J. and Barnes K. (1994) Neuropeptidases: candidate enzymes and techniques for study. Biochem. Soc. Trans. 22, 122.

Vlaskovska M., Geijer T., Nylander I., Silberring J. and Terenius L. (1994a) Changes in the prodynorphin system after chronic morphine treatments in vitro. Regulat. peptides, suppl. 1, S287.

Vlaskovska M., Nylander I., Schramm M. and Terenius L. (1994b) Effect of morphine treatment on the conversion of dynorphins to enkephalins in rat cortical neurons in primary culture and human neuroblastoma SH-SY5Y cells. Regul. Pept. 54, 311.

Vlaskovska M., Nylander I., Kasakov L., Silberring J. and Terenius L. (1995) Opiate-sensitive dynorphin conversion in SH-SY5Y human neuroblastoma cells: evidence for inhibition by dynorphin-related peptides. Analgesia, 1, 821.

Vlaskovska M., Nylander I., Schramm M., Hahne S., Kasakov L., Silberring J. and Terenius L. (1996) Opiate modulation of dynorphin conversion in primary cultures of rat cerebral cortex. Brain Res. (submitted).

Weber E., Evans C. and Barchas J. (1982) Predominance of the amino-terminal octapeptide fragment of dynorphin in rat brain regions. Nature 299, 77-79.

Zamir, N., Palkovits, M., Weber, E., Mezey, E. and Brownstein, M.J. (1984) A dynorphinergic pathway of Leu-enkephalin production in substantia nigra. Nature, 307, 643-645.

Zhu X., Robertson J., Sacks H., Dohan F., Tseng J. and Desiderio D. (1995) Opioid and tachykinin neuropeptides in prolactin-secreting human pituitary adenomas. Peptides, 16, 1097-1107.

Progress in HPLC-HPCE, Vol. 7, pp. 547-567
G. Ali Qureshi *et al.* (Eds)

APOLIPOPROTEIN E IN ALZHEIMER'S DISEASE, DEMENTIA WITH LEWYBODIES AND PARKINSON'S DISEASE.

Rianne A.J. Esselink[1], Rob A.I. de Vos[2], Istvan Vermes[3], Ernst N.H. Jansen Steur[1], Leo A.P. Ballering[4].

Department of Neurology[1] and Clinical Chemistry[3], Medisch Spectrum Twente, Regional Laboratory for Pathology[2], Enschede, and Department of Drug Safety, N.V. Organon[4], Oss, The Netherlands.

INTRODUCTION.

The apolipoprotein E (apoE) gene is a polymorfic gene with three common alleles e2,e3, and e4, coding for three isoforms: E2, E3, and E4. The gene is located on chromosome 19q13.2. The mature protein counts 299 amino acids. Apo E3 contains a single cystein at residue 112 and an arginine at position 158. E2 has cysteine residues at both positions 112 and 158, while E4 contains arginine residues at these positions. In the general population, apoE e3 is the most common allele, with a frequency of 78%, whereas e4 has an allele frequency of 15% (Davignon *et al.*, 1988). Groups with different ethnic backgrounds may have different apoE genotype frequency (Davignon *et al.*, 1988).
APOE is a plasma protein involved in lipoprotein (*ie* VLDL, LDL, HDL, chylomicrones) transport and metabolism by interaction with low density lipoprotein receptors, and the LDL receptor related protein. (Mahley, 1988; Rebeck *et al.*, 1993). The isoforms differ in their interaction with LDL receptors and lipoprotein particles.
APOE is produced and secreted in the central nervous system by astrocytes(Elshourbagy *et al.*, 1985; Pitas *et al.*, 1987; Ignatius *et al.*, 1986), and oligodendrocytes (Boyles *et*

al., 1989), and in many other tissues including the liver. It is produced in increasing amounts in the CNS after experimental brain injury (Poirer *et al.*, 1991a,b). APOE further has substantial influences in the pathogenesis of cardio-vascular diseases, Alzheimer's disease, and dementia with Lewy bodies.

APOE AND ALZHEIMER'S DISEASE.

Alzheimer's disease (AD) is the most common type of dementia. Clinically this dementia is characterized by deterioration of memory, general intellectual status, language, praxis, and/ or perception (McKhann *et al.*, 1984). It progressively influences daily life, and the patient finally becomes bedridden. The interval between the initial clinical diagnosis and death varies between 3 and 15 years. The definite diagnosis can only be made at autopsy. The main pathological hallmarks of AD are senile plaques, neurofibrillary tangles (NFT), neuropil threads, and synaptic loss. There are two main staging procedures for AD. The CERAD protocol (Mirra *et al.*, 1991) involves (1) semiquantitative assessment of senile (neuritic) plaque, and NFT frequency in various brain regions, (2) determination of an age-related plaque score based on the patient's age at death in combination with the semiquantitative measure of neuritic plaques (NP) in the neocortex, and (3) integration of the age-related plaque score with clinical information regarding the presence or absence of dementia in order to determine the level of certainty of the neuropathologic diagnosis of AD. Subjects can be classified in four AD stages: definite AD, probable AD, and possible AD, no AD (table 1.). The Braak & Braak staging (Braak and Braak, 1991) is based on the characteristic distribution pattern of neurofibrillary tangles and neuropil threads. Six stages can be distinguished: transentorhinal stages I-II, limbic stages III-IV, and the isocortical stages V-VI. ß-amyloid-protein (ßAP) is frequently present in NP's. The main structural constituent of the NFT's is hyperphosphorylated paired helical filament-tau. Tau protein is the only protein representing an integral constituent of the core structure of the paired helical filaments (Wischik *et al.*, 1995). Tau normally interacts with ß-tubulin to stabilize microtubules. Phosphorylated tau cannot bind itself to ß-tubulin, therefore the microtubular system is less stable, and neuronal degeneration ensues.

Although the association between the amount of amyloid in the brain and clinical severity of dementia is supported by some studies (Blessed *et al.*, 1968; Delaere *et al.*, 1989; Duyckearts *et al.*, 1986; Duyckearts *et al.*, 1988; Wilcock *et al.*, 1982), others could not establish a correlation between the severity of dementia and amyloid deposition density (Arriagada *et al.*,1992; Hyman *et al.*, 1993). Furthermore individuals who were not demented but, whose brains contained large numbers of senile plaques is a common feature (Arriagada and Hyman., 1990; Crystal *et al.*, 1988; Delaere *et al.*, 1990; Katzman *et al.*, 1988). Positive correlations between clinical severity of dementia and the amount of NFT were found in some studies (Arriagada *et al.*, 1992; Delaere *et al.*, 1989; Gomez-Isla *et al.*, 1996; Morris *et al.*, 1991; Price *et al.*, 1991), but not in others (Delaere *et al.*, 1989; Duyckaerts *et al.*, 1988).

The apolipoprotein E type 4 allele is overrepresented in late onset familial and sporadic Alzheimer's disease (Saunders *et al.*, 1993; Rebeck *et al.*, 1993; Mayeux *et al.*, 1993; Strittmatter *et al.*, 1993a; Corder *et al.*, 1993). Concomitant with an increase in apoE e4 frequency, the frequency of the apoE e2 allele seems to decrease. It has been suggested that the presence of the apoE e2 allele has a protective effect on the development of AD: individuals possessing this allele live longer and develop AD at a later age (Corder *et al.*, 1994), whereas the apoE e4 allele shows a "dose dependent" reduction of the age of onset of dementia. ApoE e4 homozygotes show an earlier age at the onset of dementia (75 years) than do heterozygotes (80 years), and in AD cases without an apoE e4 allele the dementia starts at a much higher age (85 years) (Corder *et al.*, 1993; Saunders *et al.*, 1993). Another study, however, did not confirm this association of apoE e4 with an earlier age of onset (Landen *et al.*, 1996). No correlation was found between e4 and the duration of dementia (Corder *et al.*, 1995a; Gomez-Isla *et al.*, 1996; Landen *et al.*, 1996; Olichney *et al.*, 1996), and the rate of cognitive deterioration (Kurz *et al.*, 1996; Asada *et al.*, 1996; Growden *et al.*, 1996). The apoE e4 frequency in AD subjects declines significantly with increasing age (Duara *et al.*, 1996; Corder *et al.*, 1994; Söbel *et al.*, 1995; Schächter *et al.*, 1994). ApoE seems to play a minor role in AD with an age at onset less than 60 and after 75 years. ApoE consequently has an age specific value for the onset of AD. Genes on chromosome 1 (Levy-Lahad *et al.*, 1995), chromosome 14 (Schellenberg *et al.*, 1992; St George-Hyslop *et al.*, 1992), and chromosome 21 (Goate *et al.*, 1991) are suspected to be risk factors for an early onset of AD.

Both sexes show an equal distribution for the three different apoE alleles in AD. (Corder *et al.*, 1994)

It is still unclear how apoE e4 affects the neuropathological changes found in AD, and

how these are related to the development of clinical AD. ApoE immunoreactivity is demonstrated in senile plaques and NFT's (Namba *et al.*, 1991; Strittmatter *et al.*, 1993a). Some *in vitro* studies have found that apoE e4 has a higher avidity for binding to ß-amyloid than apoE e3
(Strittmatter *et al.*, 1993b; Sanan *et al.*, 1994). By binding to ß-amyloid, E4 acts as a chaperone protein (Wisniewski *et al.*, 1992), leading to nonsoluble ß-amyloid deposition, and senile plaque formation (Strittmatter *et al.*, 1993a,b).

Strittmatter *et al.* (Strittmatter *et al.*, 1994) hypothesize that ApoE e3 and e2 have a higher binding avidity to the microtubular protein tau than e4. The absence of apoE e2 or apoE e3 allows increased abnormally and/or hyperphosforylated tau. In that case paired helical filaments are encountered and these are visualised in NFT's.

It has been shown that possession of the ApoE e4 allele correlates with increased amyloid deposition in the brain (Nagy *et al.*, 1995; Schmechel *et al.*, 1993; Gomez-Isla *et al.*, 1996; Rebeck *et al.*, 1993). Several studies have established a correlation between apoE e4 and NFT density (Nagy *et al.*, 1995; Polvikoski *et al.*, 1995; Ohm *et al.*, 1995). On the other hand, results of others argue against a direct role of apoE E4 in senile plaque and NFT formation (Landen *et al.*, 1996; Soininen *et al.*, 1995). Nagy *et al.* (Nagy *et al.*, 1995) demonstrated that the presence of apoE e2 provides some protection against the development of amyloid deposition and NFT's.

Although apoE e4 is associated with AD, individuals who lack apoE e4 may develop AD; apoE e4 is neither necessary nor sufficient for the development of AD. Consequently, there must be other risk factors for the development of AD.

ApoE e4 can not be regarded either as a predictor of disease. (Roses *et al.*, 1995; Corder *et al.*, 1995b) Little is known about the genotype depending risk of acquiring AD in asymptomatic individuals. A diagnostic genotyping test can be of some value when the patient is suspected for AD on clinical grounds (Roses *et al.*, 1995) although, as far as we know, there is neither a prospective study with apoE genotyping in early stages of dementia, nor a long time follow-up including neuropathological examination to prove the value of such a test.

APOE AND PARKINSON'S DISEASE.

Parkinson's disease (PD) is clinically characterized by resting tremor, rigidity, bradykinesia and impaired postural reflexes. Neuropathologically this is reflected as loss of large neuromelanin-containing neurons and gliosis in the substantia nigra and other pigmented brain stem nuclei, with Lewy bodies in some of the remaining nerve cells (Gibb *et al.*, 1990) Diagnostic accuracy during lifetime is approximately 75% (Hughes *et al.*, 1992). De Vos and Hughes found cortical Lewy bodies in all their PD cases, whether or not demented (de Vos *et al.*, 1995; Hughes *et al.*, 1993). Often a variable density of Alzheimer type pathology is found (de Vos *et al.*, 1995; Hughes *et al.*, 1993). In 44-66% of cases PD is associated with dementia (de Vos *et al.*, 1995; Hughes *et al.*, 1993). No correlation between Alzheimer type pathology and the presence of dementia in PD cases was found by Hughes *et al* (Hughes *et al.*, 1993). However our study does give a positive correlation between the limbic type of AD (Braak & Braak criteria) and dementia in PD (de Vos *et al.*, 1996). Clinically extrapyramidal sympoms are found in a significant proportion of the AD population, with or without neuropathologically established Lewy bodies (Gearing *et al.*, 1995b; Hulette *et al.*, 1995; Galasko *et al.*, 1994a; Liu *et al.*, 1997). ApoE e4 is a risk factor for AD and, as stated above, Alzheimer type pathology is often found in PD. If AD lesions in PD evolve by the same pathological mechanism as in pure AD, apoE e4 allele frequency in PD with AD lesions should be similar, and apoE might be associated with PD too. In general studies confirm that in PD the e4 frequency resembles that of the normal population (Harrington *et al.*, 1994; Hardy *et al.*, 1994; Benjamin *et al.*, 1994; Arai *et al.*, 1994; Marder *et al.*, 1994; Koller *et al.*, 1995; Whitehead *et al.*, 1996; Morris *et al.*, 1996; Nalbantoglu *et al.*, 1994; Egensperger *et al.*, 1996). Only Arai *et al.* (Arai *et al.*, 1994) found an increased e4 frequency in demented PD cases. Egensperger *et al.* (Egensperger *et al.*, 1996) found no difference in apoE e4 frequency between PD with AD lesions, PD without AD lesions, and controls, whereas apoE e4 allele frequency was significantly increased in pure AD. These data suggest that PD with AD lesions and AD represent two distinct entities. Some of the series in which the relation between apoE e4 and PD was evaluated, however, are small (n=23-24) (Hardy *et al.*, 1994; Benjamin *et al.*, 1994; Morris *et al.*, 1996), PD was not always pathologically confirmed (Arai *et al.*, 1994; Marder *et al.*, 1994; Koller *et al.*, 1995; Whitehead *et al.*, 1996) and in some of these studies no distinction was made between PD with and without dementia (Hardy *et al.*, 1994; Benjamin *et al.*, 1994). We evaluated 50 neuropathologically confirmed PD cases with and without dementia for their apoE allele frequencies and for the presence or absence of Alzheimer type pathology.

The procedure we used for apoE genotyping was as follows. ApoE genotyping was performed on formalin-fixed paraffine-embedded brain material or freshly frozen brain tissue and analysed as follows. Formalin-fixed paraffine-embedded brain tissue slices were derived from four different parts of the brain: cerebellum, frontal cortex, temporal cortex, and hippocampus.

Brain tissue was fixed for a maximum of 21 days in buffered formalin (10.0% formalin (Chemproha, Zwijndrecht, The Netherlands); 0.037% $NaH_2PO_4.2H_2O$; 0.16% $Na_2HPO_4.12H_2O$). Samples from before 1990 had been fixed in unbuffered formalin. The fixed brain tissue was routinely embedded in paraffine (Sherwood Medical Co., St. Louis, USA) at 62°C. Seven slices (10 mm each) of paraffine-embedded brain tissue were deparaffinized in an Eppendorf tube by incubation with 1 mL *m*-xylene, twice for 20 min and the pellet was subsequently washed with 96% ethanol. DNA was isolated from the dried pellet using the QIAamp tissue kit (QIAgen, Chatsworth, California) according to the manufacturer´s instructions.

The semi-nested polymerase chain reaction (PCR): for the outer PCR the following primers were used: P1 (5´-AAC AAC TGA CCC CGG TGG CG-3´)(11) and JP3 (5´-CTC GCG GAT GGC GCT GAG G-3´)(12). The amplification mixture consisted of 10 mL (±100 ng) genomic DNA, PCR Buffer (50 mM KCl, 100 mM Tris.HCl, 0.1% (w/v) gelatine, pH 8.3), 5% DMSO, 200 mM dNTP, 25 pmol of each primer, 1.5 mM $MgCl_2$, 1.25 U Taq polymerase (Perkin Elmer, Branchburg, New Jersey) in a total volume of 50 mL. The mixture was heated for 5 min at 95°C followed by 40 cycles of 1 min at 95°C, 1 min at 64°C and 2 min at 72°C, with a final elongation step of 10 min. This PCR reveals 298 bp product.

Ten mL of the outer PCR product was used in the second PCR, which was the same as the procedure described above, but with primer JP5 (5´-CGG CCA CGG CTG TCC AAG C-3´)(12) instead of primer P1. The final product has a length of 270 bp. This PCR was also used, as a single PCR, for the freshly frozen braintissue. PCR reactions were performed in a 480 DNA Thermal Cycler (Perkin Elmer).

Restriction with *CfoI*: Restriction analysis was performed on the PCR products from the single PCR on DNA isolated from freshly frozen brain tissue. For restriction with *Cfo*I 40 mL PCR product was mixed with 2 mL restriction buffer "L" (Boehringer Mannheim, Indianapolis, Indiana), 2 mL distilled water and 1 mL (10 U) *CfoI* (Boehringer Mannheim) and incubated for 4 hr at 37°C (12). The restriction pattern was read after electrophoresis of 40 mL restriction product on a 10% polyacrylamid gel (mono:bis 19:1,

Bio Rad, Hercules, California), stained afterwards with ethidium bromide.

ASO hybridisation: Two ml semi-nested PCR product (in 0.4 M NaOH, 10 mM EDTA) were denatured at 95°C for 5 min and blotted onto a nylon+ membrane (Boehringer Mannheim) followed by cross-linking with UV (254 nm) for 3 min. The blot was prehybridized (1 hr at 55°C) in hybridisation buffer (1 mM EDTA; 70 g/L Sodium Dodecyl Sulphate (SDS); 0.25 M $Na_2HPO_4.12H_2O$, pH 7.2). Hybridisation was performed with allele-specific oligonucleotides (11) Arg112: (5′-GAG GAC GTG CGC GGC CGC-3′) for the detection of allele e4, Cys112 (5′-GCG GCC GCA CAC GTC CTC-3′) and Arg158 (5′-CTG CAG AAG CGC CTG GCA G-3′) for the detection of allele e3 and Cys158 (5′-CTG CCA GGC ACT TCT GCA G-3′) for the detection of allele e2. These oligonucleotides were all 3′-biotinylated using terminal deoxynucleotide transferase (Gibco BRL, Paisley, United Kingdom), BIO-16-2′dUTP (Boehringer Mannheim) and 2′dATP (Pharmacia, Uppsala, Sweden) in a molar ratio of 1:10 at 37°C for 4 hr. Before incubation with hybridisation buffer, the probe (0.5 pmol/mL) was denatured for 5 min at 95°C. Hybridisation took place for 2 hr at 55°C. The blots were washed using 2xSSC (17.5 g/L NaCl, 8.8 g/L sodiumcitrate, pH 7.0), 1.0% (w/v) SDS, twice for 5 min at roomtemperature (RT), twice 15 min at 62°C (Arg112 and Cys112) or 60°C (Arg158 and Cys158) with 1xSSC, 1.0% (w/v) SDS, twice 5 min with 1xSSC at RT and once 1 min at RT using 100 mL washingbuffer (0.1 M maleic acid, 0.15 M NaCl) and 3 mL/L Tween-20. To avoid aspecific binding of the streptavidin-alkaline phosfatase complex (AVIDx-AP; Tropix, Bedford, Massachucets), the membrane was blocked for 30 min at RT using block buffer I (1 g casein (Sigma, St. Louis, USA) in 100 ml wash buffer) and for 15 min at RT using block buffer II (3 g casein, 1 g polyvinylpyrrolidone (PVP, Sigma) in 100 mL wash buffer). The blot was incubated for 15 min at RT using AVIDx-AP 0.05% (v/v) in block buffer I. The blot was washed four times 5 min in 3% (v/v) PBS (Phosphate Buffered Saline)/Tween 20 and subsequently twice at RT in assay buffer (0.5 mM $MgCl_2$; 1% (w/v) diethanolamine (DEA, Tropix), pH 10). Afterwards, the blot was incubated for 5 min at RT with chemiluminescent substrate CSPD (Tropix); 1% (v/v) in assay buffer and exposed to
a standard X-ray-film for 30 min.

The overall apoE allele frequencies in our series of cases with neuropathologically confirmed PD were as follows: e2, 12%; e3 69%; e4 19%; this is compares well with the frequencies in the normal population. When the group was divided in demented and non-demented cases no significant differences in apoE e4 allele frequencies were found, neither between these two groups nor compared to normal cases. In a selected subgroup of demented patients, however, in which dementia had started between 60 and 76 years of age, the apoE e4 allele frequency was significantly increased compared to patients older than 76, or younger than 60 years of age. ApoE e4 possibly influences the development of dementia in PD only in a certain agegroup as is suggested in AD. In AD dementia onset before the age of 60 is infrequent and often caused by mutations in the genes on chromosome 1 (Levy-Lahad *et al.*, 1995)), 14 (Schellenberg *et al.*, 1992; St George-Hyslop *et al.*, 1992) or 21. (Goate *et al.*, 1991) ApoE e4 plays a major role in late onset AD (>60 years) (Davignon *et al.*, 1988; Rebeck *et al.*, 1993). With increasing age the influence of apoE on the development of dementia (AD) decreases (Rebeck *et al.*, 1994; Corder *et al.*, 1996; Schächter *et al.*, 1994; Söbel *et al.*, 1995). Thus, it seems that in AD there is an influence of apoE e4 only between specific age limits. The explanantion for our finding that the apoE e4 allele frequency is significantly increased in this specific agegroup with PD and dementia could be that the same mechanisms play a role in the development of dementia in AD and PD.

In our series the presence of an apoE e4 allele significantly shortened the duration of dementia in PD, but did it not influence the age at death. If apoE e4 was present the duration was 2.1 years, without this allele 5.4 years. One explanation could be that apoE might play a role in repairing and maintaining the peripheral nervous system (Boyles *et al.*, 1989; LeBlanc *et al.*, 1990). This same mechanism might influence the central nervous system as well. After experimental brain injury apoE is produced in increased amounts in the central nervous system (Poirer *et al.*, 1991a; Stoll *et al.*, 1989). The efficiency of repair by e4 may be lower than that by e3, and therefore progression might be faster. Another explanation is the altered rate of amyloidogenesis in the presence of apoE e4 (Evans *et al.*, 1995).

ApoE e4 did not influence the age at onset and the duration of PD, duration from PD onset until onset of dementia, age at onset of dementia, or age at death neither in demented or non-demented cases, nor in the total PD group. ApoE e2 did not seem to have a protective effect on the development of dementia in PD.

Although 52% of the clinically demented PD patients met the CERAD criteria for "definite" or "probable" AD, apoE e4 was not significantly increased compared to the normal population. The amount of Alzheimer type pathology was significantly lower in the cases without dementia ("no" or "possible" AD according to the CERAD criteria). NFT pathology (according to Braak & Braak) was found to a moderate extent in our demented PD patients. Only 8% met stage V-VI of Braak & Braak, and 46% met stage III-

IV. No correlation between apoE e4 and the Braak & Braak staging could be established.

Based on these data we conclude that there is no difference in apoE e4 frequency between demented and non demented PD cases. We have to consider the possibility that apoE e4 may be is associated with a shorter duration of dementia in PD and apoE e4 probably plays a role in a specific age group (60-76 years) of demented PD patients. Because there is a significant difference in apoE e4 allele frequency between two conditions with Lewy body formation namely "dementia with Lewy body" (initial symptom cognitive impairment) and Parkinson's disease (initial symptom extrapyramidal features), one can conclude that apoE e4 does not play a role in the formation of Lewy bodies.

THE ROLE OF APOE E4 IN "DEMENTIA WITH LEWY BODIES".

Core features of "dementia with Lewy bodies" are progressive cognitive decline, fluctuation of cognition with pronounced variations in attention and alertness, visual hallucinations, and parkinsonism. Supportive features of the diagnosis are repeated falls, syncope, transient loss of consciousness, neuroleptic hypersensitivity, systematized delusions, and hallucinations in other modalities. (McKeith *et al.*, 1996) In this chapter Parkinson's disease with dementia, and AD with concommitant Lewy bodies are not regarded as "dementia with Lewy bodies" on the basis of their distinctive initial symptoms and clinical characteristics.

Neuropathologically "dementia with Lewy bodies" is characterized by Lewy bodies in the brain stem, limbic cortex, and neocortex. Lewy bodies are intracytoplasmic, spherical, eosinophilic neuronal inclusion bodies. Their major components are cytoskeletal neurofilamental proteins and ubiquitin. Lewy bodies can be visualized with hematoxylin and eosin, but ubiquitin staining has substantially raised the sensitivity for detecting Lewy bodies. Some Alzheimer type pathology (predominantly ß-amyloid deposition and diffuse plaque formation) is a common features in most, but not all, demented cases with Lewy bodies (Perry *et al.*, 1990; Hansen *et al.*, 1990; Dickson *et al.*, 1991). NFT's are only seen in a minority of cases (Ince *et al.*, 1991). Generally Alzheimer type pathology in "Dementia with Lewy bodies" generally is not severe enough to fulfil the morphological criteria for AD.

From clinical and neuropathological studies it is known that Lewy bodies are found in the brain stem and cortex in 15-25% of elderly demented patients (Kosaka *et al.*, 1984);

Perry *et al.*, 1990; Lennox *et al.*, 1989a; Hansen *et al.*, 1990; Dickson *et al.*, 1991). So "dementia with Lewy bodies" is after "pure " AD the second largest pathological subgroup of all dementias. In 21-24% of the patients with AD concomitant Lewy bodies are found at neuropathological evaluation (Gearing *et al.*, 1995a; Hulette *et al.*, 1995; Galasko *et al.*, 1994a). "Dementia with Lewy bodies" is variously referred to as "senile dementia of Lewy body type" (Perry *et al.*, 1990; McKeath *et al.*, 1994), "diffuse Lewy body disease" (Lennox *et al.*, 1989b; Dickson *et al.*, 1991), "cortical Lewy body dementia" (Gibb *et al.*, 1989), and "Lewy body variant of Alzheimer's disease"(Hansen *et al.*, 1990; Frostl *et al.*, 1993). In most of these entities the apoE e4 allele frequency is increased compared to the normal population (Harrington *et al.*, 1994; Hardy *et al.*, 1994; Kawanishi *et al.*, 1996; Galasko *et al.*, 1994b; Olichney *et al.*, 1996; St Clair *et al.*, 1994; Benjamin *et al.*, 1994; Pickering-Brown *et al.*, 1994; Lippa *et al.*, 1995; Gearing *et al.*, 1995b; Martinoli *et al.*, 1995).

As in AD the apoE e4 load is associated with an earlier age at onset of cognitive symptoms in the "Lewy body variant of AD", but this is not associated with a shorter duration of the disease (Olichney *et al.*, 1996). There is no correlation between apoE genotype and age of onset of symptoms in "senile dementia of the Lewy body type" (Perry *et al.*, 1996). Unlike in the AD population, the apoE e2 allele frequencies in "senile dementia of the Lewy body type" are not reduced compared to normal cases (Harrington *et al.*, 1994). Therefore apoE e2, thus, does not seem to have a protective effect in Lewy body dementia (Morris *et al.*, 1996).

In "dementia with Lewy bodies" no relationship between apoE genotypes and the density of Lewy bodies in the substantia nigra, or other brain regions was found, suggesting that APOE does not have a major effect on Lewy body formation (Gearing *et al.*, 1995b, Lippa *et al.*, 1995). ApoE e4 did correlate with NP, but not with NFT in the "Lewy body variant of AD" (Olichney *et al.*, 1996; Lippa *et al.*, 1995). On the other hand Hansen's observation suggest that cases with "Lewy body variant of AD" and an e4 allele have more advanced stages of Alzheimer type pathology according to the Braak & Braak criteria than cases that lack an apoE e4 allele (Hansen *et al.*, 1994). Consequently, a correlation between apoE e4 and NFT density cannot be ruled out. In "senile dementia of the Lewy body type" no relation could be established between the presence of apoE e4 and density of neocortical plaques or neocortical tangles. (Perry *et al.*, 1996). The relation between apoE e4 and neuropathology in "dementia with Lewy bodies" remains therefore obscure.

The apoE e4 allele is a risk factor in AD and probably also in "dementia with Lewy bodies" with a substantial amount of

Alzheimer type pathology (McKeith *et al.*, 1996). Both have senile plaques formation in common, but differ in NFT density. This suggests that apoE e4 is associated more with amyloid depositions than with NFT's.

ApoE e4 plays a role in "dementia with Lewy bodies", however exact mechanisms and influence on neuropathology remain unclear. The precise nosological entity of "dementia with Lewy bodies" needs still further clinical and morphological study, and it is possible that the pathogenetic role of apoE in "dementia with Lewy bodies" is related only to the Alzheimer component of this illusive entity. It is to be hoped that the new consensus criteria will elucidate more about "dementia with Lewy bodies" (McKeith *et al.*, 1996).

CONCLUSION.

ApoE e4 is associated with an increased risk for and earlier onset of Alzheimer's disease. The role of apoE in the so called "dementia with Lewy bodies" is still a matter of debate, and awaits new studies using the consensus criteria (McKeith *et al.*, 1996). In PD with or without dementia apoE plays no or only a minor role.

REFERENCES

Arai, H., Muramatsu, T., Higuchi, S., Sasaki, H., Trojanowski, J.Q. (1994). Apolipoprotein E gene in Parkinson's disease with or without dementia. *Lancet*. **344**,889.

Arriagada, P., Hyman BT. (1990). Predominance of A4 over Alz-50 positive plaques in non-demented elderly [abstract]. *Neurobiol. Aging*. **11**,324.

Arriagada, P.V., Growdon, J.H., Hedley-Whyte, E.T., Hyman, B.T. (1992). Neurofibrillary tangles but not senile plaques parallel duration and severity of Alzheimer's disease. *Neurology*. **42**,631-639.

Asada, T., Kariya, T., Yamagata, Z., Kinoshita, T., Asaka, A. (1996). ApoE epsilon 4 allele and cognitive decline in patients with Alzheimer's disease. *Neurology*. **47**,603.

Benjamin, R., Leake, A., Edwardson, J.A., McKeith, I.G., Ince, P.G., Perry, R.H., Morris, C.M. (1994). Apolipoprotein E genes in Lewy body and Parkinson's disease. *Lancet*. **343**,1565.

Blessed, G., Tomlinson, B.E., Roth, M. (1968). The association between quantitative measures of dementia and of senile change in the cerebral grey matter of elderly subjects. *Br. J. Psychiatry*. **114**,797-811.

Boyles, J.K., Zoellner, C.D., Anderson, L.J., Kosik, L.M., Pitas, R.E., Weisgraber, K.H., Hui, D.Y., Mahley, R.W., Gebicke-Haerter, P.J., Ignatius, M.J., et-al. (1989). A role for apolipoprotein E, apolipoprotein A-I, and low density lipoprotein receptors in cholesterol transport during regeneration and remyelination of the rat sciatic nerve. *J. Clin. Invest.* **83**,1015-1031.

Braak, H., Braak, E. (1991). Neuropathological stageing of Alzheimer-related changes. *Acta. Neuropathol. Berl.* **82**,239-259.

Corder, E.H., Saunders, A.M., Strittmatter, W.J., Schmechel, D.E., Gaskell, P.C., Small, G.W., Roses, A.D., Haines, J.L., Pericak-Vance, M.A. (1993). Gene dose of apolipoprotein E type 4 allele and the risk of Alzheimer's disease in late onset families. *Science.* **261**,921-923.

Corder, E.H., Saunders, A.M., Risch, N.J., et al. (1994). Protective effect of apolipoprotein E type 2 allele for late onset Alzheimer disease. *Nat. Genet.* **7**,180-184.

Corder, E.H., Saunders, A.M., Strittmatter, W.J. (1995a). Apolipoprotein E, survival in Alzheimer's disease patients, and the competing risk of death and Alzheimer's disease. *Neurology.* **45**,1323-1328.

Corder, E.H., Basun, H., Lannfelt, L., Viitanen, M., Winblad, B. (1995b). Apolipoprotein E-e4 gene dose. *Lancet.* **346**,967-968.

Corder, E.H., Lannfelt, L., Viitanen, M., Corder, L.S., Manton, K.G., Winblad, B., Basun, H. (1996). Apolipoprotein E genotype determines survival in the oldest old (85 years or older) who have good cognition. *Arch. Neurol.* **53**,418-422.

Crystal, H., Dickson, D., Fuld, P., et al. (1988). Clinico-pathological studies in dementia. Nondemented subjects with patholofically confirmed Alzheimer's disease. *Neurology.* **38**,1682-1687.

Davignon, J., Gregg, R.E., Sing, C.F. (1988). Apolipoprotein E polymorfism and atherosclerosis. *Atheriosclerosis.* **8**,1-21.

Delaere, P., Duyckaerts, C., Brion, J.P., Poulain, V., Hauw, J.J. (1989). Tau, paired helical filaments and amyloid in the neocortex: a morphometric study of 15 cases with graded intellectual status in aging and senile dementia of Alzheimer type. *Acta. Neuropathol. Berl.* 77,645-653.

Delaere, P., Duyckaerts, C., Masters, C., Beyreuther, K., Piette, F., Hauw, J.J. (1990). Large amounts of neocortical beta A4 deposits without neuritic plaques nor tangles in a psychometrically assessed, non-demented person. *Neurosci. Lett.* **116**:87-93.

Dickson, D.W., Ruan, D., Crystal, H., Mark, M.H., Davies, P., Kress, Y., Yen, S.H. (1991). Hippocampal degeneration differentiates diffuse Lewy body disease (DLBD) from Alzheimer's disease: light and electron microscopic immunocytochemistry of CA2-3 neurites specific to DLBD. *Neurology.* **41**,1402-1409.

Doherty, F.J., Laszlo, L., Lowe, J., Lennox, G., Landon, M., Mayer, R.J. (1989). Ubiquitin-protein conjugates: clinical and experimental findings. *Revis. Biol. Celular.* **20**,255-273.

Duara, R., Barker, W.W., Lopez-Alberola, R., Loewenstein, D.A., Grau, L.B., Gilchrist, D., Sevush, S., St.George-Hyslop, S. (1996). Alzheimer's disease: interaction of apolipoprotein E genotype, family history of dementia, gender, education, ethnicity, and age of onset. *Neurology.* **46**,1575-1579.

Duyckaerts, C., Delaere, P., Poulain, V., Brion, J-P., Hauw, J.J. (1988). Does amyloid precede paired helical filaments in the senile plaque? A study of 15 cases with graded intellectual status in aging and Alzheimer's disease. *Neurosci. Lett.* **91**,354-359.

Duyckaerts, C., Hauw, J.J., Bastenaire, F., et al. (1986). Laminar distribution of neocortical senile plaques in senile dementia of the Alzheimer type. *Acta. Neuropathol. Berl.* **70**,249-256.

Egensperger, R., Bancher, C., Kosel, S., Jellinger, K., Mehraein, P., Graeber, M.B. (1996). The apolipoprotein E epsilon 4 allele in Parkinson's disease with Alzheimer lesions. *Biochem. Biophys. Res. Commun.* **224**,484-486.

Elshourbagy, N.A., Liao, W.S., Mahley, R.W., Taylor, J.M. (1985). Apolipoprotein E mRNA is abundant in the brain and adrenals, as well as in the liver, and is present in other periferal tissues of rats and marmots. *Proc. Natl. Acad. Sci.* 1985;82,203-207.

Evans, K.C., Berger, E.P., Cho, C.G., Weisgraber, K.H., Lansbury, P.T. (1995). Apolipoprotein E is a kinetic but not a thermodynamic inhibitor of
amyloid formation: implications for the pathogenesis and treatment of Alzheimer disease. *Proc. Natl. Acad. Sci. USA.* **92**,763-767.

Förstl, H., Burns, A., Luthert, P., Cairns, N., Levy, R. (1993). The Lewy-body variant of Alzheimer's disease. Clinical and pathological findings. *Br. J. Psychiatry.* **162**,385-392.

Galasko, D., Hansen, L.A., Katzman, R., Wiederholt, W., Masliah, E., Terry, R., Hill, L.R., Lessin, P., Thal, L.J. (1994a). Clinical-neuropathological correlations in Alzheimer's disease and related dementias. *Arch. Neurol.* **51**,888-895.

Galasko, D., Saitoh, T., Xia, Y., Thal, L.J., Katzman, R., Hill, L.R., Hansen, L. (1994b). The apolipoprotein E allele epsilon 4 is overrepresented in patients with the Lewy body variant of Alzheimer's disease. *Neurology.* **44**,1950-1951.

Gearing, M., Mirra, S.S., Hedreen, J.C., Sumi, S.M., Hansen, L.A., Heyman, A. (1995a). The Consortium to Establish a Registry for Alzheimer's Disease (CERAD). Part X. Neuropathology confirmation of the clinical diagnosis of Alzheimer's disease. *Neurology.* **45**,461-466.

Gearing, M., Schneider, J.A., Rebeck, G.W., Hyman, B.T., Mirra, S.S. (1995b). Alzheimer's disease with and without coexisting Parkinson's disease changes: apolipoprotein E genotype and neuropathologic correlates. *Neurology.* **45**,1985-1990.

Gibb, W.R., Luthert, P.J., Janota, I., Lantos, P.L. (1989). Cortical Lewy body dementia: clinical features and classification. *J. Neurol. Neurosurg. Psychiatry.* **52**,185-192.

Gibb, W.R.G. (1990). The Lewybody. \Curr. *Opin. Neurol. Neurosurg.* **3**,346-349.

Goate, A., Chartier-Harlin, M.C., Mullan, M., Brown, J., Crawford, F., Fidani, L., Giuffra, L., Haynes, A., Irving, N., James, L., et-al. (1991). Segregation of a missense mutation in the amyloid precursor protein gene with familial Alzheimer's disease. *Nature.* **349**,704-706.

Gomez-Isla, T., West, H.L., Rebeck, G.W., Harr, S.D., Growdon, J.H., Locascio, J.J., Perls, T.T., Lipsitz, L.A., Hyman, B.T. (1996). Clinical and pathological correlates of apolipoprotein E epsilon 4 in Alzheimer's disease. *Ann. Neurol.* **39**,62-70.

Growdon, J.H., Locascio, J.J., Corkin, S., Gomez-Isla, T., Hyman, B.T. (1996). Apolipoprotein E genotype does not influence rates of cognitive decline in Alzheimer's disease. *Neurology.* **47**,444-448.

Hansen, L.A., Salmon, D., Galasko, D., Masliah, E., Katzman, R., DeTeresa, R., Thal, L., Pay, M.M., Hofstetter, R., Klauber, M., et-al. (1990). The Lewy body variant of Alzheimer's disease: a clinical and pathologic entity. *Neurology.* **40**,1-8.

Hansen, L.A., Galasko, D., Samuel, W., Xia, Y., Chen, X., Saitoh, T. (1994). Apolipoprotein-E e4 is associated with increased neurofibrillary pathology in the Lewy body variant of Alzheimer's disease. *Neurosci. Lett.* **182**,63-65.

Hardy, J., Crook, R., Prihar, G., Roberts, G., Raghavan, R., Perry, R.H. (1994). Senile dementia of the Lewy body type has an apolipoprotein E epsilon 4 allele frequency intermediate between controls and Alzheimer's disease. *Neurosci. Lett.* **182**,1-2.

Harrington, C.R., Louwagie, J., Rossau, R., Vanmechelen, E., Perry, R.H., Perry, E.K., Xuereb, J.H., Roth, M., Wischik, C.M. (1994). Influence of apolipoprotein E genotype on senile dementia of the Alzheimer and Lewy body types. Significance for etiological theories of Alzheimer's disease. *Am. J. Pathol.* **145**,1472-1484.

Heggie, P., Burdon, T., Lowe, J., Landon, M., Lennox, G., Jefferson, D., Mayer, R.J. (1989). Ubiquitin gene expression in brain and spinal cord in motor neurone disease. *Neurosci. Lett.* **102**,343-348.

Hughes, A.J., Daniel, S.E., Kilford, L., Lees, A.J. (1992). Accuracy of clinical diagnosis of idiopathic Parkinson's disease: a clinico-pathological study of 100 cases. *J. Neurol. Neurosurg. Psychiatry.* **55**,181-184.

Hughes, A.J., Daniel, S.E., Blankson, S., Lees, A.J. (1993). A clinicopathologic study of 100 cases of Parkinson's disease*Arch. Neurol.* **50**:140-148.

Hulette, C., Mirra, S., Wilkinson, W., Heyman, A., Fillenbaum, G., Clark, C. (1995). The Consortium to Establish a Registry for Alzheimer's Disease (CERAD). Part IX. A prospective cli-niconeuropathologic study of Parkinson's features in Alzheimer's disease. *Neurology.* **45**,991-1995.

Hyman, B.T., Marzloff, K., Arriagada, P.V. (1993). The lack of accumulation of senile plaques or amyloid burden in Alzheimer's disease suggests a dynamic balance between amyloid deposition and resolution. *J. Neuropathol. Exp. Neurol.* **52**,594-600.

Ignatius, M.J., Gebicke-Harter, P.J., Skene, J.H.P., Schilling, J.W., Weisgraber, K.H., Mahley, R.W., Shooter, E.M. (1986). Expression of apolipoprotein E during nerve degeneration and regeneration. *Proc. Natl. Acad. Sci. USA*. **83**,1125-1129.

Ince, P.G., Irving, D., McArthur, F., et al. (1991). Quantitative neuropathological study of Alzheimer-type pathology in the hippocampus: comparison of senile dementia of Alzheimer type, senile dementia of Lewy body type, Parkinson's disease and non-demented elderly control patients. *J. Neurol. Sci.* **106**,142-152.

Katzman, R., Terry, R., De Teresa, R., et al. (1988). Clinical, pathological, and neurochemical changes in dementia: a subgroup with preserved mental status and numerous neocortical plaques. *Ann. Neurol.* **27**,138-144.

Kawanishi, C., Suzuki, K., Odawara, T., Iseki, E., Onishi, H., Miyakawa, T., Yamada, Y., Kosaka, K., Kondo, N., Yamamoto, T. (1996). Neuropathological evaluation and apolipoprotein E gene polymorphism analysis in diffuse Lewy body disease. *J. Neurol. Sci.* **136**,140-142.

Koller, W.C., Glatt, S.L., Hubble, J.P., Paolo, A., Troster, A.I., Handler, M.S., Horvat, R.T., Martin, C., Schmidt, K., Karst, A., et-al. (1995). Apolipoprotein E genotypes in Parkinson's disease with and without dementia. *Ann. Neurol.* **37**,242-245.

Kosaka, K., Yoshimura, M., Ikeda, K., et al. (1984). Diffuse type of Lewy body disease: progressive dementia with abundant cortical Lewy bodies and senile changes of varying degree - a new disease? *Clin. Neuropathol.* **3**,185-192.

Kurz, A., Egensperger, R., Haupt, M., Lautenschlager, N., Romero, B., Graeber, M.B., Muller, U. (1996). Apolipoprotein E epsilon 4 allele, cognitive decline, and deterioration of everyday performance in Alzheimer's disease. *Neurology*. **47**,440-443.

Landen, M., Thorsell, A., Wallin, A., Blennow, K. (1996). The apolipoprotein E allele epsilon 4 does not correlate with the number of senile plaques or neurofibrillary tangles in patients with Alzheimer's disease. *J. Neurol. Neurosurg. Psychiatry.* **61**,352-356.

LeBianc, A.C., Poduslo, J.F. (1990). Regulation of apolipoprotein E gene expression after injury of the rat sciatic nerve. *J. Neurosci. Res.* **25**,162171.

Lennox, G., Lowe, J., Godwin-Austen, R.B., Landon, M., Mayer, R.J. (1989a). Diffuse Lewy body disease: an important differential diagnosis in dementia with extrapyramidal features. *Prog. Clin. Biol. Res.* **317**,121-130.

Lennox, G., Lowe, J., Landon, M., Byrne, E.J., Mayer, R.J., Godwin-Austen, R.B. (1989b). Diffuse Lewy body disease: correlative neuropathology using anti-ubiquitin immunocytochemistry. *J. Neurol. Neurosurg. Psychiatry.* **52**,1236-1247.

Lennox, G., Lowe, J., Morrell, K., Landon, M., Mayer, R.J. (1989c). Anti-ubiquitin immunocytochemistry is more sensitive than conventional techniques in the detection of diffuse Lewy body disease. *J. Neurol. Neurosurg. Psychiatry*. **52**,67-71.

Levy-Lahad, E., Wasco, W., Poorkaj, P., et-al. (1995). Candidate gene for the chromosome 1 familial Alzheimer's disease locus. *Science*. **269**,973-977.

Lippa, C.F., Smith, T.W., Saunders, A.M., Crook, R., Pulaski-Salo, D., Davies, P., Hardy, J., Roses, A.D., Dickson, D. (1995). Apolipoprotein E genotype and Lewy body disease. *Neurology*. **45**,97-103.

Lowe, J., Aldridge, F., Lennox, G., Doherty, F., Jefferson, D., Landon, M., Mayer, R.J. (1989a). Inclusion bodies in motor cortex and brainstem of patients with motor neurone disease are detected by immunocytochemical localisation of ubiquitin. *Neurosci. Lett.* **105**,7-13.

Lowe, J., Morrell, K., Lennox, G., Landon, M., Mayer, R.J. (1989b). Rosenthal fibres are based on the ubiquitination of glial filaments. *Neuropathol. Appl. Neurobiol.* **15**,45-53.

Mahley, R.W. (1988). Apolipoprotein E: cholesterol transport protein with expanding role in cell biology. *Science*. **240**,622-630.

Marder, K., Maestre, G., Cote, L., Mejia, H., Alfaro, B., Halim, A., Tang, M., Tycko, B., Mayeux, R. (1994). The apolipoprotein epsilon 4 allele in Parkinson's disease with and without dementia. *Neurology*. **44**,1330-1331.

Martinoli, M.G., Trojanowski, J.Q., Schmidt, M.L., Arnold, S.E., Fujiwara, T.M., Lee, V.M., Hurtig. H., Julien, J.P., Clark, C. (1995). Association of apolipoprotein epsilon 4 allele and neuropathologic findings in patients with dementia. *Acta. Neuropathol. Berl.* **90**,239-243.

Mayer, R.J., Lowe, J., Lennox, G., Doherty, F., Landon, M. (1989a). Intermediate filaments and ubiquitin: a new thread in the understanding of chronic neurodegenerative diseases. *Prog. Clin. Biol. Res.* **317**,809-818.

Mayer, R.J., Lowe, J., Lennox, G., Landon, M., MacLennan, K., Doherty, F.J. (1989b). Intermediate filament-ubiquitin diseases: implications for cell sanitization. *Biochem. Soc. Symp.* **55**,193-201.

Mayeux, R., Stern, Y., Ottman, R., Tatemichi, T.K., Tang, M.X., Maestre, G., Ngai, C., Tycko, B., Ginsberg, H. (1993). The apolipoprotein epsilon 4 allele in patients with Alzheimer's disease. *Ann. Neurol.* **34**,752-754.

McKeith, I.G., Fairbairn, A.F., Bothwell, R.A., Moore, P.B., Ferrier, I.N., Thompson, P., Perry, R.H. (1994). An evaluation of the predictive validity and inter-rater reliability of clinical diagnostic criteria for senile dementia of Lewy body type. *Neurology*. **44**,872-877.

McKeith, I.G., Galasko, D., Kosaka, K., et al. (1996). Consensus guidelines for the clinical and pathologic diagnosis of dementia with Lewy bodies (DLB): Report of the consortium on DLB international workshop. *Neurology*, **47**,1113-1124.

McKhann, G., Drachman, D., Folstein, M., Katzman, R., Price, D., Stadlan, E.M. (1984). Clinical diagnosis of Alzheimer's disease: report of the NINCDS-ADRDA work group under the auspices of department of health and human services task force on Alzheimer's disease. *Neurology*. **34**,939-944.

Mirra, S.S., Heyman, A., McKeel, D., Sumi, S.M., Crain, B.J., Brownlee, L.M., Vogel, F.S., Hughes, J.P., van Belle, G., Berg, L. (1991). The Consortium to Establish a Registry for Alzheimer's Disease (CERAD). Part II. Standardization of the neuropathologic assessment of Alzheimer's disease. *Neurology*. **41**,479-486.

Morris, J.C., McKeel, D.W., Storandt, M., Rubin, E.H., Price, J.L., Grant, E.A., Ball, M.J., Berg, L. (1991). Very mild Alzheimer's disease: informant-based clinical, psychometric, and pathologic distinction from normal aging. *Neurology*. **41**,469-478.

Morris, C.M., Massey, H.M., Benjamin, R., et al. (1996). Molecular biology of APO E alleles in Alzheimer's and non-Alzheimer's dementias. *J. Neural. Transm*. **36**,889-895.

Nagy, Z., Esiri, M.M., Jobst, K.A., Johnston, C., Litchfield, S., Sim, E., Smith, A.D. (1995). Influence of the apolipoprotein E genotype on amyloid deposition and neurofibrillary tangle formation in Alzheimer's disease. *Neuroscience*. **69**,757-761.

Nalbantoglu, J., Gilfix, B.M., Bertrand, P., Robitaille, Y., Gauthier, S., Rosenblatt, D.S., Poirier, J. (1994). Predictive value of apolipoprotein E genotyping in Alzheimer's disease: results of an autopsy series and an analysis of several combined studies. *Ann. Neurol*. **36**,889-895.

Namba, Y., Tomonaga, M., Kawasaki, H., Otomo, E., Ikeda, K. (1991). Apolipoprotein E immunoreactivity in cerebral amyloid deposits and neurofibrillary tangles in Alzheimer's disease and kuru plaque amyloid in Creutzfeldt-Jakob disease. *Brain. Res*. **541**,163-166.

Ohm, T.G., Kirca, M., Bohl, J., Scharnagl, H., Gross, W., Marz, W. (1995). Apolipoprotein E polymorphism influences not only cerebral senile plaque load but also Alzheimer-type neurofibrillary tangle formation. *Neuroscience*. **66**,583-587.

Olichney, J.M., Hansen, L.A., Galasko, D., Saitoh, T., Hofstetter, C.R., Katzman, R., Thal, L.J. (1996). The apolipoprotein E epsilon 4 allele is associated with increased neuritic plaques and cerebral amyloid angiopathy in Alzheimer's disease and Lewy body variant. *Neurology*. **47**,190-196.

Perry, R.H., Irving, D., Blessed, G., Fairbairn, A., Perry, E.K. (1990). Senile dementia of Lewy body type. A clinically and neuropathologically distinct form of Lewy body dementia in the elderly. *J. Neurol. Sci*. **95**,119139.

Perry, R.H., Jaros, E.B., Irving, D., et al. (1996). What is the neuropathological basis of dementia associated with Lewy bodies? In: *Dementia with Lewy bodies: clincical, pathological, and treatment issues*, R. Perry, I. McKeith and E. Perry (Eds). Cambridge University press, Cambridge, pp. 212-223.

Pickering-Brown, S.M., Mann, D.M., Bourke, J.P., Roberts, D.A., Balderson, D., Burns, A., Byrne, J., Owen, F. (1994). Apolipoprotein E4 and Alzheimer's disease pathology in Lewy body disease and in other beta-amyloid-forming diseases. *Lancet.* **343**,1155.

Pitas, R.E., Boyles, J.K., Lee, S.H., Foss, D., Mahley, R.W. (1987). Asterocytes synthesize apolipoprotein E and metabolize apolipoprotein E-containing lipoproteins. *Biochim. Biophys. Acta.* **917**,148-161.

Poirier, J., Hess, M., May, P.C., Finch, C. (1991a). Astrocytic apolipoprotein E mRNA and GFAP mRNA in hippocampus after entorhinal cortex lesioning. *Brain. Res. Mol. Brain. Res.* **11**,97-106.

Poirier, J., Hess, M., May, P.C., Finch, C.E. (1991b). Cloning of hippocampal poly(A) RNA sequences that increase after entorhinal cortex lesion in adult rat. *Brain. Res. Mol. Brain. Res.* **9**,191-195.

Polvikoski, T., Sulkava, R., Haltia, M., Kainulainen, K., Vuorio, A., et al. (1995). Apolipoprotein E, dementia, and cortical deposition of beta amyloid protein. *N. Engl. J. Med.* **333**,1242-1247.

Price, J.L., Davis, P.B., Morris, J.C., White, D.L. (1991). The distribution of tangles, plaques and related immunohistochemical markers in healthy aging and Alzheimer's disease. *Neurobiol. Aging.* **12**,295-312.

Rebeck, G.W., Reiter, J.S., Strickland, D.K., Hyman, B.T. (1993). Apolipoprotein E in sporadic Alzheimer's disease: allelic variation and receptor interactions. *Neuron.* **11**,575-580.

Rebeck, G.W., Perls, T.T., West, H.L., Sodhi, P., Lipsitz, L.A., Hyman, B.T. (1994). Reduced apolipoprotein epsilon 4 allele frequency in the oldest old Alzheimer's patients and cognitively normal individuals. *Neurology.* **44**,1513-1516.

Roses, A.D. (1995). Apolipoprotein E genotyping in the differetial diagnosis, not prediction, of Alzheimer's disease. *Ann. Neurol.* **38**,6-14.

Sanan, D.A., Weisgraber, K.H., Russell, S.J., et al. (1994). Apolipoprotein E associates with beta amyloid peptide of Alzheimer's disease to form novel monofibrils. Isoform apoE4 associates more efficiently than apoE3. *J. Clin. Invest.* **94**,860-869.

Saunders, A.M., Strittmatter, W.J., Schmechel, D., et al. (1993). Association of apolipoprotein E allele epsilon 4 with late-onset familial and sporadic Alzheimer's disease. *Neurology.* **43**,1467-1472.

Schächter, F., Faure-Delanef, L., Guenot, F., Rouger, H., Froguel, P., Lesueur-Ginot, L., Cohen, D. (1994). Genetic associations with human longevity at the APOE and ACE loci. *Nat. Genet.* **6**,29-32.

Schellenberg, G.D., Bird, T.D., Wijsman, E.M., Orr, H.T., Anderson, L., Nemens, E., White, J.A., Bonnycastle, L., Weber, J.L., Alonso, M.E., et-al. (1992a). Genetic linkage evidence for a familial Alzheimer's disease locus on chromosome 14. *Science.* **258**,668-671.

Schellenberg, G.D., Boehnke, M., Wijsman, E.M., Moore, D.K., Martin, G.M., Bird, T.D. (1992b). Genetic association and linkage analysis of the apolipoprotein CII locus and familial Alzheimer's disease. *Ann. Neurol.* **31**,223-227.

Schmechel, D.E., Saunders, A.M., Strittmatter, W.J., Crain, B.J., Hulette, C.M., Joo, S.H.,

Pericak-Vance, M.A., Goldgaber, D., Roses, A.D. (1993). Increased amyloid beta-peptide deposition in cerebral cortex as a consequence of apolipoprotein E genotype in late-onset Alzheimer disease. *Proc. Natl. Acad. Sci. USA*. **90**,9649-9653.

Söbel, E., Louhija, J., Sulkava, R., Davanipour, Z., Kontula, K., Miettinen, H., Tikkanen, M., Kainulainen, K., Tilvis, R. (1995). Lack of association of apolipoprotein E allele epsilon 4 with late-onset Alzheimer's disease among Finnish centenarians. *Neurology*. **45**,903-907.

Soininen, H., Kosunen, O., Helisalmi, S., Mannermaa, A., Paljarvi, L., Talasniemi, S., Ryynanen, M., Riekkinen, P. (1995). A severe loss of choline acetyltransferase in the frontal cortex of Alzheimer's patients carrying apolipoprotein e4 allele. *Neurosci. Lett.* **187**,79-82.

St Clair, D., Norrman, J., Perry, R., Yates, C., Wilcock, G., Brookes, A. (1994). Apolipoprotein E epsilon 4 allele frequency in patients with Lewy body dementia, Alzheimer's disease and age-matched controls.*Neurosci. Lett.* **176**,45-46.

St George-Hyslop, P., Haines, J., Rogaev, E., Mortilla, M., Vaula, G., Pericak-Vance, M., Foncin, J.F., Montesi, M., Bruni, A., Sorbi, S., et-al. (1992). Genetic evidence for a novel familial Alzheimer's disease locus on chromosome 14. *Nat. Genet.* **2**,330-334.

Stoll, G., Meuller, H.W., Trapp, B.D., Griffin, J.W. (1989). Oligodendrocytes but not astrocytes express apolipoprotein E after injury of rat optic nerve. *Glia*. **2**,170-176.

Strittmatter, W.J., Saunders, A.M., Schmechel, D., Pericak-Vance, M., Enghild, J., Salvesen, G.S., Roses, A.D. (1993a). Apolipoprotein E: high-avidity binding to beta-amyloid and increased frequency of type 4 allele in late-onset familial Alzheimer disease. *Proc. Natl. Acad. Sci USA*. **90**,1977-1981.

Strittmatter, W.J., Weisgraber, K.H., Goedert, M., Saunders, A.M., Huang, D.Y., et al. (1994). Hypothesis: microtubule instability and paired helical filament formation in the Alzheimer disease brain are related to apolipoprotein E genotype. *Exp. Neurol.* **125**,163-171.

Strittmatter, W.J., Weisgraber, K.H., Huang, D.Y., Dong, L.M., Salveson, G.S., et al. (1993b). Binding of human apolipoprotein E to synthetic amyloid beta peptide: isoform-specific effects and implications for late-onset Alzheimer disease. *Proc. Natl. Acad. Sci. USA*. **90**,8098-8102.

Vos de, R.A.I., Jansen, E.N.H., Stam, F.C., Ravid, R., Swaab, D.F. (1995). 'Lewy body disease': clinico-pathological correlations in 18 consecutive cases of Parkinson's disease with and without dementia. *Clin. Neurol. Neurosurg.* **97**,13-22.

Vos de, R.A.I., Jansen, E.N.H., Yilmazer, D., Braak, H., Braak, E. ((1996). Pathological and clinical features of Parkinson's disease with and without dementia. In: *Dementia with Lewy bodies: clincical, pathological, and treatment issues*, R. Perry, I. McKeith and E. Perry (Eds). Cambridge University press, Cambridge, pp. 255-267.

Whitehead, A.S., Bertrandy, S., Finnan, F., Butler, A., Smith, G.D., Ben-Shlomo, Y. (1996). Frequency of the apolipoprotein E epsilon 4 allele in a case-control study of early onset Parkinson's disease. *J. Neurol. Neurosurg. Psychiatry.* **61**,347-351.

Wilcock, G.K., Esiri, M.M. (1982). Plaques, tangles and dementia. A quantitative study. *J. Neurol. Sci.* **56**,343-356.

Wischik, C.M., Edwards, P.C., Harrington, C.R. (1995). The tau protein amyloidosis of Alzheimer's disease: its mechanisms, potential triggers and consequences. In: *Neurobiology of Alzheimer's disease,* D. Dawbarn, S.J. Allen, (Eds). BIOS Scientific, Oxford, pp. 89-148.

Wisniewski, T., Frangione, B. (1992). Apolipoprotein E: a pathological chaperone protein in patients with cerebral and systemic amyloid. *Neurosci. Lett.* **135**,235-238.

ABSTRACT.

The apolipoprotein E (apoE) gene is a polymorfic gene with three common alleles e2,e3, and e4. It is a plasma protein involved in lipoprotein transport and metabolism. The apolipoprotein E type 4 allele is overrepresented in late onset familial and sporadic Alzheimer's disease. It is still unclear how apoE e4 affects the neuropathological changes found in AD, and how these are related to the development of clinical AD.
Parkinson's disease is associated with dementia in 44-66% of the cases. In addition to the specific pathological characteristics of Parkinson's disease, often a variable density of Alzheimer type pathology is found. ApoE e4 is not a risk factor for Parkinson's disease in general, but it does increase the progression of dementia in Parkinson's disease, and apoE e4 might play a role in the development of dementia in Parkinson's disease in the agegroup between 60 and 76 years of age.
ApoE e4 plays a role in "dementia with Lewy bodies", especially when a substantial amount of Alzheimer pathology is present, however exact mechanisms and influence on neuropathology remain unclear.

Progress in HPLC-HPCE, Vol. 7, pp. 569-596
G. Ali Qureshi *et al.* (Eds)

CLASSICAL NEUROTRANSMITTERS AND THEIR CORRELATION WITH NEUROPEPTIDES IN CEREBROSPINAL FLUID OF PATIENTS WITH NEUROLOGICAL DISORDERS

G.Ali Qureshi
Clinical Research Center, Novum, Karolinska Institute, Huddinge University Hospital, S-141 57 Stockholm, Sweden.

INTRODUCTION

The nervous system is an organ system that coordinates sensory information and generates appropriate behavioural responses. All cells of an organism have some ability to respond to external stimuli by changes in the cell membrane but nerve cells are specialized for this task. Neurons are specialized for short and long distance transmision thoughout the body. The present knowledge suggests that the information which is transmitted through the CNS is both in the form of electrical and chemical signals. The release of neurotransmitters upon arrival of an action potential is generally beleived to depend on the fusion of the cell membrane, followed by extrusion of transmitter molecules into the synaptic cleft. The released transmitters will interact with a receptor molecules in the presynaptic membrane, and it may also bind to receptors located on the presynaptic nerve terminal (*Starke et al., 1989*). The transmitters are removed from the synaptic cleft through diffusion, uptake into the presynaptic terminal or by enzymatic degradation. Neuropeptides are normally inactivated by proteolytic enzymes, and the resulting fragments removed through diffusion. The progress in transmitter physiology has been of fundamental importance to all fields of the medical science. During the last two decades, the number of putative neurotransmitters in the CNS has increased considerably and now omprises substances with large chemical differences i.e. amines, peptides, purine, amino acids, prostagladins and leukotrienes.

Until, recent years the monoamines, acetylcholine and amino acids were thought to be the only classical neurotransmitters ,however, a large number of peptides, many of which were originally characterized in non neural tissues , have been shown to exist also in CNS (*Hökfelt, 1980, 1992*). These compounds are often referred to as neuropeptides and represent a heterogeneous group of molecules, the smallest ones built up of only two amino acids with larger polypeptides consisting of 40 or more amino acids.

The neuropeptides differ from the so-called classical transmitters since a) they function in a lower concentrations than classical transmitters, b) synthesis of peptides is directed by mRNA in perikaryon as a part of a much larger prohormone, from which the active peptide is cleaved by peptidase whereas classical transmitters are formed from dietary sources by enzymes, c) classical transmitters respond and adopt very rapidly to external stimuli and have their synthesis machinery with great capacity whereas neuropeptides respond slower and their synthesis machinery have a more limited capacity and speed, and d) molecular heterogeneity of peptides can exist, the different forms have different, sometimes opposite effects (*Copper et al., 1986*). Neuropeptides are widely distributed throughout the brain in specific nerve cells in coexistence with monoamines and /or other neuropeptides (*Hökfelt, 1992; Everitt et al., 1984; Hendry et al., 1984*). They are beleived to participate in several physiological and pathophysiological processes, including pain sensation, memory, neuroendocrine functions, regulation of release of monoamine transmitters and regulation of mood (*Teledgy, 1983; McCann et al., 1987*). The effects are brought about by pimary actions of neuropeptides or their modulation of the effects of monoamines transmitters.

Transmission of nerve impulses from one neuron to another or from a neuron to a peripheral effector cell such as muscle cells is a critical event in the nervous system. Its elucidation is essential not only for our understanding of normal neuronal function, but it is highly probable that changes in the chemical transmission process may underlie or at least are related to various disease processes in nervous system. Moreover, it is now clear that many of the common drugs used to treat various diseases in the nervous system act via interferring with the chemical transmission process (*Hökfelt et al., 1992; Meltzer et al., 1987*).

Little more than 20 years ago, it became evident that a number of endocrine cells in the body are able to synthesize and secrete more than one hormone in each cell (*Hökfelt et al., 1978; 1987*). These studies have also revealed that peptides in most instances are present in neurons which already contained , for example catecholamines or the cholinergic transmitter acetylcholine or an amino acid transmitter.

As dicussed above, the data obtained so far in general support a role of peptides in chemical transmission auxillary to classical transmitters. Several peptides such as neuropeptide Y (NPY), cholecystokinin (CCK), substance P (SP), somatostatin and vasopressin are known to have ability to act as neurotransmitters. These peptides are also known to co-localize with classical neurotransmitters within a single neuron which provide a means to transmit more complex types of signals. Studies on the relationship between low molecular weight transmitters and peptides in the CNS suggest the existence of both pre- and postsynaptic interaction between coexisting transmitters, in vivo and in vitro (*Lundberg et al., 1983; Bartfai et al., 1986*). One of the first example of coexistence in the central nervous system was the demonstration of CCK-L1 in dopamine neurons in the ventral mesencephalon, especially in the ventral tegmental area (*Hökfelt et al., 1980*). It has also been shown that small population of A10 dopamine neurons contains neurotensin (*Hökfelt et al., 1980*). On the basis of co-existence with monoamine transmitters and distinct regional distribution, the assumption has been made that neuropeptides play a role in CNS disorders (*Chan-Palay, 1988; Takeuchi et al., 1988, Heilig et al., 1990; Nemeroff, 1991*).

The separation of biologically important substances has become increasingly and critical for the advances in biological sciences. High-performance liquid chromatography (HPLC) is one of the most important arsenals for the analysis of increasingly complex materials and significant advances have been made in recent years. The speed, the resolving power and experimental flexibility that the technique HPLC engender have resulted in one of the major, multidisciplinary success in the biological sciences over the last two decades. In this chapter, CSF analysis of classical neurotransmitters such as catecholamines i.e. dopamine, noradrenaline and indoleamine serotonin, and their relation with neuropeptides, such as cholecystokinin, substance P and neuropeptide Y in various neurologic disorders have been shown.

MATERIAL AND METHODS

Patients and healthy subjects

All patients were recruited from the department of Neurology, Huddinge University Hospital, Sweden, except patients with tuberculous meningitis who were recruited from the department of Neurology , the Aga Khan University Hospital, Karachi, Pakistan. CSF samples were obtained by lumbar puncture within 4-8 hours of admission. 20 PD patients (7 females, with mean age 72±12 years), . 16 patients (6 females, mean age 62±4 years) with ischemic stroke (CVD), 20 patients (8 females,mean age, 54±6 years) with definite multiple sclerosis (MS), 18 patients (9 females, mean age, 46±5 years) with aseptic meningitis (AM) and 14 patients (5 females, mean age 51±4 years) with tuberculous meningitis (TBM) were included in this study. The diagnosis of PD was made according to established clinical criteria (*Lang et al., 1989*) . None of these patients exhibited clinical indications of sympotomatic parkinsonism or depression and all had shown clear response to L-DOPA. Among PD patients, 6 patients were included who did not receive L-DOPA and other DAergic drugs. Computerized tomography (CT scan) and/or MRI confirmed the evidence of stroke and its site. All the strokes were in territory arround the middle cerebral artery. Ten of the patients had hemiplegia and 6 had hemiparesis at the time of admission. All the patients showed normal CSF cell count while CSF protein and IgG were significantly higher (Table 1). MS was diagnosed according to Schaumacher criteria (*Schaumacher etal., 1965*). 9 of the MS patients had exacerbation, 5 had remission, while 6 had chronic progressive form of MS at the time of sampling. Exacerbation was defined as sudden appearance of new signs and symptoms or sudden worsening of previous signs and symptoms lasting more than 24 hours.. All of these patients had oligoclonal bands in their CSF. Routine CSF examination showed mononuclear pleocytosis (> 5x106 /liter) in all patients . Slightly raised CSF/serum albumin ratio reflecting low-grade blood-brain barrier damage was present in MS patients, and elevated IgG Index (*Link etal., 1977*) were observed The diagnosis of the menigitis disease was based on standard diagnosis criteria (*Futrell et al., 1992*). Aseptic meningitis was diagnosed on the basis of clinical features, culture, sensitivity tests and laboratory investigations of CSF at the time of admission at hospital. Subsequent examination of CSF was carried on the basis of clinical and therapeutic considerations.The diagnosis of tuberculous meningitis was established on clinical and CSF findings , positive cerebrospinal Gram stain and latex agglutination and positive presence of growth of pathogenic bacteria.

CSF samples were also collected from 14 healthy individual (5 females, mean age 53±5 years) with complaints of mascular tension headache. These individual were considered as healthy subjects (HS) because CSF routine analysis as well as blood complete examination, liver function test, electrolytes, ESR were within normal limits. Their general physical examination, neurological observation and CT scan of head also normal. The clinical data on the healthy subjects and patient groups are presented in Table 1. All values are expressed as mean ± Standard Error Mean (SEM).

TABLE 1. Clinical data on the healthy subjects (HS), patients with Parkinson's disease(PD),cerebrovescular disorders (CVD), multiple sclerosis (MS), aseptic (AM) and Tuberculous meningitis (TBM).

Patients	n	Females	Age(years)	CSF-albumin	CSF-IgG	IgG-index
HS	14	5	53±5	218±17	34±4	0.44±0.01
PD	20	7	72±12	332±42*	51±7**	0.43±0.01
CVD	16	6	70±10	313±31*	46±6*	0.42±0.01
MS	20	8	54±6	188±14	65±7**	1.03±0.14**
AM	18	9	46±5	412±53**	83±12***	0.68±0.06**
TBM	14	5	51±4	442±61**	88±11***	0.73±0.11**

*p<0.05,**p<0.01,***p<0.001

Lumbar puncture and routine CSF analysis

10-12 ml cerebrospinal fluid (CSF) was collected from each patient and healthy subject in a sitting position at the L4-L5 levels. Blood samples were collected by venipuncture . The basic CSF analyses included : cell counting by phase-contrast microscopy (*Sörnäs, 1967*), determination of CSF/serum albumin ratio and CSF/immunoglobuline G (IgG) index (*Tilbbling et al., 1977*) as well as isoelectric focusing for detection of oligoclonal IgG band. Serum and CSF albumin and IgG were determined using Hitachi 737 Automatic Analyzer (Naka Works, Hitachi Ltd., Tokyo, Japan. CSF and serum samples were kept at -70° C if not analysed immediately.

Strategy of chemical analysis of catecholamines and neuropeptides

The concentrations of biogenic amines in CSF, serum as well as in tissues are usually very low. Consequently, the demand for analysis of low concentrations of biogenic amines in biological samples spurred the improvement of analytical methods. Enhanced specificity and better detectability are of equal importance . Continuing progress with analytical procedures has often led to corrections of concentration values for biogenic amines determined with less sophisticated methods, since actual levels often turned out to be much lower.

Apart from chromatographic procedures, techniques have been developed which are based on the specificity of enzymic reaction, often in combination with the use of radioactive substrates as well as immunological methods. The radioenzymatic are considered to be very sensitive, the technique can be used for multiple component analysis without extra separation step and besides the technique has the disadvantage of radiation hazards.

Among the analytical assays, gas chromatography / mass spectrometry (GC/MS) is considered to be one of the most accurate and specific methods for quantitation of monoamines and their metabolites, but the technique is expensive for routine analysis and requires skillful technician for sample clean-up and derivatization. A break through in the determination of biogenic amines by rapid improvement in column liquid chromatography paired with the development of new detection methods especially electrochemical and fluorescence detection provided with an unique possibility to analysis biogenic amines routinely. Among these two detection methods, electrochemical (EC) detection system is among the most commonly used quantitative systems.The sensitivity of detection is dependent on the applied potential from +0.60 to +1.0 V, raising with increase in the applied potential. Because of base-line distrubances, resulting from endogenous compounds in biological samples, and risk of reducing the performance of the EC detector, the applied potential between +0.65 and +0.75 V is suitable which gives adequate sensitivity to accurately quantitate catecholamines and their metabolites .

On the other hand, most of the inconsistent results on the levels of monoamines, indolamines and their metabolites are caused by sample instability and degradation on storage. The cause of decomposition of most of these substances are dependent on light, room temperature and the use of an acid. Failure to standardize storage times and conditions within and between the studies may also result into invalid results.Various procedures in our laboratory show that a deproteinization with 4% 5-sulfosalicylic acid (SSA) under temperature-controlled condition in presence of sodium bisulfite to stabalize monoamines and indoleamines and their metabolites provide with 90% recovery of these substances. An addition of phosphate buffer, to deproteinized sample also avoids decomposition of these substances.

The CSF was prepared as previously described (*Qureshi et al., 1988*) and the concentration of DA, homovanillic acid (HVA), 3,4-dihydroxyphenylacetic acid (DOPAC), NA , 3-methoxy-4-hydroxyphenylglycol (MHPG), 5-hydroxytryptamine (5-HT) and 5-hydroxyindoleacetic acid (5-HIAA) was measured using previously described (*Baig et al., 1991*). Fig. 1 shows the separation of monoamines, indoleamines and their metabolites in standard and CSF sample under similar experimental conditions. Measurements were also made in serum but since no differences were found these will not be considered further.

In the years since its introduction reverse phase chromatography (RPC) has become the predominant technique for peptide separations. The versatility of RPC is reflected in its successful application to the isolation of peptides from a wide variety of sources. Many biological active peptides are often found in only small quantities and may require extensive purification. Optimum resolution of peptides by RPC generally involves a compromise between satisfactory separation of peptide components and reasonable run time. This optimization may be acheived by a systematic manupulation of both the stationary phase (i.e. type of column support, hydrophobic functionalities) and the mobile phase (i.e. pH, temperature, solvent composition, gradient rate, flow rate) . In addition, the difference in the polarities of peptides in a peptide mixture can be maximized through careful choice of ion-pairing reagents. As a rule, the actual effect on the retention time of a peptide depends strongly on both the hydrophobicity of the anionic pairing reagent and the number of positively charged groups on the peptide. The quantitation of neuropeptides require a sensitive, rapid and reliable method of detection. Ultraviolet detection is not sensitive enough and there is no selective reagent for peptide derivatization to improve detection limit. so most of the neuropeptides are quantitated with HPLC in combination with radioimmunoassay (RIA), radioreceptor assay or mass spectrometry. However, CCK-4 and CCK-8s both contain tryptophan residues which are electrochemically oxidizable and therefore this possibility is utilized in developing a method for analysis based on separation of CCK-4 and CCK-8s on a reverse phase column and the quantitation is made by a use of an electrochemical detection (*Qureshi et al., 1993*) . Fig. 2. shows the separation of CCK-4 and CCK-8s in CSF sample and commercial standard. For measurement of CCK-peptides, the chemical identity of these peptides was ensured using a novel HPLC system for micropurification (SMART) and subsequent fast atom bombardment mass spectrometry (*Qureshi et al., 1993*).

The quantitation of NPY was done according to the previously described radioimmunoassay (*Theodorsson-Norheim et al., 1985*).

SP-L1 was determined using SP2 antisera according to the previously described method (*Brodin et al., 1986*) based on RIA procedure. The limit of detection using this procedure was 0.1-0.2 pmol/l .

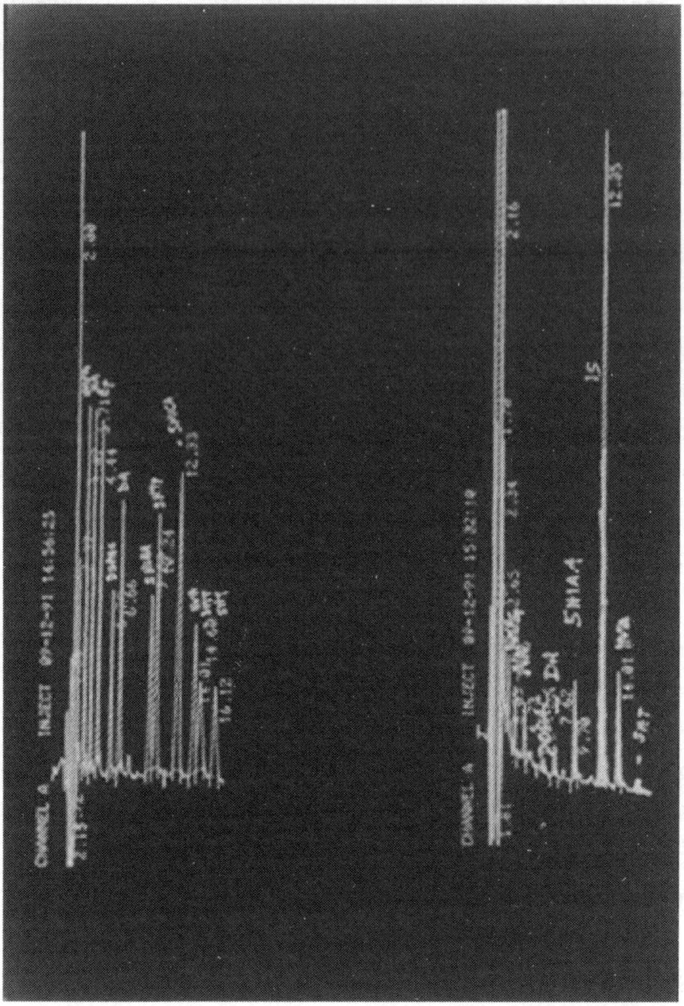

Fig. 1. HPLC separation of (A) a standard mixture containing 100 pmol/l of each substance, and (B) CSF sample from MS patient under the identical chromatographic conditions as given in the reference (Baig et al., 1991).
Abbreviation for the catecholamines, indoleamines and their metabolites: MHPG=3-methoxy,4-hydroxyphenylglycol,NA=noradrenine,EP=Epinepherine, DOPAC=3,4-dihydroxyphenylacetic acid, DA=Dopamine, 5-HIAA= 5-hydroxy indole acetic acid, 5-HTP=5-hydroxy tryptophan, 5-HICA=5-hydroxy indole carboxylic acid, HVA= homovanillic acid, 3-MT= 3-methoxy trptamine, 5-HT= serotonin.

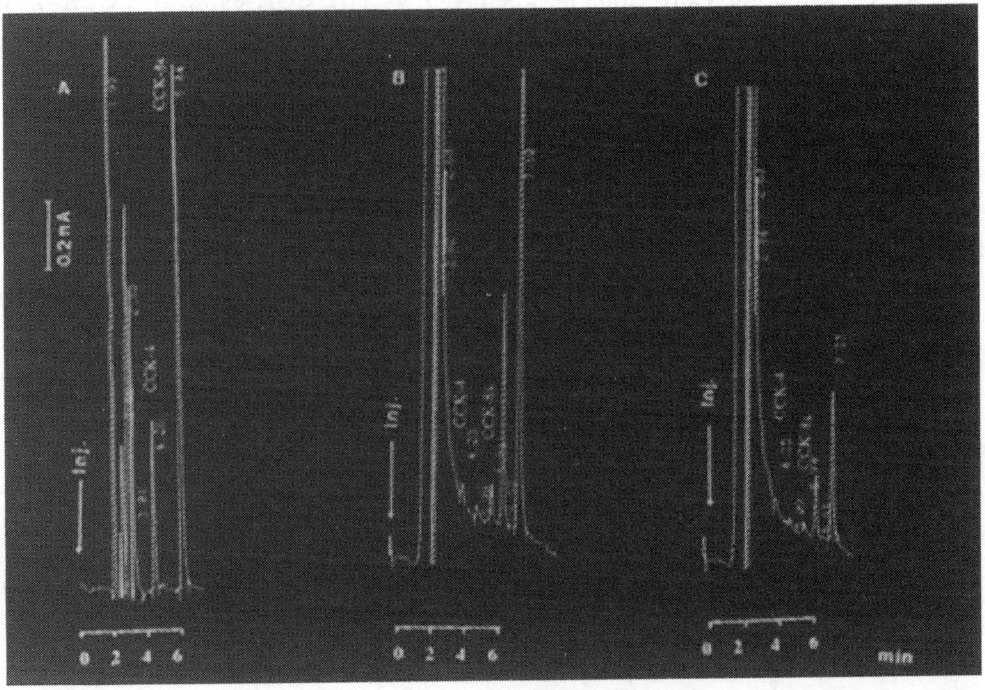

Fig. 2. RP-HPLC separation of synthetic peptides (A) CCK-4 and CCK-8s 20 and 100 pmol/l respectively and perchloric extracts of CSF sample from (B) healthy subject and a (C) patient with TB meningitis. HPLC conditions are identical as given the reference (Qureshi etal., 1993)

Data are presented as mean ±SEM and the differences in concentration of catecholamines, indoleamines, CCK-4, CCK-8s, NPY and SP were analyzed with ANOVA and group comparison were made with t-text. A p-value less than 0.05 was considered significant. Correlation between NPY vs. NA, CCK-8s vs. DA, CCK-4 vs, NA and SP vs. 5-HT were submitted to a stepwise polynomial regression analysis (*Zar, 1984*).

MONOAMINES AND THEIR METABOLITIES

The catecholamines, dopamine (DA), noradrenaline (NA) and the indoleamine, 5-hydroxytryptamine (5-HT) act as important neurotransmitters in the brain.

Abnormalities of biogenic amine metabolism in the central nervous system (CNS) have been implicated in various psychiatric and neurological disorders. (*Gottfries, 1990: Davis et al., 1991*). The estimation of monoamine and indoleamine are shown to be helpful for the diagnosis and interpretation of these disorders (*Burns et al., 1985; Roy et al., 1986*).

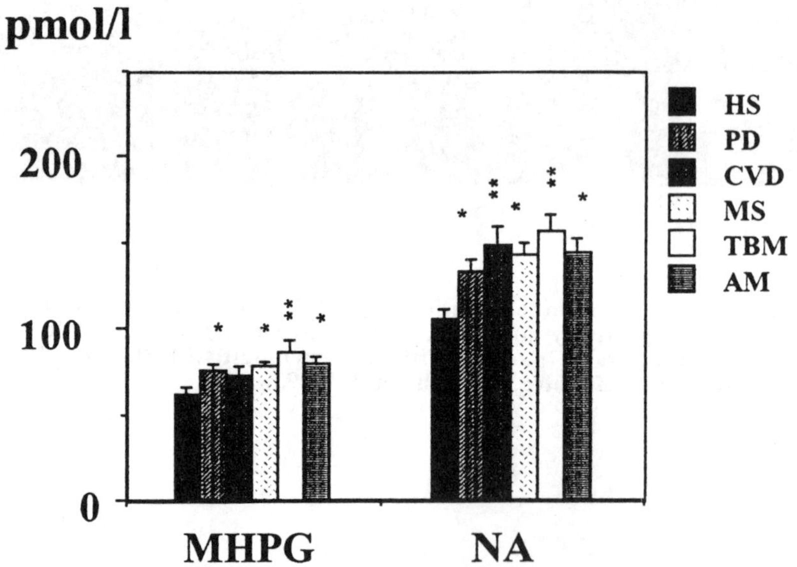

Fig. 3. CSF levels of NA and its metabolite MHPG in various neurologic patients and healthy subjects. All values are expressed as mean ± SEM. *p<0.05, **p<0.01.

DA and NA are synthesized from the amino acid tyrosine derived from food intake or protein break down. The main metabolites of DA are homovalinic acid (HVA) and 3,4-dihydroxyphenylacetic acid (DOPAC) and of NA, 3-methoxy-4-hydroxyphenylglycol (MHPG). The highest concentrations of DA are observed in basal ganglia and of NA in the hypothalamus. The major inactivation of monoamines is due to an active reuptake of the parent substances from the synaptic cleft into the presynaptic bouton. Apart from this reuptake, neurotransmitters and their metabolites in the synaptic cleft are cleared to the cerebrovenous blood. However, a small fraction diffuses into the CSF (*Björklund et al.,1984*). Our results showed (*Baig et al., 1997*) that the CSF levels of NA and its metabolite MHPG were increased in all patients with neurologic disorders (Fig. 3) whereas DA was decreased in patients with PD, MS and TBM and unchanged level were found in CVD and AM patients (Fig. 4).

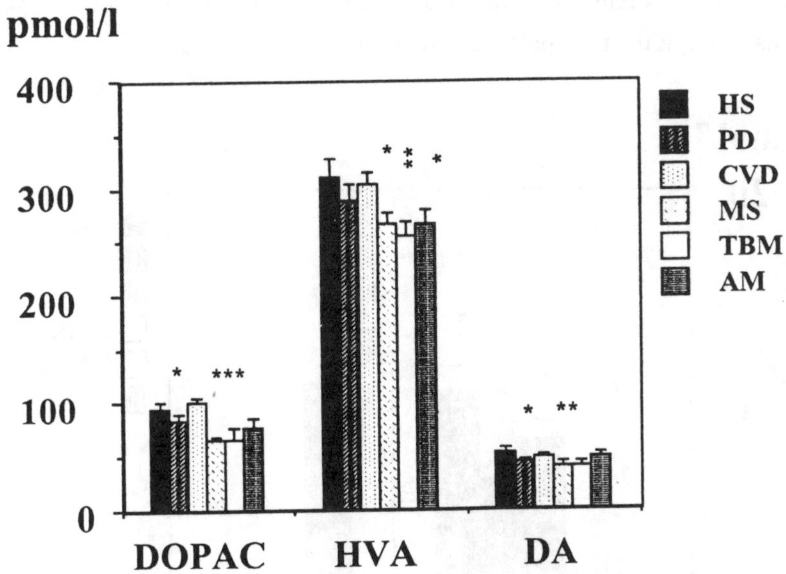

Fig. 4. CSF levels of dopamine (DA) and its metabolites DOPAC and HVA in various neurologic patients and healthy subjects. All values are expressed as mean ± SEM. *$p<0.05$, **$p<0.01$.

Its metabolites , HVA and DOPAC were decreased in all patients except CVD (Fig. 4). The elevated CSF level of NA and its metabolite MHPG in these patients indicates that the enzymes (monoamineoxidase and catechol-O-methyl transferase) converting NA into MHPG are activated under these patients. The role of CNS and influence of various neurotransmitters on the immune systems is well documented. The CNS activity could also be influenced by immune system. These findings suggest the importance of neurotransmitters in MS and PD patients since MS is an immune modulated disease where as PD is a disease which clearly shows DA deficiency in the brain. Apart from catecholamines, serotonin (5-HT) plays an important role in CNS. 5-HT is synthesized from the amino acid tryptophan (TRP) and is metabolized mainly to 5-hydroxyindoleacetic acid (5-HIAA). 5-HT neurons originate from the raphe nuclei in the brain stem and adjacent nuclear groups and, as the noradrenergic system, project widely to most areas in the brain (*Copper et al., 1991*). The highest concentration of 5-HT is found in subcortical nuclei (*Wester et al., 1990*). 5-HIAA and 5-HT are known to pass through the BBB but the rate of TRP entry in the brain is dependent on its availibility in the blood and its relationship with other large neutral amino acids through the transport system (*Fernstrom et al., 1972*).

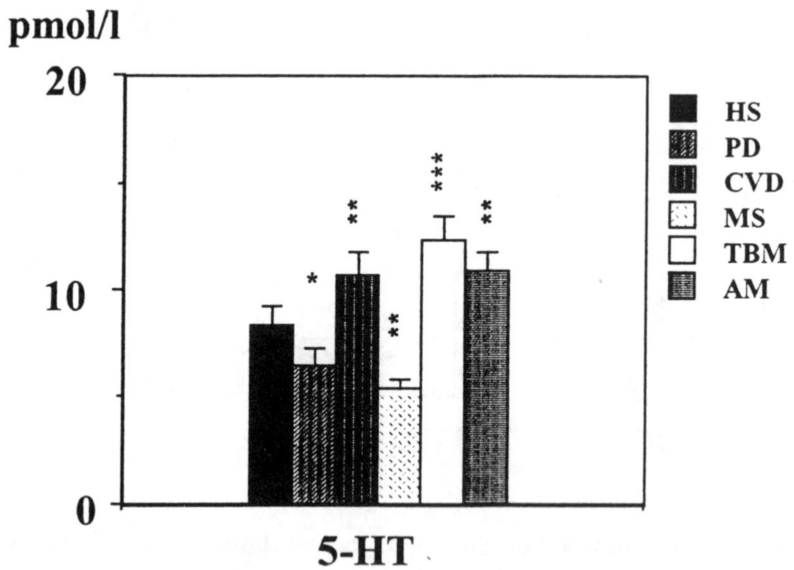

Fig. 5. CSF levels of serotonin (5-HT) in various neurologic patients and healthy subjects. The value is expressed as mean ± SEM. *p<0.05,p<0.01,***p<0.001.**

5-HT is decreased in MS, PD patients whereas increased in CVD, TBM and AM patients (Fig. 5) suggesting that under different neurologic disorders, the enzymes (tryptophan hydroxylase and aromatic amino acid decarboxylase) converting TRP into 5-HT are affected , however, the enzyme (aldehyde dehydrogenase) converting 5-HT into 5-HIAA is not affected since the level of 5-HIAA was unchanged under all these neurological conditions except in PD patients as previously shown (*Baig et al., 1997.* The level of TRP is increased in CVD, MS and AM patients and decreased in PD patients however, unchanged level were observed in TBM patients . Similar pattern was observed for 5-HIAA (Fig. 6).

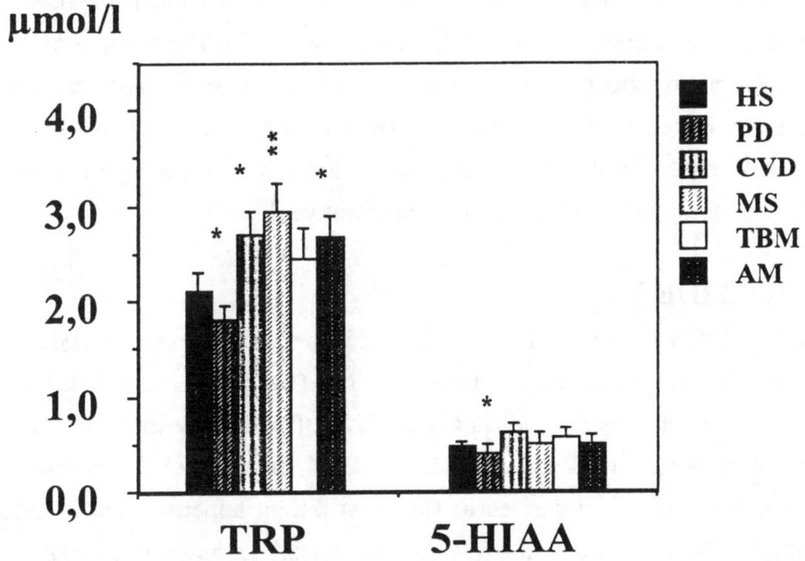

Fig. 6. CSF levels of tryptophan (TRP) and 5-hydroxy indole acetic acid (5-HIAA) in healthy subjects and patients with neurologic disorders.The values are expressed as mean±SEM.*p<0.05, **p<0.01.

The CNS activity could also be influenced by immune system. The possibility that alteration of neurotransmitter production and turn over may play a significant role for either the CNS or systamatic changes during these diseases has received little attention. Only a study in infant rats with meningitis documented an increase in fore brain NA and DA levels at the time of acute infection and evidence of persistent perturbation of monoamines neuronal transmission in adult rats surviving the disease (*Konkol et al., 1987*) . These authors hypothesized that the neurotransmitters changes may explain certain neurological saquelae, such as motor hyperactivity that are observed following meningitis.

NEUROPEPTIDES IN DEGENERATIVE DISEASES.

The presence of receptor-active opioid peptides in human cerebrospinal fluid was demonstrated more than 20 years ago (*Terenius et al., 1975*). One of the major reasons for analysing CSF peptides instead of plasma and serum is that neuropeptide do not readily cross the blood-brain barrier and CSF due to constant exchange with the nervous tissue rather than plasma can be anticipated to contain peptides which derive from the CNS (*Terenius et al., 1975; Terenius, 1982*). The changes in the CSF peptide concentration might be attributed to certain symptoms and can even be interpreted as characteristic features of the pathological conditions (*Nemeroff, 1991*). Neuropeptide Y, cholecystokinin and substance P are some of the important peptides which play vital biological roles in various degenerative disorders.

NEUROPEPTIDE Y

Neuropeptide Y (NPY) is a peptide consisting of 36 amino acid residues and has been shown to modulate a number of functions of CNS (*Heilig et al., 1990*). NPY has several physiological effects (*Dumont et al., 1992*) of which the most potent are an increase in vasoconstriction (*Grundemer et al., 1994*) and an increase in food intake (*Lee et al., 1994).* NPY is described as the most potent appetite stimulator known (*Dumont et al., 1992*) . It is found in high concentrations in several regions of the brain including nuclei of brain stem and nerve fibers surrounding cerebral vessels has been proposed to play a role in regulating cerebral blood flow (CBF) and systemic vegetative functions. Vascular inflammation resulting in spasm and thrombosis has been documented during meningitis by histopathology and angiography and is likely to lead to focal reductions of CBF (*Gado et al., 1974*). Clinical and experimental studies have further documented that autoregulation of CBF is lost during meningitis , making CBF directly dependent on cerebral perfusion pressure (*Paulsen et al., 1974*). Both occurrence of vasospasm and the loss autoregulation suggest that the regulation of cerebral vascular tone may be disturbed during meningitis. It is also demonstrated its coexistence with NA (*Hökfelt et al., 1983*). NPY plays an important role in anxiety, depression and eating disorders. Depression is a multifactorial processes (*Everitt et al., 1984*) therefore, it is of interest to study the coexistence of NPY with NA, 5-HT and DA systems irrelevant to depression since various neurologic patients such as MS and PD show sign of depression commonly.

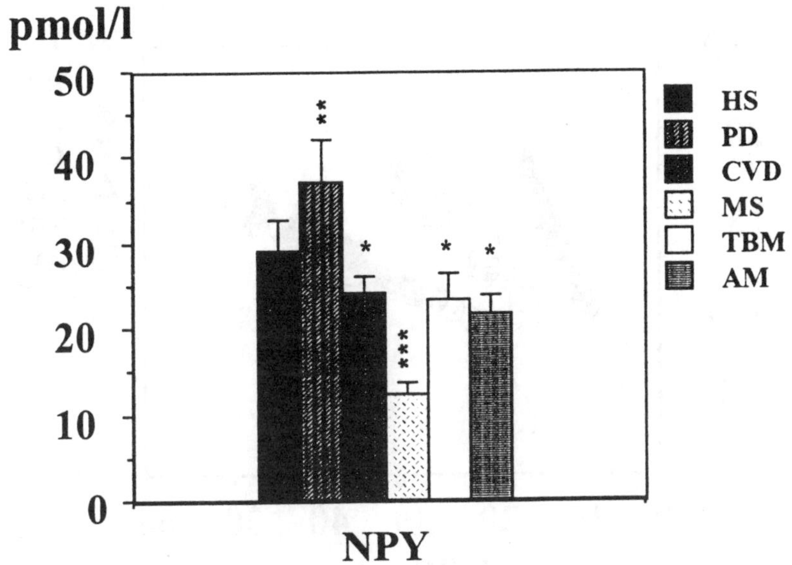

Fig. 7. CSF level of NPY in various neurologic patients and healthy subjects. The value is epressed as mean ± SEM. *p<0.05, **p<0.01, ***p<0.001.

Low levels of NPY in CSF has been reported in depressed patients, particularly in anxiety (*Widerlöv et al., 1988*). Among our patients only PD patients showed increased CSF level of NPY (Fig. 7) whereas all other groups showed significant decreased in CSF level of NPY (*Qureshi et al. 1997b*). The observed reduced NPY concentrations in CSF of MS, CVD, and meningitis patients are of interest since the infection of the CNS due to meningitis and their potential relationship to the local and systematic consequences could be the reason for this change in NPY levels in infection disorder whereas the reduction as is observed in CVD and MS patients could be attributed to the reduction in CBF. In larger vessels, NPY has been shown to exert powerful vasoconstrictive properties in vivo and vitro (*Suzuki et al., 1988*) , a finding that has led to the hypothesis that NPY may contribute to the vasospasm associated with subarachnoid hemorrhage (*Jackowski et al., 1989*). Few studies have shown decreased level of NPY in Alzheimer disease (*Alom et al., 1990*) , MS (*Qureshi et al., 1996; Maeda et al., 1994*) , unchanged levels in PD (*Bednar et al., 1996*) and increased levels in patients with subarchoid hemorrhage (*Susuki et al., 1989*).

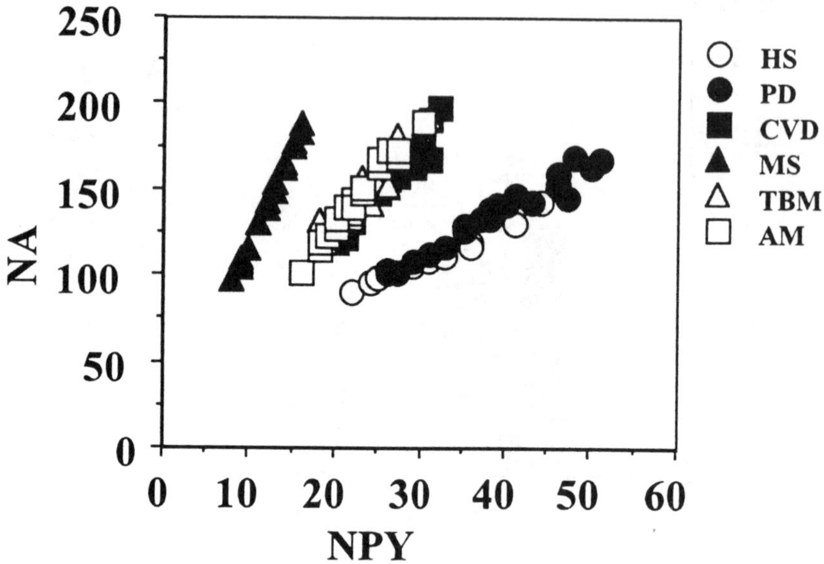

Fig. 8. Relationship between NPY and NA in CSF of HS and patients.

Treatment with a specific blocker of the NPYY1 receptor, resulted in an anxiety-type behaviour in rats (*Wahlestedt et al., 1989*). Thus, a link between anxiety and NPY seems to exist. There are several lines of evidence which suggest that NPY is a neurotransmitter or a neuromodulator since it causes contraction of cerebral arteries (*Ekland et al., 1984*). A lack of rostral caudal gradient of NPY has been demonstrated (*Berrettini et al., 1986*), however, NPY is widely distributed in the brain and spinal cord (*Allen et al., 1984*). The lack of a gradient may suggest that the spinal cord contributes to the concentration of the peptide in lumbar CSF. Similar results for other peptides including neurotensin,CCK,VIP have been reported (*Tamminga et al., 1985*). Studies in human have shown (*Allen et al., 1984*) significant concentrations of NPY at all levels of the spinal cord. Thus the lack of a CSF rostrocaudal gradient for NPY indicates a spinal cord or peripheral origin for a fraction of CSF NPY levels. The correlation between the levels of NPY and NA shows a linear relationship for HS as well as all patient groups. Among these patients , PD show the similar tendency of linearity as HS whereas other patient group do hold the linearity depending on decrease in NPY level (Fig. 8). This relationship does indicate that irrespective of decrease in NPY levels which is clearly seen in MS patients, linearity between these two values is intact

CHOLECYSTOKININ

Cholecystokinin (CCK) peptides constitutes a family of homogenous peptides, 4-58 amino acids in length, and derived from their precursor, preprocholecystokinin. CCK is a neuroactive peptide which exists in various forms and exhibits a variety of perpheral and CNS actions (*Beinfed,1983*). Besides, CCK-peptides attract considerable interest in research due to their part in termination of food intake, dopamine regulated functions and pain mechanism (*Albus, 1988; Bradweijn etal., 1984*).

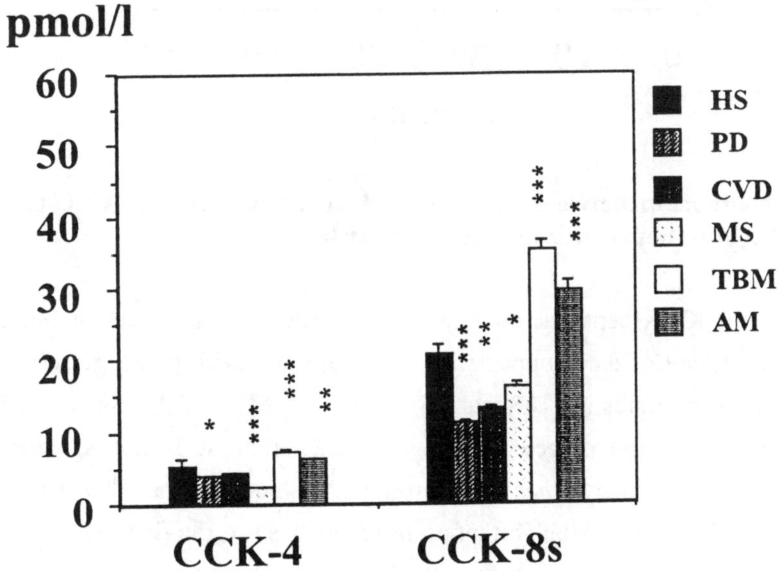

Fig. 9. CSF levels of CCK-4 and CCK-8 in various neurologic patients. All these values are expressed as mean ± SEM. *p<0.05, **p<0.01, ***p<0.001.

While some of these effects are mediated by the CCK octapeptide sulphate (CCK-8s), other CCK peptides have different effects. CCK-8 is believed to be the most prevalent form of CCK in the CNS. It appears that CCK-8s and various analogs may modulate CNS dopamine neurotransmission and have been studied as potential antipsychotic agent (*Harro et al., 1990*).

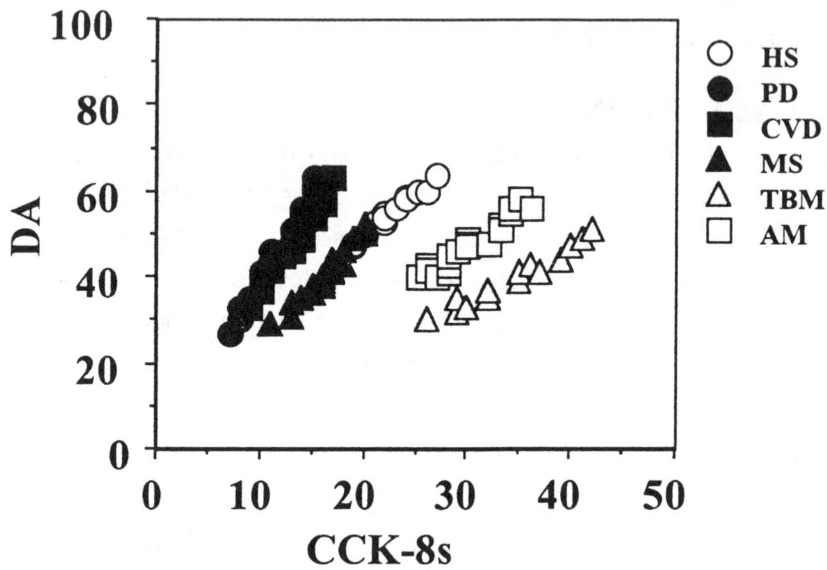

Fig. 10. Relationship between CCK-8s and dopamine (DA) in CSF of healthy subjects and in patients with neurologic disorders

Among the other CCK peptides, its tetrapeptide (CCK-4) is anxiogenic in humans and considerable amount of data supports its role in anxiety with panicogenic properties with minimal gastrointestinal effects (*Bradwejn et al., 1990*). Since CCK-ergic neurotransmition can be detected throughout the brain, it is not surprizing that interaction occur with many other neurotansmitter systems and that CCK has a variety of functional implications. Much attention has already since the first years of research on CCK in the CNS, has been drawn to the interactions between CCK and dopamine, and the role for CCK in schizopherenia has been suggested. Interactions with dopamine and with several other neurotransmitters such as GABA, serotonin and noradrenaline are also of interest with regard to anxiogenic-like effects, its involvement in cognitive processes and the CCK-ergic modulation of opioid actions, and of pain perception. Other functional role for scentral CCK (octa peptide, CCK-8s) are as a mediator of satiety responses and as a regulator of sexual and maternal behaviour and of seizure activity. As may be expected from the involvement of CCK in several of these processes, CCK has also been shown to interact with other peptides and steroid hormones

and CCK apparently has a role as a modulator of learning and memory (*Itoh et al., 1990)*. The involvement of CCK (tetrapeptide, CCK) in anxiety responses is supported by numerous reports based on animal studies as well clinical observation (*Harro et al., 1993*). Previously, we have shown (*Qureshi et al., 1996)* decreased levels of CCK-4 and CCK-8s in MS patients, however, the results in patients with PD, mood disorder, eating disorders have indicated inconsistent observations (*Albus, 1988*). Our previous study showed (*Qureshi et al., 1997b*), decreased level of CCK-4 in PD and MS patients whereas increased in TBM and AM patients. The level of CCK-8s shows decreased in PD,CVD and MS patients and increased in TBM and AM patients (Fig. 9).

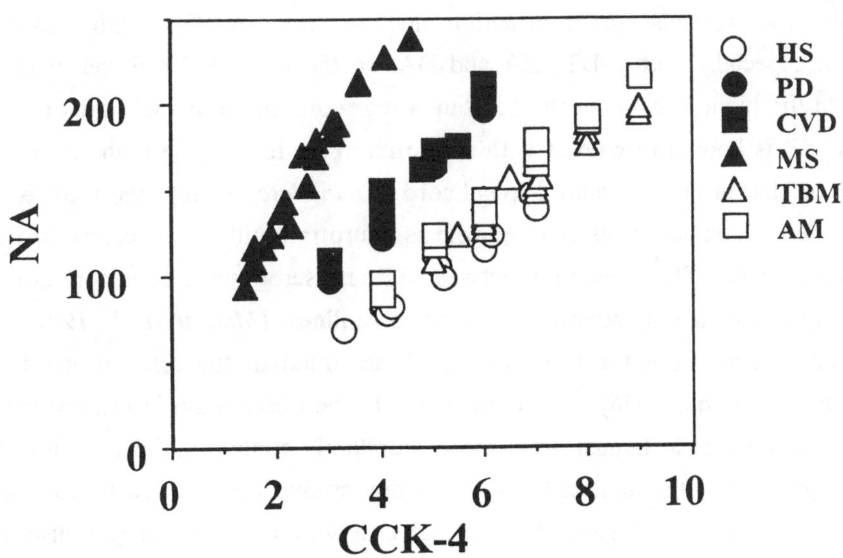

Fig. 11. Relationship between CCK-4 and NA in CSF of healthy subjects and patients with neurologic disorders.

A linear relationship was found between CSF CCK-8s and DA in healthy subjects closely followed by MS patients, whereas despite of the decreased level of DA in AM and TBM, the linearity did remain intact. The same is true for PD and CVD patients where the CSF level of CCK-8s decreased significantly but the linearity remained intact (Fig.10).

The results also show a linear relationship between CCK-4 and NA for healthy subjects closely followed by AM and TBM patients (Fig. 11). Despite of decrease in the level of CCK-4 for MS, PD and CVD patients, the relation between these two values does show a linearity indicating the dependence of these two transmitters on each other.

SUBSTANCE P

Substance P (SP) is an undecapeptide which derives from alpha, beta and gamma preprotachykinin gene transcripts and is neurotransmitter or neuromodulator of primary nociceptive afferents (*Nyberg et al., 1988*). SP is the most well known tachykinin and was for many years , the only one known to exist in mammalian tissues. SP have received great attention due to its interaction with classical transmitters, especially with 5-HT, NA and DA. In the CNS, SP is found in most regions with the highest levels in the substantia nigra and in the dorsal horns of the spinal cord. SP is known to coexist with serotonin in neurons in the raphe nuclei in the brain stem that project to ventral spinal cord (*Chan-Palay et al., 1988*). SP is an undecapeptide with an almost established role as neurotransmitter or neuromodulator (*Otsuka et al., 1993*). The interaction between SP and serotonin is of importance in view of many studies linking serotonin to depressive illness (*Meltzer et al., 1987*). In pathways descending from the raphenuclei, SP is found in the same neurons as serotonin (*Hökfelt et al., 1978*). In accordance with the observation in rat, the basal ganglia are the regions in human brain most abundantly containing SP-L1 with the substantia nigra having the highest level . Very few studies are directed to compare the levels of SP in neurodegenerative diseases, however in Huntington disease, patients showed *(Emson et al., 1980*) that SP-L1 levels in caudale nucleus and putaman are unchanged while in globus pallidus and substantia nigra are substantially reduced. Brains from patients suffering from Parkinson's disease also seems to have a decreased levels of SP in globus pallidus and substantia nigra as compared to healthy controls *(Mamborgne et al., 1983)* .Unilateral injection of SP into substantia nigra of the rat evolves dose-dependent contral lateral rotational behaviour possibily indicating activation of the ipsilateral nigro-stariatal dopamine pathway (*Olpe et al., 1977*). Furthermore, the behavioural excitation induced by a bilateral injection of SP into substantia nigra is abolished in rats with 6-OH DA lesion of the nigro-striatal dopamine pathway suggesting a possible excitotory effect of SP on DA neurons (*Kelley, 1979*).

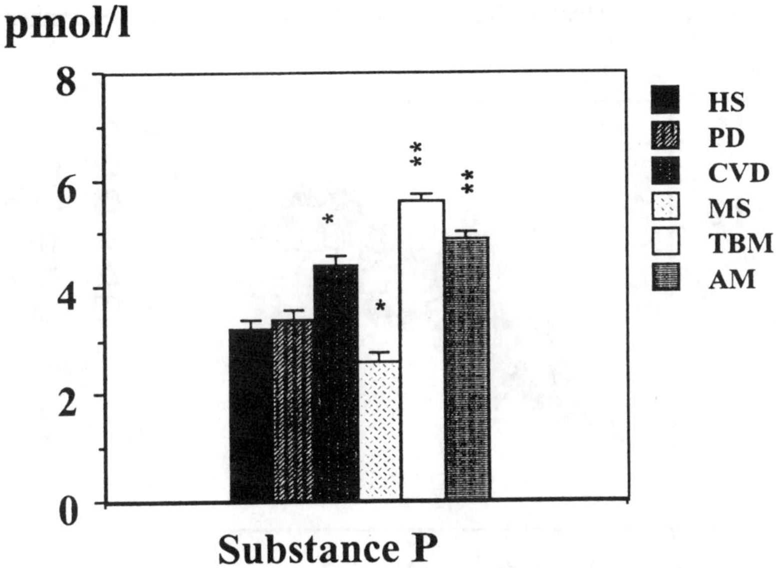

Fig. 12. CSF level of substance P in various neurologic patients. The level is expressed as mean ± SEM. *p<0.05, **p<0.01.

Very few studies are conducted to study on the levels of SP in CSF from the patients with neurologic disorders, however, since SP plays a very important role in the CNS, it is very likely that CSF would provide us with its role in various neurologic disorder. Our results (*Qureshi et al, 1997a, 1997b*) showed, increased CSF level of SP in CVD, TBM and AM patients and decreased level in MS patients (Fig. 12).

A liner relationship is found in CSF between SP and 5-HT for healthy subjects as well as in all patients groups (Fig. 13) showing clearly that irrespective of different neurologic disorders , this relationship and its linearity is intact.Similar results were shown previously where not only the linearity of SP with 5-HT was shown but the relationship between SP with NA and DA also showing relationships with linearity

(*Qureshi et al., 1997b*). A central problem relating to coexistence is if, and how, the release of coexisting transmitters can be differenially regulates. One possible mechanism may be differenes in the frequency dependence of the various transmitter pools (*Lundberg et al., 1983*).

The release of classical transmitter with low molecular weight is generally beleived to depend on the frequency by which action potentials invade the nerve terminal.

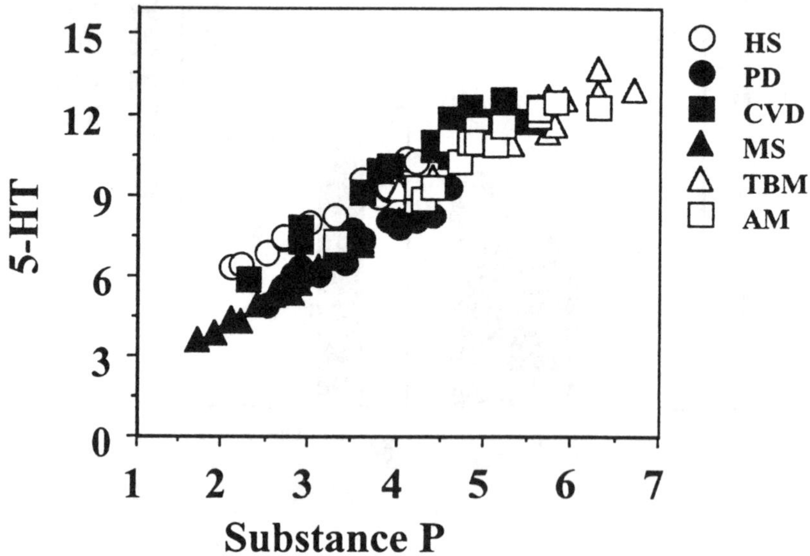

Fig. 13. Relationship between SP and 5-HT in the CSF of healthy subjects and patients with neurologic disorders.

Various studies (*Andersson et al., 1982; Bloom et al., 1984; Lundberg et al., 1986; Agoston et al., 1988*) have shown that the release of neuropeptide as calculated per pulse recurred at high frequencies compared to what usually is needed to evoke the release of classical small molecule transmitters. In addition neuropeptides were preferentially released by intermittant periods of stimulation at high frequency compared to continuous stimulation. However, it is quite clear that despite of various characterstics of these neurologic disorders, the relationship showing linearity between NA,DA and 5-HT with SP remained intact. An interesting prospective on amine/ peptide cotransmission has emerged due to the finding that chronic treatment with tricyclic antidepressand drug results in a marked increase in the tissue levels and release of SP from rat ventral spinal cord (*Brodin et al., 1984*) suggesting that the mode of action of antidepressant drugs may be related to the changes in the neuropeptide neurotransmission.

FUTURE CLINICAL IMPLICATIONS

These results undoubtedly provided uswith a possibility in understanding the interrelationship between different neurotransmitters in the brain as well as their distribution and functions. Mapping of neuropeptide genes and the studies of their linkage could provide clues to genetic factors involve in these diseases. On the basis of these results on degenerative disorders, the role of various neurotransmitters, and the role of neuropeptides could be clearly defined in order to develop effective drug therapy.

ACKNOWLEDGEMENT

The author is thankful to Dr.Shahid Baig of Aga Khan University Hospital, Karachi for kindly supplying with CSF samples from TBM patients.

REFERENCES

Agosto, D.V. and Whittaker, V.P. (1989) Characterization by size, density , osmotic fragility and immunoaffinit, of acetylcholine and vasoactiev intestinal polypeptide-containing storage particles from myentric neurones of guinea-pig. *J.Neurochem.* **52**, 1474-1480.

Albus, M. (1988) Cholecystokinin. *Prog.Neuro-Psychopharmacol & Biol.Psychiat.* **12**, S5-S21.

Allen, J.M., Gibson, S.J., Adrian, T.E., Polak,J.M. and Bloom, S.R. (1984) Neuropeptide Y in human spinal cord. *Brain Res.* **308**, 145-148.

Alom, J.,Galard,R.,Catalan,R.,Castellanos,J.M.,Schwartz,S. and Tolosa,E. (1990) cerebrospinal fluid neuropeptide Y in Alzheimer's disease. *Eur.Neurol.* **30**, 207-210.

Andersson, P-O, Bloom, S.R., Edward, A.V. and Järhult, J. (1982) Effects of stimulation of tympani in bursts of submaxillary responses in the cat. *J.Physiol..* **322**, 469-483.

Baig, S., Halawa,I. and Qureshi,G.A. (1991) HPLC as a tool in defination of abnormalities in monoamines and tryptophan metabolites in CSF from patients with neurological disorders*Biomed.Chromatogr.* **5**, 108-112.

Baig, S. and Qureshi, G.A. (1997) Catecholamines, tryptophan and their metabolites

in CSF of patients with neurologic disorders. *Biogenic Amines* (in press)
Bartfai, T., Ivenfeldt, K., Brodin, E. and Ögren, S-O (1986) Functional consequences of coexistence of classical and peptide neurotransmitters. *Prog.Brain Res.* **68**, 321-330.
Bednar, I., Södersten, P. and Qureshi, G.A. (1992) A simple HPLC method in monitoring catecholamines, tryptophan, and the metabolites in CSF of rats : a role of dopamine and noradrenaline in food intake. *J.Liq.Chromatogr.*, **15**, 3087-3096.
Bednar, I., Baig, S., Silberring, J., Stenfors, C., Södersten, P., Siden, Å. and Qureshi, G.A. (1996) Excitatory amino acids, monoamines, neuropeptide Y and Cholecystokinin in cerebrospinal fluid of Parkinson patients. *Biogenic Amine* **12(4)**, 275-284.
Beinfeld, M.C. (1983) Cholecystokinin in Central nervous system : A minireview. *Neuropeptides* **3**, 311-327.
Berrettini,W.H.,Nurnberger,J.I.Jr. and Domaggio,D.A. (1986) Neuropeptide Y immunoreactivity in human cerebrospinal fluid. *Peptides* **7**, 455-458.
Björklund, A. and Lindvall, O. (1984) Dopamine-containing systems in the CNS. In : Björklund,A. and Hökfelt, T. (eds). Handbook of Chemical neuroanatomy. Elsevier, Amsterdam, The Netherland.
Bloom, S.R., Edward, A.V. and Ghatei, M.A. (1984) Neuroendocrine response to stimulation of the splanchnic nerves in burst in the concious adrenalecotomized calf. *J.Physiol.* **346**, 519-531.
Bradwejn, J. r DeMontigny, C. (1984) Benzodiazepines antagonise cholecystokinin-induced activation of rat hippocampal neurons. *Nature* **312**, 363-364.
Brodin, E., Peterson, L-L, Ögren, S.O. and Bartfai, T. (1984) Chronic treatment with serotonin-uptake inhibitor zimelidine elevates substance P levels in rat spinal cord in vitro and in vivo. *Acta Physiol.Scand.* **122**, 209-211.
Brodin, F., Lindefors, N., dalsgaard, C-J, Theodorsson-Norheim, E. and Rosell, S. (1986) Tachykinin multiplicity in rat central nervous system as studied using antisera raised against substance P and neurokinin A. *Regul.Pept.* **13**, 253-272.
Burns, R.S., LeWitt, P.A., Ebert, M.H., Pakkenberg, H. and Kopin, J. (1985). The clinical syndrome of striatal dopamine deficincy. parkinsonism induced by 1-methyl 4-phenyl 1,2,3,6-tetra hydropyridine (MPTP). *N.Engl.J.Med.* **32**, 1418-1421.
Chan-Palay, V., Jonsson, G. and Palay, S.L. (1978) Serotonin and substance P coexist in neurons of the rat central nervous system. *Proc.Natl.Acad.Sci.* **75**, 1582-1586.
Chan-Palay, V. (1988) Galanin hyperinnervates surviving neurons of human basal nucleus of meynert in dementias of Alzheimer's and Parkinson's disease: A hypothesis for the role of galanin in accentuating cholinergic dysfunction in dementia. *J.Comp.Neurol.* **273**, 543-557.
Cooper, J.R., Bloom, F.E. and Roth, R.H. (1991) The biochemical basis of neuropharmacology. Oxford University Press, New York.
Davis, K., Kahn, R., Ko, G. and Davidson, M. (1991) Dopamine in schizophrenia :

a review and reconceptaulization. *Am.J.Phychiatry* **148**, 1474-1486.
Dumont, Y.J.C., Martel, A.., Fournier, S. Pierre, S. and Quirion, R. (1992)Neuropeptide Y and neuropeptide Y receptor subtypes in brain and peripheral tissues. *Prog. in Neurobiol.* **38**, 125-167.
Ekland,E., Edvinsson,L., Wahlestedt,C., Uddman,R., Hakanson, R. and Sundler,F. (1984) Neuropeptide Y coexists and cooperates with noraldrenaline in perivascular nerve fibers. *Regul.Pept.* **8**, 225-235.
Elliot, G.R. and Barchas, J.D. (1979) Neuroregulator : neurotransmitters and neuromodulators. *Behav.Brain Sci.* **2**, 423-424.
Emson, P.C., Arrengi, A., Clement-Jones, V., Sandberg, B.E.B. and Rossor, M. (1980). Regional distribution of methionine-enkephalin and Substance P-like immunoreactivity in normal human brain and in Huntington's disease. *Brain Res.* **199**, 147-160.
Everitt, B.J.,Hökfelt, T,Terenius, L.,Tatemoto, K.,Mutt, V. and Goldstein, M. (1984) Differential co-existence of NPY-immunoreactivity with catecholamines in the central nervous system of the rat. *Neuroscince* , **11**, 443-463.
Ferstrom, J.D. and Wurtman, R.J. (1972) Brain serotonin content : Physiological regulation of plasma neutral amino acids. *Science* **178**, 414-416.
Futrell, N., Schultz, L.R. and Malliken, C. (1992) Central nervous system disease in patients with systemic lupus erythematosus. *Neurology* **42**, 1649-1657.
Gado, M., Axley, J., Appleton; D.B. and Prensky, A.L. (1974) Angiography in the acute and post-traetment of Haemophilus Influenza meningitis. *Radiology* **10**, 429-444.
Gottfries, C. (1990) Brain monoamines and their metabolites in dementia. *Acta Neurol.Scand. (Suppl)* **129**, 8-11.
Grundermar, L. and Håkansson, R. (1994) Neuropeptide Y effector systems: perspectives for drug development. *Trends in Pharmacol.Sci.* **15**, 153-159.
Halawa, I., Baig, S. and Qureshi, G.A. (1991) Application of HPLC in defining the abnormalities in patients with aseptic meningitis. *J. Biomed. Chromatogr.*, **5**, 216-222.
Harro,J.,Kiivet,R.A., Lang,A. and Vasar,E. (1990) Rats with anxious or non-anxious type of exploratory behavior differ in their brain CCK-8 and benzodiazepine receptor characterstic. *Behav.Brain Res.* **39**, 63-71.
Harro, J., Vasar, E. and Bradwejn, J. (1993) CCK in animal and human research on anxiety. *Trends Pharmacol.Sci.* **14**, 244-249.
Heilig,M. and Widerlov,E. (1990) Neuropeptide Y: an overview of central distribution, functional aspects and possible involvement in neuropsychiatric illness. *Acta Psychiatr.Scand.* **8**, 295-314.
Hendry, S.H.C., Jones, E.G., DeFelipe, J., Schmechel, D., Brandon, C. and Emson, P.C. (1984) Neuropeptide-containing neurons of cerebral cortex are also GABAergic. *Proc.Natl.Acad.Sci.* **81**, 6526-6530.
Hökfelt, T., Ljungdahl, Å., Steinbusch, H., Verhofstad, A., Nilsson, G., Brodin, E., Pernow, B. and Goldstein, M. (1978), Immunohistochemical evidence of Substance P-like immunoreactivity in some 5-hydroxytryptamine-

containing neurons in the rat CNS, *Neurosci.* **3**, 517-538.
Hökfelt, T., Johansson O., Ljungdahl, Å., Lundberg, J.M. and Schultzberg, M. (1980) Peptidergic neurones. *Nature* **284**, 515-521.
Hökfelt, T., Lundberg, J.M., Lagercrantz, H., Tatemoto, K., Mutt, V., Lindberg, J. Terenius,L., Everitt, B.J., Fuxe, K., Agnati, L. and Goldstein,M. (1983) Occurrence of neuropeptide Y (NPY)-like immunoreactivity in catecholamine neurons in human medulla oblongata. *Neurosci. Lett.* **36** , 217-222.
Hökfelt, T., Millhorn, D., Seroogy, K., Tsuruo, Y., Ceccatelli, S., Lindh, B., Meister, B., Malander, T., Schalling, M., Bartfai, T. and Terenuis, L. (1987) Coexistence of peptides with classical transmitters. *Experientia* **43**, 768-780.
Hökfelt, T., Bean, A., Ceccatelli, S., Dagerlind, Å., Elde, R.P., Goldstein, M., Meister,B., Melander, T., Nicholas, A.P., Pelto-Huikko, M., Pieeribone, V., Schalling, M., verge, V. and Zhang, X. (1992) Neuropeptide and classical transmitters. *Arneim.Forsch./ Drug Res.* **42**, 196-201.
Itoh, S. and Lal, H. (1990) influences of Cholecystokinin and analogues on memory processes. *Drug Dev.Res.* **21**, 257-276.
Jackowski, A., Crockard, A. and Burnstock, G. (1989). Alteration in serotonin and NPY content of cerebrovascular sympathetic nerves following experimental subarchnoid hemorrhage. *J.Cereb.Blood Flow Metab.* **9**, 271-279.
Kelley, A.E. and Iversen, S.D. (1979). Substance P infusion into substantia nigra of the rat : Behaviour analysis and involvement of striatal dopamine. *Eur.J.Pharmacol.* **60**, 171-179.
Konkol, R.J., Chapman, L., Breese, G.R., Collier, H.M., Kilts, C., Vogel, R.R., Mailman, R.B. and Bendeich, E.G. (1987). Haemophilus Influenza Meningitis in the rat : Behaviour electrophysiological and biochemical consequences. *Ann.Neurol.* **21**, 353-360.
disorder in Huntington's disease. *Neurology* **44** (suppl.2), A338.
Lang, A.E. and Fahn, S. (1989) Assessment of Parkinson's disease. In: Munsat, T.L. (ed), *Quantification of Neurologi Deficit*. Butterworths, Boston, pp. 285-309.
Lee, M.C., Schiffman, S.S. and Pappas, T.N. (1994) Role of neuropeptides in the regulation of feeding behaviour: a review of cholecystokinin, bombesin, neuropeptide Y and galanin. *Neurosci.Biobehaviour Rev.* **18**, 313-323.
Link, H. and Tibbling, G. (1977) Priciples of albumin and IgG analyses in neurological disorders, III. Evaluation of IgG synthesis within the central nervous system in multiple sclerosis. *Scand.J.Clin.Lab.Invest.* **37**, 297-301
Lundberg, J.M. and Hökfelt, T. (1983) Coexistence of peptides and classical transmitters, *TIN* **6**, 325-33.
Lundberg, J.M., Rudehill, A., Sollevi, A., Fried, G. and Wallin, G. (1986) Frequency- and reserpine-dependence chemical coding of sympathetic transmission : differential release of noradrenaline and neuropeptide Y from pig spleen. *Neurosci.Lett.* **63**, 96-100.
Maeda,K., Yasuda,M.,Kaneda,H.,Maeda,S. and Yamadori,A. (1994) cerebrospinal fluid Neuropeptide Y and Somatostatin-like immunoreactivities in man. *Neuropeptide* **27**, 323-332.

Mauborgne, A., Javey-Agid, J., Legrand, J.C., Agid, Y. and Cesselin, F. (1983). Decrease of Substance P-like immunoreactivity in the substantia nigra and pallidum of Parkinsonian brains. *Brain Res.* **268,** 370-373.

McKann, S.M. and Weiner, R.I. (1987) *Integrative Neuroendocrinology: Molecular and Clinical aspects.* Karger, Basel.

Meltzer, H.Y. and Lowy, M.T. (1987) the serotonin hypothesis of depression. In: Meltzer,H.Y (Ed) *Psychopharmacology: The third generation of progress*, Raven Press, New York, pp 513-526.

Nemeroff, C.B. (1991) Neuropeptides and Schizopherenia: A critical review. In : *Advances in Neuropsychiatry and Psychopharmacology. Volume 1: Schizophrenia Research. C.A.Tamminga and S.C.Schultz (Eds) , Raven Press, New York,* pp 77-89.

Nyberg, F., Vaeroy, H. and Terenius, L. (1988). Opioid peptides and Substance P in the CSF. Regulation and significance to pain. In Olesen, J and Edvinsson, L. (eds) *Basic Mechanisms of Head ache,* Elsevier, Amsterdam, Holland, pp 241-258.

Olpe, H.R. and Koella, W.P. (1977). Rotatory behaviour in rats by intranigral application of Substance P and an eledoisin fragment. *Brain Res.* **126**, 576-579.

Otsuka, M. and Yoshioka, K. (1993) Neurotransmitter functions of mammalian tachykinins. *Physiol.Rev.* **73(2)**, 229-308.

Paulson, O.B., Brodersen, P., Hansen, E.L. and Kristensen, H.S.. (1974) Regional cerebal blood flow cerebral metabolic rate of oxygen, and CSF acid-base variable in patients with acute ningitis and with acute encephalitis. *Acta Med.Scand.* **196,** 191-198.

Qureshi, G.A and Baig, S.M. (1988) Quantitation of amino acids n biological samples by HPLC. *J.Chromatogr.* **459**, 237-244.

Qureshi,G.A.,Bednar, I., Min, Q., Södersten, P., Silberring, J., Nyberg, F. and Thörnwall, M. (1993) Quantitation and identification of two cholecystokinin peptides, CCK-4 and CCK-8s in rat brain by HPLC and fast atom bombardment mass spectroscopy. *Biomed.Chromatogr.* **7**, 251-255.

Qureshi,G.A., Baig,S., Bednar, I., Södersten, P., Forsberg, G. and Siden, Å. (1995) Increased cerebrospinal fluid concentration of nitrite in Parkinson's disease. *NeuroReport* **6,** 1642-1644.

Qureshi, G.A., Halawa, A. and Baig, S. (1996) Multiple sclerosis and Neurotransmission. *Biogenic Amines* **12(5)**, 353-376.

Qureshi, G.A. and Baig, S. (1997a) Cerebrospinal fluid levels of neuropeptides in neurologic patients. *Biochem.Mol.Biol. Int.* (in press)

Qureshi, G.A. and Baig, S. (1997b) Coexistence of SP with catecholamines in CSF of healthy and patients with various neurologic disorders. *Biochem.Mol.Biol.Int.* (in press)

Roy, A., Pickar, D., Linnoila, M., Doran, A.R. and Paul, S.M. (1986). CSF monoamine metabolite levels and dexamethasone suppression test in depression : relationship to life events. *Arch.Gen.Psychiatry* **43**, 356-360.

Schumacher G.A., Becke G., Kibler R.E., Kurland L., Kurtzke J.,McDowell F.,Nogler B., Sibley W., Tourtellotte W.W., Willmon T.(1965). Problem on the experimental trails of therapy in Multiple Sclerosis : report by a panel on the evaluation of experimental trails of therapy in Multiple Sclerosis. *Ann. N.Y. Acad. Sci.* **122**, 552-558.

Starke, K., Göthert, M. and Kilbinger, H. (1989) Modulation of neurotransmitter release by presynaptic autoreceptors. *Physiol.Rev.* **69**, 864-989.

Sörnäs, R. (1967) A new method for the cytological examination of the cerebrospinal fluid. *J.Neurol.Neurosurg.Psych.* **30**, 568-577.

Susuki, Y., Shibuya, M., Ikeyaki, I., Satoh, S., Takayasu, M. and Asano, T. (1988). Effects of NPY on canine cerebral circulation. *Eur.J.Pharmacol.* **146**, 271-277.

Susuki,Y.,Sato,S. and Susuki,S. (1989) Increased neuropeptide Y concentrations in cerebrosphinal fluid from patients with aneurysmal subarachnoid haemorrhage. *Stroke* **20**, 1680-1684.

Takeuchi, K., Uematsu, M., Ofuji, M., Morikoiyo, M. and Kaiya, H. (1988) Substance P involved in mental disorders. *Prog.Neuro-Psychopharmacol. and Biol.Psychiat.* **12**, 157-164.

Tamminga, C.A.,LeWitt,P.A. and Chase,T.N. (1985) Cholecystokinin and neurotensin gradients in human CSF. *Arch.Neurol.* **42**, 354-355.

Teledgy, G. (1987) Frontiers in Hormone Research, Vol. 15, Karger, Basel.

Terenius, L. and Wahlström, A. (1975) Morphine-like ligand for opioid receptors in human CSF. *Life Sci.* **13**, 1759-1764.

Terenius, L. (1982) Endorphin and modulation of pain: In Critchley, M. (ed) *Advance in Neurology*, **Vol. 53**, Raven Press, New York, U.S.A., pp 59-64.

Theodorsson, N.E., Hemsen, A. and Lundberg, J.M. (1985) Radioimmunoassay for neuropeptide Y (NPY) : chromatographic characterization of immunoreactivity in plasma and tissue extracts. *Scand.J.Clin,Lab.Invest.* **45**, 355-365.

Tibbling G., Link H., Öhman S. (1977). Principles of albumin and IgG analysis in neurological disorders. III. Evaluation of IgG synthesis within central nervous system in multiple sclerosis. *Scand. J. Clin. Lab. Invest.* **37,** 397-401. *J.Neurosci.***13**,2085-2104.

Wahlestedt, C., Ekman, R. and Widerlöv, E. (1989) NPY and the central nervous system: distribution effects and possible relationship to neurological and psychiatric disorders. *Prog.Neuropsychompharm.Biol.Psychiat.* **31**, 31-54.

Wester, P., Bergström, U. and Eriksson, A. (1990) Venticular cerebrospinal fluid monoamine transmitter and metabolite concentrations refect human brain neurochemistry in autopsy cases. *J.Neurochem.* **54**, 1148-1156.

Widerlöv, E.,Lindström, H.,Wahlestedt, C. and Ekman, R. (1988) NPY and PYY as possible cerebrospinal marker for major depression and schizophrenia respectively. *J.Psychiatr.Res.* **22**, 69-79.

Zar, J.H. (1984) *Biostatistical Analysis*. Eaglewood Cliffs, Prentice Hall, London.

Progress in HPLC-HPCE, Vol. 7, pp. 597-617
G. Ali Qureshi *et al.* (Eds)

CEREBROSPINAL FLUID MARKERS OF HIV-1 INFECTION

Magnus Gisslén

Department of Infectious Diseases, Sahlgrenska University Hospital, S-416 85 Göteborg, Sweden

INTRODUCTION

Human immunodeficiency virus type 1 (HIV-1) belongs to the lentiviruses, a genus of the retrovirus family, and was the first lentivirus found to cause disease in humans. Like other retroviruses, HIV-1 replicates through a DNA intermediate, where the viral RNA is transcribed with the enzyme reversed transcriptase to DNA. The viral DNA is then integrated into that of the host, whereafter it behaves as if it were a cellular gene, using the cellular system for replication. Lentivirus infections have earlier been described in the veterinary medicine; for example visna virus infection in sheep and caprine arthritis encephalitis virus infection in goats. All known lentiviruses share the features of neurotropism and neurovirulens.

The HIV-1 envelope glycoprotein gp120 has a high affinity for the cell surface protein CD4, which is expressed at high levels on T-helper lymphocytes and at lower levels on cells of the monocyte-macrophage lineage. HIV-1 enters $CD4^+$ T-helper lymphocytes, monocytes, and macrophages by fusion, a process triggered by the binding of gp120 to the CD4 receptor and involving a second b-chemokine receptor (Feng *et al*, 1996).

Generated by the lack of a proof-reading function during reversed transcription, there is a considerable inter- and intra-individual variability of molecular sequences in different HIV-1 strains (Goodenow *et al*, 1989). HIV-1 isolates differ in respect of replication rate, cell tropism, and ability to form syncytia in, for instance, cultured MT-2 cells, where syncytium-inducing (SI) are differentiated from non-syncytium-inducing (NSI) strains (Åsjö *et al*, 1986). In early, asymptomatic HIV-1 infection, blood isolates of HIV-1 are often of NSI-type with slow replication kinetics. Over time, when the viral load increases, a great diversity of HIV-1 strains develop, including HIV-1 strains of SI-type with high replicative ability. The presence of SI HIV-1 strains correlates with disease progression and reduced survival (Schuitemaker *et al*, 1992). HIV-1 strains vary markedly in their ability to productively infect different cells, such as T cells and macrophages. Macrophage-tropic HIV-1 strains are usually of NSI-type, while SI-strains only exceptionally are able to infect macrophages (Mammano *et al*, 1995). Differences in cell tropism are probably consequences of the distinction between SI and NSI strains to bind to different b-chemokine receptors.

HIV-1 strains isolated from brain tissue have the common property of macrophage tropism (Koyanagi *et al*, 1987). Although unable to form syncytia in MT-2 cells, the NSI macrophage-tropic HIV-1 strains frequently induce syncytia in microglia in the brain, i.e. multinucleated giant cells (Donaldson *et al*, 1994).

HIV-1 infection was long considered to be latent, with low viral replication in the asymptomatic stage. However, more recent studies have shown that there is actually a very rapid replication of HIV-1 in lymphoid tissue and blood also during the early infection, with as many as 10^8-10^9 virions produced and cleared in plasma every day. Likewise, there is a high CD4 lymphocyte turnover rate of about 10^9 CD4 cells destroyed and almost as many produced each day (Pantaleo *et al*, 1993; Ho *et al*, 1995; Wei *et al*, 1995). Despite the enormous production of CD4 lymphocytes, a continuous and gradual CD4 cell depletion takes place, leading to impaired cellular immunity. With time, occurrence of opportunistic infections and tumours is common.

The immunodeficiency resulting from infection with HIV-1 renders also the central nervous system (CNS) susceptible to opportunistic infections and tumours. Opportunistic diseases in the CNS usually occur in advanced stages of HIV-1 infection with severe immunodeficiency. Some opportunistic CNS diseases are listed in table 1.

Table 1. *Some CNS opportunistic diseases in HIV-1 infection and their associated infective agents.*

Disease	Associated infective agent
Toxoplasmosis	Toxoplasma gondii
Primary CNS lymphoma	Epstein-Barr virus
Progressive multifocal leucoencephalopathy (PML)	JC virus
Cytomegalovirus (CMV) encephalitis	CMV
Cryptococcal meningitis	Cryptococcus neoformans
Kaposis sarcoma	Human herpesvirus 8 (HHV-8)
Neurosyphilis	Treponema pallidum
Meningitis tuberculosa	M. tuberculosis

Besides opportunistic diseases in the CNS, HIV-1 itself frequently induces neurological deterioration in infected individuals. The most devastating consequence is AIDS dementia complex (ADC), also named HIV-1-associated cognitive/motor complex (Navia *et al*, 1986b; American Academy of Neurology AIDS Task Force, 1991). ADC predominantly develops in patients who are severely immunocompromised, with CD4 cell counts below 200 x 10^6/L.

ADC is a subcortical dementia characterized by a constellation of cognitive, motor, and behavioural symptoms, and the diagnosis is mainly established on clinical grounds.

The syndrome shows diverse advancement rates. Patients in more advanced stages of the dementia have a limited survival time, with 67% cumulative mortality in 6 months (Neaton *et al*, 1994). Although it is strongly suggested that ADC is a result of HIV-1 infection of the brain, the pathogenesis behind the neuropsychological deterioration is still largely unknown. Whether HIV-1 causes functional deterioration of the CNS in the early, otherwise asymptomatic, stage of HIV-1 infection is a controversial issue. Several studies have failed to identify differences in neuropsychological performance between asymptomatic HIV-1-infected individuals and HIV-negative controls (Janssen *et al*, 1989; McArthur *et al*, 1989), while other studies have demonstrated significant differences (Grant *et al*, 1987; Karlsen *et al*, 1992; Dunlop *et al*, 1993).

Very early in the infectious course, HIV-1 can be detected in the cerebrospinal fluid (CSF) (Chiodi *et al*, 1988; Sönnerborg *et al*, 1991; Schacker *et al*, 1996) and in the brain (Davis *et al*, 1992). Furthermore, some patients have meningoencephalitis already during seroconversion, which adds to the evidence for early spread of HIV-1 into the brain. How HIV-1 invades the brain remains, however, undefined. The generally accepted theory, called the "Trojan horse" hypothesis, suggests hematogenous dissemination of monocytes or lymphocytes infected with HIV-1 through the blood-brain barrier into the brain (Peluso *et al*, 1985). This theory is strengthened by evidence of lymphocyte traffic into the CNS during normal conditions. The normally limited traffic is dramatically increased during pathological conditions in the brain (Hickey *et al*, 1991). An alternative explanation for HIV-1 invasion of the brain is hematogenous dissemination through infiltration of the choroid plexus (Petito, 1996), which may be directly infected by HIV-1 (Falangola *et al*, 1995).

Evidence of productive infection of HIV-1 in the brain *in vivo* is confined to microglia, macrophages, and endothelial cells. The predominant, productively infected cell type is the microglial cell, a cell that like macrophages is derived from bone-marrow precursors. The microglial cells migrate early in gestation into the CNS and spread throughout both grey and white matter (Mattiace *et al*, 1991). Normally, microglial cells express CD4 antigen at low levels but they are likely to up-regulate the expression during cellular activation (Perry & Gordon, 1987). Microglial activation is a general phenomenon in HIV-1 infection, and, as in T-helper lymphocytes, binding of gp120 to the CD4 antigen is the probable mechanism of microglial infection.

A non-specific feature of microglial infection of various viruses and spirochetes is the formation of microglial nodules. In HIV-1 encephalitis, such nodules are distributed predominantly in cerebral white matter, subcortical grey matter, and brainstem, although focuses in the cortex occur relatively frequently (Budka, 1991). In HIV-1 encephalitis, the microglial nodules incorporate microglial cells, astrocytes, and multinucleated giant cells. Multinucleated giant cells, which consists of fused microglial cells and, perchance, macrophages, are pathognomonic of HIV infection (Sharer *et al*, 1985). Besides in microglial nodules, multinucleated giant cells are also found in inflammatory cell infiltrates and sometimes as isolated individual cells in perivascular locations (Dickson & Lee, 1996). Multinucleated giant cells and microglial nodules are associated with HIV-encephalitis and occur late in the infectious course. HIV-1 infection of microglial cells is frequently found also in the absence of detectable cellular histopathology (Kure *et al*, 1991).

It has been suggested that HIV-1 is able to infect endothelial cells in the cerebral capillaries and establish a noncytolytic but productive infection (Wiley *et al*, 1986; Moses *et al*, 1996). Cerebral endothelial cells, however, do not have the CD4 receptor, and the HIV-1 infection must therefore occur via a CD4-independent mechanism. Since cerebral capillary endothelial cells are part of the blood-brain barrier, endothelial cell infection might contribute to impaired blood-brain barrier function, which is a frequent finding in HIV-1 infection (McArthur *et al*, 1992).

It is also possible that astrocytes could be infected by HIV-1 *in vitro* (Tornatore *et al*, 1994) and probably also *in vivo* (Nuovo *et al*, 1994). The *in vivo* infection is, however, non-productive with restricted viral gene expression (Blumberg *et al*, 1994). As in endothelial cells, HIV-1 infection of astrocytes does not involve the CD4 receptor.

HIV-1 infection of oligodendrocytes, which form the myelin sheath, has not yet been persuasively demonstrated *in vivo*. The possibility of sporadic, low-grade infection of oligodendrocytes by HIV-1 has been suggested (Esiri *et al*, 1991), but this could not explain the frequent and often severe myelin pathology found in HIV-1 infection.

The question whether neurons are infected by HIV-1 is controversial. Nuevo et al demonstrated presence of HIV-1 nucleic acid in neurons, using polymerase chain reaction (PCR) amplification combined with in situ hybridisation (Nuovo *et al*, 1994). Whether this finding reflects a significant infection of the neuron or shows only an accumulation of nucleic acid without pathological importance is unclear. If infected by HIV-1, the neuronal infection is probably of a low grade following a restrictive pattern.

HIV-1-induced neuropathological changes with predominant alterations in cerebral white and subcortical grey matter are found in up to 90% of brains from patients with AIDS at autopsy (Navia *et al*, 1986a). The most frequent finding is a diffuse myelin pallor (Price *et al*, 1988; Budka, 1991), often accompanied by a pronounced astrogliosis (Monte de la *et al*, 1987). Myelin damage may be the cause of the white matter pallor (Esiri *et al*, 1991). Another explanation could be accumulation of serum proteins in white matter glia due to leakage of serum through an impaired blood-brain barrier (Power *et al*, 1993). Evidence of HIV-1 encephalitis, characterized by microglial nodules and multinucleated giant cells, is found at autopsy in about 20-30% of patients with AIDS (Budka, 1991; Maehlen *et al*, 1995). In addition to the more prominent white matter lesions, neocortical damage with loss of synapses and cortical neurons has been found in brains from patients with ADC (Wiley *et al*, 1991; Masliah *et al*, 1992). Neuropathological evidence of early CNS changes in HIV-1 infection was found by Gray et al who compared brains from asymptomatic HIV-1-infected and HIV-seronegative drug addicts who died from unnatural causes. They found that cerebral vasculitis, lymphocytic meningitis, myelin pallor with astrocytosis, and microglial proliferation were more frequent and severe in the HIV-1-seropositive cases (Gray *et al*, 1992).

Magnetic resonance imaging (MRI) and computed tomography (CT) scan show abnormalities in the brain in advanced cases of AIDS but they fail to demonstrate differences between asymptomatic HIV-1-infected individuals and HIV-seronegative controls. However, with cerebral proton magnetic resonance spectroscopy abnormalities in cerebral biochemistry have been demonstrated in asymptomatic HIV-1-infected patients (Wilkinson *et al*, 1997).

Antiretroviral treatment constitutes a major hope for HIV-1-infected patients. Effects on the systemic HIV-1 infection, with delayed progression and prolonged survival, have been found with several antiretroviral drugs. Especially combination therapy leads to a dramatic, long-lasting decrease in viral replication measured with HIV-1 RNA PCR in plasma (Gulick *et al*, 1996). Rapid and potent suppression of viral replication is thought to be an important prerequisite to prevent viral resistance.

Antiretroviral treatment effects on the CNS infection are scarcely studied yet. For effective inhibition of HIV-1 replication in the brain, the antiretroviral drug must penetrate the blood-brain barrier. Approximate ratios between CSF and serum concentrations of some antiretroviral drugs are presented in table 2 (Portegies, 1995). However, the CSF/plasma ratio can be misleading, since the time of sampling differs between different studies and the ratio normally increases during the first hours following the last ingested dose. Zidovudine (azt) and stavudine (d4T) are the only available antiretroviral drug with documented ability to penetrate into the CSF, but there are some evidences that lamivudine which has a low CSF/plasma ratio reaches sufficient concentrations in the CSF.

Table 2. *Approximated CSF/plasma ratios of some antiretroviral drugs (2-4 hours after last ingested dose).*

	Ratio
Reversed transcriptase inhibitors	
Zidovudine (azt)	0.6
Stavudine (d4T)	0.4
Didanosine (ddI)	0.2
Zalcitabine (ddC)	0.2
Lamivudine (3TC)	0.06
Protease inhibitors available at present	probably low

Clinically, zidovudine treatment reduces the risk of ADC development in HIV-1-infected patients (Portegies *et al*, 1989). Zidovudine treatment has also been shown to reduce HIV-1-induced brain lesions in neuropathological studies of patients with AIDS (Vago *et al*, 1993). Termination of treatment rapidly increased the incidence of HIV-1 encephalitis again (Gray *et al*, 1994; Maehlen *et al*, 1995). In addition to its prophylactic effect zidovudine also offers therapeutic benefits in the treatment of established ADC (Sidtis *et al*, 1993). There are no results available yet on the clinical effect of antiretroviral combination therapy on ADC.

ANALYSES OF THE CSF

The compartment closest to the brain accessible for analyses in living patients is the CSF. Under healthy conditions, no inflammatory activity is seen in the CSF, and the cell count is low, whereas CNS infections frequently cause increased CSF cell counts. Cell products and membrane fragments are released into the intercellular space which is in direct contact with the CSF. This process, called cell surface shedding, is increased in degenerative disorders affecting the cell membranes (Doljanski & Kapeller, 1976). Pathological events within the brain can therefore be reflected by abnormalities in the CSF.

Specific markers of HIV-1 infection

HIV-1 can be isolated from the CSF in about 80% of patients with early asymptomatic HIV-1 infection after repeated isolation attempts in each individual (Chiodi *et al*, 1992). With PCR, HIV-1 nucleic acid can be detected in almost all infected patients irrespective of current stage of disease (Sönnerborg *et al*, 1991). Until recently, quantitative measurement of HIV-1 infection has been limited to that of the HIV-1 core antigen, p24, a method shown to have a low sensitivity. Findings of p24 antigen in CSF are rare in neurologically asymptomatic stages of HIV-1 infection, and even in patients with severe dementia p24 antigen is detectable only in about 50% of the cases (Brew *et al*, 1994). Nevertheless, a correlation was found between p24 antigen levels in the CSF and severity of ADC (Royal *et al*, 1994; Singer *et al*, 1994). Recently developed PCR methodology (Mulder *et al*, 1994) can detect and quantify the number of HIV-1 RNA copies in CSF at all stages of HIV-1 infection. We found a geometric mean level of HIV-1 RNA of 3.78 (range <2.30-5.12) ^{10}log copies/mL in the CSF compared to 4.23 (range 2.46-5.09) ^{10}log copies/mL in serum in 16 HIV-1 infected patients at different stages of the disease (Gisslén *et al*, 1997c). Plasma HIV-1 RNA is a strong, CD4 cell independent predictor of a rapid progression to AIDS (Mellors *et al*, 1995), while the importance of HIV-1 RNA levels in CSF has not been fully evaluated yet. However, high CSF levels of HIV-1 RNA have been found in patients with ADC (Di Stefano *et al*, 1997).

Zidovudine treatment leads to a more prominent decrease of HIV-1 RNA levels in the CSF than in serum, while didanosine, another nucleoside analogue, does not effect the CSF levels of HIV-1 RNA significantly after 3-12 months of treatment, figure 1 (Gisslén *et al*, 1997c). In 10 patients with antiretroviral combination therapy (including zidovudine or stavudine) HIV-1 RNA could not be detected in the CSF with quantitative PCR technique (Gisslén *et al*, 1997b).

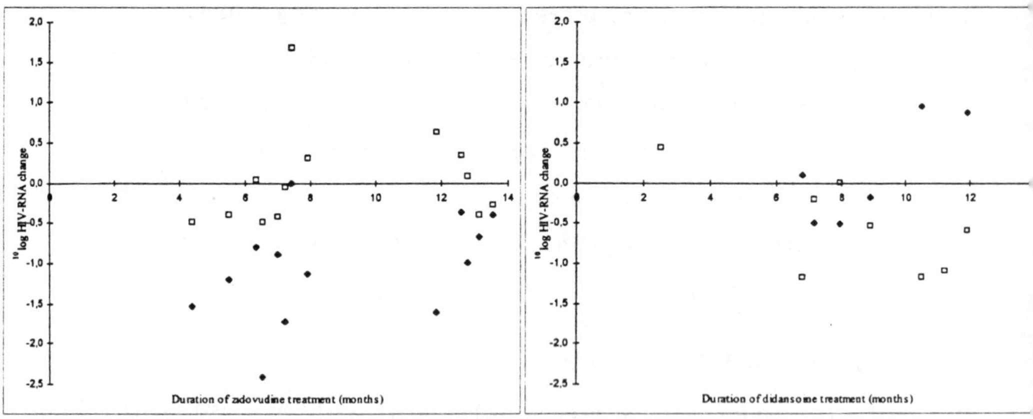

Figure 1. *Changes in CSF (˙) and serum (▒) levels of HIV-1 RNA after initiation of zidovudine or didanosine as compared with duration of treatment.*

Markers of immune stimulation

In response to CNS HIV-1 infection, the host increases its production of cytokines, such as TNF-a and IFN-g, which have been found to be increased in CSF (Griffin *et al*, 1991; Franciotta *et al*, 1992) and brain tissues (Tyor *et al*, 1992; Achim *et al*, 1993). IFN-g is capable of activating monocytes and macrophages, which increases the production of the pteridine neopterin through an activation of the enzyme GTP cyclohydrolase I, catalyzing the reaction where neopterin is formed from guanosine-5-triphosphate (GTP) (Huber *et al*, 1984). Thus, neopterin levels provide an indirect measure of IFN-g activity and activated cell-mediated immunity (Fuchs *et al*, 1988). Increased CSF levels of neopterin have been found in HIV-1 infection as well as in several other infections of the CNS (Hagberg *et al*, 1993). In parallel, IFN-g induces indoleamine-2,3-dioxygenase (IDO) activity in macrophages (Werner-Feltmyer *et al*, 1989). IDO catalyses the reaction where tryptophan is cleaved to N-formylkynurenine, which is further degraded to kynurenic and quinolinic acid (Werner *et al*, 1988). Quinolinic acid is regarded as toxic to the brain. That these mechanisms occur during HIV-1 infection is suggested by the fact that increased concentrations of neopterin and decreased concentrations of tryptophan in CSF were found before initiation of antiretroviral treatment. Furthermore, zidovudine treatment generated increased CSF tryptophan concentrations correlating with decreased CSF neopterin levels (Gisslén *et al*, 1994b). Besides neopterin, elevated levels of IFN-g, kynurenine, and quinolinic acid have been found in CSF in HIV-1-infected individuals in several studies (Heyes *et al*, 1989; Moeller *et al*, 1990; Griffin *et al*, 1991). CSF levels of neopterin and quinolinic acid have actually been positively correlated with severity of the neurological deterioration in ADC (Brew *et al*, 1990; Heyes *et al*, 1991), although no correlation was found between CSF levels of quinolinic acid and neuropathological severity of HIV-1 encephalitis (Wiley *et al*, 1992). Increased parenchymal quinolinic acid concentrations have been found in brain tissues from patients with HIV-1 encephalitis (Sei *et al*, 1995), with the highest concentrations in the basal ganglia, followed by cerebral white and cortical grey matter (Achim *et al*, 1993), which is similar to the neuropathological findings in ADC. Thus, in contrast to its beneficial effects on acute infections, chronic immune stimulation might be detrimental to the CNS in the long run.

Neopterin could be measured by commercially available radio immune assay (RIA) or enzyme linked immunosorbent assay (ELISA) methods. b2-microglobulin (b2M), a portion of the major histocompatibility complex class I antigen (MHC-I), is another marker of immune stimulation that could be quantified by ELISA. b2M is expressed in the membranes of almost all nucleated cells, with the highest concentrations on activated T-lymphocytes and macrophages. The expression of b2M is up-regulated by a variety of cytokines, including IFN-g, and increased CSF levels of b2M have been found in patients with CNS infections, HIV-1 included (Sönnerborg *et al*, 1989).

In HIV-1 infection, a gradual increase in CSF levels of neopterin and b2M could be found over time, suggesting the establishment of a chronic, progressive infection in the CNS already in the asymptomatic stages of the disease (Elovaara *et al*, 1993; Gisslén *et al*, 1994a). HIV-1 infection in the CNS is not a latent infection but an active process throughout all the infectious course in accordance with the systemic HIV-1 infection.

Cell associated markers

Glycosphingolipids are significant constituents of membrane lipids in the CNS. In contrast to proteins (Karlsson & Alling, 1984), they are not degraded in the CSF, and measurement of CSF glycosphingolipids might be used for detection of degenerative processes affecting brain cell membranes. However, although enriched in the CNS, glycosphingolipids could also be found in cells without the CNS. One have to take in account that increased passage from the blood through an impaired blood-brain barrier may result in increased CSF levels of glycosphingolipids. In contrast, some proteins are exclusively localised to certain cell types within the CNS, table 3.

Table 3. *CSF cell associated markers*

Cell/structure	CSF marker
Neurons	GM1, GD1a, GD1b, GT1b, neurofilament, neuron-specific enolase
Oligodendrocytes	Sulfatide, Galactosylceramide, GM1
Myelin	Sulfatide, Galactosylceramide
Astrocytes	GD3, GFAp
Microglia	GD3

The glycosphingolipid sulfatide (3'-sulphogalactosylceramide) is an early marker for the differentiation of oligodendrocytes and becomes highly enriched in the myelin. While sulfatide is almost absent in other brain cells, it is the major acidic glycosphingolipid in the myelin sheath (Norton & Autilio, 1966), and CSF sulfatide might be used as marker of myelin disorders. The CSF concentration of sulfatide is age-dependent and increases with ageing (Davidsson *et al*, 1991). Elevated CSF levels of sulfatide have been found in patients with vascular dementia (Fredman *et al*, 1992), where demyelination and loss of myelin lipids are constituents of the pathogenesis (Wallin *et al*, 1989).

Sulfatide is of further interest in patients with HIV-1 infection since the HIV-1 envelope glycoprotein gp120 has been shown to bind to sulfatide, which has been suggested to be a receptor for gp120 in peripheral nerves (van den Berg *et al*, 1992).

Gangliosides are sialic-acid-containing glycosphingolipids that are normal components of all human cells. They are mainly localized to the outer layer of the plasma membrane, and their concentration and composition vary in a cellular- and differentiation-specific pattern. The concentration of gangliosides is much higher in brain tissue than in any extra-neural organ and about three times higher in the cerebral cortex than in the cerebral white matter. The ganglioside pattern in the neuron is dominated by the four gangliosides GM1, GD1a, GD1b, and GT1b (Svennerholm, 1980). CSF measurements of gangliosides might be used to detect processes affecting neuronal membranes. In normal ageing, the total concentrations of gangliosides in CSF decrease while the proportion of GM1 increases with advancing age (Blennow *et al*, 1992). Alzheimers disease type I is associated with high proportions of GM1 and GD1a and low proportions of GD1b and GT1b without increased total CSF levels (Blennow *et al*, 1991).

Sulfatide and gangliosides could be quantitatively determined by thin-layer chromatography overlay technique, TLC-ELISA (Davidsson *et al*, 1991). Lipids are extracted by addition of methanol/chloroform to 0.5 mL CSF to obtain a final ratio of methanol/chloroform/CSF (4:8:3, by vol.). Sulfatide is separated from most other lipids by phase partitioning, by adding chloroform, methanol, and 0.1M KCl to obtain a solvent ratio of chloroform/methanol/water (4:2:1, by vol.), whereby sulfatide is separated into the lower phase. Aliquots of the sulfatide-containing fraction and standards of purified sulfatide are applied as 5 mm lanes on a silica gel TLC plate. After chromatography in organic solvent, the plate is dipped in polyisobutylmethacrylate solution to prevent unspecific binding and to attach the silica gel to the plate. The plate is incubated in subsequent order with anti-sulfatide mouse monoclonal antibody, alkaline phosphatase-conjugated goat antimouse-Ig antibody, and phosphatase substrate.

The sulfatide monoclonal antibody recognises sulfate attached to galactose in terminal position of the carbohydrate chain and part of the ceramide portion of the glycolipid. The antibody is therefore lipid-specific and does not react with other sulfated non-lipid molecules or glycolipids with terminal galactose like galactosylceramide (Fredman *et al*, 1988). Among lipids, the antibody reacts only with sulfatide, sulfated lactosylceramide, and seminolipid and not with any extended structures with a terminal sulfate attached to galactose.

Gangliosides could be measured in the CSF with a method similar to that quantifying sulfatide. The lipid extract is desalted by dialysis. After chromatography on TLC plates, GD1a, GD1b, and GT1b are hydrolysed with sialidase to GM1. The plate is then incubated with a cholera toxin-B subunit, and determination of the individual gangliosides GM1, GD1a, GD1b, and

GT1b is performed with an anticholera toxin-B subunit monoclonal antibody. The gangliosides are quantified by densometric scanning of plates at 620 nm.

Almost 50% of asymptomatic HIV-1-infected patients have signs of myelin disturbance measured as increased CSF sulfatide concentrations, and/or of neuronal involvement measured as increased CSF concentrations of the ganglioside GM1 (Gisslén *et al*, 1996, 1997a). Twenty-two out of 28 patients with AIDS had CSF sulfatide levels above mean plus 2 standard deviations in the HIV-seronegative controls, and some of the patients with AIDS had extremely high CSF sulfatide levels. However, in this small number of patients, no differences in CSF sulfatide concentrations were found between demented and non-demented AIDS patients.

Although oligodendrocytes, the cells that form the myelin sheath, are not significantly infected by HIV-1, interference by HIV-1 or gp120 could not be ruled out. Gp120 is exposed to the environment, and since this protein has been found to be a ligand to sulfatide (Furuta *et al*, 1994), a gp120-sulfatide interaction might influence the functions of oligodendrocytes. Parmantier et al could not demonstrate binding of gp120 to rat oligodendrocytes *in vitro*, but they found that anti-sulfatide antibodies bound to the oligodendrocyte cell membrane lead to increased intracellular calcium levels (Parmantier *et al*, 1995). Intrathecal synthesis of anti-sulfatide IgG has been associated with peripheral nerve disease in AIDS (de Gasperi *et al*, 1996).

Another possible mechanism behind myelin disturbance is damage caused by serum substances leaked into the white matter due to alterations in the blood-brain barrier function. Accumulations of serum proteins have been found in white matter in patients with AIDS, especially in those with ADC (Power *et al*, 1993), and serum substances have been shown to induce damage to oligodendrocytes *in vitro* (Compston *et al*, 1991).

The CSF concentrations and the proportion of the ganglioside GM1 were increased in about one third of HIV-1-infected patients, independent of stage of disease. The highest levels of GM1 in CSF were found in patients with CD4 cell counts below 200 x 10^6/L. Though the vast majority of HIV-1-infected patients harbour the virus in the CNS during several years of infection, only about 20% of untreated patients develop ADC. It is thus likely that not all HIV-1-infected individuals show signs of brain damage. However, 29/42 (69%) patients analysed had increased values of either sulfatide and/or GM1 in CSF. Nine of the patients (21%) had elevated CSF levels of both sulfatide and GM1, figure 2

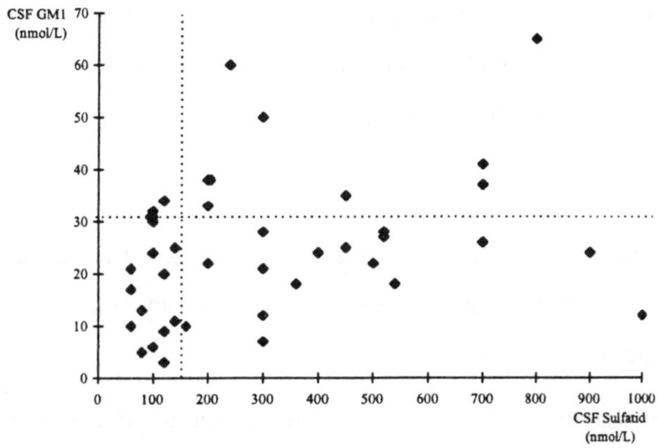

Figure 2. *CSF ganglioside GM1 and sulfatide in 42 HIV-1-infected individuals. Mean +2SD in HIV-seronegative controls is indicated; GM1 30.2 nmol/L, sulfatide 157.4 nmol/L.*

Other potential markers of neuronal degeneration are the proteins neurofilament and neuron-specific enolase which are major structural elements of neurons (Karlsson *et al*, 1993; Rosengren *et al*, 1996). Increased CSF levels of neurofilament have been found in Alzheimer dementia and amyotrophic lateral sclerosis (Rosengren *et al*, 1996) and increased CSF levels of neuron-specific enolase have been correlated to signs of brain damage on computed tomography scans in patients with ischemic cerebrovascular disorders (Persson *et al*, 1987).

Glial fibrillary acidic protein (GFAp) is the main protein constituent of the intermediate filaments of astrocytes and increased CSF levels of GFAp could be found in astrogliosis. High GFAp concentrations in the CSF have been found in CNS diseases associated with astrogliosis, for example Alzheimers disease, vascular dementia, bacterial meningitis, acute viral encephalitis, and Lyme neuroborreliosis (Rosengren *et al*, 1994; Dotevall *et al*, 1996). HIV-1 infection of the CNS is associated with astrogliosis and GFAp could be supposed to be increased in the CSF. Studies on HIV-1 infected patients regarding CSF analyses of GFAp, neurofilament and neuron-specific enolase are proceeding but no results are available yet.

No CSF marker of microglial activation or degeneration have been established to date. The ganglioside GD3 constitutes a major component of microglial glycosphingolipids but is also present in astrocytes. Comparisons between CSF levels of GD3 and GFAp might serve as markers of microglial affection.

Longitudinal studies on cell associated markers are needed to define the role of these substances in clinical practice and to ascertain their ability to serve as markers of risk for neurological deterioration in HIV-1 infection.

CONCLUSIONS

HIV-1 infects the brain early in the infectious course and establishes a chronic, progressive infection of the CNS with gradually increased intrathecal immunoactivation already at the asymptomatic stages of the disease.

One third of asymptomatic HIV-1-infected individuals and three fourths of patients with AIDS show signs of myelin disturbance, measured as increased CSF concentrations of sulfatide.

About one third of HIV-1-infected patients, independent of stage, have increased levels of the ganglioside GM1 in CSF as a sign of neuronal involvement.

CSF measurements of the ganglioside GD3, neurofilament, neuron-specific enolase, and GFAp might throw further light upon the pathogenesis behind neurological disease in HIV-1 infection.

However, the clinical role of CSF cellular markers in HIV-1 infection remains to be defined.

Intervention with zidovudine reduces the intrathecal immunoactivation and decreases the CSF viral load. Antiretroviral combination therapy dramatically reduces the CSF viral load. A finding that adds optimism to the hope of an effective treatment of HIV-1 infection.

REFERENCES

Achim, C.L., Heyes, M.P. and Wiley, C.A. (1993). Quantitation of human immunodeficiency virus, immune activation factors, and quinolinic acid in AIDS brains. *Journal of Clinical Investigation* **91**, 2769-2775.

American Academy of Neurology AIDS Task Force. (1991). Nomenclature and research case definitions for neurological manifestations of human immunodeficiency virus-type 1 (HIV-1) infection. *Neurology* **41**, 778-785.

Blennow, K., Davidsson, P., Wallin, A., Fredman, P., Gottfries, C.-G., Karlsson, I., Månsson, J.-E. and Svennerholm, L. (1991). Gangliosides in cerebrospinal fluid in probable Alzheimer's disease. *Archives of Neurology* **48**, 1032-1035.

Blennow, K., Davidsson, P., Wallin, A., Fredman, P., Gottfries, C.G., Månsson, J.-E. and Svennerholm, L. (1992). Differences in cerebrospinal fluid gangliosides between "probable Alzheimer's disease" and normal aging. *Aging Clin Exp Res* **4**, 301-306.

Blumberg, B.M., Gelbard, H.A. and Epstein, L.G. (1994). HIV-1 infection of the developing nervous system: central role of astrocytes in pathogenesis. *Virus Research* **32**, 253-267.

Brew, B., Bhalla, R., Paul, M., Gallardo, H., McArthur, J., Schwartz, M. and Price, R. (1990). Cerebrospinal fluid neopterin in human immunodeficiency virus type 1 infection. *Annals of Neurology* **28**, 556-560.

Brew, B.J., Paul, M.O., Nakajima, G., Khan, A., Gallardo, H. and Price, R.W. (1994). Cerebrospinal fluid HIV-1 p24 antigen and culture: sensitivity and specificity for AIDS-dementia complex. *Journal of Neurology, Neurosurgery and Psychiatry* **57**, 784-789.

Budka, H. (1991). Neuropathology of HIV infection. *Brain Pathology* **1**, 163-175.

Chiodi, F., Sönnerborg, A., Albert, J., Gaines, H., Norkrans, G., Hagberg, L., Åsjö, B., Strannegård, Ö. and Fenyö, E.M. (1988). Virus isolation and detection of HIV specific antibodies in cerebrospinal fluid of patients with varying clinical conditions. *Journal of the Neurological Sciences* **85**, 245-257.

Chiodi, F., Keys, B., Albert, J., Hagberg, L., Lundeberg, J., Uhlén, M., Fenyö, E.M. and Norkrans, G. (1992). HIV-1 is present in the cerebrospinal fluid of a majority of infected individuals. *Journal of Clinical Microbiology* **30**, 1768-1771.

Compston, A., Scolding, N., Wren, D. and Noble, M. (1991). The pathogenesis of demyelinating disease: insight from cell biology. *Trends in Neurosciences* **14**, 175-182.

Davidsson, P., Fredman, P., Månsson, J.-E. and Svennerholm, L. (1991). Determination of gangliosides and sulfatide in human cerebrospinal fluid with microimmunoaffinity technique. *Clinica Chimica Acta* **197**, 105-116.

Davis, L.E., Hjelle, B.L., Miller, V.E., Palmer, D.L., Llewellyn, A.L., Merlin, T.L., Young, S.A., Mills, R.G., Wachsman, W. and Wiley, C.A. (1992). Early viral brain invasion in iatrogenic human immunodeficiency virus infection. *Neurology* **42**, 1736-1739.

de Gasperi, R., Gama Sosa, M.A., Patarca, R., Battistini, S., Lamoreux, M.R., Raghavan, S., Kowall, N.W., Harrington Smith, K., Fletcher, M.A. and Kolodny, E.H. (1996). Intrathecal synthesis of anti-sulfatide IgG is associated with peripheral nerve disease in acquired immunodeficiency syndrome. *AIDS Research and Human Retroviruses* **12**, 205-211.

Di Stefano, M., Monno, L., Fiore, J.R. and Angarano, G. (1997). In vivo evidence of HIV-1 productive infection in cerebrospinal fluid of patients with HIV-1 encephalitis. *AIDS* **11**, 133-135.

Dickson, D. and Lee, S. (1996). Microglia in HIV-related CNS neuropathology: an update. *Journal of Neuro-AIDS* **1**, 57-83.

Doljanski, F. and Kapeller, M. (1976). Cell surface shedding - the phenomenon and its possible significance. *Journal of Theoretical Biology* **62**, 253-270.

Donaldson, Y.K., Bell, J.E., Holmes, E.C., Hughes, E.S., Brown, H.K. and Simmonds, P. (1994). In vivo distribution and cytopathology of variants of human immunodeficiency virus type 1 showing restricted sequence variability in the V3 loop. *Journal of Virology* **68**, 5991-6005.

Dotevall, L., Rosengren, L.E. and Hagberg, L. (1996). Increased cerebrospinal fluid levels of glial fibrillary acidic protein (gfap) in lyme neuroborreliosis. *Infection* **24**, 125-129.

Dunlop, O., Björklund, R., Abdelnoor, M. and Myrvang, B. (1993). Total reaction time: a new approach in early HIV encephalopathy? *Acta Neurologica Scandinavica* **88**, 344-348.

Elovaara, I., Nykyri, E., Poutiainen, E., Hokkanen, L., Raininko, R. and Suni, J. (1993). CSF follow-up in HIV-1 infection: intrathecal production of HIV-specific and unspecific IgG, and beta-2-microglobulin increase with duration of HIV-1 infection. *Acta Neurologica Scandinavica* **87**, 388-396.

Esiri, M.M., Morris, C.S. and Millard, P.R. (1991). Fate of oligodendrocytes in HIV-1 infection. *AIDS* **5**, 1081-1088.

Falangola, M.F., Hanly, A., Galvao Castro, B. and Petito, C.K. (1995). HIV infection of human choroid plexus: a possible mechanism of viral entry into the CNS. *Journal of Neuropathology and Experimental Neurology* **54**, 497-503.

Feng, Y., Broder, C.C., Kennedy, P.E. and Berger, E.A. (1996). HIV-1 entry cofactor: functional cDNA cloning of a seven-transmembrane G protein-coupled receptor. *Science* **272**, 872-877.

Franciotta, D.M., Melzi d'Eril, G.L., Bono, G., Brustia, R., Ruberto, G. and Pagani, I. (1992). Tumor necrosis factor alpha levels in serum and cerebrospinal fluid of patients with AIDS. *Functional Neurology* **7**, 35-38.

Fredman, P., Mattsson, L., Andersson, K., Davidsson, P., Ishizuka, I., Jeansson, S., Månsson, J.-E. and Svennerholm, L. (1988). Characterization of the binding epitope of monoclonal antibody to sulphatide. *Biochemical Journal* **251**, 17-22.

Fredman, P., Wallin, A., Blennow, K., Davidsson, P., Gottfries, C. and Svennerholm, L. (1992). Sulfatide as a biochemical marker in cerebrospinal fluid of patients with vascular dementia. *Acta Neurologica Scandinavica* **85**, 103-106.

Fuchs, D., Hausen, A., Reibnegger, G., Werner, E.R., Dierich, M.P. and Wachter, H. (1988). Neopterin as a marker for activated cell-mediated immunity: application in HIV infection. *Immunology Today* **9**, 150-155.

Furuta, Y., Eriksson, K., Svennerholm, B., Fredman, P., Horal, P., Jeansson, S., Vahlne, A., Holmgren, J. and Czerkinsky, C. (1994). Infection of vaginal and colonic epithelial cells by the human immunodeficiency virus type 1 is neutralized by antibodies raised against conserved epitopes in the envelope glycoprotein gp120. *Proceedings of the National Academy of Sciences of the United States of America* **91**, 12559-12563.

Gisslén, M., Chiodi, F., Fuchs, D., Norkrans, G., Svennerholm, B., Wachter, H. and Hagberg, L. (1994a). Markers of immune stimulation in the cerebrospinal fluid during HIV infection: A longitudinal study. *Scandinavian Journal of Infectious Diseases* **26**, 523-533.

Gisslén, M., Larsson, M., Norkrans, G., Fuchs, D., Wachter, H. and Hagberg, L. (1994b). Tryptophan concentrations increase in cerebrospinal fluid and blood after zidovudine treatment in patients with HIV type 1 infection. *AIDS Research and Human Retroviruses* **10**, 947-951.

Gisslén, M., Fredman, P., Norkrans, G. and Hagberg, L. (1996). Elevated cerebrospinal fluid sulfatide concentrations as a sign of increased metabolic turnover of myelin in HIV type 1 infection. *AIDS Research and Human Retroviruses* **12**, 149-155.

Gisslén, M., Hagberg, L., Norkrans, G., Lekman, A. and Fredman, P. (1997a). Increased cerebrospinal fluid ganglioside GM1 concentrations indicating neuronal involvement in all stages of HIV-1 infection. *Journal of NeuroVirology* **3**, 148-152.

Gisslén, M., Hagberg, L., Svennerholm, B. and Norkrans, G. (1997b). HIV-1 RNA is not detectable in the cerebrospinal fluid during antiretroviral combination therapy. *AIDS* In press.

Gisslén, M., Norkrans, G., Svennerholm, B. and Hagberg, L. (1997c). The effect on cerebrospinal fluid HIV RNA levels after initiation of zidovudine or didanosine. *Journal of Infectious Diseases* **175**, 434-437.

Goodenow, M., Huet, T., Saurin, W., Kwok, S., Sninsky, J. and Wain-Hobson, S. (1989). HIV-1 isolates are rapidly evolving quasispecies: evidence for viral mixtures and preferred nucleotide substitutions. *Journal of Acquired Immune Deficiency Syndromes* **2**, 344-352.

Grant, I., Atkinson, H., Hesselink, J., Kennedy, C., Richman, D., Spector, S. and McCutchan, A. (1987). Evidence for early central nervous system involvement in the acquired immunodeficiency syndrome (AIDS) and other human immunodeficiency virus (HIV) infections. *Annals of Internal Medicine* **107**, 828-836.

Gray, F., Lescs, M.-C., Keohane, C., Paraire, F., Marc, B., Durigon, M. and Gherardi, R. (1992). Early brain changes in HIV infection: Neuropathological study of 11 HIV seropositive, non-AIDS cases. *Journal of Neuropathology and Experimental Neurology* **51**, 177-185.

Gray, F., Belec, L., Keohane, C., De Truchis, P., Clair, B., Durigon, M., Sobel, A. and Gherardi, R. (1994). Zidovudine therapy and HIV encephalitis: a 10-year neuropathological survey. *AIDS* **8**, 489-493.

Griffin, D.E., McArthur, J.C. and Cornblath, D.R. (1991). Neopterin and interferon-gamma in serum and cerebrospinal fluid of patients with HIV-associated neurologic disease. *Neurology* **41**, 69-74.

Gulick, R.M., Mellors, J., Havlir, D., Eron, J., Gonzalez, C., McMahon, D., Richman, D., Valentine, F., Rooney, J., Jonas, L., Meibohm, A., Emini, E. and Chodakewitz, J. (1996). Potent and sustained antiretroviral activity of indinavir, zidovudine and lamivudine (abstract Th.B.931). *XI International Conference on AIDS, Vancouver, BC, July 7-12*

Hagberg, L., Dotevall, L., Norkrans, G., Larsson, M., Wachter, H. and Fuchs, D. (1993). Cerebrospinal fluid neopterin concentrations in central nervous system infection. *Journal of Infectious Diseases* **168**, 1285-1288.

Heyes, M.P., Rubinow, D., Lane, C. and Markey, S.P. (1989). Cerebrospinal fluid quinolinic acid concentrations are increased in acquired immune deficiency syndrome. *Annals of Neurology* **26**, 275-277.

Heyes, M.P., Brew, B.J., Martin, A., Price, R.W., Salazar, A.M., Sidtis, J.J., Yergey, J.A., Mouradian, M.M., Sadler, A.E., Keilp, J., Rubinow, D. and Markey, S.P. (1991). Quinolinic acid in cerebrospinal fluid and serum in HIV-1 infection. *Annals of Neurology* **29**, 202-209.

Hickey, W.F., Hsu, B.L. and Kimura, H. (1991). T-lymphocyte entry into the central nervous system. *Journal of Neuroscience Research* **28**, 254-260.

Ho, D.D., Neumann, A.U., Perelson, A.S., Chen, W., Leonard, J.M. and Markowitz, M. (1995). Rapid turnover of plasma virions and CD4 lymphocytes in HIV-1 infection. *Nature* **373**, 123-126.

Huber, C., Batchelor, R., Fuchs, D., Hause, A., Lang, A., Niederwieser, D., Reibnegger, G., Swetly, P., Troppmair, J. and Wachter, H. (1984). Immune response-associated production of neopterin release from macrophages primarily under control of interferon-gamma. *Journal of Experimental Medicine* **160**, 310-316.

Janssen, R., Saykin, A., Cannon, L., Campbell, J., Pinsky, P., Hessol, N., O'Malley, P., Lifson, A., Doll, L., Rutherford, G. and Kaplan, J. (1989). Neurological and neuropsychological manifestations of HIV-1 infection: association with AIDS-related complex but not asymptomatic HIV-1 infection. *Annals of Neurology* **26**, 592-600.

Karlsen, N., Reinvang, I. and Fröland, S. (1992). Slowed reaction time in asymptomatic HIV-positive patients. *Acta Neurologica Scandinavica* **86**, 242-246.

Karlsson, B. and Alling, C. (1984). Molecular size of myelin basic protein immunoactivity in spinal fluid. *Journal of Neuroimmunology* **6**, 141-150.

Karlsson, J.E., Wang, S., Rosengren, L.E. and Haglid, K.G. (1993). Quantitative alterations of S-100 protein and neuron specific enolase in the rat nervous system after chronic 2,5-hexanedione exposure. *Neurochemical Research* **18**, 203-208.

Koyanagi, Y., Miles, S., Mitsuyasu, R.T., Merrill, J.E., Vinters, H.V. and Chen, I.S.Y. (1987). Dual infection of the central nervous system by AIDS viruses with distinct cellular tropisms. *Science* **236**, 819-822.

Kure, K., Llena, J.F., Lyman, W.D., Soeiro, R., Weidenheim, K.M., Hirano, A. and Dickson, D.W. (1991). Human immunodeficiency virus-1 infection of the nervous system: an autopsy study of 268 adult, pediatric, and fetal brains. *Human Pathology* **22**, 700-710.

Maehlen, J., Dunlop, O., Liestoel, K., Dobloug, J.H., Goplen, A.K. and Torvik, A. (1995). Changing incidence of HIV-induced brain lesions in Oslo, 1983-1994: effects of zidovudine treatment. *AIDS* **9**, 1165-1169.

Mammano, F., Salvatori, F., Ometto, L., Panozzo, M., Chieco Bianchi, L. and De Rossi, A. (1995). Relationship between the V3 loop and the phenotypes of human immunodeficiency virus type 1 (HIV-1) isolates from children perinatally infected with HIV-1. *Journal of Virology* **69**, 82-92.

Masliah, E., Achim, C.L., Ge, N., DeTeresa, R., Terry, R. and Wiley, C.C. (1992). Spectrum of human immunodeficiency virus-associated neocortical damage. *Annals of Neurology* **32**, 321-329.

Mattiace, L.A., Brosnan, C. and Dickson, D.W. (1991). Distribution and characterization of microglia in the normal human fetal brain (abstract). *Soc Neurosci Abstr* **17**, 734.

McArthur, J.C., Cohen, B., Selnes, O., Kumar, A., Cooper, K., McArthur, J.H., Soucy, G., Cornblath, D., Chmiel, J., Wang, M.-C., Starkey, D., Ginzburg, H., Ostrow, D., Johnson, R., Phair, J. and Polk, F. (1989). Low prevalence of neurological and neuropsychological abnormalities in otherwise healthy HIV-1-infected individuals: results from the multicenter AIDS cohort study. *Annals of Neurology* **26**, 601-611.

McArthur, J.C., Nance-Sproson, T.E., Griffin, D.E., Hoover, D., Selnes, O.A., Miller, E.N., Margolick, J.B., Cohen, B.A., Farzadegan, H. and Saah, A. (1992). The diagnostic utility of elevation in cerebrospinal fluid beta2-microglobulin in HIV-1 dementia. *Neurology* **42**, 1707-1712.

Mellors, J.W., Kingsley, L.A., Rinaldo Jr, C.R., Todd, J.A., Hoo, B.S. and Kokka, R.P. (1995). Quantitation of HIV-1 RNA in plasma predicts outcome after seroconversion. *Annals of Internal Medicine* **122**, 573-579.

Moeller, A.A., Fuchs, D., Reibnegger, G., Zaudig, M., Werner, E.R. and Wachter, H. (1990). Changes of neopterin and tryptophan in serum and cerebrospinal fluid of patients in different stage of HIV-1 infection. *Pteridines* **2**, 91-93.

Monte de la, S.M., Ho, D.D., Schooley, R.T., Hirsch, M.S. and Richardson, E.P. (1987). Subacute encephalomyelitis of AIDS and its relation to HTLV-III infection. *Neurology* **37**, 562-569.

Moses, A., Stenglein, S. and Nelson, J. (1996). HIV infection of the brain microvasculature and its contribution to the AIDS dementia complex. *Journal of Neuro-AIDS* **1**, 85-99.

Mulder, J., McKinney, N., Christopherson, C., Sninsky, J. and Greenfield, L. (1994). Rapid and simple PCR assay for quantitation of human immunodeficiency virus type 1 RNA in plasma: application to acute retroviral infection. *Journal of Clinical Microbiology* **32**, 292-300.

Navia, B., Cho, E.-S., Petito, C. and Price, R. (1986a). The AIDS dementia complex: II Neuropathology. *Annals of Neurology* **19**, 525-535.

Navia, B., Jordan, B. and Price, R. (1986b). The AIDS dementia complex: I Clinical features. *Annals of Neurology* **19**, 517-524.

Neaton, J.D., Wentworth, D.N., Rhame, F., Hogan, C., Abrams, D.I. and Deyton, L. (1994). Considerations in choice of a clinical endpoint for AIDS clinical trials. Terry Beirn

Community Programs for Clinical Research on AIDS (CPCRA). *Statistics in Medicine* **13**, 2107-2125.

Norton, W.T. and Autilio, L.A. (1966). The lipid composition of purified bovine brain myelin. *Journal of Neurochemistry* **13**, 213-222.

Nuovo, G.J., Gallery, F., MacConnell, P. and Braun, A. (1994). In situ detection of polymerase chain reaction-amplified HIV-1 nucleic acids and tumor necrosis factor-alpha RNA in the central nervous system. *American Journal of Pathology* **144**, 659-666.

Pantaleo, G., Graziosi, C., Demarest, J.F., Butini, L., Montroni, M., Fox, C.H., Orenstein, J.M., Kotler, D.P. and Fauci, A.S. (1993). HIV infection is active and progressive in lymphoid tissue during the clinically latent stage of disease. *Nature* **362**, 355-358.

Parmantier, E., Monge, M., Yagello, M., Cabon, F., Demerens, C., Gluckman, J.C. and Zalc, B. (1995). HIV-1 envelope glycoprotein gp120 does not bind to galactosylceramide-expressing rat oligodendrocytes. *Virology* **206**, 1084-1091.

Peluso, R., Haase, A., Stowring, L., Edwards, M. and Ventura, P. (1985). A Trojan horse mechanism for spread of visna virus in monocytes. *Virology* **147**, 231-236.

Perry, H. and Gordon, S. (1987). Modulation of CD4 antigen on macrophages and microglia in rat brain. *Journal of Experimental Medicine* **166**, 1138-1143.

Persson, L., Hardemark, H.G., Gustafsson, J., Rundstrom, G., Mendel Hartvig, I., Esscher, T. and Pahlman, S. (1987). S-100 protein and neuron-specific enolase in cerebrospinal fluid and serum: markers of cell damage in human central nervous system. *Stroke* **18**, 911-918.

Petito, C. (1996). Ependyma and choroid plexus. *Journal of Neuro-AIDS* **1**, 101-110.

Portegies, P., Gans de, J., Lange, J., Derix, M., Speelman, H., Bakker, M., Danner, S. and Goudsmit, J. (1989). Declining incidence of AIDS dementia complex after introduction of zidovudine treatment. *British Medical Journal* **299**, 819-821.

Portegies, P. (1995). HIV-1, the brain, and combination therapy. *Lancet* **346**, 1244-1245.

Power, C., Kong, P.-A., Crawford, T., Wesselingh, S., Glass, J., McArthur, J. and Trapp, B. (1993). Cerebral white matter changes in acquired immunodeficiency syndrome dementia: Alteration of the blood-brain barrier. *Annals of Neurology* **34**, 339-350.

Price, R.W., Brew, B., Sidtis, J., Rosenblum, M., Scheck, A.C. and Cleary, P. (1988). The brain in AIDS: central nervous system HIV-1 infection and AIDS dementia complex. *Science* **239**, 586-592.

Rosengren, L.E., Wikkelsö, C. and Hagberg, L. (1994). A sensitive ELISA for glial fibrillary acidic protein; application in CSF of adults. *Journal of Neuroscience Methods* **51**, 197-204.

Rosengren, L.E., Karlsson, J.E., Karlsson, J.O., Persson, L.I. and Wikkelso, C. (1996). Patients with amyotrophic lateral sclerosis and other neurodegenerative diseases have increased levels of neurofilament protein in CSF. *Journal of Neurochemistry* **67**, 2013-2018.

Royal, W., Selnes, O.A., Concha, M., Nance Sproson, T.E. and McArthur, J.C. (1994). Cerebrospinal fluid human immunodeficiency virus type 1 (HIV-1) p24 antigen levels in HIV-1-related dementia. *Annals of Neurology* **36**, 32-39.

Schacker, T., Collier, A.C., Hughes, J., Shea, T. and Corey, L. (1996). Clinical and epidemiologic features of primary HIV infection. *Annals of Internal Medicine* **125**, 257-264.

Schuitemaker, H., Koot, M., Kootstra, N.A., Dercksen, M.W., de Goede, R.E., van Steenwijk, R.P., Lange, J.M.A., Schattenkerk, J.E., Miedema, F. and Tersmette, M. (1992). Biological phenotype of human immunodeficiency virus type 1 clones at different stages of infection: progression of disease is associated with a shift from monocytotropic to T-cell tropic virus populations. *Journal of Virology* **66**, 1354-1360.

Sei, S., Saito, K., Stewart, S.K., Crowley, J.S., Brouwers, P., Kleiner, D.E., Katz, D.A., Pizzo, P.A. and Heyes, M.P. (1995). Increased human immunodeficiency virus (HIV) type 1 DNA content and quinolinic acid concentration in brain tissues from patients with HIV encephalopathy. *Journal of Infectious Diseases* **172**, 638-647.

Sharer, L.R., Cho, E.S. and Epstein, L.G. (1985). Multinucleated giant cells and HTLV-III in AIDS encephalopathy. *Human Pathology* **16**, 760.

Sidtis, J., Gatsonis, C., Price, R., Singer, E., Collier, A., Richman, D., Hirsch, M., Schaerf, F., Fischl, M., Kieburtz, K., Simpson, D., Koch, M., Feinberg, J. and Dafni, U. (1993). Zidovudine treatment of the AIDS dementia complex: Results of a placebo-controlled trial. *Annals of Neurology* **33**, 343-349.

Singer, E.J., Syndulko, K., Fahy Chandon, B.N., Shapshak, P., Resnick, L., Schmid, P., Conrad, A.J. and Tourtellotte, W.W. (1994). Cerebrospinal fluid p24 antigen levels and intrathecal immunoglobulin G synthesis are associated with cognitive disease severity in HIV-1. *AIDS* **8**, 197-204.

Svennerholm, L. (1980). *Structure and biology of cell membrane gangliosides. 43rd Nobel Symp., Stockholm 1978.* Ouchterlony, Ö. and Holmgren, J. (eds) Cholera and related diarrheas. pp. 80-87 Basel, Karger.

Sönnerborg, A., von Stedingk, L.-V., Hansson, L.-O. and Strannegård, Ö. (1989). Elevated neopterin and beta2-microglobulin levels in blood and cerebrospinal fluid occur early in HIV-1 infection. *AIDS* **3**, 277-283.

Sönnerborg, A., Johansson, B. and Strannegård, Ö. (1991). Detection of HIV-1 DNA and infectious virus in cerebrospinal fluid. *AIDS Research and Human Retroviruses* **7**, 369-373.

Tornatore, C., Meuers, K., Atwood, W., Conant, K. and Major, E. (1994). Temporal patterns of human immunodeficiency virus type 1 transcripts in human fetal astrocytes. *Journal of Virology* **68**, 93-102.

Tyor, W.R., Glass, J.D., Griffin, J.W., Becker, P.S., McArthur, J.C., Bezman, L. and Griffin, D.E. (1992). Cytokine expression in the brain during the acquired immunodeficiency syndrome. *Annals of Neurology* **31**, 349-360.

Vago, L., Castagna, A., Lazzarin, A., Trabattoni, G., Cinque, P. and Costanzi, G. (1993). Reduced frequency of HIV-induced brain lesions in AIDS patients treated with zidovudine. *Journal of Acquired Immune Deficiency Syndromes* **6**, 42-45.

Wallin, A., Gottfries, C., Karlsson, I. and Svennerholm, L. (1989). Decreased myelin lipids in Alzheimer's disease and vascular dementia. *Acta Neurologica Scandinavica* **80**, 319-323.

van den Berg, L., Sadiq, S., Lederman, S. and Latov, N. (1992). The gp120 glycoprotein of HIV-1 binds to sulfatide and to the myelin associated glycoprotein. *Journal of Neuroscience Research* **33**, 513-518.

Wei, X., Ghosh, S.K., Taylor, M.E., Johnson, V.A., Emini, E.A., Deutsch, P., Lifson, J.D., Bonhoeffer, S., Nowak, M.A. and Hahn, B.H. (1995). Viral dynamics in human immunodeficiency virus type 1 infection. *Nature* **373**, 117-122.

Werner, E., Fuchs, D., Hausen, A., Jaeger, H., Reibnegger, G., Werner-Felmayer, G., Dierich, M. and Wachter, H. (1988). Tryptophan degradation in patients infected by human immunodeficiency virus. *Biological Chemistry Hoppe-Seyler* **369**, 337-340.

Werner-Feltmyer, G., Werner, E., Fuchs, D., Hausen, A., Reibnegger, G. and Wachter, H. (1989). Tumor necrosis factor-alfa and lipopolysaccharide enhance interferon-induced tryptophan degradation and pteridine synthesis in human cells. *Biological Chemistry Hoppe-Seyler* **370**, 1063-1069.

Wiley, C.A., Schrier, R.D., Nelson, J.A., Lampert, P.W. and Oldstone, M.B.A. (1986). Cellular localization of human immunodeficiency virus infection within the brains of acquired immune deficiency patients. *Proceedings of the National Academy of Sciences of the United States of America* **83**, 7089-7093.

Wiley, C.C., Masliah, E., Morey, M., Lemere, C., DeTeresa, R., Grafe, M., Hansen, L. and Terry, R. (1991). Neocortical damage during HIV infection. *Annals of Neurology* **29**, 651-657.

Wiley, C.A., Achim, C.L., Schrier, R.D., Heyes, M.P., McCutchan, J.A. and Grant, I. (1992). Relationship of cerebrospinal fluid immune activation associated factors to HIV encephalitis. *AIDS* **6**, 1299-1307.

Wilkinson, I.D., Miller, R.F., Miszkiel, K.A., Paley, M.N., Hall-Craggs, M.A., Baldeweg, T., Williams, I.G., Carter, S., Newman, S.P., Kendall, B.E., Catalan, J., Chinn, R.J. and Harrison, M.J. (1997). Cerebral proton magnetic resonance spectroscopy in asymptomatic HIV infection. *AIDS* **11**, 289-295.

Åsjö, B., Morfeldt-Månson, L., Albert, J., Biberfeld, G., Karlsson, A., Lidman, K. and Fenyö, E.M. (1986). Replicative capacity of human immunodeficiency virus from patients with varying severity of HIV infection. *Lancet* **2**, 660-662.

Progress in HPLC-HPCE, Vol. 7, pp. 619-647
G. Ali Qureshi *et al.* (Eds)

Brain lipid levels in neurodegenerative disorders

Magnus Söderberg and Pavel J. Sindelar

Division of Pathology (IMPI) and Clinical Research Center,Karolinska Institute, S-141 86 Huddinge, Sweden

INTRODUCTION

In the field of neurodegenerative diseases a large number of investigations have been performed on various protein components and neurotransmitters. Since the methodological development of molecular biology, an increasing number of studies have also been performed on genetic aberrations. Nowhere near as much effort has been put into research on brain lipid modifications in these diseases. Developments of new methods for isolation and quantification of lipids, improved chromatography materials and the establishment of gas and high performance liquid chromatography (GC and HPLC) have made it possible to investigate lipid composition in greater detail and to identify subtle lipid changes in connection with pathological conditions. Our understanding of lipid biosynthesis and the role of lipids in various physiological mechanisms has also increased and has thus led to a more relevant interpretation of the effects of lipid changes. Consequently, there are many exciting prospects when lipid levels are investigated in different neurodegenerative disorders.

Neuronal membranes are highly specialised structures involved in receiving, processing, transporting and transmitting information. These functions are not only dependent on receptors, ion channels and enzymes, but also on a delicate balance in the lipid composition of these membranes. Lipids provide the membranes with suitable stability, fluidity and permeability. They are also required for the proper function of integral membrane proteins and serve as precursors for many metabolically active compounds. In fact, the brain is characterised by a high content of neutral lipids, glycolipids and phospholipids (Sastry, 1985). The largest neutral lipid component is cholesterol, making up 30% of the total lipid weight. The phospholipids contain relatively high concentrations of phosphatidylethanolamine (PE), phosphatidylserine (PS) and phosphatidylinositol (PI), and the concentration of PE-plasmalogen is the highest among all organs. The brain also has the highest concentrations of proteins covalently modified with lipids; acylation with myristic or palmitic acid, glycosyl phosphatidylinositol-anchoring and isoprenylation with farnesol or geranylgeraniol (Shultz *et al*, 1988; Ferguson *et al*, 1988; Sinenski and Lutz, 1992). The main function of these lipid modifications is to enable specific interactions with membranes, once again emphasising the importance of an appropriate membrane lipid environment.

Interest is growing concerning the involvement of lipids in neurodegenerative disorders. The finding of the possible role for apolipoprotein E in the development of Alzheimer's disease (AD) (Schmechel *et al.*, 1993), indicating disturbances in lipid trafficking and metabolism, make studies of lipid constituents imperative in the future studies of this disorder. The pathological accumulation of prion proteins in scrapie-infected cells has recently been showed to be accompanied by a decrease in the fluidity of the plasma membrane (Wong *et al.*, 1996), which is likely to be caused by a change in lipid composition.

This chapter focuses on phospholipids and neutral lipids in two major neurodegenerative conditions, Alzheimer's disease and prion disorders. Though the aetiologies of these diseases are different, there are several similarities between them. The similarities in lipid modifications are pronounced and imply that the pathogenic mechanisms could be the same in both conditions. Furthermore, the lipid changes described are quite specific and involve mainly mevalonate pathway lipids. Thus, general membrane destruction is not reflected in these results, but rather that specific metabolic changes result in these modifications. It is therefore our conclusion that lipids play a major role in the development of both Alzheimer's and prion diseases and that further investigations will reveal new insights into the mechanisms of neurodegeneration.

MATERIALS AND METHODS

We have used different extraction and quantification procedures on brain tissue from rats, mice and from man and have obtained satisfactory results. The methods have mostly been primarily developed for use on liver tissue and some modifications have been made to make them equally useful on nervous tissue. Since the brain is very rich in lipids on tissue gram basis, it is preferable to use 10% homogenates instead of the more usual 20%. This both saves material and results in lower losses during washing of the organic lipid-containing phase.

We have used several different extraction and separation methods over the years. The ones that are presented here are the ones presently in use at our laboratory. They have proven to be both speedy and reliable over thousands of analyses. At the present time, phospholipids are separated using HPLC, because of the sensitivity and versatility of this method. However, a portion of the phospholipid results presented has been obtained by thin layer chromatography and a section concerning this technique has therefore also been included.

Homogenisation

The tissue homogenates are prepared in isotonic fluids and we have mostly used 0.25 *M* sucrose or 0.15 *M* KCl with good results. The latter is preferable when peroxidation products are to be determined, since sucrose interferes with the detection of malondialdehyde, a product of lipid peroxidation. Brain tissue contains relatively little connective tissue, which should simplify homogenisation of the samples. However, to effectively disrupt the tissue for lipid extraction, it is necessary to use an extensive homogenisation procedure. The best is to cut the tissue in small pieces in an isotonic water-based solvent and then run it through a blender. We use an Ultra-Turrax blender set on 24000 rpm for 2 min, for a 1 g sample. The volume was then adjusted to obtain a 10% homogenate. This was stored at -20°C until further analysis.

Internal standards

Internal standards should be used whenever possible, since they are required for both quantification and calculation of yield. There are several different approaches to the question of internal standards.

When measuring cholesterol we have tried using radioactive cholesterol as well as sterols with a similar chemical structure to cholesterol, which are not normally present in brain tissue. Radioactive compounds have the advantage of behaving exactly in the same manner as the quantified compound, but after separation determination of radioactivity is required, with either a radioactive flow detector or a scintillation counter. The other method is to use a similar sterol. We have used ergosterol, which is easily detectable on HPLC and behaves very much like cholesterol during the extraction procedures. Ergosterol has a disadvantage in its tendency to spontaneously degrade if it is stored in dissolved form. It is therefore necessary to prepare a fresh standard before each experiment. Both methods are useful, but over time we have come to use the ergosterol approach, since it gives the most reproducible results and is also slightly more expedient.

Ubiquinone is present in different forms in different species of animals and plants. The difference is the length of the polyisoprene chain attached to the benzoquinone ring. In rat and mouse most ubiquinone is present in a form with nine isoprene units in the side chain (Q9), while in man the form with ten isoprene units (Q10) is the dominating one. Forms with detectable amounts of shorter ubiquinones are not found in these species. Thus, we are using a short ubiquinone, ubiquinone-6 (Q6) as an internal standard. It is easily detectable on HPLC and behaves similarly to the longer forms during extraction and separation.

The polyisoprenes dolichol and dolichyl phosphate are also present in forms with different chain lengths. In both rodents and man the forms with 17 to 21 isoprene units dominate. In this case, dolichol-23 and dolichyl-23 phosphate are used as internal standards, respectively. The standards are obtained by isolation of dolichol-23 from bovine pituitary glands (Radominska-Pyrek *et al*, 1979), with subsequent phosphorylation to dolichyl-23 phosphate (Danilov and Chojnacki, 1981).

When investigating phospholipids, it is much more difficult to find a proper internal standard. This is due to the much more heterogeneous nature of these compounds. The differences in electrical charge due to their different head groups make it impossible use a single substance that is adequately similar to be a relevant internal standard for the entire group. However, repeated extractions of the non-lipid residue have shown that a single extraction is quite efficient and yields recoveries around 95-100%.

Lipid extraction

Depending on the nature of the lipids of interest, different extraction procedures should be used to obtain maximal recovery. For complete extraction of dolichol and dolichyl phosphate, alkaline hydrolysis has to be performed. This treatment, however, destroys the phospholipids and most of the ubiquinone. Thus, extraction of ubiquinone and phospholipids is performed separately from the extraction of the polyprenols (dolichol and dolichyl phosphate).

All solvents are supplemented with 5 μM of butylated hydroxytoluene to prevent lipid peroxidation and mixtures of solventes are presented as volume to volume (v/v).

For most purposes, 2 ml of 10% brain homogenate was enough for the determination of the levels of lipids. When cholesterol and ubiquinone were to be extracted, 20 μg of Q6 and 200 μg of freshly dissolved ergosterol were added to the homogenate and vortexed vigorously in a 50 ml Pyrex glass tube. After this, 6 ml of chloroform and 6 ml of methanol were added to obtain a chloroform/methanol/water (C/M/W) ratio of 1/1/0.3. The lipids were then extracted with magnetic stirring at 40°C for 60 min. The tubes were spun at 1200 *g* for 15 min and the supernatant was transferred to a new tube. The tissue residue was reextracted with 5 ml of chloroform/methanol (C/M) 2/1 at 40°C for 60 min, after which 1 ml of methanol was added. The tubes were again centrifuged at 1200 *g* for 15 min. The supernatants from the two extractions were combined and the non-lipid residue was saved for determination of DNA levels (Burton, 1956). The combined supernatants were adjusted to a C/M/W ratio of 3/2/1 to obtain phase separation. The lower phase was retained and evaporated under N_2. The lipid residue was redissolved in 2 ml of chloroform and was named Fraction I.

For the determination of dolichol and dolichyl phosphate, 1 ml of 10 % brain homogenate was extracted in tubes with screw caps. The homogenate was then supplemented with 10 µg dolichol-23 and 1 µg dolichyl-23 phosphate as internal standards. To perform alkaline hydrolysis, 3 ml of methanol and 1.5 ml of 60% KOH was added, after which the tubes were capped tightly and heated at 100°C for 60 min. After cooling, the hydrolysate was extracted twice with 5 ml of diethyl ether and vigorous shaking. The pooled ether extracts were washed with 10 ml of 5% acetic acid and the upper phase, containing the lipids, was retained and subsequently evaporated under N_2. The lipid residue was redissolved in 400 µl C/M (2/1) and was named Fraction II.

Acid hydrolysis, which releases oligosaccharide bound dolichyl phosphate but not dolichyl phosphate covalently bound to proteins, was performed in some experiments. Internal standards of dolichol and dolichyl phosphate were added to 2 ml of 10% tissue homogenate, which was adjusted to C/M/W (1/1/0.3) with the addition of 6 ml of chloroform and 6 ml of methanol. HCl was added to a concentration of 0.1 M. The sample was heated to 55°C for 45 min, and finally heated in a boiling water bath for 10 min. After cooling, pH was adjusted to 6.0 with NaOH. The sample was then adjusted to a C/M/W ratio of 3/2/1 by the addition of 3 ml of chloroform and 1 ml of water. Centrifugation was performed to achieve complete phase separation, after which the lower phase was removed. The protein residue was reextracted with 14 ml of C/M/W (1/1/0.3) at 40°C for 60 min. This extract was then adjusted to contain C/M/W (3/2/1). The lower phases were combined, washed four times with theoretical upper phase (C/M/W, 3/48/47) and evaporated under N_2. The lipid residue was redissolved in 400 ml C/M (2/1) and further treated as Fraction II.

Lipid separation

Fraction I, containing cholesterol, ubiquinone and phospholipids, was placed on a Silica-SepPak cartridge (Millipore), which had been preequilibrated in chloroform. The cartridge was then washed with 25 ml of chloroform to remove the neutral lipids; cholesterol, ubiquinone and also dolichol. Phospholipids were then removed from the Silica-SepPak cartridge with 30 ml of methanol and the solvent was evaporated under N_2. The phospholipids were redissolved in 1 ml of chloroform and stored under N_2 at -20°C until further analysis with HPLC or TLC. The chloroform fraction from the Silica-SepPak cartridge was evaporated under N_2. The residue was dissolved in 100 µl of C/M (2/1) and was saved for further HPLC analysis.

Fraction II, containing dolichol and dolichyl phosphate, was supplemented with 15 ml of methanol/water (M/W) 98/2, containing 20 m*M* of H_3PO_4. This mixture was applied to a C_{18}-SepPak cartridge that had first been washed with 10 ml of M/W (98/2) containing 20 m*M* of

H_3PO_4. The eluate, containing cholesterol, was discarded. To remove H_3PO_4 from the C_{18}-SepPak cartridge, it was washed with 10 ml of M/W (98/2). Dolichol and dolichyl phosphate were then eluted with 20 ml of C/M (2/1). This eluate was supplemented with ammonia to give a final concentration of 0,5%. This mixture was placed on a Silica-SepPak cartridge, which had first been washed with 10 ml of C/M (2/1) containing 0,5% of ammonia, and dolichol was removed with 20 ml of C/M (2/1) containing 0,5% of ammonia. C/M/W (1/1/0.3; 30 ml) was subsequently applied to the Silica-SepPak cartridge to remove dolichyl phosphate. Both eluates were evaporated under N_2. Dolichol was then dissolved in 100 µl of C/M (2/1), while dolichyl phosphate was dissolved in 100 µl of methanol/2-propanol/hexane (3/2/1), containing 2m*M* tert-butylammonium dihydrogen phosphate. These samples were then used in HPLC analysis.

HPLC analysis of neutral lipids

HPLC analyses of neutral lipids were performed on a Hewlett-Packard Hypersil ODS 3 µm (pore size) reversed-phase column, attached to a dual pump HPLC.

For the analysis of cholesterol and ubiquinone, a convex gradient was used, from the initial M/W 9/1 in pump system A, to methanol/2-propanol/hexane (M/P/H) 2/1/1 in pump system B, with a flow rate of 1.5 ml/min and a program time of 50 min. The retention times of the different substances were determined after running standard samples of ubiquinones -6, -9 and -10, as well as ergosterol and cholesterol. The absorption of the eluates was monitored at 210 nm and the peak areas were integrated on a Shimadzu Chromatopac C-R4A.

For the separation of dolichol, pump system A contained methanol/2-propanol/water (M/P/W) 40/60/5 and pump system B contained M/P/H (3/2/1). A convex gradient was used, from 100% system A solvent to 100% system B solvent over 25 min at a flow rate of 2 ml/min. The same conditions were used for the separation of dolichyl phosphate, except that the solutions contained 2m*M* of tert-butylammonium dihydrogen phosphate. Absorption was monitored at 210 nm (Fig 1) and peaks were integrated using automated methods as before.

A fraction of the dolichol is present in α-unsaturated form. To determine this percentage, dolichol peaks were collected after HPLC and were evaporated under N_2. The lipids were redissolved in 100 µl of C/M (2/1) and run on a straight phase HPLC system with a silica column, using 14% diethyl ether in hexane as the solvent (Ekström *et al*, 1987).

HPLC analysis of phospholipids

All major phospholipid classes were separated by HPLC (Dugan *et al*, 1986). An aliquot of the phospholipid fraction was used to determine total lipid phosphorus (Valtersson and Dallner, 1982). A fraction corresponding to 1.5 µmol of lipid phosphorus was evaporated under N_2 and redissolved in 30 µl of hexane/2-propanol (H/P) 3/2 and was then injected into a

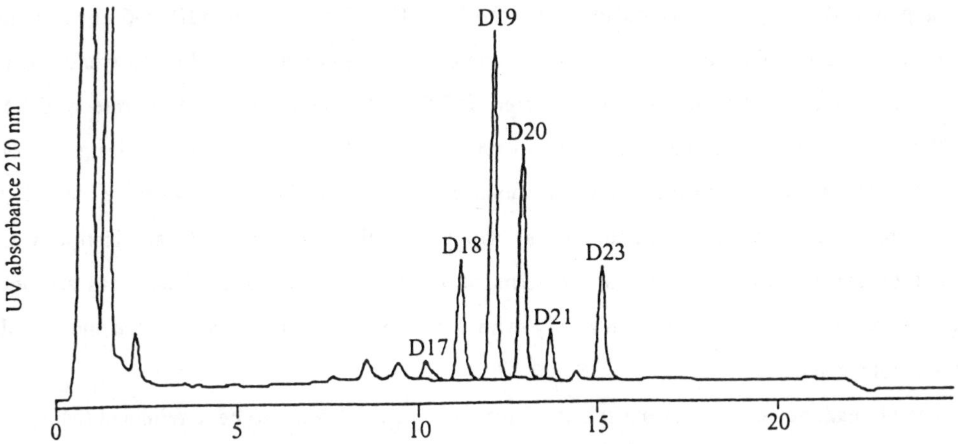

Fig 1. Appearance of dolichols on HPLC, monitored at 210 nm, from the frontal cortex of a 80-year-old man without neuropathological disorders. The indicated peaks are the naturally occurring dolichols with 17-21 isoprene units (D17-D21), as well as the added internal standard dolichol-23 (D23). Numbers on the horizontal axis indicate minutes of HPLC run time.

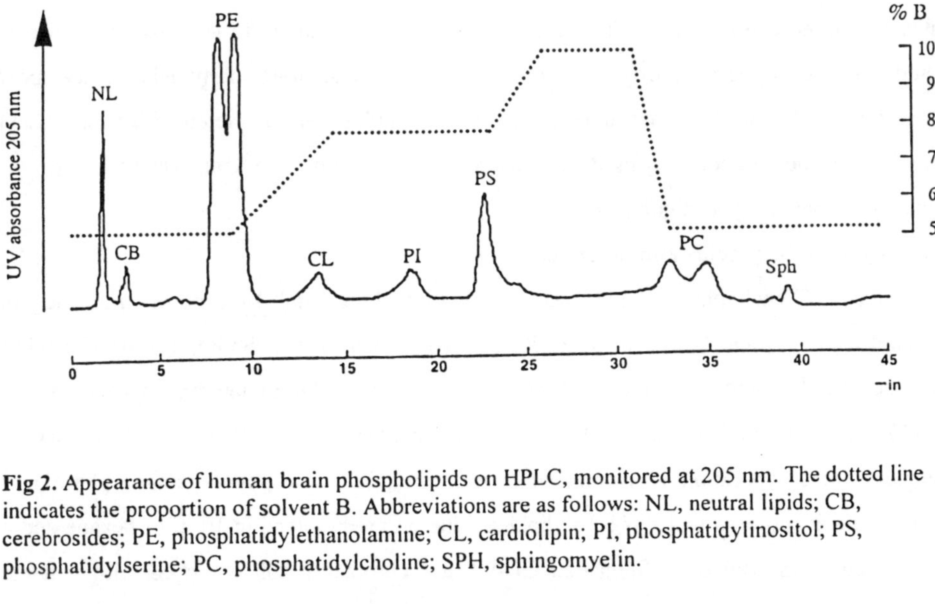

Fig 2. Appearance of human brain phospholipids on HPLC, monitored at 205 nm. The dotted line indicates the proportion of solvent B. Abbreviations are as follows: NL, neutral lipids; CB, cerebrosides; PE, phosphatidylethanolamine; CL, cardiolipin; PI, phosphatidylinositol; PS, phosphatidylserine; PC, phosphatidylcholine; SPH, sphingomyelin.

dual pump HPLC system. The column was a Zorbax SIL (25 cm x 4.6 mm ID, 5-6 μm particle size) from Dupont Co., maintained at 34°C. Pump system A contained H/P (3/2), while pump system B contained hexane/2-propanol/water (H/P/W) 56.7/37.8/5.5. The flow rate was 1.5 ml/min and a stepwise gradient from 50% to 100% solvent B was used, over a total run time of 45 min. The absorption of the eluate was monitored at 205 nm (Fig 2). Peaks were identified by comparison with chromatograms of known phospholipid standards. Quantification was made by collection of the peaks and determination of lipid phosphorus. Peak areas are not useful for quantification, since they vary with the fatty acid composition of the individual phospholipid species.

The plasmalogen content of phosphatidylethanolamine (PE) can also be determined (Murphy *et al*, 1993). An aliquot of PE was dried under N_2 and subjected to the vapours from conc. HCl for a few minutes. This converts PE-plasmalogen (alkenylacyl-PE) to lyso-PE, while alkylacyl-PE and diacyl-PE remain intact. The products were separated on the same HPLC system as above except that the flow rate was 1.8 ml/min and that a steeper gradient was used. Quantification was performed by determining lipid phosphorus.

TLC analysis of phospholipids

The solvent was evaporated from the phospholipid fraction. The lipids were redissolved in 200 μl of C/M (2/1) and 20 μl of the solution was applied to a Silica gel 60 plate (Merck, Darmstadt, Germany) for two-dimensional TLC. The plates were developed with chloroform/methanol/acetic acid/water (106/50/12/6) in the first direction and with chloroform/acetone/methanol/acetic acid/water (50/20/10/10/5) in the second direction. The plates were dried and sprayed with a water solution of Rhodamin 6G to visualize spots. Iodine should not be used, as it destroys double bounds on polyunsaturated fatty acids. Silica spots were scraped off, placed in glass tubes and the phospholipids were extracted with 5 ml of C/M (2/1). Half of the solution was used to measure lipid phosphorus content, while the rest was used to determine fatty acid composition.

GC analysis of fatty acids and aldehydes

An aliquot of each phospholipid class was dissolved in 1 ml 14% boron trifluoride in methanol under N_2 and heated to 90°C for 30 min (Morrison and Smith, 1964). The tubes were cooled to 0°C and the resultant fatty acyl methyl esters (FAME) and dimethyl acetals (DMA) were extracted by adding 4 ml of cold n-pentane and 1 ml of cold 5 *M* NaOH. Complete phase separation was obtained by centrifugation. The upper pentane phase was removed and evaporated under N_2. The lipid residue was redissolved in 10 μl of hexane and a 0.2 μl aliquot was run on a Shimadzu GC-9A gas chromatograph equipped with a flame ionisation detector. Hydrogen was used as the carrier gas and the column was a fused silica capillary with cyanosiloxane 60 as the stationary phase (30 m x 0.25 mm ID, df 0.18 μm). The retention times of different FAME and DMA were determined by comparison with known standards. A standard run has to be performed every day, since minor changes in the status of the column may have dramatic effects on the retention times. This makes it impossible to give an exact temperature gradient that will always work on this type of GC system. Generally, the start temperature was 130°C, which was increased to a final temperature of 190°C over a time of 40 min. The final temperature was held for 5 min.

Quantification was achieved by integrating the peak areas on a Shimadzu Chromatopac C-R3A.

α-Tocopherol, reduced and oxidised ubiquinone

These substances were quantified immediately after homogenisation of the brain tissue. A rapid extraction procedure was used to avoid oxidation (Åberg *et al*, 1992). Ubiquinone-6 and δ-tocopherol were added as internal standards to 1 ml of 10% brain tissue homogenate. The sample was supplemented with 18 ml of methanol and 12 ml of petroleum ether and thoroughly mixed by extensive vortexing. The upper phase was removed, evaporated under N_2, redissolved in 100 μl of C/M (2/1) and labelled Fraction III. An aliquot of about 30 μl from Fraction III was injected onto the same HPLC system as for the determination of cholesterol and total ubiquinone. The retention times of the reduced forms of Q9 and Q10 were determined after reduction of standards with sodium borohydride. For the determination of α-tocopherol, 20 μl from Fraction III was injected onto an HPLC equipped with an L-ECD-6A EC detector from Shimadzu. The column used was the same as for the separation of neutral lipids, i.e. a Hewlett-Packard Hypersil ODS 3 μm. The system was isocratic, consisting of M/W (9/1) in pump system A and methanol/2-propanol 4/1 in pump system B. Both systems contained 20 m*M* $LiClO_4$. The flow was 85% from pump system A and 15% from pump system B, with a flow rate of 1.5 ml/min, while the run time was 12 min. Quantification was performed by integration of peak areas.

RESULTS

Lipid changes during ageing

Fresh autopsy material was used to study the lipid compositions of different regions of the human brain from the age of 33 to 92 (Söderberg *et al*, 1990). Controls demonstrated that *post mortem* changes involving the lipids investigated were very limited and did not influence the results obtained. The total content of phospholipid on gram wet weight brain tissue basis is decreased by only 5-10% at about 90 years of age (Table 1).

Table 1. *Phospholipid contents of various regions of the human brain.*

	Content (mg/g wet weight) at age (years)			
Region	33-36(n=7)	54-57(n=8)	69-72(n=8)	89-92(n=7)
Gray matter	18.2 ± 1.7	17.8 ± 1.2	17.4 ± 1.9	16.4 ± 1.6*
White matter	35.1 ± 2.3	34.7 ± 4.3	29.9 ± 4.2*	28.2 ± 3.4*
Nucleus caudatus	21.2 ± 1.6	20.8 ± 1.8	20.2 ± 2.7	18.4 ± 2.8*
Hippocampus	23.6 ± 1.8	20.5 ± 1.9*	18.5 ± 2.1*	17.2 ± 2.4*
Pons	32.3 ± 2.1	31.0 ± 2.6	28.2 ± 3.1*	26.6 ± 3.6*
Cerebellum	23.3 ± 3.3	22.3 ± 1.9	22.0 ± 1.6	20.2 ± 1.8
Medulla oblongata	34.5 ± 3.4	33.3 ± 4.7	32.9 ± 2.3	28.4 ± 1.6*

The gray and white matter in all analyses was taken from the frontal lobe.
The values represent means ± SD.
*Significant decrease ($p<0,05$) as compared with the youngest age group.

The phospholipid compositions of the regions studied are quite specific but remain unchanged during ageing. For instance, large differences are seen in the ratio between the two major phospholipids, phosphatidylethanolamine (PE) and phosphatidylcholine (PC) (Table 2).

Table 2. *Phosphatidylethanolamine/phosphatidylcholine and dolichol-18/dolichol-20 ratios in different regions of the brain.*

Region	PE/PC ratio	D18/D20 ratio
Gray matter	0.77 ± 0.06	0.40 ± 0.03
White matter	1.71 ± 0.15	1.39 ± 0.08
Nucleus caudatus	0.80 ± 0.09	0.91 ± 0.05
Hippocampus	0.93 ± 0.11	0.39 ± 0.03
Pons	1.68 ± 0.14	0.47 ± 0.05
Cerebellum	1.41 ± 0.10	1.27 ± 0.08
Medulla oblongata	1.43 ± 0.09	0.67 ± 0.04

The values represent means ± SD.

Among the lipid products of the mevalonate pathway, dolichol levels showed the most extensive changes with age. A drastic increase in these levels was observed in all regions examined, with a several-fold elevation during the time period analysed (Fig. 3). Interestingly, the dolichol chain-length pattern was region-specific. For example, the dolichol-18 (D18)/dolichol-20(D20) ratio was 0.40 in grey matter and 1.39 in white matter (Table 2). The polyisoprenoid chain-length patterns were constant throughout the life span in all regions examined.

In contrast to dolichol, dolichyl-P concentrations were quite constant in all regions of the brain (Table 3). A moderate increase, not exceeding 10-20%, could be observed during the time span studied.

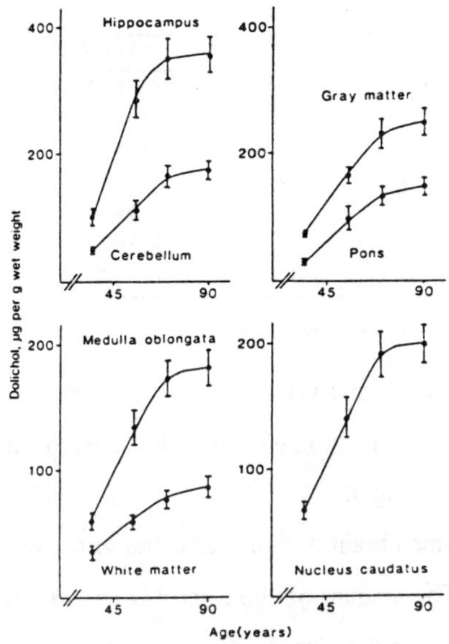

Fig 3. Dolichol content of various regions of the brain during aging. The various time points represent age groups 33-36, 54-57, 69-72 and 89-92 years old. The values are the means ± SD's, with eight individuals in each age group.

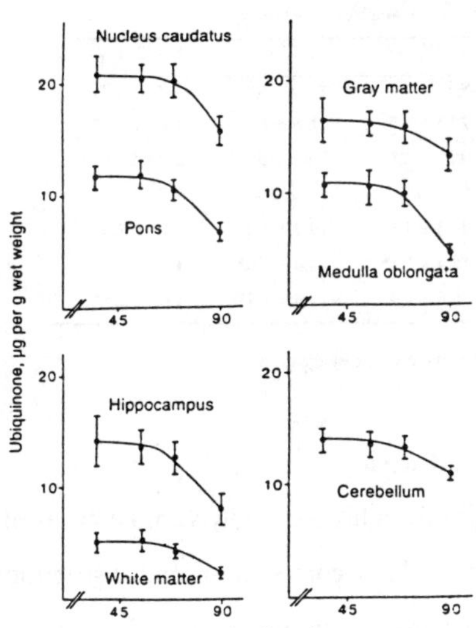

Fig 4. Ubiquinone content of various regions of the brain during aging. The various time points represent age groups 33-36, 54-57, 69-72 and 89-92 years old. The values are the means ± SD's

Table 3. *Content of dolichyl phosphate in various regions of the human brain.*

	Content (µg/g wet weight) at age (years)			
Region	33-36(n=7)	54-57(n=8)	69-72(n=8)	89-92(n=7)
Gray matter	19.4 ± 1.8	20.7 ± 1.5	21.4 ± 2.4	25.8 ± 2.5*
White matter	8.2 ± 0.8	7.9 ± 1.0	9.1 ± 0.7*	9.8 ± 0.9*
Nucleus caudatus	20.4 ± 1.4	22.8 ± 2.2*	21.8 ± 1.8	26.9 ± 2.4*
Hippocampus	24.1 ± 2.9	26.2 ± 2.6	25.3 ± 2.1	28.8 ± 3.7*
Pons	11.1 ± 0.9	12.0 ± 1.7	13.8 ± 0.9*	12.4 ± 1.5
Cerebellum	23.9 ± 3.1	23.8 ± 2.1	25.1 ± 3.0	28.9 ± 2.4*
Medulla oblongata	8.9 ± 0.8	9.3 ± 1.0	10.4 ± 1.6*	9.9 ± 0.9*

The values represent means ± SD.
*Significant difference ($p<0.05$) as compared with the youngest age group.

Measurements of the levels of total Q10 (reduced and oxidised) showed decreases in the contents of this lipid in all regions studied, with the most pronounced decrease occurring between the ages of 70 and 90 (Fig 4).

One-third of the lipid in human brain is cholesterol and virtually all the cholesterol in brain is found in unesterified form. The amount of cholesterol is particularly high in regions which are enriched in myelin. In light of its functional importance, one might expect relative stability in brain levels of cholesterol. Therefore, the behaviour of this lipid during ageing is somewhat surprising. In certain regions, the cholesterol concentration is entirely unchanged , while in other regions the levels decrease by as much as 40% with ageing (Table 4).

Table 4. *Content of cholesterol in various regions of the human brain.*

	Content (mg/g wet weight) at age (years)			
Region	33-36(n=7)	54-57(n=8)	69-72(n=8)	89-92(n=7)
Gray matter	10.2 ± 1.2	9.1 ± 0.9	8.6 ± 1.0*	7.7 ± 0.6*
White matter	22.4 ± 2.9	20.5 ± 4.7	18.6 ± 2.1*	16.4 ± 1.7*
Nucleus caudatus	10.9 ± 0.9	9.3 ± 1.2*	7.6 ± 0.9*	6.7 ± 0.5*
Hippocampus	8.1 ± 0.7	8.3 ± 1.0	7.9 ± 0.5	7.2 ± 0.8*
Pons	20.7 ± 2.9	21.3 ± 3.1	19.7 ± 2.0	20.1 ± 2.6
Cerebellum	17.5 ± 2.0	15.4 ± 1.1*	14.0 ± 1.6*	12.8 ± 1.3*
Medulla oblongata	17.8 ± 2.0	16.6 ± 1.4	15.4 ± 1.9*	14.0 ± 1.6*

The values represent means ± SD.
*Significant difference ($p<0.05$) as compared with the youngest age group.

Lipid changes in Alzheimer's disease

Ten different regions of brains from histologically verified cases of Alzheimer's disease (AD) were investigated with respect to lipid composition and distribution (Söderberg *et al*, 1992; Guan *et al*, 1994). Such regional investigation is absolutely necessary, since the disease affects certain regions selectively and, at least morphologically, does not affect other regions.

The total phospholipid content in most regions of AD brains was the same as that of the controls on gram tissue basis, but in the nucleus caudatus and white matter some decrease was seen (Table 5).

TABLE 5. *Phospholipid contents of different regions of the brains of control and AD patients.*

	Content (mg/g wet weight)	
Region	Control (n=9)	AD (n=9)
Frontal cortex	16.4 ± 1.0	16.1 ± 3.5
Precentral cortex	18.2 ± 3.2	17.5 ± 2.5
Temporal cortex	16.7 ± 2.7	15.7 ± 3.1
Occipital cortex	19.3 ± 3.9	18.0 ± 2.7
Frontal white matter	28.5 ± 4.2	22.7 ± 3.7*
Nucleus caudatus	19.8 ± 1.8	15.9 ± 2.2*
Hippocampus	17.0 ± 1.6	15.4 ± 1.9
Pons	28.7 ± 4.3	27.2 ± 3.3
Cerebellum	25.2 ± 3.3	23.1 ± 2.9
Medulla oblongata	31.3 ± 5.3	30.8 ± 3.8

The values represent means ± SD.
*Significant difference ($p < 0.05$) as compared with controls, using Student's t test.

No major modifications of the two largest components PE and PC could be detected in the Alzheimer group (Fig 5), apart from regional increases in PC and decreases of PE. The level of lipid-extractable PI was increased in regions affected by the disease, such as frontal and temporal cortex and hippocampus. Those regions rich in myelin exhibited a lowered sphingomyelin content. The regions investigated varied in cardiolipin content, which accounted for 1-2% of the total phospholipid (Table 6). The amount, calculated on a phospholipid basis, did not change in Alzheimer cases, neither in the frontal and temporal cortex (which are affected by the disease), nor in the occipital cortex (which does not appear to be involved in the disease process).

TABLE 6. *Total phospholipid and cardiolipin contents of various regions of the brain in AD.*

Region	Phospholipid (µmol P/mg protein)	Cardiolipin (nmol P/mg protein)	Cardiolipin (% of total phospholipid P)
Frontal cortex			
Control	0.397±0.055	4.46±0.46	1.12±0.22
AD	0.404±0.068	3.96±0.35	0.99±0.15
Temporal cortex			
Control	0.460±0.074	3.38±0.27	0.75±0.16
AD	0.487±0.060	3.27±0.22	0.68±0.09
Occipital cortex			
Control	0.389±0.044	5.60±0.43	1.45±0.18
AD	0.440±0.056	6.62±0.81	1.47±0.20

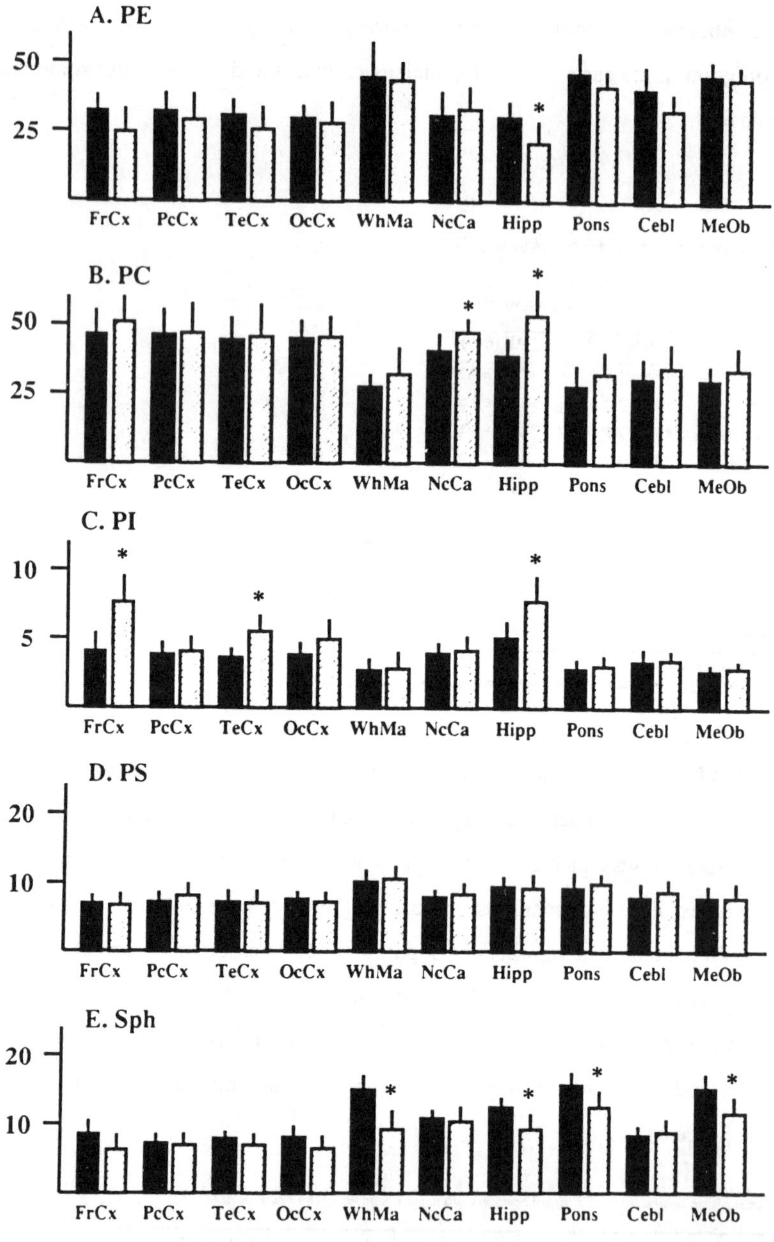

Fig 5. Phospholipid compositions of various regions of control and Alzheimer brains. These results were obtained using thin layer chromatography (TLC) techniques. Solid bars represent the controls, striped bars represent Alzheimer cases. Numbers on the vertical axis are % of the total lipid phosphorus. **A.** Phosphatidylethanolamine (PE). **B.** Phosphatidylcholine (PC). **C.** Phosphatidylinositol (PI). **D.** Phosphatidylserine (PS). **E.** Sphingomyelin (Sph). The abbreviations used on the horisontal axis are as follows: FrCx, frontal cortex; PcCx, precentral cortex; TeCx, temporal cortex; OcCx, occipital cortex; WhMa, frontal white matter; NcCa, nucleus caudatus; Hipp, hippocampus; Cebl, cerebellum; MeOb, medulla oblongata. The values are the means of nine experiments. The vertical bars represent the SD's. Asterisks represent a significant difference ($p<0.05$), using Student's t test.

The values represent the means ± SD for 6 cases.
P = inorganic ortophosphate after release from phospholipids.

As pointed out earlier, the dolichol content of brain increases during ageing. Most surprisingly, the amounts of dolichol in all regions of the AD brain are decreased in comparison with age-matched controls (Table 7). Significantly, this decrease was not restricted to those regions which are most heavily affected by the disease, but could be observed in most investigated parts of the brain.

TABLE 7. *Dolichol and dolichyl phosphate contents of different regions of the brains of control and AD patients.*

	Dolichol content (μg/g wet weight)		Dolichyl phosphate content (μg/g wet weight)	
Region	Control (n=9)	AD (n=9)	Control (n=9)	AD (n=9)
Frontal cortex	199 ± 53	111 ± 35*	21.2 ± 4.1	32.2 ± 8.4*
Precentral cortex	167 ± 42	142 ± 39	17.7 ± 1.2	32.5 ± 7.5*
Temporal cortex	204 ± 41	122 ± 37*	16.1 ± 1.7	31.9 ± 6.2*
Occipital cortex	178 ± 29	165 ± 31	19.5 ± 2.0	41.1 ± 5.5*
Frontal white matter	84 ± 7	52 ± 17*	8.1 ± 1.7	8.3 ± 3.8
Nucleus caudatus	182 ± 41	131 ± 40*	22.4 ± 1.9	37.0 ± 5.1*
Hippocampus	270 ± 87	151 ± 20*	25.9 ± 3.4	33.5 ± 6.5*
Pons	134 ± 26	111 ± 22	12.4 ± 1.5	25.2 ± 4.6*
Cerebellum	167 ± 27	97 ± 34*	25.2 ± 2.6	27.7 ± 3.6
Medulla oblongata	152 ± 33	106 ± 19*	9.7 ± 1.2	20.2 ± 4.9*

The values represent means ± SD.
*Significant difference ($p < 0.05$) as compared with controls, using Student's *t* test.

The decrease in dolichol content was not associated with any modification in the polyisoprenoid chain-length pattern. About 5-6% of human brain polyprenes are normally found in the α-unsaturated form and this value was decreased somewhat in the diseased state (not shown).

The most prominent lipid modification observed in Alzheimer´s disease involved dolichyl phosphate. The total content of this compound was increased substantially in most regions, and in some regions by as much as 100% (Table 7). When extraction was performed using chloroform/methanol, less than 10% of this lipid was extracted (Table 8). Upon acid hydrolysis, which removes oligosaccharide chains from dolichyl phosphate, the large majority of this lipid intermediate could be extracted both in controls and in AD. This experiment demonstrates that in the brain oligosaccharide synthesis is going on continuously and that the whole dolichyl phosphate pool is involved in this process, since glycosylated species of dolichol phosphate are short-lived. After alkaline hydrolysis, only a small additional amount of this intermediate was liberated, in contrast to the liver, where the portion of dolichyl phosphate released by alkali makes up one-third of the total (Thelin *et al.*, 1991). This indicates that dolichyl phosphate is not attached covalently to proteins in the human brain to any larger extent.

TABLE 8. *Amounts of dolichyl phosphate obtained using different preparative procedures.*

Region	Dolichyl phosphate amount (μg/g wet weight) No hydrolysis	Acid hydrolysis	Alkaline hydrolysis
Frontal cortex			
Control (n=5)	2.5 ± 0.7	21.5 ± 2.5	22.2 ± 3.8
AD (n=5)	2.9 ± 0.6	29.8 ± 4.7	30.7 ± 7.1
Frontal white matter			
Control (n=5)	0.4 ± 0.1	7.7 ± 0.9	8.2 ± 1.7
AD (n=5)	0.4 ± 0.1	8.0 ± 0.7	8.5 ± 2.6

The values represent means ± SD.

During ageing the amount of ubiquinone decreases, but in the AD state, ubiquinone content increased significantly in most regions investigated. The elevation is as large as 100% (Table 9), while the levels of cholesterol remain unchanged.

TABLE 9. *Contents of cholesterol and ubiquinone in different regions of control and AD brains.*

	Cholesterol content (mg/g wet weight)		Ubiquinone content (μg/g wet weight)	
Region	Control (n=9)	AD (n=9)	Control (n=9)	AD (n=9)
Frontal cortex	8.2 ± 2.4	8.1 ± 2.4	13.1 ± 0.7	16.9 ± 1.8*
Precentral cortex	8.7 ± 2.1	8.5 ± 1.9	14.7 ± 1.8	17.9 ± 2.3
Temporal cortex	9.1 ± 1.9	8.7 ± 2.5	14.3 ± 1.5	16.8 ± 2.8
Occipital cortex	10.7 ± 3.1	9.5 ± 2.2	16.0 ± 0.9	15.8 ± 2.1
Frontal white matter	20.8 ± 8.3	17.3 ± 6.3	2.6 ± 0.4	4.2 ± 1.7*
Nucleus caudatus	8.9 ± 3.0	7.2 ± 2.0	14.9 ± 1.3	22.4 ± 3.2*
Hippocampus	10.1 ± 2.4	8.3 ± 3.4	8.7 ± 0.8	11.8 ± 1.6*
Pons	20.3 ± 7.9	18.1 ± 6.8	7.2 ± 1.2	16.1 ± 4.9*
Cerebellum	15.6 ± 4.5	13.5 ± 2.7	9.7 ± 2.9	14.8 ± 3.2*
Medulla oblongata	22.0 ± 8.2	21.1 ± 5.3	5.0 ± 1.3	9.4 ± 2.0*

The values represent means ± SD.
*Significant difference ($p < 0.05$) as compared with controls, using Student's t test.

Fatty acid composition of phospholipids in Alzheimer's disease

It is generally accepted that the glycerol backbone of phospholipids has a different turnover rate than the fatty acyl moieties and it is a common finding that the latter moieties can be considerably modified without changing the overall phospholipid composition. There are no major changes in the amounts of the individual phospholipids in the AD brain, but it was of obvious interest to also analyse the fatty acid (FA) compositions of the phospholipids.

Phosphatidylethanolamine (PE) and phosphatidylcholine (PC) constitute 80% of the total phospholipid in the brain and, for this reason, these lipids were selected for analysis (Söderberg *et al*, 1991). In addition, cardiolipin was also studied in detail (Guan *et al*, 1994). The FA compositions of PE and PC are region-specific, with as much as ten-fold differences being observed when different areas were compared. During ageing, between 33 and 92 years of age, no significant changes in these compositions are seen.

Upon comparison of the distribution patterns of the three main groups of FA, the Alzheimer brains displayed deviations from the controls. Generally, the amounts of monounsaturated FA were about the same, but there was a clear tendency towards an increase in total saturated and a decrease in polyunsaturated FA levels (Fig 6). In contrast to the controls, however, the variations between different AD patients were quite considerable, possibly reflecting different developmental stages of the disease.

The saturated FA, 14:0, 16:0 and 18:0, in PE were increased significantly in connection with AD (Table 10), not only in the frontal grey matter and hippocampus, which are major targets of the disease, but also in white matter and pons, which, at least morphologically, do not exhibit extensive cellular damage. The major monounsaturated component of PE, 18:1, which may account for as much as 50% of the total FA content, remained unchanged during the disease process. The levels of the three major polyunsaturated components, 20:4, 22:4 and 22:6, all of which are richly represented in PE in all regions, were all decreased extensively in the four brain regions of the AD cases.

TABLE 10. *Fatty acid composition of phosphatidylethanolamine in AD.*

	Fatty acid (% of total)							
Region	14:0	16:0	18:0	18:1	20:1	20:4	22:4	22:6
Frontal cortex								
Control	0.7	9.4	35.0	10.6	1.0	10.7	9.7	23.5
AD	2.5*	12.6*	42.7*	12.5	1.1	6.9*	5.6*	12.6*
Frontal white matter								
Control	0.4	7.7	10.5	44.6	6.9	8.0	15.8	4.6
AD	2.4*	11.3*	17.4*	44.9	6.2	3.7*	8.8*	2.0*
Hippocampus								
Control	0.6	8.6	25.6	21.3	1.8	12.4	11.9	16.9
AD	2.5*	14.1*	33.0*	23.2	1.8	8.1*	7.0*	7.9*
Pons								
Control	0.9	8.7	9.5	44.0	10.8	8.6	10.6	5.7
AD	1.0	9.1	16.7*	46.4	9.5	4.9*	6.2*	4.1*

The values are the means of measurements from 8-10 different individuals. SD's were 3-30% of the means.
The mean age ± SD was 80.0 ± 7.9 for AD and 79.2 ± 11.6 for controls.
*These values are significantly different from the controls ($p<0.05$), using Wilcoxon-Mann-Whitney test.

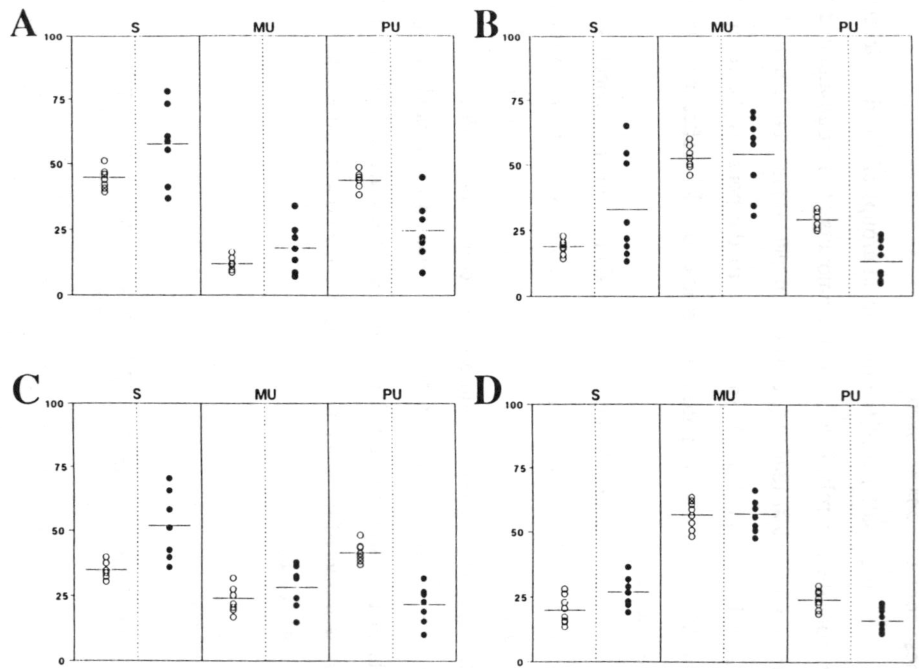

Fig 6. Distribution of saturated and unsaturated fatty acids in phosphatidylethanolamine of various regions of the brain in Alzheimer's disease. **A.** Frontal cortex. **B.** Frontal white matter. **C.** Hippocampus. **D.** Pons. The individual values are plotted separately. Open circles, controls; filled circles, Alzheimer's disease. Abbreviations are: S, saturated fatty acids; MU, monounsaturated fatty acids; PU, polyunsaturated fatty acids. The horizontal bars represent the means. The numbers on the vertical axis represent percentage of total fatty acids.

The FA composition of PC differs qualitatively from that of PE. Only 2-3% of the total FA in PC is polyunsaturated, the rest being saturated and, in part, monounsaturated. This composition in various regions of the control brains was very similar to that found in the diseased brains and no major modifications could be observed (Table 11).

TABLE 11. *Fatty acid composition of phosphatidylcholine in AD.*

	Fatty acid (% of total)							
Region	14:0	16:0	18:0	18:1	20:1	20:4	22:4	22:6
Frontal cortex								
Control	3.5	51.2	12.3	27.0	0.4	2.6	0.4	1.4
AD	2.8	53.8	14.0	25.1	*tr.*	2.0*	*tr.*	0.8*
Frontal white matter								
Control	2.3	31.7	16.7	42.6	1.4	2.2	0.6	*tr.*
AD	1.8	33.8	19.8	40.8	1.0	1.3*	0.4	*tr.*
Hippocampus								
Control	2.7	43.7	16.1	31.2	*tr.*	2.9	*tr.*	0.9
AD	2.1	46.9	17.6	29.0	*tr.*	2.1*	*tr.*	0.6
Pons								
Control	0.7	33.5	22.5	37.0	2.2	0.7	*tr.*	0.5
AD	0.9	35.4	19.0	38.7	1.9	0.8	*tr.*	0.6

The values are the means of measurements from 8-10 different individuals. SD's were 2-15% of the means.
The mean age ± SD was 80.0 ± 7.9 for AD and 79.2 ± 11.6 for controls.
Traces amount are identified as *tr.*
*These values are significantly different from the controls ($p<0.05$), using Wilcoxon-Mann-Whitney test.

As described previously, no changes in cardiolipin content were detected in the brain samples from AD patients. The total amounts of saturated FA (14:0, 16:0 and 18:0), monounsaturated FA (16:1 and 18:1), and polyunsaturated FA (18:2, 20:4, 22:4 and 22:6) in cardiolipin were about the same in the different brain regions investigated (Table 12). In the affected brain regions there was a moderate decrease, 10-25%, in the levels of 20:4 and 22:6. As a result, the total PUFA in cardiolipin decreased by 9% (Fig 7), which is not nearly as large as the decrease observed in the case of PE. Most significantly, in the occipital cortex, which is not a primary target of this disease, the FA pattern in cardiolipin was not modified at all.

Lipid changes in scrapie-infected mice

An animal model of prion disease, the scrapie-infected mouse brain, was investigated with respect to both phospholipid and neutral lipid compositions (Guan *et al*, 1996). This model is very interesting from the point of view of neurodegenerative diseases, since the process is

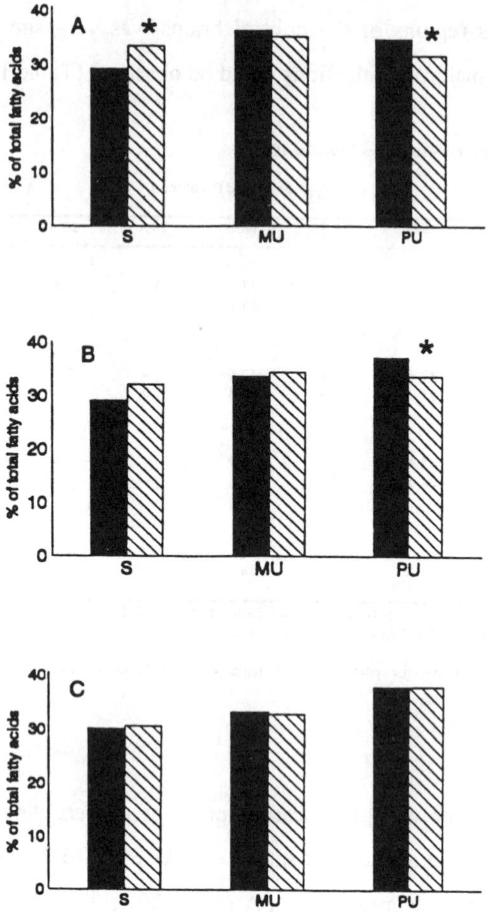

Fig 7. Distribution of saturated and unsaturated fatty acids in cardiolipin of various regions of the brain in Alzheimer´s disease. **A.** Frontal cortex. **B.** Temporal cortex. **C.** Occipital cortex. Abbreviations are: S, saturated fatty acids; MU, monounsaturated fatty acids; PU, polyunsaturated fatty acids. Solid bars represent the controls, striped bars represent Alzheimer cases. The values are the means of measurements on six cases. SD´s were 1-4% of these means. Asterisks represent a significant difference ($p<0.05$), using Student´s t test.

initiated by the injection of the scrapie agent and the disease can thereafter be continuously monitored up to the terminal stage.

The total amount of dolichol in the brains of terminally ill mice was decreased by 30% (Fig 8*A*), in contrast to the phosphorylated form, which increased by 30% (Fig 8*B*). Most significantly, in the later stages of the disease, at 120 days, before clinical symptoms had developed, the dolichol and dolichyl phosphate amounts were already modified. The isoprenoid chain-length patterns for both dolichol and dolichyl phosphate were very similar during the entire process.

Ubiquinone is present in two forms with different chain lengths in mouse brain. Q9 is present in the largest amounts, but about 25% of the total is Q10. Both forms of this redox lipid were increased greatly in amount at the terminal stage, when the amounts were 2.5-fold those found in control brains (Table 13). Utilisation of this lipid as an antioxidant by brain tissue is indicated by the fact that only 20-25% of the total was found in reduced form, both in control and diseased brain. The other lipid soluble antioxidant, α-tocopherol, was also elevated in terminally ill mice by about 50%.

TABLE 12. *Fatty acid composition of cardiolipin in AD.*

	Fatty acid (% of total)								
Region	14:0	16:0	16:1	18:0	18:1	18:2	20:4	22:4	22:6
Frontal cortex									
Control	2.3	9.7	3.2	17.3	32.9	7.6	11.6	3.7	11.7
AD	2.2	11.4*	3.5	19.9	31.6	7.0	10.4	3.5	10.6
Temporal cortex									
Control	1.5	7.5	2.8	20.1	30.8	6.8	12.7	4.2	13.4
AD	2.4	8.7	3.5	20.9	30.9	7.2	11.4	3.4	11.6*
Occipital cortex									
Control	1.5	8.4	0.9	19.7	32.3	10.6	8.0	4.9	14.4
AD	1.4	8.8	1.1	19.5	31.7	10.1	7.6	5.8	14.1

The values are the means of measurements from 6 different individuals. SD's were 3-8% of the means.
*These values are significantly different from the controls ($p<0.05$), using group *t*-test.

As in other animal species, cholesterol accounts for a large portion of total brain lipids in the mouse. In contrast to the other mevalonate pathway lipids, cholesterol levels remained unaltered, even in the terminal phase of the disease (not shown).

The phospholipid pattern, including the smaller components, remained unchanged between the time of infection and 140 days after infection (Table 14).

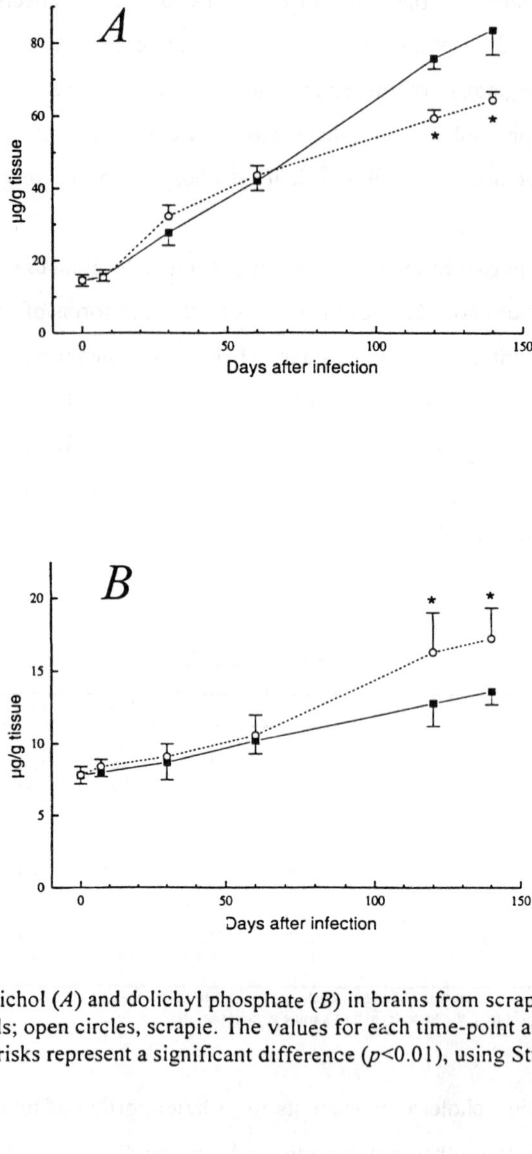

Fig 8. Contents of dolichol (*A*) and dolichyl phosphate (*B*) in brains from scrapie-infected mice. Filled squares, controls; open circles, scrapie. The values for each time-point are means ± SD (bars) for five animals. Asterisks represent a significant difference (p<0.01), using Student's *t* test.

TABLE 13. *Ubiquinone and α-tocopherol contents in brains of terminally ill scrapie-infected mice.*

Region	Control	Scrapie
Total Q (μg/g.w.w.)	40.3 ± 2.0	99.9 ± 10.5*
Q10/Total Q	0.261 ± 0.021	0.322 ± 0.025
Reduced Q/Total Q	0.212 ± 0.012	0.235 ± 0.025
Reduced Q9/Total Q9	0.214 ± 0.015	0.266 ± 0.027*
Reduced Q10/Total Q10	0.205 ± 0.006	0.176 ± 0.003
Total reduced Q (μg/g.w.w.)	8.5 ± 0.3	23.5 ± 2.1*
α-tocopherol (μg/g.w.w.)	10.1 ± 1.1	15.5 ± 2.2*

The values represent the means ± SD of measurements from 5 mice.
Q is short for ubiquinone.
*These values are significantly different from the controls ($p<0.05$), using Student's *t* test.

TABLE 14. *Phospholipid composition in brains from scrapie-infected mice*

	% of total phospholipid phosphorus						
Days post inoculation	PE alkenylacyl	PE diacyl[§]	CL	PI	PS	PC	SPH
5 days							
Scrapie	18.5	21.6	2.4	3.5	9.9	40.7	3.6
Control	19.1	21.0	2.4	3.4	8.7	41.9	3.6
30 days							
Scrapie	19.3	20.0	2.6	3.5	9.4	41.9	3.4
Control	19.7	22.4	2.4	3.4	8.4	40.3	3.4
60 days							
Scrapie	18.9	20.4	2.7	3.2	9.7	41.3	3.8
Control	19.0	21.9	2.6	3.2	9.2	40.3	3.5
120 days							
Scrapie	21.1	20.5	2.6	3.2	9.4	39.8	3.8
Control	21.0	20.8	2.6	3.1	9.4	39.5	3.6
140 days							
Scrapie	20.5	20.8	2.5	3.4	8.6	40.7	3.5
Control	19.8	20.9	2.4	3.4	9.6	40.6	3.4

The values represent the means ± SD of measurements from 5 mice. SD's were 2-7% of the means.
The abbreviations are: PE, phosphatidylethanolamine; CL, cardiolipin; PI, phosphatidylinositol; PS, phosphatidylserine; PC, phosphatidylcholine; and SPH, sphingomyelin.
[§]Includes alkylacyl-PE (usually less than 1% of total phospholipid phosphorus).

GC analysis was employed to measure the FA and aldehyde compositions of the isolated individual phospholipids. In this neurodegenerative condition, both the FA and aldehyde compositions of the individual phospholipids remained unchanged (Table 15).

TABLE 15. *Fatty acid and aldehyde composition of phosphatidylethanolamine in scrapie-infected mice*

	Fatty acid (% of total fatty acids)							Aldehyde (% of total aldehydes)			
Days post inoculation	16:0	18:0	18:1	20:1	20:4	22:4	22:6	16:0	18:0	18:1ω9	18:1ω7
5 days											
Scrapie	8.4	23.1	16.7	2.8	13.9	6.2	28.9	34.3	41.1	12.6	12.0
Control	8.3	24.8	15.0	2.7	14.1	5.9	29.2	36.3	41.4	12.8	9.5
30 days											
Scrapie	7.8	22.8	16.7	3.7	12.8	6.1	30.2	28.9	42.6	12.9	15.7
Control	9.1	23.5	17.1	3.2	12.5	5.4	31.9	29.9	42.7	12.8	14.9
60 days											
Scrapie	7.4	22.2	17.5	3.6	12.2	5.6	31.8	24.8	39.1	12.5	23.6
Control	6.9	21.9	18.8	4.4	12.3	6.1	29.7	24.9	39.2	12.8	23.1
120 days											
Scrapie	7.3	19.0	22.7	5.5	12.6	5.5	27.0	25.8	34.7	12.7	26.8
Control	7.0	19.9	22.4	6.4	10.9	5.6	27.9	24.8	37.7	14.2	23.3
140 days											
Scrapie	7.6	20.6	20.8	4.8	12.4	5.5	28.3	25.5	39.5	12.5	22.9
Control	7.0	21.6	17.8	4.8	12.3	5.9	30.7	24.4	37.2	13.0	24.9

The values represent the means ± SD of measurements from 5 mice. SD's were 3-10% of the means.

DISCUSSION

Many properties of the brain are governed by the amounts, types and distributions of the various lipids present. Investigations to date have been focused on the two major lipid components, phospholipids and cholesterol. Our investigations emphasise the importance of other products of the mevalonate pathway. The level of dolichol, which is known to affect the biophysical properties of membranes, increases several-fold during ageing, whereas its phosphorylated counterpart, the intermediate in glycoprotein synthesis, is modified to only a very limited extent. The decrease in ubiquinone content indicates that the amount of lipid-soluble antioxidant available to the brain is diminished, which is very likely to be deleterious, since protection from free radicals and reactive oxygen metabolites may be inadequate. Several lines of evidence point to the possible importance of peroxidation in the pathogenesis of AD (Subbarao *et al.*, 1990; Mattson and Goodman, 1995). An important finding is that the ubiquinone amounts are increased in all parts of the AD brain. Since the number of mitochondria is not changed in the diseased brain (Sumpter *et al*, 1986), this elevation does not indicate an increase or enlargement of the mitochondrial respiratory system. The most plausible explanation for this finding is that the antioxidant level is elevated as a response to an increased production of free radicals. Thus, the elevation in ubiquinone would be involved in cellular antioxidant defence systems. In this context, the question arises as to what the level of lipid peroxidation in these brains is and whether oxidative damage to protein or DNA occurs. The decrease in polyunsaturated fatty acid content is a strong indication that lipid peroxidation occurs. The increase in carbonyl content of proteins in Alzheimer´s disease (Smith *et al*, 1996) also supports the oxidative damage theory. So far, no extensive analysis has been performed on peroxidative changes of DNA.

The largest product of the mevalonate pathway is cholesterol. Its absolute concentration is dependent on the brain region investigated. The concentrations of this lipid were neither modified in ageing nor in AD.

Changes in the phospholipid composition of AD brains are relatively limited. This, however, does not mean that these changes are unimportant. The slight regional decrease in PE and increase in PC, as well as the decrease in sphingomyelin, could be explained as being secondary to the loss of certain neuronal structures. The increase in free phosphatidylinositol is more intriguing, as it could imply modifications in the phosphatidylinositol-anchoring of membrane proteins. Previously, the possibility was raised that alterations in mitochondrial function are elicited by changes in cardiolipin content and structure (Pettegrew *et al*, 1990). However, the levels and structure of cardiolipin are not modified in the brain in connection with AD and thus, disturbances in mitochondrial function cannot be explained by changes in this lipid.

At the present stage, it is not yet clear whether the changes in the fatty acid pattern of the major phospholipids observed here are characteristic for AD or are also found in other neurodegenerative conditions. On the other hand, the modifications described differentiate this disease from ageing. The substantial changes in the saturated to polyunsaturated fatty acid ratio in the major phospholipid, phosphatidylethanolamine, very probably influences the function of various membranes in the brain by modifying their fluidity and permeability. In addition, the decrease in arachidonic acid content is likely to be deleterious from the point of view of production of important metabolites, such as prostaglandins and leukotrienes. The fact that neither phosphatidylcholine nor cardiolipin displayed any significant changes in fatty acid compositions further emphasise the specificity of the changes in the fatty acid composition of phosphatidylethanolamine.

During development of scrapie in mouse brain, the levels of mevalonate pathway lipids changed greatly, with dolichol amounts being decreased, while dolichyl phosphate and ubiquinone levels are increased. These modifications, together with the unaltered concentrations of cholesterol, are probably important factors in changing the functional status of brain membranes. The highly significant increase in ubiquinone content was paralleled by an elevation in α-tocopherol, suggesting that the brain cells had an increased requirement of antioxidants. By far the largest amounts of brain lipids, phospholipids and cholesterol, as well as the FA compositions of the phospholipids remain completely unaltered during the entire disease period. This finding emphasises the specific nature of the modifications in lipid metabolism occurring in prion infection and indicate that cell damage is not caused by lysosomal disruption elicited by accumulation of the prion proteins in this organelle.

The findings in this study suggest that this is an important field of research in investigating the pathogenesis of AD. The biosynthetic enzymes of the mevalonate pathway, both prior to the branching point and those subsequently participating in the synthesis of dolichol, dolichyl phosphate and the side-chain of ubiquinone, are of particular interest because of the modified levels of these lipids. The increased levels of dolichyl phosphate and the highly glycosylated form of this lipid even in the diseased state are indicative of an increased rate of protein glycosylation, possibly of a pathological nature. The β-amyloid precursor protein (APP) contains two N-glycosidically linked oligosaccharide chains (Påhlsson *et al*, 1992). It would be of considerable interest to characterise the structures and compositions of the oligosaccharide chains associated with dolichyl phosphate and compare them with those of APP. This could give an idea as to what extent abnormal sugar sequences are synthesised in connection with this disease.

Recent reports indicate that administration of certain drugs leads to modifications in the lipid compositions of various organs, including increased ubiquinone contents (Åberg *et al*, 1994). In this way, there is a possibility that we can further improve the antioxidant status of the brain and thereby slow the progress of Alzheimer´s disease. Consequently, it will be necessary to analyse the effects of a number of drugs on this biosynthetic pathway and to determine conditions which cause favourable changes by selectively affecting individual pathways.

In order to find an experimental model, we have turned here to another group of neurodegenerative diseases, which has been the object of increasing interest. Prion diseases occur both in humans and animals and are available in several experimental systems. One has access to infected brains from rats, mice and hamster, transgenic mice and infected neuroblastoma cells. It is remarkable that many of the properties of prion diseases exhibit great similarities to those observed in AD. Both of these conditions affect the brain selectively and lead to dementia. Furthermore, the agents which are considered to be responsible for these diseases have similar appearances. In AD, the amyloid precursor protein, APP, is found in the cell membrane of normal neuronal cells. Upon cleavage, a portion re-enters the cell and, under appropriate physical conditions, causes cell toxicity (Yankner *et al*, 1990). The prion protein is also associated with the cell membrane of normal neurons, re-enters the cell and becomes pathogenic upon conformational change (Taraboulos *et al*, 1992). There are forms of prion disease which possess amyloid plaques, as well as neurofibrillary tangles (Watanabe and Duchen, 1993). Neither of these diseases modifies the phospholipid composition or cholesterol content of brain, while in both cases dolichol is decreased, whereas both dolichyl phosphate and ubiquinone are increased. In a few percent of both diseases mutations in the prion and the preamyloid gene, respectively, have been observed (Collinge and Palmer, 1993; Selkoe, 1996). For this reason, it appears quite reasonable to conduct metabolic labelling, enzyme studies and regulatory studies in the prion material and then compare the results obtained to the conditions found in AD, in order to identify possible common denominators.

During the past decade, great effort has been invested in identifying factors responsible for the development of neurodegenerative diseases and, particularly, of Alzheimer's disease. In spite of the fact that a number of clearly defined functional modifications have been reported, our knowledge about the aetiology is still far from complete. An important explanation for the relatively slow progress in understanding this disease is the lack of suitable experimental systems. Several promising models have been presented, especially those involving transgenic animals (Price *et al*, 1996). Transgenic mice will be most valuable for the analysis of branch-point enzymes of the mevalonate pathway and to establish whether the kinetic properties of these enzymes are modified and/or whether up- and down-regulations are elicited at the transcriptional and/or translational levels. Parallel studies of lipid half-lives using brain slices could efficiently complement these analyses. Subfractionation to determine both the localisation and redox state of ubiquinone would be advantageous. These systems could also be used for studies on lipid peroxidation.

Scrapie-infected neuroblastoma cells is a very good system for isoprenylation studies and for investigating the role of G-proteins in the propagation of the scrapie agent. It is also very possible that a large number of covalent modifications of proteins involving lipids have not yet been identified. Such modifications could have causative roles in disease processes. Considering the development in this field, there are good reasons to believe that our knowledge about lipid changes in connection with neurodegenerative diseases will increase greatly in the near future and that close connections between lipid modifications and the pathological processes will be found.

CONCLUSIONS

Some lipid modifications in the AD brain occur in all regions, while others are seen only in those regions which are damaged by the disease. The modifications in phospholipid contents are modest, while changes in the levels of mevalonate pathway lipids are more pronounced and of different character than those found during ageing. Changes in the levels of mevalonate pathway lipids are also observed in scrapie-infected mice. It appears that changes in lipid compositions and distributions could be important events in the development of AD and prion disorders and may contribute to the functional disturbances connected with these conditions. Thus, advances in lipid analysis and biochemistry are of great importance in future investigations of neurodegenerative disorders.

REFERENCES

Åberg, F., Appelkvist, E. L., Dallner, G. and Ernster, L. (1992) Distribution and redox state of ubiquinones in rat and human tissues. *Arch. Biochem. Biophys.* **295**, 230-234.

Åberg, F., Zhang, Y., Appelkvist, E. L. and Dallner, G. (1994) Effects of clofibrate, phthalates and probucol on ubiquinone levels. *Chem. Biol. Interact.* **91**, 1-14.

Burton, K (1956) A study of the conditions and mechanism of the diphenylamine reaction for the colorimetric estimation of deoxyribonucleic acid. *Biochem. J.* **62**, 315-323.

Collinge, J. and Palmer, M. S. (1993) Prion diseases in humans and their relevance to other neurodegenerative diseases. *Dementia* **4**, 178-185.

Danilov, L. L. and Chojnacki, T. (1981) A simple procedure for preparing dolichyl monophosphate by the use of $POCl_3$. *FEBS Lett.* **131**, 310-312.

Dugan, L. L., Demediuk, P., Pendley II, C. E. and Horrocks, L. A. (1986) Separation of phospholipids by high-performance liquid chromatography: all major classes including ethanolamine and choline plasmalogens and most minor classes, including lysophosphatidylethanolamine. *J. Chromatogr.* **378**, 317-327.

Ekström, T. J., Chojnacki, T. and Dallner, G. (1987) The α-saturation and terminal events in dolichol biosynthesis. *J. Biol. Chem.* **262**, 10460-10468.

Ferguson, M. A. J., Williams, A. F. and Ruoslahti, E. (1988) Cell surface anchoring of proteins via glycosyl-phosphatidylinositol structures. *Ann. Rev. Biochem.* **57**, 285-320.

Guan, Z. Z., Söderberg, M., Sindelar, P. and Edlund, C. (1994) Content and fatty acid composition of cardiolipin in the brain of patients with Alzheimer's disease. *Neurochem. Int.* **25**, 295-300.

Guan, Z. Z., Söderberg, M., Sindelar, P., Prusiner, S. B., Kristensson, K. och Dallner, G. (1996) Specific lipid changes in the brains of scrapie-infected mice. *J. Neurochem.* **66**, 277-285.

Mattson, M. P. and Goodman, Y. (1995) Different amyloidogenic peptides share a similar mechanism of neurotoxicity involving reactive oxygen species and calcium. *Brain Res.* **676**, 219-224.

Morrison, W. R. and Smith, L. M. (1964) Preparation of fatty acid methyl esters and dimethylacetals from lipid with boron fluoride-methanol. *J. Lipid Res.* **5**, 600-608.

Murphy, E. J., Stephens, R., Jurkowitz-Alexander, M. and Horrocks, L. A. (1993) Acidic hydrolysis of plasmalogens followed by high-performance liquid chromatography. *Lipids* **28**, 565-568.

Påhlsson, P., Shakin-Eshleman, S. H. and Spitalnik, S. L. (1992) N-linked glycosylation of ß-amyloid precursor protein. *Biochem. Biophys. Res. Commun.* **189**, 1667-1673.

Pettegrew, J. W., Panchalingam, K., Strychor, S. and Branthoover, G. (1990). Analysis of membrane phospholipids in Alzheimer 's disease brain by ^{31}P NMR. *Soc. Neurosci. Abstr.* **16**, 241.13.

Price, D. L., Becher, M. W., Wong, P. C., Borchelt, D. R., Lee, M. K. and Sisodia, S. S. (1996) Inherited neurodegenerative diseases and transgenic models. *Brain Pathol.* **6**, 467-480.

Radominska-Pyrek, S., Chojnacki, T. and Pyrek, S. (1979) Fully unsaturated decaprenol from bovine pituitary glands. *Biochem. Biophys. Res. Commun.* **86**, 395-401.

Sastry, P.S. (1985) Lipids of nervous tissue: composition and metabolism. *Prog. Lipid. Res.* **24**, 69-176.

Schmechel, D. E., Saunders, A. M., Strittmatter, W. J., Crain, B. J., Hulette, C. M., Joo, S. H., Pericak-Vance, M. A., Goldgaber, D. and Roses, A. D. (1993) Increased amyloid ß-peptide deposition in cerebral cortex as a consequence of apolipoprotein E genotype in late-onset Alzheimer disease. *Proc. Natl. Acad. Sci. U. S. A.* **90**, 9649-9653.

Selkoe, D. J. (1996) Amyloid β-protein and the genetics of Alzheimer's disease. *J. Biol. Chem.* **271**, 18295-18298.

Shultz, A. M., Henderson, L. E. and Oroszlan, S. (1988) Fatty acylation of proteins. *Ann. Rev. Cell Biol.* **4**, 611-647.

Sinenski, M. and Lutz, R. J. (1992) The prenylation of proteins. *BioEssays* **14**, 25-31.

Smith, M. A., Perry, G., Richey, P. L., Sayre, L. M., Anderson, V. E., Beal, M. F. and Kowall, N. (1996) Oxidative damage in Alzheimer's. *Nature* **382**, 120-121.

Söderberg, M., Edlund, C., Alafuzoff, I., Kristensson, K. and Dallner, G. (1992) Lipid composition in different regions of the brain in Alzheimer's disease/senile dementia of Alzheimer's type. *J. Neurochem.* **59**, 1646-1653

Söderberg, M., Edlund, C., Kristensson, K. and Dallner, G. (1990) Lipid compositions of different regions of the human brain during aging. *J. Neurochem.* **54**, 415-423.

Söderberg, M., Edlund, C., Kristensson, K. and Dallner, G. (1991) Fatty acid composition of brain phospholipids in aging and in Alzheimer's disease. *Lipids* **26**, 421-425.

Subbarao, K. V., Richardson, J. S. and Ang, L. C. (1990) Autopsy samples of Alzheimer's cortex show increased peroxidation *in vitro*. *J. Neurochem.* **55**, 342-345.

Sumpter, P. Q., Mann, D. M., Davies, C. A., Yates, P. O., Snowden, J. S. and Neary, D. (1986) An ultrastructural analysis of the effects of accumulation of neurofibrillary tangle in pyramidal neurons of the cerebral cortex in Alzheimer's disease. *Neuropathol. Appl. Neurobiol.* **12**, 305-319.

Taraboulos, A., Raeber, A. J., Borchelt, D. R., Serban, D. and Prusiner, S. B. (1992) Synthesis and trafficking of prion proteins in cultured cells. *Mol. Biol. Cell* **3**, 851-863.

Thelin, A., Löw, P., Chojnacki, T. and Dallner, G. (1991) Covalent binding of dolichyl-P to proteins in rat liver. *Eur. J. Biochem.* **195**, 755-761.

Valtersson, C. and Dallner, G. (1982) Compartmentalization of phosphatidylethanolamine in microsomal membranes from rat liver. *J. Lipid Res.* **23**, 868-875.

Watanabe, R. and Duchen, L. W. (1993) Cerebral amyloid in human prion disease. *Neuropathol. Appl. Neurobiol.* **19**, 253-260.

Yankner, B. A., Duffy, L. K. and Kirschner, D. A. (1990) Neurotrophic and neurotoxic effects of amyloid *β* protein: Reversal by tachykinin neuropeptides. *Science* **250**, 279-282.

Progress in HPLC-HPCE, Vol. 7, pp. 649-672
G. Ali Qureshi *et al.* (Eds)

NOVEL FLUOROMETRY OF NORSALSOLINOL, AN ENDOGENOUS 1,2,3,4-TETRAHYDRO-6,7-DIHYDROXYISOQUINOLINE FROM DOPAMINE

Masanori Yoshioka[a],Akira Shirahata[b],Zenzo Tamura[c], Simone Parvez[d] and Hasan Parvez[d]

[a]*Faculty of Pharmaceutical Sciences, Setsunan University, 45-1 Nagaotoge-cho, Hirakata, Osaka 573-01, Japan,* [b]*Faculty of Pharmaceutical Sciences, Josai University, 1-1 Keyakidia, Sakato, Saitama 350-02, Japan,* [c]*Faculty of Pharmaceutical Sciences, The University of Tokyo, Tokyo 113, Japan, and* [d]*Neuropharmacology Unit, University of Paris XI, 91400 Orsay, France.*

INTRODUCTION

Tetrahydroisoquinolines(THQs) are condensation products of catecholamines and their analogs with aldehydes by enzymatic or non-enzymatic reaction (Cohen, 1976). THQs found in body fluids are considered to be related with some diseases such as diabetes, aging, Parkinsonism, phenylketonuria and alcoholism (Collins, 1980; Lasala and Coscia, 1979). It is well known that they are present as alkaloids in many natural products (Kametani, 1977). We noticed that a THQ was generated during incubation of epinephrine with ascorbic acid for radioimunoassay and supposed that this kind of condensation of catecholamines with nascent formaldehyde from ascorbic acid would ocurr in physiological condition (Shirahata *et al.*, 1982). Recently, an important report that 1-methyl-4-phenyl-1,2,3,6-tetrahydropyridine (MPTP) in narcotic patients gave rise to Parkinson's syndrome emphasized THQs as endogenous MTPTs (Yoshida *et al.*, 1990).

There are several methods to determine THQs by gas chromatography (GC)/mass spectrometry (MS) which are accurate, but inconvenient (Niwa *et al.*,

1992). High performance liquid chromatography(HPLC) with an electrochemical detector is also good for practical use (Riggin and Kissinger, 1977). These methods, however, are not selective enough to discriminate THQs from various ingredient in physiological samples. During sampling and pretreatment in these methods, they did not take care of artificial condensation of catecholamines with contaminated aldehydes to give rise to THQs. Further, despite that dopamine is an important transmitter in the brain, its THQ derivative has not been analyzed so far.

In this paper, we attempted to develop a determination method of 1,2,3,4-tetrahydro-6,7-dihydroxyisoquinoline(DA-Fp, so called norsalsolinol), one of THQs from dopamine, by HPLC with a new fluorescent reaction. After syntheses of several THQs, their separation and fluorescent reaction were examined and optimized for biological samples. Structures of THQs concerned in this paper are shown in Table 1. Two types of probable condensation products are classified into type p and o, naming after para or ortho substitution of an aldehyde to the catechol moiety (Shirahata *et al.*, 1997).

Table 1 Abbreviation and Structures of Tetrahydroisoquinolines Studied in this Paper.

THQ	R_1	R_2	R_3	Structure
DA-Fp	H	H	H	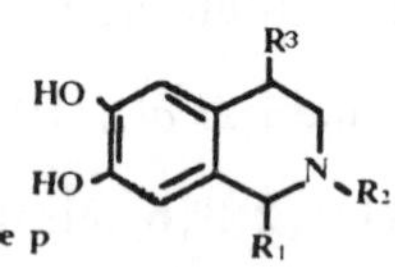 type p
En-Fp	H	CH_3	H	
DA-Ap	CH_3	H	H	
NE-Fp	H	H	OH	
E-Fp	H	CH_3	OH	
NE-Fo	H	H	OH	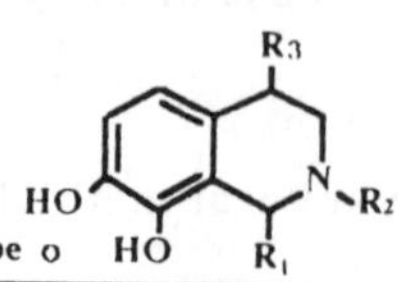 type o
E-Fo	H	CH_3	OH	

EXPERIMENTAL

Materials *l*-Phenylephrine・HCl and 1,2,3,4-tetrahydro-6,7-dihydroxy-2-methylisoquinoline(En-Fp)・HCl were obtained from Sigma(St Louis). *dl*-Epinephrine, *dl*-and *l*-norepinephrine, and dopamine・HCl were obtained from nacalai tesque(Kyoto). *l*-Epinephrine was obained from Merck(Darmstadt). Sodium metaperiodate was obtained from Kanto Chemical Co.(Tokyo). 1,2,3,4-Tetrahydro-6-hydroxy -2-methylisoquinoline was kindly donated by Dr. Toshihiko Okamoto of Faculty of Pharmaceutical Sciences, University of Tokyo. Amberlite XAD-4(nonionic porous polystyrene polymer) obtained from Rohm & Haas(Philadelphia) was ground into fine grains of 150-200 mesh. Boric acid gel was obtained from Aldrich Chemical Co.(Milwaukee). Hitachi gel 3011C(macroreticular carboxylate-anion exchange resin, 10 μm in mean diameter) was kindly donated by Mr. Akira Narita, Nissei Sangyo Co., Tokyo. Alumina of neutral activity in grade 1 was obtained from Woelm(Eschwege), washed with 6 N HCl at 100℃ until the aquaous phase became colorless, subsequently with water until the washing was neutral, heated at 115℃ for an hour, and stored in a desiccator. All the other chemicals of reagent grade were commercially available.

Measurement of Chemical Characteristics All melting points were measured on a Yanaco MP-S3 melting point apparatus(Yanagimoto, Kyoto) and were uncorrected. ^{1}H-NMR spectra were measured with a FMN-FX 100 NMR spectrometer(JEOL, Tokyo). The chemical shifts were expressed in ppm using trimethylsilylpropanesulfonic acid sodium salt as an internal standard. Fluorescence spectra were measured by a MPF-2A fluorescence spectrophotometer(Hitachi, Tokyo).

Syntheses 1,2,3,4-Tetrahydro-6,7-dihydroxy-isoquinoline(DA-Fp) Hydrochloride

One gram of dopamine · HCl and 0.5 ml of 35% formaldehyde were mixed with 5 ml of methanol and heated at 45℃. The reaction mixture was concentrated to dryness *in vacuo*. The residue was dissolved in a small aliquot of 2 N HCl. The solution was added acetone to give white crystals. The crystals were recrystallized with 2 N HCl and acetone. The pure crystals obtained were 250 mg; mp 256℃ dec(262℃ reported by Buck, 1934). ^{1}H-NMR(D_2O) d: 2.96(2H, t, j=6.8 Hz, -CH_2-), 3.48(2H, j=6.8 Hz, -CH_2-), 4.22(2H, s, -CH_2-), 6.67(1H, s, Ar-H), 6.71(1H, s, Ar-H). Anal. Calcd for $C_9H_{12}O_2NCl$: C, 53.63; H, 6.00; N, 6.95. Found: C, 53.50; H, 6.02; N, 6.81.

1,2,3,4-Tetrahydro-4,6,7-trihydroxy-isoquinololine(NE-Fp) Hydrochloride

The synthesis was carrided out according to the method described by Sarges (1974). One gram of *dl*-norepinephrine · HCl in 1.5 ml of water was added with 0.4 ml of 35% formaldehyde dropwise. After stirring additionally 1 hour, 10 ml of acetone was added and the precipitated solid was recrystallized with water-acetone 7 times. The crystals obtained were 180 mg; mp 172℃ dec(172℃ reported by Sarges, 1974). Anal. Calcd for $C_9H_{12}O_3NCl \cdot H_2O$: C, 45.86; H, 5.99; N, 5.97. Found: C, 45.80; H, 5.94; N, 5.82. ^{1}H-NMR(D_2O) data were completely the same as described by Sarges (1974).

1,2,3,4-Tetrahydro-4,7,8-trihydroxy-isoquinoline(NE-Fo) Hydrochloride

The synthesis was carried out according to the similar manner to 1,2,3,4-tetrahydro-4,6,7-trihydroxy-2-methylisoquinoline(E-Fo) as described in our previous paper (Shirahata *et al.*, 1982). One gram of *dl*-norepinephrine · HCl was dissolved in 4 ml of 6 N HCl and added to 100 ml of 0.1 M tris buffer(pH, 7.0)-10 mM ascorbic acid solution containing 1 ml of 35% formaldehyde. After stirring at room temparature for an hour, the solution was added 50 g of NaCl. The solution was passed through a column(1 cm X 20 cm) of Amberlite XAD-4 equilibrated with 20% NaCl. The adsorbed product was eluted under a linear gradient of 20-0% NaCl in 0.01 N HCl. The eluate was monitored by a UV detector at 280 nm. The UV absorbing fractions were pooled and applied to a column of boric acid gel in ten sample volumes. The column was washed with twenty column volumes of water and the product was eluted with 0.025 N HCl. The fractions of UV absorbing product were pooled and concentrated to dryness *in vacuo*. The residue was recrystallized from methanol-ether three times. The white crystals obtained were 110 mg; mp 190℃(196-198℃ reported by Sarges, 1974). Anal. Calcd for $C_9H_{12}O_3NCl$: C, 49.67; H, 5.56; N, 6.44. Found: C, 49.40; H, 5.59; N, 6.33. ^{1}H-NMR(D_2O) data were completely the same as described by Sarges (1974).

1,2,3,4-Tetrahydro-6,7-dihydroxy-1-mehtylisoquinoline(DA-Ap) Hydrochloride

Dopamine·HCl(0.5 g) was condensed with acetaldehyde according to the method described by King *et al.*(1974). The product was recrystallized from methanol-ether three times. The white crystals obtained were 180 mg; mp 231℃(231℃ reported by King *et al.*, 1974). Anal. Calcd for $C_{10}H_{14}O_2NCl$: C, 55.60; H, 6.54; N, 6.50. Found: C, 55.61; H, 6.58; N, 6.34. ^{1}H-NMR(D_2O) data were completely the same as described by King *et al.* (1974).

Fluorescent Reaction of DA-Fp with Sodium Periodate

One-tenth ml of an aquaeous solution of DA-Fp at a given concentration was distributed into a siliconized test tube. The solution was added with 3.9 ml of 1 mM $NaIO_4$ in 0.1 M phosphate (pH 8.0) and stood at 37℃ for 15 min. The fluorescence spectrum was measured with the fluorescence spectrophotometer.

Fluorescence Determination of DA-Fp by HPLC

A column(500 mm X 2 mm, I. D.) of Hitachi gel 3011C was maintained at 45℃. An eluent consisted of 0.1 M acetate, 0.03 M disodium hydrogen phosphate and 0.05 M NaCl(pH 4.25) and was flowed through the column at 0.70 ml/min. The inlet pressure of the column was 50-55 kg/cm^2. After the column, the eluate was mixed with a fluorescence reagent solution of 10 mM $NaIO_4$ at a flow rate of 0.16 ml/min and with the other buffer solution(pH 9.2) consisting of 0.5 M disodium hydrogen phosphate and 0.4 M Na_2CO_3 at a flow rate of 0.25 ml/min. The mixed solution was flowed through a reaction coil (20 m X 0.25 mm, I.D.) at 45℃. The reaction mixture was monitored by a F4T4/BL fluorescence detector(SLM-Aminco, Rochester) with a 7-60 filter(Corning Inc., New York) for excitation and a Wratten 2A filter(Eastman Kodak Co., Rochester) for emission.

Sampling of Biological Materials.

Urine was collected from the bladder of a Japanese rabbit weighing 3 kg by catheterization and adjusted to pH 1.5 with 6 N HCl. The acidified urine was added *l*-phenylephrine which final concentration was made 1%. The urine was stored at -20℃ until analysis. Blood was intravenously collected from the ear of the rabbit. The blood was added with one-tenth volume of a solution of 2% EDTA containing 10% *l*-phenylephrine and centrifuged at 3500 g and the plasma was stored at -20℃. The other plasma was obtained from the rabbit, after L-DOPA(25 mg/Kg body weight) was intravenously administrated into the ear of the rabbit. Human urine excreted was treated and stored in the same way as the rabbit urine. A brain was taken out from male Wistar rats weighing 400-500 g immediately after being sacrificed. The brain was frozen on dry ice, weighted(around 2.0 g) and added with a five fold wet weight of 1% *l*-phenylephrine in 0.4 M $HClO_4$. Then, it was homogenized in a glass homogenizer and centrifuged at 10000 g at 4℃ for 30 min. The supernatant was stored at -20℃.

Pretreatment of Sample

To the stored sample(5 ml of urine, 2 ml of plasma or one brain supernatant) was added 800 pmol of En-Fp as an internal standard in 100 μl of 0.1 M HCl. The sample solution was adjusted to pH 1.5 with 6 M HCl and hydrolyzed at 80℃ for 20 min. To the hydrolyzate was added 1.0 ml of 10% $Na_2S_2O_5$ and 5 ml of 0.5 M tris-HCl buffer(pH 8.5) containing 0.02 M EDTA. The solution was re-adjusted to pH 8.5 with 4 M NH_4OH. It was added with 100 mg of the washed alumina and stirred for 10 min. The suspension was filtered through a No. 2 filter paper(Toyo Roshi Co., Tokyo). The residue on the filter paper was washed with 5 ml of water 3 times. Then, the analyte on the residue was eluted with 0.5 ml of 0.5 M acetic acid. The eluate was analyzed by the HPLC-fluorescent reaction above.

RESULTS

Fluorescent reaction of DA-Fp We thought that there were two ways to find a specific fluorescent reaction of DA-Fp. One was direct oxidation of DA-Fp (Corrodi and Jonsson, 1967), and the other was a condensation of ethylenediamine with the catechol moeity of DA-Fp as also described for catecholamines (Nagatsu, 1973).

First, we tried to examine the oxidation reactions using various reagents such as $NaIO_4$, $NaIO_3$, $NaBrO_3$, H_2O_2 and $K_3[Fe(CN)_6]$ at 1 mM each in test tubes as described in Experimental. Only $NaIO_4$ among them showed a strong fluorescence with the emission maximum at 404 nm and the excitation maximum at 362 nm. The other oxidation reagents showed less than one-fiftieth of the fluorescent intensity of $NaIO_4$ under any conditions such as pH 5, 7 or 9 at 25, 60 or 90℃. On the other hand, the condensation reaction of ethylenediamine at 25% at 60℃ for 30 min showed a weak fluorescence with the emission maximum at 500 nm and the excitation maximum at 400 nm. The fluorescence intensity with ethytlenediamine was one-thirtieth of the one with $NaIO_4$. Thus, we selected $NaIO_4$ as the best reagent and pursued its quantitative conditions in the followings.

As shown in Fig. 1, fluorescence intensity of the derivative from DA-Fp attained a plateau at pH 8.0-8.5. A phosphate buffer at 0.1 M was suitable, as a glycine or tris buffer at 0.1 M interacted with the fluorescence reagent so that the fluorescent reaction with DA-Fp did not occur. A citrate buffer at pH 7 also seemed to interact with it.

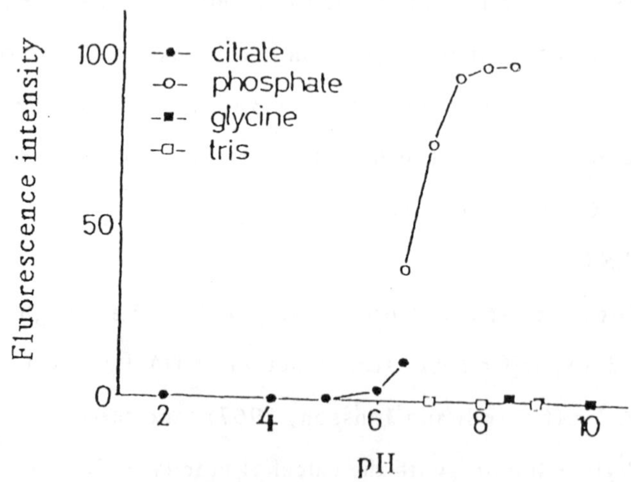

Fig. 1 Dependence of Fluorescent Reaction upon pH of Various Buffers

Half μM DA-Fp was reacted with 1 mM $NaIO_4$ at 37℃ for 15 min.

Time courses of the fluorescent reaction were drawn using four solutions at various concentrations of $NaIO_4$ as shown in Fig. 2. An 1-mM solution alon produced stable and the highest fluorescence, hence we decided to use 1 mM $NaIO_4$. When 0.5 μM DA-Fp was reacted at pH 8.0 with 1 mM $NaIO_4$, the fluorescence intensity attained the plateau in 20 min at 18℃, in 5 min at 37℃ and in 2 min at 45℃.

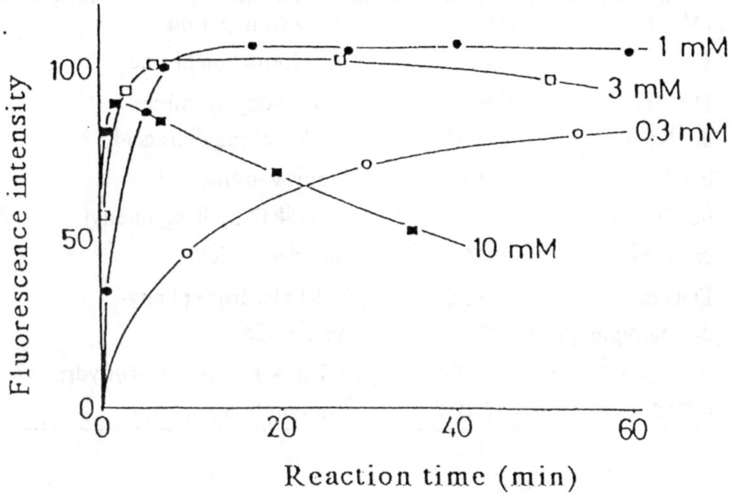

Fig. 2 Time Courses of Fluorescent Reactions of Various Concentrations of $NaIO_4$
Half μM DA-Fp was reacted at pH 8.0 at 37℃.

In the mannual reaction, the reaction at 37℃ for 15 min was made a custom. Now that, we came up against a superable problem of specificity of the fluorescent reaction. Fluorescence intensities of various related compounds reacted with 1 mM $NaIO_4$ under the same condition as above described are compared as shown in Table 2. Fortunately, the reaction was rather specific for DA-Fp, although En-Fp and DA-Ap showed weak fluorescence. The other THQs, catecholamines and related compounds did a little. From these characteristcs, the fluorescent reaction of DA-Fp was different from the known reaction. It was good that a working curve of DA-Fp was linear from 2 pmol to 2 nmol per tube as well as that of En-Fp or DA-Ap.

Table 2 Relative Fluorescence Intensity (RFI) of Derivatives from THQs and Related Compounds Reacted with $NaIO_4$.

Compound	RFI	Compound	RFI
DA-Fp	100	DL-Metanephrine	0.14
En-Fp	18	DL-Normetanephrine	0.14
DA-Ap	6.5	3-Methoxytyramine	0.01
DL-NE-Fp	0.54	DL-Vanilmandelic acid	0.89
DL-E-Fp	0.16	Homovanillic acid	0.09
DL-NE-Fo	0.36	DL-3,4-Dihydroxyphenyl-mandelic acid	0.75
DL-E-Fo	0.01		
Dopamine	0.30	3,4-Dihydroxypheny-acetic acid	0.18
D-*l*-Norepinephrine	0.27		
D-*l*-Epinephrine	0.08	1,2,3,4-Tetrahydro-6-hydroxy-2-mehtylisoquinoline	0.01
L-DOPA	0.01		

HPLC-on-line fluorescent reaction with $NaIO_4$ The mannual method, however, was limited to pure samples. It was expected that there were many kinds of catecholamines and related compounds in biological samples. In order to increase selectivity of the determination, the fluorescent reaction should be on-line coupled with HPLC. We examined kinds of columns, separation and reaction conditions. Finally, we adopted the system as described in Experimental. As the column, a reversed phase column of octadecyl silica gel (Devolosil 5 or 10, Nomura Chemical Co., Seto, or LS-410, Tosoh, Tokyo), an adsoption column of porous polystyrene polymer beads(Htachi gel 3010 or 3011, Nissei Sangyo Co., Tokyo) and a strong anion exchange column of porous polystyrene beads (Hitachi gel 3011P, Nissei Sangyo Co.) were tested in vain. The weak anion exchange column of Hitachi gel 3011C showed a good resolution of THQs and catecholamines as shown in Fig.3. In this system, DA-Fp was well separated from DA-Ap and En-Fp as an internal standard, and dopamine, norepinephrine and epinephrine were also separated as shown with arrows.

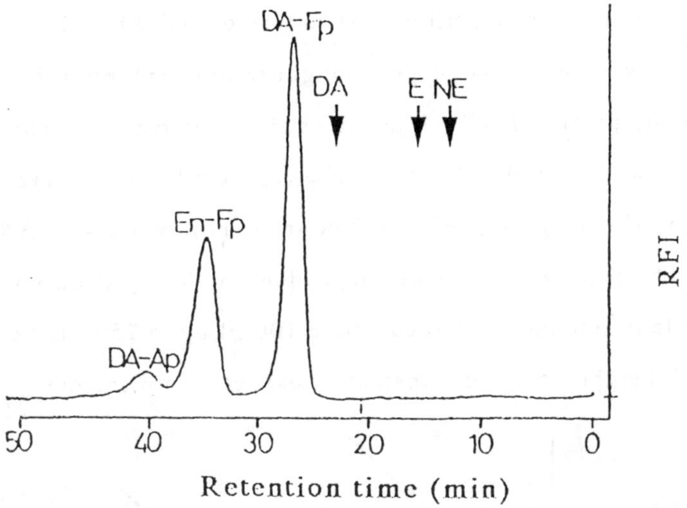

Fig. 3 Chromatogram of a Mixture of 200 pmol each of DA-Fp, En-Fp and DA-Ap
The arrows show retention times of catecholamines.

As an eluent, the acetate and phosphate buffer did not interact with $NaIO_4$ and was adopted. At pH higher than 4.0, the resolution of DA-Fp from dopamine was more than 1.0. The elution time, however, was too long to practical analysis. In order to make it short, NaCl added to the buffer showed higher in efficiency than methanol. The flow rate of the eluent and column temparature were set 0.7 ml/min and 45℃, taking into consideration the analytical time and stability of the column. Next, the fluorescent reaction after the column was based on the mannual method and checked, again. The concentration of $NaIO_4$ became 1.5 mM after the reaction coil, when it was pumped at the minimum constant flow rate. The concentration of the mixture was within the range from 0.5 to 3.0 mM for the stable derivatizaton of DA-Fp. Its peak height was the highest at pH 9.2 of the phosphate-carbonate buffer, when its pH was changed to be from 8.0 to 9.5 after the mixing coil. When the coil length was changed from 5 to 30 m, its peak height reached a plateau at 20 m which was used always. Thus, the HPLC-on line derivatization system was optimized as described in Experimental. A working curve of peak height ratio of DA-Fp to En-Fp was linear from 2 to 800 pmol of DA-Fp injected as shown in Fig. 4. The minimum detectable amount was 2 pmol (S/N=5). The repeatability of the peak height ratio of 200 pmol of DA-Fp injected at a 2 hour interval was CV=1.5 %(N=5) within a day. The peak height was not dependent on the injection volume from 5 to 100 μl, while it gradually declined above 100 μl up to 250 μl checked, although its peak height ratio was constant as expected in principle.

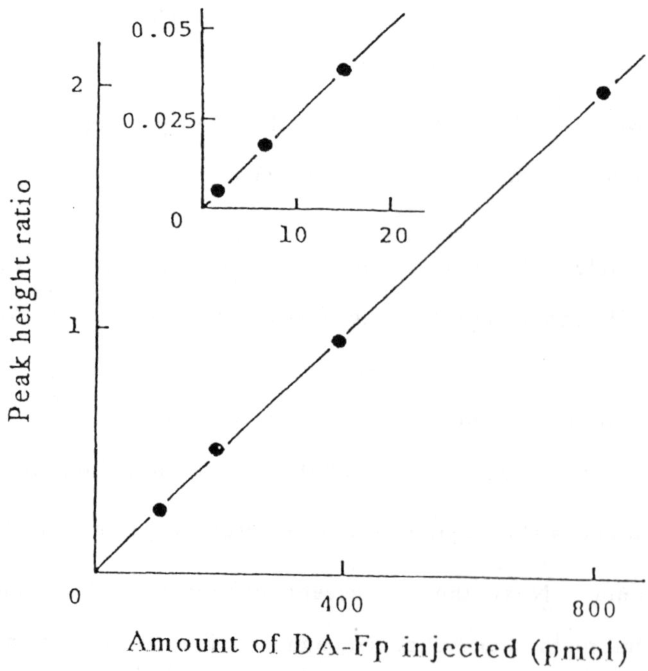

Fig. 4 A Working Curve of DA-Fp Measured by HPLC

The injected amount of En-Fp was 800 pmol.

Pretreatment of sample Although the HPLC-fluorescent derivatization system was highly selective for DA-Fp, it was still essential to clean-up the biological samples. As DA-Fp was a kind of catecholamine derivatives, its pretreatment before HPLC was done according to that of catecholamines in the biological samples as described in Experimental. It consisted of hydrolysis of conjugated DA-Fp and adsoption of freed DA-Fp on the alumina. The peak height of DA-Fp of the urine of the rabbit administered L-DOPA attained a plateau at 10 min of the hydrolysis as shown in catecholamine conjugates, which suggested the presence of the conjugate form of almost half of DA-Fp together with its free form. In addition to these proper treatments, we elaborated two means to suppress the artificial conversion of catecholamines present in the biological samples into THQs. One was an addition of excess *l*-phenylephrine which produced a THQ with concomitant formaldehyde in the sample and reagent solutions as described by Kaito *et al.* (1975). When a human urine sample was supplemented with 2 x 10^{-4} M dopamine as a final concentration, where the artificial production of DA-Fp was 1.2% in yield after incubation at 37℃ for 3 hours, the peak height of DA-Fp sharply declined by the addition of 0.1 ml of 5% *l*-phenylephrine and was compatible with that of the sample without the supplementation of dopamine by the addition of more than 0.5 ml of 5% *l*-phenylephrine(Data not shown). The other was an addition of 10% $Na_2S_2O_5$ which worked on not only anti-oxidation, but also suppression of the artificial production of DA-Fp from dopamine in the sample. When the urine sample supplemented with 2 x 10^{-4} M dopamine was pretreated, the peak height of DA-Fp gradually declined depending on the added volume of 10% $Na_2S_2O_5$ down to that of the sample without the supplementation by the addition of 1 ml of 10% $Na_2S_2O_5$. Thus, the pretreatment was decided as described in Experimental. By this treatment, addition recovery curves of DA-Fp were linear up to 500 pmol added to the urine(one of chromatogram as shown in Fig. 5), 300 pmol to the plasma and 200 pmol to the brain as shown in Fig. 6. Both the overall recoveries of DA-Fp and En-Fp were around 60%. DA-Fp in the normal plasma was not detected.

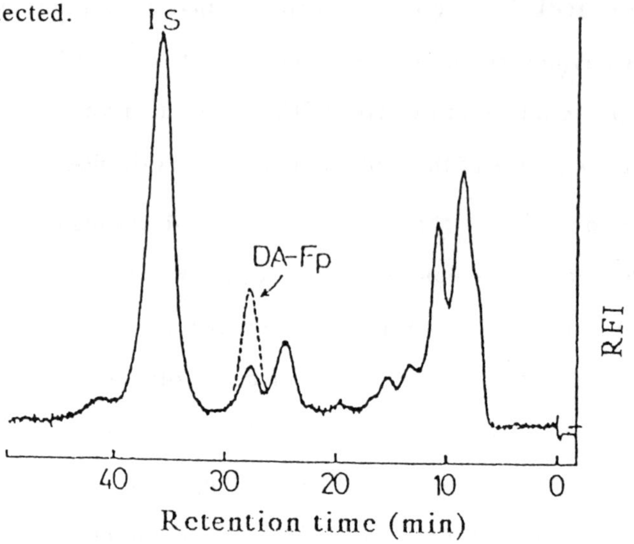

Fig. 5 Chromatogram of a Rabbit Urine

The DA-Fp peak increased as shown by a dotted line, when 20 pmol of standard DA-Fp was injected with the same urine sample.

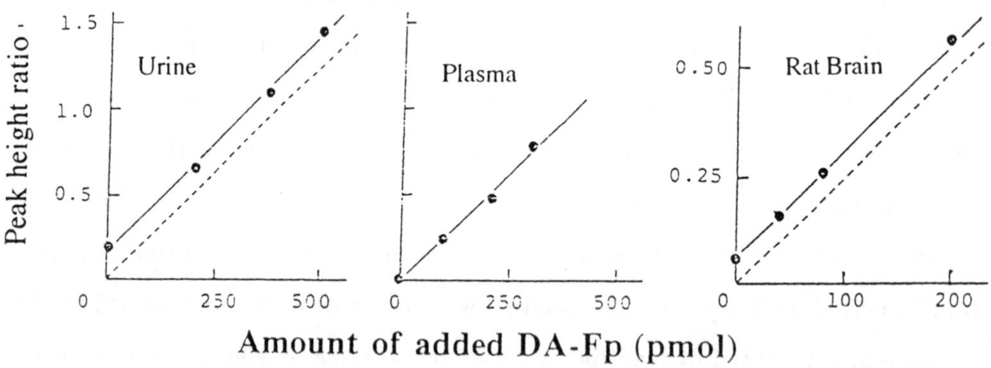

Fig. 6 Working Curves of the Biological Samples Added with DA-Fp

Five ml of urine, 2 ml of plasma or 2 g of rat brain was added 800 pmol of the internal standard and a given amount of DA-Fp.

Determination of biological samples At last, we were able to analyze the biological samples. As shown in the chromatogram of a rabbit urine of Fig. 5, the peak of DA-Fp appeard after several unknown peaks. When the standard DA-Fp was added to the urine as described in the addition recovery test, the peak height increased to make the retention time sure. In this way, DA-Fp excretion was measured after the administration of L-DOPA as shown in Fig. 7. Its excretion increased very highly in an hour and gradually decreased in a few hours. Its peak, however, was not observed in the plasma, which meant its plasma level less than 5 pmol/ml.

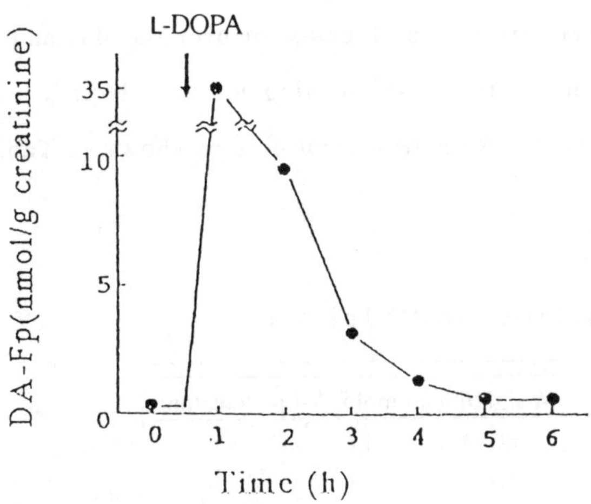

Fig. 7 Time Courses of Excretion of DA-Fp in Urine of a Rabbit Administrated with L-DOPA

The rabbit was injected intravenously in the ear with 25 mg/kg L-DOPA. The urine samples were collectectedjust before the administration and each hour thereafter.

In a human urine, DA-Fp excretion was found as shown in a chromatogram of Fig. 8.

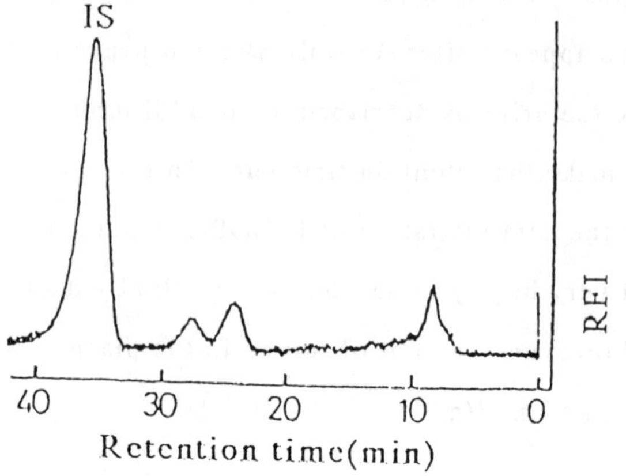

Fig. 8 Chromatogram of a Normal Human Urine Sample

Eight subjects were starved from 21 o'clock on the previous day and the urines excreted at the second time in the next morning were collected. The mean excretion value of DA-Fp was 2.8 nmol/ g creatinine as shown in Table 3.

Table 3 Urinary Excretion of DA-Fp in Men

Subject	Age/year	Weight/kg	nmol DA-Fp/g creatinine
HN	36	60	4.6
AH	23	70	1.0
YM	23	49	3.4
JS	40	54	4.1
AS	27	76	2.0
YN	26	55	1.0
TT	41	53	5.3
TT	32	70	0.8
		Mean	2.8 ± 1.7

We doubted whether the excretion was natural or artificial in the bladder. One of the urines with or without the supplementation of 2 x 10^{-4} M dopamine was incubated at its own pH 6.8 or pH 7.8 adjusted with 4 M NH_4OH at 37℃. In the presence of dopamine, the artificial conversion into DA-Fp at pH 7.8 linearly increased up to 1.2% in yield in 3 hours, while that at pH 6.8 did up to 0.2%. Without the supplementation, DA-Fp was not detected even after 5 hour incubation. It was unlike that DA-Fp was not the artificial product in the bladder. Further, DA-Fp was found in the normal rat brain as shown in Fig. 9. The mean amount was 8 ± 2pmol/ g wet weight (N=3). In order to confirm the reliabilty of the obtained values of above biological samples, we re-chromatographed fractions from each peak. The fluorescent derivative of the peak was collected and injected into the other column(500 mm X 2 mm, I.D.) of

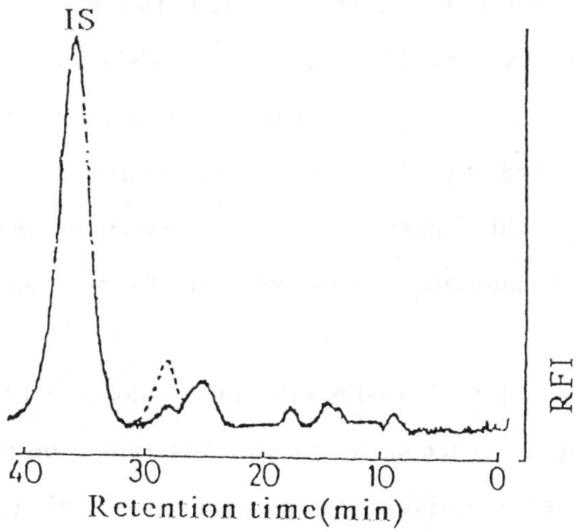

Fig. 9 Chromatogam of a Rat Brain Sample

The DA-Fp peak increased as shown by a dotted line, when 20 pmol of standard DA-Fp was injected with the same urine sample.

Hitachi gel 3011C at 45℃. An eluent consisted of 0.05 M citrate buffer (pH 3.8) and methanol(4:1, v/v) and its flow rate was set at 0.7 ml/min. The eluent was monitored by the same fluorescence detector. All the DA-Fp peak was completely baseline-separated and showed a single symmetric peak. A working curve was linear from 5 to 200 pmol of DA-Fp injected. All the values measured were compatible with those obtained by the first HPLC-fluorescent derivatization.

DISCUSSION

In the syntheses, we obtained all the standard THQs except DA-Fo. When dopamine was condensed with acetaldehyde, type p and o were produced (King *et al.*, 1974). It was easily confirmed that both types of THQs were produced from norepinephrine with formaldehyde in this paper and from epinephrine in our previous paper (Shirahata *et al.*, 1982), although type p was abundant. Under the same synthetic condition, we could not detect type o from dopamine with formaldehyde by HPLC and thin layer chromatography using various mobile phases. This fact suggested that type o from dopamine was very minor even in biological materials. Thus, we aimed to focus on the analysis of DA-Fp.

The fluorescent reaction of DA-Fp with $NaIO_4$ was found to be a new reaction, because its fluorescence maximum was different from that of 3, 4-dihydro-6,7-dihydroxyisoquinoline showing an emission maximum at 480 nm by an excitation maximum at 410 nm or that of 6,7-dihydroxyisoquinoline showing an emission maximum at 490 nm by an excitation maximum at 330 nm (Corrodi and Jonsson, 1967, Anton and Sayre, 1964). Kaito *et al.* reported that the THQ from *l*-phenylephrine was converted into a fluorescent derivative with

$K_3Fe(CN)_6$ ten times higher in efficiency than $NaIO_4$ (1975). In our condition, $K_3Fe(CN)_6$ was not effective to DA-Fp. The fluorescence spectrum of DA-Fp reacted with $NaIO_4$ was different from that of THQ from *l*-phenylephrine with $K_3Fe(CN)_6$. Then, we tried to purify the derivative by HPLCs, but did not succeed, because the yield of the purification was too little. Judging from our several experiments, it is to say at least that 1- and 2-position of the fluorescent derivative from DA-Fp are not substituted, because each fluorescent derivative from DA-Fp, En-Fp or DA-Ap showed a different retention time in the re-chromatography. Further, 3- and 4-position of the fluorescent derivative were also not substituted, because [3-^{14}C]-[4-^{3}H]-DA-Fp, we synthesized, gave the fluorescent derivative with the same radioactivity ratio of [3-^{14}C]/ [4-^{3}H] purified by the re-chromatography as the ratio of the starting material. During the purification, the derivative did not adsorb onto the boric gel, which suggested that the catechol moiety of the derivative was not intact. A larger scale of the purification is necessary to identify the fluorescent derivative.

We found that normal distilled water also contained formaldehyde from the environment, unless the water was kept in a glass-stoppered flask as described in our previous paper (Shirahata *et al.*, 1982). This kind of artifacts of DA-Fp was completely prevented in our sampling and pretreatment. These conditions will be useful for analyses of catecholamines and related compounds. Much more, the present method is applicable to determine dopamine which is converted to DA-Fp.

Thus, DA-Fp, we measured, is natural in organisms. Its source, however, is either dietary or endogenous. The latter possibilty is evident from that the excretion of DA-Fp increased after administration of L-DOPA into the rabbit. The urinary excretion of DA-Ap, so called salsolinol, after L-DOPA administration in man was firstly reported by Sandler *et al.* (1973). DA-Fp, however, was not analyzed in their experiments. In our experiments, DA-Fp was detected after the adminstration of L-DOPA, but DA-Ap was not. It was thought that the excretion level of DA-Ap was under the detction limit. Sandler *et al.* injected into patients ethanol to produce acetoaldehyde immediately after the L-DOPA administration (1973). It was assumed that acetoaldehyde concentration was so low in a normal body. The possible dietary source was not ruled out in the normal subjects as shown in Table 3, because natural plants contained THQs as described (Kametani, 1977, Sandler *et al.*, 1973). It is really to say that the values of DA-Fp we obtained was sum of both origins at different ratio. The plasma level of DA-Fp is assumed to be very low, as the plasma level of dopamine was (Imai *et al.*, 1970). Further, DA-Fp excretion into human urine was less than one-hundredth of dopamine excretion. The origins of formaldehye are sevaral substances such as S-adenosylmethionine (Axelrod and Daly, 1965), 5-methyltetrahydrofolate (Lauwers *et al.*, 1975), air (Shirahata *et al.*, 1982), smoking (Sprince *et al.*, 1979), methanol (Axelrod and Daly, 1965), methylenechloride(Ahmed and Anders, 1978), and possible methylamine derivatives from drugs. The present determination method will be promising to study on these biosyntheses of DA-Fp.

Further more, metabolism of DA-Fp is a keen problem, because the oxidized products should be toxic as MTPT. It was already described that THQs including DA-Fp were substrates to catechol-O-methyltransferase (Smissman *et al.*, 1976) and monoamine oxidase (Naoi *et al.*, 1989). 1,2,3,4-Tetrahydroisoquinoline and 1,2,3,4-tetrahydro- 2- methylisoquinoline were detected in Parkinsoninan and normal human brains (Kohno *et al.*, 1986), and were supposed to be main neurotoxins inducing Parkinsonian syndrome (Niwa *et al.*, 1987). Finally, dopamine-derived 1,2,3,4-tetrahydroisoquinolines were cytotoxic to a dopaminergic cell model, clonal rat pheochromocytoma PC12h cell (Maruyama *et al.*, 1994). There appear several unknown peaks in the chromatogram of Fig. 5. 8.and 9. They may be clues to find such products by the present HPLC. Thus, the present determination method of DA-Fp will be determinant to elucidate etiology of Pakinsonism.

Summary

Supposing that endogenous 1,2,3,4-tetrahydro-6,7-dihyroxyisoquinoline (DA-Fp) should be a potential marker involved in etiology of various diseases such as Parkinsonism, we attempted to develop a fluorescence determination method of DA-Fp. It was synthesized by condensation of dopamine with formaldehyde according to the known method. As a fluorescence reagent to DA-Fp, periodate was screened out among various oxidation reagents. The optimal reaction condition was obtained with 0.25 mM $NaIO_4$ in 0.1 M phosphate buffer (pH 8.0) at 37℃ for 15 min. From the fluorescence spectrum of the derivative, it was found to be a new reaction specific for DA-Fp. This reaction was on-line coupled with high performance liquid chromatography (HPLC), which enabled us highly sensitive determination of DA-Fp. A working curve was linear from 2 pmol to 800 pmol of DA-Fp per injection. For determination of DA-Fp in biological materials, the pretreatment before HPLC was optimized by hydrolysis of its conjugate and suppression of the artifact with *l*-phenylephrine. Urinary excretion of DA-Fp in men was measured by the present method. In a rabbit administrated with L-DOPA, the urinary excretion of DA-Fp increased endogenously. An amount of DA-Fp was determined in rat brains. The significance of DA-Fp in these biological materials was discussed and evaluated. Thus, it was concluded that the present method would be useful for study on tetrahydroisoquinolines involved in many diseases.

REFERENCES

Ahmed A. E. and Anders M. W. (1978). Metabolism of dihalomethanes to formaldehyde and inorganic halide- 2 Studies on the mechanism of the reaction. *Biochem. Pharmcol.*, **27**, 2021-2025.

Anton A. H. and Sayre D. F. (1964). The distribution of dopamine and DOPA in various animals and a method for their deternination in diverse biological material. *J. Pharmacol. Exp. Therap.*, **145**, 326-336.

Axelrod J. and Daly J. (1965). Pituitary gland: enzymatic formation of methanol from S-adenosylmethionine. *Science*, **150**, 892-893.

Buck J. S. (1934). Some substituted tetrahydroisoquinoline hydrochlorides. *J. Am. Chem. Soc.*, **56**, 1769-1771.

Cohen G. (1976). Alkaloid products in the metabolism of alcohol and biogenic amines. *Biochem. Pharmacol.*, **25**, 1123-1128.

Collins M. A. (1980). Neuroamine condensations in human subjects. *Adv. Exp. Med. Biol.*, **126**, 87-102.

Corrodi H. and Jonsson G. (1967). The formaldehyde fluorescence method for the histochemical demonstration of biogenic monoamines. A review of the methodology. *J. Histochem. Cytochem.*, **15**, 65-78.

Imai K., Sugiura M. and Tamura Z. (1970). The presence of conjugated dopamine in normal human plasma. *Chem. Pharm. Bull.*, **18**, 2134.

Kaito T., Kasuya K., Sagara K. and Yoshida T. (1975). Studies on fluorometric analysis of phenol derivatives. 5 Fluorometric determination of phenylephrine hydrochloride. *Yakugaku Zasshi.*, **95**, 985-991.

Kametani T. (1977). *The Total Synthesis of Natural Products*, ApSimon J.(Ed.). John Wiley & Sons, New York, pp.1-272.

King G. S., Goodwin B. L. and Sandler M. (1974). Isosalsolinol formation: a secondary reaction in the Pictet-Spengler condensation. *J. Pharm. Pharmac.*, **26**, 476-478.

Kohno M., Ohta S. and Hirobe M. (1986). Tetrahydroisoquinoline and 1-methyl-tetrahydroisoquinoline as novel endogenous amines in rat brain. *Biochem. Biophys. Res. Comm.*, **140**, 448-454.

Lasala J. M. and Coscia C. J. (1979). Accumulation of a tetrahydroisoquinoline in phenylketonuria. *Science*, **203**, 283-284.

Lauwers W., Leysen J., Verhoeven H., Laduron P. and Claeys M. (1975). Identification of alkaloids; the condensation products of biogenic amines with formaldehyde, enzymatically formed from 5-methyltetrahydrofolic acid. *Biomed. Mass Spectrom.*, **2**, 15-22.

Maruyama W., Takahashi T., Minami M., Takahashi A., Dostert P., Nagatsu T. and Naoi M. (1994). Cytotoxicity of dopamine-derived 6,7-dihydroxy-1,2,3,4-tetrahydroxyisoquinolines. *Adv. Neurol.*, **60**, 224-230.

Nagatsu T. (1973). *Biochemistry of Catecholamines*, University of Tokyo Press, Tokyo, pp. 219-222.

Naoi M., Matsumura S., Parvez H., Takahashi T., Hirata Y., Minami M. and Nagatsu T. (1989). Oxidation of N-methyl-1,2,3,4-tetrahydroisoquinoline into the N-methyl-isoquinolinium ion by monoamine oxidase. *J. Neurochem.*, **52**, 653-655.

Niwa T., Maruyama W., Nakahara D., Takeda N., Yoshimizu H., Tatematsu A., Takahashi A., Dostart, P., Naoi M. and Nagatsu T. (1992). Endogenou synthesis of N-methylsalsolinol, an analogue of 1-methyl-4-phenyl-1,2,3,6-tetrahydropyridine, in rat brain during *in vivo* microdialysis with salsolinol, as demonstrated by gas chromatography-mass spectrometry. *J. Chromatogr.*, **578**, 109-115.

Niwa T., Takeda N., Kaneda N., Hashizume Y. and Nagatsu T. (1987). Presence of tetrahydroisoquinoline and 2-methyl-tetrahydroquinoline in Parkinsonian and normal human brains. *Biochem. Biophys. Res. Comm.*, **144**, 1084-1089.

Riggin R. M. and Kissinger P. T. (1977). Determination of tetrahydroisoquinoline alkaloids in biological materials with high performance liquid chromatography. *Analytical Chem.*, **49**, 530-532.

Sandler M., Carter S. B., Hunter K. R. and Stern G. M. (1973). Tetrahydroisoquinoline alkaloids: *in vivo* Metabolites of L-Dopa in man. *Nature*, **241**, 439-443.

Sarges R. (1974). The synthesis of tetrahydroisoquinolines related to catecholamines. *J. Heterocyclic. Chem.*, **11**, 599-601.

Shirahata A., Yoshioka M. and Tamura Z. (1982). Unexpected conversion of epinephrine into tetrahydroisoquinolines in a solution containing ascorbic acid. *Chem. Pharm. Bull.*, **30**, 1352-1357.

Shirahata A., Yoshioka M. and Tamura Z. (1997). Fluorometric determination of 1,2,3,4-tetrahydro-6,7-dihydroisoquinoline in biological materials by HPLC. *Chem. Pharm. Bull.*, **45**, in press.

Smissman E. E., Reid J. R. and Walsh D. A. (1976). Synthesis and biological activity of 2- and 4-substituted 6,7-dihydroxy-1,2,3,4-tetrahydroisoquinolines. *J. Med. Chem.*, **19**, 127-131.

Sprince H., Parker C. M. and Smith G. G. (1979). Comparison of protection by L-ascorbic acid, L-cysteine, and adrenergic-blocking agents against acetalaldehyde, acrolein, and formaldehyde toxicity: implications in smoking. *Agents and Actions*, **9**, 407-414.

Yoshida M., Niwa T. and Nagatsu T. (1990). Parkinsonism in monkeys produced by chronic administration of an endogenous substance of the brain, tetrahydroisoquinoline: the behavioral and biochemical changes. *Neurosci. Lett.*, **119**, 109-113.

Progress in HPLC-HPCE, Vol. 7, pp. 673-721
G. Ali Qureshi *et al.* (Eds)

BIOPTERIN AND NEOPTERIN AS NEUROCHEMICAL MARKERS OF NEUROLOGICAL DISEASES

HIROSHI KUZUYA,[1] HIDEO SAKAMOTO,[1] MAKOTO HIBIYA,[3] KAN SHIMPO,[2] HIDEHIKO BEPPU[2] and TOSHIHARU NAGATSU[3]

[1]*Division of Molecular Biology,* [2]*Fujita Memorial Division of Pharmacognosy, and* [3]*Division of Molecular Genetics II Neurochemistry, Institute for Comprehensive Medical Science, Fujita Health University, Toyoake, Aichi 470-1192, Japan*

INTRODUCTION

Biopterin is a metabolite of (6*R*)-5,6,7,8-tetrahydrobiopterin, which is required as an essential cofactor in biosynthesis of catecholamine and indoleamine neurotransmitters in the central nervous system and in biosynthesis of nitric oxide in many species and tissues(Mayer and Werner, 1995). There is a close relationship between a deficiency of biopterin cofactor and neurological disturbances such as atypical phenylketonuria (Kaufman *et al.*, 1978; Niederwieser *et al.*, 1979) and Parkinson disease (Nagatsu *et al.*, 1981; Lovenberg *et al.*, 1979). Its concentrations in body fluids and tissues reflect various physiological conditions (Hausen *et al.*, 1989; Nagatsu *et al.*, 1989; Nakahara *et al.*, 1989). (6*R*)-5,6,7,8-Tetrahydrobiopterin is synthesized from guanosine triphosphate (GTP) via 7,8-dihydroneopterin triphosphate; GTP → 7,8-dihydroneopterin triphosphate → 6-pyruvoyl tetrahydropterin → tetrahydrobiopterin.

Determination of biopterin and neopterin is an essential part of biochemical and clinical studies of these neurological diseases. This article describes three assay methods of biopterin and neopterin, i.e., high-performance liquid chromatography (HPLC) with fluorimetric detection of oxidized form by either pre-column or post-column oxidation, radioimmunoassay, and enzyme-linked immunosorbent assay, and discusses some diseases associated with changes in biopterin and neopterin concentrations.

CHEMISTRY OF BIOPTERIN

The term "pterin", which originally meant a factor in the pigments of butterfly wings, is now used for the natural pteridines that have the structure of 2-amino-4-hydroxypteridine. As shown in Figure 1, pterin in body fluids has the structure of the favored tautomeric oxoform, 2-amino-4-oxo-3,4-dihydropteridine. There are two groups of natural pterin compounds: one group is the "conjugated" pterins: folic acid derivatives, which have a methylamino benzoyl glutamic acid side chain at 6-position, the other group is the "unconjugated" pterins which have a side chain with less carbons at 6-position. Biopterin and neopterin belongs to unconjugated pterins.

A well-established and physiologically important function of unconjugated pterins is the role as the cofactor, (6*R*)- L-*erythro*-6-(1,2-dihydroxypropyl) pterin (tetrahydrobiopterin), of the pterin-requiring monooxygenases which catalyze the hydroxylation of aromatic amino acids (Kaufman and Fischer, 1974).

The active cofactor form of biopterin is the reduced, tetrahydro form, and in tissues biopterin is present mostly in the tetrahydro form. Kaufman (1963) first isolated L-*erythro*-7,8-dihydrobiopterin from the rat liver as the cofactor of phenylalanine hydroxylase (phenylalanine 4-monooxygenase) and found that 7,8-dihydrobiopterin is reduced by dihydrofolate reductase to 5,6,7,8-tetrahydrobiopterin which acts as a direct electron donor with the cofactor activity. Chemical reduction of L-*erythro*-biopterin to the tetrahydro forms introduces another center of asymmetry at position 6 of the pteridine ring, and results in the racemic mixtures of diastereoisomers [(6*RS*)- or (6DL)-tetrahydrobiopterin], whereas enzymatic reduction of L-*erythro*-7,8-dihydrobiopterin produces only one *R*-form of the 6-diastereoisomers (Matsuura *et al.*, 1985). The racemic (6*RS*)-mixtures can be separated by liquid chromatography (Bailey *et al.*,1978).

The absolute configuration of the tetrahydrobiopterin at the C-6 chiral center produced stereospecifically with reference to the C-6 chiral center from 7,8-dihydrobiopterin or 7,8-dihydro-6-methylpterin by the action of dihydrofolate reductase was determined to be (6*R*)-L-*erythro*-5,6,7,8-tetrahydrobiopterin, or (6*S*)-5,6,7,8-tetrahydro-6-methylpterin. Although the enzymatically reduced tetrahydro-6-methylpterin and tetrahydrobiopterin have the same absolute configuration at the C-6 chiral center, the symbol for

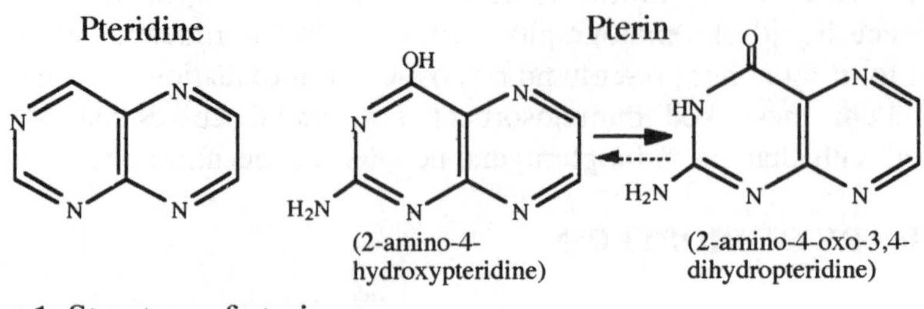

Figure 1. Structure of pterins.

the tetrahydrobiopterin becomes (6*R*) according to the *R,S* naming system, owing to the presence of the OH group at the 1' position of the side chain (Matsuura *et al.*,1985).

In the hydroxylation of aromatic amino acids by pterin-requiring monooxygenases, the cofactor, i.e., (6*R*)-L-*erythro*-5,6,7,8-tetrahydrobiopterin, is simultaneously oxidized to quinonoid L-*erythro*-7,8-dihydrobiopterin via 4α-hydroxy-tetrahydrobiopterin. Quinonoid dihydrobiopterin is reduced back to (6*R*)-L-*erythro*-tetrahydrobiopterin by dihydrobiopterin reductase (DHPR) in tissues. Quinonoid dihydrobiopterin is also spontaneously converted to 7,8-dihydrobiopterin and then further oxidized to biopterin (Fig. 2).

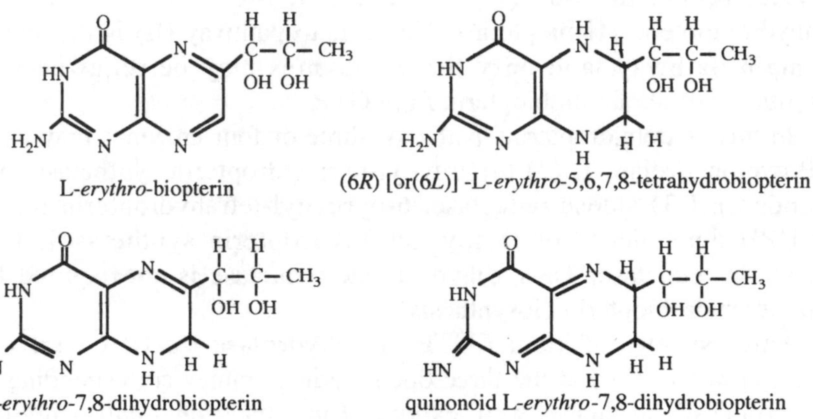

Figure 2. Structure of four forms of L-*erythro*-biopterin.

BIOSYNTHESIS OF TETRAHYDROBIOPTERIN

(6*R*)-L-*erythro*-5,6,7,8-tetrahydrobiopterin, the cofactor for pterin-requiring monooxygenases, is synthesized from guanosine triphosphate (GTP) in mammals. Historically, it had been assumed that tetrahydrobiopterin is synthesized from GTP by the "dihydropterin" pathway as the biosynthesis of tetrahydrofolate: guanosine triphosphate (GTP)−(1) → D-*erythro*-7,8-dihydroneopterin triphosphate−(2) → sepiapterin (a natural dihydropterin)−(3) → L-*erythro*-7,8-dihydrobiopterin−(4) → (6*R*)-L-*erythro*-5,6,7,8-tetrahydrobiopterin. Thus, four enzymes were assumed to exist for the biosynthesis: (1) GTP cyclohydrolase I; (2) the enzyme that converts D-*erythro*-7,8-dihydroneopterin triphosphate to sepiapterin; (3) sepiapterin reductase; and (4) dihydrofolate reductase.

However, it is now confirmed that the *de novo* biosynthetic pathway of (6*R*)-tetrahydrobiopterin is the "tetrahydrobiopterin" pathway (Fig. 3) (Milstien and Kaufman, 1985; Milstien and Kaufman, 1986; Heintel *et al.*, 1985; Switchenko *et al.*, 1984). GTP−(1) → D-*erythro*-7,8-dihydroneopterin triphosphate (H2neopterin-PPP)−(2) → 6-pyruvoyl-tetrahydropterin (6-pyruvoyl-H4pterin) ["compound X or dyspropterin" as termed by Masada *et al.*, 1985). Akino's group (Masada *et al.*, 1985) proposed two pathways in the conversion of 6-pyruvoyl-tetrahydropterin to tetrahydrobiopterin: (I) 6-lactoyl-tetrahydropterin [6-lactoyl-H4pterin, 6-(1'-oxo-2'-hydroxypropyl)-tetrahydropterin, or 5,6-dihydrosepiapterin] [lower left-hand side (I) of Figure 3, the 1'-keto pathway], or (II) 6-(1'-hydroxy-2'-oxopropyl)-tetrahydropterin [lower, right-hand side (II) of Figure 3, the 2'-keto pathway], to tetrahydrobiopterin (H4biopterin). The 2'-keto pathway (II) is thought to be the main pathway, and only three enzymes may be required for the biosynthesis of tetrahydrobiopterin from GTP.

In this "tetrahydropterin" pathway, three or four enzymes may exist:(1) GTP cyclohydrolase I; (2) 6-pyruvoyl-tetrahydropterin synthetase (Mg^{2+} -dependent); [(3) aldose reductase, 6-pyruvoyl-tetrahydropterin reductase (NADPH-dependent), or lactoyl-tetrahydropterin synthetase]; and (4) sepiapterin reductase. Thus, dihydrofolate reductase is not required for *de novo* tetrahydrobiopterin biosynthesis.

Three species of human GTP cyclohydrolase I cDNA were cloned (Togari *et al.*, 1992) and the three open reading frames corresponding to all reported cDNA sequences were expressed in *Escherichia coli*. Only one of the recombinant protein had the activity of GTP cyclohydrolase I (Gütlich *et al.*, 1994). Human genes of the enzyme were characterized (Ichinose *et al.*, 1995). Mutations of the GTP cyclohydrolase I gene were discovered in autosomal recessive GTP cyclohydrolase I deficiency and in autosomal dominant hereditary progressive dystonia with marked diurnal fluctuation (HPD) or DOPA-responsive dystonia (DRD) (also termed as Segawa's disease) (Ichinose *et al.*, 1994; Ichinose *et al.*, 1995).

GTP cyclohydrolase I is the first and rate-limiting enzyme for tetrahydrobiopterin biosynthesis. A potent inhibitor of GTP cyclohydrolase I, 2,4-diamino-6-hydroxypyrimidine, decreased the contents of tetrahydrobiopterin, dopamine, and norepinephrine in the brain (Suzuki *et al.*, 1988).

A regulatory protein for the control of GTP cyclohydrolase I activity was found in rat liver (Harada *et al.*, 1993) and its cDNA was cloned (Milstein *et al.*, 1996; Yoneyama *et al.*, 1997). The protein was initially called p35, but now termed as GTP cyclohydrolase I feedback regulatory protein,

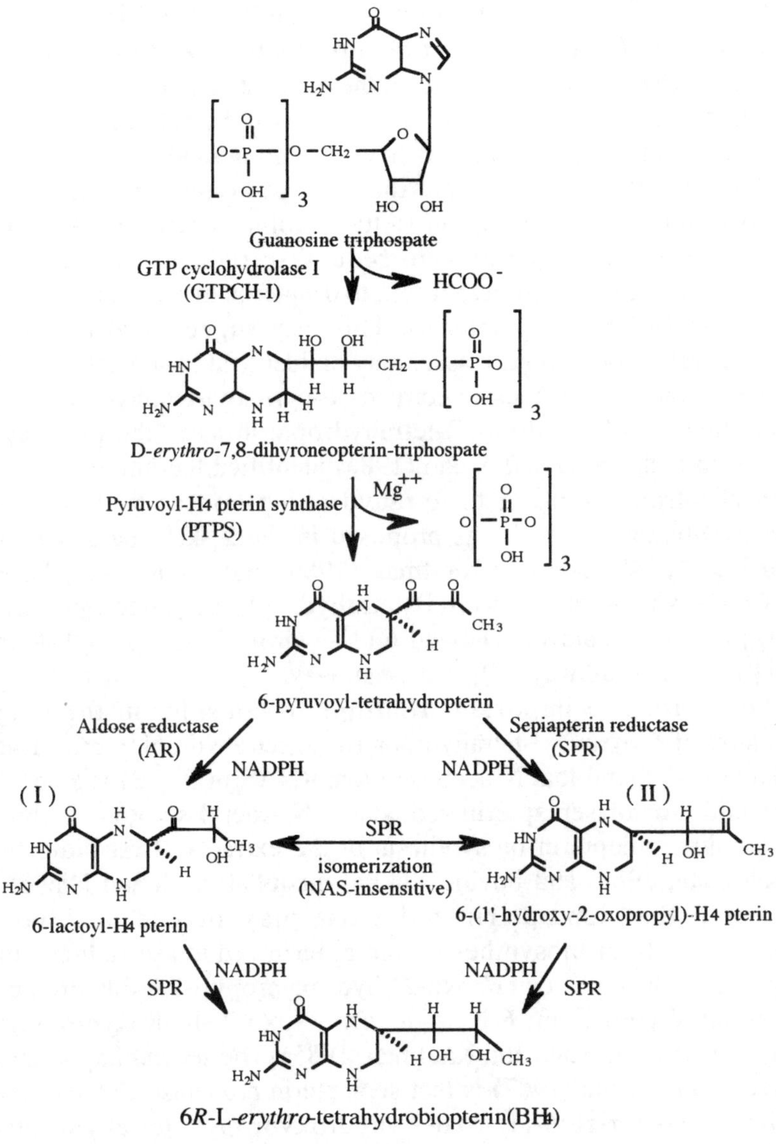

Figure 3. The probable biosynthetic pathway of (6R)-L-erythro-tetrahydrobiopterin (H4biopterin) from guanosine triphoshate (GTP): (I) the 1'-keto pathway with 6-lactoryl-tetrahydropterin (6-lactoyl-H4pterin) as the intermediate, (II) the 2'-kato pathway with 6-(1'-hydroxy-2'-oxopropyl)-tetrahydropterin [6-(1'hydroxy-2-oxopropyl)-H4pterin] as the intermediate. The isomerizasion reaction by SPR is N-acetylserotonin (NAS)-insensive(Katoh and Sueoka, 1987).

GFRP(see Fig. 7). GFRP mediates feedback inhibition of GTP cyclohydrolase I activity by tetrahydrobiopterin and also mediates the stimulatory effect of phenylalanine on the enzyme activity.

Milstien and Kaufman(1985) first proposed the pathway (I) in Figure 3. They found that enzymes from rat liver and brain, and from bovine adrenal medulla can catalyze the formation of 6-pyruvoyl-, 6-lactoyl-, and 6-dihydroxypropyl-tetrahydropterin (tetrahydrobiopterin). Sueoka and Katoh (1985) reported that sepiapterin reductase reduces various carbonyl compounds including 6-pyruvoyl-tetrahydropterin in addition to sepiapterin (6-lactoyl-7,8-dihydrosepiapterin). This may suggest that the enzyme 6-pyruvoyltetrahydropterin reductase, may be identical to sepiapterin reductase. However, sepiapterin reductase converts 6-pyruvoyl-tetrahydropterin mainly to 6-(1'-hydroxy-2'-oxopropyl)-tetrahydropterin as in the pathway (II) of Figure 3. In fact, Smith and Nichol(1986) identified the intermediate from 6-pyruvoyl-tetrahydropterin to tetrahydrobiopterin as 6-(1'-hydroxy-2'-oxopropyl)-tetrahydropterin, as proposed in the 2'-keto pathway (pathway (II) of Fig. 3). Milstien and Kaufman (1986) and Smith(1987) have found both 6-(1'-oxo-2'hydroxypropyl)-tetrahydropterin (6-lactoyl-tetrahydropterin) [in 1'-keto pathway (I)l and 6-(1'-hydroxy-2'-oxopropyl)-tetrahydropterin [in 2'-keto pathway (II)] in the pathway.

There are two important findings to consider in the enzymatic conversion of 6-pyruvoyl-tetrahydropterin to tetrahydrobiopterin. Katoh and Sueoka (1987) found that N-acetylserotonin(NAS in Fig. 3) is a specific and potent inhibitor of sepiapterin reductase. N-Acetyl-serotonin inhibited *de novo* tetrahydrobiopterin biosynthesis in the extracts of rat brain (Milstien and Kaufman, 1985) and bovine adrenal medulla(Smith and Nichol, 1986), suggesting that sepiapterin reductase may indeed be involved in tetrahydrobiopterin biosynthesis. Sepiapterin reductase reduces not only sepiapterin [6-lactoyl-, or (1'-oxo-2'-hydroxypropyl)-7,8-dihydropterin] but also the tetrahydro form, 6-lactoyl-, or (1'-oxo-2'-hydroxypropyl)5,6,7,8-tetrahydropterin.(Milstien and Kaufman, 1985). The second important finding by Katoh and Sueoka (1987) is that sepiapterin reductase also has isomerase activity to isomerize 6-(1'-oxo-2'-hydroxypropyl)-tetrahydropterin (6-lactoyl-tetrahydropterin) to 6-(1'-hydroxy-2'-oxopropyl)-tetrahydropterin. The C'-keto structure of the product is confirmed by ^{1}H nuclear magnetic resonance (NMR) spectroscopy. This activity is N-acetylserotonin insensitive, as indicated in Figure 3. These findings may suggest a possibility that lactoyl-tetrahydropterin synthetase first reduces 6-pyruvoyl-tetrahydropterin to 6-lactoyl-tetrahydropterin, which is then isomerized to 6-(1'-hydroxy-2'-oxopropyl)-tetrahydropterin and is further reduced to tetrahydrobiopterin. However, sepiapterin reductase alone may synthesize tetrahydrobiopterin from 6-pyruvoyl-tetrahydropterin via 6-(1'-hydroxy-2'-oxopropyl)-tetrahydropterin.

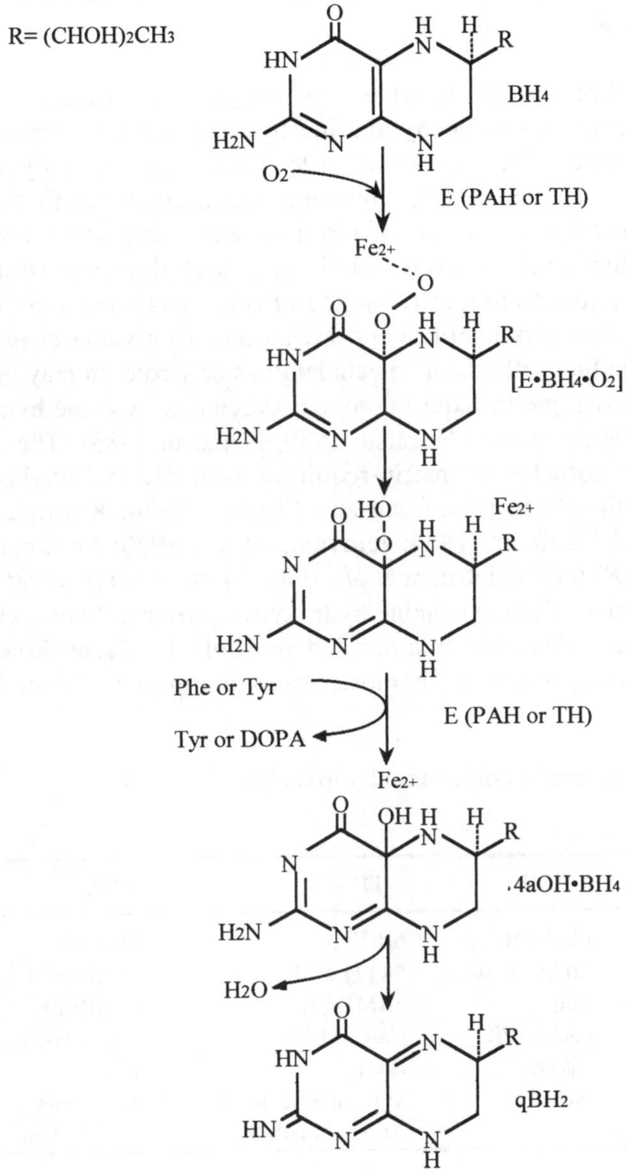

Figure 4. The mechanism of reaction of phenylalanine hydroxylase and tyrosine hydroxylase. BH4, (6*R*)-L-*erythro*-tetrahydrobiopterin; qBH2, quinonoid dihydrobiopterin; E, enzyme; PAH, phenylalanine hydroxylase; TH, tyrosine hydroxylase. (Dix *et al.*, 1987)

PHYSIOLOGICAL ROLE OF TETRAHYDROBIOPTERIN

TETRAHYDROBIOPTERIN AS COFACTOR OF AROMATIC AMINO ACID HYDROXYLASES

There are three pterin-requiring monooxygenases which are aromatic amino acid hydroxylases, phenylalanine hydroxylase (Kaufman, 1963), tyrosine hydroxylase (Nagatsu *et al.*, 1964) and tryptophan hydroxylase (Lovenberg *et al.*, 1967). The latter two enzymes catalyze the first and rate-limiting steps of the biosynthesis of physiologically important monoamine neurotransmitters and hormones such as catecholamines (dopamine, noradrenaline, and adrenaline)(Nagatsu *et al.*, 1964) and indoleamines [serotonin (5-hydroxytryptamine) and melatonin] (Ichiyama *et al.*, 1970). Regulation of the biosynthesis of catecholamines or serotonin may be mainly mediated by the two pterin-requiring monooxygenases, tyrosine hydroxylase and tryptophan hydroxylase (Nagatsu, 1980; Nagatsu, 1985). The naturally occurring pterin cofactor for pterin-requiring aromatic hydroxylases is L-*erythro*-tetrahydropterin (Kaufman and Fischer, 1974; Kaufman, 1963; Brennemam and Kaufman, 1964; Ichiyama *et al.*, 1970; Friedman *et al.*, 1972) and the (6*R*) form (Matsuura *et al.*, 1980; Matsuura *et al.*, 1985).

The properties of phenylalanine hydroxylase, tyrosine hydroxylase, and tryptophan hydroxylase are summarized in Table 1. These three pterin-requiring monooxygenases are homotetramers composed of four identical

Table 1.
Pterin-requiring aromatic amino acid hydroxylase

	PAH	TH	TPH
Cofactor	6R-BPH4	6R-BH4	6R-BH4
Subunit	50 kD × 4	58 kD × 4	50 kD × 4
Activator	Phe	cAMP-PK	cAMP-PK
	cAMP-PK	Ca-CaM-PK	Ca-CaM-PK
Distribution	Liver,	Brain	Brain
	Kidney	Sympathetic nerves	Mast cells
		Adrenal medulla	Pineal gland

Abbreviations: PAH, phenylalanine hydroxylase; TH, tryosine hydroxylase; TPH, tryptophan hydroxylase; BH4, tetrahydrobiopterin; cAMP-PK, cyclic AMP-dependent protein kinase; Ca-CaM-PK, calcium-calmodulin-dependent protein kinase.

subunits of *Mr* 50-60 kD, and require not only a tetrahydropterin cofactor but also Fe^{2+} (one atom per subunit). The structures of tyrosine hydroxylase of the rat (Grima *et al.*, 1985) and human (Grima *et al.*, 1987; Kaneda *et al.*, 1987; Kobayashi *et al.*, 1987), phenylalanine hydroxylase of the rat (Robson *et al.*, 1984) and human (Kwok *et al.*, 1985), and tryptophan hydroxylase of the rabbit (Grenett *et al.*, 1987) are determined based on the nucleotide sequences by cDNA cloning. These three enzymes may contain the regulatory domain in the N-terminal region and the catalytic domain in the C-terminal region. The structures of the catalytic domain in the C-terminal regions have high homology in the amino acid sequences, whereas the structures of the regulatory domain in the N-terminal regions are considerably different (Ledley *et al.*, 1985). Moreover, four different forms of human tyrosine hydroxylase mRNA are produced by alternative RNA splicing from a single gene (Grima *et al.*, 1987; Kaneda *et al.*, 1987). The N-terminal regulatory domain is the inhibitory region that acts as an inhibitory factor on the C-terminal catalytic domain and determines the substrate specificity. Phosphorylation of a serine in the N-terminal regulatory domain by various protein kinases such as cyclic adenosine monophosphate (cAMP)-dependent protein kinase and Ca^{2+}-calmodulin-dependent protein kinase II activates the enzymes.

The mechanism by which the tetrahydrobiopterin-requiring monooxygenase, phenylalanine hydroxylase (Lazarus *et al.*, 1981) and tyrosine hydroxylase (Dix *et al.*, 1987) activates dioxygen for substrate hydroxylation may be via oxygenation of its tetrahydropterin cofactor to a 4α-carbinolamine concurrent with the hydroxylation of phenylalanine to tyrosine or tyrosine to 3,4-dihydroxyphenylalanine (DOPA). The 4α-carbinolamine dehydrates by 4α-carbinolamine dehydratase to the final observed product of enzyme turnover, quinonoid dihydrobiopterin (Fig. 4). The reaction mechanism of tryptophan hydroxylase may also be similar to that of phenylalanine hydroxylase or tyrosine hydroxylase.

Since the amino acid sequence of N-terminal domain of tyrosine hydroxylase is longer than that of phenylalanine hydroxylase or tryptophan hydroxylase, the regulatory mechanism of tyrosine hydroxylase by phosphorylation or some other mechanism appears to be more complex than that of phenylalanine hydroxylase or tryptophan hydroxylase.

Relations between the biosynthetic pathway of tetrahydrobiopterin and the biosynthesis of catecholamines (dopamine, noradrenaline, and adrenaline) or serotonin mediated by three pterin-requiring monooxygenases, i.e., phenylalanine hydroxylase, tyrosine hydroxylase, and tryptophan hydroxylase, are illustrated in Fig. 5.

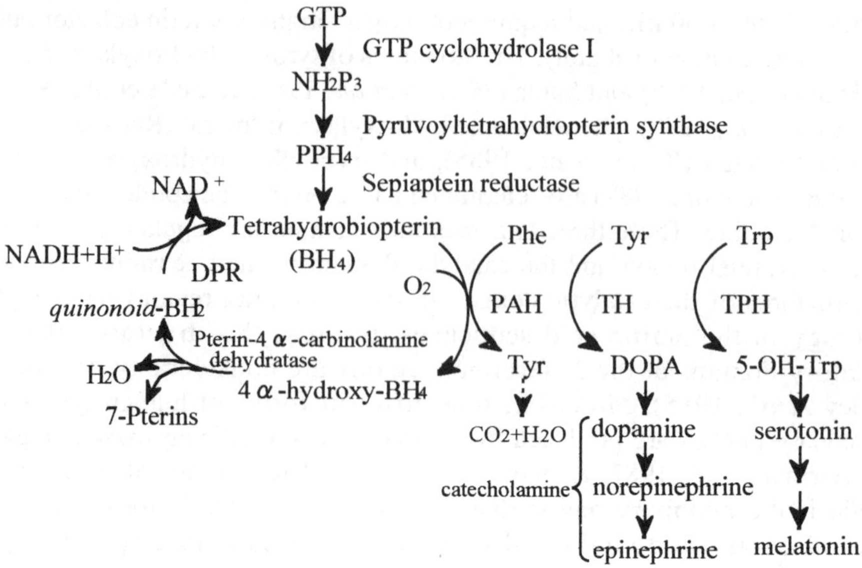

Figure 5. Tetrahydrobiopterin as cofactor of aromatic amino acid hydroxylases. PAH, phenylalanine hydroxylase; TH, tryosine hydroxylase; TPH, tryptophan hydroxylase.

REGULATION OF AROMATIC AMINO ACID HYDROXYLASES BY TETRAHYDROBIOPTERIN UNDER PHYSIOLOGICAL CONDITIONS

The intracellular concentration of tetrahydrobiopterin under physiological conditions is assumed not to be saturating toward aromatic amino acid hydroxylases, and therefore the concentration of the tetrahydrobiopterin cofactor could be a regulatory factor for the biosynthesis of catecholamines and serotonin for tyrosine hydroxylase and tryptophan hydroxylase reactions.

One approach to examine whether or not tyrosine hydroxylase or tryptophan hydroxylase is saturated with the tetrahydrobiopterin cofactor in vivo is to measure the increase in DOPA or 5-hydroxytryptophan formation in the presence of an aromatic L-amine acid decarboxylase inhibitor after addition of tetrahydrobiopterin at the cellular or animal level.

When tetrahydrobiopterin is added to the incubation medium containing an aromatic L-amino acid decarboxylase inhibitor and tissue slices either of the brain striatum region containing tyrosine hydroxylase (Hirata *et al.*, 1983) or of the brain raphe region containing tryptophan hydroxylase (Sawada *et al.*, 1986a; Sawada and Nagatsu, 1988) the accumulation of DOPA or 5-hydroxytryptophan is stimulated about 3-4-fold. The rate of uptake of the natural (6*R*) form or unnatural (6*S*) form of tetrahydrobiopterin into the raphe tissue slices is similar, but the formation of 5-hydroxytryptophan is stimulated maximally (350%) at 42 μM (6*R*)-tetrahydrobiopterin, but the (6*S*) form has no significant effect. Only (6*R*)-tetrahydrobiopterin, but not (6*S*) form stimulated the formation of tyrosine from phenylalanine in the tissue slices of the rat liver (Koizumi *et al.*, 1987). It is noteworthy that the (6*S*) isomer showed comparable *in vitro* cofactor activity as the natural (6*R*) isomer toward tyrosine hydroxylase (Oka *et al.*, 1981).

However, the stimulatory effect of (6*R*)-tetrahydrobiopterin in an animal model is somewhat controversial (6*R*)-Tetrahydrobiopterin at 40 mg/kg body weight, together with 3-hydroxybenzyl hydrazine hydrochloride (NSD-1015), a decarboxylase inhibitor, 1 h after the intraperitoneal injection, increased the DOPA accumulation under inhibition of DOPA decarboxylation in the adrenals by 2.5-fold, whereas the (6*S*) isomer had no significant effect (Matsuura *et al.*, 1986).

The *in vivo* effect of peripheral administration of tetrahydrobiopterin on the tyrosine hydroxylase or tryptophan hydroxylase in the brain is more difficult to prove and controversial. Peripherally administered (6*R*)- or (6*S*)-tetrahydrobiopterin and a synthetic cofactor, 6-methyltetrahydropterin, do enter the brain through the blood-brain barrier (Kapatos and Kaufman, 1981). The natural (6*R*) form of tetrahydrobiopterin may be more effective than the unnatural (6*S*) form. However, the transfer into the brain from blood of tetrahydrobiopterin does not appear to stimulate significantly tyrosine hydroxylase or tryptophan hydroxylase in the brain at least under physiological conditions. When (6*RS*)-tetrahydrobiopterin is administered intravenously (20 mg/kg) to rhesus monkeys, the cofactor levels increase with peak levels at 90-180 min, but no apparent effect on synthesis of dopamine, noradrenaline, or serotonin is observed (Miller *et al.*, 1986). On the other hand, the intracerebroventricular injection of (6*RS*)-L-*erythro*-tetrahydrobiopterin enhanced the tyrosine hydroxylation in rat striatum *in vivo* (Kettler *et al.*, 1974). Intracerebroventricularly administered (6*R*)-L-*erythro*-tetrahydrobiopterin increased the accumulation of DOPA and 5-hydroxytryptophan, respectively, suggesting that (6*RS*)-tetrahydrobiopterin concentration may not reach saturation levels within dopaminergic and serotonergic neurons in the brain (Miwa *et al.*, 1985). In an experiment in which DOPA accumulation was measured *in vivo* using microdialysis in the striatum, the intraperitoneal administration of (6*R*)-tetrahydrobiopterin (50 mg/kg) increased the accumulation of DOPA under the inhibition of aromatic L-amino acid decarboxylase in the striatum both of transgenic mice carring human tyrosine hydroxylase gene and of non-transgenic mice by about 4-fold. The results suggest that the peripherally administered 6*R*-

tetrahydrobiopterin activates *in vivo* tyrosine hydroxylase activity in the nigrostritiatal dopamine neurons in both wild and human tyrosine hydroxylase transgenic mice (Nagatsu *et al.*, 1993).

Stimulation of dopamine synthesis after peripheral administration of tetrahydrobiopterin (20 mg/kg) was proved by positron emmision tomography (PET) into striatum of the monkey brain using L-[beta-^{11}C]DOPA. Interestingly, further stimulation was observed by simultaneous peripheral administration of L-tyrosine with L-[beta-^{11}D]DOPA. These results suggest that the low dose loading of tyrosine facilitates the activity of tetrahydrobiopterin on the presynaptic dopamine biosynthesis. (Tsukada *et al.*, 1996).

Intracerebral distribution of tyrosine hydroxylase and tryptophan hydroxylase parallels the distribution of the tetrahydrobiopterin in both rat (Bullard *et al.*, 1978; Levine *et al.*, 1979) and in human brains (Nagatsu *et al.*, 1986b; Sawada *et al.*, 1987), especially in the striatum. Tetrahydrobiopterin, GTP cyclohydrolase I, and tyrosine hydroxylase are localized in the rat striatum, (Levine *et al.*, 1981). Coexistence of tyrosine hydroxylase and GTP cyclohydrolase I in the nigrostratal dopamine neurons was proved by a double-labeling immunofluorescence method with a confocal laser-scanning microscope (Nagatsu, I. *et al.*, 1997). In human brains, biopterin and GTP cyclohydrolase I are mainly localized in dopaminergic, noradrenergic, and serotonergic regions, however, the distribution of neopterin which is derived from D-*erythro*-7,8-dihydroneopterin triphosphate, an intermediate of tetrahydrobiopterin biosynthesls from GTP, has no correlation with that of biopterin (Nagatsu *et al.*,1986; Sawada *et al.*, 1987; Sawada and Nagatsu, 1987). Neopterin is found only in humans and monkeys, and its physiological significance is unknown (Fig. 6).

Among various psychotropic drugs, D-amphetamine, especially after 10-day infusion, resulted in parallel reductions in biopterin, dopamine, and serotonin in the rat striatum.(Lee and Mandell, 1985).

Injection of manganese into the rat striatum causes a rapid fall in the biopterin and dopamine content. The fall in dopamine levels may result from a rapid replacement of the amine from its storage sites by manganese followed by a decrease in the rate of dopamine synthesis caused by the drop in biopterin cofactor levels (Lista *et al.*, 1986).

In the adrenal gland, reserpine increased tyrosine hydroxylase protein threefold, but cellular tyrosine hydroxylase activity measured in tissue slices increased sixfold. This differences may be due to increased biopterin cofactor levels induced by reserpine (Togari *et al.*, 1985).

Repeated subcutaneous injection of thyrotropin-releasing hormone (TRH) and its analog, butyrolactone-carbonyl-L-prolinamide citrate (DN

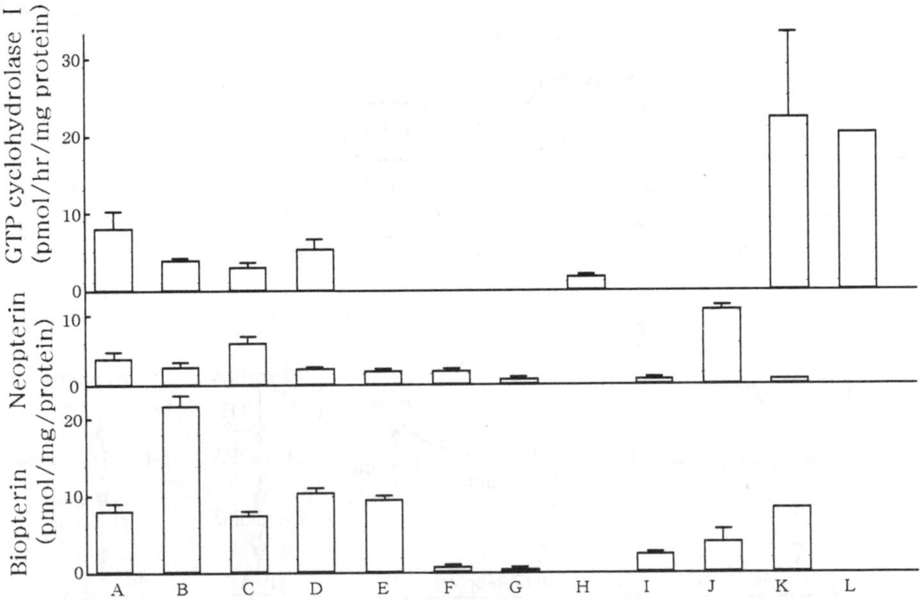

Figure 6. Distribution of GTP cyclohydrolase I, neopterin and biopterin in the brain regions of humans. A: Substantla nigra, B: Coudate nucleus, C: Putamen, D: Globus pallidus, E: Nucleus occumbens, F: Amygdata, G: Hippocampus, H: Frontol cortex, I: Hypothalamus, J: Locus coeruleus, K: Rophe nucleus, L: Reticular formation.

1417) reduced biopterin levels in the raphe nucleus 70%, but cellular tryptophan hydroxylase activity in raphe tissue slices was reduced by 30% . This difference was found to be due to induction or activation of tryptophan hydroxylase (Sawada *et al.*, 1987).

These results suggest that tetrahydrobiopterin concentrations could be a regulatory factor for the activities of catecholamines and serotonin, but the rate-limiting role still remains unclear.

Although the main physiological function of tetrahydrobiopterin is considered to be due to its role as the cofactor of pterin-requiring amino acid hydroxylases, there are still unknown physiological roles. The distribution of tetrahydrobiopterin in various tissues other than catecholamine- and indoleamine-containing cells suggests some ubiquitous function of tetrahydrobiopterin. It should be noted that tetrahydrobiopterin is also an essential cofactor for nitric oxide synthetase (Tayeh and Marletta 1989; Know et al., 1989).

Relationships between metabolism of tetrahydrobiopterin and catecholamines and/or serotonin, and their pathological metabolites used as biochemical markers in the differential diagnosis are illustrated in Fig. 7.

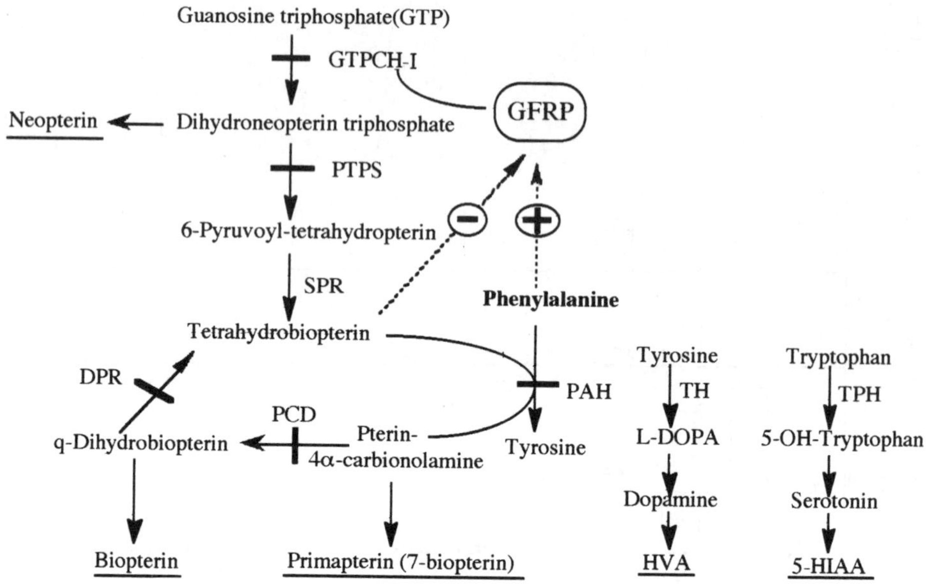

Figure 7. Biosynthesis and regeneration of tetrahydrobiopterin including hydroxylation of aromatic amino acids. Pathological metabolites used as specific markers in the differential diagnosis are marked in squares. (Blau *et al.*, 1996). GTPCH-I, GTP cyclohydrolase I; PTPS, 6-pyruvoyl - tetrahydroptrin synthase; SPR, sepiapterin reductase; PCD, pterin-4α-carbinolamine dehydratase; DPR, dihydropteridine reductase; PAH, phenylalanine hydroxylase; TH, tyrosine hydroxylase; TPH, tryptophan hydroxylase; HVA, homovanillic acid; 5-HIAA, 5-hydroxyindoleacetic acid; GFRP, GTPCH-I feedback regulatory protein.

PHYSIOLOGICAL ROLE ON NEUROTRANSMITTER RELEASE

(6*R)*-Tetrahydrobiopterin causes the release of dopamine, serotonin and glutamic acid *in vitro* in rat striatum by dialytic infusion of (6*R*)-tetrahydrobiopterin (Fig. 8) (Koshimura *et al.*, 1990; Mataga, *et al.*, 1991) and acetylcholin in the hippocampus (Ohue *et al.*, 1991; Ohue *et al.*, 1992). The release of neurotransmitters by (6*R*)-tetrahydrobiopterin is not dependent on the stimulation of biosynthesis as cofactor of tyrosine hydroxylase. The function of catecholaminergic system probably mediates the release of glutamate and acetylcholine.

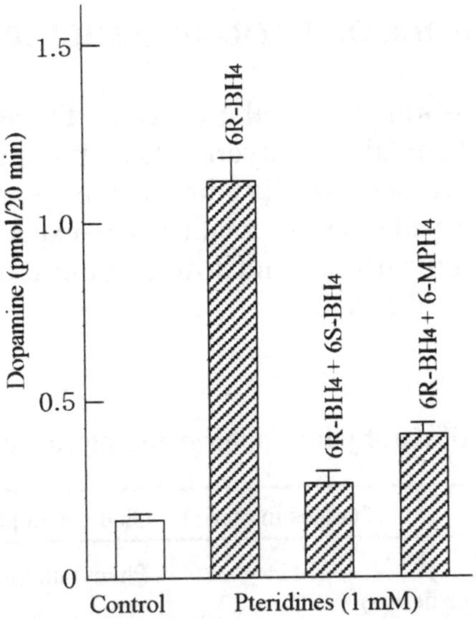

Figure 8. Effects of co-infusion of *6*S-BH4 or 6-MPH4 with 6*R*-BH4 on dopamine levels in dialysates. All rats were intraperitoneally injected with α-methyl-*p*-tyrosine (250 mg/kg) for inhibiting tyrosine hydroxylase. Twenty minutes after the injection, 6*R*-BH4 alone or in combination with either 6*S*-BH4 or 6-MPH4 was added to the perfusion fluid. Dopamine levels in pteridine-treated groups are the maximal value after administration of pteridines. (Miwa *et al.*, 1992)

REGULATION OF NITRIC OXIDE SYNTHETASE ACTIVITY BY TETRAHYDROBIOPTERIN

(6*R)*-Tetrahydrobiopterin is an essential cofactor of nitric oxide synthetase (NOS) (Kwon *et al.*, 1989; Tahey and Marlleta, 1989; Stuehr *et al*., 1989). The pteridine was thought to have a function in NOS similar to its role in pterin-dependent aromatic amino acid hydroxylases. However, the pteridine is apparently not involved directly in the NOS-catalyzed, L-arginie-independent reduction of molecular oxigen, but the possibility exists that the enzyme-bound cofactor is redox active. Binding of the pteridine is essential for formation of active dimer (Tzeng *et al.*,1996; Klatt *et al.*, 1995, 1996; Nathan and Xie, 1994). Km values for (6*R)*-tetrahydrobiopterin of aromatic amino acid hydroxylases are relatively high (10 - 100 μM) as compared with relatively low concentrations to obtain one-half maximal activity of NOS (0.03 - 0.1 μM). (Werner-Felmayer and Gross, 1996). Thus NOS require about 1000-fold less tetrahydrobiopterin for its activation than aromatic amino acid hydroxylases. (Kinoshita *et al.*, 1997).

PATHOLOGICAL ROLE OF TETRAHYDROBIOPTERIN

Since tetrahydrobiopterin is the cofactor for aromatic amino acid hydroxylases, i.e., phenylalanine hydroxylase, tyrosine hydroxylase, and tryptophan hydroxylase, and also for nitric oxide synthetase, deficiency of tetrahydrobiopterin in pathological conditions is expected to result in high phenylalanine concentrations, and low catecholamine and serotonin concentrations in tissues (Table 2).

Table 2.
Pathological changes of tetrahydrobiopterin and pterin-related enzymes

Diseases	Changes in pterins	Changes in pterin-related enzymes
Classical phenylketonuria (phenylalanine hydroxylase defect)	BH4→	Phenylalanine hydroxylase ↓
Dihydropteridine reductase defect	BH4 ↓	Dihydropteridine reductase ↓
GTP cyclohydrolase I deficiency	NP ↓ BP ↓	GTP cyclohydrolase I ↓
Hereditary progressive dystonia(HPD) /DOPA responsive dystonia(DRD) /Segawa's disease	BP ↓ (brain: nigrostriatum)	GTP cyclohydrolase I ↓ Tyrosine hydroxylase ↓ (brain: nigrostriatum)
6-Pyruvoyltetrahydropterin synthetase defect	BP ↓ NP ↑	6-Pyruvoyltetrahydropterin syntase ↓ Tyrosine hydroxylase ↓ (brain: nigrostriatum)
Parkinson's disease	BP ↓ (brain: nigrostriatum)	GTP cyclohydrolase I ↓ Tyrosine hydroxylase ↓
Depression	CSF BP ↓ ? plasma BP ↑ urine BP ↑	Tryptophan hydroxylase ↓ ?
AIDS	BP ↓ NP ↑	
Cancer	NP ↑	

Abbreviations: BH4, tetrahydrobiopterin; BP, (total) biopterin; NP, (total) neopterin; AIDS, acquired immunodeficiency syndrome; →, no change; ↑, increase; ↓, decrease.

Phenylketonuria

Phenylketonuria is a hereditary disease due to phenylalanine hydroxylase abnormality (McKusick 26160) (Kaufman, 1985). Phenylalanine hydroxylase, which produces tyrosine from phenylalanine, is an obligatory enzyme for complete degradation of phenylalanine, and the enzyme defect causes accumulation of phenylalanine in the blood (hyperphenylalaninemia), and the contents of such abnormal metabolites as phenylpyruvic acid are increased and excreted in urine. Accumulation of phenylalanine or the metabolites during development damages the brain and results in mental deficiency. Phenylketonuria can be prevented by a phenylalanine-restricted diet when hyperphenylalaninemia is diagnosed immediately after birth.

In contrast to phenylalanine hydroxylase abnormality, which is also called "classical or typical phenylketonuria," phenylketonuria due to tetrahydrobiopterin deficiency (Table 3) is called "atypical phenylketonuria." Tetrahydrobiopterin deficiency also causes deficiency of catecholamines and serotonin. Hence, the symptoms are more severe than those of classical phenylketonuria (phenylalanine hydroxylase deficiency) and cannot be treated by mere restriction of phenylalanine in diet. Instead, symptoms of atypical phenylketonuria are diminished by administration of tetrahydrobiopterin together with L-DOPA and L-5-hydroxytryptophan to supplement catecholamines and serotonin. Four enzyme abnormalities of tetrahydrobiopterin biosynthesis have been found. Dihydropteridine reductase

Table 3.
Biochemical differentiation of phenylketonuria (hyperphenylalaninemia)

Abnormal enzyme	Response to low Phe diet	Biopsy or culture	Phe load	BH4 load	Pterin (urine or plasma)	Catecholamines and serotonin (urine, plasma, CSF)
PAH	(+)	Liver	Plasma BH4 ↑	Plasma Phe→	Normal	Slightly low
DPR	(−)	Liver Erythrocyte Leukocytes Amnionic cells	Plasma BH4 ↑	Plasma Phe ↓	BH4 ↓ BH2 ↑	Low
PTPS	(−)	Erythrocytes	Plasma BH4→	Plasma Phe ↓	NP ↑ BP ↓ NP/BP ↑	Low
GTPCH-I	(−)	Liver Leukocytes	Plasma BH4→	Plasma Phe ↓	NP ↓ BP ↓	Low
[HPD/DRD] GTPCH-I	(−)	Leukocytes	Plasma BH4→	Plasma Phe ↓	NP→ BP→	Dopamine in striatum Low

Abbreviations: PAH,phenylalanine hydroxylase; TH, tyrosine hydroxylase; TPH, tryptophan hydroxylase; DPR, dihydropteridine reductase; PTPS, 6-pyruvoyltetrahydropterin synthetase; GTPCH-I, GTP cyclohydrolase I; BP, biopterin; BH4, tetrahydrobiopterin; BH2, 7,8-dihydroneopterin; Phe, phenylalanine; →, no change; ↑, increase; ↓, decrease.

deficiency (McKusick 26163) (Kaufman *et al.*, 1975) causes not only hyperphenylalaninemia, but also low homovanillic acid (a main metabolite of dopamine) and low 5-hydroxyindoleacetic acid (a main metabolize of serotonin) in cerebrospinal fluid. Low dihydropteridine reductase activity can be proved in cultured skin fibroblasts and leukocytes of patients (Butler *et al.*, 1978). Administration of tetrahydrobiopterin or a synthetic tetrahydropterin such as (6*RS*)-methyltetrahydropterin is expected to be a promising supplementary therapy. An effective dose of tetrahydrobiopterin of 300-500 μmol/kg day (70-120 mg/kg day) is suggested (Kaufman *et al.*, 1978). Tetrahydrobiopterin (10-20 mg/kg) usually lowers the plasma phenylalanine. But in some patients tetrahydrobiopterin loading in amounts larger than 20 mg/kg is required (Endes, 1987).

Some phenylketonuric patients have defects in tetrahydrobiopterin biosynthesis (McKusick 26164, 26169) (Kaufman *et al.*, 1978).

GTP cyclohydrolase I deficiency blocks tetrahydrobiopterin biosynthesis at the first step of conversion of GTP to D-*erythro*-7,8-dihydroneopterin triphosphate (Niederwieser *et al.*, 1982; Niederwieser *et al.*, 1984), and these patients are unable to produce any pterins. Thus, both neopterin and biopterin are decreased in urine.

6-Pyruvoyl tetrahydrobiopterin synthetase (phosphate-eliminating enzyme) deficiency (Niederwieser *et al.*, 1985) and probably 6-pyruvoyl tetrahydropterin reductase deficiency (Dhondt *et al.*, 1985) are characterized by hyperphenylalaninemia, a high excretion of neopterin and a low biopterin concentration in urine.

Parkinson's Disease

Parkinson's disease is pathologically characterized by selective cell loss of the nigrostriatal dopaminergic neurons in the brain. The specific biochemical changes are decreases in dopamine, tyrosine hydroxylase, biopterin (Nagatsu *et al.*, 1981) and GTP cyclohydrolase I (Nagatsu *et al.*, 1986a) in the striatum. Biopterin concentrations are also reduced in the cerebrospinal fluid (Lovenberg *et al.*, 1979) and serum (Yamaguchi *et al.*, 1983). The decreases in biopterin and GTP cyclohydrolase I in the striatum may be due to the cell loss of dopaminergic neurons. However, in one case of juvenile parkinsonism, tyrosine hydroxylase in the striatum was decreased with normal tyrosine hydroxylase in the substantia nigra and without cell loss (Narabayashi *et al.*, 1986). Narabayashi *et al.*, (1982) reported that oral administration of tetrahydrobiopterin is effective for the treatment of parkinsonian symptoms. This may suggest that tetrahydrobiopterin can stimulate residual tyrosine hydroxylase in the striatum of parkinsonian patients.

Alzheimer-Type Senile Dementia

Tetrahydrobiopterin levels of cerebrospinal fluid decline by approximately a third over a 40-year span, probably reflecting loss of aminergic neurons in the brain (Williams *et al.*, 1980). The serum biopterin levels are also significantly reduced at older ages in age-matched controls (Aziz *et al.*, 1983). Therefore, biopterin cofactor level may decrease with age. The concentrations of total biopterin in the brains of patients with Alzheimer-type senile dementia were significantly reduced in the putamen and substantia nigra (Sawada *et al.*, 1987). These results suggest some disturbance in tetrahydrobiopterin metabolism in Alzheimer-type senile dementia and that the reduction in biogenic amines may be an event independent of the cholinergic dysfunction. The efficacy of tetrahydro-biopterin to improve clinical symptoms of Alzheimer-type senile dementia remains for further study.

Depression

There has been a monoamine hypothesis on endogenous depression; the reduction in serotonin and/or noradrenaline may be the cause of some types of endogenous depression. Based on this hypothesis, tetrahydrobiopterin administration to stimulate tryptophan hydroxylase or tyrosine hydroxylase was tried for the treatment of endogenous depression, and successful treatment with concomitant increase in 5-hydroxyindoleacetic acid and homovanillic acid and biopterin in cerebrospinal fluid was reported (Curtius *et al.*, 1983; Ito *et al.*, 1984). Brain tetrahydrobiopterin synthesis in postmortem cortex samples of four patients with a history of severe depression but without histopathological evidence for Alzheimer-type senile dementia was found to be significantly reduced in the depressed subjects (Blair *et al.*, 1984). The results suggest decreased tetrahydrobiopterin biosynthesis in the brain in depression. However, tetrahydrobiopterin levels in cerebrospinal fluid in patients with depressed patients were reported not to be different from those in normal patients (Kellner *et al.*, 1983). Plasma biopterin levels were found to be significantly elevated in depressive phase of patients with depression, but in remission phase the values did not differ from normal controls (Hashimoto *et al.*, 1987). Urinary excretion of biopterin in severely depressed patients was also reported to be increased, although the small number of patients examined in remission had reduced biopterin output (Duch *et al.*, 1984). Increased biopterin in plasma and urine in depressed patients seems to be contradictory to assumed decrease in biosynthesis of tetrahydrobiopterin in the brain. However, though the origin of plasma biopterin is not determined yet, increased excretion of cortisol and biopterin from adrenal cortex could be assumed in depression patients.

Measurements of biopterin in blood plasma of patients with depression or its related disorders are shown in Tables 4 to 7 (Nagatsu *et al.*, 1990).

Table 4.
Total biopterin concentration in blood plasma of depressive patients

Patient	n	Age	Biopterin (pmol/ml)
healthy subject	12	43.1±16.0	14.1±2.3
depresive patient	12	46.2±13.9	25.6±5.7*
anxiety (+)	4	50.8±13.6	32.5±4.3*
anxiety (−)	8	43.9±14.5	22.2±3.6*
period of depression			
0 - 6 months	4		22.0±3.1*
6 - 12 months	3		23.2±1.4*
over than 12 months	5		29.9±5.9*
remmision phase of dedepression	10	43.0±9.3	13.9±3.1

**p* <0.01, significantly different from control healthy subject.

Table 5.
Total biopterin concentration in blood plasma of patients with emotinal disorder

Patient	n	Hamilton score	Biopterin (pmol/ml)
healthy subject	12		14.1±2.3
monopolar depression	8		
depressive phase		22.9±2.9	25.4±5.8*
remission phase		6.8±2.0	12.1±3.4
bipolar depression	4		
hypomania phase			35.5±6.8*
depressive phase		25.3±3.0	28.6±5.2*
remission phase		8.3±2.1	13.4±2.5
hypomania	1		
hypomania phase			35.0
remission phase			14.0

**p* <0.01, significantly different from control healthy subject.

Table 6.
Total biopterin concentration in blood plasma of the same indivisual depressive patient

Patient	n	Hamilton score	Biopterin (pmol/ml)
healthy subject	12		14.1±2.3
depressive patient	6		
depressive phase		24.5±5.3	27.2±6.0*
remission phase		5.8±2.2	13.6±1.9

*p <0.01, significantly different from control healthy subject.

Table 7.
Total biopterin and tetrahydrobiopterin concentrations in the blood plasma of patients with emotional disorder

Patient	n	BP (pmol/ml)	BH4 (pmol/ml)	BP/BH4 (pmol/ml)
healthy subject	8	23.0±2.9	16.6±2.0	0.71±0.10
monopolar depression	8	42.7±3.6*	10.6±3.0*	0.25±0.08*
bipolar depression	4			
depressive phase		37.6±3.2*	7.1±3.5*	0.20±0.10*
hypomania phase		55.2±4.2*	23.0±2.6	0.42±0.05*

*p <0.01, significantly different from control healthy subject. BP: biopterin, BH4: tetrahydrobiopterin.

Infantile Autism

The levels of pterin compounds in the lumbar cerebrospinal fiuid (CSF) of ten neurologically normal and twenty autistic children were measured using HPLC system with fluorimetric detection by post-column sodium nitrite oxidation (Tani and Ohno, 1994a; Tani *et al*., 1994b). Among pterin compounds 7.8-dihydroneopterin (NH2) and 6*R*-5,6,7,8-tetrahydrobiopterin (*R*-BH4) levels in autistic children were significantly reduced to 66.1 and 41.5%, respectively, of those found in the controls. The autistic children were divided into three subgroups according to their CSF homovanillic acid (HVA) level and age: the group with an elevated HVA level, the younger (<7 year old) group with a normal HVA level, and the older (>7 year old) group with a normal HVA level. In comparison, the further reduction was observed in NH2 and *R*-BH4 levels in the younger group and in the 7,8-dihydropterin (PH2) level in the group with an elevated HVA level. Thus, these results suggest that the endogenous biosynthesis of *R*-BH4 in the brain may be reduced in autistic children as compared with that in neurologically normal children (Fig. 9)(Tani *et al*., 1994b) .

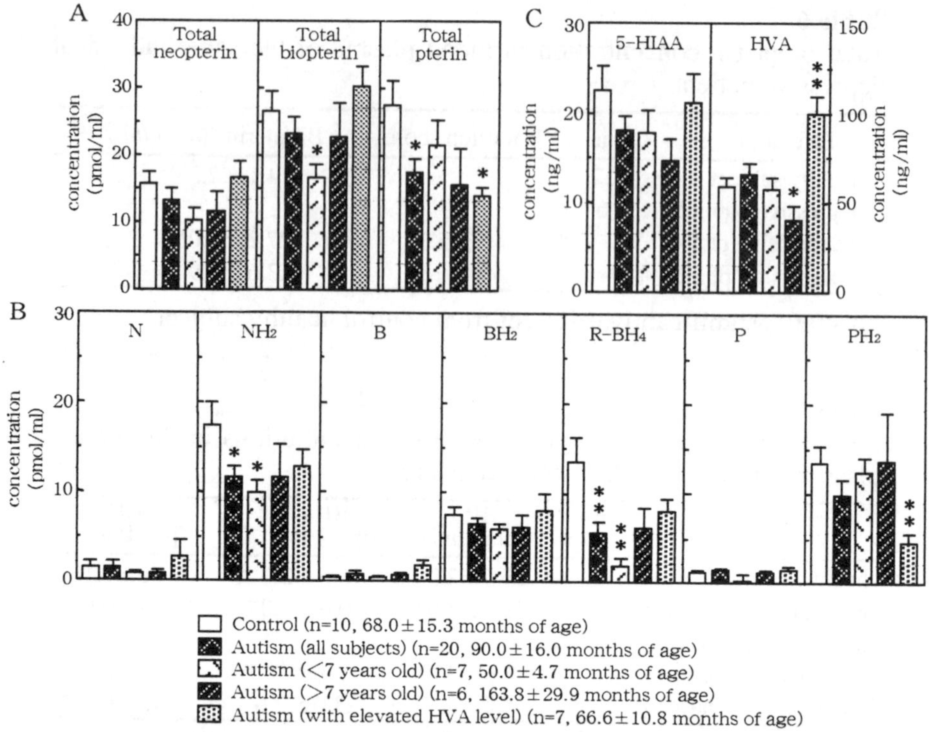

Figure 9. CSF levels of total forms of pterin compounds (A), individual pterin compounds: N, Neopterin; NH2, dihydroneopterin; B, biopterin; BH2, dihydrobiopterin; R-BH4, 6*R*- tetrahydrobiopterin; P, pterin; PH2, dihydropterin (B), and monoamine metabolites, 5-HIAA, 5-hydroxyindole-acetic acid; HVA, homovanillic acid (C) in control and autistic groups. Each collmn represents the mean ± S.E.M. Statistically significant differences (unpaired two-tailed Student *t*-test) between control and autistic groups are indicated $*p<0.05$, $**p<0.01$. (Tani *et al.*,1994)

Dystonia

In hereditary progressive dystonia (HPD)/DOPA responsive dystonia (DRD) (Segawa's disease) that is an autosomal dominant disease caused by mutations of GTP cyclohydrolase I as discovered by Ichinose *et al.* (1994), only tyrosine hydroxylase activity and dopamine content in the nigrostrital dopaminergic neurons are decreased, causing DOPA responsive dystonia (Table 8).

Infantile Meningitis

Cerebrospinal fluid (CSF) neopterin levels in children with neurologic

Table 8.
GTP cyclohydrolase I activity in HPD patients, carriers and controls

Sample	Sex	Number	GTPCH-I (pmol/mg protein/h)
HPD patients	total	7	1.9±0.4* (0.3-3.6)
	female	6	1.8±0.5
	male	1	2.4
HPD carriers	male	2	6.7, 6.9
Healthy control	total	13	18.8±2.9 (9.0-46.1)
	female	8	15.3±1.6
	male	5	24.3±6.9

Data are represented as mean ± SE. The numbers in paretheses indicate the range of values. GTPCH-I: GTP cyclohydrolase I. (Ichinose *et al.*, 1994)
*$p < 0.01$ compared to total control value.

diseases were measured and shown to be an useful marker of inflammatory central nervous diseases (Nakazawa, 1996). They were higher in patients with aseptic and bacterial menigitis than in those with non-infectious and non-neurologic diseases. The levels in bacterial menigitis were higher compared with those in aseptic meningitis (Fig. 10).

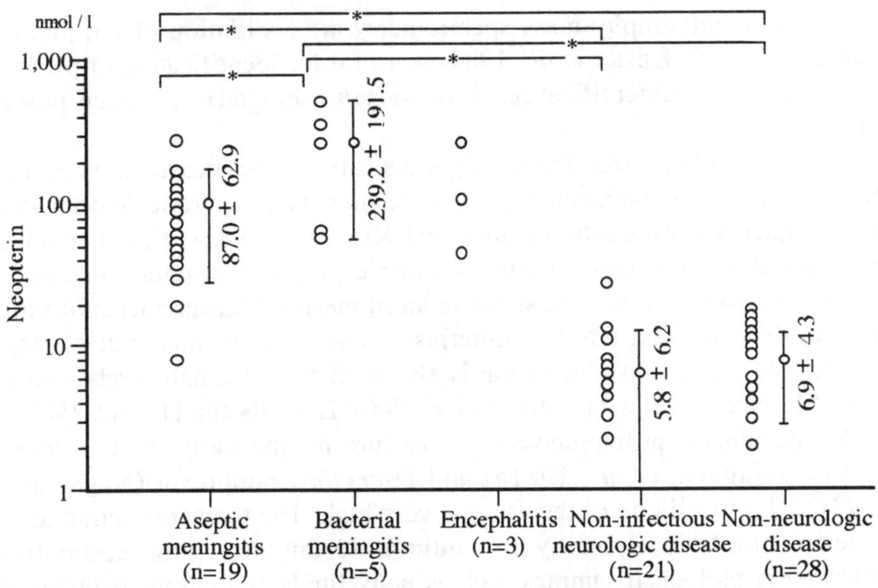

Figure 10. Neopterin concentration in cerebrospinal fluid (CSF) of patients with neurologic diseases.CSF neopterin levels were higher in patients with aseptic and bacterial meningitis than in those with non-infectious and non-neurologic diseases. The levels in bacterial meningitis were higher compared with those in aseptic meningitis. ∗ $p< 0.0001$ (Nakazawa, 1996).

ASSAY OF BIOPTERIN

Tissue concentration of total biopterin, including the four forms (tetrahydro-, quinonoid dihydro-,7,8-dihydrobiopterin and biopterin), has been measured mainly by the bioassay, based on the growth-stimulating effect for *Crithidia fasciculata,* a trypanosoma flagellate, which specifically requires a 6-substituted pterin (Fukushima *et al.*,1978). This bioassay is sensitive (less than 1 ng/ml) and specific for 6-hydroxyalkylpterins such as L-*erythro*-biopterin or L-*erythro*-neopterin *(Crithidia* factors) which have growth-stimulating effect, but it is nonspecific, time consuming and requires technical skill.

A radioenzymatic method uses *Pseudomonas* or rat liver phenylalanine hydroxylase, which is dependent on a tetrahydropterin, with radioactive phenylalanine as substrate (Guroff *et al.*,1967). Any tetrahydropterins including (6*R*)-L-*erythro*-tetrahydrobiopterin, which have the cofactor activity, are measured.

A method using high-performance liquid chromatography (HPLC) with fluorescence detection is applicable for the analysis of pterins in urine, blood, and tissues.with a sensitivity of femtomole level (Fukushima and Nixon, 1980).

Gas chromatography-mass spectrometry after sililation of the pterins (Ebels *et al.*, 1979; Kuster *et al.*, 1984) is useful for identification of pterins, and was applied to identification of 6-L-*erythro*-biopterin in sheep pineal gland.

Reduced and oxidized forms of pterins can also be measured by HPLC with electrochemical detection, especially using a dual-electrode detector in a parallel-adjacent configuration (Lunte and Kissinger, 1983). This method is highly sensitive, but requires careful sample preparation under anaerobic conditions to prevent loss of unstable reduced pterins. Measurements of fully oxidized, dihydro- and tetrahydropterins, using in series dual-cell electrochemical and fluorescence detection is also used for urine and cerebrospinal fluid (CSF) pterin analysis (Howells *et al.*,1986; Howells and Hyland, 1987).

We developed an immunoassay procedure for the assays of L-*erythro*-biopterin (Nagatsu *et al.*, 1981a) and D-*erythro*-neopterin (Nagatsu *et al.*,1984). The specificity of the assay is very high. The total concentration of biopterin (biopterin, 7,8-dihydro-, quinonoid dihydro-, and tetrahydrobiopterins) in biological samples such as body fluids and tissues is obtained by iodine oxidation in acidic conditions before the radioimmunoassay, whereas iodine oxidation in alkaline conditions gives the concentration only of the former two. This radioimmunoassay is simple, sensitive (100 fmol), and highly specific for the natural biopterin cofactor, L-*erythro*-biopterin, and therefore it is suitable for physiological and clinical studies. We also developed a polarization fluoroimmunoassay for routine determination of L-*erythro*-biopterin and D-*erythro*-neopterin in human urine (Sawada *et al.*, 1984). This method employs fluorescein-labeled biopterin and neopterin. Although the sensitivity (50 pmol) is lower than the radioimmunoassays (100 fmol) (Nagatsu *et al.*, 1981a; Nagatsu *et al.*, 1984), biopterin and neopterin in human urine samples can be measured rapidly.

A rapid and sensitive reversed-phase ion-pair liquid chromatographic system with fluorimetric detection by post-column sodium nitrite oxidation was developed for measuring six pterin compounds (6*R*-5, 6, 7, 8-tetrahydrobiopterin, 6*S*-5,6,7,8-tetrahydrobiopterin, 7,8-dihydrobiopterin, biopterin, pterin, and D-neopterin) (Tani and Ohno, 1994). The coefficients of variation for these pterins were 0.705-3.714%, and the minimum detectable amount was ca. 10-20 pg at a signal-to-noise ratio of 3. A linear detector response was also verified. This system can be applied to analyses not only of various rat tissues but also of dialysates collected by brain microdialysis.

A method using high-performance liquid chromatography (HPLC) with fluorescence detection, an enzyme immunosorbent assay (ELISA), and a reversed-phase ion-pair liquid chromatographic system, and radioimmunoassay for measurement of pterins are described in this review.

REVERSE-PHASE HIGH-PERFORMANCE LIQUID CHROMATOGRAPHY
(Fukushima, T and Nixon, J. C.(1980). *Analytical Biochemistry* 102,176-188)

Principle

This HPLC fluorescence method consists of an initial iodine oxidation of the reduced forms (tetrahydro-, quinonoid dihydro-, and 7,8-dihydrobiopterin) to highly fluorescent biopterin, a partial purification on a Dowex 50 column, and subsequent HPLC analysis with fluorescence detection. Using iodine in HCl, quinonoid dihydro-, 7,8-dihydro-, and tetrahydrobiopterin are oxidized to biopterin. When iodine oxidation is carried out in NaOH, only 7,8-dihydrobiopterin is oxidized to biopterin. Thus, both the total and reduced forms (tetrahydro- and quinonoid dihydrobiopterin) can be differentially measured.

Procedure

Assay for total biopterin

Total biopterin (H_4-biopterin + QH_2-biopterin + H_2-biopterin + oxidized biopterin) can be measured as follows. A tissue is homogenized in 10 vol of 0.1 M phosphoric acid. The homogenate (4.5 ml) is mixed with 2 M TCA

(0.5 ml) and then iodine solution (0.5% I_2/1% KI in 0.2 M TCA) is added to excess. The amount of iodine is determined by a persistence of iodine color in the homogenate for the period of oxidation. The homogenization step through the addition of iodine is carried out at 0 to 5°C. The oxidation mixture is allowed to stand 1 h at room temperature under reduced light, and then the sample is centrifuged at 3000 g for 10 min. Two milliliters of the supernatant is applied to a Dowex 50 column (200 to 400 mesh [X12], H^+) with a column bed volume of 0.2 ml (5×10 mm) packed in a Pasteur pipet. After the column is washed with 5 ml of water, pteridines are eluted with 2.8 ml of 1M NH_4OH. The eluate is neutralized with 0.2 ml of glacial acetic acid. Up to 200 μl of the neutralized eluate is applied directly to an analytical HPLC column, Most rat tissues can be analyzed by this procedure. When a higher concentration of pteridines in the eluate is required, a tissue can be homogenized in 3 vol of 0.1 M phosphoric acid.(Fig. 11)

Assay for alkaline-stable biopterin

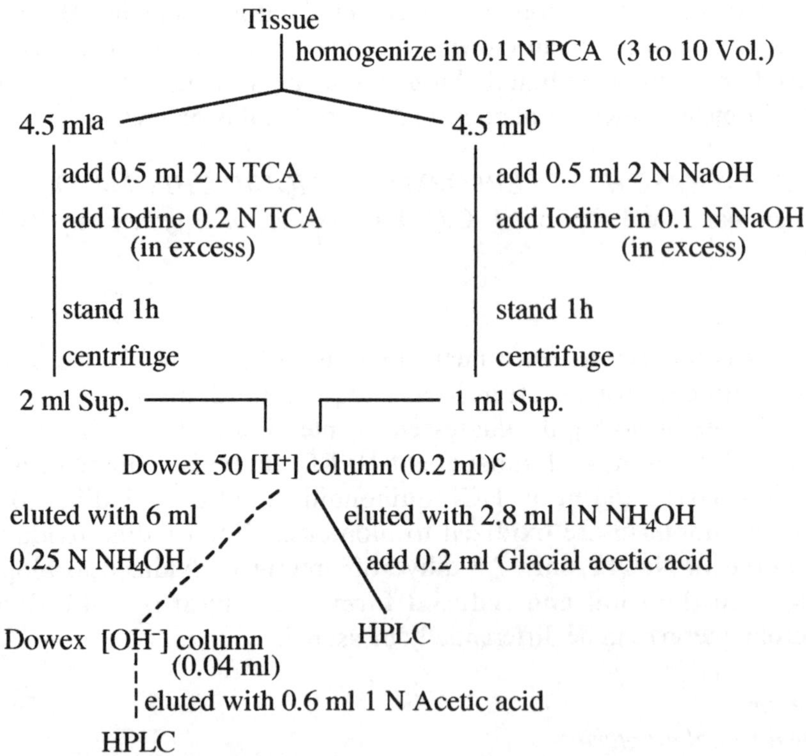

Figure 11. A standard assay procedure for tissue biopterin. a: Oxidation in acid, b: Oxidation in alkali, c: When the amount of biopterin is small, the Dowex 1 column can be used to concentrate. (Fukushima and Nixon, 1980)

Alkaline-stable biopterin (H_2-biopterin + oxidized biopterin) can also be measured. Tissue homogenate (4.5 ml) in 0.1 M phosphoric acid is made alkaline by addition of 2 M NaOH (0.5 ml). The mixture is oxidized by the addition of an alkaline iodine solution which has been prepared by mixing 9 volumes of 1% iodine in 2% KI and 1 vol of 1 M NaOH. The amount of iodine required for complete oxidation of reduced pteridines is two to three times of that used for acid oxidation of most tissues. Since the tissue homogenates are viscous under the alkaline conditions, a glass rod or spatula is used for mixing. After standing for 60 min, the sample is mixed with 0.6 ml of 2 M TCA. Iodine color should develop at this step. Any precipitate formed is removed by centrifugation, and supernatant (1 ml) is applied to a Dowex 50 column as described above. Since QH_2-biopterin behaved similarly to H_4-biopterin, the amount of H_4-biopterin plus QH_2-biopterin can be estimated from the difference between the total (acid oxidized biopterin level) and the alkaline-stable biopterin. (Fig. 11)

HPLC

A model 6000A solvent delivery system, a Model U6K injector, and a Model 440 absorbance detector were obtained from Waters Associates, Inc. A FS-970 continuous flow fluorometer (Schoeffel Instrument Corp.) was used with an excitation wave length set at 350 nm and emission cutoff below 410 nm. Both a Whatman 10 μ ODS reverse-phase column, Partisil 10 (4.6 mm × 25 cm), and a μBondapak C_{18} column (3.9 mm × 30 cm) from Waters Associates were used separately for pteridine analysis. The HPLC separation was achieved using an isocratic solvent system of 5% or 20% methanol in water. Compounds were quantitated by their peak height or peak area in comparison with standards. Peak area was measured by a Hewelett Packard 9100B Programable Calculator.

Biopterin dissolved in distilled water (1 μg/ml) was diluted for HPLC studies. This concentrated solution was stored at -20°C and was stable for several months under these conditions. The concentrated standard was diluted into an expected concentration range for biopterin within the sample. Other pteridine standards were prepared and stored similarly.

Separation of pteridines

The pteridines were dissolved in water and subjected to HPLCis presented in Table 9. The relative elution volume for each pteridine from Whatman Partisil 10 ODS, 4.6 mm × 25 cm; elution: methanol/water, 5% or 20% v/v at 0.8 ml/min. Data are expressed as relative elution volumes to the elution volume of biopterin with 5% methanol.

Table 9.
Elution volumes of various pteridines

Compound	Methanol (%) 5	20
Biopterin #	(100)	58
Threobiopterin #	118	59
H2-Biopterin §	123 *	—
Neopterin #	63	46
Threoneopterin #	72	48
H2-Neopterin §	76 *	—
6-Carboxypterin	23	23
6-Hydroxymethylpterin	124	59
Xanthopterin	86 *	—
Isoxanthopterin	115	59
Pterin	161	65
6-Formylpterin	151 *	65
6-Formyldihydropterin §	—	79
Sepiapterin §	—	113
Deoxysepiapterin §	—	169
Lumazine	102	—
6-Hydroxylumazine	42	—
7-Hydroxylumazine	23	—

Biopterin was eluted with 8 ml of 5% methanol. # D and L isomers of the compound showed the same retention volumes. § Detected by UV absorption at 254 nm. *Broad peak. (Fukushima and Nixon, 1980)

Total biopterin contents in rat tissues

Biopterin concentrations in various rat tissues are presented in Tables 10 and 11. Liver was homogenized in 10 vol and brain was homogenized in 3 vol of 0.1N phosphoric acid. The reduced forms of biopterin was added to each homogenate and the sample was then oxidized in acid or alkali (Table 10). Male Sprague-Dawley rats (150 to 250 g) were used. Homogenates of liver, blood, and brain were oxidized in 0.2N TCA. Other tissues were oxidized in 0.1N HCl (Table 11).

Table 10.
Recovery of H2- and H4-biopterin from homogenates of rat liver and brain

Tissue	Oxidation conditions	Test No.	Biopterin added (ng/ml homogenate) Dihydro	Tetrahydro	Biopterin found (ng/ml)	Recovery (%)
Liver	Acid	1	0	0	159	—
		2	143	0	272	79
		3	0	128	269	86
	Alkali	1	0	0	12	—
		2	143	0	136	87
		3	0	128	24	9
Brain	Acid	1	0	0	18	—
		2	14.3	0	29.8	83
		3	0	12.8	28.8	84
	Alkali	1	0	0	1.4	—
		2	14.3	0	13.1	82
		3	0	12.8	3	13

(Fukushima and Nixon, 1980)

Table 11.
Distribution of total biopterin in rat tissues

Tissue	Biopterin (μg/g wet wt)
Pineal gland #	12.5 ± 2.0
Liver	1.6 ± 0.4
Adrenal	0.95 ± 0.10
Blood §	0.12 ± 0.05
Bone marrow(red)	1.2
Spleen	0.9
Lung	0.35
Intestine(small)	0.2
Kidney *	0.16
Brain	0.09
Muscle	0.08
Eye	0.08
Pancreas	0.07
Heart	0.07
Urine @	48/day/rat

Calculated on the assumption that each gland weight 1.5 mg. § younger rats may show higher values. * yields several unidentified fluorescent peaks in addition to biopterin and pterin. @ mean of 6-day analyses of two rats. (Fukushima and Nixon, 1980)

HIGHLY SENSITIVE , SPECIFIC ENZYME-LINKED IMMUNOSORBENT ASSAY OF NEOPTERIN AND BIOPTERIN IN BIOLOGICAL SAMPLES

(Ogiwara, S., Kiuchi, K., Nagatsu, T., Teradaira, R., Nagatsu, I., Fujita, K. and Sugimoto, T. (1992). *Clinical Chemistry* 38, 1954 - 1958)

Principle

Since either biopterin or neopterin is a hapten without antigenicity, L-*erythro*-biopterinyl-capronyl-bovine serum albumin is synthesized and used as antigen to obtain monospecific polyclonal antibodies toward L-*erythro*-biopterin or D-*erythro*-neopterin. A radiolabeled ligand is prepared by synthesizing L-*erythro*-biopterinyl- or D-*erythro*-neopterinyl-[^{125}I]tyramine. The antibody can stereospecifically recognize L-*erythro*-biopterin or D-*erythro*-neopterin.

Procedure

Binding the hapten to the microtiter plate

Each well of a microtiter plate was rinsed three times with 200 µl of 0.01 mol/l phosphate-buffered saline containing Tween 20 (0.5 ml/l) after each step of the following binding and immunoassay processes. The entire procedures were carried out at 24 °C unless otherwise indicated.

To each well of a microtiter plate we added 150 µl of 0.1 g/l poly-L-lysine solution and kept this at 4 °C for 10 h. The plate was incubated with 100 g/l glutaraldehyde reagent (200 µl/well) for 1 h; with the N^2-(3-aminopropyl) derivative of neopterin (Nagatsu, Matsuura, and Sugiura, 1989),10 µmol (100 µl), for 1 h; with 10 g/l sodium borohydride reagent (200 µl) for 0.5 h; and finally with 100 ml/l normal sheep serum (200 µl) for 1 h. The plate can be stored at 4°C for several weeks without changing its characteristics.

The microtiter plate for the immunoassay of biopterin was similarly prepared by using the N^2-(3-aminopropyl) derivative (Niederwieser *et al.*, 1979) of biopterin.

Immunoassay of neopterin and biopterin

To assay neopterin in serum, we first reacted the sample with a fixed amount of anti-neopterin antiserum. The amount of excess antiserum was determined with the prepared microtiter plate as described below.

We incubated 150 µl of 2000-fold-diluted anti-neopterin antiserum and 20 µl of the sample solution [or 0.01 mol/l phosphate-buffered saline, pH 7.4, containing neopterin standards (from 0 to 100 pmol)] for 1 h. We then transferred 150 µl of the incubated mixture to a well on the plate and let this stand for 1 h. After washing the well (see above), we added 150 µl of goat anti-rabbit IgG-horseradish peroxidase conjugate diluted 2000-fold with 0.01 mol/L phosphate-buffered saline, pH 7.4, and incubated for 1 h. We then added 200 µl of 5 g/l *o*-phenylenediamine solution containing 80 µl of 10 mol/l H_2O_2. After 10 min, we stopped the reaction by adding 50 µl of 6 mol/l H_2SO_4 to the well, and measured the absorbance at 490 nm with an NJ2000 immuno reader (Japan Intermed Co., Tokyo, Japan).

The determination of biopterin was carried out similarly with a microtiter plate containing bound biopterin derivative (Niederwieser *et al.*, 1979), and with use of the anti-biopterin antiserum and biopterin standards (from 0 to 100 pmol).

HPLC

Analytical HPLC to determine neopterin and biopterin in human serum was carried out with a JASCO 880-PU (Japan Spectroscopic Co., Tokyo, Japan), a JASCO FP-210 fluorescence detector (excitation, 355 nm; emission, 450 nm), an SIC Chromatocorder 12 (from GL Science Co., Tokyo, Japan), a 4.6 mm (i.d.) x 250 mm Develosil reversed-phase column (Nomura Chemical Co., Aichi, Japan), and ammonium phosphate buffer (30 mmol/l, pH 3.5) as the elution solvent at a flow rate of 1 ml/min.

Sensitivity and specificity

The measurable amounts of neopterin and biopterin in the present immunoassay ranged from ~ 0.1 pmol (30 pg) to 100 pmol (3 ng) per well for each. The percent cross-reaction of various pterin ligands to the anti-neopterin antiserum and anti-biopterin antiserum (Table 12) indicates a high specificity of the present immunoassay.

Table 12.
Cross-Reaction of Pterin Ligands with Antisera against Neopterin and Biopterin

Ligand	Anti-neopterin antiserum	Anti-biopterin antiserum
D-erythro-Neopterin	100	0.002
L-erythro-Neopterin	0.02	0.13
D-threo-Neopterin	0.2	0.02
L-erythro-Biopterin	0.002	100
D-erythro-Biopterin	3	0.003
L-threo-Biopterin	0.002	0.3
D-threo-Biopterin	0.02	4
Pterin-6-carboxylic acid	0.0006	0.0007
Pterin	0.0001	0.003

Cross-reactivity, %

Physiological data and reliability

To prove the reliability of the present method, we also measured the concentrations of neopterin and biopterin in human serum by HPLC-fluorometry. The values determined by the present immunoassay agreed well with the values obtained by HPLC-fluorometry (Table 13).

Table 13.
Concentrations of neopterin and biopterin in human serum determined by the present immunoassay and HPLC

Neopterin		Biopterin	
Immuno assay	HPLC	Immuno assay	HPLC
8.2	8.1	3.5	3.7
6.1	6.5	3.2	3.6
9.5	9.7	2.7	3.1
8.0	7.6	4.9	4.8
8.9	9.0	5.0	4.8
8.5	8.2	4.9	5.0
7.5	7.3	4.0	3.9

Each value is an average of two experiments (SD < 0.3). Concentration, nmol/l

ASSAY BY POST-COLUMN SODIUM NITRITE OXIDATION
(Tani, Y. and Ohno, T. (1993). *Journal of Chromatography* 617, 249 - 255.)

Principle

Pterins are directly determined by means of post-column sodium nitrite ($NaNO_2$) oxidation using a reversed-phase ion-pair LC system.

Procedure

LC equipment

A schematic diagram of the LC system (Shimadzu, Japan) is shown in Figure 12. It consisted of two pumps (LC-6A), a system controller (SCL-6B), an autoinjector (SIL-6B), a Cosmosil 5C18 reversed-phase column (250 × 4.6 mm i.d., Nacalai Tesque, Japan) with a Cosmosil 5C18 guard column (50 × 4.6 mm i.d., Nacalai Tesque) coupled with a fluorimetric detector (RF-535). The excitation wavelength was 350 nm, and the emission wavelength was 440 nm. The temperature of the column and reaction coil were set at 40°C and 80°C, respectively, using two column ovens (CTO-6A, Shimadzu). The concentration of pterins was determined by calculating the areas under the curves using an integrator (C-R4AX, Shimadzu).

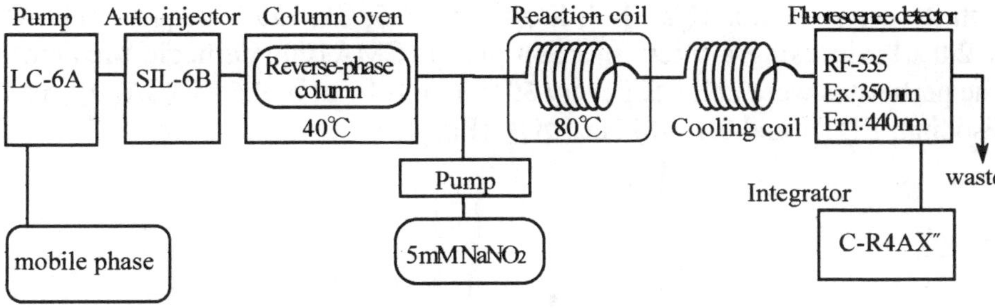

Figure 12. Post-column LC-FD system. (Tani and Ohno, 1994)

Mobile phase

The mobile phase was 0.1 M sodium phosphate buffer (pH 3.0) containing 5% (v/v) methanol, 3 mM sodium octylsulphate, 0.1 mM disodium EDTA and 0.1 mM ascorbic acid (to prevent oxidation). The flow-rate was maintained at 1.0 ml/min. The mobile phase was filtered through a 0.45-μm membrane filter (Millipore) and then degassed under vacuum before use.

Oxidation of reduced pterins

Reduced pterins were oxidized according to the method of Suzuki and Owada (1991) with some modifications. After separating the pterins on the reversed-phase column, 7, 8-dihydrobiopterin (BH_2), *R*-and *S*-THPP, which are not naturally fluorescent, were oxidized to their fluorescent compounds by $NaNO_2$ (5 mM, flow-rate, 1.0 ml/min) in the reaction coil (80°C)

Preparation of standard solutions

Standard solutions of *R*-THBP, *S*-THBP, BH_2, biopterin, pterin and neopterin (1.0 mg/ml each) were prepared in 0.5 M perchloric acid (PCA) containing 0.1 mM disodium EDTA and 0.1 mM $Na_2S_2O_3$.

Animal studies

The tissues were homogenized in 1.0-3.0 ml of 0.5 M PCA containing 0.1 mM disodium EDTA and 0.1 mM $Na_2S_2O_3$. After centrifugation (1700g, 15 min), the supernatant was filtered through a 0.45-μm membrane filter and then directly injected into the LC system in 10-100 μl volumes. The samples were prepared and the pterins were assayed within 24 h.

The microdialysis probe (CMA12, Carnegie Medicin), covered by a dialysis membrane 3.0 mm in length and 0.5 mm in diameter, was perfused with Ringer's solution (Na^+, 147; K^+, 4; Ca^{2+}, 4.5; Cl^-, 155.5 mequiv., pH 6.4) at 2.0 μl/min using a microinjection pump (CMA100, Carnegie Medicin). The perfusates were collected every 30 min into 10 μl of 0.5 M PCA, 0.1 mM disodium EDTA and 0.1 mM $Na_2S_2O_3$. (Fig. 13)

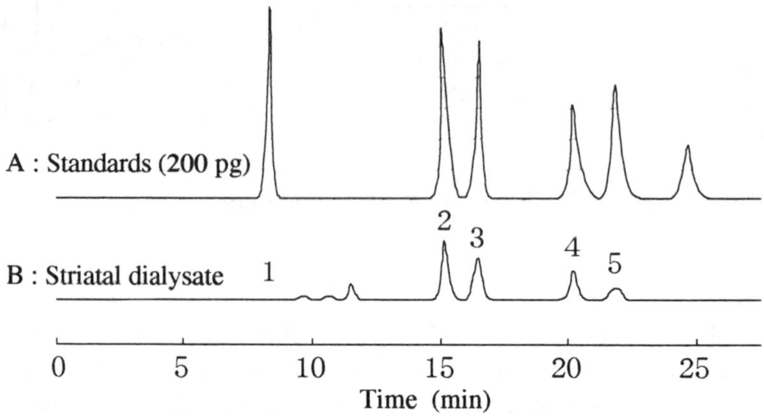

Figure 13. Reversed-phase ion-pair LC-FD separation of neopterin, biopterin, pterin, BH_2, R- BH_4 and *S*- BH_4. (A) standerd mixture (500 pq each) ; (B) rat diencephalon (50 μl inlected), (C) rat liver (10 μl injcted) . Peaks: 1 = neopterin, 2 = biopterin, 3 = pterin, 4 = dihydrobiopterin (BH_2), 5 = 6*R*- tetrahydrobiopterin (*R* - BH_4), 6 = 6*S*-tetrahydrobiopterin (*S*- BH_4) (Tani and Ohno, 1994)

Concentration of pterins in rat tissues

The concentrations of pterins in various tissues are shown in Table 14.

Table 14.
Distribution of pterin compounds in various brain regions, the kidney and the liver of young adult (eight weeks of age) Fischer-344 rats

Tissue	Neopterin	Biopterin (ng/g tissue)	Pterin (ng/g tissue)	BH_2 (ng/g tissue)	R-BH_4 (ng/g tissue)	S-BH_4
Cortex	N.D.	0.55±0.37	10.4±0.3	16.4±0.7	62.9±2.3	N.D.
Hippocampus	N.D.	3.76±0.50	69.1±6.8	50.9±2.1	102.8±4.9	N.D.
Striatum	N.D.	6.15±0.47	60.1±2.2	104.1±9.6	695.7±98.4	N.D.
Diencephalon	N.D.	1.32±0.30	32.5±2.1	31.2±1.6	176.8±6.8	N.D.
Cerebellum	N.D.	0.72±0.24	61.2±14.6	21.9±0.4	89.3±8.9	N.D.
Brainstem	N.D.	1.79±0.17	45.7±6.8	32.8±0.9	110.0±3.6	N.D.
Kidney	N.D.	1.33±0.31	90.6±16.1	101.8±30.9	235.8±22.9	N.D.
Liver	N.D.	16.67±2.37	357.1±46.7	261.7±21.0	2883.2±172	N.D.

N.D., not detectable. (Tani and Ohno, 1994)

Pterin compound levels in rat striatal dialysates

A representative chromatogram obtained after injection of a dialysate from rat striatum is shown in Fig. 14. Small amounts of *R*-THBP, BH_2, biopterin and pterin were detected in intrastriatal dialysate, but neopterin and *S*-THBP were not. Preliminary experiments showed that ascorbic acid (final concentration, 0.86 mM) prevents the oxidation of reduced pterins in a sampling tube.

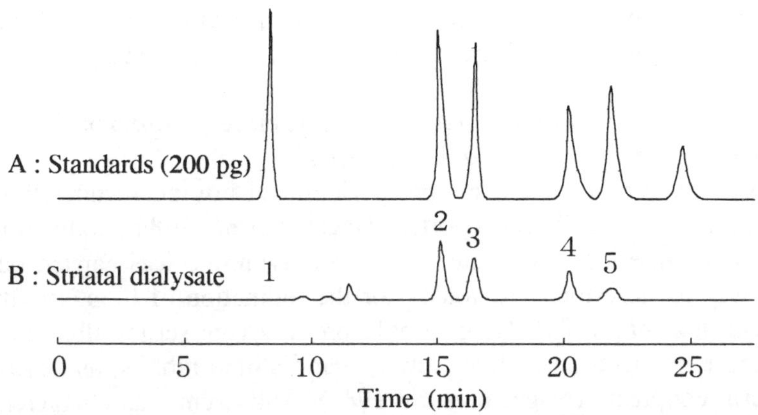

Figure 14. Reversed-phase ion-pair LC-FD separation of pterin compounds. (A) standard mixture (200 pg each), (B) rat striatal dialysate (50 μl injected). The perfusate was collected at 2.0 μl/min every 30 min in a small tube containing 10 μl of 0.5 M PCA, 0.1 mM disodium EDTA, 0.1 mM $Na_2S_2O_3$ and 6 mM ascorbic acid. (Tani and Ohno, 1994)

A SENSITIVE ASSAY OF GTP CYCLOHYDROLASE I ACTIVITY IN RAT AND HUMAN TISSUE USING RADIOIMMUNOASSAY OF NEOPTERIN

(Sawada, M., Horikoshi, T., Masada, M., Akino, M., Sugimoto, T., Matuura, S. and Nagatsu, T. (1986). *Analytical Biochemistry* 154, 361-366.)

Principle

D-*erythro*-7,8-Dihydroneopterin triphosphate formed from GTP by GTP cyclohydrolase I was oxidized by iodine and dephosphorylated by alkaline phosphatase to D-*erythro*-neopterin, and quantified by a radioimmunoassay for D-*erythro*-neopterin. This method was highly sensitive and required only 0.2 mg of rat liver tissues for the measurement of the activity. It was reproducible and can be applied for the simultaneous assay of many samples. The activity of GTP cyclohydrolase I was measured in several rat tissues.

Procedure

Human brains and rat tissues

Rat tissues were obtained from adult male Wistar rats, and used either immediately after decapitation or 2-4 h postmortem for the assay or stored frozen at - 80℃. Ten control and six parkinsonian human brains were obtained from patients. Control patients had no history of neurological disorders and had not received drugs known to affect monoamine metabolism. The brains were removed between 2 and 4 h postmortem. Caudate nucleus was dissected and frozen on dry ice and stored at - 80℃.

Preparation of the ^{125}I-labeled neopterinyltyramine, neopterinylcaproyl-bovine serum albumin conjugate and antiserum

Neopterinyltyramine and neopterinylcaproyl-bovine serum albumin were synthesized as previously reported (Nagatsu *et al.*, 1984). Iodination of neopterinyltyramine into ^{125}I-labeled neopterinyltyramine was carried out in the same way as described previously for the iodination of biopterinyltyramine (Nagatsu *et al.*, 1984). Neopterinylcaproyl-bovine serum albumin was used for the immunization of five New Zealand albino rabbits, as described for biopterin conjugate (Nagatsu *et al.*, 1981). Antiserum was characterized as previously reported (Nagatsu *et al.*, 1984).

Preparation of enzyme source

Human brains or rat tissues were homogenized in 3 vol of cold 0.1 M Tris-HCl buffer, pH 7.8, containing 2.5 mM EDTA, 0.3 M KCl, and 10% glycerol. After centrifugation at 39,000g for 30 min, 450 μl of the supernatant was applied to 5× 60-mm Sephadex G-25 column and eluted with the same buffer. The first 450 μl of elute was discarded and the next 450 μl was collected to be used as an enzyme source. This desalting procedure was necessary to protect the product, D-*erythro*-neopterin triphosphate, from further enzymatic metabolism.

Assay procedure of GTP cyclohydrolase I

The reaction mixture (350 μl) contained 0.6 mM GTP, 2.5 mM EDTA, 0.3 M KCl, 10% glycerol in 0.1 M Tris-HCl buffer, pH 7.8, and 100 μl of enzyme preparation (Sephadex elute). The reaction proceeded at 37℃ for 60 min in the dark and was terminated by the addition of 20 μl of 1M HCl solution. The mixture was oxidized by the addition of 10 μl of a solution containing 8% iodine and 16% KI in the dark for 10 min and the oxidation was terminated by the addition of 10 μl of 8% ascorbic acid. Neopterin triphosphate formed was dephosphorylated by the addition of 30 μl of 1M NaOH followed by 3.5 units of alkaline phosphatase and subsequent incubation for 60 min at

37℃ in the dark. The reaction was terminated by the addition of 50 μl of 1M HCl. After centrifugation, neopterin was purified by a Dowex 50W column and the elute was lyophilized as previously reported (Nagatsu *et al.*, 1984).

Preparation of samples for radioimmunoassay and radioimmunoassay procedure

Preparation of samples and radioimmunoassay procedure were performed as described previously (Nagatsu *et al.*, 1984).

HPLC-fluorometry

Dried materials were dissolved in 300 μl of distilled water and each 100 μl was injected on a HPLC column. HPLC was performed using a Shimadzu LC-3A pump, a Shimadzu SIL-1A sample injector, and a reversed-phase column (4.6 × 250 mm) of 10 μm μBondapak C18 (Waters Associates). A Shimadzu RF-500LCA spectrofluorometer with the highest fluorometer setting was used for continuous fluorescence measurement of the column elute at 440 nm with excitation at 350 nm. The limit of sensitivity of neopterin under this condition was 1 pmol. The carrier buffer for the HPLC was 10 mM sodium phosphate buffer, pH 7.0, with a flow rate of 1.5 ml/min.

GTP cyclohydrolase I activity

The activity of GTP cyclohydrolase I in rat tissues by this method is shown in Table 15.

Table 15.
Distribution of GTP cyclohydrolase I activity in rat tissues

Tissue	n	GTPCH-I (pmol/mg protein / h)
Brain		
Cerebellum	5	4.4±1.1
Cortex	5	10.1±2.2
Hippocampus	5	5.5±0.9
Hypothalamus	5	13.8±1.4
Midbrain	5	12.9±1.2
Pons medulla	5	8.4±0.2
Striatum	6	16.3±0.8
Thalamus	5	9.8±0.4
Adrenal	5	19.0±2.5
Kidney	5	5.0±1.4
Liver	6	107.0±6.5

Abbreviations: GTPCH-I, GTP cyclohydrolase I (Mean ± S.E.).

Correlation between this assay and HPLC-fluorimetric assay

The correlation between this assay and HPLC-fluorimetric assay is satisfactory; the correlation coefficient (γ) between the two methods was 0.95 (Fig. 15).

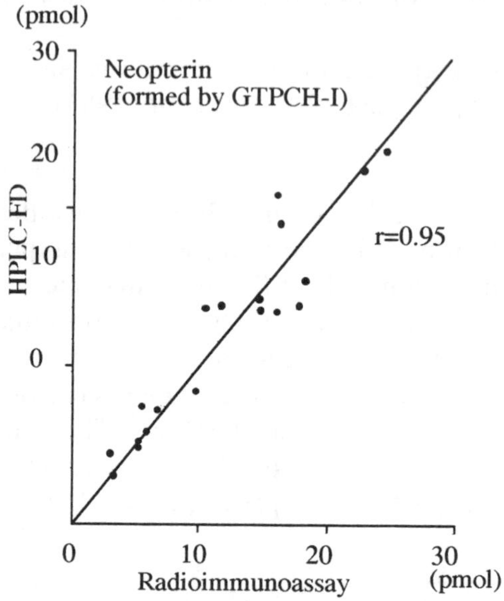

Figure 15. Correlation between neopterin formations as determined by radio-immunoassay and HPLC-fluorometric detection.
The line of best fit is shown.

CONCLUSION

The current status of physiological and pathological neurochemistry of biopterin has been reviewed in this article. The regulatory role of tetrahydrobiopterin as the cofactor of pterin-requiring aromatic amino acid monooxygenases for the biosynthesis of catecholamines and indoleamines as important neurotransmitters and hormones is clear, and our knowledge has been increasing rapidly. The clinical application of (6*R*)-L-erythro-5,6,7,8-tetrahydrobiopterin is promising. However, there may be still unknown physiological roles of biopterin and the related pterins. The role of (6*R*)-L-*erythro*-tetrahydrobiopterin as a cofactor of nitric oxide synthase should be significant, since NO is an important neurotransmitter and intercellular chemical mediator. Elucidation of such unknown roles of tetrahydrobiopterin and related pterins in biochemistry and molecular biology may be important for a better understanding of pathological roles of biopterin in clinical biochemistry.

REFERENCES

Aziz, A.A., Leeming, R.J. and Blair, J.A. (1983). Tetrahydrobiopterin metabolism in senile dementia of Alzheimer type. *J. Neurol. Neurosurg. Psychiat.* **46**, 410-413.

Bailey, S. W. and Ayling, J. E. (1978). Separation and properties of the 6-diastereoisomers of L-erythro-tetrahydrobiopterin and their reactivities with phenylalanine hydroxylase. *J. Biol. Chem.* **253**, 1598-1605.

Blair, J.A., Barford, P.A., Morar, C., Pheasant, A.E., Hamon, C.G., Whitburn, S.B., Leeming R.J., Reynolds, G.P. and Coppen, A. (1984). Tetrahydrobiopterin metabolism in depression. *Lancet* **2**, 163.

Blau, N., Barnes, I. and Dhondt, J.L. (1996). International detabese of tetrahydrobiopterin deficiencies. *J. Inher. Dis.* **19**, 8-14.

Brenneman, A.R. and Kaufman, S. (1964). The role of tetrahydrobiopterines in the enzymatic conversion of tyrosine to 3,4-dihydroxyphenylalanine. *Biochem. Biophys. Res. Commun.* **17**, 177-181.

Bullard, W.P., Guthrie, P.B., Russo, P.V. and Mandell, A.J. (1978). Regional and subcellular distribution and some factors in the regulation of reduced pterins in rat brain. *J. Pharmacol.Exp.Therap.* **206**, 4-20.

Butler, I.J., Koslow, S.H., Krumholz, A., Holtzman, N. A.and Kaufman, S. (1978). A disorder of biogenic amines in dihydropteridine reductase deficiency. *Annal. Neurology* **3**, 224-230.

Curtius, H.C., Niederwieser, A., Levine, R.A., Lovenberg, W., Woggon, B.and Angst, J. (1983). Successful treatment of depression with tetrahydrobiopterin. *Lancet* **1**, 657-658.

Dewey, V. C. and Kider, G. W. (1971). Assay of unconjugated pteridines. In: D. B. McCormick and L. D. Wright, Eds., Methods in Enzymology, Vol. 18, Part B, Academic Press, New York, p. 618.

Dhondt, J.L., Cotton, R.G.and Danks, D.M. (1985) Liver enzyme activities in hyperphenylalaninaemia due to a defective synthesis of tetrahydrobiopterin. *J. Inheri. Met. Dis.* **8**, 47-47.

Dix, T. A., Kuhn, D. M. and Benkovic, S. J. (1987). Mechanism of oxygen activation by tyrosine hydroxylase. *Biochemistry* **26**, 3354-3361.

Duch, D.S., Woolf, J.H., Nichol, C.A., Davidson, J.R. and Garbutt, J.C. (1984). Urinary excretion of biopterin and neopterin in psychiatric disorders. *Psychiatry Res.* **11**, 83-89.

Ebels, I., de Moree, A., Hus-Citharel, A. and Moszkowska, A. (1979). A survey of some active sheeppineal fractions and a discussion on the possible significance of pteridines in those fractions in vitro and in vivo assays. *J. Neural Transm.* **44**, 97-116.

Endes, W., Ibel, H., Kierat, L,. Balu, N. and Curtius, H.-Ch. (1987).

Tetrahydrobiopterin and "non-responsive" dihydropteridine reductase deficiency. *Lancet* **2**(223), 8552.

Friedman, P.A., Kappelman, A.H. and Kaufman, S. (1972). Partial purification and characterization of tryptophan hydroxylase from rabbit hind brain. *J. Biol. Chem.* **247**, 4165-4173.

Fukushima, T., Kobayashi, K., Eto, I. and Shiota, T. A. (1978). Differential microdetermination for the various forms of biopterin. *Anal. Biochem.* **89**, 71-79.

Fukushima, T. and Nixon, J. C. (1980). Analysis of reduced forms of biopterin in biological tissues and fluids. *Anal. Biochem.* **102**, 176-188.

Fujishiro, K., Hagihara, M., Takahashi, A. and Nagatsu, T. (1990). Concentrations of neopterin andbiopterin in the cerebrospinal fluid of patients with pakinson's disease. *Biochem. Med. Metabol. Biol.* **44**, 97-100.

Grenett, H. E., Ledley, F. D., Reed, L. L. and Woo, S. L. (1987). Full-length cDNA for rabbit tryptophan hydroxylase, functional domains and evolution of aromatic amino acid hydroxylases. *Proc. Natl. Acad. Sci. USA.* **84**, 5530-5534. 38.

Grima, B., Lamouroux, A., Blanot, F., Biguet, N. F. and Mallet, J. (1985). Complate coding sequence of rat tyrosine hydroxylase mRNA. *Proc. Natl. Acid. Sci. USA.* **82**, 617-621.

Grima, B., Lamouroux, A., Boni, C., Julien, J.-F., Javoy-Agid, F. and Mallet, J. (1987). A single humangene encoding multiple tyrosine hydroxylases with different predicted functional characteristics. *Nature.* **326**, 707-711.

Guroff, G., Rhoads, C. A. and Abramowitz, A. A. (1967). simple radioisotope assay for phenylalaninehydroxylase cofactor. *Anal. Biochem.* **21**, 273-278.

Gutlich, M., Jaeger, E., Rucknagel, K.P., Werner, T., Wolfgang, R., Ziegler, I. and Bacher, A. (1994). Human GTP cyclohydrolase I: only one out of there cDNA isoforms gives rise to the active enzyme. *Biochem. J.* **302**, 215-221.

Harada, T., Kagamiyama, H. and Hayakeyama, K. (1993). Feedback regulation mechanisms for the control of GTP cyclohydrolase I activity. *Science* **260**, 1507-1510.

Hashimoto, R., Ozaki, N., Ohta, T., Kasahara, Y., Kaneda, N. and Nagatsu, T. (1987). Total biopterin levels of plasma in patients with depression. *Neuropsychobiology* **17**, 176-1777.

Hausen, A., Fuchs, D. and Reibnegger G. (1989)Neopterin in clinical use . *Pteridines* **1**, 3-10.

Heintel, D., Leimbacher, W., Redweik, U., Zagalak, B. and Curtius, H. C. (1985). Purification and properties of the phosphate eliminating enzyme involved in the biosynthesis of BH4 in man. *Biochem. Biophys. Res.*

Commun. **127**, 213-219.

Hirata, Y., Togari, A. and Nagatsu, T. (1983). Studies on tyrosine hydroxylase system in rat brain slices using high-performance liquid chromatography with electrochemical detection. *J. Neurochem.* **40**, 1585-1589.

Howells, D.W., Smith, I. and Hyland, K. (1986). Estimation of tetrahydrobiopterin and other pterins incerebrospinal fluid using reversed-phase high-performance liquid chromatography with electrochemical and fluorescence detection. *J. Chromator.* **381**, 285-294.

Howells, D.W. and Hyland, K. (1987). Direct analysis of tetrahydrobiopterin in cerebrospinal fluid byhigh-performance liquid chromatography with redox electrochemistry, prevention of autoxidation during storage and analysis. *Clin. Chim. Acta* **167**, 23-30.

Ichinose, H., Ohye, T., Takahashi, E., Seki, N., Hori, T., Segawa, M., Nomura, Y., Endo, K., Tanaka, H., Tsuji, S., Fujita, K. and Nagatsu, T. (1994). Hereditary progressive dystonia with marked diurnal fluctuation caused by mutations in the GTP cyclohydrolase I gene. *Nature Genet.* **8**, 234-242.

Ichinose, H, Ohye, T, Matsuda,Y, Hori, T, Blau, N, Burline, A, Rouse, B, Matalan, R, Fujita, K and Nagatsu, T (1995) Characterization of mouse and human GTP cyclohydrolase I genes. *J. Biol. Chem.* **270**, 10062-10072.

Ichinose, H., Ohye, T., Matsuda,Y., Hori, T., Blau, N., Burline, A., Rouse, B., Matalan, R., Fujita, K. and Nagatsu, T. (1995). Characterization of mouse and human GTP cyclohydrolase I genes. *J. Biol. Chem.* **270**, 10062-10071.

Ichiyama, A., Nakamura, S.,. Nishizuka, Y. and Hayaishi, O. (1970). Enzymatic studies on the biosynthesis of serotonin in mammalian brain. *J. Biol. Chem.* **245**, 1699-1709.

Imazumi, K., Sasaki, T., Takahashi, K. and Takai, Y. (1994). Identification of a Rabphilin-3A-interacting protein as GTP cyclohydrolase I in PC12 calls. *Biochem. Biophys. Res. Commun.* **205**, 1409-1416.

Ito, T., Fujita, K., Matsuura, S. and Nagatsu, T. (1988). Treatment of depression with (6R)-tetrahydrobiopterin, the natural cofactor of tyrosine hydroxylase and tryptophan hydroxylase. *Biog. Amin.* **5**, 489-493.

Kaneda, N., Kobayashi, K., Ichinose, H., Kishi, F., Nakazawa, A., Kurosawa, Y., Fujita, K. and Nagatsu, T. (1987). Isolation of a novel cDNA clone for human tyrosine hydroxylase. Alternative RNA splicing produces four kinds of mRNA from a single gene. *Biochem. Biophys. Res. Commun.* **146**, 971-975.

Kapatos, G. and Kaufman, S. (1981). Peripherally administered reduced pterins do enter the brain. *Science.* **212**, 955-956. 30. Katoh, S., Sueoka,

T. and Yamada, S. (1982). Direct inhibition of brain sepiapterin reductase by a catecholamine and an indoleamine. *Biochem. Biophys. Res. Commun.* **105**, 75-81.

Katoh, S. and Sueoka, T. (1987). Isomerization of 6-lactoyl tetrahydropterin by sepiapterin reductase. *J. Biochem.* **101**, 275-278.

Kaufman, S. (1963). The structure of the phenylalanine hydroxylation cofactor. *Proc. Natl. Acad. Sci. USA* **50**, 1085-1093.

Kaufman, S, and Fisher, D. B. (1974). In: O. Hayaishi. Ed., Molecular Mechanism of Oxygen Activation, Academic Pjress. New York, p. 285.

Kaufman, S., Holtzman, N.A., Milstien, S., Butler, L.J. and Krumholz, A. (1975) Phenylketonuria due to a deficiency of dihydropteridine reductase. *New Engl. J. Med.* **293**, 785-790.

Kaufman, S., Berlow, S., Summer, G.K., Milstien, S., Schulman, J.D., Orloff, S., Spielberg S. and Pueschel, S. (1978). Hyperphenylalaninemia due to a deficiency of biopterin. A variant form of phenylketonuria. *New Engl. J. Med.* **299**, 673-679.

Kaufman, S., Berlow, S., Summer, G.K., Milstien, S., Schulman, J.D., Orloff, S., Spielberg, S. and Pueschel, S. (1978). Hyperphenylalaninemia due to a deficiency of biopterin. A variant form of phenylketonuria. *New Engl. J. Med.* **299**, 673-679

Kaufman, S. (1985). Hyperphenylalaninaemia caused by defects in biopterin metabolism. *J. Inheri. Met. Dis.* **8**, 20-27.

Kaufman, S. (1987). The enzymmatic conversion of phenylalanine to tyrosine. *J. Biol. Chem.* **226**, 511-524.

Kellner, C.H., Rakita, R.M., Rubinow, D.A., Gold, P.W., Ballenger, J.C. and Post, R.M. (1983). Tetrahydrobiopterin levels in cerebrospinal fluid of affectively ill patients. *Lancet* **2**, 55-56.

Kettler, R., Bartholini, G. and Pletscher, A. (1974). In vivo enhancement of tyrosine hydroxylation in rat striatum by tetrahydrobiopterin. *Nature* **249**, 476-478.

Kinoshita, H., Tsutsui, M., Milstien, S. and Katusic, Z.S. (1997). Tetrahydrobiopterin, nitric oxide and regulation of cerebral arterial tone. *Progr.Neurobiol.* **52**, 295-302.

Klatt, P., Schmidt, K., Lehner, D., Glatter, O., Bachinger, H P. and Mayer, B. (1995). Structural analysis of porcine brain nitric oxide synthase reveals a role for tetrahydrobiopterin and L-arginine in the formation of an SDS-resistant dimer. *EMBO J.* **14**, 3684-3695.

Klatt, P., Pfeiffer, S., List, B. M., Lehner, D., Glatter, O., Bachinger, H. P., Werner, E. R., Schmidt, K. and Mayer, B. (1996). Characterization of heme-deficient neuronal nitric oxide synthase reveals a role for heme in subunit dimerization and binding of the amino acid substrate and tetrahydrobiopterin. *J. Biol. Chem.* **271**, 7336-7342.

Koizumi, S., Sugimoto, T., Matsuura, S. and Nagatsu, T. (1987). Effects of (6R)- and (6S)-tetrahydrobiopterin on phenylalanine hydroxylration in rat liver tissue slices. *Biog. Amin.* **4**, 487-491.

Kobayashi, K., Kaneda, N., Ichinose, H., Kishi, F., Nakazawa, A., Kurosawa, Y., Fujita, K. and Nagatsu, T. (1987). Isolation of a full-length cDNA clone encoding human tyrosine hydroxylase type 3. *Nucleic Acids Res.* **15**, 6733.

Koshimura, K., Miwa, S, Lee., K, Fujiwara., M, and Watanabe, Y. (1990) Enhancement of dopamine release in vivo from the rat striatum by dialytic perfusion of 6R-L-ertthro- 5, 6, 7, 8-tetrahydrobiopterin. *J. Neurochem.* **54**, 1391-1397.

Kuster, T., Matasovic, A. and Niederwieser, A. (1984). Application of gas chromatography-mass spectrometry to the study of biopterin metabolism in man. Detection of biolumazine and 2'-deoxysepialumazine. *J. Chromatogr.* **290**, 303-310.

Kwok, S. C., Ledley, F. D., DiLella, A. G., Robson, K. J. and Woo, S. L. (1985). Nucleotide sequence of a full-length complementary DNA clone and amino acid sequence of human phenylalanine hydroxylase. *Biochemistry* **24**, 556-561.

Kwon, N.S., Nathan, C.F. and Stuehr, D.J. (1989). Reduced biopterin as a cofactor in the generation of nitrogen oxides by murine macrophages. *J. Biol. Chem.* **264**, 20496-20501.

Lamouroux, A. Blanot, F., Biguet, N. F. and Mallet, J. (1985). Complete coding sequence of rat tyrosine hydroxylase mRNA. *Proc. Natl. Acad.Sci. USA*. **82**, 617-621.

Lazarus, R. A., Dietrich, R. F., Wallick, D. E. and Benkovic, S. J. (1981). On the mechanism of action of phenylalanine hydroxylase. *Biochemistry* **20**, 6834-6841.

Ledley, F. D., DiLella, A. G., Kwok, S. C. M. and Woo, S. L.C. (1985). Homology between phenylalanine and tyrosine hydroxylases reveals common structural and functional domains. *Biochemistry* **24**, 3389-3394.

Lee, E.H. and Mandell, A.J. (1985). Relationships between drug-induced changes in tetrahydrobiopterin and biogenic amine concentrations in rat brain. *J. Pharmacol. Exp. Therap.* **234**, 141-146.

Levine, R.A., Kuhn, D.M. and Lovenberg, W. (1979). The regional distribution of hydroxylase cofactor in rat brain. *J. Neurochem.* **32**, 1575-1578.

Levine, R.A., Miller, L.P. and Lovenberg, W. (1981). Tetrahydrobiopterin in striatum: localization in dopamine nerve terminals and role in catecholamine synthesis. *Science* **214**, 919-921.

Lista, A., Abarca, J,. Ramos, C. and Daniels, A.J. (1986). Rat striatal dopamine and tetrahydrobiopterin content following an intrastriatal injection of manganese chloride. *Life Sci.* **38**, 2121-2127.

Lovenberg, W., Jequier, E. and Sjoerdsma, A. (1967). Tryptophan hydroxylation, measurement in pineal gland, brainstem, and carcinoid tumor. *Science* **155**, 217-219.

Lovenberg, W., Levine, R.A., Robinson, D.S., Ebert, M., Williams, A,C. and Calne, D.B. (1979). Hydroxylase cofactor activity in cerebrospinal fluid of normal subjects and patients with Parkinson's disease.*Science* **204**, 624-626.

Lovenberg, W., Levine, R.A., Robinson, D.S., Ebert, M., Williams, A.C. and Calne, D.B. (1979). Hydroxylase cofactor activity in cerebrospinal fluid of normal subjects and patients with Parkinson's disease. *Science* **204**, 624-626.

Lunte, C.E. and Kissinger, P.T. (1983). Determination of pterins in biological samples by liquid chromatography/electrochemistry with a dual-electrode detector. *Anal. Chem.* **55**, 1458-1462.

Masada, M., Akino, M., Sueoka, T. and Katoh, S. (1985). Dyspropterin, an intermediate formed fromdihydroneopterin triphosphate in the biosynthetic pathway of tetrahydrobiopterin. *Biochim. Biophys. Acta* **840**, 235-244.

Mataga, N., Imanura, K. and Watanabe, Y. (1991). 6R-tetrahydrobiopterin perfusion enhances dopamine, serotonin, and glutamate outputs in dialysate from rat striatum and frontal cortex. *Brain Res.* **551**, 64-71.

Matsuura, S., Sugimoto, T., Hasegawa, H., Imaizumi, S. and Ichiyama, A. (1980). Studies biologically active pteridines. III. The absolute configuration at the C-6 chiral center of tetrahydrobiopterin cofactor and related compounds. *J. Biochem.* **87**, 951-957.

Matsuura, S., Sugimoto, T., Masada, S., Murata, Y, and Iwasaki, H. (1985). Stereochemistry of biopterin cofactor and facile methods for the determination of the stereochemistry of a biologically active 5, 6, 7, 8 -tetrahydrobiopterin. *J, Biochem.* **98**, 1341-1348

Matsuura, S., Murata, S., Sugimoto, T., Sawada, M. and Nagatsu, T. (1986). Preparation and cofactor activity of (6*S*)- tetrahydrobiopterin. *Chem, Express* **1**, 403.

Mayer, B. and Werner, E.R. (1995). Inseach of a function for tetrahydrobiopterin un the biosynthesis of nitric oxide. *Naunyn-schmiedeberg's Arch. Pharmacol.* **351**, 453-463.

Miller, L., Insel, T., Scheinin, M., Aloi, J., Murphy, D.L., Linnoila, M. and Lovenberg, W. (1986). Tetrahydrobiopterin administration to rhesus macaques. Its appearance in CSF and effect on neurotransmitter synthesis. *Neurochem. Res.* **11**, 291-298.

Milstien, S. and Kaufman, S. (1983). Tetrahydro-sepiapterin is an intermediate in tetrahydrobiopterin biosynthesis. *Biochem. Biophys. Res. Commun.* **115**, 888-893.

Milstien, S. and Kaufman, S. (1985). Biosynthesis of tetrahydrobiopterin, conversion of dihydroneopterintriphosphate to tetrahydropterin intermediates.*Biochem. Biophys. Res. Commun.* **128**, 1099-1107.

Milstien, S. and Kaufman, S. (1986). In: Chemistry and Biology of Pteridines, B, A. Cooper and V. M. Whitehead, Eds., Walter de Gruyter, Berlin, p. 169.

Milstien, S., Jaffe, H., Kowlessur, D. and Bonner, T.I. (1996). Purification and cloning of the GTP cyclohydrorase I feedback regulatory protein, GFRP. *J. Biol. Chem.* **271**, 19743-19751.

Miwa, S., Watanabe, Y. and Hayaishi, O. (1985).6R-L-erythro-5,6,7,8-tetrahydrobiopterin as a regulator of dopamine and serotonin biosynthesis in the rat brain. *Arch. Biochem. Biophys*. **239**, 234-241.

Miwa, S., Watanabe, Y., Koshimura, K. and Masaki, T. (1992). *Folia Pharmacol. Japon* **100**, 367-381.

Nagatsu, T., Levitt, M. and Udenfriend, S. (1964). Conversion of L-tyrosine ton 3, 4-dihydroxyphenylalanine by cell-free preparations of brain and sympathetically innervate tissues. *Biochem. Biophys. Res. Commun.* **14**, 543-549.

Nagatsu, T. (1980). In: Advances in Experimental Medicine: A Centenary Tribute to Claude Bernard, H. Parvez and S. Parvez, Eds.. Elsevier. Amsterdam, p. 527.

Nagatsu, T., Yamaguchi, T., Kato, T., Sugimoto, T., Matsuura, S., Akino, M., Tsushima, S., Nakazawa, N. and Ogawa, H. (1981a). Radioimmunoassay for biopterin in body fluids and tissues. *Anal. Biochem.* **110**, 182-189.

Nagatsu, T., Yamaguchi, T., Kato, T., Sugimoto, T., Matsuura, S., Akino, M., Nagatsu, I., Iizuka, R., and Narabayashi, H. (1981b). Biopterin in human brain and urine from controls and Parkinsonian patients: application of a new radioimmunoassay. *Clin Chim Acta* **109**, 305-311.

Nagatsu, T., Sawada, M., Yamaguchi, T., Sugimoto, T., Matsuura, S., Akino, M., Nakazawa, N. and Ogawa, H. (1984). Radioimmunoassay for neopterin in body fluids and tissues. *Anal. Biochem.* **141**, 472-480.

Nagatsu, T. (1985). Biopterin cofactor and monoamine-syntherising monooxygenases. In: Selected Topics in Neurochemistry, N. N. Osborne, Ed., Pergamon Press, New York, p. 325.

Nagatsu, T., Horikoshi, T., Sawada, M., Nagatsu, I., Kondo, T., Iizuka, R. and Narabayashi, H. (1986a). Biosynthesis of tetrahydrobiopterin in purbeinoohian human brain : activity of GTP cyclohydrolase I. In:

Advances in Neurolgy. Vol. 45. M. D. Yahr and K. J. Bergmann, Eds., Raven Press, New York, p. 223-226.

Nagatsu, T., Sawada, M., Akino, M., Masada, M., Sugimoto, T. and Matsuura, S. (1986b). Distribation of GTP cyclohydrolase I, neopterin, and biopterin in the human brain. In: *Chemistry and Biology of Peritidines*, B. A. Cooper and V. M. Whitehead, (Eds.), Walter de Gruyter, Berlin, pp. 223-226.

Nagatsu, T., Matsuura S. and Sugimoto T. (1989). Physiological and clinical chemistry of biopterin. *Med. Res. Rev*. **9**, 25-44.

Nagatsu, T., Hashimoto, R., Ohta, T. and Kasahara, Y. (1990). Depression and the abnormal metabolism. *Pharma Media* **8**, 21-24.

Nagatsu, T. (1991) Genes for human catecholamine-synthesizing enzymes. *Neurosci. Res.* **12**, 315-345.

Nagatsu, T., Nakahara, D., Kobayashi, K., Morita, S., Sawada, H., Mizuguchi, T. and Kiuchi, K. (1994) Peripherally administered (6R)-tetrahydrobiopterin increases in vivo tyrosine hydroxylase sctivity activity in the striatum measurwd by microdialysis both in normal mice and in transgenic mice carrying human tyrosine hydroxylase. *Neurosci. Lett.* **182**, 44-46.

Nagatsu, I., Arai, R., Sasaki, M., Yamayaki, Y., Takeuchi, T., Karasawa, N. and Nagatsu, T. (1997). Immunohistochemical colocalization of GTP cyclohydrolase I in the nigrostriatal system with tyrosine hydrolase. *Neurosci. Lett.* **224**, 185-188.

Nakahara, D., Ozaki, N., Miura, Y., Miura, H. and Nagatsu, T. (1989). Increased dopamine and serotonin metabolism in rat nucleus accumbens produced by intracranial self-stimulation of medial forebrain bundle as measured by in vivo microdialysis. *Brain Res*. **495**, 178-181.

Nakazawa, T. (1996). Cerebrospinal fluid neopterin levels in children with neurologic disease. (Japanese). *Brain Develop.* **28**, 291-298.

Narabayashi, H., Kondo, Y., Nagatsu, T., Sugimoto. T. and.Matsuura, S. (1982). *Proc. Jpn. Acad.*, **58** (Ser. B), 283.

Narabayashi, H., Yokochi, M., Iizuka, R. and Nagatsu, T. (1986). In: *Handbook of Clinical Neurology*, Vol. 5, P. J. Vinken, G. W. Bruyn, H. L. Klawans, (Eds.), Elsevier, Amsterdam, p. 153.

Nathan, C. and Xie, Q.W. (1994). Nitric oxide synthases: roles, tolls, and controls. *Cell* **78**, 915-918.

Nicol, C. A., Smith, G. K. and Duch, D.S. (1985) Biosynthesis and metabolism of tetrahydrobiopterin and molybdorterin. *Ann. Rex. Biochem*. **54**, 729-164.

Niederwieser, A., Curtius, H.C., Bettoni, O., Bieri, J., Schircks, B., Viscontini, M. and Schaub J. (1979). Atypical phenylketonuria caused by 7,8-dihydrobiopterin synthetase deficiency. *Lancet* **1**, 131-133.

Niederwieser, A., Curtius, H.C., Wang, M. and Leupold, D. (1982). Atypical phenylketonuria with defective biopterin metabolism. Monotherapy with tetrahydrobiopterin or sepiapterin, screening and study of biosynthesis in man. *Eur. J. Pediatr.* **138**, 110-112.

Niederwieser, A., Blau, N., Wang, M., Joller, P., Atares, M. and Cardesa-Garcia, J. (1984) GTP cyclohydrolase I deficiency, a new enzyme defect causing hyperphenylalaninemia with neopterin, biopterin, dopamine, and serotonin deficiencies and muscular hypotonia. *Eur. J. Pediatr.* **141**, 208-214.

Niederwieser, A., Leimbacher, W., Curtius, H.C., Ponzone, A., Rey, F. and Leupold, D. (1985) Atypical phenylketonuria with "dihydrobiopterin synthetase" deficiency: absence of phosphate-eliminating enzyme activity demonstrated in liver. *Eur. J. Pediatr.* **144**, 13-16.

Oka, K., Kato, T., Sugimoto, T., Matsuura, S.and Nagatsu, T. (1981). Kinetic properties of tyrosine hydroxylase with natural tetrahydrobiopterin as cofactor. *Biochim. Biophys. Acta* **661**, 45-53.

Ohue, T., Koshimura, K., Lee, K., Watanabe, Y. and Miwa, S. (1991). A novel action of 6R-L-*erythro*-5,6,7,8-tetrahydrobiopterin, a cafactor for hydroxylases of phenylalanine, tyrosine and tryptophan: enhancement of acetylcholine release in vivo in the rat hippocampus. *Neurosci. Lett.* **128**, 93-96.

Ohue, T, Koshimura, K, Akiyama, Y, Watanabe, Y, and Miwa, S, (1992) Monoamine-mediated dnhancement of acetylcholine in rat hippocampus by 6R-L-erythro- 5, 6, 7, 8-tetrahydrobiopterin. *Brain Res.* **570**, 173-179.

Robson, K. J., Beattie, W., James, R. J., Cotton, R. C., Morgan, F. J. and Woo, S. L. (1984). Sequence comparison of rat liver phenylalanine hydroxylase and its cDNA clones. *Biochemistry* **23**, 5671-5675.

Sawada, M., Yamaguchi, T., Sugimoto, T., Matsuura, S. and Nagatsu, T. (1984). Polarization fluoroimmunoassay of biopterin and neopterin in human urine. *Clin. Chim Acta* **138**, 275-282.

Sawada. M., Sugimoto, T., Matsuura, S. and Nagatsu T. (1986a). 6R-tetrahydrobiopterin increases the activity of tryptophan hydroxylase in rat raphe slices. *J. Neurochem.* **47**, 1544-1547.

Sawada, M., Horikoshi, T., Masada, M., Akino, M., Sugimoto, T., Matuura, S. and Nagatsu, T. (1986b). *Anal. Biochem.* 154, 361-366.

Sawada, M., Hirata, Y., Arai, H., Iizuka, R. and Nagatsu, T. (1987). Tyrosine hydroxylase, tryptophan hydroxylase, biopterin, and neopterin in the brains of normal controls and patients with senile dementia of Alzheimer type. *J. Neurochem.* **48**, 760-764.

Sawada, M., Hirata, Y., Minami, M. and Nagatsu, T. (1987). Effects of thyrotropin releasing hormoneand its analogue, DN 1417, on biopterin concentration and tryptophan hydroxylation in rat rapheslices. *Life Sci.* **41**, 2733-2737.

Sawada, M. and Nagatsu, T. (1987) Distrilution of pterins in biological tissues. In: *Unconjugated pterins in neurobiolozy: basic and clinical aspects*, Lovenberg, W. and Levine, R. A. (Eds.), Taylor and Francis, London, pp. 131-156.

Sawada, M. and Nagatsu, T. (1988) Tryptophan hydroxylase activity in brain slices. *Int. J. Biochem.* **20**, 1033-1038.

Smith, G. K. and Nichol, C. A. (1983). In: *Biochenlical and Clinical Aspects of Pteridines*. Vol. 2, Walter de Gruyter, Berlin, p. 123.

Smith, G. K. and Nichol, C. A. (1986). Synthesis, utilization, and structure of the tetrahydropterin intermediates in the bovine adrenal medullary de novo biosynthesis of tetrahydrobiopterin. *J. Biol. Chem.* **261**, 2725-2737.

Smith, G. K. (1987). On the role of sepiapterin reductase in the biosynthesis of tetrahydrobiopterin. *Arch. Biochem. Biophys.* **255**, 254-266.

Stuehr, D.J., Kwon, N.S., Gross, S.S., Thiel, B.A., Levi, R. and Nathan, C.F. (1989). Synthesis of nitrogen oxides from L-arginine by macrophage cytosol: requirement for inducible and contitutive components. *Biochem. Biophys. Res. Commun.* **161**, 420-426.

Sueoka, T. and Katoh, S. (1985). Carbonyl reductase activity of sepiapterin reductase from rat erythrocytes. *Biochim. Biophys. Acta* **843**, 193-198.

Suzuki, S., Watanabe, Y., Tsubokura, S., Kagamiyama, H. and Hayashi, O. (1988). Decrease in tetrahydrobiopterin content and neorotransmitter amine biosynthesis in rat brain by an inhibitor of guanosine triphosphate cyclohydrolase. *Brain Res.* **446**, 1-10.

Suzuki, K. and Owada, M. (1991). A simple and sensitive method for the determination of pterins in cerebrospinal fluid. Clinical usefulness for management of tetrahydrobiopterin deficiency. *J. Inher. Metab. Dis.* **14**, 825-830.

Suzuki, K. and Owada, M. (1991). A simple and sensitive method for the determination of pterins in cerebrospinl fluid. Clinical for management of tetrahydrobiopterin deficiency. *J. Inher. Metab. Dis.* **14**, 825-830.

Switchenko, A.C., Primus, J.P. and Brown, G.M. (1984). Intermediates in the enzymic synthesis of tetrahydrobiopterin in Drosophila melanogaster. *Biochem. Biophys. Res. Commun.* **120**, 754-760.

Tani, Y. and Ohno, T. (1993). Analysis of 6R- and 6S-tetrahydrobiopterin and other pterins by reversed-phase ion-pair liquid chromatography with fluorimetric detection by post-column sodium nitrite oxidation. *J. Chromatogr.* **617**, 249-255.

Tani, Y., Fernell, E., Watanabe, Y., Kanai, T. and Landgstom, B. (1994).

Decrease in 6*R*-5,6,7,8-tetrahydrobiopterin content in cerebrospinal fluid of autistic patients. *Neurosci.Lett.* **181**, 169-172.

Tayeh, M, A, and Marletta, M, A, (1989). Macrophage oxidation of L-arginine to nitric oxide, nitrite, and nitrate. *J. Biol. Chem.* **264**, 19654-19658.

Togari, A., Kojima, K. and Nagatsu, T. (1985). Molecular and physiological properties of tyrosine hydroxylase induced by reserpine in rat adrenal gland. *Life Sci.* **37**, 1605-1611.

Togari, A., Ichinose, H., Matsumoto, S., Fujita, K. and Nagatsu, T. (1992). Multiple mRNA forms of human GTP cyclohydrolase I.*Biochem. Biophys.Res.Commun.* **187**, 359-365.

Tsukada, H., Lindner, K. J., Hartvig, P., Tani, X., Valtysson, J., Bjurling, P., Kihlbery, T., Westerberg, G., Watanabe, Y. and Langstrom, B. (1996). Effcet of 6*R*-L-erythro-5, 6, 7, 8-tetrahydrobiopterin and infusion of L-tyrosine on the in vivo L-[beta-^{11}C] DOPA dioposition in the momkey brain. *Brain. Red.* **713**. 92-98.

Tzeng, E., Shears II. L. L., Robbins., P. D., Pitt, B. R., Geller, D. A., Watkins. S. C., Simmons, R. L. and Billiar, T. R. (1996). Vascular gene transfer of the human inducible nitric oxide synthase characterization of activity and effects on myointimal hyperplasia. *Mol. Med.* **2**, 211-225.

Werner-Felmayer, G. and Gross, S. S. (1996). Analysis of tetrahydrobiopterin and its role in nitric oxide synthesis. In: *Methods in Nitric Oxide Research*. Feelisch, M. and Stamler, J. S. (Eds). John Wiley and Sons, New York, pp. 271-299.

Williams, A., Ballenger, J., Levine, R., Lovenberg, W. and Calne, D. (1980). Aging and CSF hydroxylase cofactor. *Neurology* **30**, 1244-1246.

Woolf, J.H., Nichole, C.A.and Duch, D.S. (1983). Determination of biopterin and other pterins in tissues and body fluids by high-performance liquid chromatography. *J. Chromatogr.* **274**, 398-402.

Yamaguchi, T., Nagatsu, T., Sugimoto, T., Matsuura, S., Kondo, T., Iizuka, R. and Narabayashi, H. (1983). Effects of tyrosine administration on serum biopterin in normal controls and patients with Parkinson's disease. *Science* **219**, 75-77.

Yoneyama, T., Brewer, J.M. and Hatakeyama, K. (1997). GTP cyclohydrolase I feedback regulatory Protein is a pentamer of identical subunits. *J. Biol. Chem.* **272**, 9690-9696.

Progress in HPLC-HPCE, Vol. 7, pp. 723-779
G. Ali Qureshi *et al.* (Eds)

NEUROPEPTIDE ANALYSIS IN HUMAN PITUITARY TUMORS

Xuegong Zhu[1] and Dominic M. Desiderio[1,2,3]*
[1]The Charles B. Stout Neuroscience Mass Spectrometry Laboratory, [2]Department of Neurology, [3]Department of Biochemistry, The University of Tennessee, Memphis,800 Madison Avenue, Memphis, TN 38163, U.S.A.

Introduction.

This chapter describes the purification and analysis of neuropeptides in human pituitary tumors and controls (post-mortem pituitaries were used as controls). The focus is on two opioid neuropeptides, methionine enkephalin (ME) and beta-endorphin (BE), and one tachykinin neuropeptide, substance P (SP). The purification and analysis strategies described in this chapter are also readily exportable to the study of many other different types of compounds.

It has been hypothesized that an altered metabolism of neuropeptides is associated with human anterior pituitary tumor formation (Zhu, 1993). That alteration may be any defect within the multiple-step enzyme-mediated metabolic pathways of a neuropeptide. Because the tissue concentration of a neuropeptide is the manifestation of a balance between the production and degradation of a neuropeptide, one way to test the above hypothesis is to compare accurately the concentration of the neuropeptide in pituitary tumors versus the controls.

The characterization of ME, BE, and SP in human pituitary tissues (controls and tumors) includes the extraction and purification of these neuropeptides, their qualitative (determination of the amino acid sequence) and quantitative analysis, and a comparison of the quantitative data between controls and tumors to test the working hypothesis. These quantitative data are necessary to determine whether the concentrations of these neuropeptides, which reflect the summation of the production and degradation of these neuropeptides, correlates with pituitary tumor types, and to determine whether any of the neuropeptides is associated with a certain type of pituitary tumor. Because SP exists in the human pituitary at the femtomole level, and ME and BE exist at the picomole to nanomole levels, radioimmunoassay (RIA) was used for SP quantification and mass spectrometry

(MS) for the qualitative and quantitative analysis of ME and BE. The primary objective of this study was to optimize the molecular specificity of the ME and BE quantification.

Figure 1 illustrates the overall analytical strategy that was used in this research program. Each step will be discussed below in greater detail. The efficiency of neuropeptide extraction and purification procedures will be discussed. Because post-mortem pituitaries were used as the controls, the post-mortem stability of neuropeptides will be discussed. The effects of peptidase inhibitors on peptide recovery will be discussed. A solid-phase extraction (SPE) method, which is a rapid, simple, and efficient way of producing the peptide-rich fractions from a tissue extract, will be discussed. Gradient reversed-phase high performance liquid chromatography (RP-HPLC) followed by isocratic RP-HPLC, which separates neuropeptides contained in the SPE peptide-rich fractions, will also be discussed.

There are several commonly used detectors to measure HPLC-purified neuropeptides, including radioreceptorassay (RRA), RIA, immunofluorescence, enzyme-linked immunosorbent assay (ELISA), and MS. Of all of these detectors, only MS provides amino acid sequence information of a neuropeptide. The others either rely on receptor-ligand binding or antigen-antibody binding, and in varying degrees they compromise the structural specificity of the analysis.

MS, especially tandem mass spectrometry (MS/MS), has the potential of providing the optimum molecular specificity for the qualitative and quantitative analysis of a neuropeptide. An MS/MS instrument contains two mass spectrometers, MS-1 and MS-2. In an MS/MS experiment, the protonated molecule ion, (M+H)+, of a neuropeptide is selected in MS-1, and, after collision-induced decomposition (CID) in a collision cell, all of the fragment ions that derive from that (M+H)+ ion are collected in MS-2. The resulting product-ion spectrum contains most, if not all, of the amino acid sequence-determining ions, which can be used to identify the neuropeptide with structural specificity. Figure 2 schematically represents the steps involved in the generation of that product-ion spectrum. In Figure 2, the hypothetical peptide is ionized to form its corresponding (M+H)+ ion; CID generates fragment ions, including the amino acid sequence-determining fragment ions. Those fragment ions are collected in the product-ion spectrum. That spectrum also provides a basis for the quantification of the neuropeptide, in which the transition of the (M+H)+ ion to one of its amino acid sequence-determining fragment ion is monitored to establish an important precursor-product relationship. The qualitative and quantitative analyses of endogenous neuropeptides by MS

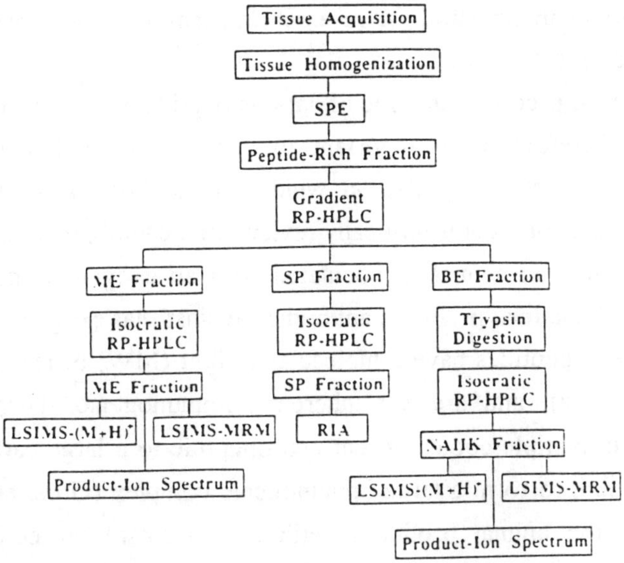

Figure 1. General scheme used for peptide quantification.

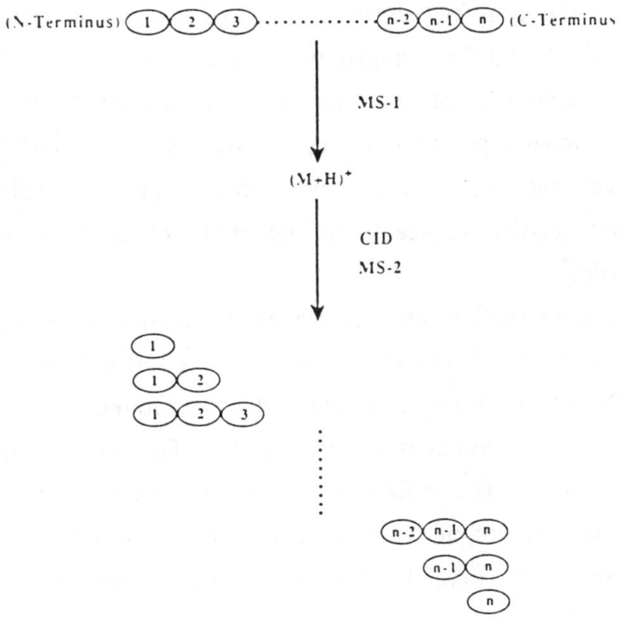

Figure 2. A hypothetical peptide, the formation of its (M+H)+, and the use of collision-induced dissociation (CID) and MS/MS to produce, and to collect, respectively, the N-terminus-containing and the C-terminus-containing amino acid sequence-determining fragment ions.

provide the important information for research in many different areas, including basic and clinical sciences.

RIA data are the most predominant type of data on peptide measurements that have been published. Therefore, it is important to understand the limitations of an immune-based method such as RIA, in which the antibody against the target peptide has the central role. Let us critically review the general procedure of raising such antibodies. When an animal is inoculated with an exogenous molecule, the immune system of the animal produces antibodies against that exogenous molecule. However, most neuropeptides have a molecular weight (MW) of less than several thousand Daltons (Da), and are not inherently immunogenic. Therefore, it is necessary to attach multiple copies of that neuropeptide to a large carrier protein, such as thyroglobulin, to form a large immunogenic complex (Gros, et al., 1978). When inoculated, the animal produces antibodies against that complex. That heterogeneous group of antibodies consists of not only antibodies against the target neuropeptide but also those antibodies against parts of carrier protein or against a combination of the target neuropeptide and parts of the carrier protein. Nonetheless, RIA has a high sensitivity of detecting a neuropeptide (down to the lower femtomole level) and provides important preliminary data for the analysis of a neuropeptide. In this chapter, we review the RIA measurement of SP as well as the MS analysis of ME and BE in human pituitary tumors and controls. Those three neuropeptides are examples of neuropeptides that participate in many different neuronal and hormonal regulatory processes (Wagner and Chavkin, 1995). As mentioned above, the basic principles involved in the analysis of these neuropeptides are readily applicable to the study of many other biologically important molecules.

During the peptide extraction and purification processes, a substantial loss [for example, in the rat pituitary, 57% ME and 44% SP are lost after the first RP-HPLC step (Zhu and Desiderio, 1993)] is unavoidable. Deuterated ME and BE internal standards are used as to compensate for that loss. The ratio of the endogenous peptide to its internal standard is determined by MS. The amount of ME or BE is calculated from the amount of the internal standard that has been added and a corresponding calibration curve. Because these internal standards are exactly the same molecules, except for a very small mass difference, and because they are added before tissue homogenization, the quantification data for ME and BE represent the accurate tissue concentration of these two neuropeptides.

Opioid and Tachykinin Neuropeptides in the Anterior Pituitary.

This section reviews briefly the biosynthesis of ME, BE, and SP. Like many other neuropeptides, these three neuropeptides are polyprotein in nature; namely, they are translated as part of a large precursor protein. Through enzyme-mediated post-translational processing steps, the neuropeptides are generated as bioactive forms. Because such processing involves more than one step, there may be many intermediate-sized precursor proteins present in a biological matrix. That complex mixture requires a sophisticated analytical strategy to measure accurately those neuropeptides.

Methionine Enkephalin.

Proenkephalinhuman, 1-267.

The post-translational processing of proenkephalinhuman, 1-267 to its component neuropeptides has been reviewed extensively (Udenfriend and Kilpatrick, 1984). The intact proenkephalinhuman, 1-267 has not been isolated from biological tissues; rather, its amino acid sequence has been deduced from its cDNA data, and is shown in Figure 3. Proenkephalinhuman, 1-267 contains four copies of ME (YGGFM), one copy of LE (YGGFL), and one copy each of two C-terminal-extended MEs, the octapeptide ME-RGL and the heptapeptide ME-RF (Numa, 1984). Analyses of ME-containing opioid peptide-containing proteins (OPCPs) have been performed in the anterior pituitary, brain, and the chromaffin cells of the bovine adrenal medulla, which contains a high concentration of ME. According to our working hypothesis, a defect in the production of ME may be associated with human pituitary tumor formation. Therefore, in addition to ME, it is important to study the specific ME-containing OPCPs in the human pituitary, because that information may help to pinpoint the exact pathway of ME processing in which a defect is associated with human pituitary tumor formation. That topic is part of a separate research project in our laboratory, and will not be discussed here.

Localization of Methionine Enkephalin in the Anterior Pituitary.

In the study of localization of neuropeptides in the biological tissues, immunohistochemistry and other immune-based techniques are commonly used, and thus the terms ME-like immunoreactivity (ME-li), BE-like immunoreactivity (BE-li), and SP-like immunoreactivity (SP-li) will be used in this chapter to indicate the fact that the amino acid sequence of each endogenous neuropeptide was not determined.

ME was the first opioid neuropeptide found in high concentrations in such biological tissues as the adrenal medulla, brain areas, spinal cord, sympathetic ganglia, and peripheral neurons (Clement-Jones and Besser, 1983; Hughes, et al., 1975; Watson, et al., 1978). However, the localization of ME in the pituitary gland has been scattered and may be controversial. Different researchers localized ME-li in different anterior pituitary cell types: although some experiments demonstrated the presence of ME-li in the rat pituitary somatotrophs, others showed the presence of ME-li in the human pituitary thyrotrophs (Duka, et al., 1978; Tramu and Leonardelli, 1979; Weber, et al., 1978).

Beta-Endorphin

Proopiomelanocortin (POMC)human, 1-267.

The intact BE precursor protein, proopiomelanocortin (POMC)human, 1-267, has also not been isolated from any biological tissue. Its sequence has been deduced from its cDNA sequence (Numa, 1984), and is shown in Figure 4. The POMC sequence contains several biologically active peptides such as BE, adrenocorticotropic hormone (ACTH), melanocyte-stimulating hormone (MSH), and luteinizing hormone (LPH), all of which have been isolated from biological tissues. In addition, recent reports showed the presence of BE-containing OPCPs (e.g., a protein of ca 10 kDa) (Yan, et al., submitted).

Localization of Beta-Endorphin in the Anterior Pituitary.

The localization of BE in the pituitary has been well-studied. The pituitary gland is one of the major sites of POMC synthesis; POMC gives rise to BE along with several other neuropeptides (Millan and Herz, 1985; Loh, 1992). The pituitary gland consists of three sections: the anterior, intermediate, and posterior lobes. The corticotrophs in the anterior lobe produce POMC peptides. After birth, the human intermediate lobe disappears, leaving behind only a scattered thin layer of intermediate lobe gland cells, i.e., melanocytes. BE has been quantified in bovine (Zhu and Desiderio, 1994), rat (Zhu and Desiderio, 1993), and human (Dass, et al., 1991; Desiderio, et al., 1993) pituitaries. In many species, BE-li was localized in the corticotrophs in the anterior pituitary and melanocytes in the intermediate pituitary, but not in the posterior pituitary (Bloom, et al., 1977). There is a difference in POMC processing in the anterior pituitary and intermediate pituitary: in the anterior pituitary, beta-LPH and BE-li exist in equimolar amounts; in the intermediate pituitary, beta-LPH is entirely processed to BE-li (Millan and Herz, 1985).

MARFLTLCTWLLLLGPGLLATVRAECSQDCATCSYRLVRP
ADINFLACVMECEGKLPSLKIWETCKELLQLSKPELPQDG
TSTLRENSKPEESHLLAKRYGGFMKRYGGFMKKMDELYPM
EPEEEANGSEILAKRYGGFMKKDAEEDDSLANSSDLLKEL
LETGDNRERSHHQDGSDNEEEVSKRYGGFMRGLKRSPQLE
DEAKELQKRYGGFMRRVGRPEWWMDYQKRYGGFLKRFAEA
LPSDEEGESYSKEVPEMEKRYGGFMRF

Figure 3. cDNA-derived amino acid sequence of preproenkephalinhuman, 1-267 (Numa, 1984). The ME sequence is underlined each time that it occurs.

MRPSCCSRSGALLLALLLQASMEVRGWCLESSQCQDLTTE
SNLLECIRACKPDLSAETPMFPGNGDEQPLTENPRKYVMG
HFRWDRFGRRNSSSSGSSGAGQKREDVSAGEDCGPLPEGG
PEPRSDGAKPGPREGKRSYSMEHFRWGKPVGKKRRPVKVY
PNGAEDESAEAFPLEFKRELTGQRLREGDGPDGPADDGAG
AQADLEHSLLVAAEKKDEGPYRMEHFRWGSPPKDKRYGGF
MTSEKSQTPLVTLFKNAIIKNAYKKGE

Figure 4. cDNA-derived amino acid sequence of POMChuman,1-267 (Numa, 1984), in which the BE1-31 sequence is underlined.

MKILVALAVFFLVSTQLFAEEIGANDDLNYWSDWYDSDQI
KEELPEPFEHLLQRIARRPKPQQFFGLMGKRDADSSIEKQ
VALLKALYGHGQISHKRHKTDSFVGLMGKRALNSVAYERS
AMQNYERRR

Figure 5. cDNA-derived amino acid sequence of bPPThuman (Harmar, et al., 1986), in which the SP sequence is underlined.

Substance P

Preprotachykinins.

SP belongs to a peptide family, the tachykinins, which shares a common carboxyl-terminal amino acid sequence, Phe-X-Gly-Leu-Met-NH2, where X is an aliphatic or aromatic amino acid, generally either Phe or Val (Maggio, 1988). In mammalian tissues, two genes encode all of the tachykinins. The preprotachykinin A (PPT-A) gene encodes SP, neurokinin A, neuropeptide K, and neuropeptide g (Krause, et al., 1989). The PPT-B gene encodes neurokinin B (Kotani, et al., 1986). Figure 5 shows the cDNA-derived amino acid sequence of one of the three human SP precursor proteins, bPPT. aPPT and gPPT are similar to bPPT, except for the deletions of the amino acid sequence at some points. Most of the human SP is produced from aPPT and bPPT (Harmar, et al., 1986).

Localization of Substance P in the Anterior Pituitary.

SP has been localized in neuronal fibers and gland cells in pituitaries. By using immunohistochemical methods, the presence of SP-li fibers was demonstrated in the anterior and the posterior lobe of the rat pituitary (Mikkelsen, et al., 1989). In the posterior pituitary, SP-li fibers varied in size, and were located near the capillaries and in the neuropil; in the anterior pituitary, the SP-li fibers distributed around the gland cells. The presence of SP-li fibers in the anterior pituitary has also been demonstrated in monkeys (Ju and Liu, 1989; Liu and Ju, 1988) and dogs (Ju and Liu, 1990; Ju and Zhang, 1990). There were species differences in the abundance and distribution pattern of SP-li fibers in the anterior pituitary (Ju, et al., 1991). For example, SP-li fibers were abundant in the anterior pituitary in monkeys, moderate in dogs, and fewer in rats. In monkeys, SP-li fibers were concentrated in the dorsoposterior quadrant of the anterior pituitary, whereas in dogs and rats they were distributed fairly evenly. The SP-li fibers in the anterior pituitary did not appear to arise from the posterior pituitary, because a few fibers were seen to penetrate from the periphery of the gland into a more central part, but no fibers were seen to penetrate from the intermediate or posterior lobe (Mikkelsen, et al., 1989).

SP has also been localized in gland cells of the anterior pituitary. In the rat pituitary, mammotrophs and gonadotrophs were stained by an anti-SP serum, whereas somatotrophs, corticotrophs, and thyrotrophs were not stained (Morel, et al., 1982). However, in guinea pigs, by using a different immunocytochemical technique, SP-li was localized in a sub-population of thyrotrophs (DePalatis, et al.,

1982). More recently, RIA localized SP-li in somatotrophs and thyrotrophs of the rat anterior pituitary and in thyrotrophs of the human anterior pituitary (Roth and Krause, 1990).

Neuropeptides and Anterior Pituitary Tumors.

The pituitary is the most important relay station for many neuroendocrinological processes, and is highly regulated by neuropeptidergic systems such as the opioid and tachykinin neuropeptides. A pituitary adenoma often affects the neuroendocrinological status of the body by disturbing one or more hormonal regulation pathways, or in some cases of non-secreting pituitary tumors, causes hypopituitarism, in which all pituitary hormones decrease in concentration. Furthermore, opioid and tachykinin neuropeptides have been shown to be important regulators in the pituitary functions.

Human Pituitary Tumors.

Among all the intracranial tumors, there are about 10% pituitary tumors, the majority of which are pituitary adenomas, which are a disease of relatively high incidence and can be found in up to 23% of unselected adult autopsies (Asa, et al., 1984; Kovacs and Horvath, 1987). Morphologically, all pituitary adenomas are benign epithelial lesions derived from and composed of hormone-secreting cells of the anterior pituitary (Asa and Kovacs, 1992). Some of the pituitary adenomas are slow-growing tumors whose expansion is limited to the sella turcica, and these pituitary tumors are termed as microadenomas (less than ten mm in diameter), whereas others are rapid-growing and spread into neighboring tissues, and are termed as macroadenomas (more than ten mm in diameter) (Asa and Kovacs, 1992).

Pituitary adenomas often consist of one or more types of pituitary hormone-secreting cells. The patient who bears such a tumor often shows symptoms that are caused by an oversecretion of one or more pituitary hormones, such as acromegaly (growth hormone), amenorrhea-galactorrhea syndrome [prolactin (PRL)], or Cushing's disease (ACTH). Pituitary adenomas are classified by histological, immunohistochemical, and microscopic criteria. For example, if a pituitary adenoma secretes PRL, then it is classified as a PRL-secreting tumor. However, a significant portion of pituitary adenomas (25-30%) do not show any over-secretion of pituitary hormones. These tumors do not disturb the patient's neuroendocrinological status, but they produce symptoms such as headache, visual loss, or hypopituitarism, which are caused by the invading effects of these fast-growing non-secreting pituitary tumors (Klibanski, 1987).

Models of Pituitary Tumor Formation.

The molecular mechanism of pituitary tumor formation has not been established unequivocally, but researchers have proposed several theories, the major two of which are the hypothalamic dysregulation theory and the pituitary tumor de novo formation (Molitch, 1987).

The first theory suggests that a pituitary tumor forms because of an imbalanced secretion of hypothalamic-stimulating versus -inhibiting factors. Under physiological conditions, the pituitary secretion activities are regulated by various hypothalamic-stimulating factors such as growth hormone-releasing hormone (GRH), gonadotropin-releasing hormone (GnRH), corticotropin-releasing hormone (CRH), and thyrotropin-releasing hormone (TRH), and by hypothalamic-inhibiting factors such as somatostatin and dopamine. These hypothalamic factors are secreted from the hypothalamus to the portal blood and are carried to the pituitary gland. If the hypothalamus secretes too much of a stimulating factor or too little of a inhibiting factor, then a specific type of pituitary cell initiates excessive cellular activities, which lead to hyperplasia and subsequent tumor formation (Melmed, et al., 1983).

According to the second theory, pituitary tumors are caused by anomalies inside the pituitary gland (Melmed, et al., 1983). One description of this theory is that an abnormal paracrine or autocrine regulation in the pituitary causes tumor formation. The analysis of the immunoreactive cell distribution pattern has demonstrated that pituitary tumors and non-tumor pituitary both develop by the multicentric proliferation of cell clones (Herzog and Schlote, 1992).

Approximately 80 - 90% of all pituitary tumors can be explained by the de novo theory. Whether the remaining 10 - 20% can be explained by the theory of hypothalamic dysregulation is still controversial (Molitch, 1987). Therefore, the currently available evidence weighs towards the theory of de novo formation. That theory will be explored further in the following section.

Neuropeptides and Pituitary Tumor Formation.

Considerable experimental evidence has shown that opioid neuropeptides, such as ME and BE, and tachykinin neuropeptides, such as SP, play an important role in the regulation of physiological functions of the anterior pituitary. These peptides exist in the anterior pituitary. They regulate the secretion of the pituitary hormones

PRL, GH, luteinizing hormone (LH), follicle-stimulating hormone (FSH), ACTH, and thyrotropin (TSH), which participate in body growth, metabolism, and cellular communications. ME, BE, and SP, among other neuropeptides, are increasingly accepted as autocrine or paracrine agents in the anterior pituitary. For example, SP has been shown to stimulate PRL secretion (Kato, et al., 1976; Rivier, et al., 1977; Vijayan and McCann, 1979) but to inhibit ACTH (Larsen, et al., 1993) and LH secretion (Kalra, et al., 1992; Picanco-Diniz, et al., 1990); ME has been shown to stimulate PRL (Dupont, et al., 1980; Grossman and Rees, 1983) and GH secretion (Grossman and Rees, 1983; Millan and Herz, 1985); and BE has been shown to stimulate PRL (Dupont, et al., 1980; Grossman and Rees, 1983) secretion but to inhibit ACTH (Grossman and Rees, 1983) and LH secretion (Grossman and Rees, 1983; Leadem and Kalra, 1985).

Figure 6 is a scheme that illustrates the localization of ME, BE, and SP in specific pituitary cell types and their possible regulatory effects on the pituitary hormone secretion. This figure was drawn based on the weighted trend that is demonstrated in the literature. Although the data are not exhaustive, they do provide strong evidence that these neuropeptides play important functional roles in the anterior pituitary. Those data also support the conclusion that these neuropeptides are autocrine or paracrine agents in the anterior pituitary.

Unlike endocrine control, in which hormones are secreted into the blood and travel to the distant target tissues, an autocrine or paracrine control is a mechanism in which neuropeptides or other molecules diffuse across the intercellular space to affect the neighboring target tissues. In primitive organisms and the early development of higher organisms, where the circulatory system is absent, the autocrine or paracrine control may be an important mechanism of intercellular communication (Peillon, et al., 1991). Therefore, a malfunction of autocrine or paracrine control by neuropeptides, such as ME, BE, or SP, will contribute to the formation of pituitary tumors.

Evidence has shown that various neuropeptides exist in pituitary tumors and non-tumor pituitaries, but that pituitary tumors contain higher concentrations of stimulatory neuropeptides (Peillon, et al., 1991). On the other hand, as inhibitory agents, opioid peptides, such as ME, have been shown to regulate early developmental events and neoplasia by interacting with the zeta-opioid receptor. This regulatory role has been reported on a variety of eukaryotic and prokaryotic cells that undergo embryogenesis (Zagon, et al., 1991, 1992, 1993).

Researchers have thus hypothesized (Zhu, 1993) that an overproduction or underproduction of neuropeptides is associated with pituitary tumor formation, and that an altered production of neuropeptides may be a manifestation of a defect in any of the enzyme-mediated steps that are involved in the metabolism of these neuropeptides, which may include the transcription of specific mRNAs, translation of neuropeptide precursor proteins, and post-translational processing of those precursor proteins to different-sized precursor proteins, to the functional neuropeptides, and to inactive metabolites. The research program described in this chapter involved an accurate comparison of the concentrations of ME, BE, and SP in human pituitary tumors versus the controls.

Extraction and HPLC Purification of Opioid and Tachykinin Neuropeptides.

Post-Mortem Stability of Opioid and Tachykinin Neuropeptides.

In this research program, the post-mortem human pituitaries were used as the controls. Although the post-mortem stability of neuropeptides is neither a major problem in small animal studies nor in human post-surgical studies, because those tissues and fluids can be acquired immediately, stability is a significant factor when using human post-mortem pituitary as controls, where an appropriate period of post-mortem delay time must be considered (Perry and Perry, 1983). Many articles reported that the post-mortem stability of several neuropeptides extended over several days (Arai, et al., 1987; Beal, et al., 1986; Beal and Martin, 1986; Emson, et al., 1980; McGregor and Bloom, 1983). Those conclusions were questioned, and experiments were designed to investigate the post-mortem stability of neuropeptides, using the rat pituitary as model (Zhu and Desiderio, 1993).

Spokes-Koch Cooling Curve.

The post-mortem stability of neuropeptides was studied using the rat pituitary as the model. The Spokes-Koch post-mortem human brain time-temperature cooling curve (Figure 7) (Spokes and Koch, 1978) was used. In brief, temperatures were measured in the ventral thalamus region (brain core temperature) and below the parietal cortex surface (superficial temperature) in two cadavers. The body temperature at death (non-febrile) was assumed to be 37°C, and the probe temperature was monitored after placing the cadaver in a 4°C mortuary refrigerator (one cadaver entered three hours, and one four hours, after death). The resulting brain-cooling curves were obtained, and were replotted (See below) in this study.

In situ Post-Mortem Incubation of the Rat Pituitary.

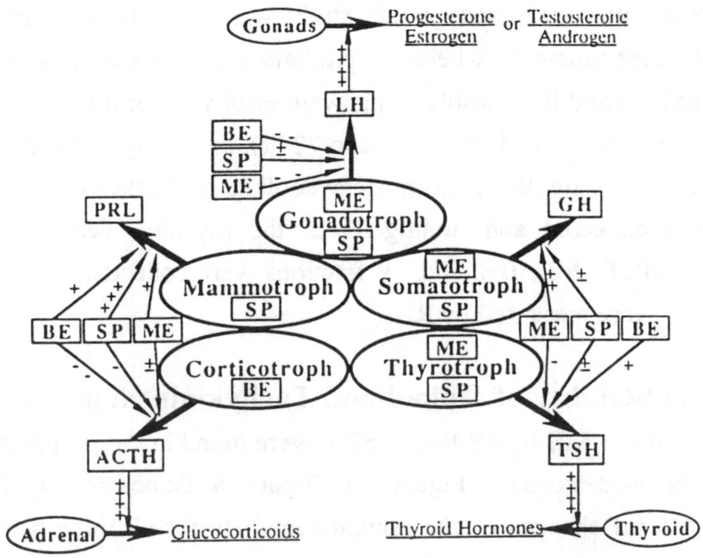

Figure 6. Regulatory roles of ME, SP, and BE in hormone secretions in the anterior pituitary, as discussed in the text. The five large ovals in the center symbolize the five major types of gland cells in the anterior pituitary; the three small ovals symbolize the peripheral glands. All rectangles symbolize the peptides that are present in the anterior pituitary. Peptides inside the large ovals indicate the reported localization of those peptides in those specific pituitary gland cell types. Underlined names are hormones that are secreted by peripheral glands. Thick arrows symbolize secretion processes; thin arrows symbolize regulatory processes. A plus sign (+) indicates a stimulatory effect; a minus sign (-) indicates an inhibitory effect; both signs (±) indicate that the literature is controversial on whether the process is stimulatory or inhibitory. The number of + or - signs indicates how strongly the current literature supports that regulatory process.

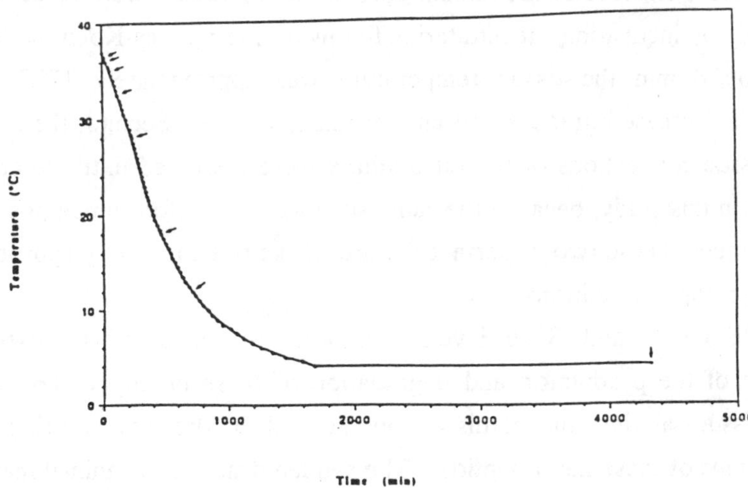

Figure 7. Post-mortem time-temperature cooling curve used for rat pituitaries (Spokes and Koch, 1978). Arrows indicate the incubation times used in this study. Individual dots indicate the temperature changes (n = 66). With permission (Zhu and Desiderio, 1993).

The experimental protocol used in this study was: animals were sacrificed by decapitation (except time = 0, where the pituitary was dissected directly from the anesthetized animal) and their whole heads were incubated for a series of times (15 min, 30 min, 1 hr, 2 hr, 4 hr, 8 hr, 12 hr, and 72 hr), according to the Spokes-Koch cooling curve, to imitate the post-mortem cooling of human brain (Figure 7); pituitaries were dissected and homogenized; the peptides were extracted and purified by RP-HPLC; ME, BE, and SP fractions were collected individually; and ME, BE, and SP were measured by RIA.

Post-Mortem Stability of Opioid and Tachykinin Neuropeptides.

Significant changes of ME-li, BE-li, and SP-li were found in the rat pituitary within the first 12 hr post-mortem (Figure 8). Figure 8 demonstrates clearly that significant and rapid changes of the immunoreactivity for all three neuropeptides occur in the earlier, previously uninvestigated, period (<1 hr). For example, the ME-li data (Figure 8A) demonstrate four separate regions where a statistically significant change in ME-li occurs: a rapid increase from 0 to 15 min, a rapid decrease from 15 to 60 min, a slow increase from 60 to 240 min, and a stable phase from 240 to 720 min. Similar concentration changes were also demonstrated for BE-li (Figure 8B) and SP-li (Figure 8C).

Before rationalizing these post-mortem changes of neuropeptide concentrations, it is important to indicate two features of the experimental design in this study. First, the temperature changes of the human post-mortem pituitary were imitated closely in this study by incubating rat pituitaries following the Spokes-Koch model. Thus, from 0 to 30 min the tissue temperature was approximately 37°C, and the temperature decreased gradually to and stabilized at 4 °C. Second, the local post-mortem tissue connections of the rat pituitary were preserved until the end of the incubation in this study, because the same situation occurs for human post-mortem pituitary tissues. These two important features make this study very appropriate for the human autopsy conditions.

The ME-li, BE-li, and SP-li levels measured in this study represented the summation of the production and degradation of these neuropeptides in the rat pituitary tissues at the time of dissection. Several factors may contribute to the concentrations of these neuropeptides. The sudden death of the animal may trigger a rapid mobilization of the neuropeptides from the surrounding tissues into the pituitary through the neuronal connections that were preserved in this study, resulting in the early rapid increase of the neuropeptides (0 to 15 min for ME-li, 0

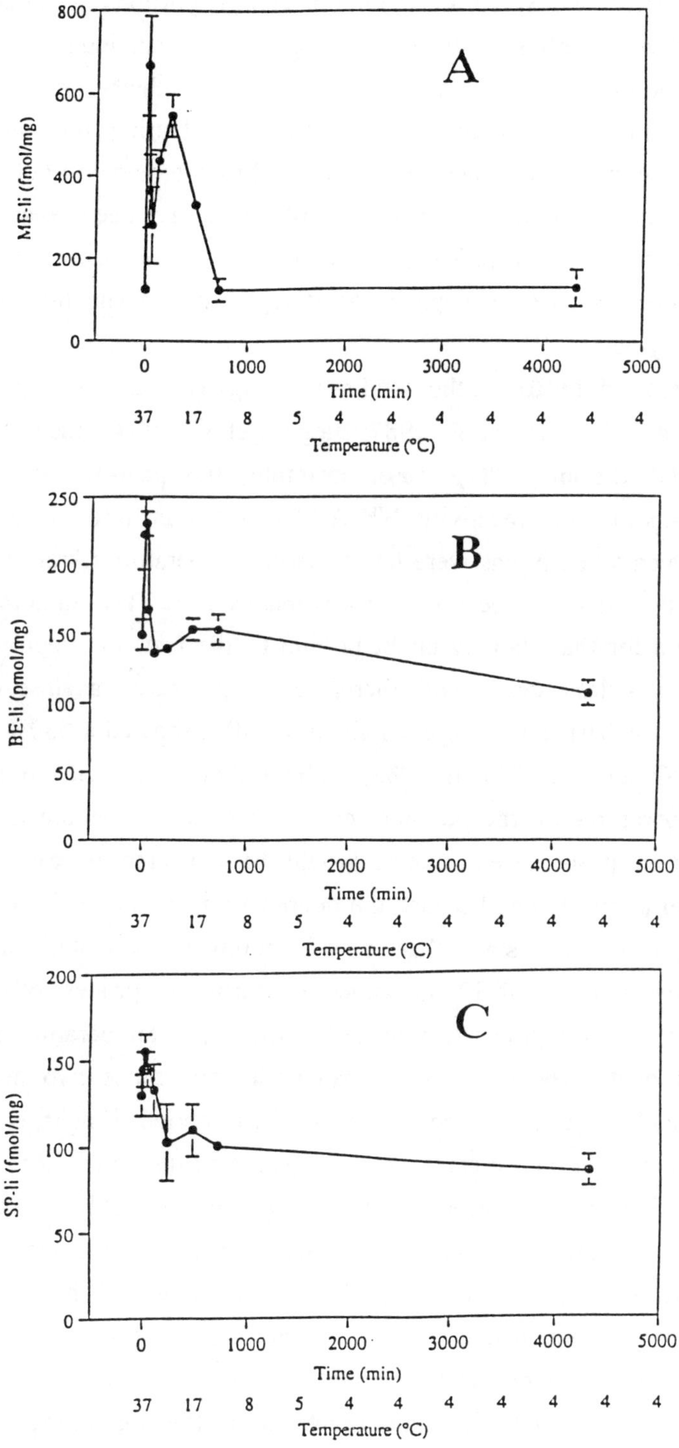

Figure 8. Plot of ME-li (A), BE-li (B), and SP-li (C) versus post-mortem time. With permission (Zhu and Desiderio, 1993).

to 30 min for BE-li and SP-li). Furthermore, various portions of the neuropeptide precursor molecules within the neuropeptidergic systems continue to undergo post-mortem metabolism (de novo synthesis, processing of higher molecular weight neuropeptide-containing proteins, or degradation). During the early time of incubation, these metabolic processes are favored by a temperature of 37°C. From 15-60 min, the neuropeptide precursor molecules may have been exhausted and the degradation of these neuropeptides may predominate, presumably from the action of various aminopeptidases, enkephalinase, dipeptidylcarboxypeptidases, and other proteolytic enzymes.

These conclusions differ from other studies that suggested that neuropeptides were stable for up to 72 hr (Arai, et al., 1987; Biggins, et al., 1984; Emson, et al., 1980; McGregor and Bloom, 1983). For example, the post-mortem stability of neurokinin A-like immunoreactivity (NK A-li) was demonstrated for at least 48 hr after death, even when brains were left at room temperature (Arai, et al., 1987). After 48 hr, the NK A-li decreased approximately 50%. The authors speculated that the reason for that stability might be due to the fact that neuropeptides are packaged in vesicles, and would therefore be protected against cytoplasmic peptidase activity. That post-mortem stability of SP-li showed a >30% loss at 4°C and 28°C(McGregor and Bloom, 1983). The authors state that the effect of the different temperatures on the post-mortem content of the peptide is unexpected because lysis in post-mortem tissues would be expected to cause a gradual degradation of peptides, and because the degradation should be slower at a lower temperature. They also state that it is theoretically possible that a more immunoreactive fragment of SP was generated during the process of degradation, and that more of this fragment is produced at the higher temperature. ME and SP were both studied in post-mortem human brain, and those two neuropeptides showed no significant change in content over a 72-hr period (Emson, et al., 1980).

Therefore, the data in this research do not agree with the conclusions of many of those previous reports. However, the differences can be readily rationalized. One explanation for that difference is that, in our study, more samples were taken during the initial rapid temperature decrease of the cooling curve, and therefore, a greater level of accuracy is obtained in monitoring that important time-frame, when the neuropeptide metabolism occurs at the greatest rate.

The advantages of using the rat pituitary to study the post-mortem stability of neuropeptides are: (1) the ready availability of rats, and the easy temperature control of post-mortem incubation; (2) it is an appropriate reference for human post-mortem pituitary (the study of the post-mortem stability of neuropeptides in human pituitaries is impossible within the first few hours following death; but, it is possible to incubate the rat pituitary as it remains within the decapitated head, surrounding tissues, and connections); and (3) human and rat pituitaries are similar, such as similar biologic mechanisms and biochemical pathways. Also, HPLC purification preceded the measurement of neuropeptide immunoreactivity as a means to minimize any cross-reactivity problems. The disadvantage of using the rat pituitary is species differences (for example, the human pituitary lacks the intermediate lobe).

The conclusions from this post-mortem study in the rat pituitary are important in evaluating the data from human pituitary tissues (see below). There is no delay time for the post-surgical pituitary tumor tissues. However, the post-mortem delay time must be observed, and can vary from 4-40 hr, a time-frame that overlaps the stable phase in the post-mortem rat pituitary for BE and SP. Care must be taken for post-mortem tissues between 4 to 10 hr for the ME data, because in rat post-mortem studies, ME-li did not reach the stable phase until approximately 10 hr.

In our research program, ten post-mortem human pituitaries were used as the control group for the characterization of ME, BE, and SP in human pituitary controls and tumors (see below). The post-mortem delay ranges from 8 to 18 hr, mostly from 12 to 16 hr. Based on the above conclusions, ME, BE, and SP are relatively stable during this time-frame and are not significantly different from the time zero when the post-surgical pituitary tumors were acquired. It is appropriate to use this group as the control from the post-mortem stability point of view. The average values of neuropeptides of this group were used as the basis for the analysis of the quantitative data obtained from human pituitary tumors.

Peptidase Inhibitors and Neuropeptide Analysis.

Another problem in the characterization of neuropeptides in human pituitary tissues is that the concentration of neuropeptides is usually low, especially in some pituitary tumors (Desiderio, et al., 1993). The absolute amount of neuropeptides in some tumors is also very low, because some tumor samples are less than one mg in wet weight. Therefore, it is imperative to optimize the tissue processing and peptide extraction procedures to maximize the amount of neuropeptide that can be collected for peptide characterization after the two-step RP-HPLC purification. One important aspect of optimizing the neuropeptide recovery is to minimize the metabolism of the neuropeptides during the process of tissue homogenization, when neuropeptides come into close contact with various extracellular and intracellular peptidases (Thorsett and Wyvratt, 1987; Turner, et al., 1985). Several possible ways exist to control the metabolism of peptides during tissue processing. An extreme in pH and/or temperature is normally used to minimize the hydrolytic and enzymatic degradation of peptides. Often, one or more enzyme inhibitors are added to the biological matrix in the belief that they help to preserve peptides within the tissue homogenate (Causon and McDowall, 1992). For studying the effects of peptidase inhibitors on the neuropeptide recovery, either acetic acid (1.0 M) or 2-amino-2-hydroxymethyl-1,3-propanediol-hydrochloric acid (Tris-HCl) buffer (pH 7.5, 0.05 M) was used to extract peptides from bovine pituitary tissues in the presence or absence of one of several peptidase inhibitors (Table 1) (Zhu and Desiderio, 1994). Following RP-HPLC purification, the three neuropeptides were measured by RIA.

In the absence of inhibitors, the highest ME-li and BE-li recovery occurred when the tissue was homogenized in acetic acid (1.0 M) and incubated at 4 °C (Figure 9). The SP-li data are inconclusive in that the SP-li recovery is not significantly different between using acetic acid (4 °C) and using Tris-HCl (4 °C). However, other studies have concluded that either acetic acid (McGregor and Bloom, 1983) or formic acid (Igwe, et al., 1988) is an efficient medium for the extraction of SP from biological tissues.

In the presence of inhibitors and in the acetic acid homogenate, no effect on neuropeptide recovery was noted for any inhibitor studied (Data not shown) (Zhu and Desiderio, 1994). In the Tris-HCl homogenate, where the neutral pH permits optimal enzyme activity, the following observation was made (Zhu and Desiderio, 1994). For BE-li and SP-li, all inhibitors had little or no effect on the neuropeptide recovery at either 4 °C or 37°C (Data not shown). However, the ME-li data (Figure 10) showed the effect of one inhibitor in improving the neuropeptide recovery in Tris-HCl buffer (37 °C). In the presence of 200 µg bestatin (in a 10 ml homogenate), ME-li approached that recovery found in an acetic acid homogenate incubated at 4°C (Figure 10 versus panel B in Figure 9), and the degradation of ME was virtually stopped.

The phenomenon that bestatin stopped any enzymatic degradation of ME in the bovine pituitary homogenate is very interesting. Other researchers have shown that different neuropeptide-inactivating enzymes are found in different tissues that degrade the same peptide, and that the relative importance of these enzymes in inactivating this peptide may be different (Turner, et al., 1985). In this present

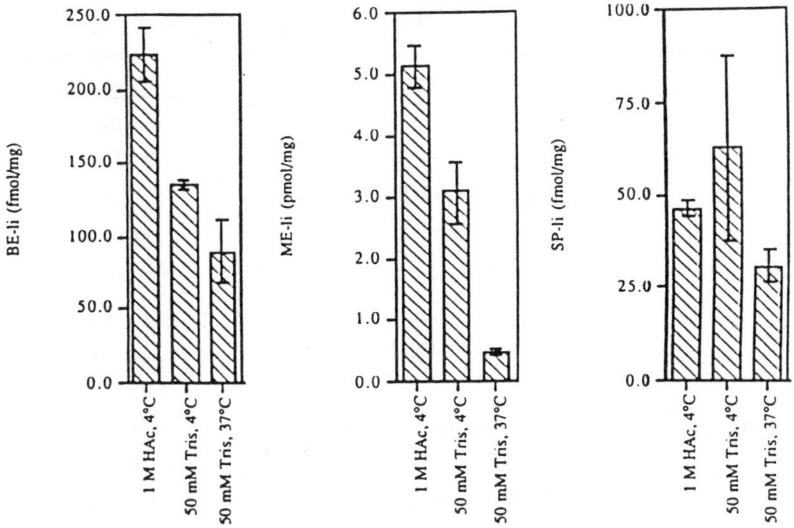

Figure 9. RIA measurements of the three peptides in the absence of inhibitors. The bovine pituitary tissue was homogenized in either 1 M acetic acid or 50 mM Tris-HCl (pH = 7.5). The homogenate in acetic acid was incubated for 5 hr at 4 °C. The homogenate in Tris buffer was either incubated for 0 hr or 1 hr at 37 °C, followed by the addition of 1 M acetic acid (v:v = 1:2) and incubated for 5 hr at 4 °C. With permission (Zhu and Desiderio, 1994).

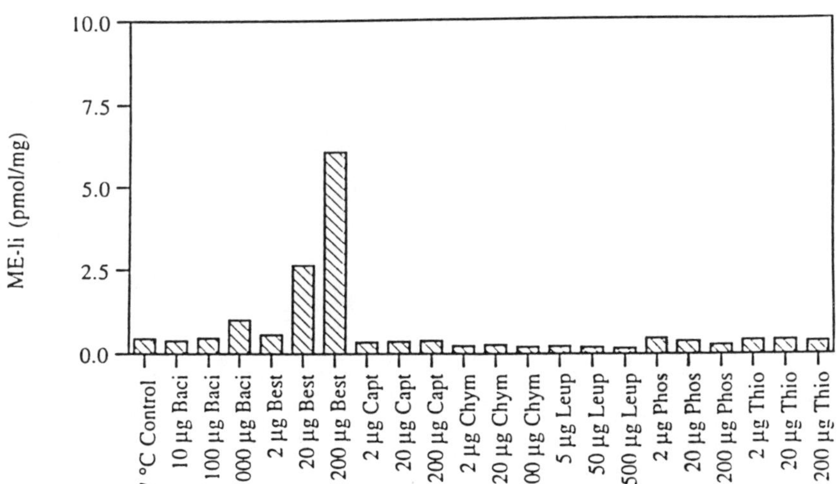

Figure 10. ME-li when the bovine pituitary tissue was homogenized in 4 °C 50 mM Tris buffer (pH = 7.5) and incubated with a peptidase inhibitor for 1 hr at 37 °C. In controls, no inhibitor was added.

Table 1 List of Inhibitors and Their Specificity.

Inhibitor	Specificity	Ref
Bacitracin	Peptidases	(Chang and Catrecasas, 1983)
Bestatin	Aminopeptidases	(Thorsett and Wyvratt, 1987)
Captopril	Angiotensin converting enzyme	(Thorsett and Wyvratt, 1987)
Leupeptin	Serine and thiol- peptidases	(Umezawa, 1976; Frommer, 1979)
Phosphoramidon	Metalloendopeptidases	(Thorsett and Wyvratt, 1987)
Thiorphan	Endopeptidase 24.11	(Thorsett and Wyvratt, 1987)

study, a bestatin-sensitive aminopeptidase seems to be the primary inactivating enzyme for ME in the bovine pituitary. Similarly, a bestatin-sensitive aminopeptidase was found in canine (Kusmierz and Desiderio, 1992) and human (Kaneko, et al., 1993) lumbar cerebrospinal fluid (CSF). Other researchers have also observed a bestatin-sensitive aminopeptidase activity to be the primary enkephalin inactivator in the ileum (Geary, et al., 1982).

Many researchers boil tissue homogenates to quickly inactivate enzymes. However, controlled experiments have shown that cold acetic acid (0 °C) is much more effective than heated acetic acid (90 °C) in SP recovery from biological tissues (Igwe, et al., 1988).

We conclude that peptidase inhibitors are not necessary in neuropeptide analysis if acetic acid and a low temperature are both used for the tissue processing. Therefore, during the peptide extraction from human pituitary controls and tumors, no inhibitors were added, and the homogenate media was 1.0 M acetic acid and the temperature was 4 °C. All of the extraction procedures, from tissue homogenization to protein precipitation to SPE until the peptide-rich fraction was lyophilized, were completed within one day to minimize the exposure of peptides to any peptidases or other degrading agents.

Solid-Phase Extraction of Opioid and Tachykinin Neuropeptides.

The supernatant obtained after the centrifugation of the tissue homogenate contains a host of various different biomolecules, such as proteins, peptides, carbohydrates, lipids, and others. SPE, by using a pre-constructed mini-cartridge packed with a reversed-phase material, is a fast and efficient way of isolating peptides from other molecules in that supernatant. The packing material inside the SPE cartridge is a solid silica or polymer, onto which is attached a long-chain hydrocarbon (usually C-18). Peptides are electrically-charged molecules in an aqueous solution. They can form ion-pairs with buffer molecules. For example, in triethylamine-formic acid (TEAF) buffer, pH 3.0, the carboxyl group of a peptide may form an ion-pair with a protonated triethylamine molecule, and the amino group of a peptide may form an ion-pair with a formate ion. The formation of ion-pairs increases the relative hydrophobicity of a peptide, and that peptide can now interact with the hydrocarbon chains inside the SPE cartridge. In our research program, conditions are chosen so that peptide ion-pairs are retained preferentially in the SPE column. Other relatively more hydrophilic molecules such as carbohydrates and salts are eluted by a low concentration of trifluoroacetic acid (TFA; 0.1%). The peptide ion-pairs that have been retained in the SPE cartridge are eluted with a buffer containing high concentrations (50-100%) of an organic modifier, such as acetonitrile.

Two experimental parameters must be observed in an SPE. The first parameter is the capacity of the SPE cartridge. Our experience has shown that the SPE cartridge we used (Sep-Pak Classic by Millipore-Waters, Bedford, MA, USA) contains picomole equivalents of binding sites, and is suitable for the study of human pituitary tissues. The second parameter is the flow-rate of the loading and eluting of the sample. Our experiments with tritium-labeled ME and SP showed that the best recovery (over 92%) was obtained when the flow-rate was 75 µl/min (Higa and Desiderio, 1989). Generally the recovery of neuropeptides from a tissue extract is in the range of 80 to 85% when the flow-rate is 1.0 ml/min (Zhu, 1993).

Reversed-Phase HPLC Purification of Opioid and Tachykinin Neuropeptides.

RP-HPLC is an effective method to purify peptides that are contained in the peptide-rich fraction that is eluted from the SPE before their analyses (Desiderio, 1996). Like the SPE described above, RP-HPLC is based on the interaction between peptide ion-pairs and the hydrophobic surface of silica gels or the other particles that are packed within the analytical HPLC column. The strength of the hydrophobic interaction is different for each different peptide. Therefore, when eluted by an organic modifier such as acetonitrile, different peptides are separated from each other to varying degrees. A gradient HPLC effectively separates peptides over a wide range of relative hydrophobicity and MW. However, an HPLC fraction obtained from a biological extract does not contain only the target peptide (Kusmierz, et al., 1990). Thus, it is important that HPLC fractions are further analyzed by an appropriate method.

As a first step of purification, the peptide-rich fraction obtained from the SPE is separated by a gradient RP-HPLC, using a mBondapak octadecylsilyl (ODS) analytical column (Millipore-Waters, Bedford, MA, USA). For example, Figure 11 shows a typical gradient RP-HPLC chromatogram that is obtained from a human pituitary tissue sample. The arrows indicate the retention times (obtained in a separate experiment to avoid any column contamination) of synthetic ME, BE, and SP, and the corresponding horizontal brackets indicate the fractions that are collected for ME, BE, and SP.

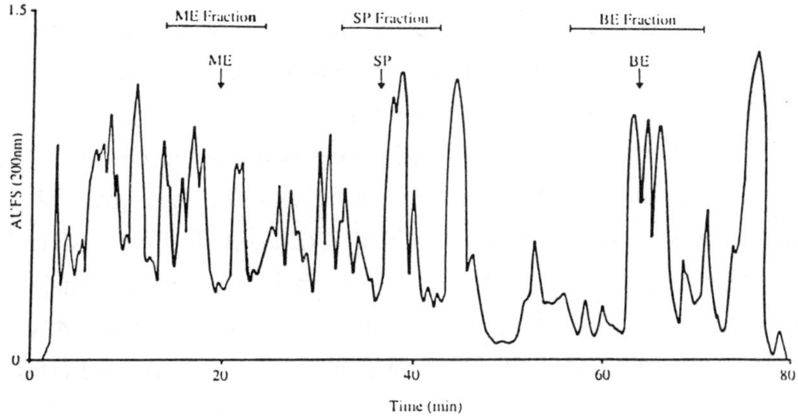

Figure 11. Gradient RP-HPLC chromatogram of the peptide-rich extract obtained from a human pituitary sample. The arrows indicate the retention time of synthetic peptides, and the brackets indicate the peptide fractions that were collected for analysis. With permission (Zhu, et al., 1995).

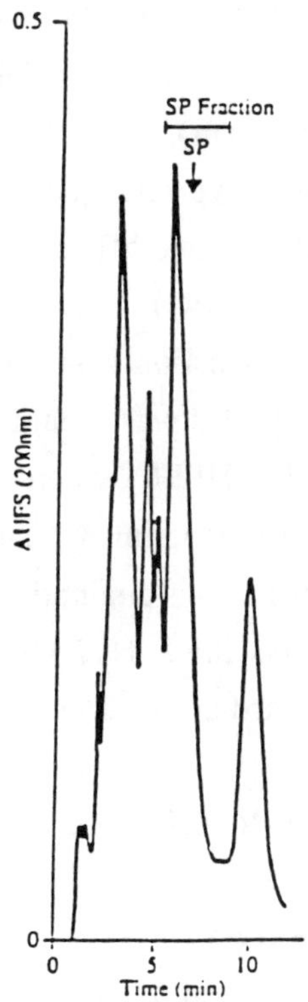

Figure 12. Isocratic RP-HPLC chromatogram of SP. The mobile phase was 18% acetonitrile in 40 mM TEAF buffer (pH 3.0). The arrow indicates the retention time of synthetic SP. The bracket indicates the fraction that was collected for SP quantification. With permission (Zhu, et al., 1995).

To increase the detection sensitivity and specificity, it is preferred that the fractions of ME, BE, and SP collected from the gradient HPLC be purified further by an isocratic RP-HPLC, using a polymer-based analytical column (Polymer Laboratories, Amherst, MA, USA). Another reason that a subsequent RP-HPLC is performed is that, although SP and ME can be analyzed intact by RIA and MS, respectively, BE is trypsinolyzed to form NAIIK (Zhu, et al., 1995), which has very favorable MS properties (Dass, et al., 1991) and is used for the qualitative and quantitative analysis of BE. Trypsinolysis may or may not by required, depending on the mass range and sensitivity of the MS instrument that is used (Dass, et al., 1991).

Figure 12 shows the isocratic RP-HPLC chromatogram of the SP fraction from the gradient RP-HPLC. The arrow in Figure 12 indicates the retention time of synthetic SP on the same column (obtained in a separate experiment to avoid column contamination, and the horizontal bracket indicates the fractions that were collected for SP quantification by RIA.

Figure 13 shows the isocratic RP-HPLC chromatogram of the ME fraction from the gradient RP-HPLC. The arrow indicates the retention time of synthetic ME (obtained in a separate experiment), and the horizontal bracket indicates the fraction collected for the qualitative and quantitative analysis of ME. The inset shows the (M+H)+ ion for endogenous ME at m/z 574 and for the corresponding internal standard at m/z 579 (see below).

After the trypsinolysis of the BE fraction from the gradient RP-HPLC (Zhu, et al., 1995), the mixture of tryptic peptides was separated on isocratic RP-HPLC. Figure 14 shows an example of that RP-HPLC, where the arrow indicates the retention time of synthetic NAIIK (separate experiment), and the horizontal bracket indicates the fraction that was collected for the qualitative and quantitative analysis of the endogenous BE. The inset shows the (M+H)+ ion for NAIIK at m/z 558 and for the corresponding internal standard at m/z 562.

Analytical Methods Used to Characterize Opioid and Tachykinin Neuropeptides.

Radioimmunoassay of Substance P.

RIA is a widely used detection method for the analysis of peptides. It is based on an interaction of a peptide and its antibodies. Theoretically, RIA possesses a high level of molecular specificity and detection sensitivity (Yalow, 1980). There is a general agreement about the detection sensitivity of RIA measurement of peptides, because

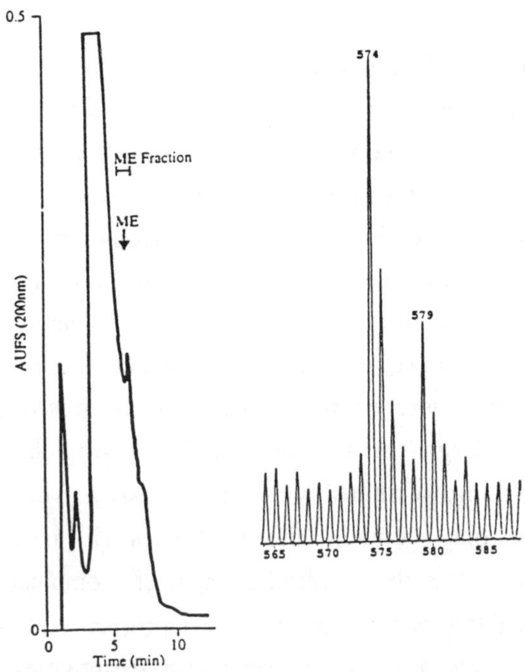

Figure 13. Isocratic RP-HPLC chromatogram of ME. The mobile phase was 15% acetonitrile in 40 mM TEAF buffer (pH 3.0). The arrow indicates the retention time of synthetic ME. The bracket indicates the fraction that was collected for ME quantification and amino acid sequence-determining ion analysis of ME. The inset shows the limited mass range scan, which contains the MH+ ion of ME at m/z 574 and of (2H5)ME at m/z 579. With permission (Zhu, et al., 1995)

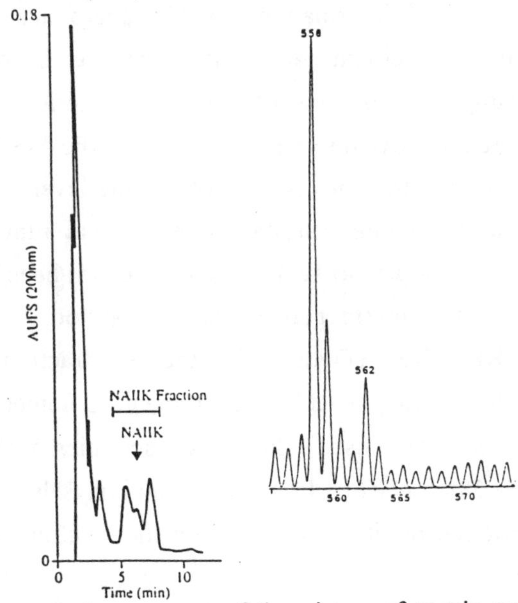

Figure 14. Isocratic RP-HPLC chromatogram of the mixture of tryptic peptides produced by trypsin treatment of the BE fraction collected from the gradient RP-HPLC separation. The mobile phase was 2.2% acetonitrile in 40 mM TEAF buffer (pH 3.0). The arrow indicates the retention time of synthetic NAIIK. The bracket indicates the NAIIK fraction that was collected for BE quantification and amino acid sequence-determining ion analysis of NAIIK. The inset shows the limited mass range scan, which contains the (M+H)+ ion of NAIIK at m/z 558 and of (2H4)NAIIK at m/z 562. With permission (Zhu, et al., 1995).

the detection limit is at the low femtomole level. However, the molecular specificity of RIA is questionable (Desiderio, 1984). For one thing, an RIA measurement does not have any direct connection to the amino acid sequence of the target peptide; rather, it detects the percentage of radio-labeled peptide that has displaced the endogenous peptide in an antigen-antibody interaction, in which the determining factor is the three-dimensional shape and interacting surface of an antigen and antibody. Additionally, many biologically important neuropeptides have a MW of less than 4000 Da, and thus lack any inherent immunogenicity. Therefore, the antibody against these neuropeptides must be raised by attaching multiple copies of them to a larger carrier protein such as thyroglobulin. That immunogenic complex is injected into an animal (Gros, et al., 1978). The antiserum that is raised in this way not only contains the antibodies (if any) against the original neuropeptide, but also the antibodies against combinations of the original neuropeptide and different parts of the carrier protein. Although many researchers who use RIA provide a list of possible cross-reactivities of an antibody with numerous other known molecules, that list cannot be exhaustive, and many unknown endogenous species may also interfere with an RIA measurement.

Nonetheless, RIA is an important tool in neuroscience research, and has provided a tremendous amount of information on the anatomical localization of many neuropeptides. In many circumstances, RIA may be the only practical analytical method. For example, in the research program described here, SP exists in the human pituitary at only low fmole levels, whereas the MS detection threshold for many endogenous neuropeptides is at fmole-pmole levels. Therefore, RIA is used to measure SP in the human pituitary controls and tumors. It is necessary to indicate the fact that the amino acid sequence has not been confirmed by denoting the data as SP-li, and to interpret such data with caution.

After isocratic RP-HPLC (Figure 12), the SP fractions were collected and reconstituted in buffer for an RIA measurement. Commercial RIA kits (IncStar, Stillwater, MN, USA) were used. The cross-reactivities of the anti-SP antiserum in the kit were <0.002% for ME, LE, eledoisin, and physalemin, and 0.008% for BE. The SP-li data for ten pituitary controls and thirty pituitary tumors are shown in Table 2.

Mass Spectrometric Analysis of Methionine Enkephalin and Beta-Endorphin.

Qualitative Analysis of Methionine Enkephalin and Beta-Endorphin.

Table 2. RIA Measurements of SP-li in Human Pituitary Controls and Tumors.

Control	SP-li (fmol/mg)	Tumor	SP-li (fmol/mg)
1	105	1	89.0
2	92	2	20.1
3	103	3	14.5
4	62.8	4	15.3
5	90.8	5	9.6
6	47.6	6	9.6
7	95.9	7	6.3
8	141	8	9.0
9	136	9	439
10	120	10	333
		11	142
		12	36.9
		13	12.1
		14	97.5
		15	9.6
		16	9.4
		17	9.4
		18	3.5
		19	2.3
		20	5.6
		21	1.7
		22	2.4
		23	5.1
		24	11.0
		25	5.8
		26	22.0
		27	20.7
		28	47.1
		29	220
		30	294

In the analysis of endogenous neuropeptides, it is extremely important to first confirm the amino acid sequence of those neuropeptides. MS/MS is a powerful technique to provide such sequence information. For the qualitative analysis of a neuropeptide, the MS/MS is set so that MS-1 selects the (M+H)+ of the target neuropeptide (m/z of 574 for ME or m/z of 558 for NAIIK), and, after CID, MS-2 collects the product ions that derived from that (M+H)+ ion. That product-ion spectrum contains amino acid sequence-determining fragment ions of the endogenous neuropeptide. By comparing such product-ion spectrum with that obtained from corresponding synthetic peptide, the amino acid sequence of the endogenous neuropeptide is established unequivocally.

Figure 15 shows the product-ion spectrum of the (M+H)+ ion from synthetic ME. The standard nomenclature of ions (Biemann, 1990; Roepstorff and Fohlman, 1988) is used to rationalize the genesis of the amino acid sequence-determining fragment ions from the fragmentation of the ME peptide chain, as shown in Figure 16. Figure 17 contains the product-ion spectrum of the (M+H)+ ion of the endogenous ME that was extracted from one human pituitary controls, and Figure 18 contains the product-ion spectrum of the (M+H)+ ion of the endogenous ME from a human pituitary tumor. The product-ion spectra in Figure 17 and 18 are both similar to the spectrum in Figure 15. Those data confirm the amino acid sequence of the endogenous ME obtained from that human pituitary control and tumor.

In our research program, the analysis of BE is accomplished by analyzing NAIIK after trypsinolysis of BE fractions obtained in the gradient RP-HPLC (Zhu, et al., 1995). A survey of the protein sequences in published databases indicates that the pentapeptide sequence NAIIK is unique to BE in mammals (Yan, et al., 1993). Because of the uniqueness of NAIIK to BE, the qualitative analysis of NAIIK is performed to ensure the molecular specificity of the BE detection. Like ME, the first step in BE analysis is to confirm the amino acid sequence of NAIIK in the isocratic RP-HPLC fractions. A product-ion spectrum of the (M+H)+ ion of synthetic NAIIK has been obtained (Figure 19). The same standard ion nomenclature (Biemann, 1990; Roepstorff and Fohlman, 1988) is used to rationalize the genesis of the amino acid sequence-determining fragment ions from the (M+H)+ ion of NAIIK and is shown in Figure 20. Figure 21 contains the product-ion spectrum of the (M+H)+ ion of NAIIK, which was generated by trypsinolysis of the BE fraction extracted from a human pituitary control, and Figure 22 the product-ion spectrum from the (M+H)+ ion of NAIIK from a human

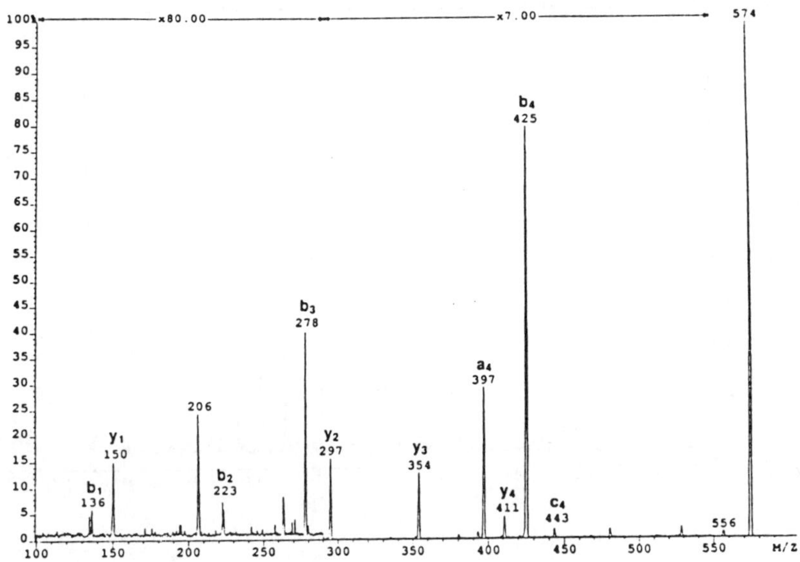

Figure 11. Gradient RP-HPLC chromatogram of the peptide-rich extract obtained from a human pituitary sample. The arrows indicate the retention time of synthetic peptides, and the brackets indicate the peptide fractions that were collected for analysis. With permission (Zhu, et al., 1995).

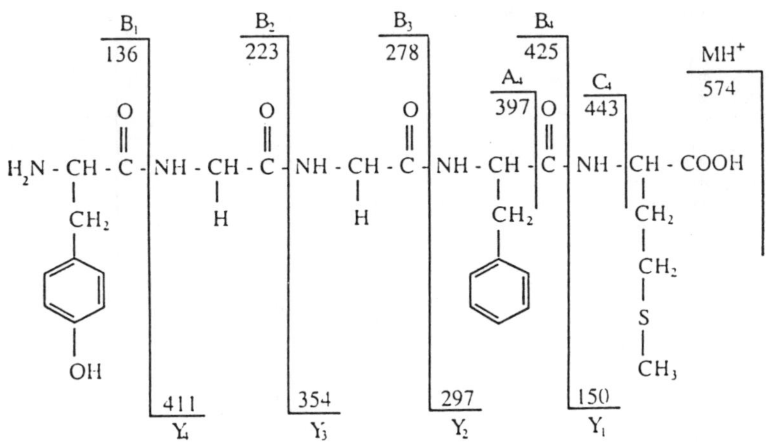

Figure 12. Isocratic RP-HPLC chromatogram of SP. The mobile phase was 18% acetonitrile in 40 mM TEAF buffer (pH 3.0). The arrow indicates the retention time of synthetic SP. The bracket indicates the fraction that was collected for SP quantification. With permission (Zhu, et al., 1995).

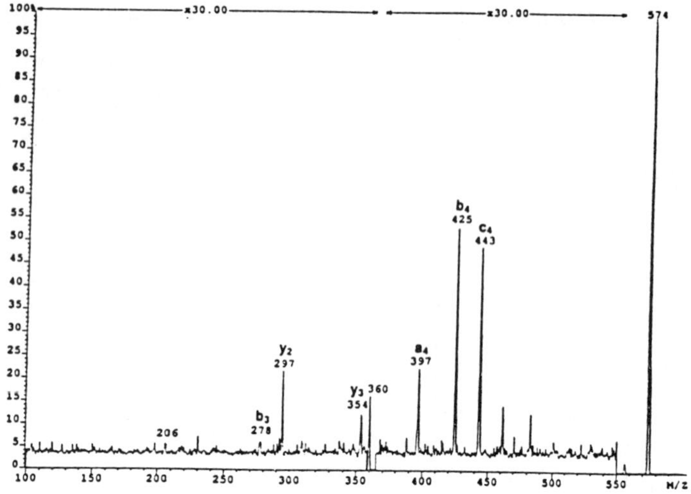

Figure 17. Product-ion spectrum of the (M+H)+ ion of native ME extracted from a human pituitary control sample.

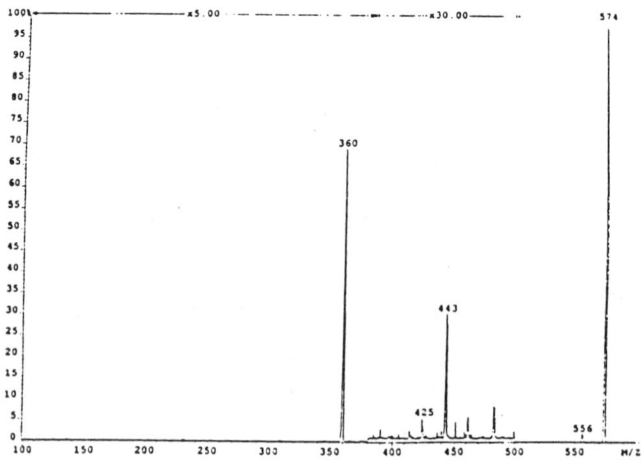

Figure 18. Product-ion spectrum of the (M+H)+ ion of native ME extracted from a human pituitary tumor sample. With permission (Zhu, et al., 1995).

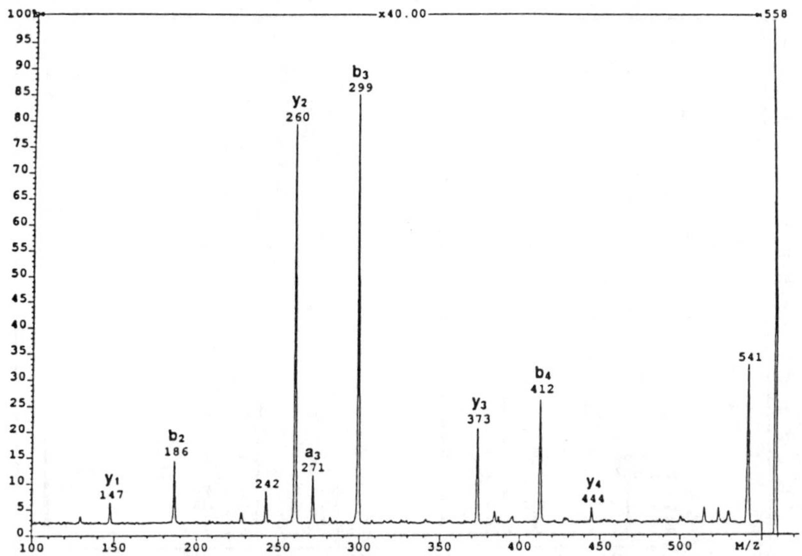

Figure 19. Product-ion spectrum of synthetic NAIIK. The (M+H)+ at m/z 558 of synthetic NAIIK was selected, and this spectrum was collected in the Product-Q mode. With permission (Zhu, et al., 1995)

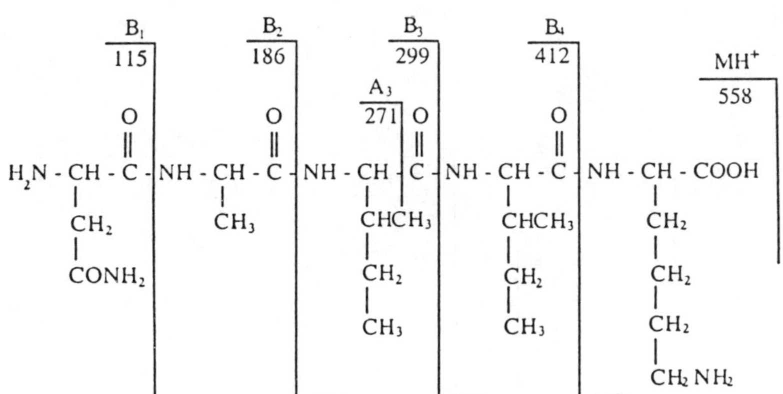

Figure 20. Peptide bond fragmentation pattern of NAIIK. The standard nomenclature (Biemann, 1990; Roepstorff and Fohlman, 1988) of peptide bond fragmentation is used to designate the amino acid sequence-determining fragment ions. With permission (Zhu, et al., 1995).

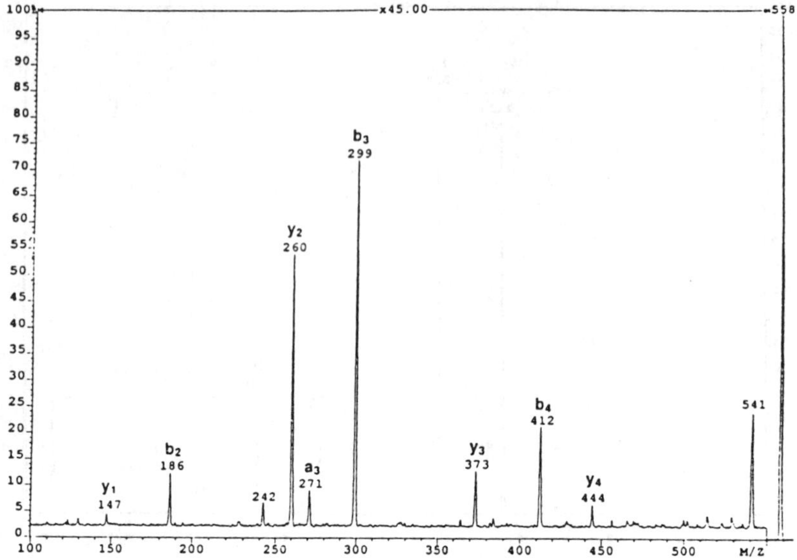

Figure 21. Product-ion spectrum of the (M+H)+ ion of native NAIIK extracted from a human pituitary control sample.

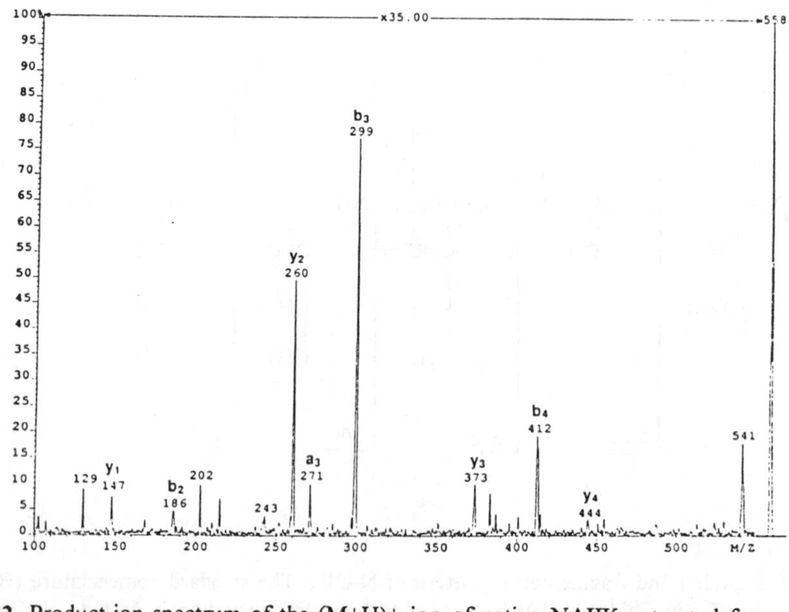

Figure 22. Product-ion spectrum of the (M+H)+ ion of native NAIIK extracted from a human pituitary tumor sample. With permission (Zhu, et al., 1995)

pituitary tumor. Figures 21 and 22 are both qualitatively similar to the spectrum contained in Figure 19. Therefore, the amino acid sequence of NAIIK obtained from trypsinolysis of the endogenous BE extracted from human pituitary controls and tumors was confirmed by the presence of those amino acid sequence-determining fragment ions.

Quantitative Analysis of Methionine Enkephalin and Beta-Endorphin.

There are two commonly used types of MS techniques to quantify neuropeptides, the (M+H)+ ion measurement and multiple reaction monitoring (MRM). In the first technique, the intensity of the (M+H)+ ion of the neuropeptide is recorded; in the second technique, the ion current corresponding to the transition of the (M+H)+ ion of the neuropeptide to one of its amino acid sequence-determining ion is recorded. Both methods use an internal standard that is the same molecule as the target neuropeptide, except for the replacement of several hydrogens of the neuropeptide by deuteriums. The ion current due to the internal standard can be compared to that current of the target neuropeptide, and the amount of the endogenous neuropeptide can be calculated from a calibration curve.

Internal Standard.

For the quantification of an endogenous neuropeptide by MS, an internal standard is added to the tissue before homogenization. The internal standard added is the same as the target neuropeptide, except that several hydrogens have been replaced by the stable-isotope deuterium; e.g., ((2H5)Phe)ME. In a mass spectrum, ((2H5)Phe)ME and ME both produce Phe-containing ions, but that differ by 5-Da; those ions can be analyzed simultaneously. That internal standard has the same chemical and physical characteristics as the target neuropeptide, and thus can compensate readily for any loss during tissue homogenization, SPE, and RP-HPLC. One example was mentioned above that there were a substantial loss of neuropeptides during extraction and purification processes. Furthermore, in the long process of extracting and purifying the endogenous ME, the side chain of the methionine residue may be air-oxidized to form ME-sulfoxide, which is separated from the authentic ME in RP-HPLC and lost in the final collection of ME fractions. However, the deuterated ME experiences the same oxidation process at the same rate as the authentic ME does. Therefore, by measuring the ratio of signal intensities of endogenous ME to deuterated ME, the amount of the endogenous

ME can be calculated, based on the amount of deuterated ME that has been added to the tissue before homogenization and using a corresponding calibration curve.

In our research program, we used ((2H5)Phe)ME (Kusmierz, et al., 1990) and ((2H4)Ile)BE (Dass, et al., 1991) as internal standards for the quantitative analysis of endogenous ME and BE, respectively, in the human pituitary controls and tumors. Specifically, one mg ((2H5)Phe)ME and five mg ((2H4)Ile22)BEhuman, 1-31 were added to each tissue before homogenization (Zhu, 1993; Zhu, et al., 1995). The homogenate was incubated (4 hr; 4 °C) to ensure that the endogenous and exogenous peptides experience the same environment and to precipitate proteins and cellular debris. After an aliquot (5%) was taken for total protein determination, the homogenate was subjected to centrifugation, SPE, and RP-HPLC for the extraction and purification of ME and BE, and for subsequent qualitative and quantitative analyses.

(M+H)+ Measurements.

As mentioned above, one technique of neuropeptide quantification by MS is to measure the intensity of the (M+H)+ ions. There are advantages in using (M+H)+ information to quantify a neuropeptide. One advantage is that it relates the neuropeptide measurement to the MW of the neuropeptide. Also, by limiting the scanning to a small mass range that flanks the (M+H)+ ion, sufficient detection sensitivity can be achieved for many neuropeptides.

The ME and NAIIK fractions from isocratic RP-HPLC (Figures 13 and 14, respectively) were analyzed first by liquid secondary-ion mass spectrometry (LSIMS) in the magnet-scanning mode to demonstrate the presence of the appropriate (M+H)+ ions of these two neuropeptides and of their corresponding internal standards. The (M+H)+ ion data for ME and (2H5)ME were obtained by scanning the magnet across the limited m/z range of 550-635, and the (M+H)+ ion data for NAIIK and (2H4)NAIIK was obtained by scanning the magnet across the limited m/z range of 550-635.

Opioid and Tachykinin Neuropeptides in the Anterior Pituitary.

Methionine Enkephalin.

1. Proenkephalinhte for both neuropeptides was two seconds per decade. The ratio of the (M+H)+ ion of each neuropeptide to its corresponding internal standard was used to determine the amount of the endogenous neuropeptide from a calibration curve (see below).

To obtain a calibration curve (Figure 23) (Zhu, 1993) for quantifying ME in the human pituitary tissues, one set of standard solutions was prepared. The solutions contained 0.25, 0.5, 1, 2, 4, 8, or 16 mg ME and a fixed amount of 1 mg (2H5)ME. All these solutions were prepared in 50 ml 0.1% acetic acid in water-methanol (1:1, v:v). The ratio of intensity of m/z 574 to 579 was plotted against each ME amount. To obtain a calibration curve (Figure 24) (Zhu, 1993) for quantifying endogenous BE in the human pituitary tissues, one set of standard solutions was prepared. Each solution contained 0.5, 1, 2, 4, 8, 16, or 32 mg BE and a fixed amount of 5 mg (2H4)BE. The entire content of each solution was treated individually with trypsin (Zhu, 1993; Zhu, et al., 1995), and the tryptic peptide NAIIK (and deuterated NAIIK) was purified by isocratic RP-HPLC. The lyophilized fractions were reconstituted into 50 ml 0.1% TFA. The ratio of the intensity of m/z 558 to 562 was plotted against each amount of BE.

Table 3 collects the ME and BE measurements by the (M+H)+ quantification method. All measurements are expressed as pmol/mg total protein. For some control pituitaries, BE measurements are not available because the samples were lost in the experiments.

Tandem MS Measurements.

Although the (M+H)+ quantification of neuropeptides is relatively straightforward and easy to perform, that method does have some limitations. The (M+H)+ ions of peptides are usually not the most abundant ions in the MS spectra. They are sometimes overwhelmed by contaminating ions from the oligomerization of matrix molecules such glycerol, or from matrix-peptide interactions that form various adducts (Dass and Desiderio, 1988). Thus, the (M+H)+ quantification method often presents a poor signal-to-noise ratio, and therefore poor sensitivity and molecular specificity. The MRM mode of MS/MS, however, provides the highest sensitivity and molecular specificity for the quantification of neuropeptides, because the specific transition of the (M+H)+ ion of the peptide to one of its unique product ions is monitored in MRM (Desiderio, 1984, 1991, 1996). When an internal standard is used to quantify an endogenous neuropeptide, two ion-channels are used to monitor simultaneously the transition of the (M+H)+ ion of the neuropeptide to one of its product ions and the transition of the (M+H)+ ion of the internal standard to one of its product ions. Figure 25 shows the precursor-product ion relationship (amino acid sequence; m/z value) that was used for the MRM quantification of ME and BE (via NAIIK). Data for each neuropeptide and its internal standard are shown. Specifically, in the quantification of ME, Channel 1

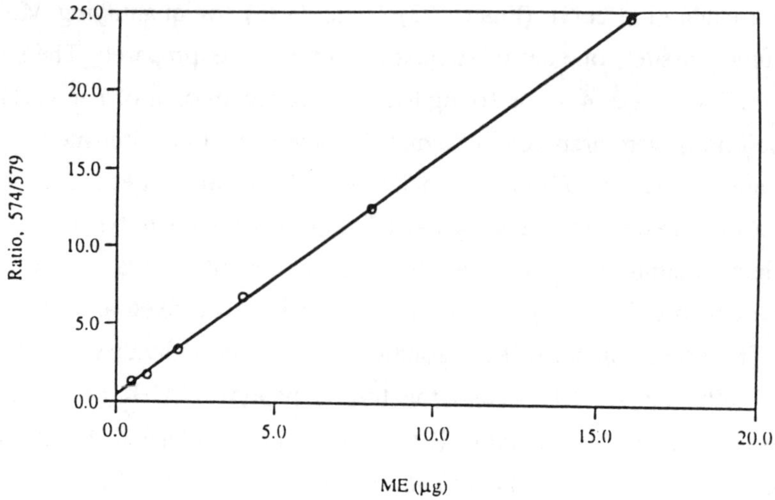

Figure 23. Calibration curve for quantification of ME by (M+H)+ measurement.

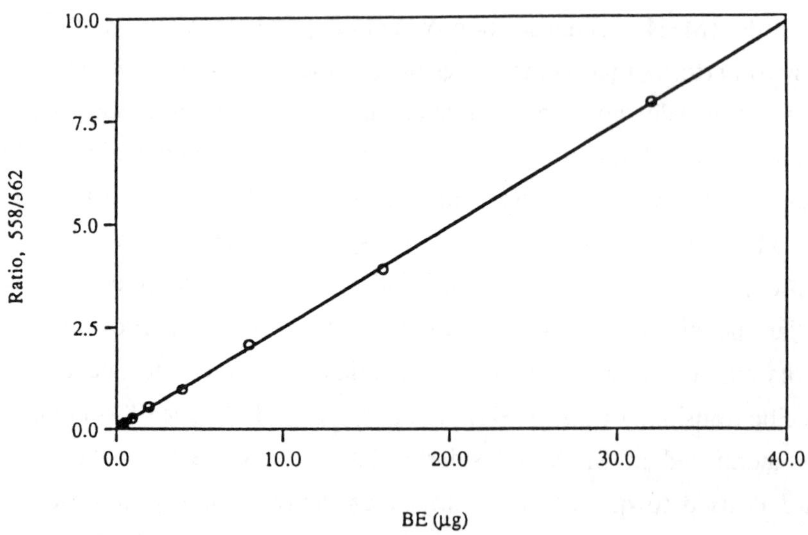

Figure 24. Calibration curve for quantification of BE by (M+H)+ measurement.

Beta-Endorphin (BE)

Channel	Precursor	Product
1	$NAIIK^+$, m/z 558	NAI^+, m/z 299
2	$(^2H_4)NAIIK^+$, m/z 562	$(^2H_4)NAI^+$, m/z 303

Methionine Enkephalin (ME)

Channel	Precursor	Product
1	$YGGFM^+$, m/z 574	$YGGF^+$, m/z 425
2	$(^2H_5)YGGFM^+$, m/z 579	$(^2H_5)YGGF^+$, m/z 430

Figure 25. Schematic representation of the precursor-product ion relationships that were monitored during MRM for the quantification of endogenous BE (top) and of ME (bottom). See text for details.

Table 3. BE and ME Measurements in Human Pituitary Controls and Tumors by LSIMS-(M+H)$^+$.

-, not measured. n.d., not detectable.

Control	BE (pmol/mg)	ME (pmol/mg)	Tumor	BE (pmol/mg)	ME (pmol/mg)
1	-	18.5	1	377	414
2	-	15.7	2	109	366
3	-	16.9	3	65.1	224
4	-	18.5	4	34.2	116
5	228	20.1	5	31.6	172
6	655	26.6	6	17.1	143
7	704	30.0	7	33.6	102
8	882	35.2	8	21.9	82.3
9	481	20.1	9	191	53.3
10	424	26.5	10	27.0	41.2
			11	14.3	36.6
			12	17.5	470
			13	129	159
			14	320	45.1
			15	14.8	52.5
			16	13.2	84.4
			17	24.4	55.0
			18	7.4	40.3
			19	3.9	16.2
			20	6.4	34.4
			21	12.8	12.9
			22	5.9	18.3
			23	45.6	37.8
			24	450	115
			25	31.8	62.7
			26	573	280
			27	297	395
			28	4180	772
			29	367	n.d.
			30	343	n.d.

monitors the transition of (YGGFM+H)+ (m/z 574) to (YGGF+H)+ (m/z 425), and Channel 2 monitors the transition of ((2H5)YGGFM+H)+ (m/z 579) to ((2H5)YGGF+H)+ (m/z 430). In the quantification of BE, Channel 1 monitors the transition of (NAIIK+H)+ (m/z 558) to (NAI+H)+ (m/z 299), and Channel 2 monitors the transition of ((2H4)NAIIK+H)+ (m/z 562) to ((2H5)NAI+H)+ (m/z 303). The ratio of intensities in Channel 1 to Channel 2 is calculated. The amount of the endogenous ME or BE is calculated from that ratio and the corresponding calibration curve (see below). To obtain calibration curves for the MRM quantification of ME (Figure 26) and BE (Figure 27), the procedure described in the (M+H)+ measurement is followed, except that the MS instrument is set to the MRM mode, and the ion current that corresponds to the transition of the (M+H)+ ion to one of its fragment ions is recorded. The ratio of the ion currents that are recorded in Channel 1 to Channel 2 is plotted against the amount of ME or BE that has been added to the peptide mixture. Table 4 collects the ME and BE measurements by the MRM quantification method. All measurements are expressed as pmol/mg total protein.

Human Pituitary Quantitative Measurements.

As an example of interpreting the data that are obtained from the quantitative analysis of the endogenous neuropeptides by RIA and MS, this section summarizes and discusses the measurements of SP-li, ME, and BE in human pituitary controls and tumors with the corresponding clinical correlation. Because MRM measurements have a higher molecular specificity than (M+H)+ measurements, the MRM data for ME and BE are used for discussion. SP-li measurements are used in the discussion because they are the only available SP measurements.

Human Pituitary Controls.

Table 5 summarizes the quantitative data of SP-li, ME, and BE in ten human pituitary controls. The post-mortem time ranges from 8 to 18 hr (13.4 ± 1.0 hr), mostly from 12 to 16 hr. As concluded from the post-mortem stability of neuropeptide study in the rat pituitary (Zhu and Desiderio, 1993), these endogenous neuropeptides are stable during this time-frame; thus, it is appropriate to use the data in Table 5 as controls to analyze the data derived for the human pituitary tumors.

Human Pituitary Tumors.

By comparing the neuropeptide measurement data in human pituitary controls versus tumors, the coefficient of variation of the measurements of all three

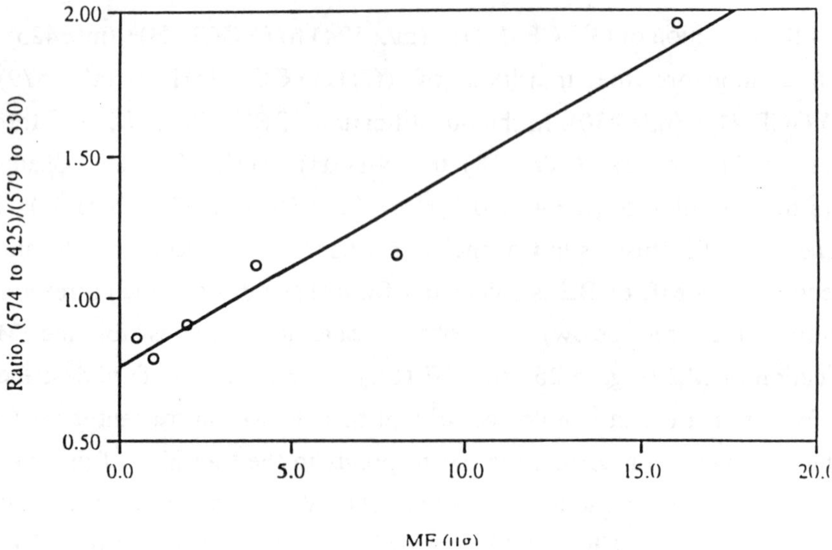

Figure 26. Calibration curve for quantification of ME by MRM.

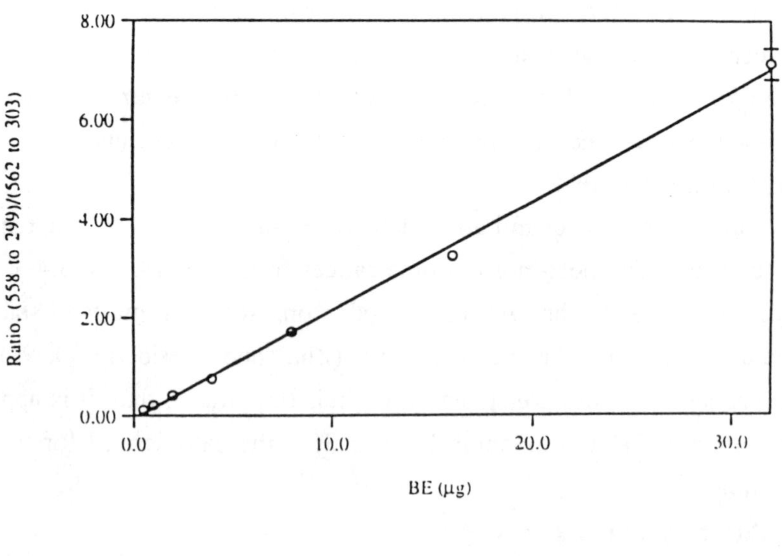

Figure 27. Calibration curve for quantification of BE by MRM.

Table 4. BE and ME Measurements in Human Pituitary Controls and Tumors by LSIMS-MRM.
-, not measured. n.d., not detectable.

Control	BE (pmol/mg)	ME (pmol/mg)	Tumor	BE (pmol/mg)	ME (pmol/mg)
1	-	56.4	1	1530	1740
2	-	67.3	2	272	1320
3	-	53.9	3	187	934
4	-	87.7	4	57.3	682
5	333	101	5	95.2	1060
6	671	92.5	6	97.4	747
7	816	67.8	7	63.5	612
8	1040	76.5	8	25.6	173
9	398	108	9	351	69.9
10	308	55.4	10	35.8	87.5
			11	16.9	124
			12	18.1	53.1
			13	249	181
			14	366	56.9
			15	18.7	71.1
			16	13.8	99.5
			17	24.9	38.2
			18	9.6	40.0
			19	5.2	26.6
			20	10.1	35.9
			21	23.5	20.6
			22	10.0	36.6
			23	61.6	24.0
			24	467	112
			25	67.4	377
			26	885	2040
			27	666	2140
			28	6840	5460
			29	507	n.d.
			30	598	n.d.

Table 5. Neuropeptide Measurements in Post-Mortem Human Pituitary Controls.

Ave., mean ± s.e.m.; -, not measured; n.a., not available; PM, post-mortem.

Sample	Age	Sex	PM (hr)	SP-li (fmol/mg)	BE (pmol/mg)	ME (pmol/mg)
1	49	M	8	105	-	56.4
2	60	M	16	92.0	-	67.3
3	22	M	14	103	-	53.9
4	26	M	18	62.8	-	87.7
5	26	M	16	90.8	333	101
6	24	M	14	47.6	671	92.5
7	43	M	15	95.9	816	67.8
8	57	M	12	141	1040	76.5
9	49	M	9	136	398	108
10	52	M	12	120	308	55.4
Ave.	40.8 ± 4.7		13.4 ± 1.0	99.4 ± 9.2	594 ± 121	76.6 ± 6.2

neuropeptides is much larger in the tumor group than that in the control group. That variation is not surprising, considering the fact that the tumor group consists of a heterogeneous population of several different types of pituitary tumors. Therefore, such a variation of measurements could be attributed to their biological heterogeneity. Clinically, this group contains several different types of tumors, and they are classified based on the symptoms of the patient and pathology reports, which also includes the immunohistological staining of the tumor specimen by the antisera against several different pituitary hormones. For example, if a patient had an elevated serum PRL level, and the tumor specimen was stained by an anti-PRL antiserum, then the tumor is designated a PRL-secreting tumor. On the other hand, if the patient did not have any abnormal hormone levels, and the staining of antisera to all pituitary hormones is negative, then the tumor is designated as a non-secreting tumor.

Although the human pituitary tumor specimens we used included several different types of tumors, such as ACTH-secreting, GH-secreting, PRL-secreting, and non-secreting tumors, we discuss here only two types of tumors, namely, PRL-secreting and non-secreting tumors, because other types of tumors do not have the minimum number samples required for statistical analysis.

Prolactin-Secreting Pituitary Tumors.

Table 6 summarizes the clinical profiles of a group of PRL-secreting human pituitary tumors, and Table 7 summarizes the corresponding SP-li, ME, and BE measurements. The presence of a PRL-secreting tumor in a patient results in an increased level of PRL in the blood. Clinically, a serum PRL-li of 3-23 ng/ml is considered normal. The serum level of PRL-li in Table 7 ranges from 41 ng/ml to 880 ng/ml. This group contains microadenomas (less than 10 mm in diameter) and macroadenomas (more than 10 mm in diameter). There is no correlation between the tumor size and the serum PRL-li level in the patient ($p>0.05$).

In tumors DM and ES (Table 7), the concentration of SP-li is substantially lower than the control group (Table 5). There are two possible explanations for the decreased SP-li in the PRL-secreting tumor. If SP is localized in the human mammotroph as it is in the rat (see Figure 6), then SP may be part of a feedback mechanism. For example, when serum PRL decreases, SP increases to stimulate the production of PRL; when PRL increases, SP decreases so that less PRL will be produced. Alternatively, if SP is not localized in the human mammotroph, then we are merely comparing the average SP-li concentration in all cell types of the

Table 6. Diagnostic Data of PRL-secreting Human Pituitary Tumors
m, month; w, week; y, year.

ID	Age	Sex	Tumor Size (mm)	Serum PRL-li (ng/ml)	Diagnosis	Parlodel[1]
DM	69	M	30 x 20 x 20	472	PRL-secreting pituitary tumor Sepsis Staphylococcus aureus	No
ES	38	M	35 x 35 x 35	94	PRL-secreting pituitary tumor Acromegaly persistent Hypopituitarism	No
EM	30	M	6 x 5 x 5	n/a	PRL-secreting pituitary tumor	Yes 1 w prior
CR	36	F	15 x 10 x 7	103	PRL-secreting pituitary tumor Postoperative diabetes	Yes 2 m prior
DW	35	F	12 x 12 x 12	113	PRL-secreting pituitary tumor Goiter and nodules of thyroid	Yes >3 y prior
GE	36	F	10 x 9 x 7	197	PRL-secreting pituitary tumor Retrobulbar neuritis Multiple sclerosis	Yes 6 m prior
SE	22	F	25 x 25 x 25	n/a	PRL-secreting pituitary tumor	Yes >3 y prior
DA	29	F	14 x 17 x 18	880	PRL-secreting pituitary tumor Intratumoral hemorrhage	Yes 8 y prior
AP	22	F	7 x 7 x 7	271	PRL-secreting pituitary tumor Rathke's cleft cyst of pituitary	Yes 1 w prior
LV	40	F	3 x 4 x 4	109	PRL-secreting pituitary tumor Hypothyroidism	No
AJ	20	F	3 x 3 x 3	41	PRL-secreting pituitary tumor Hypothyroidism	No

[1] The length of time between the stopping of the Parlodel treatment and the surgery is noted.

Table 7. Quantitative Data of PRL-secreting Human Pituitary Tumors

Avg., mean ± s.e.m.; n.d., not detectable.

ID	SP-li (fmol/mg)	BE (pmol/mg)	ME (pmol/mg)
Male (Control)	99.4 ± 9.2	594 ± 121	76.6 ± 6.2
Male			
DM	9.6	18.7	71.1
ES	2.3	5.2	26.6
Male (On Parlodel)			
EM	20.7	666	2140
Female (On Parlodel)			
CR	89.0	1530	1740
DW	15.3	57.3	682
GE	9.6	97.4	747
SE	6.3	63.5	612
DA	5.8	67.4	377
AP	22.0	885	2040
Avg.	24.7 ± 13.1	450 ± 253	1030 ± 278
Female (Hypothyroidism)			
LV	220	507	n.d.
AJ	294	598	n.d.

anterior pituitary to that in a single cell type (mammotroph) in which SP does not localize preferentially.

Also in tumors DM and ES (Table 7), the concentration of BE is lower than in the control group; that difference is consistent with previous reports that indicated that BE production was decreased in human pituitary tumors (Desiderio, et al., 1993).

In an evaluation of the patients' medical history (Table 6) of those PRL-secreting tumors, a drug named Parlodel (chemical name: bromocriptine mesylate) may have caused an overproduction of ME and BE, because most of the patients in this group had received Parlodel treatment. Whenever a patient has galactorrhea, which is a symptom of PRL-secreting pituitary tumors, Parlodel is prescribed. Especially in microadenomas, Parlodel treatment is preferred to the surgical removal of the tumor. For macroadenomas, transsphenoidal surgery is preferred. Apparently ME and BE were much increased in patients recently treated with Parlodel. In the male group, patient EM, who had recently received Parlodel treatment, had much higher concentrations of ME and BE in the tumor than patients DM and ES, who never received Parlodel treatment. Similarly, in female group, patients who received Parlodel treatment near surgery (CR and AP) had higher concentrations of ME and BE in their tumors than those patients who received Parlodel treatment a longer time ago (DW, GE, SE, and DA).

Parlodel is an ergot derivative with potent dopamine receptor agonist activity, and inhibits PRL secretion by activating post-synaptic dopamine receptors (D2) (O'Halloran, et al., 1991). In rat experiments, Parlodel treatment was shown to increase the ME concentration in the anterior pituitary (George and Kertesz, 1985; Papas, et al., 1988). For SP, the chronic treatment of Parlodel increased SP mRNA but not SP-li in the rat anterior pituitary (O'Halloran, et al., 1991). The chronic treatment of human subjects with Parlodel was also reported to increase plasma BE-li (Alfoldi, et al., 1991). Our data, which indicate that patients who had Parlodel treatment showed increased concentrations of ME, BE, and SP-li, are consistent with these reports. There have been reports that Parlodel not only relieved the symptoms caused by high levels of PRL, but that it also shrank the tumors (Bevan, et al., 1992; Zgliczynski, et al., 1992a, 1992b). One possible explanation of the tumor-shrinking effects of Parlodel is that Parlodel may increase the level of ME and BE in the anterior pituitary. ME or BE, or both, acts upon certain growth-related opioid receptors. One such receptor, the zeta-receptor, has been described (Zagon, et al., 1991, 1992, 1993) and has been found to be important in early development and tumorigenesis. Our data demonstrated an association between Parlodel treatment and the ME and BE concentrations in the PRL-secreting tumors. For example, EM, CR, and AP received a recent treatment of Parlodel (within two months before the transsphenoidal surgery), and all showed the highest concentrations of ME and BE.

Two female PRL-secreting tumor samples (LV and AJ) were considered separately from other female PRL-secreting tumors because they were obtained from patients with hypothyroidism (Table 6). It was reported that thyroid hormones influenced the ME concentration in the pituitary (Figure 6). For example, in PTU-induced hypothyroidism, the ME-li decreased dramatically in the rat anterior pituitary (Tang and Man, 1988). Convincing reports exist on the regulation of SP content in the anterior pituitary by the thyroid status. For example, a thyroid hormone regulation of SP-li in the rat anterior pituitary was reported: hypothyroidism increased and thyroxin (T4)-induced hyperthyroidism decreased the SP-li concentration (Aronin, et al., 1984). This finding has been confirmed by others (Jonassen, et al., 1987; Jones, et al., 1989). The SP-li increase induced by hypothyroidism was accompanied by an increase in PPT mRNA (Jonassen, et al., 1987). The effects of the thyroid status on SP-li levels in the anterior pituitary were independent of a direct link from the hypothalamus, because they were also demonstrated in anterior pituitaries that were implanted under the renal capsule of rats (Aronin, et al., 1988). Different from ME and SP, BE levels in the anterior pituitary did not change with the thyroid status. For example, in the PTU-induced hypothyroidism, BE-li in the rat anterior pituitary did not change (Tang and Man, 1988). In tumors LV and AJ, ME was non-detectable by the quantification methods in this research program, implying that the pituitary ME concentration in these two tumors was at least one thousand-times lower than the controls, because our detection sensitivity of ME by MS was approximately hundreds of femtomoles (Tolun, et al., 1987). Also, in patients LV and AJ, SP-li increased substantially compared to that in other female patients. Therefore, our data on the peptide concentrations in human hypothyroidism are consistent with those reported by others.

Non-Secreting Pituitary Tumors.

Tables 8 and 9 contain the diagnostic and neuropeptide quantification data, respectively, of a group of non-secreting human pituitary tumors. In this group of tumors (Table 9), the amount of SP-li and BE in four tumors from males decreased significantly compared to the control group ($p < 0.05$). The present observation

Table 8. Diagnostic Data of Non-secreting Human Pituitary Tumors
l.s./s.s., large sellar or suprasellar.

I.D.	Age	Sex	Tumor Size (mm)	Diagnosis
MP	46	F	l.s./s.s.	Non-secreting pituitary tumor
				Confusion
				Arthritis (on medication)
GW	60	M	30 x 25 x 20	Non-secreting pituitary tumor
				Confusion
				Arthritis (on medication)
				Seizure
				Cardiovascular diseases
				others
EH	77	F	30 x 30 x 30	Non-secreting pituitary tumor
JW	49	M	25 x 25 x 25	Non-secreting pituitary tumor
				Severe depression
				On Co-Dependency/Drug &
				Alcohol Treatment program
KD	46	F	18 x 18 x 18	Non-secreting pituitary tumor
				Severe depression
				Schizo-affective disorder
MC	67	M	15 x 15 x 15	Non-secreting pituitary tumor
				Panhypopituitarism
TW	76	M	30 x 40 x 35	Non-secreting pituitary tumor
				Hypopituitarism
HW	79	F	l.s./s.s.	Non-secreting pituitary tumor
				Cerebrovascular infarction
TS	38	M	30 x 20 x 20	Non-secreting pituitary tumor
				Aseptic meningitis
LB	51	F	30 x 30 x 30	Non-secreting pituitary tumor
				Hypertension
DG	35	F	30 x 30 x 30	Non-secreting pituitary tumor
EL	50	M	25 x 25 x 25	Non-secreting pituitary tumor
JT	64	F	30 x 15 x 20	Non-secreting pituitary tumor
				Panhypopituitarism
GP	37	F	14 x 14 x 14	Non-secreting pituitary tumor

Table 9. Quantitative Data of Non-secreting Human Pituitary Tumors

ID	SP-li (fmol/mg)	BE (pmol/mg)	ME (pmol/mg)
Male (Control)	99.4 ± 9.2	594 ± 121	76.6 ± 6.2
Male			
MC	142	16.9	124
TW	36.9	18.1	53.1
TS	4.9	13.8	99.5
EL	2.4	10.0	36.6
Avg.	46.4 ± 32.7	14.7 ± 1.8	78.3 ± 20.2
Male (On arthritis medication)			
GW	9.6	95.2	1060.4
Male (Severe depression)			
JW	439	351	69.9
Female			
EH	9.0	25.6	173
HW	12.1	249	181
LB	3.5	9.6	40.0
DG	1.7	23.5	20.6
JT	5.1	61.6	24.0
GP	11.0	467	112
Avg.	7.1 ± 1.7	139 ± 74.9	91.8 ± 30.2
Female (On arthritis medication)			
MP	20.1	272	1320
Female (Severe depression)			
KD	333	35.8	87.5

that non-secreting pituitary tumors contain lower BE is consistent with our previous publication (Desiderio, et al., 1993). The SP-li data are new to this research program. On the other hand, ME did not show any change from the controls, but in that earlier publication a decreased ME was noted. This phenomenon implied that an imbalance occurred between the POMC and proenkephalin systems in these tumors, and that there was a more substantial loss of POMC peptides. This decrease may be due to a down-regulation of the synthesis, an increased level of degradation of the POMC neuropeptides, or both factors in these tumor samples.

In non-secreting tumors GW and MP (Table 9), the much-elevated ME level may be related to the medication given for arthritis in those two patients. Generally, drugs for arthritis treatment have analgesic and antiinflammatory effects. There have been no reports on how these drugs affect the ME levels in the anterior pituitary. These tumors also showed decreased SP-li and BE concentrations.

In non-secreting tumors JW and KD (Table 9), the SP-li concentration increased substantially. Both of these tumors were obtained from patients with severe depression. One of them (KD) also had a schizo-affective disorder. There has been no reported relationship of SP in the pituitary to any mental diseases such as severe depression, although there have been some reports that showed an increased SP-li in certain brain areas in patients with mental disorders (Malek-Ahmadi, 1992; Takeuchi, et al., 1988). Our limited data indicate a relationship between severe depression and increased SP-li levels in the anterior pituitary. Similar to other non-secreting tumors, these two tumors also had a decreased BE concentration but unchanged ME concentration.

Taken together, our data demonstrated a significantly decreased BE concentration in human pituitary tumors, except in those PRL-secreting tumors from patients who were recently treated by Parlodel. Our data also demonstrated a significantly decreased SP-li concentration in these male tumors except one tumor obtained from hypothyroidism and one tumor obtained from severe depression patients. The ME concentration, however, did not show any significantly changes in most of the tumors, but increased in the tumors from the patients who were on Parlodel or arthritis medications. These data collectively indicate an association of the tachykinin and POMC systems with human anterior pituitary tumors. The proenkephalin system may also be associated with the hormone-secreting tumors, although more samples are needed to be able to test for statistical significance.

Conclusions.

This chapter reviews the strategy for extracting, purifying, and analyzing neuropeptides in human pituitary tissues. The analysis of two opioid neuropeptides and one tachykinin neuropeptide in human pituitary tumors versus pituitary controls is used as an example to describe each step that is involved in that strategy. It has been shown that neuropeptides are relatively stable from 8 to 72 hr post-mortem, and especially from 12 to 18 hr, which is the time-frame when most post-mortem human tissues are acquired for research. The data described in this chapter also demonstrated that peptidase inhibitors are not necessary as long as the tissue is homogenized in acidic conditions and the initial processing is done promptly and at a low temperature (4 °C). SPE combined with RP-HPLC is a very efficient way to purify neuropeptides extracted from tissues. RIA is a widely-used quantitative technique for neuropeptides, and it possesses a high sensitivity (femtomole level). However, RIA does not elucidate the amino acid sequence of the target neuropeptide. MS, especially MS/MS, greatly increases molecular specificity. It not only provides a way of confirming the identity of the target neuropeptide by demonstrating the existence of amino acid sequence-determining fragment ions (product-ion spectra) before the quantification, but it relates the quantification process directly to the sequence information of the target neuropeptide (MRM mode). The use of a deuterated peptide for the quantitative analysis by MS provides an accurate measurement of the real tissue concentration of a neuropeptide, because it compensates for any loss of the target neuropeptide during extraction and purification processes.

Acknowledgments.

The authors acknowledge the financial support of the U.S. National Institutes of Health (GM 26666), and each scientific collaborator who contributed to the research reviewed.

References

Alfoldi, A., Simko, K., Fekete, M. I., and Farsang, C. (1991). Adrenergic and dopaminergic regulation of circulating beta-endorphin-like immunoreactivity in hypertension. Acta Physiol. Hung. 78: 99-106.

Arai, H., Emson, P. C., and Carrasco, L. H. (1987). Huntington's disease: changes in tachykinin content in postmortem brains. Ann. Neurol. 22: 587-594.

Aronin, N., Coslovsky, R., and Chase, K. (1988). Hypothyroidism increases substance P concentrations in the heterotopic anterior pituitary. Endocrinol. 122: 2911-2914.

Aronin, N., Morency, K., Leeman, S. E., Braverman, L. E., and Coslovsky, R. (1984). Regulation by thyroid hormone of the concentration of substance P in the rat anterior pituitary. Endocrinol. 114: 2138-2142.

Asa, S. L., and Kovacs, K. (1992). Pituitary pathology in acromegaly. Endocrinol. Metab. Clin. N. Amer. 21: 553-574.

Asa, S. L., Horvath, E., Kovacs, K., and Ezrin, C. (1984). Cytology of the Normal Pituitary and Pituitary Tumors. In Pituitary Tumors. Odell, W. D., and Nelson, D. H. (Eds.); Futura. Mount Kisco, New York. pp. 71-143.

Beal, M. F., and Martin, J. B. (1986). Neuropeptides in neurological disease. Ann. Neurol. 20: 547-565.

Beal, M. F., Mazurek, M. F., Lorenz, L. J., Chattha, G. K., Ellison, D. W., and Martin, J. B. (1986). An examination of neuropeptide Y postmortem stability in an animal model simulating human autopsy conditions. Neurosci. Lett. 64: 69-74.

Bevan, J. S., Webster, J., Burke, C. W., and Scanlon, M. F. (1992). Dopamine agonists and pituitary tumor shrinkage. Endocrine Rev. 13: 220-240.

Biemann, K. (1990). Nomenclature for peptide fragment ions (positive ions) in mass spectrometry. Meth. Enzymol. 193: 886-887.

Biggins, J. A., McDermott, J. R., Smith, A. I., Candy, J. M., Perry, R. H., Bateman, D. E., and Edwardson, J. A. (1984). Chromatographic characterization of neuropeptides in post mortem human brain. J. Neurol. Sci. 66: 47-58.

Bloom, F., Battenberg, E., Rossier, J., Ling, N., Leppaluoto, J., Vargo, T. M., and Guillemin, R. (1977). Endorphins are located in the intermediate and anterior lobes of the pituitary gland, not in the neurohypophysis. Life Sci. 20: 43-48.

Causon, R. C., and McDowall, R. D. (1992). Sample pretreatment techniques for the bioanalysis of peptides. J. Contr. Rel. 21: 37-48.

Chang, K. J., and Cuatrecasas, P. (1983). Receptors and second messengers. In Brain Peptides. Krieger, D. T. (Ed.); Wiley-Interscience. New York. pp. 565-593.

Clement-Jones, V., and Besser, G. M. (1983). Clinical perspective in opioid peptides. Br. Med. Bull. 39: 95-100.

Dass, C., and Desiderio, D. M. (1988). Particle beam induced interaction between peptides and liquid matrices. Anal. Chem. 60: 2723-2729.

Dass, C., Kusmierz, J. J., and Desiderio, D. M. (1991). Mass spectrometric quantification of endogenous ß-endorphin. Biol. Mass Spectrom. 20: 130-138.

DePalatis, L. R., Fiorindo, R. P., and Ho, R. H. (1982). Substance P immunoreactivity in the anterior pituitary gland of the guinea pig. Endocrinol. 110: 282-284.

Desiderio, D. M. (1984). Analysis of neuropeptides by liquid chromatography and mass spectrometry. Elsevier. New York. 235 pp.

Desiderio, D. M. (1991). Mass spectrometry of biologically important neuropeptides. In Mass Spectrometry of Peptides. Desiderio, D. M. (Ed.); CRC Press. Boca Raton, FL. pp. 367-400.

Desiderio, D. M. (1996). Mass spectrometric quantification of neuropeptides. In Methods in Molecular Biology, Vol. 61. Chapman, J. R. (Ed.); Humana Press, Totowa, N.J. pp. 57-65.

Desiderio, D. M., Kusmierz, J. J., Zhu, X., Dass, C., Hilton, D., Robertson, J. T., and Sacks, H. S. (1993). Mass spectrometric analysis of opioid and tachykinin neuropeptides in non-secreting and ACTH-secreting human pituitary adenomas. Biol. Mass Spectrom.22: 89-97.

Duka, T., Höllt, V., Przewlocki, R., and Wesche, D. (1978). Distribution of methionine- and leucine-enkephalin within the rat pituitary gland measured by highly specific radioimmunoassays. Biochem. Biophys. Res. Commun. 85: 1119-1127.

Dupont, A., Barden, N., Cusan, L., Merand, Y., Labrie, F., and Vaudry, H. (1980). ß-endorphin and met-enkephalins: their distribution, modulation by estrogens and haloperidol, and role in neuroendocrine control. Fed. Proc. 39: 2544-2550.

Emson, P. C., Arregui, A., Clement-Jones, V., Sandberg, B. E. B., and Rossor, M. (1980). Regional distribution of methionine-enkephalin and substance P-like immunoreactivity in normal human brain and in Huntington's disease. Brain Res. 147: 147-160.

Frommer, W. (1979). J. Med. Plant Res. 35: 195.

Geary, L. E., Wiley, K. S., Scott, W. L., and Cohen, M. L. (1982). Degradation of exogenous enkephalin in the guinea-pig ileum: relative importance of aminopeptidase, enkephalinase and angiotensin converting enzyme activity. J. Pharmacol. Exp. Ther. 221(1): 104-111.

George, S. R., and Kertesz, M. (1985). Dopamine receptors regulate met-enkephalin content in pituitary. Brain Res. 334: 187-189.

Gros, C., Pradelles, P., Rouget, C., Bepoldin, O., Dray, F., Fournie-Zaluski, M. C., Roques, B. P., Pollard, H., Llorens-Cortes, C., and Schwartz, J. C. (1978). Radioimmunoassay of methionine- and leucine-enkephalins in regions of rat brain andcomparison with endorphins estimated by a radioreceptorassay. J. Neurochem. 31: 29-45.

Grossman, A., and Rees, L. H. (1983). The neuroendocrinology of opioid peptides. Br. Med. Bull. 39: 83-88.

Harmar, A. J., Pierotti, A. R., Armstrong, A., Chapman, K., Chahl, J. S., Going, J.,Edwards, C. R. W., Rosie, R., and Pascall, J. (1986). Structure and post-translational processing of the preprotachykinins. Substance P and Neurokinins, 21-23.

Herzog, T., and Schlote, W. (1992). Pituitary adenomas: tumor cells growth by cluster formation. Pattern analysis based on immunohistochemistry. Acta Neuropathol. 84: 509-515.

Higa, T., and Desiderio, D. M. (1989). Optimizing recovery of peptides from an octadecylsilyl (ODS) cartridge. Int. J. Peptide Protein Res. 33: 250-255.

Hughes, J., Smith, T. W., Kosterlitz, H. W., Fothergill, L. H., Morgan, B. A., and Morris, H. (1975). Identification of two related pentapeptides from the brain with potent opiate agonist activity. Nature 258: 577-579.

Igwe, O. J., Felice, L. J., Seybold, V. S., and Larson, A. A. (1988). Optimization of high-performance liquid chromatography-radioimmunoassay protocols for the analysis of substance P and some of its metabolic fragments. J. Chromatogr. 432: 113-126.

Jonassen, J. A., Mulliken-Kilpatrick, D., McAdam, A., and Leeman, S. E. (1987). Thyroid hormone status regulates preprotachykinin: a gene expression in male rat anterior pituitary. Endocrinol. 121: 1555-1561.

Jones, P. M., Ghatei, M. A., Steel, J., O'Halloran, D., Gon, G., Legon, S., Burrin, J. M., Leonhardt, U., Polak, J. M., and R, B. S. (1989). Evidence for neuropeptide Y synthesis in the rat anterior pituitary and the influence of thyroid hormone status: comparison with vasoactive intestinal peptide, substance P, and neurotensin. Endocrinol. 125: 334-341.

Ju, G., and Liu, S. (1989). Substance P-immunoreactive nerve fibers in the pars distalis of the anterior pituitary of macaques. J. Chem. Neuroana. 2: 349-360.

Ju, G., and Liu, S. (1990). Substance P-like immunoreactive nerve fibers in the pars distalis of the anterior pituitary in the dog. Cell Tissue Res. 261: 323-331.

Ju, G., and Zhang, X. (1990). An electron microscopic study on substance P-like immunoreactive nerve fibers in the pars distalis of the adenohypophysis in the dog. Neurosci. 38: 503-513.

Ju, G., Liu, S., and Zhang, X. (1991). Peptidergic innervation of the pars distalis of the adenohypophysis. Neuroendocrinol. 53: 41-44.

Kalra, P. S., Sahu, A., Bonavera, J. J., and Kalra, S. P. (1992). Diverse effects of tachykinins on luteinizing hormone release in male rats: mechanism of action. Endocrinol. 131: 1195-1201.

Kaneko, T., Wood, G., Crouch, W. L., and Desiderio, D. M. (1993). Assay of preproenkephalin A processing enzymes in human cerebrospinal fluid. Anal. Biochem.215: 45-53.

Kato, Y., Chihara, K., Ohgo, S., Iwasaki, Y., Abe, H., and Imura, H. (1976). Growth hormone and prolactin release by substance P in rats. Life. Sci. 19: 441-446.

Klibanski, A. (1987). Nonsecreting pituitary tumors. Endocrinol. Metabol. Clin. 16: 793-804.

Kotani, H., Hoshimaru, M., Nawa, H., and Nakanishi, S. (1986). Structure and gene organization of bovine neuromedin K precursor. Proc. Natl. Acad. Sci. USA 83: 7074-7078.

Kovacs, K., and Horvath, E. (1987). Pathology of pituitary tumors. Endocrinol. Met. Clin. 16: 529-551.

Krause, J. E., MacDonald, M. R., and Takeda, Y. (1989). The polyprotein nature of substance P precursors. BioEssays 10: 62-69.

Kusmierz, J. J., and Desiderio, D. M. (1992). Characterization of an aminopeptidase in cerebrospinal fluid. J. Chromatogr. 574: 189-196.

Kusmierz, J. J., Sumrada, R., and Desiderio, D. M. (1990). Fast atom bombardment mass spectrometric quantitative analysis of methionine-enkephalin in human pituitary tissues. Anal. Chem. 62: 2395-2400.

Larsen, P. J., Jessop, D., Patel, H., Lightman, S. L., and Chowdrey, H. S. (1993). Substance P inhibits the release of anterior pituitary adrenocorticotrophin via a central mechanism involving corticotrophin-releasing factor-containing neurons in the hypothalamic paraventricular nucleus. J. Neuroendocrinol. 5: 99-105.

Leadem, C. A., and Kalra, S. P. (1985). Effects of endogenous opioid peptides and opiates on luteinizing hormone and prolactin secretion in ovariectomized rats. Neuroendocrinol. 41: 342-352.

Liu, S. J., and Ju, G. (1988). Substance P-like immunoreactive nerve fibers in the pars distalis of the adenohypophysis of macaque monkeys. Neurosci. Lett. 94: 1-4.

Loh, Y. P. (1992). Molecular mechanisms of ß-endorphin biosynthesis. Biochem.Pharmacol. 44: 843-849.

Maggio, J. E. (1988). Tachykinins. Annu. Rev. Neurosci. 11: 13-28.

Malek-Ahmadi, P. (1992). Substance P and neuropsychiatric disorders: an overview. Neurosci. Biobehav. Rev. 16: 365-369.

McGregor, G. P., and Bloom, S. R. (1983). Radioimmunoassay of substance P and its stability in tissue. Life Sci. 32: 655-662.

Melmed, S., Braunstein, G. D., and Horvath, E. (1983). Pathophysiology of acromegaly. Endocrinol. Rev. 4: 271-290.

Mikkelsen, J. D., Larsen, P. J., Møller, M., Vilhardt, H., and Sœrmark, T. (1989). Substance P in the median eminence and pituitary of the rat: Demonstration of immunoreactive fibers and specific binding sites. Neuroendocrinol. 50: 100-108.

Millan, M. J., and Herz, A. (1985). The endocrinology of the opioids. Int. Rev. Neurobiol. 26: 1-83.

Molitch, M. E. (1987). Pathogenesis of pituitary tumors. Endocrinol. Metabol. Clin. 16: 503-527.

Morel, G., Chayvialle, J. A., Kerdelhue, B., and Dubois, P. M. (1982). Ultrastructural evidence for endogenous substance-P-like immunoreactivity in the rat pituitary gland. Neuroendocrinol. 35: 86-92.

Numa, S. (1984). Opioid Peptide Precursors and Their Genes. In The Peptides--Analysis, Synthesis, Biology, Vol. 6. Udenfriend, S., and Meienhofer, J. (Eds.); Academic Press, Inc. Orlando, FL. pp. 1-23.

O'Halloran, D. J., Jones, P. M., Ghatei, M. A., and Bloom, S. R. (1991). Rat anterior pituitary neuropeptides following chronic prolactin manipulation: a combined radioimmunoassay and mRNA study. J. Endocrinol. 131: 411-419.

Papas, S., Shin, S. H., and Obonsawin, M. C. (1988). Biphasic changes in anterior pituitary met-enkephalin concentration following reserpine treatment. Neuroendocrinol. 47: 149-153.

Peillon, F., LeDafniet, M., Pagesy, P., Li, J. Y., Benlot, C., Brandi, A. M., and Joubert, D. (1991). Neuropeptides of anterior pituitary origin. Path. Res. Pract. 187: 577-580.

Perry, E. K., and Perry, R. H. (1983). Human brain neurochemistry - some postmortem problems. Life Sci. 33: 1733-1743.

Picanco-Diniz, D. L. W., Valenca, M. M., Franci, C. R., and Antunes-Rodrigues, J. (1990). Role of substance P in the medial preoptic area in the regulation of gonadotropin and prolactin secretion in normal or orchidectomized rats. Neuroendocrinol. 51: 675-682.

Rivier, C., Brown, M., and Vale, W. (1977). Effect of neurotensin, substance P and morphine sulfate on the secretion of prolactin and growth hormone in the rat. Endocrinol.100: 751-754.

Rivier, C., Brown, M., and Vale, W. (1977). Effect of neurotensin, substance P and morphine sulfate on the secretion of prolactin and growth hormone in the rat. Endocrinol.100: 751-754.

Roepstorff, P., and Fohlman, J. (1988). Proposal for a common nomenclature for sequence ions in mass spectra of peptides. Biomed. Mass Spectrom. 11: 601.

Roth, K. A., and Krause, J. E. (1990). Substance-P is present in a subset of thyrotrophs in the human pituitary. J. Clin. Endocrinol. 71: 1089-1095.

Spokes, E. G. S., and Koch, D. J. (1978). Post-mortem stability of dopamine, glutamine decarboxylase and choline acetyltransferase in the mouse brain under conditions simulating the handling of human autopsy material. J. Neurochem. 31: 381-383.

Takeuchi, K., Uematsu, M., Ofuji, M., Morikiyo, M., and Kaiya, H. (1988). Substance P involved in mental disorders. Prog. Neuro-Psychopharmacol. & Bio. Psychiat. 12: S157-S164.

Tang, F., and Man, S. Y. (1988). Hypothyroidism and pituitary contents of immunoactive met-enkephalin and ß-endorphin in male rats of different ages. Proc. Soc. Exp. Biol. Med. 188: 82-86.

Thorsett, E. D., and Wyvratt, M. J. (1987). Inhibition of zinc peptidases that hydrolyse neuropeptides. In Neuropeptides and Their Peptidases. Turner, A. J. (Ed.); Ellis Horwood. Leeds. pp. 229-292.

Tolun, E., Dass, C., and Desiderio, D. M. (1987). Trace level measurement of enkephalin peptides at the attomole/femtomole level by FAB-MS. Rapid Commun. Mass Spectrom. 1: 77-79.

Tramu, G., and Leonardelli, J. (1979). Immunohistochemical localization of enkephalins in median eminence and adenohypophysis. Brain Res. 168: 457-471.

Turner, A. J., Matsas, R., and Kenny, A. J. (1985). Are there neuropeptide-specific peptidases? Biochem. Pharmacol. 34(9): 1347-1356.

Udenfriend, S., and Kilpatrick, D. L. (1984). Proenkephalin and the Products of Its Processing: Chemistry and Biology. In The Peptides--Analysis, Synthesis, Biology, Vol. 6. Udenfriend, S., and Meienhofer, J. (Eds.); Academic Press, Inc. Orlando, FL. pp. 26-68.

Umezawa, H. (1976). Structure and activities of protease inhibitors of microbial origin. Meth. Enzymol. 45: 678-679.

Vijayan, E., and McCann, S. M. (1979). In vivo and in vitro effects of substance P and neurotensin on gonadotropin and prolactin release. Endocrinol. 105: 64-68.

Wagner, J. J., and Chavkin, C. I. (1995). Neuropharmacology of endogenous opioid peptides. In Psychopharmacology: the Fourth Generation of Progress. Bloom, F. E., and Kupfer, D. J. (Eds.); Raven Press. New York. pp. 579-529.

Watson, S. J., Akil, H., Richards, C. W., and Barchas, J. D. (1978). Evidence for two separate opiate peptide neuronal systems. Nature 275: 226-228.

Weber, E., Voigt, K. H., and Martin, R. (1978). Pituitary somatotrophs contain [Met]enkephalin-like immunoreactivity. Proc. Natl. Acad. Sci. USA 75: 6134-6138.

Yalow, R. S. (1980). Radioimmunoassay. Ann. Rev. Biophys. Bioeng. 9: 327-350.

Yan, L., Fridland, G., Tseng, J., and Desiderio, D. M. (1993). Evidence of an intermediate-sized beta-endorphin-containing presursor molecule in bovine pituitary by mass spectrometry. Biochem. Biophys. Res. Commun. 196: 521-526.

Yan, L., Zhu, X., Tseng, J., and Desiderio, D. M. (1996). Beta-endorphin-containing proteins in human pituitary. Peptide. Submitted.

Zagon, I. S., Gibo, D. M., and McLaghlin, P. J. (1991). Zeta (z), a growth-related opioid receptor in developing rat cerebellum: identification and characterization. Brain Res. 551: 28-35.

Zagon, I. S., Gibo, D. M., and McLaughlin, P. J. (1992). Ontogeny of zeta (z), the opioid growth factor receptor, in the rat brain. Brain Res. 596: 149-156.

Zagon, I. S., Goodman, S. R., and McLaughlin, P. J. (1993). Zeta (z), the opioid growth receptor: identification and characterization of binding subunits. Brain Res. 605: 50-56.

Zgliczynski, W., Zgliczynski, S., Makowska, A., Slowinska-Srzednicka, J., Janik, J., Jeske, W., and Walecki, J. (1992a). New long-acting bromocriptine (Parlodel MR and Parlodel LAR) in the treatment of pituitary tumours with hyperprolactinomia. Endokrynologia Polska 43: 234-241.

Zgliczynski, W., Zgliczynski, S., Makowska, A., Slowinska-Srzednicka, J., Janik, J., Jeske, W., and Walecki, J. (1992b). Mixed pituitary tumours-effects of bromocriptine treatment: Parlodel MR and Parlodel LAR. Endokrynologia Polska 43: 242-249.

Zhu, X. (1993). Characterization of methionine enkephalin, substance P, and beta-endorphin in human pituitary controls and tumors. The University of Tennessee, Memphis. Ph.D. Dissertation.

Zhu, X., and Desiderio, D. M. (1993). Methionine enkephalin-like immunoreactivity, substance P-like immunoreactivity and ß-endorphin-like immunoreactivity post-mortem stability in rat pituitary. J. Chromatogr. 616: 175-187.

Zhu, X., and Desiderio, D. M. (1994). Peptidase inhibitors are not necessary in the analysis of neuropeptides. Anal. Lett. 27: 2267-2280.

Zhu, X., Robertson, J. T., Sacks, H. S., Dohan, F. C., Tseng, J., and Desiderio, D. M. (1995). Opioid and tachykinin neuropeptides in prolactin-secreting pituitary adenomas. Peptides 16: 1097-1107.

Progress in HPLC-HPCE, Vol. 7, pp. 781-816
G. Ali Qureshi *et al.* (Eds)

STUDIES ON NEURONAL APOPTOSIS USING A COMET ASSAY AND CULTURED CELLS

WAKAKO MARUYAMA[1] and MAKOTO NAOI[2*]

[1]*Laboratory of Biochemistry and Metabolism, Department of Basic Gerontology, National Institute for Longevity Sciences, Obu 474, Japan, and*
[2]*Department of Biosciences, Nagoya Institute of Technology, Nagoya 466, Japan,*

I. INTRODUCTION

Necrosis and apoptosis are common types of neuronal cell death based on morphological observation. Necrosis is induced by cell injury or destruction of cell homeostasis and in acute pathological process the cells swell and lyse and the intracellular contents spill into the extracellular space. The necrosis is followed by an inflammatory tissue reaction with appearance of macrophages and microglia in the brain. On the other hand, apoptosis is morphologically characterized by shrinking and fragmentation of nucleus and cytoplasm. The cells are rapidly phagocyted by the neighbors or macrophages or microglia, and the contents of the cells don't leak and there is no inflammation reaction. Apoptosis is observed in "normal" cell death and is induced by activation of a suicide cell death program. The programmed cell death which is morphologically defined as apoptosis controls shape and size of living bodies in normal and pathological processes (Kerr *et al.*, 1972; Martz and Howell, 1989; Oppenheim, 1991; Williams, 1991; Johnson and Deckwerth, 1993). It is involved in keeping homeostasis by taking a balance between cell replication and cell death (Raff, 1992; Raff *et al.*, 1993), and it requires synthesis of protein and mRNA for the process of cell death, so apoptosis has been neglected as a death process of postmitotic neurons in adult human brains. However, programmed cell death is now confirmed to be induced not only in the normal evolution and development of the neurons (Chu-Wang and Oppenheim, 1978; Clark, 1990; Cunningham, 1982), but also in some neurodegenerative diseases, even in genetic diseases (Busciglio and Yankner, 1995). Programmed cell death is associated with endogenous endonuclease activity (McCall and Cohen, 1991) and commonly detected by analysis of extracted genomic DNA and the appearance of ladder of nucleosomal DNA in agarose gel. This method cannot distinct individual cells undergoing apoptosis. Apoptosis can be detected by histological observation also. Typical morphological features of apoptosis are nuclear chromatin condensation, compactness of cytoplasmic organelles and appearance of pedunculated protuberances on the cell surface (Wyllie *et* , 1984). However, apoptotic cell appearance is limited to only a few minutes, which makes it difficult to distinguish cells undergoing apoptosis (Kerr *et al.*, 1987). A method of *in situ* labeling of DNA breaks in nuclei is commonly applied to tissue sections. This method is based on the specific binding of terminal deoxynucleotidyl transferase (TdT) to 3'-OH ends of DNA, ensuing a synthesis of nuclear polydeoxynucleotide polymers (Cavrieli *et al.*,

1992). After proteolysis, TdT is used to incorporate biotinylated deoxyuridine at sites of DNA breaks. The signal was amplified by avidin-peroxidase to be detected by light microscope.

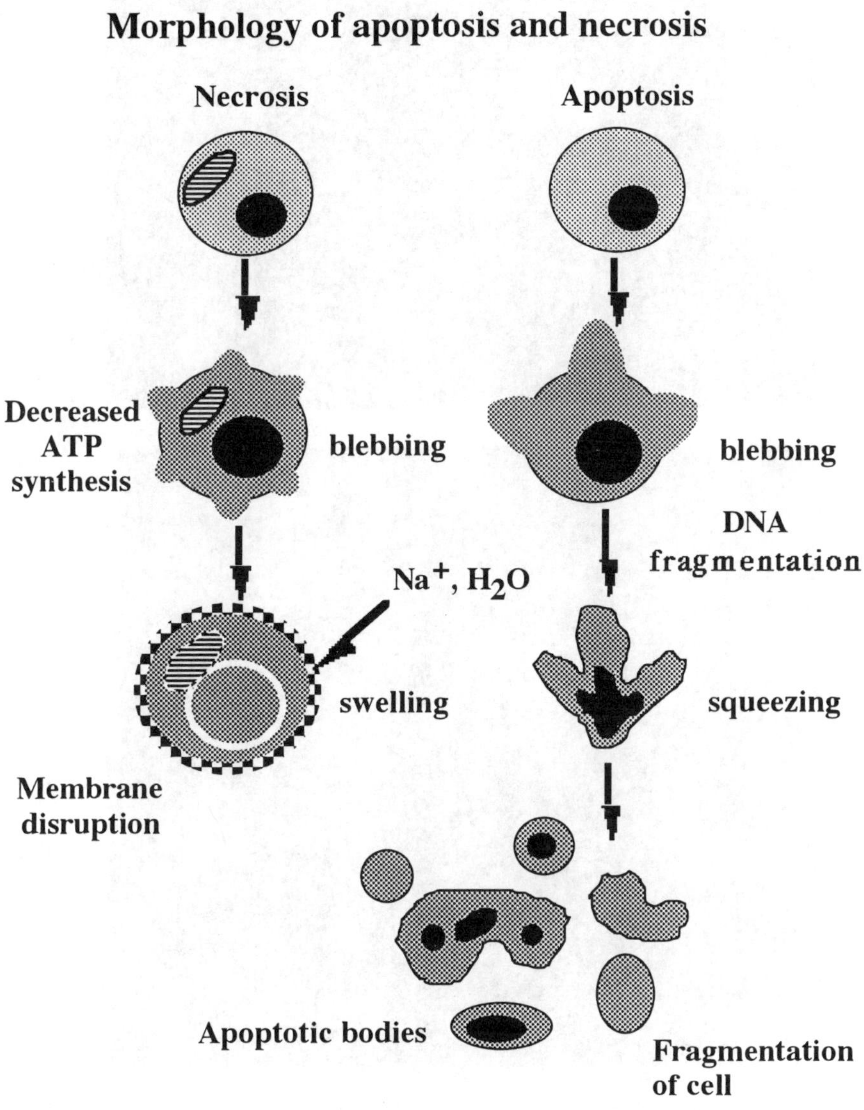

Figure 1. Two mechanisms of cell death; apoptosis and necrosis.

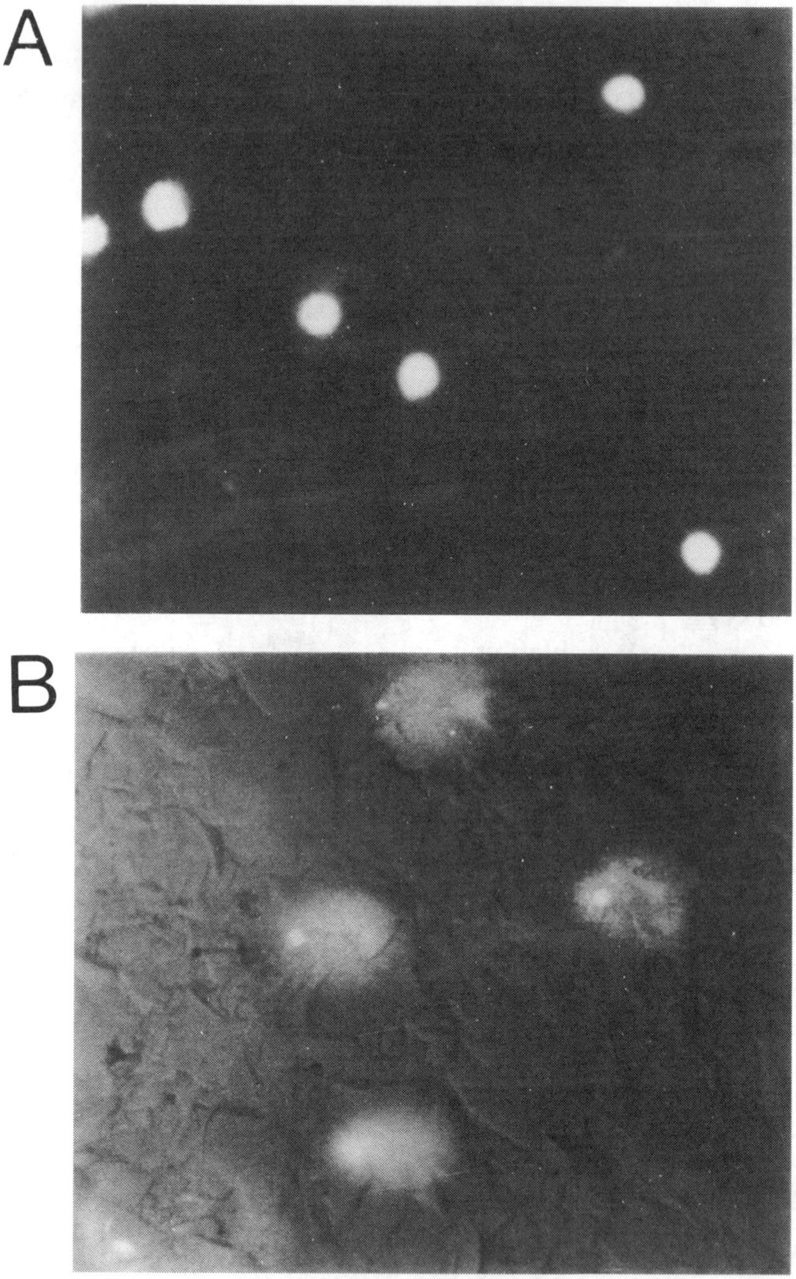

Figure 2. SH-SY5Y cells incubated with 1 mM of *N*-methyl(*R*)salsolinol for 3 hours (B) or control cells (A) were embedded in agarose and lysed by alkaline solution, then applied to electrophoresis. Cells treated with *N*-methyl(*R*)salsolinol shows typical comet image with DNA head and tail.

The TdT-mediated dUTP-biotin nick end labeling (TUNEL) is now generally applied to detect apoptotic DNA damage. These methods are useful for morphological analyses of apoptosis in tissues and cultured cells, however, they are not applicable for quantitative analysis of the cell death process.

As a quantitative assay of DNA breaking, a single cell gel electrophoresis assay (comet assay) has been used in the fields of cancer research (*see reviews*: McKelvey-Martin *et al.*, 1993; Fairbairn *et al.*, 1995). In the comet assay, a small number of cells are embedded in agarose, lysed, subjected to electrophoresis and visualized by staining with a fluorescent DNA-binding dye. When cells contain DNA strand breaks, the electric current pulls the negatively charged broken and relaxed DNA from nucleus toward the anode. In undamaged cells, most of the supercoiled DNA remains within the nuclear cage, whereas a damaged cell shows a "comet" image with a fluorescent head and a tail contained fragments and relaxed supercoils of DNA (Figure 1). It is an adaptation of previous methods, such as nucleoside sedimentation (Cook and Brazell, 1976), and the halo assay (Roti Roti and Wright, 1987). This comet assay has been applied to examine the DNA damage induced by oxidative stress (Anderson *et al.*, 1994; Kruszewski *et al.*, 1994), mutagens (Hartmann *et al.*, 1995), and radiation (Olive *et al.*, 1993).

DNA fragmentation in cells undergoing apoptosis is measured either by flow cytometry after staining with DNA binding dyes (Nicoletti *et al.*, 1991; Lowe *et al.*, 1993), or by agarose gel electrophoresis of oligonucleosomal subunits produced by specific endonucleases after chromatin cleavage (Zychlinsky *et al.*, 1991). The comet assay is much simpler and time-saving method to measure DNA fragmentation in apoptotic cells than these methods (Gopalakrishna and Khar, 1995).

Recently we successfully applied this assay for the quantitative studies on the apoptosis caused by an endogenous neurotoxin, 1(*R*), 2(*N*)-dimethyl-6,7-dihydroxy-1,2,3,4-tetrahydroisoquinoline, *N*-methyl(*R*)salsolinol. As shown in Figure 2 and 7, this isoquinoline enantio-selectively induced apoptosis in dopaminergic neuroblastoma SH-SY5Y cells. Only in the cells incubated with this toxin, DNA fragments migrate in a form of tail from intact DNA head, which looks like a "comet" (Maruyama *et al,* 1997b). This review summarizes the principle and protocol of the comet assay and discusses its future application to the molecular studies of the pathogenesis of the neurodegenerative diseases.

II. PRINCIPLE OF A SINGLE CELL GEL ELECTROPHORESIS ASSAY

Single-strand breaks are commonly measured by procedures, such as treatment at high pH or alkaline unwinding, alkaline elution and alkaline sucrose sedimentation (*see reviews*; Ahnström, 1988; Whitaker *et al.*, 1991). The principle is that the DNA base-pairing is disrupted by alkaline and the discontinuities in single strand can be detected. At neutral pH, the long duplex molecules maintain the continuities regardless of occasional single-strand breaks, which, however, release supercoiling and induce a relaxed structure of DNA. The release of supercoiling in neutral conditions or in combination with unwindings in alkaline can be detected. Rydberg and Johanson (1978) observed that after X-ray irradiation the nuclei of alkaline-lysed cells became increasingly diffuse in a dose-dependent way. The cells were imbedded in agarose on microscope slides and lysed under mild alkaline conditions. After neutralization the cells were stained with acridine orange and the extent of DNA damage was quantitatively measured by

Table 1.
Application of the comet assay

Tissue	Cell line		References
Human tissues			
Blood	Lymphocytes		Ahuja *et al.*, 1993; Betti *et al.*, 1994
	T-cells		Arlett *et al.*, 1993
	Granulocytes		Vijayalaxmi *et al.*, 1993
Epithelia	Lens		Kleiman and Spectir, 1993
	Mucosa		Mckelvey *et al.*, 1992
Fibrolasts	Skin-derived		Arlett *et al.*, 1993
	Foreskine-derived		Singh *et al.*, 1991
Adenocarcinoma			Olive *et al.*, 1993b
Lymphoma			Östling *et al.*, 1987
Human cell culture			
Neuroblastoma	SH-SY5Y		Maruyama *et al.*, 1997
Glioma	U-87		Olive *et al.* 1994a
Blood	B-Cell	Raji	Fairbarin *et al.*,1994b
		TK6	Evans *et al.*, 1993
	T-cell	HUT-78	Shafer *et al.*, 1994
Carcinoma	Cervix	SiHa	Olive and Banath, 1993
		Hela	Collins *et al.*, 1993
	Colon	WiDr	Olive *et al.*, 1991
		HT-29	Olive *et al.*, 1994b
Melanoma	MeWo		Mükker *et al.*, 1994
	Ht-144		Olive *et al.* 1994b
Fibroblast	IBR3		Olive and Banath, 1993
Breast keratinocyte	NHEK-267-1		Ward *et al.*, 1993
Animal tissues			
Brain	Mouse		Fairbarin *et al.*, 1994c; Tice *et al.*, 1990
	Sheep		Fairbarin *et al.*, 1994a
Lymphocytes	Mouse		Plappert *et al.*, 1994
	Rat		Betti *et al.*, 1993
	Dog		Kreja *et al.*, 1993
Splenocytes	Mouse		Fairbarin *et al.*, 1994b
Thymocytes	Mouse		Fairbarin *et al.*, 1994c
Bone marrow	Mouse		Olive *et al.*, 1994b
Kidney	Mouse		Fairbarin *et al.*, 1994c
	Rat		Latinwo *et al.*, 1994
Liver	Mouse		Olive *et al.*, 1994b
	Rat		Betti *et al.*, 1993
Mucosal epithelia	Rat		Betti *et al.*, 1993

After neutralization the cells were stained with acridine orange and the extent of DNA damage was quantitatively measured by comparing the ratio of green fluorescence of double-stranded DNA to red of single stranded DNA. It was the first quantitation of DNA in a single cell. This method was improved by Östling and Johanson (1984), as the comet assay by use of an electrophoretic microgel technique. Cells are imbedded in agarose on microscopic slides, lysed, subjected to electrophoresis and stained with a DNA binding dye to detect DNA. Intact DNA in undamaged cells appears as a spherical mass ("head"), whereas in the cells exposed to toxins, ionizing radiation, or oxidative stress, damaged DNA can be seen streaming from the head as a "tail" in the direction of the anode, forming a "comet" image. Östling and Johanson observed that the extent of DNA liberated from the head of the comet by electrophoresis was a function of dose of irradiation.

Under original conditions reported by Östling and Johanson, the cells are lysed in neutral conditions by detergents and salts, which are not potent enough to remove all the protein. The comet images observed by this neutral method represent a released "halo" of DNA by loss of DNA supercoiling. By electophoresis some of DNA stained with acridine orange migrate as a tail in the direction of the anode, which indicates the presence of single-strand breaks. As a modification, under more stringent conditions for lysis more than 95% protein is dissolved and the broken duplex complex is allowed to migrate. This method visualizes the continuous long duplex complex and detects DNA double-strand breaks (Olive 1989 and Olive *et al.*, 1991). More recently lysis in the strong alkaline conditions with detergents were used to dissolve DNA-base pairing completely, and the discontinuities in the single strands can be detected (Singh *et al.*, 1988). Their condition to lyse the cells in high alkaline reveals alkaline labile sites and frank breaks, and staining with ethidium bromide increases the sensitivity to detect DNA migration from the nucleus. A comet image of a head to a tail are visualized fluorometrically, and the length and the fluorescent intensity are related to the number of DNA strand breaks.

This method can detect DNA breaks in individual cells, and not only the extent of DNA damage, but the distribution of cells with various degree of the DNA damage. The heterogeneous response of the cells to cytotoxic treatment indicates the involvement of endogenous factors, especially in the case of clinical samples, such as lymphocytes or tumor cells. This method has began to be applied to experimental conditions to elucidate the mechanism of genotoxicity and DNA repairing.

The sensitivity of the comet assay was estimated by incubation of cells irradiated by UV with exogenous UV endonuclease and by examination of the consequent breaks both by alkaline comet assay and by alkaline unwinding. As few as 1 DNA breaks per 10^{10} dalton are detected in the comet assay (Gedik *et al.*, 1992). It is much more sensitive than the histopathological detection of apoptotic cells by TUNEL method or by electro-microscopic observation.

III PROTOCOL FOR THE COMET ASSAY

The most commonly used conditions are based on those reported by Singh *et al* (1988). The principle of the assay is shown in Figure 2. A high pH (> 12.3) is generally used to facilitate denaturation, unwinding, expression of single strand breaks and DNA breaks that become visible by alkaline treatment.

In the field of cancer research, virtually any kind of cells from clinical samples, such as lymphocytes, tumor cells and biopsied cells, and cultured cells from primary culture to clonal cultured cells have been applied to the comet assay, as summarized in Table 1. In the field of neurochemistry the comet assay was used for the brain tissues from scrapie-infected sheep brain (Fairbarin *et al.*, 1994). However, this assay had never been applied to cultured neural cells until our application to SH-SY5Y cells in relation to apoptosis induced by neurotoxins (Maruyama *et al.*, 1997b).

Preparation of a single cell suspension

As an example of cell preparation from cultured cells, SH-SY5Y cells are cultured with Cosmedium-001 (CosmoBio, Tokyo, Japan) supplemented by 5% Nakashibetsu precolostrum newborn calf serum (Mitsubishi Kasei, Tokyo, Japan) in an atmosphere of 95% air-5% CO_2. To recover the cells, the cells are dissociated with trypsin, gathered, washed with Cosmedium-serum solution, then with phosphate-buffered saline (PBS) twice. In general, mild trypsinization is preferable to a scraping and mechanical disaggregation.

The culture of monolayer cells on fully frosted microscope slide followed by *in situ* treatment and analyzed by the comet assay was proposed by Singh *et al.* (1991). However, this monolayer method has never been used for neural cells.

was homogenized in 2 ml of PBS with a mortar and pestle, and used for the comet assay. Singh *et al.* (1995) isolated rat brains, immediately washed with PBS containing 200 μM *N*-*t*-butyl-a-phenylnitrone, minced with a pair of scissors to prepare pieces of approx. 1 mm^3. The pieces were dispersed into single-cell suspension using a 5000-μl pipetman. A small volume (10 μl, approx. 10 000 cells) of the cell suspension was mixed with 0.5% agarose.

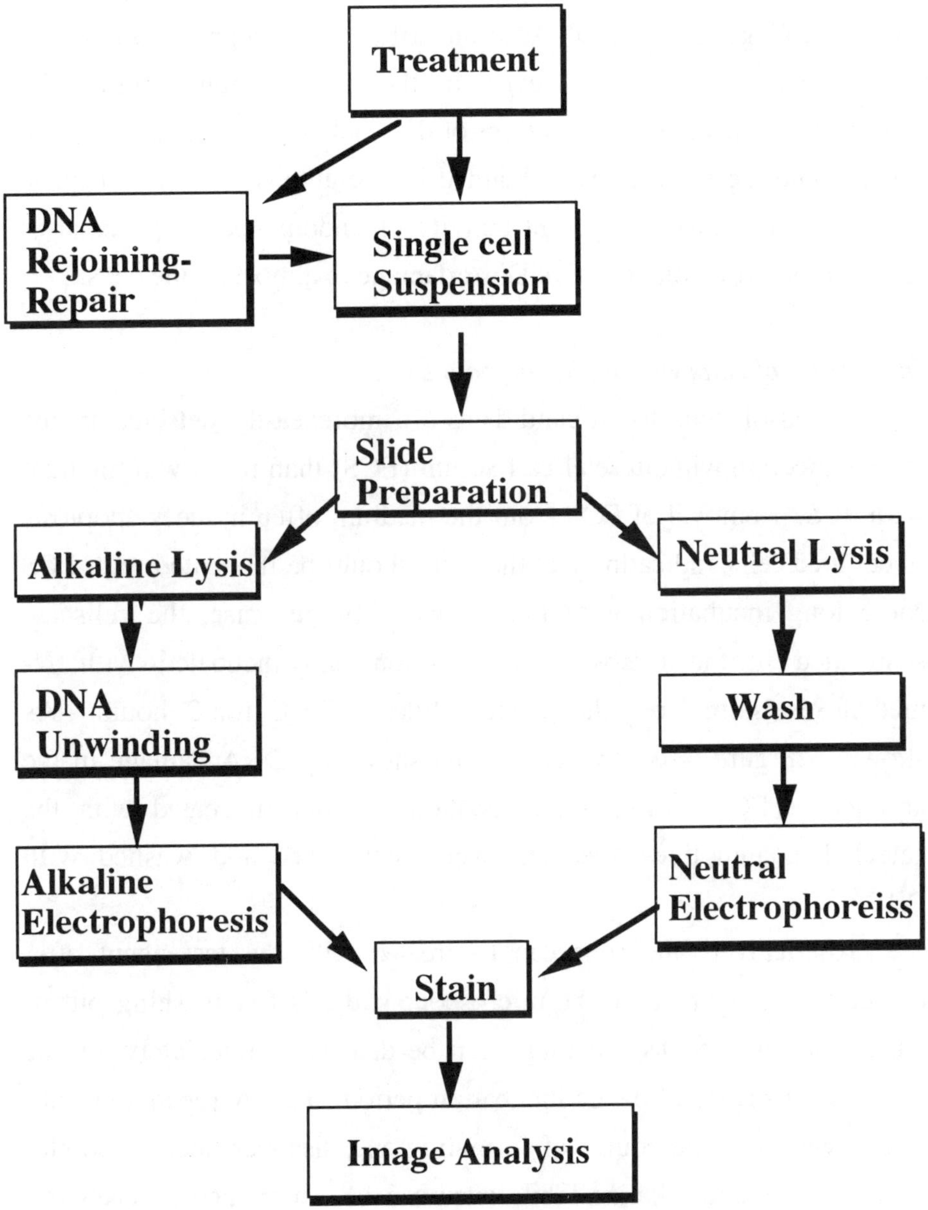

Figure 3. Protocol of the comet assay. The cells are embedded in agarose gel and subjected to neutral or alkaline lysis with detergents and high concentration of salts.

From tumors the cells are prepared by mincing and incubation with trypsin, collagenase and DNAase and filtering through 50 μm nylon mesh (Olive, 1989b). However, in the case of human brain the postmortem time and the conditions of the final stage of the patients are critical, and the brain tissues obtained by the autopsy are not adequate for this assay, since the potent activity of endonucleases, proteinases and lipases cause non-selective DNA damage in almost all the cells.

Incubation of cells with toxic compounds

The effects of neurotoxin candidates are more easily detected in the culture medium without fetal calf serum (FCS) than in growth medium with FCS. Removal of FCS from the medium often induces apoptosis in cultured cells, indicating that the cells should be incubated with FCS for a long incubation with the toxins. In our case, the cells are suspended in the Krebs-Ringer solution and incubated with *N*-methyl(*R*)salsolinol or other isoquinolines at 37°C for 3 hours. As shown in Figure 2, control cells don't show any DNA damage in the absence of FCS during the incubation After incubated with the catechol isoquinolines, the cells were centrifuged and washed with PBS.

Alternatively the cells can be treated with the test agent after embedding in agarose on the microscope slide. After washing out of the test agent, the DNA damage can be detected immediately, or the assay is performed after the incubation period to allow repair to occur. In this way the time course of formation and disappearance of single-strand breaks and alkaline-labile sites can be determined (McKelvey-Martin *et al.*, 1993).

This method is further devised to examine the involvement of the intracellular mechanism to DNA damage. In neutral conditions, cells embedded sparsely in agarose are exposed to various salt solutions to sequentially remove non-histone and then histone proteins (Olive and Banath, 1995). In alkaline conditions, cells embedded in agarose are lysed in alkaline as described below and incubated with the test agents, then subjected to electrophoresis (Kasamatsu *et al.*, 1996). This method can distinguish whether DNA damage is induced by a pure chemical event or induced by enzymatically activated toxins or so-called apoptotic mechanism in the cells.

An application of this assay to *in vivo* experiments has been proposed. Animals are treated with test compounds and after a short exposure period (e.g. 1-72 hours), cells are explanted, checked for viability and immediately subjected to the comet assay. This method has been applied for toxicokinetic studies with ingested compounds by analysis of the gastrointestinal tract (Pool-Zobel *et al.*, 1992). As another application to *in vivo* experiments, lymphocytes are prepared from humans 24 hours after exercise (Hartmann *et al.*, 1994). It was shown that strenuous exercise caused DNA migration by the comet assay. This method has never been applied to the brain tissues, because the viability of the nerve cells cannot be preserved during the isolation procedures. If possible, it may be a very convenient method to detect apoptotic cells in animal models exposed to neurotoxins or oxidative stress.

Factors affecting the cell experiments

As reported by Olive *et al.* (1990), cell cycle stages don't affect induction of DNA damage in alkaline comet assay, suggesting that this method can be applied to asynchronous culture cells and primary cells where the cell-cycle cannot be controlled. When the comet assay is performed under neutral conditions cell-cycle differences are observed. DNA from S phase cells shows much less damage that from cells in other phases of the cell cycle (Olive *et al.*, 1991).

Preparation of slides

The cells are suspended in low melting point agarose at a final concentration of 0.5 - 1.0% at 35 - 45°C. Two basic types of casted gel prevail; the sandwich gel and the single layer gel.

In the former method the cells are contained in the middle layer of three distinct layers of agarose. Fully frosted or roughened microscope slides are covered with 100 μl of 0.5% normal melting agarose in PBS at 45°C, immediately covered with a No. 1 coverslip and then kept at 4°C for 10 min to allow the agarose to solidity. The first layer is used to promote even and firm attachment of the second and third layers. Cell suspension (1 000 to 500 000 cells per slide) in 10 μl of PBS are mixed with 75 μl of 0.5% low melting point agarose at 37°C. After gently removing the coverslip the cell suspension is rapidly pipetted onto the first agarose layer, spread using a coverslip and maintained at 4°C for 110 min. After removal of the coverslip the third layer of 0.5% low melting point agarose (75 μl) at 37°C is added, spread using a coverslip and allowed to solidity at 4°C.

The single layer gel method is a simpler alternative, and the first and third layer of agarose can be omitted. The cells (50 000 cells) are suspended in 100 μl of PBS free from calcium and magnesium, 20 μl of which is mixed with 140 μl of 1% low-melting agarose in PBS. The mixture (100 μl) is applied on a frosted microscope slide, and allowed to stand at 4°C for 10 min. This method is simple and time-saving, when one can skillfully smear the gel on the slide.

Neutral and alkaline lysis and electrophoresis

Neutral and alkaline lysis conditions are used for double and single strand break detection, respectively. In general alkaline lysis is more commonly used than the neutral. The neutral lysis buffer is a mixture

0.5% - 2.5% sodium dodecylsulphate (SDS) or sodium laurylsulphate (SLS) and 10 - 30 mM 2Na EDTA. pH 8.3, in distilled water or in 10 mM Tris buffer. The incubation is at 50°C for 4 hours (Olive *et al.*, 1994a). As a modification, the cells are lysed in 30 mM EDTA, 0.5% SDS and 0.25 mg/ml proteinase K at pH 8.0 at 4°C for 1 hour, followed by lysis at 37°C for 12 - 16 hours, then rinsed in 90 mM Tris-HCl - 90 mM boric acid- 2 mM EDTA buffer (pH 8.0). The cells prepared rat brains are lysed in ice-cold lysis solution (2.5 M NaCl, 1% sodium N-lauroyl sarcosine, 100 mM EDTA 2Na, 10 mM Tris buffer to a final pH of 10 and 1% Triton X-100) at 0°C for 1 hour. Then, slides are treated with DNAase-free proteinase K (Boehringer-Mannheim, Indianapolis, IN) for 1 hour at 37°C (Singh *et al.*, 1995). The lysed cells are electrophoresed for 25 min at 0.55V/cm in 90 mM Tris - 2 mM EDTA - 90 mM boric acid buffer (pH 8.5) (Olive *et al.*, 1991) or 45 mM Tris - 45 mM boric acid - 1 mM EDTA (pH 8.0) at 1V/cm and 12°C (Kent *et al.*, 1995).

In alkaline lysis, the slide is subjected to alkaline lysis at 4°C for 1 hour (minimum) in 10 mM Tris buffer, containing 2.5 M NaCl, 100 mM EDTA 2Na, 1% sarcosine and 1% Triton X-100, pH 10.0, to which dimethylsulfoxide (DMSO) is added to 10% just before use. The lysis may be prolonged to overnight at 4°C.

The slides are placed in a horizontal slab of an electrophoresis system (*i.e.*, Horizon 11-14 Gel Casting System, Life Technologies, Gaithersburg, MD; Hoefer Scientific, San Francisco, CA), side by side, avoiding spaces and with the agarose end facing the anode, then the tank is filled with an alkaline electrophoresis buffer, 300 mM NaOH solution containing 1 mM EDTA 2Na, and the slides are left at 4°C for 20 min to allow unwinding of the DNA and remove salts and detergents before electrophoresis. Electrophoresis is carried out at 4°C or at room temperature for 20 min with 25 V. The current is adjusted to 300 mV by raising or lowering the buffer level in the tank. The voltage and time of electrophoresis are related to the level of DNA damage expressed in the cells and the salt concentration of phoresis buffer. For microscopic observation only a few mm migration is enough to detect the comet formation, and the slides may be subjected to electrophoresis for a very short runs (5 - 30 min) and with low voltages (0.5 - 5.0 V/cm). This is especially the case with the assay of cells prepared primarily from tissues.

Staining

The slides are washed twice with 0.2 M Tris-HCl buffer, pH 7.5, to be neutralized. Then, DNA is stained with 50 - 100 μl of ethidium bromide (2- 20 μg/ml in distilled water) or 4',6-diamidino-2-phenylindole (DAPI) solution (1 - 5 μg/ml). A sensitive fluorescence dye YOYO-1 (Molecular Probes, Eugene, OR) is also used (25 μl of 1 μM solution). Staining with propidium iodide (2.5 μg/ml) for 60 min is commonly used for image analysis.

Analysis of comet images

In intact cells DNA is stained as a fluorescent core in compact spherical mass. In damaged cells broken ends of the negatively charged DNA migrate in the electric field towards the anode, forming a comet image. The migration is a function of both the size of the DNA and the number of broken ends. Tail length increases with the amount of breaking (Singh *et al.*, 1988), when the damage is relatively mild. In severely damaged cells, the length was found to be constant over a range of DNA break frequencies of 1 -14 per 10^{10} dalton, and the fluorescence intensity to increase markedly with increased frequency of the breaks (Gedik *et al.*, 1992).

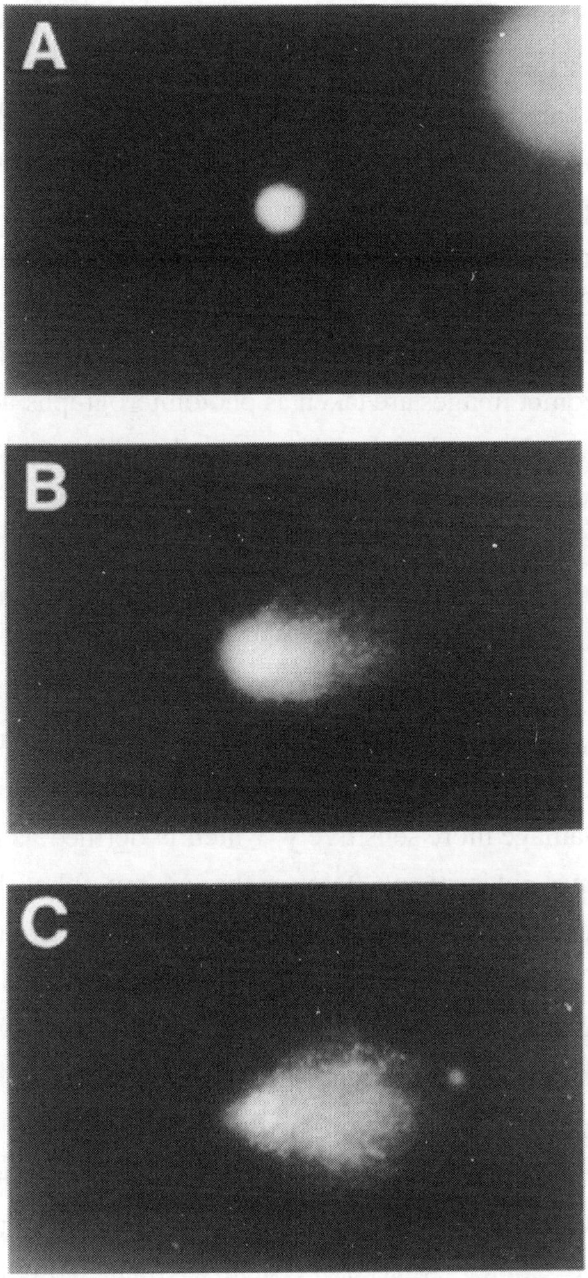

Figure 4. Cells showing comet images with different degrees of DNA damage. A, control cell ; B, cell with moderate DNA damage; C, with severe DNA damage.

The comet images are detected fluorometrically using an epi-fluorescence microscope with a 100 W mercury lamp, and an excitation filter of 515-560 nm (green light) or 580 nm and a barrier filter of 590 nm for ethidium bromide or propidium iodide staining, respectively. For slides stained with YOYO-1, excitation at 490 nm with an emission filter at 515 nm and dichromic filter at 500 nm wavelength are used. For detection of DNA stained with DAPI, the fluorescence is measured with a 435 nm barrier filter and 365 nm excitation filter (dichroic mirror 400 nm). The comet images are taken as photomicrographs at 400 time magnification, or observed on a TV screen through a video camera attached to a microscope at 200 time magnification.

The extent of DNA damage quantitatively is measured in 25 - 200 randomly selected cells per population being tested, using the tail length (Tice *et al.*, 1990), total comet length (Deeley *et al.*, 1991), tail inertia (Hellman *et al.*, 1995), or the ratio of head diameter to tail length (Olive, 1989a). Recently a parameter, "tail moment" is used to quantify DNA damage more sensitively, which is defined as a function of the fluorescence intensity of DNA in the tail multiplied by the tail length, or the percentage of migrated DNA (Olive *et al.*, 1990 ; Ashby *et al.*, 1995). These parameters were confirmed to be linear to DNA damage, and the comet moment was proved to be linear over a larger range than other parameters (Kent *et al.*, 1995). Figure 4 shows sample photos to measure the comet length.

The comet images are also analyzed by a fluorescence image processing system (Olive *et al.*, 1990). The systems commercially available are a Comet Analysis system (Kinetic Imaging Ltd., Liverpool, U. K.), Colourmorph Comet Assay (Perceptive Instruments, Halstead, U, K.), Fenestra Comet (Confocal Technologics, U. K.). Comet images are also captured using

OPTIMAS (Data Cell Ltd., Maidenhead, U. K.), an image analysis software package and digitized to 256 gray values using a PCVision + 512 x 512 oixel frame grabber (Kent *et al.*, 1995).

IV. NEUROTOXIN HYPOTHESIS OF THE PATHOGENESIS OF PARKINSON'S DISEASE

Parkinson's disease is now one of the most common neurological disorders among the aged, and the major pathological features are selective cell death of dopamine neurons in the substantia nigra and locus ceruleus. The mechanism of cell death in parkinsonian brains has be considered to be due to inhibition of ATP synthesis (Mizuno *et al.*, 1989; Schapira *et al.*, 1990), or to oxidative stress (Yoritaka *et al.*, 1996). On the other hand, there has been a increasing number of papers reporting the occurrence of apoptosis in parkinsonian brains, as shown by electro-microscopic observation (Anglade *et al.*, 1997) or by staining with TUNEL method (Mochizuki *et al.*, 1996). These results suggest that apoptosis may be another mechanism underlying the selective cell death of dopamine neurons in parkinsonian brain. Biochemical changes in parkinsonian brain are characterized by reduction of dopamine and the activity of a rate-limiting enzyme in catecholamine synthesis, tyrosine hydroxylase [tyrosine, tetrahydropteridine: oxygen oxidoreductase (3-hydroxylating), EC 1.4.16.2]. The reduction is now considered to be secondary to the cell death of dopaminergic neurons in the nigro-striatal system.

The induction of parkinsonism in humans by 1-methyl-4-phenyl-1,2,3,6-tetrahydropyridine (MPTP) (Langston *et al.*, 1983) made the "neurotoxins" to be one of the most possible hypothesis of the pathogenesis. MPTP is oxidized into a more potent dopaminergic neurotoxin, 1-methyl-4-phenylpyridinium ion (MPP^+), by type B monoamine oxidase [monoamine: oxidoreductase (deaminating), EC 1.4.3.4] (Chiba *et al.*, 1984). The results using MPTP-induced animal models of Parkinson's disease suggest common characteristics of a dopaminergic neurotoxin; transport through the blood-brain-barrier or *in situ* synthesis in the brain, metabolism into a more toxic compound, uptake into dopamine neurons by a dopamine transporter, accumulation in mitochondria or binding with neuromelanin (*see a review,* Tipton and Singer, 1993). The final mechanism of the cell death may be either ATP depletion by inhibition of mitochondrial respiratory chain (Mizuno *et al.*, 1987) or generation of reactive oxygen species (Chiueh *et al.*, 1993). Recently, MPTP or MPP^+ were reported to induce apoptosis in cultured cerebellar granule neurons (Dipasquale *et al.*, 1991), cultured rat ventral mesencephalic and striatal cells (Mochizuki *et al.*, 1994) and in pheochromocytoma PC12 cells (Mutoh *et al.*, 1994).

The finding of MPTP suggests that endogenous or xenocompound, uptake into dopamine neurons by a dopamine transporter, accumulation in mitochondria or binding with neuromelanin (*see a review,* Tipton and Singer, 1993). The final mechanism of the cell death may be either ATP depletion by inhibition of mitochondrial respiratory chain (Mizuno *et al.*, 1987) or generation of reactive oxygen species (Chiueh *et al.*, 1993). Recently, MPTP or MPP^+ were reported to induce apoptosis in cultured cerebellar granule neurons (Dipasquale *et al.*, 1991), cultured rat ventral mesencephalic and striatal cells (Mochizuki *et al.*, 1994) and in pheochromocytoma PC12 cells (Mutoh *et al.*, 1994).

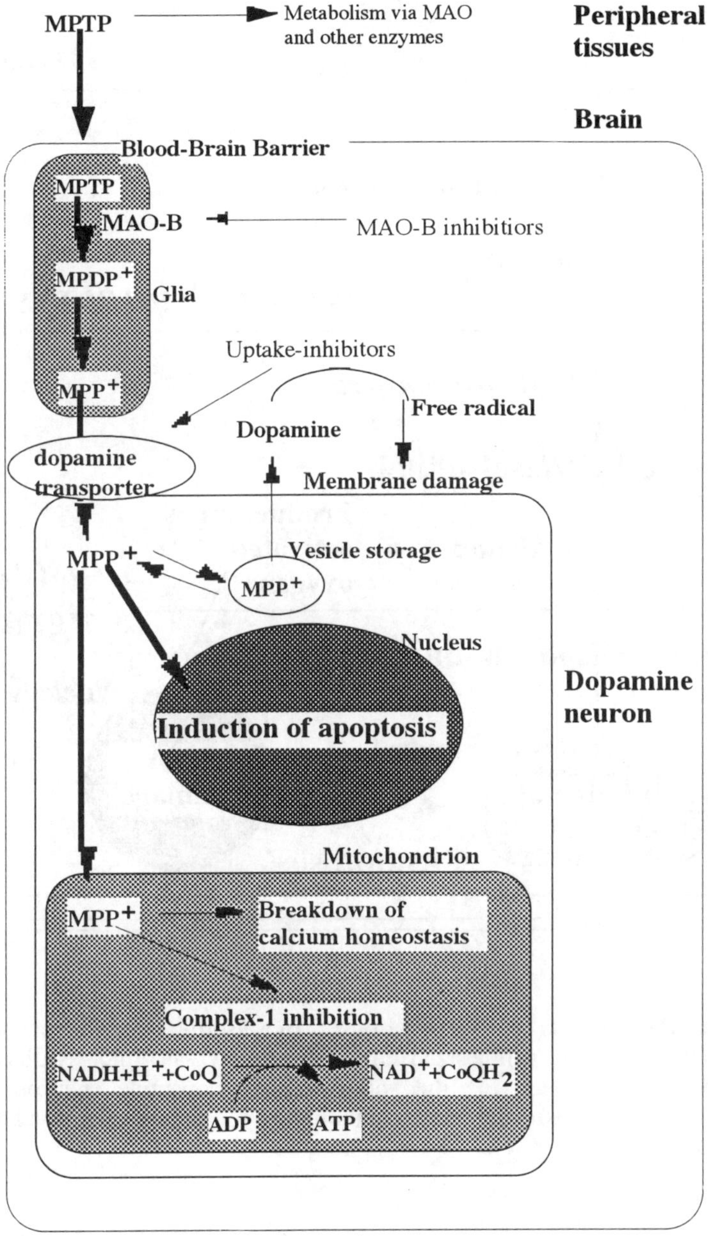

Figure 5. Mechanism of neurotoxicity of MPTP to dopamine neurons. MPTP is transported into the brain and metabolized into MPP^+ by type B monoamine oxidase (MAO-B). MPP^+ is concentrated in the mitochondria and inhibits complex I. Oxidation of MPTP and released dopamine generate reactive oxygen species and finally apoptosis is induced.

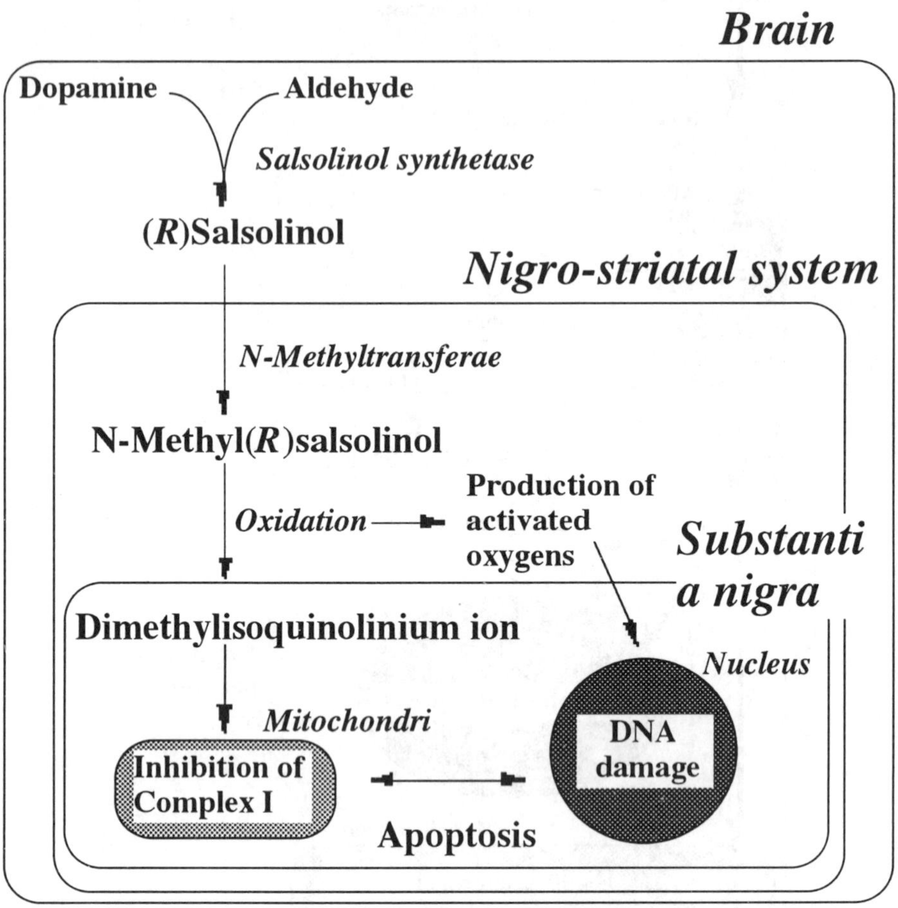

Figure 6. Biosynthesis and distribution of (*R*)salsolinol derivatives in human brains. (*R*)Salsolinol is synthesized enzymatically by salsolinol synthase in the brain and N-methylated in the nigro-striatal dopamine neurons. *N*-Methyl(*R*)salsolinol is oxidized and isoquinolinium ion produced is accumulated in the substantia nigra.

The finding of MPTP suggests that endogenous or xenobiotic neurotoxins may elicit Parkinson's disease. As a neurotoxin candidates, three groups of endogenous alkaloids have been proposed; 1-methyl-6,7-dihydroxy-1,2,3,4-tetrahydroisoquinoline [salsolinol] (Naoi *et al.*, 1995b, 1996a, b, c), 1,2,3,4-tetrahydroisoquinoline

(Nagatsu and Yoshida, 1988), and β-carbolines (Collins *et al.*, 1992, 1996; Matsubara *et al.*, 1992; Matsubara 1996).

We speculate that dopaminergic neurotoxins should have chemical structure similar to dopamine and MPTP to obtain the specific cytotoxicity to dopamine neurons, and we examined derivatives of dopamine-derived salsolinol. Salsolinol has been considered to be synthesized as a racemic form by a non-enzymatic Pictet-Spengler reaction, which is supported by the occurrence of racemic salsolinol in food and wine (Strolin Benedetti *et al.*, 1989).

However, recently we found that in the human brain only the (*R*)-enantiomer is present (Maruyama *et al.*, 1996a and b, 1997a). The enzymatic biosynthesis of (*R*)salsolinol had been suggested by its dominant presence in human urine (Dostert *et al.*, 1989 and 1990). It was proved by the isolation of a novel enzyme catalyzing condensation of dopamine with acetaldehyde (Naoi *et al.*, 1996d). (*R*)Salsolinol was further *N*-methylated into 1(*R*), 2(*N*)-dimethyl-6,7-dihydroxy-1,2,3,4-tetrahydroisoquinoline [*N*-methyl(*R*)salsolinol], as proved by *in vivo* microdialysis in the rat brain (Maruyama *et al.*, 1992). *N*-Methyl(*R*)salsolinol is oxidized into 1,2-dimethyl-6,7-dihydroxy-isoquinolinium ion by non-enzymatic (Maruyama *et al.*, 1995a) or enzymatic oxidation (Naoi *et al.*, 1995a). These catechol isoquinolines are detected in human brains, *N*-methyl(*R*)salsolinol accumulates in the nigra-striatal system and catechol isoquinolinium ion is detected only in the substantia nigra, whereas (*R*)salsolinol is distributed in whole brain regions (Maruyama *et al.,* 1997a).

Recently we succeeded in preparation of an animal model of Parkinson's disease by injection of *N*-methyl(*R*)salsolinol into the rat striatum (Naoi *et al.*, 1996b). The behavioral, biochemical and histopathological changes were very similar to those observed in Parkinson's disease. After continuous infusion of *N*-

methyl(*R*)salsolinol into the striatum, dopamine neurons in the substantia nigra were selectively depleted without necrotic tissue reaction. These results suggest that an apoptotic process may account for the cell death of dopamine neurons caused by the neurotoxin.

V. APPLICATION OF COMET ASSAY FOR NEURONAL CELL DEATH CAUSED BY ENDOGENOUS NEUROTOXIN

Figure 6 shows the histogram of the length of comet image of the cells treated with (*R*)- and (*S*)enantiomers of salsolinol and *N*-methylsalsolinol, and catechol isoquinolinium. The mean head-tail distances of control and cells incubated with (*R*)- and (*S*)salsolinol or catechol isoquinolinium ion were distributed between 10 to 25 μm (mean ± SD, 11.2 ± 0.02), which is consistent with intact nuclei with undamaged DNA, whereas those of the cells incubated with *N*-methyl(*R*)salsolinol were larger than 45 μm and did not overlap with that of the former cell groups. The tail-length of the *N*-methyl(*S*)salsolinol-treated cells were rather variable giving values in the range 10 to 45 μm. This histogram indicates that at least some of the cells in this group were positive for DNA damage. The tail-length of 45 μm or greater were taken to indicate extensive apoptosis. MPP^+ did not induce DNA damage in the cells under the experimental conditions used.

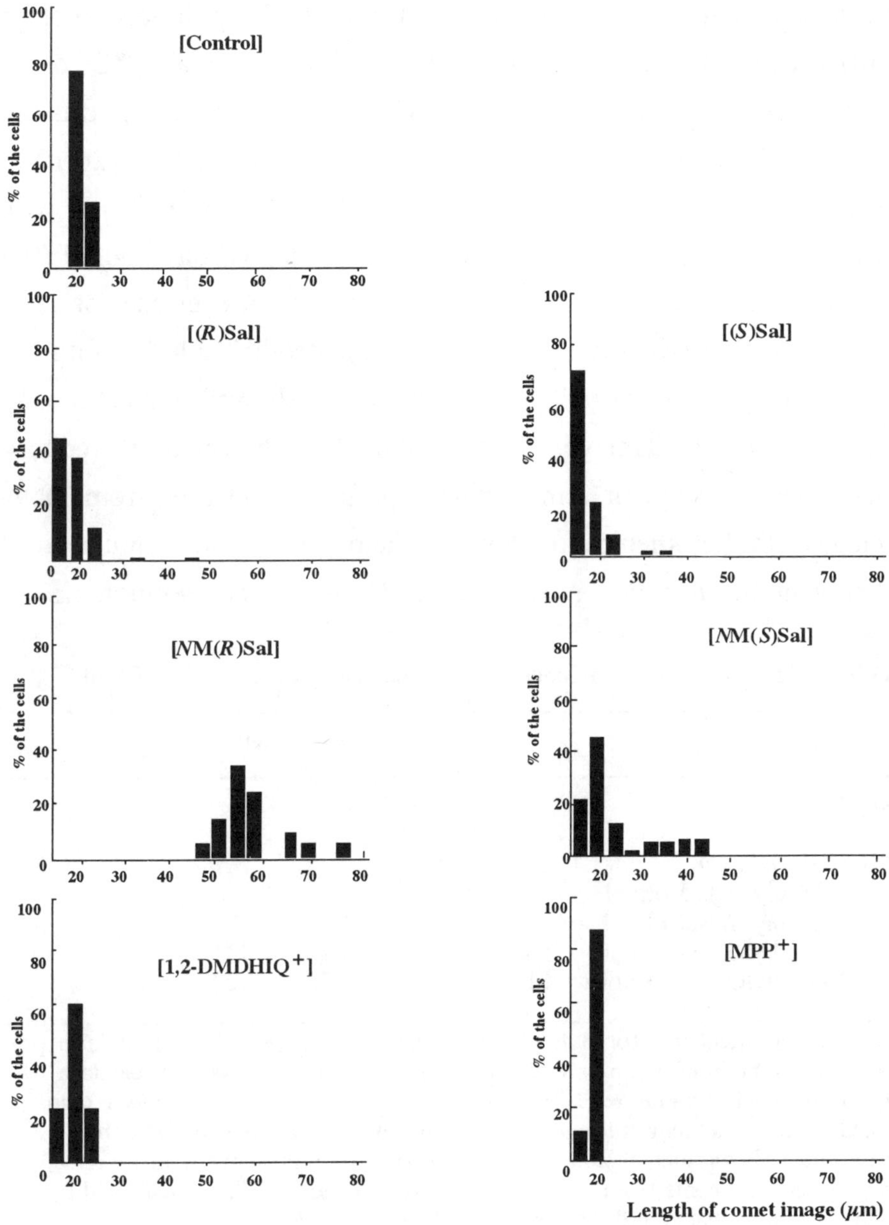

Figure 7. Frequency distribution of DNA migration distance in SH-SY5Y cells incubated with catechol isoquinolines. The cells were incubated with 1 mM of the (*R*)- and (*S*)-enantiomers of Sal and *N*MSal, or DMDHIQ$^+$, and the length of comet image (from comet head to the tail) was measured. The distribution of DNA-damaged cells was expressed as the percentage of the total 200 cells. Each column represents the mean value of four experiments.

The mechanism of apoptosis induced by *N*-methyl(*R*)salsolinol is clarified by the comet assay. The cells were incubated at 37°C for 20 min with reduced glutathione (100 μM, final concentration), catalase (0.5 mg) and superoxide dismutase (2 000 units), clorgyline (20 μM), deprenyl (20 μM) and semicarbazide (100 μM), or 60 min with cycloheximide (5 μM). Then, the effect of *N*-methyl(*R*)salsolinol (0.5 mM) was examined after a 3 hour incubation. The effects of anti-oxidants and anti-oxidative enzymes are presented in Table 2. With 0.5 mM *N*-methyl(*R*)-salsolinol about 18% cells showed typical comet images of DNA damage. Pre-treatment with catalase, reduced glutathione, deprenyl or semicarbazide protected the cells from DNA damage. On the other hand, clorgyline and superoxide dismutase did not prevent the DNA damage induced by *N*-methyl(*R*)salsolinol.

Table 2. The effects of antioxidants on DNA damage induced by *N*M(*R*)Sal.

	Comet positive cells (%)
Control	0
Cells incubated with	
N-Methyl(*R*)salsolinol	17.2 ± 0.3
N-Methyl(*R*)salsolinol + SOD	15.2 ± 1.27
N-Methyl(*R*)salsolinol + Catalase	4.22 ± 1.93*
N-Methyl(*R*)salsolinol + GSH	7.39 ± 0.42*

The cells were incubated for 3 hours with 0.5 mM *N*M(R)Sal after 20 min pre-incubation in medium alone or with superoxide dismutase (2000 U), catalase (0.5 mg) and reduced glutathione (100 μM). The percents of cells with DNA damage, defined by the comet length longer than 45 μm, were measured by the comet assay with a total cell number of 200. Each value and bar represent the mean ± SD of triplicate measurements from three separate experiments. * Significantly different from the cells incubated with *N*-methyl(*R*)salsolinol alone ($p < 0.05$).

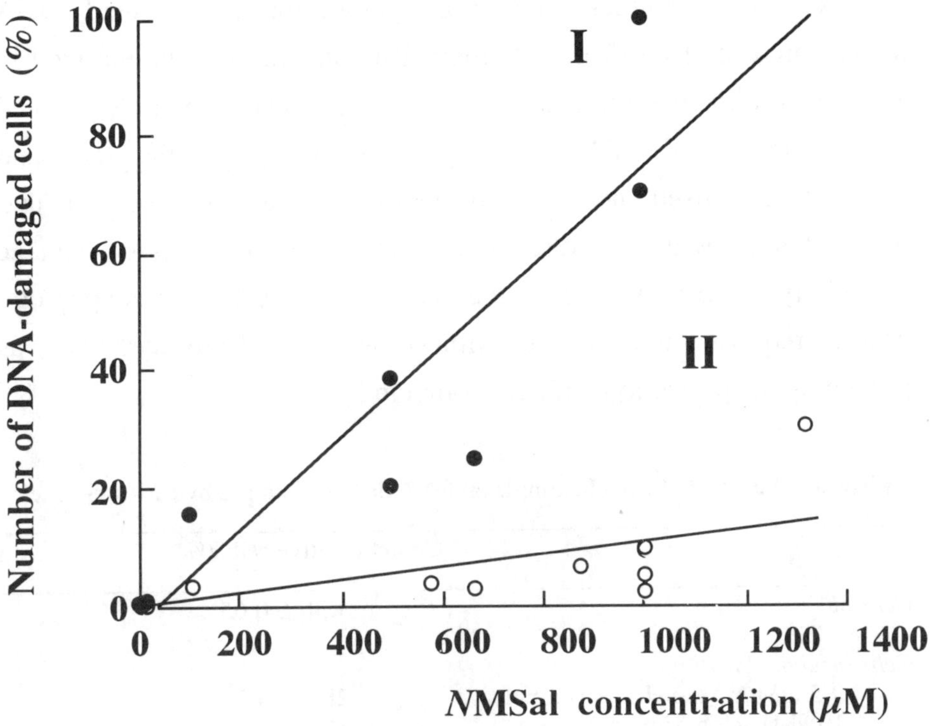

Figure 7. The number of DNA-damaged cells by various concentration of *N*-methylsalsolinol (*N*MSal). (*R*)-enantiomer of *N*MSal was more potent than (*S*)-enantiomer. SH-SY5Y cells were incubated with *N*M(*R*)Sal (I, filled circle) or NM(*S*)Sal (II, open circle) for 3 hours and DNA damage was determined by comet assay. The DNA-damage cells were defined by the comet length longer than 45 μm. Each circle represents mean of duplicate experiments.

The effects of the concentration were studied using the (*R*)- and (*S*)enantiomers of *N*-methylsalsolinol. Figure 7 shows that the induction of the DNA damage was dependent on the concentration ($p < 0.01$) for both the enantiomers. With 1 mM *N*-methyl(*R*)salsolinol almost all the cells were positive for DNA damage. *N*-Methyl(*S*)salsolinol was less potent in inducing DNA damage, and at 1 mM only about 10% cells were positive.

A protein synthesis inhibitor cycloheximide prevented DNA damage induced by 0.5 mM *N*-methyl(*R*)salsolinol as summarized in Table 3. Pre-incubation of the cells with 5 μM cycloheximide reduced the occurrence of DNA damage. Allowing for the fact that cycloheximide itself induced some DNA damage, it can be seen that pre-treatment with cycloheximide prevented the DNA damage essentially completely. This suggests that the DNA damage is apoptotic process requiring the protein synthesis and also RNA synthesis, as proved by the protection with actinomycin D.

Table 3. The effect of cycloheximide on DNA damage induced by *N*M(*R*)Sal.

	Comet positive cells (%)
Control	1.30 ± 0.65
Cells incubated with	
0.5 mM *N*-Methyl(*R*)salsolinol	28.1 ± 3.7
10 μM cycloheximide	10.7 ± 1.6*
N-Methyl(*R*)salsolinol + cycloheximide	6.40 ± 4.20

The cells were pre-incubated with 5 μM cycloheximide for 1 hour before addition of 0.5 mM *N*-Methyl(*R*)salsolinol. The number of the cells with DNA damage, defined by the comet length longer than 45 μm, was assessed by comet assay with a total cell number of 200. Each value and bar represent the mean ± SD of triplicate measurements from three separate experiments. * Significantly different from the cells incubated with *N*-methyl(*R*)salsolinol alone ($p < 0.05$).

CONCLUSION

The comet assay is very sensitive and a small number of cells are required for the quantitative analysis of the apoptosis, without use of radioactive labeling. This method has been used in the field of cancer research to predict the efficiency of radio- and chemotherapy. For the first time we use this assay for the screening the neurotoxins and for the study on the cytotoxic mechanism. As shown here, the comet assay may be the most convenient and reproducible method for the research of the program cell death in the neuronal cells. As an example, we found that only *N*-methylsalsolinol can induce apoptosis in SH-SY5Y cells and the cell death program is initiated by an intracellular mechanism, which is potentiated by the (*R*)enantiomer and involves the generation of reactive oxygen species. One of the most important benefits of this assay is the assessing heterogeneity in DNA damage within a mixed population of cells. By use of synchronized cells, the effects of the cell cycle on the program cell death induced can also be examined by choosing the neutral lysis conditions.

At present there are several difficulties to apply this assay for the brain tissues from autopsied samples or animal models. The alkaline lysis damages DNA into fragments with very small molecular weight. If we can find milder conditions for lysis and electrophoresis, this method will take place of the hitherto-known assay methods, such as TUNEL assay or electromicroscopic observation.

Acknowledgment

This work was supported by a Grant-In-Aid for Scientific Research in Priority Areas and for Exploratory Research from the Ministry of Education, Science and Culture (M.N.), and the Grant for Longevity Sciences (9C-03) (M.N.) and (9C-02) (W. M.) from the Ministry of Health and Welfare , Foundation for Comprehensive Research on Aging and Health (W.M.) of Japan.

References

Ahnström G. (1988) Techniques to measure DNA single strand breaks in cells. *J. Radiat. Biol.* **54**, 697-707.

Ahuja Y.R., Anuradha G., Jaiswal M., Rajeswari N., Raju K.N. and Jain S.N. (1993) Single cell gel electrophoresis on leukocytes of patients with precancerous and cancerous lesions of the cervix. *Am. J. Hum. Gen.* **53**, 1515.

Anderson D., Yu T.-W., Phillips B.J. and Schmezer P. (1994) The effect of various antioxidants and other modifying agents on oxygen-radical-generated DNA damage in human lymphocytes in the COMET assay. *Mut. Res.* **307**, 261-271.

Anglade P., Vyas S., Javoy-Agid F., Herrero M.T., Michel P.P., Marquez J., Mouatt-Prigent A., Ruberg M., Hirsh E.C. and Agid Y. (1997) Apoptosis and autophagy in nigral neurons of patients with Parkinson's disease. *Histol. Histopathol*. **12**, 25-31.

Arlett C.F., Lowe J.E., Harcourt S.A., Waugh A.P.W., Cole M.J., Roza L., Diffey B.L., Mori T., Nikaido O. and Green M.H.L. (1993) Hypersensitivity of human lymphocytes to UV-B and solar irradiation. *Cancer Res.* **53**, 609-614.

Ashby J., Tinwell H., Lefevre P.A. and Browne M.A. (1995) The single cell gel electrophoresis assay for induced DNA damage (comet assay): measurement of tail length and moment. *Mutagenesis* **10**, 85-90.

Betti C., Barale R. and Pool-Zobel B.L. (1993) Comparative studies on cytotoxic and genotoxic effects of two organic mercury compounds in lymphocytes and gastric mucous cells of Sprague-Dawley rats. *Env. Mol. Mutagen* **22**, 172-180.

Betti C., Davini T., Giannessi L., Loprieno N. and Barale R. (1994) Microgel electrophoresis assay (comet test) and SCE analysis in human lymphocytes from 100 normal subjects. *Mutation Res.* **307**, 232-233.

Busciglio J. and Yankner B.A. (1995) Apoptosis and increased generation of reactive oxygen species in Down's syndrome neurons *in vitro*. *Nature* **378**, 776-779.

Cavrieli Y., Sherman Y. and Ben-Sasson A. (1992) Identification of programmed cell death in situ via specific labeling of nuclear DNA fragmentation. *J. Cell Biol.* **119**, 493-501.

Chiba K., Trevor A.J. and Castagnoli N. Jr. (1984) Metabolism of the neurotoxic amine, MPTP, by brain monoamine oxidase. *Biochem. Biophys. Res. Commun*. **120**, 574-578

Chiueh C.C., Miyake H. and Peng M.-T. (1993) Role of dopamine autoxidation, hydroxyl radical generation and calcium overload in underlying mechanisms involved in MPTP-induced parkinsonism. *Adv. Neurol.* **60**, 251-258.

Chu-Wang I.W. and Oppenheim R.W. (1978) Cell death of motorneurons in the chick embryo spinal cord. *J. Comp. Neurol.* **177**, 33-58.

Clark P.G.H. (1990) Developmental cell death: morphological diversity and multiple mechanisms. *Anat. Embryol.* **181**, 195-213.

Cook P.R. and Brazell L.A. (1976) Conformational constraints in nuclear DNA. *J. Cell Sci.* **22**, 287-302.

Collins A.R., Duthie S.J. and Dobson V.L. (1993) Direct enzymic detection of endogenous oxidative base damage in human lymphocyte DNA. *Carcinogenesis* **14**, 1733-1735.

Collins M.A., Neafsey E.J., Matsubara K., Cobuzzi R. Jr. and Rollema H. (1992) Indole-N-methylated β-carbolinium ions as potential brain-bioactivated neurotoxins. *Brain Res* **570**, 154-160.

Collins M.A., Neafsey E/J. and Matsubara K. (1996) β-Carbolines: metabolism and neurotoxicity. *Biogenic Amines* **12**, 171-180.

Cunningham T.J. (1982) Naturally-occurring neuron death and its regulation by developing neural pathways. *Int. Rev. Cytol.* **74**, 163-185.

Deeley J.O.T., Mitchell S., Sanders A. and Moore J.L. (1991) Microgel electrophoresis: a predictive test of radiosensitivity. *Int. J. Radiat. Biol.* **60**, 947.

Dipasquale B., Marini A. M. and Youle R. J. (1991) Apoptosis and DNA degradation induced by 1-methyl-4-phenylpyridinium in neurons. *Biochem. Biophys. Res. Commun.* **181**, 1442-1448.

Dostert P., Strolin Benedetti M., Dordain G. and Vernay D. (1989) Enantiometric composition of urinary salsolinol in Parkinsonian patients after Madopar. *J .Neural Transm. [P-D Sect]* **1**, 269-278

Dostert P., Strolin Benedetti M., Bellotti V., Allievi C. and Dordain G. (1990) Biosynthesis of salsolinol, a tetrahydroisoquinoline alkaloid, in healthy subjects. *J. Neural Transm. [GenSect]* **81**, 215-223.

Evans H.H., Ricanati M., Horng M-F., Jiang Q., Mencl J. and Olive P.L. (1993) DNA double-strand break rejoining deficiency in TK6 and other human B-lymphoblast cell lines. *Radiat. Res.* **134**, 307-315

Fairbarin D.W., Carnahan K.G., Thwaits R.N., Grigsby R.V., Holyoak G.R. and O'Neill K.L. (1994a) Detection of apoptosis induced DNA cleavage in scrapie -infected sheep brain. *FEMS Microbiol. Lett.* **115**, 341-346.

Fairbairn D.W., Mayers D. and O`Neill K.L. (1994b) Detection of DNA damaging agents in environmental water samples. *Bull. Environ. Contam. Tox*, **52**, 687-690.

Fairbarin D.W., Reyes W.A., Grigsby R.V. and O`Neill K.L. (1994c) Laser scanning microscopic analysis of DNA damage in frozen tissues. *Cancer Lett.* **76**, 127-132.

Fairbarin D.W., Olive P.L. and O'Neill K. L. (1995) The comet assay: a comprehensive review. *Mutation Res* **339**, 37-59.

Gedik C.M., Ewen S.W.B. and Collins A.R. (1992) Single cell gel electrophoresis applied to the analysis of UV-C damage and its repair in human cells. *Int. J. Radiat. Biol.* **62**, 313-320.

Gopalakrishna P. and Khar A. (1995) Comet assay to measure DNA damage in apoptotic cells. *J. Biochem. Biophys. Methods* **30**, 69-73.

Hartley A., Stone J.M., Heronm C., Cooper J.M. and Shapira A. H. V. (1994) Complex I inhibitors induce dose-dependent apoptosis in PC12 cells: Relevance to Parkinson's disease. *J. Neurochem.* **63**, 1987-1990.

Hartmann A., Plappert U., Raaddatz K, Grünert-Fuchs M. and Speit G. (1994) Does physical activity induce DNA damage? *Mutagenesis* **9**, 269-272.

Hartmann A., Herkommer K., Blück M. and Speit G. (1995) DNA-damaging effect of cyclophoaphamide on human blood cells in vivo and in vitro studies with the single-cell gel test (Comet assay). *Envir. Molec. Mutagen.* **25**, 180-187.

Hellman B., Vaghef H. and Boström B. (1995) The concepts of tail moment and tail inertia in the single cell gel electrophoresis assay. *Mutation Res.* **336**, 123-131.

Johnson E.M. and Deckwerth T.L. (1993) Molecular mechanism of developmental neuronal death. *Annu. Rev. Neurosci.* **16**, 31-46.

Kasamatsu T., Kohda K. and Kawasaki Y. (1996) Synergetic cytotoxicity of bleomycin and polyhydric alcohols; DNA strand breakage evaluated by "comet assay". *Biol. Pharm. Bull.* **19**, 632-635.

Kent C.R.H., Eady J.J., Ross G.M. and Steel G.G. (1995) The comet moment as a measure of DNA damage in the comet assay. *Int. J. Radiat. Biol.*, **67**, 655-660.

Kerr J.F.R., Wyllie, A.H. and Currie A.R. (1972) Apoptosis: basic biological phenomenon with wide-ranging implications in tissue kinetics. *Br. J. Cancer* **26**, 239-257.

Kerr J.F.R., Wyllie A.H. and Currie A.R. (1987) Apoptosis. In; Perspectives on mammalian cell death, C.S. Potten, Ed., Oxford University Press, Oxford, pp. 93-128.

Kleiman N. J. and Spector A. (1993) DNA single strand breaks in human lens epithelial cells from patients with cataract. *Current Eye Res.* **12**, 423-431.

Kreja L., Plappert U. and Nothdurft W. (1993) In vitro study in the radiosensitivity of canine hemopoietic and stroma cells with the single cell gel (comet) assay. *Exp. Hematol.* **21**, 1051.

Kruszewski M., Green M.H.L., Lowe J.E. and Szumiel I. (1994) DNA strand breakage, cytotoxicity and mutagenicity of hydrogen peroxide treatment at 4°C and 37°C in L5178Y sublines. *Mut. Res.* **308**, 233-241.

Langston J. W., Ballard P. Tetrud J. W. and Irwin I. (1983) Chronic Parkinsonism in humans due to a product of meperidine-analog synthesis. *Science* **219**, 979-980.

Latinwo, L.W., Ikediobi C.O. and Singh N.P. (1994) Effects of low levels of ionizing particles on rat kidney DNA. *FASEB J.* **8**, A906.

Lowe S.W., Schmitt E.M., Smith S.W., Osborne B.A. and Jachs T. (1993) p53 is required for radiation-induced apoptosis in mouse thymocytes. *Nature* **362**, 847-849.

Martz E, and Howell D.M. (1989) CTL: virus control cells first and cytolytic cell second? *Immunol. Today* **10**, 79- 86.

Maruyama W., Nakahara D., Ota M., Takahashi T., Takahashi A., Nagatsu T. and Naoi M. (1992) N-Methylation of dopamine-derived 6,7-dihydroxy-1,2,3,4-tetrahydroisoquinoline, (*R*)-salsolinol, in rat brains: In vivo microdialysis study. *J. Neurochem.* **59**, 395-400.

Maruyama W., Dostert P., Matsubara K. and Naoi M. (1995a) N-Methyl (*R*) salsolinol produces hydroxyl radicals: Involvement to neurotoxicity. *Free Radic. Biol. Med.* **19**, 67-75.

Maruyama W., Dostert P. and Naoi M. (1995b) Dopamine-derived 1-methyl-6,7-dihydroxyisoquinolines as hydroxyl radical promoters and scavengers in the rat brain: In vivo and in vitro studies. *J. Neurochem.* **64**, 2635-2643.

Maruyama W., Abe T., Tohgi H., Dostert P. and Naoi M (1996a) A dopaminergic neurotoxin, (*R*)-N-methylsalsolinol, increases in parkinsonian CSF. *Ann. Neurol.* **40**, 119-122.

Maruyama W., Narabayashi H., Dostert P. and Naoi M (1996b) Stereospecifc occurrence of a parkinsonism-inducing catechol isoquinoline, *N*-methyl (*R*)salsolinol, in the human intraventricular fluid. *J. Neural Transm.* **103**, 1069-1076.

Maruyama W., Sobue G., Matsubara K., Hashizume Y., Dostert P. and Naoi M. (1997a) A dopaminergic neurotoxin, 1(*R*), 2(*N*)-dimethyl-6,7-dihydroxy-

1,2,3,4-tetrahydroisoquinoline, *N*-methyl(*R*)salsolinol, and its oxidation product, 1, 2(*N*)-dimethyl-6,7-dihydroxyisoquinolinium ion, accumulate in the nigro-striatal system of the human brain. *Neurosci. Lett.* **223**, 61-64.

Maruyama W., Naoi M., Kasamatsu T, Hashizume Y., Takahashi T., Kohda K. and Dostert P. (1997b) An endogenous dopaminergic neurotoxin, *N*-methyl-(*R*)-salsolinol induces DNA damage in human dopaminergic neuroblastoma SH-SY5Y cells. *J. Neurochem.* **69**, 322-329.

Matsubara K., Neafsey E.J. and Collins M.A. (1992) Novel S-adenosylmethionine-dependent indole-N-methylation of β-carbolines in brain particulate fractions. *J .Neurochem.* **59**, 511-518.

Matsubara K. (1996) Occurrence of neurotoxic β-carbolinium cations in mammalian central nervous system. *Biogenic Amines* **12**, 161-169.

McCall C.A. and Cohen J.J. (1991) Programmed cell death in terminally differentiating keratincytes: role of endogenous endonucleases. *J. Invest. Dermatol.* **97**, 365-370.

Mckelvey V.J., Butler M. and Stewart L.H. (1992) Analysis of DNA content and integrity in cells extracted from bladder washing and voided urine specimens, in bladder cancer patients using the comet assay. *Mutation Res.* **271**,163.

McKelvey-Martin V.J., Green M.H.L., Schmerzer P., Pool-Zobel B.L., De Meo M.P. and Collins A. (1993) The single cell gel electrophoresis assay (comet assay): A European review. *Mutation Res.* **288**, 47-63.

Mizuno Y., Saitoh T. and Sone N. (1987) Inhibition of mitochondrial NADH-ubiquinone oxidoreductase activity by 1-methyl-4-phenyl-pyridinium ion. *Biochem. Biophys. Res. Commun.* **143**, 294-299.

Mizuno Y., Ohta S., Tanaka M. Takamiya S., Suzuki K., Sato T., Oya H., Ozawa T. and Kagawa Y. (1989) Deficiencies in Complex I subunits of the respiratory chain in Parkinson's disease. *Biochem. Biophys. Res. Commun.* **163**, 1450-1455.

Mochizuki H., Nakamura N., Nishi K. and Mizuno Y. (1994) Apoptosis induced by 1-methyl-4-phenyl-pyridinium ion (MPP^+) in ventral mesencephalic-striatal co-culture in rat. *Neurosci. Lett.* **170**, 191-194.

Mochizuki H., Goto G., Mori H. and Mizuno Y. (1996) Histochemical detection of apoptosis in Parkinson's disease. *J. Neurol. Sci.* **137**, 120-123.

Müller M.U., Bauch T. , Streffer C., Niedereichholz F. and Böcker W. (1994) Comet assay studies of radiation-induced DNA damage and repair in various tumor cell lines. *Int. J. Radiat. Biol.* **65**, 315-319.

Mutoh T., Tokuda A., Marini A.M. and Fujiki, N. (1994) 1-Methyl-4-phenylpyridinium kills differentiated PC12 cells with a concomitant change in protein phosphorylation. *Brain Res.* **661**, 51-55.

Nagatsu T and Yoshida M (1988) An endogenous substance of the brain, tetrahydroisoquinoline, produces parkinsonism in primates with decreased dopamine, tyrosine hydroxylase and biopterin in the nigrostriatal regions. *Neurosci. Lett.* **87**, 178-182.

Naoi M., Maruyama W., Zhang Y. H., Takahashi T., Deng Y. and Dostert P. (1995a) Enzymatic oxidation of the dopaminergic neurotoxin, 1(*R*), 2(N)-dimethyl-6,7-dihydroxy-1,2,3,4-tetrahydroisoquinoline, into 1,2-dimethyl-6,7-dihydroxyisoquinolinium ion. *Life Sci.* **57**, 1061-1066.

Naoi M., Maruyama W. and Dostert P. (1995b) Dopamine-derived 6,7-dihydroxy-1,2,3,4-tetrahydroisoquinoline; oxidation and neurotoxicity. *Progr. Brain Res.* **106**, 227-239.

Naoi M., Maruyama W., Dostert P., Hashizume Y., Nakahara D., Takahashi T. and Ota M. (1996a) Dopamine-derived salsolinol, induced parkinsonism in rat: Biochemical, pathological and behavioral studies. *Brain Res.* **709**, 285-295.

Naoi M., Maruyama W., Dostert P. and Hashizume Y. (1996b) Animal model of Parkinson's disease induced by naturally-occurring 1(*R*), 2(N)-dimethyl-6,7-dihydroxy-1,2,3,4-tetrahydroisoquinoline. *Biogenic Amines* **12,** 135-147.

Naoi M., Maruyama W., Dostert P. and Hashizume, Y. (1996c) N-Methyl-(*R*)salsolinol as a dopaminergic neurotoxin: From an animal model to an early marker of Parkinson's disease. *J. Neural Transm. Suppl.* **50**, 89-105.

Naoi M., Maruyama W., Dostert P., Kohda, K. and Kaiya T. (1996d) A novel enzyme enantio-selectively synthesizes (*R*)salsolinol, a precursor of a dopaminergic neurotoxin, *N*-methyl(*R*)salsolinol. *Neurosci. Lett.* **212**, 183-186.

Nicoletti L., Migliorati G., Pagliacci M.C., Urignani F. and Ricardi C. (1991) A rapid and simple method for measuring thymocyte apoptosis by propidium iodide staining and flow cytometry. *J. Immunol. Methods* **139**, 271-279.

Olive P.L. (1989a) Cell proliferation as a requirement for development of the contact effect in Chinese hamster V79 speroids. *Radiation Res.* **117**, 79-92.

Olive P.L. (1989b) Distribution, oxygenation and clonogenecity of macrophages in an murine tumor. *Cancer Commun,* **2**, 93-100.

Olive P.L., Banath J.P. and Durand R.E. (1990) Heterogeneity in radiation-induced DNA damage and repair in tumor and normal cells measured using the 'comet' assay. *RDA. Res.* **122**, 86-94.

Olive P.L. and Banta J.P. (1991) DNA double strand breaks measured in individual cells subjected to gel electrophoresis. *Cancer Res.* **51**, 4671-4676.

Olive P.L., Wlodek D. and Banáth J.P. (1991) DNA double-strand breaks measured in individual cells subjected to gel electrophoresis. *Cancer Res.* **51**, 4671-4674.

Olive P.L., Frizzier G. and Banta J.P. (1993) Radiation-induced apoptosis measured in TKO human B cells using the comet assay. *Rad. Res.* **136**, 130-136.

Olive P.L. and Banath J.P. (1993) Induction and rejoining of radiation-induced DNA single-strand breaks: "tail moment" as a function in the cell cycle. *Mutation Res.* **294**, 275-283.

Olive P.L., Durand R.E., LeRiche J., Olicotto L.A. and Jackson S.M. (1993) Gel electrophoresis of individual cells to quantify hypoxic fraction in human breast cancers. *Cancer Res.* **53**, 733-736.

Olive P.L., Nanath J.P. and McPhall H.S. (1994a) Lack of a correlation between radiosensitivity and DNA double- strand break induction or rejoining in six human tumor cell lines. *Cancer Res.* **545**, 3939-3946.

Olive P.L., Vikse C.M. and Durand R.E. (1994b) Hypoxic fractions measured in murine tumors and normal tissues using the comet assay. *Inter. J. Radiat. Oncol. Biol. Phys.* **29**, 487-491.

Olive P.L. and Banath J.B. (1995) Radiation-induced DNA double-strand breaks produced in histone-depleted tumor cell nuclei measured using the neutral comet assay. *Rad. Res.* **142**, 144-152.

Oppenheim R.W. (1991) Cell death during development of the nervous system. *Annu. Res. Neurosci.*, **14**, 453-501.

Östling O. and Johanson K.J. (1984) Microelectrophoresis study of radiation-induced DNA damage in individual mammalian cell. *Biochem. Biophys. Res. Commun.* **123**, 291-298.

Östling O., Johanson K.J., Blomquist E. and Hagelqvist E. (1987) DNA damage in clinical radiation therapy studied by microelectrophoresis in single tumor cells. *Acta Oncol.* **26**, 45-48.

Pool-Zobel B.L., Lanbertz R., Knoll M. and Schmerzer P. (1992) Detection of genotoxicity in cells of the gastrointestinal tract. *Gum Abstr. Mutagenesis* **7**, 166.

Plappert U., Barthel E., Raddatz K. and Seidel H.J. (1994) Early effects of benzene exposure in mice. Hematological versus genotoxic effects. *Arch. Toxicol.* **68**, 284-290.

Raff M.C. (1992) Social controls on cell survival and cell death. *Nature* **356**, 397-400.

Raff M.C., Barres B.A., Burne J.F., Coles H.S., Ishizaki Y. and Jacobson M.D. (1993) Programmed cell death and control of cell survival: Lessons from the nervous system. *Science* **262**, 695-700.

Roti Roti J.L. and Wright W.D. (1987) Visualization of DNA loops in nucleotides from HeLa cells; assays from DNA damage and repair. *Cytometry* **8**, 461-467.

Rydberg B. and Johanson K.J. (1978) Estimation of DNA strand breaks in single mammalian cells. in: P.C. Hanawalt, E.C. Friedberg and C.F. Fox '(Eds.) DNA Repair Mechanisms, Academic Press, New York, pp. 465-468.

Schapira A.H.V., Cooper J.M., Dexter D., Jenner P., Clark J.B. and Marsden C.D. (1990) Mitochondrial complex I deficiency in Parkinson's disease. *J. Neurochem.* **54**, 823-827.

Shafer D.A., Xie Y. and Falek A. (1994) Detection of opiate-enhanced increases on DNA damage, HPRT mutants and the mutation frequency in human HUT-78 cells. *Env. Mol. Mutagen.* **23**, 37-44.

Singh N.P., McCoy M.T., Tice R.R. and Schneider E.L. (1988) A simple technique for quantitation of low levels of DNA damage in individual cells. *Exp. Cell Res.* **175**, 184-191.

Singh N.P., Lai H. and Khan A. (1995) Ethanol-induced single-strand DNA breaks in rat brain cells. *Mut. Res.* **345**, 191-196.

Strolin Benedetti M., Bellotti V., Pianezola E., Moro E., Carminati P. and Dostert P. (1989) Ratio of the R and S enantiomers of salsolinol in food and human urine. *J. Neural Transm.* **77**, 47-53.

Tice R.R., Andrews P. and Singh N. P. (1990) The single cell gel assay: a sensitive technique for evaluating intercellular differences in DNA damage and repair. *Basic Life Sciences.* **53**, 291-301.

Tipton K.F. and Singer T.P. (1993) Advances in our understanding of the mechanism of the neurotoxicity of MPTP and related compounds. *J. Neurochem.* **61**, 1191-1206.

Vijayalaxmi G.S., Strauss H. and Tice R.R. (1993) An analysis of gamma-ray-induced DNA damage in human blood leukocytes, lymphocytes and granulocytes. *Mutation Res.* **292**, 123-128.

Walkinshaw G. and Waters C.M. (1995) Induction of apoptosis in catecholaminergic PC12 cells by L-DOPA. Implication for the treatment of Parkinson's disease. *J. Clin. Invest.* **95**, 2458-2464.

Ward J.A., Olive P.L., Burr A.H. and Rosin M.P. (1993) A sensitivity to oxidative stress is linked to chromosome-11 but is not due to a difference in single-strand DNA breakage or repair. *Mutation Res.* **294**, 299-308.

Williams G.T. (1991) Programmed cell death: apoptosis and oncogenesis. *Cell* **65**, 1097-1098.

Whitaker S.J., Powell s.N. and McMillan T.J. (1991) Molecular assays of radiation-induced DNA damage. *Eur. J.Cancer* **27**, 922-928.

Wolvetang E.J., Johnson K.L., Knauer K., Ralph S. J. and Linnane A.W. (1994) Mitochondrial respiratory chain inhibitors induce apoptosis. *FEBS Lett.* **339**, 40-44.

Wyllie A.H., Morris R.G., Smith A.L. and Dunlop D. (1984) Chromatin cleavage in apoptosis: association with condensed chromatin morphology and dependence on macromolecular system. *J. Pathol.* **142**, 67-77.

Yoritaka A., Hattori N., Uchida K., Tanaka M., Stadtman E.R. and Mizuno Y. (1996) Immunohistochemical detection of 4-hydroxynonenal protein adducts in Parkinson's disease. *Proc. Natl. Acad. Sci. USA.* **93**, 2696-2701.

Zychlinsky A., Zheng L.M., Liu C.C. and Young J.D.E. (1991) Cytolytic lymphocytes inducce both apoptosis and necrosis in target cells. *J. Immunol.* **146**, 393-400.

Progress in HPLC-HPCE, Vol. 7, pp. 817-837
G. Ali Qureshi *et al.* (Eds)

Three Dimensional HPLC-EC Analyses of Regional Brain Neurotransmitter Disturbances in Alzheimer's Disease With and Without Lewy Bodies.

PHILIP J. LANGLAIS

Department of Neurosciences, Univ. of California at San Diego, School of Medicine, Neurology Research Service, Veterans Affairs Medical Center, San Diego, CA, Department of Psychology, San Diego State University, San Diego, CA, 92182, U.S.A.

INTRODUCTION

An extensive body of literature relates disturbances in neurotransmitters to neurodegenerative, neuropsychiatric, and behavioral disorders in both humans and experimental animals. Although the potential role of the neuroactive compounds in several different disorders has been established, their routine analysis has not been widely implemented in the diagnosis or treatment of these disorders. The major problems underlying this failure to utilize neurotransmitter analyses are the economic and technical limitations of current methodologies. A critical limiting factor is the difficulty in adequately measuring most of the known metabolically related compounds of a particular neurotransmitter. Without a relatively complete picture of the synthetic and metabolic pathway, it is difficult to determine the nature of the disturbances contributing to alterations in a specific neurotransmitter. In addition, most methodologies are not only limited in the number of compounds that can be simultaneously measured and identified within a tissue sample, they are virtually unable to detect and identify unusual neurotransmitter-related compounds. A further difficulty with this field is the fact that the concentrations and physiological effects of a particular neurotransmitter are modified by other neurotransmitters in the same or parallel metabolic pathways. One solution to this problem is the simultaneous determination of all the compounds within the metabolic pathways of several neurotransmitters. This approach appears straighforward but has been severely hampered by the relatively poor sensitivity and separation capabilities of most analytical instruments.

High performance liquid chromatography (HPLC) coupled with amperometric-electrochemical (EC) detectors have been widely used for analyses of neurotransmitters in a variety of tissues (Kissinger et al., 1981; Mefford, 1981). These flat-plate, single-electrode amperometric detectors have been used in dual detector configurations to yield qualitative information by simultaneously recording chromatograms at different potentials to observe different segments of a compound's current/voltage curve (Kissinger, 1982; Blank, 1976). However, only a small fraction of all molecules is converted on the surface of the amperometric detector and much of the compound remains unreacted and available to react at a higher potential electrode downstream. This feature imposes an inherent limitation on the resolution achievable by series amperometric detector systems.

Another constant source of analytical error in HPLC-EC assays using a single electrochemical detector is the inability to detect coeluting compounds. The development of gradient chromatographic separations and the construction of analytical columns with a larger number of plates has reduced but by no means eliminated quantitative errors caused by coeluting compounds. These single and dual EC detection systems also suffer from a minimum of qualitative information, i.e., they are capable of identifying only a handful of compounds within a single assay. These analytical problems place significant diagnostic and theoretical limitations of studies conducted on 'raw extracts' of very complex tissue such as human post-mortem brain. Crude perchloric acid extracts of frozen postmortem brain contain most if not all of the compounds in the synthetic and metabolic pathways of neurotransmitters. Studies directed at detecting disturbances in these pathways are ideally conducted on these crude extracts which may contain over 300-400 small compounds, many of which are detectable by electrochemical sensors.

More recently, analytical instruments which combine multiple series coulometric detectors with gradient liquid chromatography have been used to succesfully separate and identify large number of neurotransmitter substances within crude extracts of brain tissue or cerebrospinal fluid (Langlais et al., 1988; Langlais and Mair, 1990; Langlais et al., 1991; Langlais et al., 1993a,b; Matson et al., 1984). The coulometric detectors used in these analytical instruments and assays have overcome many of the inherent limitations imposed by amperometric detectors. The flow through design of these newer coulometric electrodes allows for a theoretical 100% conversion and detection vs the 2-5% detection of amperometric sensors. These features greatly enhance their sensitivity and permit detection of femtogram (10^{-15}) amounts of neurotransmitter substances. Their insensitivity to electrode poisoning and the nearly complete reaction rate of the flow-through coulometric detectors have been explored and successfully utilized in the simultaneous analysis of large numbers of neurotransmitters and related substances.

One instrument in particular, the Coulometric Electrochemical Array System (CEAS, ESA Inc.) has been used by neuroscientists for over 10 years. The CEAS is a fully automated, computer driven instrument which incorporates multiple coulometric sensors (6-12) that are arranged in series. The theoretical and practical applications of the CEAS have been described in detail (Achworth and Bowers, 1997; Cullison and

Gamache, 1997). A key feature of the CEAS instrument is its ability to permit detection of coelution errors and the ability to identify compounds based on their current/voltage characteristics as well as their retention properties. The bandspreading, dead volume, and pressure characteristics of the coulometric cells used in tne CEAS are such that there is little loss of chromatographic separation nor any significant difference in registration time between the first and last sensor in the array. Each electrode can be operated at incrementally increasing or decreasing applied voltages. Since each compound is completely reacted at the preceding electrode, all of the remaining unreacted compound is available for detection at the subsequent electrode operating at a different voltage. Peaks are located in terms of the sensor from which a response is recorded and the elution/retention time. A peak can be then be plotted along an applied voltage axis (E), a elution time axis (t), and a response axis (V). The first sensor in the array is designated as T_1 and subsequent recording sensors are designated as T_2-T_n. Peaks can be characterized and identified by the ratio of the response recorded on the dominant sensor, i.e., sensor with maximal response, to that recorded on the preceding and following sensors. Typically, compounds show a dominant detector response with a leading and following response. The ratio of response across three electrodes is highly characteristic of a particular compound. For example, 3-methoxy-4-hydroxyphenylglycol (MHPH) under certain chromatographic conditions yields a ratio of .009 preceding/dominant and .113 following/dominant. Signatures obtained from the direct injection of crude brain extracts or CSF can be compared to signatures obtained from authentic standards and used to verify the purity and determine the identity of compounds. If a correct signature ratio is not obtained, this suggests that a coeluting contaminant may be present. An example of the use of signature ratios to verify the 'purity' of the peaks is shown in Table 1. These data were obtained from the analyses of a crude perchloric acid extract of putamen of a post-mortem human brain. The chromatographic conditions were similar to those used in the analyses of Alzheimer and control brain tissues described in the following section.

Unknown peaks are initially identified by matching their retention time to those obtained from standards. The signature of the sample peak is then compared to the signature of the standard by computing a ratio accuracy (sample peak signature/standard peak signature). Peaks with a ratio accuracy greater than 0.90 are considered free of any contaminants. Peaks with a ratio betwee 0.75-0.90 are more carefully examined for evidence of leading or trailing 'humps' or other indications of coeluting compounds. Peaks with a ratio less than 0.75 are considered to be contaminated. As shown in the Table 1, some compounds, e.g., norepinephrine and epinephrine, react completely on one sensor and have signatures of 0.000. In this case, the presence of a signal on adjacent sensors and a signature ratio greater than 0.000

indicates the presence of a coeluting substance. As described earlier, the serial arrangement of sensors provides another dimension, i.e., voltage, by which coeluting compounds can be separated. As shown in Table 1, methionine and L-DOPA have similar retention times but react at quite different sensors. The dominant sensor for methionine is 16 (900 mV) whereas sensor 4 (180 mV) is the dominant sensor for L-DOPA.

The CEAS technology has been applied to several interesting diagnostic and pathophysiological questions in neuropsychiatry. The following sections describe the application of this methodology to study the biochemical alterations regarding a recently described variant of Alzheimer's disease. These questions required a rather careful and detailed analyses of several neurotransmitter systems in several regions of postmortem brain The results obtained have clarified the biochemical disturbances of these variants of AD and suggested a more careful analysis of current therapies used in AD patients with parkinsonian features.

NEUROTRANSMITTER DISTURBANCES IN AD & LBV

There are several pathophysiological substrates implicated in the cognitive, affective, and motor disturbances observed in dementing illnesses among the elderly. Among the best documented are significant reductions in biochemical markers of cholinergic activity in Alzheimer's disease (AD) (Bowen et al., 1976; Rossor et al., 1984) and of dopaminergic activity in Parkinson's disease (PD) (Hornykiewicz, 1973).

In the past several years, 30-40% of clinically diagnosed and pathologically confirmed cases of AD at the San Diego Alzheimer's Disease Research Center (ADRC) have been shown to contain cortical and subcortical Lewy Bodies (LBs) (Hansen et al., 1990). Lewy bodies are intracytoplasmic eosinophilic neuronal inclusions that have been historically linked with Parkinson's disease (PD). Although PD was originally described as a movement disorder characterized by tremor, flexed posture and shuffling gait, many neurologist now recognize that PD is often accompanied by dementia. Neuropathologically the link between PD and Lewy bodies has also become muddled by the observation of Lewy bodies in AD and plaques and tangles in PD. Furthermore, recent studies have demonstrated that a high percentage of PD cases have neocortical Lewy bodies (reviewed in Hansen and Galasko, 1992). These findings make the clinical and pathological distinctions between PD and AD less clear than in the past.

An important question is whether these AD cases with subcortical and cortical Lewy represent individuals who have both AD and PD neurodegenerative changes. AD and LBV do share certain neuropathological features. Neocortical loss of synaptic density (Masliah et al., 1993) and the number of neuritic plaques within neocortex (Samuel et al., 1997) are comparable in these two disorders. Clinically, these LBV cases have all of the features diagnostic of AD but also demonstrate more severe deficits of attention, fluency (letter but not category) and visuospatial processing than the pure AD cases. In addition, features associated with PD such as masked facies and bradykinesia are uncommonly mild, none have resting tremor, and about one-half have an action tremor. Neuropathologically, these cases are also different from pure AD. Although these cases contain the usual density of neocortical senile plaques, the plaques are diffuse and there are fewer neocortical neurofibrillary tangles than pure AD cases, and the neuronal loss in LC, substantia nigra (SN) and substantia innominata (SI) is far greater than observed in the pure AD cases. These cases differ neuropathologically from typical PD in that the pattern and distribution of neuronal loss in the substantia nigra (SN) is unusually mild compared to that observed in typical PD (Gibb and Lees, 1991).

On these grounds, it was suggested that the LBV cases represent a clinically and pathologically distinct subset of AD and labelled the Lewy Body variant (LBV) of AD (Hansen et al., 1990). This view has been challenged by another investigator who proposes that the LBV cases are similar to "diffuse Lewy Body disease" (DLBD) and not a variant of AD (Dickson et al., 1987). Another group has argued that these LBV cases probably represent PD with additional involvement of cerebral cortex and the presence of AD changes (Fearnley et al., 1991). They point out that without a careful quantitative cell count in the various regions of SN, the actual loss of neurons may have been underestimated. A third group has described yet another disorder, i.e., senile dementia of Lewy body type (SDLT), in a group of psychiatric cases in which mild extrapyramidal features were apparent but most did not display the classical triad of tremor, rigidity and akinesia (Perry et al., 1990). These cases also resemble the LBV cases in that they contain Lewy body formations in cortex and basal ganglia but at densities well below those typically observed in diffuse Lewy body disease. However, in contrast to the LBV cases, these SDLT cases do not meet the clinical criteria for AD and often present initially with a confused state, often fluctuating and associated with visual hallucinations and behavioral disturbances.

Understanding the pathological-behavioral relationships in this group of patients has significant implications to both their treatment and diagnosis. Are the motor disturbances related to dopamine deficiency in basal ganglia or to some unknown metabolic defect? Why are the mild extrapyramidal disturbances of these patients is relatively unresponsive to levodopa treatment? Are these motor disturbances unrelated to a dopamine deficiency? Is the greater decline in cognitive functions due to a greater disruption of neocortical cholinergic activity or to additional loss of other neurotransmitter systems implicated in attention and cognitive processing? To provide answers to these questions and the issue of whether LBV is simply the coexistence of PD and AD, a fairly extensive neurochemical study was conducted. This study has been reported in detail elsewhere (Langlais et al., 1993) and is briefly presented here to illustrate the growing need for multiple transmitter analyses in the study of neurodegenerative disorders.

All subjects were diagnosed clinically (McKhann et al., 1984) and neuropathologically (Khachaturian, 1985) as having AD. The LBV cases were identified neuropathologically by the presence of Lewy bodies in neocortex and substantia nigra. The AD and LBV cases were matched for age of onset, sex, and postmortem interval. The age of onset, duration of illness, and age at death were similar in the AD and LBV groups. A similar proportion of patients in these groups had reached a floor on the Mini-Mental Status Exam and Blessed Information-Memory-Concentration (IMC) test (Blessed et al., 1968) within the year prior to death, i.e., 2/6 for AD group and 3/7 for the LBV group. Extrapyramidal symptoms (EPS) developed early in 5 of the 7 LBV cases, at a stage when they were relatively mildly demented and preceding neuroleptic exposure. By contrast, none of the pure AD cases had EPS relatively early in the course of illness. Two pure AD cases did develop EPS but only after neuroleptic treatment. Post-mortem tissue from control subjects was obtained from the Brain Tissue Resource Center, McLean Hospital/Harvard Medical School and general autopsies conducted at the UCSD Medical Center.

The levels of dopamine, norepinephrine, serotonin, kynurerine, their precursors tyrosine, DOPA, tryptophan, 5-hydroxytryptophan, and metabolites homovanillic acid (HVA), 3-methoxy,4-hyroxphenylglycol (MHPG), and 5-hydroxyindoleacetic acid (5-HIAA) were measured in frozen post-mortem tissue from caudate, putamen, and Brodman areas 8 (frontal cortex), 21 (temporal cortex), and 39/40 (posterior parietal cortex) with a single chromatographic assay (Langlais et al., 1991) using a gradient, series coulometric-electrochemical detector HPLC system (Model CEAS, ESA Inc.). Mobile phase A consisted of 0.1M sodium phosphate, pH 3.35 with sodium dodecyl sulfate (10 mg/L) and 5 mM nitrilotriacetic acid. Mobile phase B consisted of 50% methanol/water with 50 mg/L sodium docedyl sulfate at pH 3.45 and 5 mM nitrilotriacetic acid. Separations were accomplished at a flow rate of 1.0 ml/min, starting with 94%A/6%B for the first six mins, then increasing in a linear gradient to 60%A/40%B at 20 mins. At 21 mins, the gradient increased linearly to a final 10%A/90%B at 28 mins. The sensors were set from 0 to 900 mV in 60 mV increments. Two 8-x 0.45-cm, 3 mm particle size, C_{18} ESA HR80 columns were used in series at a temperature of 35°C. In addition, the level of tyrosine hydroxylase (TYR-OH) enzyme was measured with an immunoblotting technique (Girault et al., 1989) and the activity of choline acetyltransferase (ChAT) with a radioenzymatic assay (Fonnum et al., 1969).

FINDINGS

Severe reductions of biochemical markers of dopamine activity were observed in the basal ganglia of the LBV cases, suggesting a biochemical resemblance to typical PD. In the LBV brains, dopamine levels were reduced to 5% of the control value in the putamen (Figure 1) and to 19% of controls in the caudate (Figure 2). Previous studies have shown that the loss of dopamine in idiopathic PD is likewise greater in the putamen (<10% of control value) than caudate (10-20% of control value) (Fahn et al., 1971; Kish et al., 1988). The observations of a significant reduction of the dopamine precursor, DOPA, the major dopamine metabolite, HVA (Table 2), and TOH enzyme and normal levels of tyrosine within the putamen of the LBV cases are also consistent with a severe functional disruption of nigral cells and loss of terminals. These findings demonstrate that dopaminergic innervation of the basal ganglia of the LBV cases is as severely impaired as that observed in typical PD. Furthermore, reductions of 5-HT (Figure 1) and its metabolite 5-HIAA (Table 2) were observed in both caudate and putamen of the LBV cases that are similar to those observed in typical PD brains (Scatton et al., 1983; Arai et al., 1984; Hornykiewicz, 1982; Dubois et al., 1983; Birkmayer et al., 1974; Fahn et al., 1971).

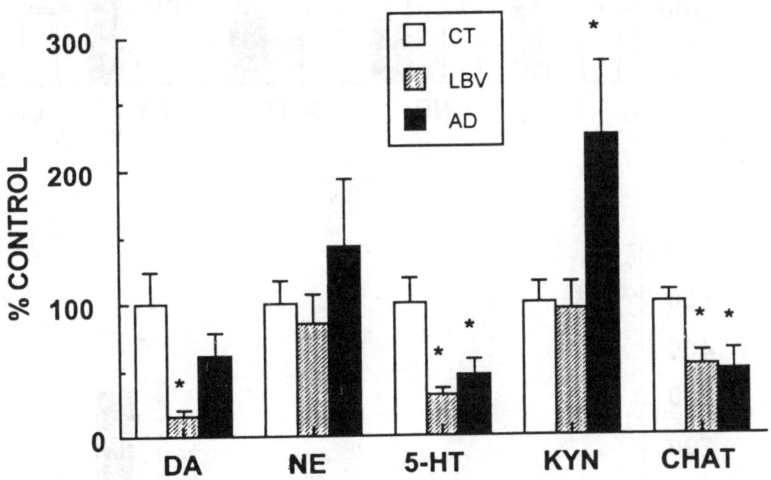

Figure 1. Concentration of dopamine (DA), norepinephrine (NE), serotonin (5-HT), kynurenine (KYN), and the activity of choline acetyltransferase (CHAT) in putamen of subjects diagnosed pathologically as pure Alzheimer disease (AD), Lewy Body variant of AD (LBV) and normal control subjects (CT). Values are the mean ± S.E.M. *p<.05 compared to controls, post hoc Scheffe test; ‡p<.05 compared to controls and AD, post hoc Scheffe test.

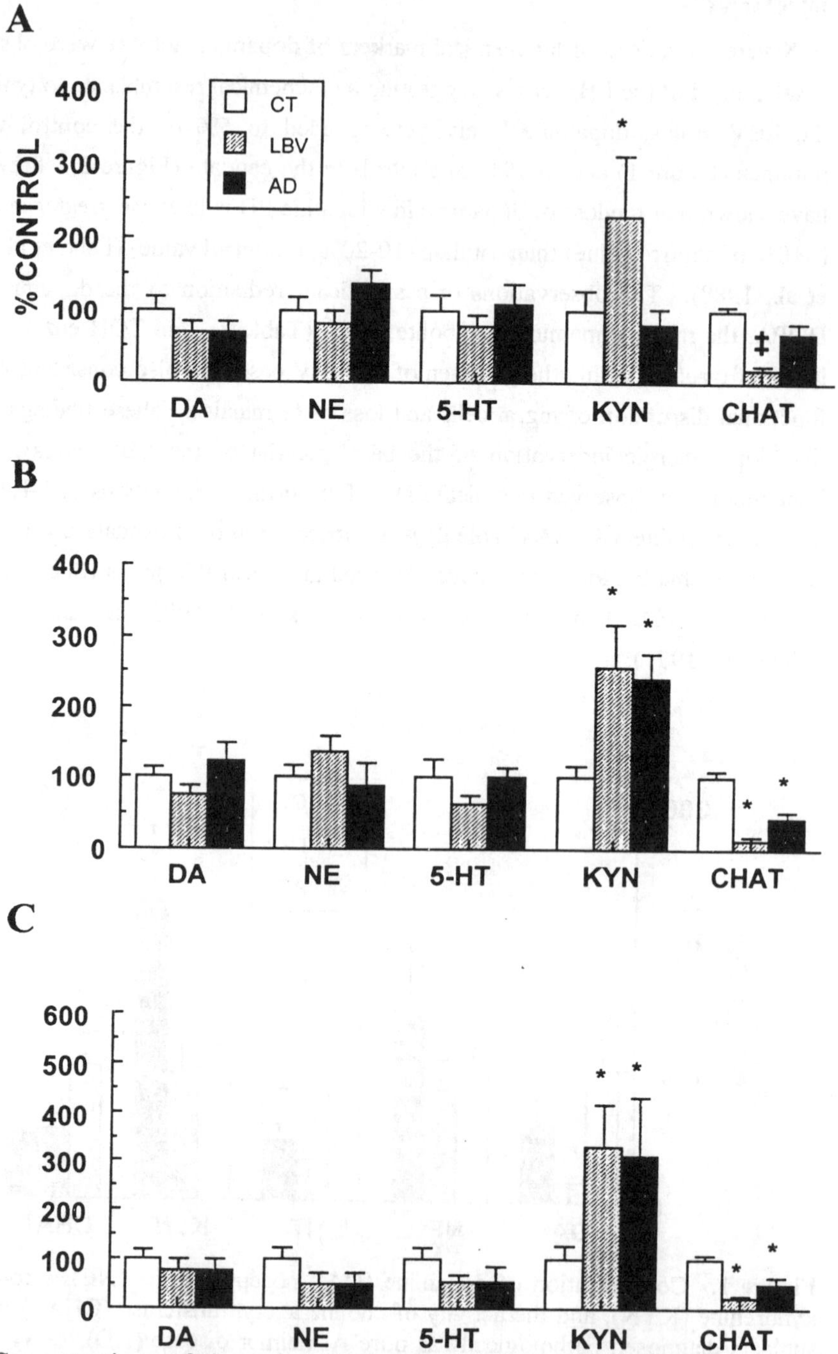

Figure 2. Concentration of dopamine (DA), norepinephrine (NE), serotonin (5-HT), kynurenine (KYN), and the activity of choline acetyltransferase (CHAT) in frontal (A), parietal (B), and temporal (C) neocortex of subjects diagnosed pathologically as pure Alzheimer disease (AD), Lewy Body variant of AD (LBV) and normal control subjects (CT).

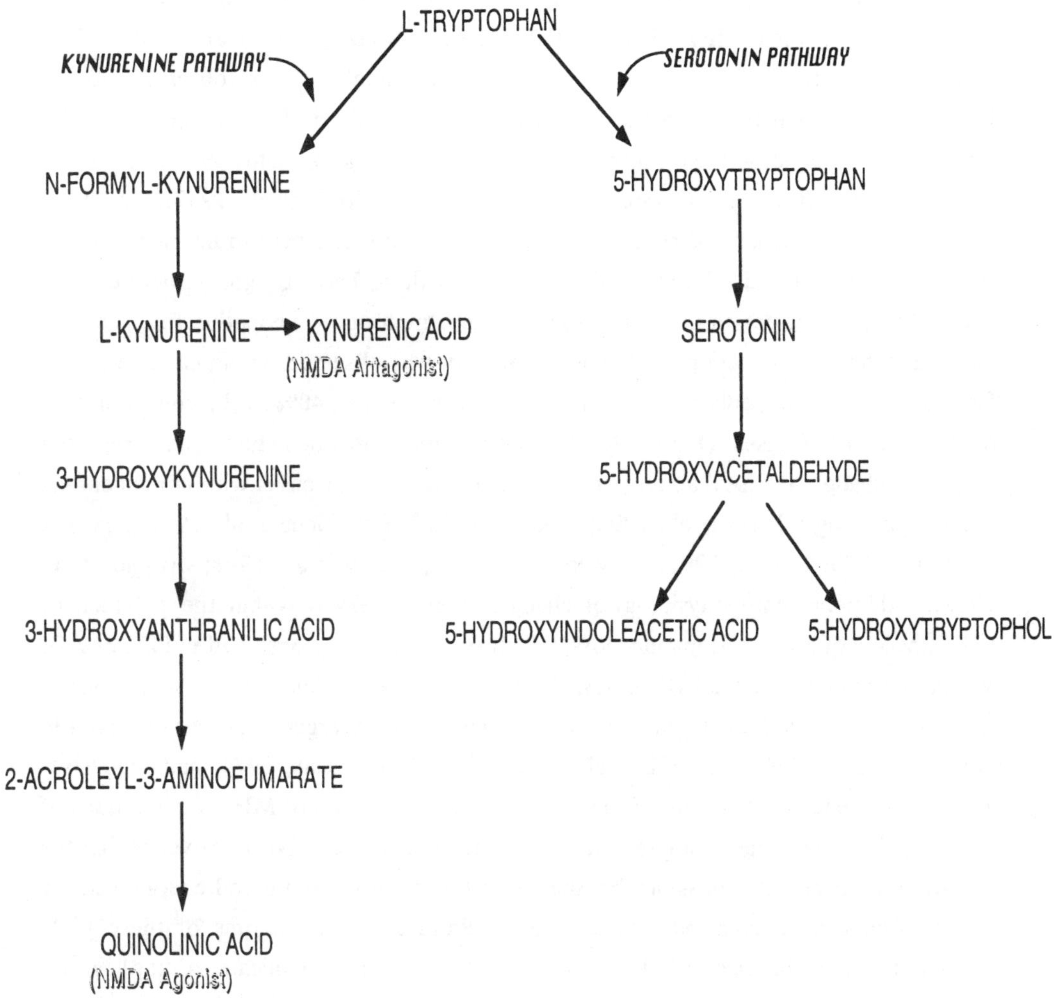

Figure 3. Schematic representation of the metabolism of tryptophan along the serotonin and kynurenine pathways

In the only other study of brain dopamine levels in cases equivalent to LBV, i.e., senile dementia of the Lewy body type (SDLT), which contains AD pathology, nigral degeneration and neocortical Lewy bodies (Perry et al., 1990) the dopamine level in caudate was only moderately reduced (42% of control levels) and congruent with the mild EPS observed in those patients. The severe dopamine depletion but relatively mild extrapyramidal symptoms (EPS) observed in these LBV cases suggest that additional pathophysiological changes exist in LBV which attenuate parkinsonism, particularly resting tremor. One possible site of damage is the subthalamic nucleus which when destroyed can reverse MPTP-induced PD in monkeys (Bergman et al., 1990). This hypothesis seems unlikely since the subthalamic nucleus was observed to be histologically unremarkable in the few LBV cases examined (Langlais et al., 1993). Damage to other structures within the extrapyramidal motor pathways, such as the globus pallidus (Bravo and Cooper, 1959) and thalamus (Kelly et al., 1987), have been shown to reduce tremor and rigidity in PD. Acetylcholine is a neurotransmitter used in both interneuronal and efferent projections within these basal ganglia structures, and anticholinergic medication improves some of the symptoms of idiopathic PD. On this point, a significant reduction in cholinergic activity was observed within caudate of our LBV cases, averaging only 52% of controls and was less than 40% of the control in four of the seven LBV cases (Figure 2). In contrast, most previous studies have reported very little change in ChAT activity in caudate of PD cases or cases with combined AD and PD pathology (Perry et al., 1990; Gaspar et al., 1980; Rinne et al., 1991; Sirvio et al., 1989; Aubert et al., 1992; Bowen et al., 1976; Perry et al., 1991; Araujo et al., 1988). Additional pathophysiological changes were observed within the putamen of LBV, the site of greatest dopamine loss, which are not typically observed in PD nor AD. Within the putamen of the LBV cases, there was a severe reduction of norepinephrine (to 4% of control value) (Figure 1). This decline is much larger than that observed in typical PD (Hornykiewicz, 1982; Fahn et al, 1971; Rinne et al., 1974) and suggests a loss of noradrenergic terminals. The presence of a normal level of MHPG in putamen of these LBV cases (Table 2) suggests a compensatory increase in NE turnover within the surviving terminals. It is possible that the reduced cholinergic and noradrenergic activity within basal ganglia contribute to the milder Parkinsonian symptoms of these LBV cases, as some theories of the pathophysiology of PD implicate a cholinergic-dopaminergic imbalance. Unfortunately, this hypothesis could not be tested because of a lack of longitudinal and detailed measures of EPS and other neurologic features. Further studies are needed to conduct similar biochemical analyses of basal ganglia structures in postmortem brain from patients with LBV, PD, and AD in whom careful assessments of motor disturbances, i.e., bradykinesia, rigidity, resting tremor, masked facies, postural instability, shuffling gait, stooped posture, and hypophonic speech are available. Comparison of the distribution and severity of these features with the magnitude of change in dopamine, norepinephrine, and acetylcholine markers within caudate and putamen should help to determine if these transmitter disturbances contribute to the milder Parkinson's symptoms of LBV. As mentioned previously, the

LBV cases studied at our ADRC are different from pure AD in that they have more severe cognitive and memory deficits than AD patients (Hansen et al., 1990). The loss of neocortical ChAT activity is greater in these LBV cases than in pure AD cases, particularly in frontal cortex (Figure 2), and may partially contribute to the suspected greater intellectual decline in these LBV cases. The loss of neocortical ChAT activity in our LBV cases was also greater than that observed in most previous studies of pure AD (Perry et al., 1985, 1990, 1991; Aubert et al., 1992; Bowen et al., 1976; Araujo et al., 1988; Rossor et al., 1984; D'Amato et al., 1987; Whitehouse et al., 1988; Crow et al., 1984). However, the severe reduction of ChAT in the LBV cases closely resemble many but not all studies of patients with combined AD and PD pathology (Perry et al., 1985, 1990, 1991; Aubert et al., 1992; D'Amato et al., 1987; Whitehouse et al., 1988) and diffuse Lewy body disease (Dickson et al., 1987). Furthermore, the levels of ChAT in the LBV cases are closer to those reported in severely demented PD cases than in most cases of pure AD (Aubert et al., 1992; Sirvio et al., 1989). In many of these PD cases, AD pathology has been reported in conjunction with the PD lesions. However, in the LBV cases, the severe ChAT depletion appears unrelated to the severity of Alzheimer pathology since neocortical plaque and tangle counts in LBV cases were similar or lower than in pure AD cases (Langlais et al., 1993). These observations suggest that the more severe reduction of ChAT activity may be more related to the presence of Lewy bodies than to numbers of plaques and tangles.

Very little is known about neocortical levels of the monoamines and their role in the behavioral and cognitive impairments of LBV, AD or PD. Several studies have failed to detect any significant differences in neocortical dopamine or homovanillic acid (HVA) levels of control and PD brains, perhaps due to the very low levels of dopamine and their high variability in postmortem brain. One study, however, has reported a significant reduction of dopamine (39% of control) in frontal cortex from PD cases (Scatton et al., 1983). In this study, and a recently published study (Ogawa et al., 1992), dopamine concentration in frontal cortex of LBV cases was reduced but these changes were not significant. However, a reduction of HVA (56% of control) was detected in frontal cortex of our LBV cases (Table 2) of a similar degree to that reported in PD (62% of control [Scatton et al., 1983]; 58% of control [Perry et al., 1991]) and in cases of combined AD/PD (Perry et al., 1991). In unmedicated PD patients, Ogawa et al., 1992 found even larger reductions (34% of control) of HVA in frontal cortex. The HVA concentration in neocortical regions from the LBV cases was also lower than in the current group of AD cases, suggesting some difference in neocortical dopamine utilization between the LBV and pure AD cases. It is clear, however, that more studies involving a larger number of subjects are needed to

TABLE 1. *Signature ratios and ratio accuracies of compounds in extract of putamen from human postmortem brain.*

COMPOUND	CONC. (ng)	R.T. (mins)	RT ERROR	RATIO ACCURACY	
				Preced/Dom.	Follow/Dom
MHPG	7.05	2.76	0.008	5/6=0.495	7/6=0.877
HGA	1.49	2.96	0.008		2/1=0.000
GUNOSINE	1267.22	4.63	0.017	12/13=0.741	14/13=0.886
NE	3.55	4.83	0.017	1/2=0.000	3/2=0.000
L-DOPA	4.38	5.56	0.009	3/4=0.00	5/4=0.892
METHONINE	2044.09	5.70	0.012	15/16=0.799	
4HPLA	58.22	6.93	0.006	9/10=0.930	11/10=0.730
3OH-ANTH	0.31	8.53	0.033	1/2=0.000	3/2=0.000
EPI	0.14	9.72	0.058	1/2=0.000	3/2=0.000
TYROSINE	3745.63	9.76	0.025	10/11=0.955	12/11=0.986
4HPAC	37.42	10.78	0.000	10/11=0.547	12/11=0.913
5-HIAA	4.06	11.39	0.017	2/3=0.000	4/3=0.871
HVA	83.01	13.86	0.110	5/6=0.986	7/6=0.801
3-OMDOPA	0.15	14.33	0.004	5/6=0.867	7/6=0.921
5-HTP	0.56	14.78	0.003	3/4=0.846	5/4=0.000
DA	9.88	15.80	0.053	1/2=0.000	3/2=0.000
MN	0.29	16.25	0.167	6/7=0.564	8/7=0.000
KYN	3.54	17.65	0.117	12/13=0.780	14/13=0.926
5-HT	2.21	21.60	0.103	2/3=0.000	4/3=0.799
3-MT	0.41	22.53	0.242	5/6=0.000	7/6=0.000
TRYPTOPAN	75.14	26.28	0.058	11/12=0.881	13/12=0.00

Abbreviations: DA-dopamine; EPI-epinephrine; HGA-homogentisic acid; 5-HIAA-5-hydroxyindoleacetic acid; HPAC-hydroxyphenylacetic acid; 4-HPLA-4-hydroxyphenyllactic acid; 5-HT-serotonin; 5-HTP-5-hydroxytryptophan; HVA-homovanillic acid;KYN-kynerunine; MHPD-3-methoxy-4-hydroxyphenylethylglycol; MN-metanephrine; 3OH-ANTH-3hydroxyanthranilic acid;3-MT-3-methoxytyramine; NE-norepinephrine; 3-OMDOPA-3-o-methyl-DOPA.

TABLE 2. *Concentration of monoamine metabolites and precursors in postmortem brain regions.*

REGION/ GROUP	HVA	MHPG	5-HIAA	TYR	DOPA	5-HTP
CAUDATE						
CT (N=6)	32.04±3.6	0.19±.03	1.11±.20	346±37	0.30±.07	0.07±.01
LBV (N=7)	11.38±2.3[a]	0.21±.04	0.33±.07 [a]	400±63	0.19±.07	0.06±.02
AD (N=6)	20.20±4.5	0.40±.03[c]	0.50±.13 [a]	435±28	0.25±.09	0.08±.02
PUTAMEN						
CT N=8)	50.54±7.5	0.35±.06	1.75±.16	296±52	2.12±.49	0.23±.08
LBV (N=7)	12.55±1.9[b]	0.32±.06	0.75±.14 a	376±18	0.69±.18[a]	0.08±.05[a]
AD (N=6)	38.09±2.7	0.39±.04	0.70±.14 [a]	391±24	1.62±.28	0.25±.10
FRONTAL						
CT (N=9)	2.32±0.4	0.17±.04	0.14±.03	339±34	0.11±.02	0.02±.01
LBV (N=6)	1.29±0.2	0.29±.12	0.11±.02	360±67	0.11±.03	0.03±.01
AD (N=6)	1.57±0.3	0.44±.10	0.15±.04	327±47	0.12±.03	0.03±.01
PARIETAL						
CT (N=9)	2.07±0.3	0.69±.12	0.08±.02	616±80	0.37±.04	0.10±.02
LBV (N=7)	1.73±0.4	0.86±.10	0.05±.01	673±79	0.38±.06	0.22±.07
AD (N=6)	2.49±0.5	0.62±.17	0.08±.01	654±83	0.48±..06	0.11±.04
TEMPORAL						
CT (N=12)	1.64±0.3	0.21±.02	0.09±.02	392±50	0.13±.02	0.03±.02
LBV (N=7)	1.10±0.1	0.20±.05	0.05±.01	423±115	0.10±.01	0.04±.03
AD (N=6)	1.53±0.2	0.14±.01	0.05±.02	654±83	0.08±.01	0.10±.03

[a] p<.01, compared to controls, post hoc Scheffe test;[b] p<.01, compared to controls and AD, post hoc Scheffe test. All statistically significant differences remained significant after Bonferroni correction for multiple comparisons. Values represent the mean ± S.E.M in ng/mg protein. HVA = homovanillic acid; MHPG = 3-methoxy-4-hydroxyphenylglycol; 5-HIAA = 5-hydroxyindoleacetic acid; TYR= tyrosine; DOPA = 3,4 Dihydroxyphenylalanine; 5-HTP = 5-hydroxytryptophan; CT = controls matched for age, sex, and postmortem interval to the LBV and AD groups; LBV = Lewy body variant; AD = Alzheimer's disease.

determine the integrity of the mesocortical dopamine systems of LBV as well as PD and AD.

We also observed reductions of serotonin (5-HT) concentration in frontal, temporal, and parietal neocortex of the LBV group (Figure 3). The high variance and small number of cases examined probably prevented the demonstration of statistically significant group differences. The absence of PD cases in our study also precluded examining whether these reductions in neocortical 5-HT in the LBV cases resemble typical PD. Two published studies have reported a significant reduction of 5-HT (49-55% of control) and increased 5-hydroxyindoleacetic acid/serotonin (5-HIAA/5-HT) ratio in frontal cortex of PD (Scatton et al., 1983; Adolfsson et al., 1979), suggesting enhanced 5-HT turnover in this region in PD. A third study found no significant changes in 5-HT concentration in frontal cortex of PD either on or off L-DOPA (Ogawa et al., 1992). There was no evidence for this pattern in the frontal cortex of our LBV group since the reductions of 5-HT and 5-HIAA were comparable, and 5-HIAA/5-HT (9.0) was similar to controls (8.4). However, the reduced 5-HT content and near normal levels of 5-HIAA in temporal cortex (Table 2) suggest a compensatory increase in 5-HT turnover in this area of the LBV group. A similar increase in the 5-HT/5-HIAA ratio was also observed in temporal cortex of the AD group. One previous study (Adolfsson et al., 1979) also observed normal 5-HT levels whereas another study reported a significant decline (39% of control) of 5-HT in frontal cortex of AD patients (D'Amato et al., 1987). A third study (Arai et al., 1984) observed a significant change of 5-HT in frontal but not temporal or parietal cortex and no significant change in 5-HIAA concentration in any of the three regions. It appears, therefore, that changes in serotonergic activity in the LBV and AD groups are highly variable and less severe within neocortical areas compared to basal ganglia structures. Correlational studies are needed to determine if these changes in serotonin activity within cortical regions are related to severity of neurological features and/or neuropsychometric measures of cognition and memory.

A significant elevation of kynurenine was also observed in temporal and parietal cortex of AD and LBV groups (Figure 4) and in the caudate of the AD group (Figure 2). These observations are significant because they suggest a possible mechanism for the loss of neurons and neurotransmitters in AD and LBV. Within the central nervous system, tryptophan is converted to either 5-HT or kynurenine through separate metabolic pathways (Figure 4). Kynurenine is further metabolized to both quinolinic acid, an n-methyl-d-aspartate (NMDA) receptor agonist (Moroni et al., 1988) and kynurenic acid, an NMDA receptor antagonist (Foster et al., 1984). Increased concentrations of kynurenine and quinolinic acid and decreased levels of kynurenic acid could result in over activation of NMDA receptors, a mechanism implicated in accelerated neuronal death (Olney, 1989).

This type of excitotoxic mechanism involving increased kynurenine and decreased kynurenic acid concentrations has been implicated in the pathogenesis of Huntington's disease (Beal et al., 1990). Our observation of a significant increase of kynurenine/5-HT + 5-HIAA in temporal and parietal cortex of the LBV group (Figure 5) is suggestive of a shift in the utilization of tryptophan from the 5-HT to kyrunenine metabolic pathways in LBV. Levels of the immediate serotonin precursor, 5-HTP, were not significantly altered in these regions (Table 2) but the values were very low and thus may not be a sensitive marker of activity within this pathway. In contrast, reduced levels of kynurenine and kynurenic acid have been reported in frontal cortex of pure PD cases (Ogawa et al., 1992).

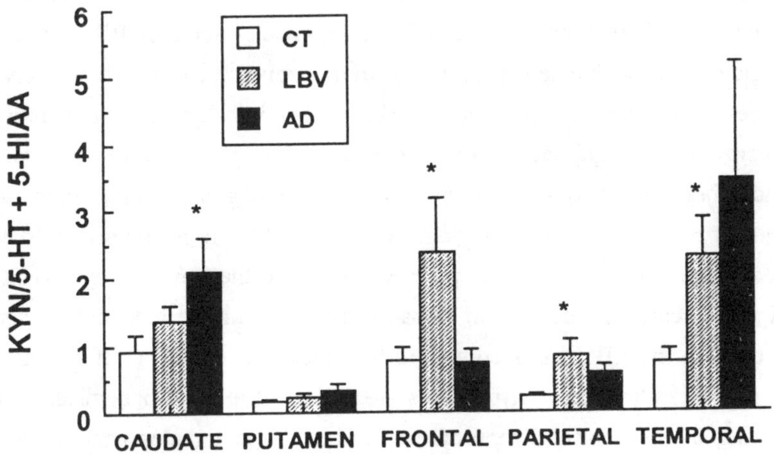

Figure 4. Ratio of the concentration of kynurenine to the combined concentrations of serotonin and its metabolite (KYN/5-HT + 5-HIAA) in basal ganglia and neocortical regions of subjects diagnosed pathologically as pure Alzheimer disease (AD), Lewy Body variant of AD (LBV) and expressed as percentage of value obtained from normal control subjects. Values are the mean ± S.E.M. *p<.05 compared to controls, post hoc Scheffe test.

Measurement of kynurenic acid and/or quinolinic acid in brain tissue are needed to provide a direct assessment of the hypothesis that these tryptophan-derived NMDA receptor ligands are significantly altered in brain regions of LBV and AD and may be strongly correlated with neuronal loss.

SUMMARY

Understanding the relationship between Lewy bodies and disturbances in neocortical and basal ganglia neurotransmitter activity will have a significant impact on the treatment of progressive dementing illnesses. As described above, results from our laboratories suggest that more severe disruptions of cholinergic, noradrenergic, and dopaminergic neurotransmission exist in a relatively large proportion of clinically and neuropathologically diagnosed cases of AD. These cases of AD have diffuse cortical and subcortical Lewy body and severe depletion of dopamine within basal ganglia characteristic of Parkinson's disease. Moreover these cases have additional loss of noradrenergic activity with basal ganglia not typically seen in PD and a decline in cholinergic activity within neocortex that is more pronounced than that observed in pure AD. These observations raise the possibility that imbalances in the interactions of dopaminergic-cholinergic and noradrenergic-cholinergic control of cortical and subcortical brain systems may be involved in the cognitive, mnemonic, and motor disturbances of these progressive dementing subjects. It is important, therefore, to examine a large number of cases of pure AD, PD, and the Lewy Body variant of AD to determine if greater disruptions in dopaminergic, noradrenergic, and or serotonergic activity exist in the LBV cases compared to typical PD and AD. If so, these findings would suggest alternative hypotheses regarding the neurochemical disturbances participating in the clinical symptoms and guide in the development of different and more effective pharmacologic treatments.

ACKNOWLEDGEMENTS

Supported by funds from the VA Merit Program, NIH/DRR grants S10 RRO 4654 and AG05131, and PHS grant MH NS 31862. This work was conducted in collaboration with Drs. Douglas Galasko, Larry Hansen, Leon Thal, and Eliezer Masliah and with the technical assistance of Michael Alford, Andy Chen, and Mark Wardlow.

REFERENCES

Ackworth, I.N. and Bowers, M. (1997). An introduction to HPLC-based electrochemical detection: From single electrode to multi-electrode arrays. In:

Coulometric Electrode Array Detectors for HPLC, I.N. Acworth, M. Naoi, H. Parvez, and S. Parvez (Eds). VSP, Utrecht, The Netherlands, pp. 3-50.

Adolfsson, R., Gottfries, C.G., Roos, B.E. and Winblad, B. (1979) Changes in the brain catecholamines in patients with dementia of Alzheimer type. *Brit.J. Psychiat.* **135**, 216-223.

Arai, H., Kosaka, K.and Iizuka, R. (1984). Changes of biogenic amines and their metabolites in postmortem brains from patients with Alzheimers-type dementia. *J. Neurochem.* **43**, 387-393.

Aubert, I., Araujo, D.M., Cecyre, D., Robitaille, Y., Gauthier, S. and Quiron, R. (1992). Comparative alterations of nicotinic and muscarinic binding sites in Alzheimer's and Parkinson's diseases. *J. Neurochem.* **58**, 529-541.

Araujo, D.M., Lapchak, P.A., Robitaille, Y., Gauthier, S. and Quirion, R. (1988). Differential alteration of various cholinergic markers in cortical and subcortical regions of human brain in Alzheimer's Disease. *J. Neurochem.* **50**, 1914-1923.

Beal, M.F., Matson, W.R., Swartz, K.J., Gamache, P. and Bird, E.D. (1990). Kynurenine pathway measurements in Huntington's disease striatum: Evidence for reduced formation of kynurenic acid. *J. Neurochem.* **55**, 1327-1339.

Bergman, H., Wichman, T. and DeLong, M.R. (1990). Reversal of experimental parkinsonism by lesions of the subthalamic nucleus. *Science* **249**, 1436-1438.

Birkmayer, W., Danielczyk, W., Neumayer, E. and Riederer, P. (1974). Nucleus ruber and L-Dopa psychosis: Biochemical post-mortem findings. *J. Neural. Trans.* **35**, 93-116.

Blank, C.L. (1976). Dual electrochemical detector for liquid chromatography. *J. Chromatogr.* **117**, 35-46.

Blessed, G., Tomlinson, B.E. and Roth, M. (1968). The association between quantitative measures of dementia and of senile change in the cerebral grey matter of elderly subjects. *Brit. J. Psychiat.* **114**, 797-811.

Bowen, D.M., Smith, C.B., White, P.and Davidson, A.N. (1976). Neurotransmitter-related enzymes and indices of hypoxia in senile dementia and other abiotrophies. *Brain* **99**, 459-496.

Bravo, G.J. and Cooper, I.S. (1959). A clinical and radiological correlation of the lesions produced by chemopallidectomy and thalamectomy. *J. Neurol. Neurosurg. Psychiat.* **22**, 1-10.

Cullison, J.K. and Gamache, P.H. (1997). Qualitative utility of coulometric array detection: Theoretical and applied aspects. In: *Coulometric Electrode Array Detectors for HPLC*, I.N. Acworth, M. Naoi, H. Parvez, and S. Parvez (Eds). VSP, Utrecht, The Netherlands, pp. 51-74.

D'Amato, R.J., Zweig, R.M., Whitehouse, P.J., Wenk, G.L., Singer, H.S., Mayeux, R., Price, D.L. and Snyder, S.H. (1987). Aminergic systems in Alzheimer's disease and Parkinson's disease. *Ann. Neurol.* **22**, 229-236.

Dickson, D.W., Davies, P., Mayeux, R., Crystal, H., Horoupian, D.S., Thompson, A. and Goldman, J.E. (1987). Diffuse Lewy body disease: Neuropathological and biochemical studies of six patients. *Acta Neuropathol.* (Berl) **75**, 8-15.

Dubois, B., Ruberg, M., Javoy-Agid, F., Ploska, A., and Agid, Y. (1983). A subcortico-cortical cholinergic system is affected in Parkinson's disease. *Brain Res.* **288**, 213-218.

Fahn, S., Libsch, L.R. and Cutler, R.W. (1971). Monoamines in the human neostriatum: Topographic distribution in normals and in Parkinson's Disease and their role in akinesia, rigidity, chorea, and tremor. *J. Neurol. Sci.* **14**, 427-455.

Fearnley, J.M., Revesz, T., Brooks, D.J., Frackowiak, R.S. and Lees, A.J. (1991). Diffuse Lewy body disease presenting with a supranuclear palsy. *J. Neurol. Neurosurg. Psychiat.* **54**, 159-161.

Fonnum, F. (1969). Radiochemical microassays for the determination of choline acetyltransferase and acetylcholinesterase activities. *J. Biochem.* **115**, 465-472.

Foster, A.C., Vezzani, A., French, E.D.and Schwarcz, R. (1984). Kynurenic acid blocks neurotoxicity and seizures induced in rats by the related brain metabolite quinolinic acid. *Neurosci. Lett.* **48**, 273-278.

Gaspar, P., Javoy-Avid, F., Ploska, A.and Agid, Y. (1980). Regional distribution of neurotransmitter synthesizing enzymes in the basal ganglia of human brain. *J. Neurochem.* **34**, :278-283.

Gibb, W.R.G. and Lees, A.J. (1991). Anatomy, pigmentation, ventral and dorsal subpopulations of the substantia nigra, and differential cell death in Parkinson's disease. *J. Neurol. Neurosurg. Psychiat.* **54**, 388-395.

Girault, J.-A., Raisman-Vozari, R., Agid, Y. and Greengard, P. (1989). Striatal phosphoproteins in Parkinson disease and progressive supranuclear palsy. *Proc. Natl. Acad. Sci.USA* **86**, 2493-2497.

Hansen, L.A. and Galasko, D. (1992). Lewy body disease. *Current Opinions Neurol. & Neurosurg.* **5**, 889-894.

Hansen, L., Salmon, D., Galasko, D,. Masliah,E., Katzman, R., DeTeresa, R., Thal, L., Pay, M.M., Hofstetter, R., Klauber, M., Rice, V., Butters, N. and Alford, M. (1990). The Lewy body variant of Alzheimer's Disease: A clinical and pathologic entity. *Neurology* **40**, 1-8.

Hornykiewicz, O. (1973). Dopamine in the basal ganglia: Its role and therapeutic implications (including the clinical use of L-DOPA). *Brit. Med. Bull.* **29**, 172-178.

Hornykiewicz, O. (1982). Brain neurotransmitter changes in Parkinson's Disease. In: *Movement Disorders,* C. D. Marsden and S. Fahn (Eds). Butterworth Scientific, London, pp.41-58.

Kelly, P.J., Ahlskog, J.E., Goerss, S.J., Daube, J.R., Duffy, J.R. and Kall, B.A. (1987). Computer-assisted stereotactic ventralis lateralis thalamotomy with microelectrode recording control in patients with Parkinson's disease. *Mayo Clin. Proc.* **62**, 655-664.

Khachaturian, Z.S. (1985). Diagnosis of Alzheimer's disease. *Arch. Neurol*. **42,** 1097-1105.

Kish, S.J., Shannak, K., and Hornykiewicz, O. (1988). Uneven pattern of dopamine loss in the striatum of patients with idiopathic Parkinson's Disease. *N. Engl. J. Med.* **318**, 876-880.

Kissinger, P.T. (1977). Amperometric and coulometric detectors for high performance liquid chromatography. *Anal. Chem.* **49**, 447-456.

Kissinger, P.T., Bruntlett, C.S., and Shoup, R.E. (1981). Neurochemical applications of liquid chromatogrpahy with electrochemical detection. *Life. Sci.* **28**, 455-465.

Langlais, P.J., Mair, R.G., Anderson, C.D., and McEntee, W.J. (1988). Long-lasting changes in regional brain amino acids and monoamines in recovered pyrithiamine treated rats. *Neurochem. Res.* **13**, 1199-1206.

Langlais, P.J., and Mair, R.G. (1990) Protective effects of the gluamate antagonist MK-801 on pyrithiamine induced lesions and amino acid changes in rat brain. *J. Neurosci.* **10**, 1664-1674.

Langlais, P.J., Connor, D. and Thal, L. (1993a). Comparison of the efrects of a single and combined neurotoxic lesion lesions of the nucleus basalis magnocellularis and dorsal noradrenergic buncle on learning and memory in the rat. *Behav. Brain Res.* **54**, 81-90.

Langlais, P.J., Thal, L.J., Hansen, L., Galasko, D., Alford, M.and Masliah, E. (1993b). Neurotransmitters in basal ganglia and cortex of Alzheimer's disease with and without Lewy bodies. *Neurology* **43**, 1927-1934.

Langlais, P.J., Wardlow, M.L.and Yamamoto, H. (1991). Changes in CSF neurotransmitters in Infantile Spasms. *Ped. Neurol.* **7**, 440-445.

Masliah, E., Mallory, M., DeTeresa, R., Alford, M., and Hansen, L. (1993). Differing patterns of aberrant neuronal sprouting in Alzheimer's disease with and without Lewy bodies. *Brain Res.* **617**, 258-266.

Matson, W.R., Langlais, P.J., Volicer, L., Gamache, P.H., Bird, E., and Mark, K.A. (1984). n-Electrode three-dimensional liquid chromatography with electrochemical detection for determination of neurotransmitters. *Clin. Chem.* **30**, 1477-1488.

McKhann, G., Drachman, D., Folstein, M.F., Katzman, R., Price, D. and Stadlan, E.M. (1984). Clinical diagnosis of Alzheimer's disease: Report of the NINCDS-ADRDA work group. *Neurology* **34**, 939-944.

Mefford, I.N. (1981). Application of high performance liquid chromatography with electrochemical detection to neurochemial analysis: Measurement of catecholamines, serotonin and metabolites in rat brain. *J. Neurosci. Methods* **3**, 207-224.

Moroni, F.P., Russi, P., Lombardi, G., Beni, M., Carla, V. (1988) Presence of kynurenic acid in mammalian brain. *J. Neurochem.* **51**, 177-180.

Ogawa, T., Matson, W.R., Beal, M.F., Myers, R.H., Bird, E.D., Milbury, P.and Saso, S. (1992). Kynurenine pathway abnormalities in Parkinson's disease. *Neurology* **42**, 1702-1706.

Olney, J.W. (1989). Excitatory amino acids and neuropsychiatric disorders. *Biol. Psychiat.* **26**, 505-525.

Perry, E.K., Curtis, M., Dick, D.J., Candy, J.M., Ratack, J., Bloxham, C.A., Blessed, G., Fairbairn, A., Tomlinson, B.E., and Perry, R.H. (1985). Cholinergic correlates of cognitive impairment in Parkinson's disease: Comparisons with Alzheimer's Disease. *J. Neurol. Neurosurg. Psychiat.* **48**, 413-421.

Perry, E.K., Marshall, E., Perry, R.H., Irving, D., Smith, C.J., Blessed, G. and Tomlinson, B.E. (1990). Cholinergic and dopaminergic activities in Senile Dementia of Lewy Body Type. *Alzheimer Dis. Assoc. Disord.* **4**, 87-95.

Perry, E.K., McKeith, I., Thompson, P., Marshall, E., Kerwin, J., Jabeen, S., Edwardson, J.A., Ince, P., Blessed, G., Irving, D., Perry, R.H. (1991). Topography, extent, and clinical relevance of neurochemical deficits in Dementia of Lewy Body Type, Parkinson's Disease, and Alzheimer's Disease. *Ann N. Y. Acad. Sci.* **640**, 197-202.

Rinne, U.K., Sonninen, V., Riekkinen, P.and Laaksonen, H. (1974). Dopaminergic nervous transmission in Parkinson's Disease. *Medical Biol .* **52**, 208-217.

Rinne, J.O., Laihinin, A., Lonberg, P., Marjamaki, P. and Rinne, U.K. (1991). A post-mortem study on striatal dopamine receptors in Parkinson's disease. *Brain Res.* **556**, 117-122.

Rossor, M.N., Iversen, L.L., Reynolds, G.P., Mountjoy, C.Q.and Roth, M. (1984). Neurochemical characteristics of early and late onset types of Alzheimer's disease. *Brit. Med. J.* **288**, 961-964.

Samuel, W., Alford, M., Hofstetter R., and Hansen, L. (1997). Dementia with Lewy Bodies versus pure Alzheimer's disease: Differences in cognition, neuropathology, cholinergic dysfunction, and synapse density. *J. Neuropathol. Exp. Neurol.* **56**, 499-508.

Scatton, B., Javoy-Agid, F., Rouquier, L., Dubois, B. and Agid, Y. (1983). Reduction of cortical dopamine, noradrenaline, serotonin and their metabolites in Parkinson's disease. *Brain Res.* **275**, 321-328.

Sirvio, J., Rinne, J.O., Valjakka, A., Rinne, U.K., Riekkinen, P.J., Paljarvi, L. (1989). Different forms of brain acetylcholinesterase and muscarinic binding in Parkinson's disease. *J. Neurol. Sci.* **90**, 23-32.

Whitehouse, P.J., Martino, A.M., Marcus, K.A., Zweig, R.M., Singer, H.S., Price, D.L. and Kellar, K.J. (1988). Reductions in acetylcholine and nicotine binding in several degenerative diseases. *Arch. Neurol.* **45**, 722-724.

Progress in HPLC-HPCE, Vol. 7, pp. 839-852
G. Ali Qureshi *et al.* (Eds)

HPLC as a tool in quantitation of CSF biopterin, neopterin, homovanillic acid, and 5-hydroxyindoleacetic acid in neurological patients

M. Candito[a], T. Nagatsu[b], J. Durant[c], P. Dellamonica[c], M. Chatel[d], E. Van Obberghen[a]

[a]Laboratoire de Biochimie, Hôpital Pasteur, 30 av. de la Voie Romaine, 06002 Nice Cedex 1, France, [b]Institute for Comprehensive Medical Science, Fujita Health University School of Medecine, Toyoake 470-11, Japan, [c]Service des Maladies Infectieuses, Hôpital de l'Archet, 06202 Nice Cedex 3, France, [d]Service de Neurologie, Hôpital Pasteur, 30 av. de la Voie Romaine, 06002 Nice Cedex 1, France

Introduction

Cerebrospinal fluid (CSF) is the biological fluid which, in defined conditions, provides a reflection of brain syntheses. High performance liquid chromatography (HPLC) is a useful tool for quantitation of certain cerebral neurotransmitters:

- upstream, by assays of tetrahydrobiopterin (BH_4), the common cofactor for phenylalanine, tyrosine and tryptophan hydroxylases, which catalyze the synthesis of dopamine, noradrenaline, adrenaline and serotonin. HPLC also allows assay of dihydroneopterin (NH_2) and its fully oxidized form: neopterin (N), the breakdown product of the BH_4 precursor: dihydroneopterin triphosphate (Nagatsu et al. 1989). BH_4 is also the cofactor for nitric acid synthase which produces nitric oxide, a candidate neurotransmitter (Werner-Felmayer et al. 1993). Cerebral BH_4 is thought to reflect the rate of neurotransmitter synthesis, which is decreased in both the brain and CSF of patients with Parkinson's disease (Yamaguchi et al. 1983; Furukawa et al. 1992). *In vitro*, NH_2 and N are released from macrophages stimulated by interferon gamma, a product of activated T cells (Huber et al. 1984). Some authors have proposed CSF total neopterin as a biochemical marker of central nervous system inflammation (Furukawa et al. 1992), and simultaneous determination of BH_4, N and NH_2 in CSF should be useful in clinical chemistry.

- downstream, by assays of the metabolites of neurotransmitters. The main metabolites found in detectable amounts in CSF are: 3-methoxy-4-hydroxyphenylglycol (MHPG), derived from noradrenaline (NA); 5-hydroxyindole-3-acetic acid (5-HIAA), derived from 5-hydroxytryptamine or serotonin (5-HT), and homovanillic acid (HVA), derived from dopamine (DA).

Assays of amine precursors or metabolites in CSF are important for investigations on the pathogenesis of central nervous system (CNS) pathologies. Numerous methods have been developed for metabolite assays, including fluorimetry, gas chromatography-mass spectrometry (GC-MS), and radioimmunoassays. Rapid and sensitive liquid chromatography coupled with amperometric or fluorometric detection has been successfully employed without complicated extraction or derivatization procedures, and requires only small volumes of CSF.

In an initial series of investigations, CSF samples from patients with neurological disorders were assayed to determine BH_4, NH_2 and N concentrations. A number of assay procedures described in the literature (Howells et al. 1986; Howells and Hyland 1987) were adapted for our system (Candito et al. 1994). Neopterin can be assayed alone by the same simple method described for plasma.

5-HIAA and HVA were also measured in CSF samples of the same patients and in HIV-1 infected patients (Candito et al. 1995). Brain biogenic amine abnormalities have been observed in various psychiatric disorders such as dementia of the Alzheimer type (DAT), and the neurological complications frequently observed in AIDS include the development of progressive dementia (Navia et al. 1986a, b; Carne et al. 1985), whose well-documented clinical features include memory alterations, slowed psychomotor function, and affective and emotional disorders. This association of cognitive, motor, and behavioral symptoms has been termed the AIDS dementia complex (ADC) (Sidtis et al. 1991).

As BH_4 and NH_2 are difficult to analyze owing to their instability, CSF samples were collected under special conditions to preserve these reduced pterins. Liquid chromatographic assays and detection of the various substances measured also required particular precautions, as discussed below.

Standard N, 5-HIAA, and MHPG were from Sigma (Coger, Paris, France); HVA was from Merck (Elvetec, Nice, France).

Tetrahydrobiopterin, dihydroneopterin and neopterin assays

BH_4, NH_2 and N were assayed by HPLC. (6R)-BH_4 and NH_2 standards were obtained from Dr. Schircks Labs (Jona, Switzerland). Standard solutions were prepared in millimolar concentrations. To BH_4 and NH_2 solutions in 0.010 mM HCl were added 1 mg/ml diethylenetriaminepentaacetic acid (DETAPAC) (Sigma) and 1 mg/ml dithioerythritol (DTE) (Sigma). The solutions were stable for 2 months at -80° C.

BH_4 was detected by electrochemical detection and N by fluorimetric detection. NH_2 was measured with the same HPLC system by fluorescence detection after electrochemical oxidation at a sufficiently high potential according to the modified method of Howells et al. (1986). As illustrated in Fig. 1, BH_4 was detected by an ESA

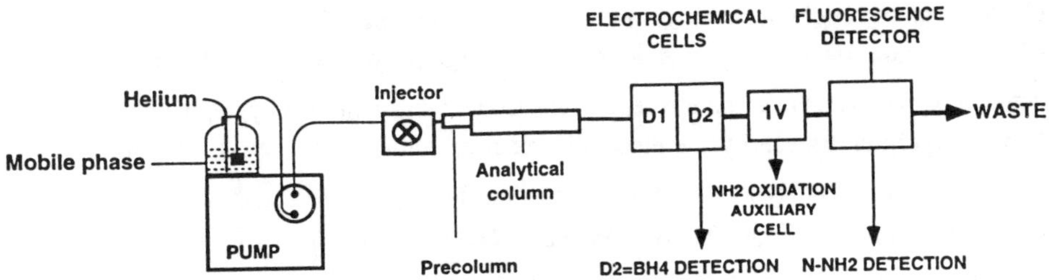

Fig. 1. Systamatic diagram of HPLC system

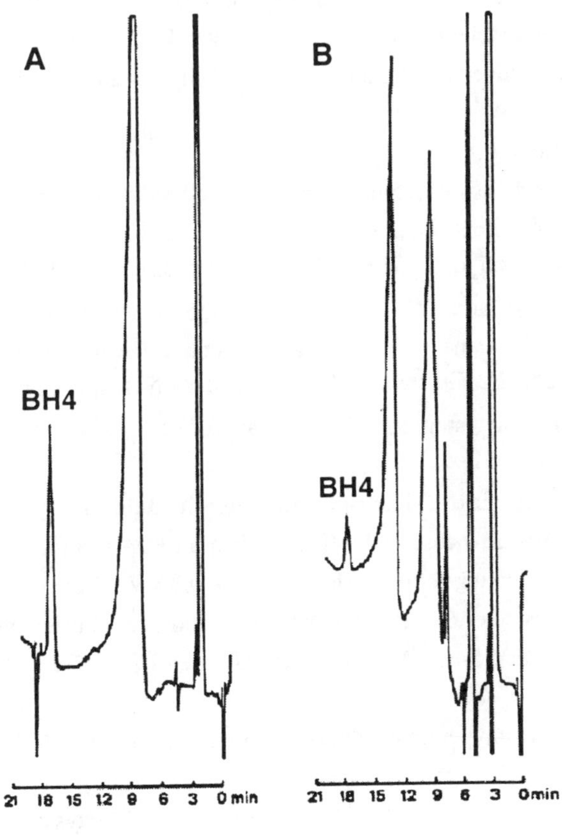

Fig. 2. Chromatogram of (A) 100 nM BH4 standard and (B) CSF sample

Coulochem electrochemical detector (Touzart and Matignon, Paris, France); electrode 1: +0.50 V (gain 10 x 99); electrode 2: -0.10 V (gain 10 x 99). As NH_2 cannot be detected by fluorescence, the effluent of electrode 2 was oxidized by application of +1.0 V by a conditioning cell by the electrochemical detector. Native N and N from NH_2 (separated on the analytical column) were detected by fluorescence with a sensitive fluorimeter (Jasco, Prolabo, Paris, France). The excitation filter selected the band at 348 nm, and a set of high transmission filters centered at 444 nm allowed emission selection.

The mobile phase consisted of 6.8 g/l sodium acetate and 1.05 g/l citric acid (pH 5.20) containing 20 mg/l disodium ethylenediaminetetraacetic acid (EDTA) (Sigma) and 150 mg/l DTE. The mobile phase was filtered through a 0.2 μM filter (Elvetec, Nice, France) and degassed before and during use by continuous bubbling with dry helium gas. The column of Ultrasphere ODS 5 μm (250 x 4.6 mm ID) was obtained from Beckman (Gagny, France).

In order to obtain reproducible results, the chromatographic system was equilibrated overnight, and the mobile phase was run to waste. Prior to injection of the samples, 150 mg/l DTE was added to the mobile phase, which was always run to waste.

Neopterin assays

Neopterin can be measured alone using the method described for plasma assays (Candito et al., 1993), but requires precautions for CSF sample collection (see below) to avoid NH_2 oxidation to N. The mobile phase consisted of a 98% phosphate buffer (0.015 M, pH 6.4) and 2% methanol with the same column. N was detected fluorimetrically at the same excitation and emission wavelengths as for simultaneous assays of NH_2 and N. A stable 0.5 mM standard N suspension was prepared as previously described and stored for 8 months at -20° C.

Homovanillic acid and 5-hydroxyindoleacetic acid assays

HVA and 5-HIAA were measured by HPLC using an ESA Coulochem electrochemical detector with the following settings: electrode 1: + 0.30 V, electrode 2: +0.44 V. The mobile phase consisted of 6.8 g/l monopotassium phosphate buffer (pH 4.0)-methanol (77:13, v/v). HVA and 5-HIAA were separated on an Ultrasphere ODS 5 μm column (250 x 4.6 mm ID).

HVA and MHPG hemipiperazate solutions were prepared in distilled water; they were stable for 2 months and 1 month respectively at +4° C in the dark. 5-HIAA solution (0.5 mM) was prepared in 0.01 M HCl and was stable for 15 days at 4° C.

Patients were hospitalized in the Neurology Unit for a variety of pathologies: (a) inflammatory disease: Guillain-Barré syndrome, meningoradiculitis, spinal arachnoiditis, lupus-like polyneuritis (7 patients); (b) multiple sclerosis (2 patients); (c)

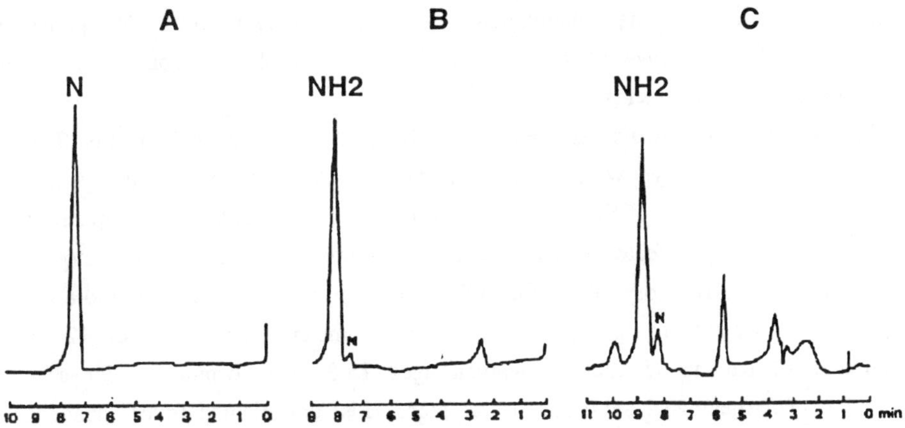

Fig. 3. Chromatogram of (A) 25 nM N,standard, (B) 25 nM NH2 standard and (C) CSF sample.

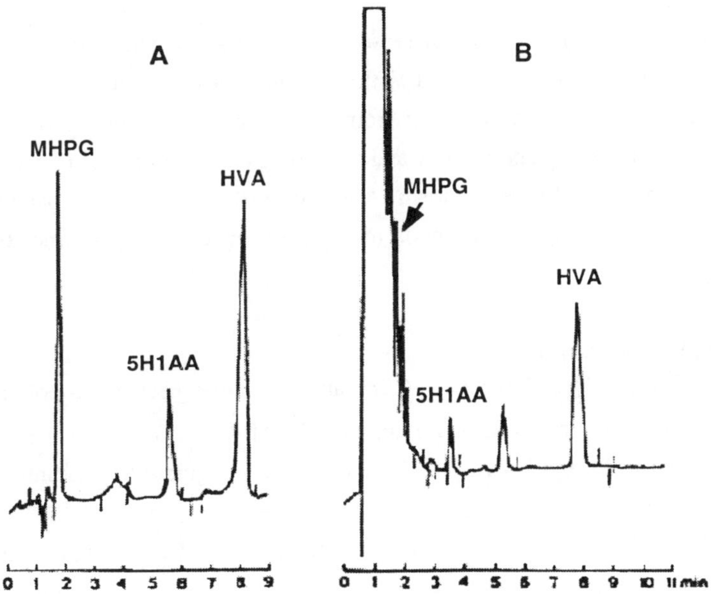

Fig. 4. Chromatogam of (A) standard mixture of catecholamines and (B) CSF sample.

infectious diseases: viral meningitis, syphilis, meningomyelitis (4 patients), subarachnoid hemorrhage (1 patient) ; AIDS with neurological complications but without dementia (2 patients).

5-HIAA and HVA were also assayed in the CSF of 11 HIV-1-infected patients (7 men, 4 women aged 24-66 yr) with a clinical diagnosis of AIDS dementia complex (ADC) according to the criteria of Navia et al. (1986) (group 1). Results were compared with levels in 11 HIV-1-infected patients (6 men, 5 women aged 25-64 yr) without any clinical signs of ADC hospitalized for a CNS pathology such as cerebral toxoplasmosis (group 2). Results for groups 1 and 2 were compared with those of 12 HIV-seronegative patients (5 men, 7 women aged 15-32 yr) hospitalized for a non-Parkinsonian meningeal syndrome (viral meningitis, unexplained fever, headaches, meningoradiculitis, or Guillain-Barré syndrome) who served as controls (group 3).

Group 1 patients were all stage IV-C1 of the 1987 Center for Disease Control and Prevention (CDC) classification, except for one who was stage II. Group 2 patients were CDC stage IV-C1 or C2.

None of the patients or controls had any clinical, radiologic, or ophthalmologic signs of intracranial hypertension. None of the controls were receiving any treatment at the time of lumbar puncture, but some of the HIV-seropositive patients (groups 1 and 2) were taking benzodiazepines, phenobarbitol, ciprofloxane, sulphamides, or an antiviral agent. One patient in group 2 was taking fluoxetine, which blocks presynaptic serotonin capture.

For the other assays and for ethical reasons, CSF was not obtained from healthy controls, but results were compared with concentrations assayed in other patients without neurological disease suffering from a variety of pathologies: (a) neoplastic diseases: medulloblastoma, pulmonary and gynecological malignancies (3 patients); (b) hormonal disorders: hypopituitarism and hyperthyroidism (2 patients); (c) cerebrovascular accidents (4 patients); (d) diffuse pyramidal syndrome and facial paralysis (2 patients).

CSF sample collection

To measure BH_4, NH_2 and N levels, CSF samples were collected into polypropylene tubes containing 1 mg/ml DTE and 1 mg/ml DETAPAC according to the procedure described by Howells and Hyland (1987) and frozen in liquid nitrogen until analyzed (within one month).

5-HIAA and HVA levels can be assayed on the same CSF samples, but these precautions are not necessary when these metabolites are to be assayed alone. MHPG, HVA and 5-HIAA are stable for at least 72 hr at 0-4° C. Addition of perchloric acid does not affect stability. Samples stored at -80° C are stable for up to 3 months (Mena et al. 1984). Freezing does not cause any significant change.

After centrifugation, the CSF samples were injected directly into the chromatographic apparatus.

CSF BH_4, NH_2, N, 5-HIAA and HVA in neurological diseases

Fig. 2A shows the chromatogram of a 100 nM BH4 standard and Fig. 2B the chromatogram of a CSF sample (patient 12). Fig. 3A is the chromatogram of a 25 nM standard of N while Fig. 3B corresponds to 25 nM NH_2 and Fig. 3C to a CSF sample. A very small amount of the NH_2 standard (less than 4%) was in the form of N under our conditions. The NH_2 peak disappeared completely when the 1 V potential was not applied at the outlet of electrode 1 by the auxiliary cell. Fig. 4A is the chromatogram of a standard mixture of MHPG, 5-HIAA, and HVA; Fig. 4B is the chromatogram of a CSF sample in which 5-HIAA and HVA were assayed. MHPG, a norepinephrine metabolite, could not always be separated from the other chromatographic peaks in CSF samples. As peak heights are strongly influenced by the pH, chromatography must be performed with a mobile phase that cannot be recycled, and calibration must be repeated. Under these conditions, the intra-assay variation was 9% and the interassay variation was 12%, calculated with a 50 nM concentration.

The results are summarized in Tables 1 and 2. In the group without any neurological disorder, the two patients with a diffuse pyramidal syndrome and facial paralysis had low BH_4 concentrations (35 and 16.5 nmol/l) and low N values (1.5 and 1.7 nmol/l); their N/BH_4 ratios were also low (0.04 and 0.1 nmol/l). Similar results were obtained for the three patients with neoplastic diseases. The N/BH_4 ratio was elevated in some of the patients who had had a cerebrovascular accident. In the literature, normal BH_4 values in CSF range from 18 to 24 nmol/l (Williams et al. 1980); normal values for N are below 3 nmol/l (Griffin et al. 1991).

In inflammatory neurological diseases, BH_4 was sometimes increased and the N/BH_4 ratio remained low. BH_4, N and NH_2 were all low in the two patients with multiple sclerosis; this concurs with the low values reported previously for patients in remission (Fredrikson et al. 1987), which was the case for our patients. N was considerably elevated in meningomyelitis; a dramatic increase was also observed in an AIDS patient with neurological complications. N is considered a marker of neurological diseases in CSF, and a marker of neurological complication in AIDS (Griffin et al. 1991).

In subarachnoid hemorrhage, N and NH_2 were elevated while BH_4 was very low, resulting in a high N/BH_4 ratio. Increased concentrations of N in CSF have previously been reported in both subarachnoid hemorrhage and patients with cerebral ischemia (Mathiesen et al. 1990), as observed in some of our patients who had suffered cerebrovascular accidents. The N/BH_4 ratio (Table 2) reflects the variations in the two parameters (Fujishiro et al. 1990).

In seven sufficiently large CSF samples, N was also measured in parallel by the method described for plasma N determination. Results differed by less than 5%.
The HVA and 5-HIAA concentrations (Tables 1 and 2) concurred with the normal values reported in the literature (Blennow et al. 1993) (5-HIAA 104 ± 27 nmol/l; HVA 230 ± 35 nmol/l). One patient with inflammatory disease (polyneuritis) exhibited decreased HVA and 5-HIAA concentrations, but this was probably due to alcohol-related deficiencies. HVA and 5-HIAA concentrations were higher in infectious diseases, menigomyelitis, and AIDS with neurological complications.

Correlations

As expected, a highly significant correlation was observed between N and NH_2 ($p < 0.001$) using single regression. NH_2 was also correlated with BH_4 ($p = 0.04$). Surprisingly, this was not the case for N and BH_4 (not significant). Despite the low number of samples tested, HVA (the product of dopamine breakdown) was weakly correlated with BH_4 ($p = 0.06$), the cofactor of tyrosine hydroxylase in dopamine biosynthesis, and with N ($p = 0.05$). HVA was also correlated with 5-HIAA (the product of serotonin breakdown) ($p = 0.04$). This might reflect the fact that both dopamine and serotonin are correlated with BH_4 concentrations in the neurological disorders investigated. Such correlations have been described in the literature (Bottiglieri et al. 1992).

CSF HVA and 5-HIAA in the AIDS dementia complex

CSF HVA and 5-HIAA values in the HIV-seronegative controls (group 3) were similar to those observed previously in a group of patients without any encephalic pathology (HVA, 230 ± 35 nmol/l; 5-HIAA, 103.6 ± 26 nmol/l) (Candito et al. 1994) and to the values reported in the literature for healthy individuals (HVA, 253 ± 109 nmol/l; 5-HIAA, 125 ± 54 nmol/l) (see above) (Table 3). In contrast, the SD were greater. CSF HVA levels were slightly decreased in group 2 and were even lower in group 1, but because of the wide distribution of results, the difference was not statistically significant compared with results for the controls (group 3). ($p = 0.08$). Mean 5-HIAA was slightly elevated in group 2 compared with the control mean, but it was lower in group 1, although again in a non-significant manner. A correlation was observed between 5-HIAA and HVA values in CSF in all three groups, and the three CSF HVA: 5-HIAA ratios were similar.

Several investigators have reported decreased CSF HVA in patients with DAT (dementia of Alzheimer type) (Parnetti et al. 1987). Others have reported that CSF HVA levels were only lower than in controls and in DAT patients with myoclonus (Kaye et al. 1988). None of our patients had myoclonus, and mean CSF 5-HIAA was decreased only in the HIV-positive patients with ADC (group 1), although in the previously mentioned studies CSF 5-HIAA was lower in DAT patients or DAT patients with myoclonus. A correlation between CSF 5-HIAA and HVA concentrations has been described in controls and in patients with personality disorders (Eriksson et al. 1991) and we also observed a weak correlation in CSF levels for several neurological diseases in this study.

HPLC permits separation of molecules with very similar structures but slightly different solubilities on columns of increasing separation capacity. Electrochemical and fluorimetric detection are very sensitive, and increase the specificity of chromatography. While such techniques are also applied to studies by microdialysis, their use in this case is still primarily restricted to animal experiments and involves numerous methodological biases.

CSF analyses are currently the only available means to estimate cerebral biosyntheses. Owing to the concentration gradient of the substances studied in CSF (Furukawa et al. 1992), a strict CSF sampling protocol is obligatory. Moreover, catabolites such as HVA and 5-HIAA, which are both lipophilic and hydrophilic, can originate partially from plasma, and thus in part be a peripheral reflection, as demonstrated for MHPG (Filser et al. 1986). Nevertheless, these analyses may provide preclinical indications for the initiation of treatment as well as allow evaluation of treatment efficacy and help elucidate the physiopathological mechanisms involved.

CSF BH_4 is decreased in dystonia (Lewitt et al. 1988) and in Alzheimer's disease, and as a biopterin cofactor in Parkinson's disease (Nagatsu et al. 1984); catecholamine-related enzymes and the biopterin cofactor are decreased in postmortem parkinsonian patients.

CSF total neopterin is markedly elevated during the exacerbation or acute stages of patients with neurosarcoidosis, neuro-Behçets disease and meningitis (Furukawa et al. 1992), in multiple sclerosis (Fredrikson et al. 1987), and in Lyme neuroborreliosis (Dotevall et al. 1990).

CSF biopterin, 5-HIAA and HVA levels are higher in aseptic meningitis, especially in the acute phase (Komori et al. 1994). In depression, 5-HIAA is decreased in patients with low red-cell folate levels (Bottiglieri et al. 1992).

For investigations on cerebral biogenic amine synthesis, HPLC with fluorimetric and amperometric detection allows quantitation of CSF BH_4, the pterin cofactor (hydroxylases have a rate-limiting effect on neurotransmitter synthesis), and other pterins such as NH_2 and N. HPLC also allows quantitation of 5-HIAA derived from serotonin and HVA derived from dopamine.

The reduced pterins BH_4 and NH_2 can be measured simultaneously with N when the samples are assayed in a reducing medium. Increased concentrations of BH_4 may be

accompanied by high levels of N and NH_2 as well as raised HVA and 5-HIAA concentrations reflecting the degree of dopaminergic and serotoninergic activation.
In this study, the BH_4 concentrations in patients with neurological disorders were higher than the normal values reported in the literature. The highest concentrations were observed in patients with syphilis, lupus-like disease, and meningoradiculitis. An AIDS patient with neurological complications also showed a dramatic elevation in BH_4 and an increased N/BH_4 ratio which may reflect neurological complication. The decreased levels of CSF HVA and 5-HIAA observed in the AIDS dementia complex might reflect lower dopaminergic and serotoninergic activity. Although these biochemical abnormalities are not disease-specific, aberrant cerebral dopaminergic and serotonin-dopamine neurotransmission could be associated with neurological diseases, and this could have an effect on therapy.

References

Blennow K, Wallin A, Gottfries CG et al. (1993) Cerebrospinal fluid monoamine metabolites in 114 healthy individuals 18-88 years of age. Eur Neuropsychopharmacol 3: 55-61

Bottiglieri T, Hyland K, Laundy M et al.(1992) Folate deficiency, biopterin and monoamine metabolism in depression. Psychol Med 22 (4): 871-876

Candito M, Cavenel C, Gugenheim J et al.(1993) Simple column liquid chromatographic assay for serum neopterin. J Chromatogr 614: 164-168

Candito M, Durant J, Gurle V, Chambon P, Dellamonica P. (1995) Cerebrospinal fluid 5-hydroxyindoleacetic acid and homovanillic acid in AIDS dementia complex. AIDS 9 (9): 1100-1101

Candito M, Nagatsu T, Chambon P, Chatel M. (1994) High-performance liquid chromatographic measurement of cerebrospinal fluid tetrahydrobiopterin, neopterin, homovanillic acid and 5-hydroxyindoleacetic acid in neurological diseases. J Chromatog B: Biomed Applic 657: 61-66

Carne CA, Tedder RS, Smith A et al. (1985) Acute encephalopathy coincident with seroconversion for anti-HTLV III. Lancet 2:1206-1208

Dotevall L, Fuchs D, Reibnegger G, Wachter H, Hagberg L. (1990) Cerebrospinal fluid and serum neopterin levels in patients with Lyme neuroborreliosis. Infection 18 (4): 210-214

Erikson E, Westberg P, Alling C et al. (1991) Cerebrospinal fluid levels of monoamine metabolites in panic disorder. Psychiatry Res 36: 243-251

Filser JG, Müller WE, Beckmann H. (1986) Should plasma or urinary MHPG be measured in psychiatric research? A critical comment. Br J Psychiat. 148: 95-97

Fredrikson S, Link H, Eneroth P. (1987) CSF neopterin as a marker of disease activity in multiple sclerosis. Acta Neurol Scand 75: 352-355

Fujishiro KI, Hagihara A, Takahashi A, Nagatsu T. (1994) Concentration of biopterin in the cerebrospinal fluid of patients with Parkinson's disease. Biomed Metabol Biol 44: 97-100

Furukawa Y, Nishi K, Kondo T, Tanabe K, Mizuno Y. (1992) Significance of CSF total neopterin and biopterin in inflammatory neurological diseases. J Neurol Sci 111: 65-72

Griffin DE, McArthur JC, Cornblath DR. (1991) Neopterin and interferon-gamma in serum and cerebrospinal fluid of patients with HIV-associated neurologic disease. Neurology 41: 69-74

Howells DW and Hyland K. (1987) Direct analysis of tetrahydrobiopterin in cerebrospinal fluid by high-performance liquid chromatography with redox electrochemistry: prevention of autoxidation during storage and analysis. Clin Chimica Acta 167: 23-30

Howells DW, Smith I, Hyland K. (1986) Estimation of tetrahydrobiopterin and other pterins in cerebrospinal fluid using reversed-phase high-performance liquid chromatography with electrochemical and fluorescence detection. J Chromatogr 381: 285-294

Huber C, Fuchs D, Niederweiser A. (1984) Neopterin, eine neue biochemischer Marker zur klinischen erfassung zellulären Immunreaktion. Klin Wochenschrift 62: 103-113

Kaye JA, May C, Daly E et al.(1988) Cerebrospinal fluid monoamine markers are decreased in dementia of the Alzheimer type with extrapyramidal features. Neurology 38: 554-557

Komori H, Matsuishi T, Yamada S, Ichikawa K, Amamoto M, Yamashita F. (1994) Elevated biopterin and homovanillic acid levels in cerebrospinal fluid from children with aseptic meningitis. J Child Neurol 9 (1): 22-25

LeWitt PA, Miller LP, Levine RA et al.(1988) Pterin abnormalities in dystonia: A metabolic marker with therapeutic implications. Adv Neurol 50: 193-201

Mathiesen T, Fuchs D, Wachter H, von Holst H. (1990) Increased CSF neopterin levels in subarachnoid hemorrhage. J Neurosurg 73: 69-71

Mena MA, Aguado EG, de Yebenes JG. (1984) Monoamine metabolites in human cerebrospinal fluid. HPLC/ED method. Acta Neurol Scand 69: 218-225

Nagatsu T, Matsura S, Sugimoto. (1989) Physiological and clinical chemistry of biopterin. Med Res Rev 9: 25-44

Nagatsu T, Yamaguchi T, Rahman MK et al. (1984) Catecholamine-related enzymes and the biopterin cofactor in Parkinson's disease and related extrapyramidal diseases. Adv Neurol 40: 467-473

Navia BA, Cho ES, Petito CK, Price RW. (1986) The AIDS dementia complex: II. Neuropathology. Ann Neurol 19b : 525-535

Navia BA, Jordan BD, Price RW. (1986) The AIDS dementia complex: I. Clinical features. Ann Neurol 19a: 517-524

Parnetti L, Gottfries J, Karlsson I, Langstrom G, Gottfries CG, Svennerholm L. (1987) Monoamines and their metabolites in cerebrospinal fluid of patients with senile dementia of Alzheimer type using high performance liquid chromatography and gas chromatography-mass spectrometry. Acta Psychiatr Scand 75: 542-548

Sidtis JJ, Price RW. (1991) Early HIV-1 infection and the AIDS dementia complex. Neurology 40: 323-326

Tomiyama K, Noguchi M, Koshikawa N, Kobayashi M.(1993) YM-09151-2 but not 1-sulpiride induces transient dopamine release in rat striatum via tetrodotoxin-insensitive mechanism. J Neurochem 60 (5): 1690-1695

Werner-Felmayer G, Werner ER, Fuchs D, Haussen A, Reibnegger G, Schmidt K, Weiss G, Watcher H.(1993) Pteridine biosynthesis in human endothelial cells. Impact on nitric oxide-mediated formation of cyclic GMC. J Biol Chem 268: 1842-1846

Williams AC, Levine RA, Chase TN, Lovenberg, Calne B.(1980) CSF hydroxylase cofactor levels in some neurological diseases. J Neurol Neurosurg Psychiat 43: 735-738

Yamaguchi T, Nagatsu T, Sugimoto T et al. (1983) Effects of tyrosine administration in normal controls and patients with Parkinson's disease. Science 219: 75-77

TABLE I - CSF values (in nmol/l) in patients without neurological disease

Patient n°	Sex	Age (yr)	BH_4	N	NH_2	N/BH_4	5-HIAA	HVA
P1	F	32	17	1.7	22	0.10	180	143
P2	M	51	35	1.5	15	0.04		
P3	M	56	43	2.1	80	0.05		
P4	F	39	49	2.3	15	0.05		
P5	F	25	34	1.3	36	0.04	230	415
P6	M	43	37	3.1		0.08		
P7	M	74	29	4.2	24	0.14		
P8	M	44	33	5	42	0.15		

P1: headaches
P2: diffuse pyramidal syndrome
P3 : cerebral arteritis
P4 : gynecologic neoplasm
P5 : medulloblastoma
P6 : pulmonary neoplasm
P7 : hypopituitarism
P8 : hyperthyroidism

F : female
M : male

TABLE II - CSF values (in nmol/l) in patients with neurological diseases

Patient n°	Sex	Age (yr)	BH_4	N	NH_2	N/BH_4	5-HIAA	HVA
P1	F	51	63	4.4	75	0.07		
P2	F	58	31	1.6	44	0.05		
P3	M	72	75	4.7		0.06	143	250
P4	M	80	25	2.2		0.09	149	169
P5	M	69	12	1.1	22	0.09		
P6	M	18	30	3.6	18	0.12	90	230
P7	F	25	46	2.2		0.04	109	330
P8	M	29	41	2.1		0.05		
P9	M	29	74	4.8	24	0.06	49	154
P10	M	45	12	1.1	22	0.09		
P11	F	50	20	2.3	49	0.01	160	77
P12	M	54	22	4.7	29	0.21	200	346
P13	M	59	21	4.1	23	0.19		
P14	F	77	31	6	37	0.09	109	210
P15	F	31	34	5.2	37	0.15		
P16	M	23	16	3	12	0.19		
P17	F	31	32	2.5	16	0.08	99	160
P18	F	69	7.5	19	66	2.5	230	140
P19	M	63	21	3.8	21	0.18		
P20	F	35	53	24	90	0.45		
P21	M	35	25	2.5	50	0.08	49	130
P22	M	31	59	102	268	1.73	200	346

Inflammatory diseases:
P1 syphilis
P2 syphilis
P3 meningoradiculitis
P4 myelinic neuropathy
P5 polyneuritis
P6 Guillain-Barré syndrome
P7 demyelinization
P8 postanorexic encephalopathy
P9 lupus-like disease

P10, P11: alcoholic patients
P12, P13, P14, P15: cerebrovascular diseases
P16, P17: multiple sclerosis
P18: subarachnoid hemorrhage
P19: viral meningitis
P20: meningomyelitis
P21: AIDS but normal CD4/CD8 ratio indicating sufficient cellular immunity
P22: AIDS with neurologic involvement

TABLE III. Cerebrospinal fluid 5-HIAA and HVA concentrations (nmol/L) in controls and in HIV-infected patients with and without the AIDS dementia complex (ADC)

	HVA	5-HIAA	HVA/5-HIAA ratio
Controls*	245 ± 102	117 ± 61	1.69 ± 0.52
(n = 12)	74-430	24-235	0.50-2.95
HIV+ patients without ADC	156 ± 107	156 ± 98	1.33 ± 1.2
(n = 11)	18-340	32-328	0.56-4.2
HIV+ patients with ADC	91 ± 71	76 ± 49	1.44 ± 0.49
(n = 11)	32-230	30-170	0.2-3.0

*Controls were seronegative for HIV

The patient receiving fluoxetine was in Group 2; his CSF 5-HIAA and HVA values were respectively 194 (surprisingly not increased compared with the mean value) and 330 nmol/L (among the highest concentrations).

List of Contributors

Ian N.Acworth
ESA Inc.,
22, Alpha Road,
Chemsford, USA.

K.Akita
Department of Pharmacology,
Faculty of Pharmaceutical Sciences,
The Health University of Hokkaido
Ishikari-Tobetsu, Hokkaido, Japan.

Bruce A.Bailey
ESA Inc.,
22, Alpha Road,
Chemsford, USA.

Leo A.P.Ballering
Dpartment of Drug safety,
N.V.Organon, Oss,
The Netherlands.

Margherita S.Benedetti
Preclinical Development & Human
Pharmacology, Zambon Group,
Milan, Italy.

Hidehiko Beppu
Fujita Memorial Division of
Pharmacognosy, Fujita University,
Toyoake, Japan.

R.Bertolotti
CNRS, Molecular Genetics,
Department of Hepato-
Gastroenterology,
Faculty of Medicine, University of
Nice,Nice, France.

G.Baier-Bitterlich
Institute of Medical and Biochemistry,
The University of Innsbruck,
Innsbruck,
Austria.

P.Caudy
TSP Corporation France,
Les Ulis, France.

J.-F.Cloix
Universite d'Orleans, P.O.Box 6759,
Orleans, Cedex, France.

Dominic M.Desiderio
The Charles B.Stout Neuroscience
Mass Spectrometry Laboratory, The
University of Tennessee, Memphis,
USA.

T.Endo
Department of Pharmacology,
Faculty of Pharmaceutical Sciences,
The Health University of Hokkaido
Ishikari-Tobetsu, Hokkaido, Japan.

R.A.J.Esselink
Department of Neurology,
Medisch Spectrum Twente, Enschede,
The Netherlands.

D.Fuchs
Institute of Medical and Biochemistry,
The University of Innsbruch,
Innsbruck,
Austria.

T.Gasse,
Department of Neurology,
The University of Innsbruck,
Austria.

M.Gisslen
Department of Infectious Diseases,
Sahlgrenska University Hospital,
Gothenburg, Sweden.

Gary A.Gudelsky
College of Pharmacy,
The University of Cincinnati Medical
Center, Cincinnati, Ohio, USA.

N.Hamaue
Department of Pharmacology,
Faculty of Pharmaceutical Sciences,
The Health University of Hokkaido
Ishikari-Tobetsu, Hokkaido, Japan.

O.Helary
Universite d'Orleans, P.O.Box 6759,
Orleans, Cedex, France.

T.K.Hever
Universite d'Orleans, P.O.Box 6759,
Orleans, Cedex, France.

Makoto Hibiya
Division of Molecular Genetics II
Neurochemistry, Institute for
Comprehensive Medical Science,
Fujita Health University,
Toyoake, Japan.

M.Hirafuji
Department of Pharmacology,
Faculty of Pharmaceutical Sciences,
The Health University of Hokkaido
Ishikari-Tobetsu, Hokkaido, Japan.

S.Kimura
Department of Clinical Pharmacology
and Toxicology, Faculty of
Pharmaceutical Sciences,The Health
University of Hokkaido, Ishikari-
Tobetsu, Hokkaido, Japan.

Hiroshi Kuzuya
Division of Molecular Biology,
Institute for Comprehensive Medical
Science, Fujita Health University,
Toyoake, Japan.

Philip J.Langlais
Department of Neuosciences,
The University of California at San
Diego, San Diego, USA.

Friedrich Leblhuber
Department of Gerontology,
Landesnerveklinik Wagner-Jauregg,
Linz, Austria.

Kazutaka Maeyama
Department of Pharmacology,
ehime University School of Medicine,
Ehime, Japan.

Trimothy J.Maher
Massachusetts College of Pharmacy,
179, Longwood Avenue, Boston,
USA.

Wakako Maruyama
Laoratory of Biochemistry and
Metabolism, Department of Basic
Gerontology, National Institute for
Longevity Sciences, Aichi, Japan.

Kazuo Matsubara,
Department of Pharmacy,
Asahikawa Medical College,
Asahikawa, Japan.

Masuru Minami
Department of Pharmacology,
Faculty of Pharmaceutical Sciences,
The Health University of Hokkaido
Ishikari-Tobetsu, Hokkaido, Japan.

Yohio Monma
Department of Pharmacology,
Faculty of Pharmaceutical Sciences,
The Health University of Hokkaido,
Ishikari-Tobetsu, Hokkaido, Japan.

Ken Morii
1st Department of Pharmaclogy,
Hokkaido University School of
Medicine, Sapporo, Japan.

T.Nagatsu
Institute for Comprehensive Medical
Sciences, Fujita Health University
School of Medicine, Toyoake, Japan.

Nishio Nakamara
Department of Laboratory Technology,
The College of Medical Technology,
Hakkaido Univrsity; Sapporo Japan.

Makoto Naoi
Department of Biosciences,
Nagoya Institute of Technology,
Nagoya, Japan.

A.Nicotra
Dipartimento di Biologia,
Animale e dell' Uomo,
Universita di Roma,
Roma, Italy.

J.Nikolic
Institute of Biochemistry,
Faculty of Medicine,
The University of Nis,
Nis, Yoguslavia.

Ingrid Nylander,
Department of Pharmaceutical Biosciences,
Division of Pharmacology,
Uppsala university,
Uppsala, Sweden.

Hasan Parvez
Neuropharmacology Unit,
University Paris XI, Orsay, Cedex,
France.

Simone Parvez
Neuroendocrinologie &
Neuropharmacologie du development,
Universite de Riem, Riem, France.

George Perry
Institute of Pathology,
Case Western Reserve University,
Cleveland, Ohio, USA.

J.Pichon
universite d' Orleans,
Orleans, Cedex, France.

G.Ali Qureshi
Clinical Research Center, Novum,
Huddinge University Hospital,
Karolinska Institute,
Stockholm, Sweden.

Hideya Saito
Department of Pharmacology,
Faculty of Pharmaceutical Sciences,
The Health University of Hokkaido
Ishikari-Tobetsu, Hokkaido, Japan.

Hideo Sakamoto
Division of Molecular Biology,
Institute for Comprehensive Medical Science, Fujita Health University,
Toyoake, Japan.

Kan Shimpo
Fujita Memorial Division of Phamacognosy.
Fujita Health University,
Toyoake, Japan.

Akiva Shirahata
Faculty of Pharmaceutical Science,
Josai University, Sakato,Saitama,
Japan.

Jerzy Silberring
laboratory for High Resolution Mass Spectrometry, Faculty of Chemistry,
Jagiellonian University, Krakow,
Poland.

Pavel J.Sindelar
Clinical Research Center, Novum,
Huddinge University Hospital,
Karolinska Institute.
Huddinge, Sweden.

Mark A.Smith
Institute of Pathology,
Case Western Reserve University,
Cleveland, Ohio, USA.

Stacy E.Stephans
Program in Basic & Clinical Neurosciences, Department of Psychiatry,Case Western Reserve University,Cleveland, Ohio, USA.

Ernst N.H.Jansen Stuer
Department of Neurology,
Medisc Spectrum Twente,
Enschede, The Netherlands.

Magnus Söderberg
Division of Pathology (IMPI),
Huddinge University Hospital,
Karolinska Institute,
Huddinge, Sweden.

Zenzo Tamura
Faculty of Pharmaceutical Sciences,
The University of Tokyo,
Tokyo, Japan.

Keath Tipton,
Department of Biochemistry,
Trinity College,
Dublin, Ireland.

Hiroko Togachi
1st Department of Pharmacology,
Hokkaido University School of Medicine,
Sapporo, Japan.

Kenichi Ueno
Department of Pharmacology,
Faculty of Pharmaceutical Sciences,
The Health University of Hokkaido
Ishikari-Tobetsu, Hokkaido, Japan.

Barbara Sperner-Unterweger
Department of Psychiatry,
University of Innsbruck,
Innsbruck, Austria.

Istvan Vermes
Department of Clinical Chemistry,Medisch Spectrum Twente,
Enschede, The Netherlands.

Victor Raul Villanueva
Institut de Chimie des Substances Naturelles, CNRS,
Gif-sur-Yvette, Cedex, France.

Rob A.I.de Vos
Medisch Spectrum Twente,
Regional Laboratory for Pathology,
Enschede, The Netherlands.

Helmut Wachter
Institute of Medical and Biochemistry,
The Universty of Innsbruck,
Innsburck, Austria.

Takehiko Watanabe
department of Pharmacology I,
Tohoka University School of Medicine,
Sendai, Japan.

Bernhard Widner
Institute of Medical and Biochemistry,
The Universty of Innsbruck,
Innsburck, Austria.

Bryan K.Yamamoto
Program in Basic & Clinical Neurosciences, Department of Psychiatry,
Case Western Reserve University,
Cleveland, Ohio, USA.

Mitsuhiro Yoshioka
1st Department of Pharmacology,
Hokkaido University School of Medicine,Sapporo, Japan.

Masanori Yoshioka
Faulty of Pharmaceutical Sciences,
Setsunan University, Hirakata,
Osaka, Japan.

Xuegong Zhu
The Charles B.Stout Neuroscience
Mass Spectrometry Laboratory, The
University of Tennessee, Memphis,
USA.